T0325819

Introduction to Mobile Robot Control

Introduction to Mobile Robot Control

Spyros G. Tzafestas
School of Electrical and Computer Engineering
National Technical University of Athens
Athens, Greece

AMSTERDAM • BOSTON • HEIDELBERG • LONDON NEW YORK • OXFORD
PARIS • SAN DIEGO • SAN FRANCISCO • SINGAPORE • SYDNEY • TOKYO

Elsevier
32 Jamestown Road, London NW1 7BY
225 Wyman Street, Waltham, MA 02451, USA

First edition 2014

Notices
Knowledge and best practice in this field are constantly changing. As new research and experience broaden our understanding, changes in research methods, professional practices, or medical treatment may become necessary.

Practitioners and researchers must always rely on their own experience and knowledge in evaluating and using any information, methods, compounds, or experiments described herein.

In using such information or methods they should be mindful of their own safety and the safety of others, including parties for whom they have a professional responsibility. To the fullest extent of the law, neither the Publisher nor the authors, contributors,or editors, assume any liability for any injury and/or damage to persons or property as a matter of products liability, negligence or otherwise, or from any use or operation of any methods, products, instructions, or ideas contained in the material herein.

British Library Cataloguing-in-Publication Data
A catalogue record for this book is available from the British Library

Library of Congress Cataloging-in-Publication Data
A catalog record for this book is available from the Library of Congress

ISBN: 978-0-12-417049-0

This book has been manufactured using Print On Demand technology. Each copy is produced to order and is limited to black ink. The online version of this book will show color figures where appropriate.

Dedication

To the Robotics Teacher and Learner

For the things we have to learn before we can do them, we learn by doing them.

Aristotle

The second most important job in the world, second only to being a parent, is being a good teacher.

S.G. Ellis

In learning you will teach, and in teaching you will learn.

Phil Collins

Contents

Preface

Robotics has been a dominant contributor to the development of the human society over the years. It is a field that needs the synergy of a variety of scientific areas such as mechanical engineering, electrical-electronic engineering, control engineering, computer engineering, sensor engineering, and others. Robots and other automated machines have to live together with people. In this symbiosis, human needs and preferences should be predominantly respected, incorporated, and implemented. To this end, modern robots, especially wheeled or legged mobile robots, incorporate and realize in a purposeful and profitable way the *perception–action cycle* principle borrowed from biological systems and human cognitive and adaptation capabilities.

The objective of this book is to present in a cohesive way a set of fundamental conceptual and methodological elements, developed over the years for nonholonomic and omnidirectional wheeled mobile robots. The core of the book (Chapters 5 through 10) is devoted to the analysis and design of several mobile robot controllers that include basic Lyapunov-based controllers, invariant manifold–based controllers, affine model–based controllers, model reference adaptive controllers, sliding-mode and Lyapunov-based robust controllers, neural controllers, fuzzy-logic controllers, vision-based controllers, and mobile manipulator controllers. The topics of mobile robot drives, kinematics, dynamics, and sensing are covered in the first four chapters. The topics of path planning, motion planning, task planning, localization, and mapping are covered in Chapters 11 and 12, including most fundamental concepts and techniques at a detail compatible with the purpose and size of the book. Chapter 13 provides a selection of experimental results obtained by many of the methods studied in this book. These results were drawn from the research literature and include some of the author's results. Chapter 14 provides a conceptual overview of some generic systemic and software architectures developed for implementing integrated intelligent control of mobile robots. Finally, Chapter 15 provides a tour to the applications of mobile robots in the factory and society at an encyclopedic level.

For the convenience of the reader, the first section of each chapter involves the required mathematical, mechanics, control, and fixed-robot background concepts that are used in the chapter. The book is actually complementary to most books in the field, in the sense that it provides a solid model-based analysis and design of a large repertory of mobile robot control schemes, not covered in other books.

This book is suitable for senior undergraduate and graduate instructional courses on general and mobile robotics. It can also be used as an introductory reference book by researchers and practitioners in the field that need a consolidated methodological source for their work.

I am grateful to all publishers and authors for granting their permission to include in the book the requested illustrations and experimental plots.

Spyros. G. Tzafestas
Athens, April 2013

List of acknowledged authors and collaborators

The work on mobile robotics of the following researchers and collaborators, which was used and compiled for producing the present book, is acknowledged.

Authors		
Abdelkader H.H.	Edwards G.	Mühlenfeld A.
Adams M.D.	Engedy I.	Muir P.F.
Adom A.	Espiau B.	Myers K.
Afshar A.	Ettelt E.	Narayankar I.
Agüero C.	Euderde E.	Nayar S.
Aguilar L.T.	Everett H.R.	Nesnas A.
Ahmadabadi M.N.	Fetter Lages W.	Okamoto Jr J.
Aicardi M.	Fierro R.	Olunloyo V.O.S
Albus J.S.	Fischer C.	Orebäck A.
Alsina P.J.	Fourquet J.Y.	Oriolo G.
Al-Zorkany M.	Fraser G.	Osmic N.
Amato N.	Fu L.C.	Ostrowski J.P.
Andreff N.	Fukuda T.	Padois V.
Ang Jr M.H.	Furtwängler R.	Papadopoulos E.
Araujo A.	Gallina P.	Pascoal A.
Araujo H.	Ganguly P.	Petrinic T.
Arkin R.C.	Gans N.R.	Petrovic I.
Arney T.	Garrido S.	Pimenta L.C.A.
Asheriad M.	Gasparetto A.	Poulakakis J.
Ashoorizad M.	Gerkey B.	Pourboghrat F.
Asmore M.	Geva S.	Puiu D.
Awad H.A.	Geyer C.	Qian J.
Ayari I.	Gholipour A.	Rachid A.
Ayomoh M.K.O.	Gilioli M.	Ramaswamy P.S.A.
Baek S.H.	Giordano P.R.	Rekleitis I.
Bailey T.	Gordon N.J.	Reyhanoglu M.
Baker S.	Grassi Jr V.	Rives P.
Balakrishnan S.N.	Grassi Jr Y.	Rizon M.
Balestrino A.	Grigorescu S.	Roy N.
	Grosu V.	
Barnes N.	Hager G.	Ruiz-Ayùcar J.
Barreto J.P.	Handeck U.D.	Ruspini E.
Bar-Shalom Y.	Hemerly E.M.	Sablantnög S.
Barzamini R.	Hong D.	Safadi H.

Bastin G.
Bayar G.
Beattle B.
Bench-Capon T.J.M.
Bian H.J.
Bicchi A.
Blanco D.
Boles W.W.
Bonert M.
Borenstein J.
Bornstedt B.
Bretl T.
Bright G.
Brooks R.A.
Brun Y.
Buss M.
Byrne J.C.
Campion G.
Campos J.
Cañas J.M.
Cardenas S.
Carreli R.
Casalino G.
Castillo O.
Ceyer C.M.
Chang C.F.
Chang K.C.
Chatti A.
Chaumette F.
Chen Y.
Cherubini A.
Chirikjian G.S.
Chiron P.
Choi Y.H.
Chong C.Y.
Christensen H.L.
Chung J.H.
Chung W.
Cocias T.
Coenen F.P.
Corke P.
Coste-Manière E.
Cote C.
Cuerra P.N.
Cwa D.
D'andrea R.
D'andrea-Novel B.
Dailey M.N.
Horvath G.
Howard A.
Huang C.I.
Hutchinson S.
Ivanjko E.
Izumi K.
Jarvis R.A.
Jiang P.
Kalmar-Nagy T.
Karlsson M.P.
Kerry M.
Khatib O.
Kim B.M
Kim J.H
Kim M.
Kimoto K.
Koku A.B
Konolige K.
Konukseven E.I.
Koren Y.
Kraetzschmar G.
Kragic D.
Kramer J.
Krüger D.
Kumar V.
Kunitake Y.
Lacevic B.
Laengle T.
Lages W.F.
Lapierre L.
Latombe J.C.
Lavalle S.M.
Lee S.
Lee T.K.
Leonard J.L.
Lewis F.L.
Liang S.
Liang Z.
Lietmann T.
Limsoonthrakul S.
Lindström M.
Lozano-Perez T.
Lueth T.C.
Ma J.
Macesanu G.
Maeyama S.
Mamat M.
Martin F.
Saffioti A.
Saich J.
Salmond D.J.
Samson C.
Saridis G.N.
Sarkar N.
Sato K.
Scheutz M.
Schmidt D.C.
Schmidt G.
Schneider S.
Scholl K.U.
Selmic R.
Sethian J.A.
Shave M.J.R.
Sheu P.G.Y.
Sidek N.
Simmons R.
Smith A.F.M.
Soetanto D.
Soria C.M.
Southall B.
Spinu V.
Spletzer J.
Steinbauer G.
Su J.
Suga Y.
Tadijne M.
Tang J.
Tayebi A.
Taylor C.J.
Taylor T.
Thrun S.
Tian Y.
Tsakiris D.
Tsiotras P.
Ullman M.
Utz H.
Velagic J.
Vidal Calleja T.A.
Wang C.
Wang L.C.
Wang Y.
Wotawa A.
Xie W.
Xue Q
Yacacob S.
Yamamoto T.
Yamamoto Y.

Daniilidis K.	Martinet P.	Yang F.
Das A.K	Matellán V.	Yang J.M
Das T.	Medeiros A.A.D.	Yang M.
De Oliveira V.M.	Mehrandezh M.	Yaugham R.
De Pieri E.R.	Melchiori G.	Yong L.S.
Dehgam S.M.	Meystel A.M.	Youm Y.
DeLuca A.	Mezouar Y.	Yun X.
DeVilliers M.	Micaelli A.	Yuta S.
		Zefran M.
DeVon D.	Milios E.	Zelinsky A.
Diaz B.M.	Moldoveanu F.	Zhang Q.
Ding F.G.	Montemerlo M.	Zhang Y.
Doroftei I.		
Du J.	Morales B.	Zouzdani J.
Dudek G.	Moreno L.	
Durrant-Whyte H.F.	Moret E.N.	
Collaborators		
Deliparaschos K.M.	Moustris G.	Tang J.
Fukuda T.	Rigatos G.	Tzafestas C.S.
	Sgouros N.M.	
Katevas N.	Shiraishi Y.	Tzafestas E.S.
Krikochoritis T.	Skoundrianos E.	Watanabe K.
Melfi A.	Stamou G.	Zavlangas P.

Principal symbols and acronyms

t, k	Continuous, discrete time
l, s, D, d	Linear distance (length)
$\mathbf{p},\mathbf{d},\mathbf{x}$	Position vector
$\mathbf{R}$	Rotation matrix
$\mathbf{R}_x, \mathbf{R}_y, \mathbf{R}_z$	Rotation matrix w.r.t. axis x, y, z
$\mathbf{n},\mathbf{o},\mathbf{a}$	Normal, orientation, and approach unit vectors
$\mathbf{A},\mathbf{T}$	Homogeneous (4×4) matrix
$\mathbf{q}$	Generalized variable (linear, angular)
$\mathbf{v},\upsilon$	Linear velocity vector
$\omega,\dot{\theta}$	Angular velocity vector
$\mathbf{J}(\mathbf{q})$	Jacobian matrix
$\mathbf{J}^{-1}(\mathbf{q})$	Inverse of $\mathbf{J}(\mathbf{q})$
$\mathbf{J}^{\dagger}(\mathbf{q})$	Generalized inverse (pseudoinverse) of $\mathbf{J}(\mathbf{q})$
$D-H$	Denavit–Hartenberg
$\det(\cdot)$	Determinant
$\mathbf{F},\tau,(\mathbf{N})$	Force, torque vector
L	Lagrangian function
K, P	Kinetic energy and potential energy
$\mathbf{D}(\mathbf{q})$	Inertial matrix
$\mathbf{g}(\mathbf{q})$	Gravity term
HRI	Human–robot interface
GHRI	Graphical human–robot interface
IC	Intelligent control
ICA	Intelligent control architecture
NL	Natural language
NL-HRI	Natural language HRI
UI	User interface
$\mathbf{C}(\mathbf{q},\dot{\mathbf{q}})\dot{\mathbf{q}}$	Centrifugal/Coriolis term
$Oxyz$	Coordinate frame
WMR	Wheeled mobile robot
MM	Mobile manipulator
LS	Least squares
ϕ, ψ	WMR direction, steering angles
DOF	Degree of freedom
COG	Center of gravity
COM	Center of mass
$\mathbf{M}(\mathbf{q})$	Nonholonomic constraint

CS	Configuration space
GPS	Global positioning system
l_f	Lens focal length
$\mathbf{K}_p, \mathbf{K}_v$	Position, velocity gain matrix
KF, EKF	Kalman filter, extended Kalman filter
$SLAM$	Simultaneous localization and mapping
$\hat{\mathbf{x}},\hat{\boldsymbol{\theta}}$	Estimate of $\mathbf{x},\boldsymbol{\theta}$
$\tilde{\mathbf{x}},\tilde{\boldsymbol{\theta}}$	Error of the estimate $\hat{\mathbf{x}},\hat{\boldsymbol{\theta}}$
$\boldsymbol{\Sigma}_x, \boldsymbol{\Sigma}_\theta$	Covariance matrix of $\tilde{\mathbf{x}},\tilde{\boldsymbol{\theta}}$
AI	Artificial intelligence
RF	Radio frequency
CAD	Computer-aided design
$\bar{x}(s)$	Laplace transform of $x(t)$
$G(s), \bar{g}(s)$	Transfer function
ω_n	Natural angular frequency
ζ	Damping factor
P, PD	Proportional, proportional plus derivative
PI, PID	Proportional plus integral (plus derivative)
$MRAC$	Model reference adaptive control
SMC	Sliding mode control
$\mathbf{J}_{im}$	Image Jacobian
$VRS(VRC)$	Visual robot servoing (control)
$FL(FC)$	Fuzzy logic (fuzzy control)
NN	Neural network
BP	Back propagation
RFN	Radial basis function neural network (RBF-NN)
MLP	Multi-layer perception
NF	Neurofuzzy
MB	Model-based
tg, tan	Trigonometric tangent of an angle
tg^{-1}, arctan	Inverse of tg, tan

Quotations about robotics

Aristotle *If every tool, when ordered, or even of its own accord, could do the work that benefits it, just as the creations of Daedalus move by themselves. . ., then there would be no need of apprentices for the master workers or of slaves for their lords.*

Allen Newel *From where I stand it is easy to see the science lurking in robotics. It lies in the welding of intelligence to energy. That is, it lies in intelligent perception and intelligent control of motion.*

Rod Grupen *At bottom, robotics is about us. It is the discipline of emulating our lives, of wondering how we work.*

Rob Spencer *Got a dirty, dangerous, dull job? Let a robot do it and keep your workers safe.*

Marvin Minsky *We wanted to solve robot problems and needed some vision, action, reasoning, planning, and so on. . .. Eventually, robots will make everything.*

David Hanson *Making realistic robots is going to polarize the market. You will have people who love it and some people who will really be disturbed.*

John McCarthy *Every aspect of learning or any other feature of intelligence can in principle be so precisely described that a machine can be made to simulate it. No robot has ever been designed that is ever aware of what is doing; but most of the time, we aren't either.*

John McDermott *To be useful, a system has to do more than just correctly perform some task.*

Chuck Gosdzinski *The top two awards don't even go to robots.*

1 Mobile Robots: General Concepts

1.1 Introduction

Mobile robots are robots that can move from one place to another autonomously, that is, without assistance from external human operators. Unlike the majority of industrial robots that can move only in a specific workspace, mobile robots have the special feature of moving around freely within a predefined workspace to achieve their desired goals. This mobility capability makes them suitable for a large repertory of applications in structured and unstructured environments. Ground mobile robots are distinguished in *wheeled mobile robots* (WMRs) and *legged mobile robots* (LMRs) Mobile robots also include *unmanned aerial vehicles* (UAVs), and *autonomous underwater vehicles* (AUVs). WMRs are very popular because they are appropriate for typical applications with relatively low mechanical complexity and energy consumption. Legged robots are suitable for tasks in non-standard environments, stairs, heaps of rubble, etc. Typically, systems with two, three, four, or six legs are of general interest but many other possibilities also exist. Single-leg robots find rare applications because they can only move by hopping. Mobile robots also include mobile manipulators (wheeled or legged robots equipped with one or more light manipulators to perform various tasks).

The objective of this chapter is to present the fundamental general concepts of WMRs. In particular, the chapter

- provides a list of the main historical landmarks of general robotics and mobile robots;
- discusses the locomotion issues of ground (wheeled, legged) mobile robots;
- investigates the wheel and drive types of mobile robots (nonholonomic, omnidirectional); and
- introduces the concepts of mobile robot degree of mobility, degree of steerability, and maneuverability.

1.2 Definition and History of Robots

1.2.1 What Is a Robot?

The term "*robot*" (*robota*) was used for the first time in 1921 by the Czech writer *Karel Capek* and means slave servant or forced labor. In science and technology there is not a global or unique definition of a robot. When *Joseph Engelberger*, the father of modern robotics, was asked to define a robot he said: "I can't define a

Introduction to Mobile Robot Control. DOI: http://dx.doi.org/10.1016/B978-0-12-417049-0.00001-8

robot but I know one when I see one." The *Robotics Institute of America* (RIA) defines an industrial robot as "a reprogrammable multi-functional manipulator designed to move materials, parts, tools, or specialized devices through variable programmed motions for the performance of a variety of tasks which also acquire information from the environment and move intelligently in response." This definition does not capture mobile robots.

The definition adopted in the *European Standard EN775/1992* is as follows: "Manipulating industrial robot is an automatically controlled reprogrammable multi-purpose, manipulative machine with several degrees of freedom (DOF), which may be either fixed in place or mobile for use in industrial automation applications.

Ronald Arkin says: "An intelligent robot is a machine able to extract information from its environment and use knowledge about its work to move safely in a meaningful and purposive manner."

Rodney Brooks says: "To me a robot is something that has some physical effect on the world, but it does it based on how it senses the world and how the world changes around it."

In summary, a robot is referred in the literature as a machine that performs an intelligent connection between perception and action. An autonomous robot is programmed to work without human intervention, and with the aid of embodied artificial intelligence can perform and live within its environment. Today's mobile robots can move around safely in cluttered surroundings, understand natural speech, recognize real objects, locate themselves, plan paths, and generally think by themselves. Intelligent mobile robot design employs the methodologies and technologies of intelligent, cognitive, and behavior-based control. Mobile robots must maximize flexibility of performance subject to minimal input dictionary and minimal computational complexity.

1.2.2 Robot History

The history of robots can be divided in two general periods [1,2]:

- Ancient and preindustrial period
- Industrial and robosapien period

1.2.2.1 Ancient and Preindustrial Period

The first robot in the worldwide history (around 2500–3000 BC) is the Greek mythodological mechanical creature called *Talos* ("Τάλως") [3]. This name is attributed both to a human being (the son of Daedalus' sister Perdika) and a mechanical artificial entity constructed by Hephaestus, under the order of Zeus, with bronze body and a single vein from the neck up to the ankle, where a copper nail blocked it out. Talos was gifted by Zeus to Europe who afterward gave him to her son Minos to guard Crete. Talos died when the Argonaut Poas removed the copper nail from his heel. This resulted in the spilling out of the *ichor* ("the blood of the immortals") flowing in the Poas' vein. The name Talos was given to Asteroid 5786 discovered by Robert McNaught, on September 31, 1991 at Siding Spring Observatory in Coonabarabran, New South Wales (Australia). Around 350 BC the friend of Plato Archytas of Tarentum constructed a mechanical bird ("*pigeon*") which was propelled by steam. This represents one of the earlier historic studies of flight or airplane model:

Around 270 BC: Ktesibios ("Κτησίβιος") has discovered the water clock that involves movable parts, and wrote his book "*About Pneumatics*" (Περί Πνευματικής) where he has shown that air is a material entity.
Around 200 BC: Chinese artisans design and construct mechanical automata, such as orchestra, etc.
Around 100 AD: Heron of Alexandria designs and constructs several regulating mechanisms, such as the odometer, the steam boiler (*aclopyle*), the automatic opening of temples, and automatic distribution of wine.
Around 1200 AD: The Arab author Al Jazari writes "*Automata*" which is one of the most important texts in the study of the history of technology and engineering.
Around 1490: Leonardo Da Vinci constructs a device that looks as an armored knight. This seems to be the first humanoid robot in Western civilization.
Around 1520: Hans Bullman (Nuernberg, Germany) builds the first real *android* in robot history imitating people (e.g., playing musical instruments).
1818: Mary Shelley writes the famous novel *Frankenstein* based on an artificial life creature (robot) developed by Dr. Frankenstein. All robots in this novel turned eventually against human kind in a frightening way.
1921: The Chzech dramatist *Karel Capek* coins the term *robota* (robot) in his play named "Rossum's Universal Robots" meaning compulsory or slavery work.
1940: The science fiction writer Isaac Asimov used for the first time the terms "robot" and "robotics." In 1942, he wrote "*Runaround*" a story that involved his three laws of robotics (known as *Asimov's laws*).

1.2.2.2 Industrial and Robosapien Period

This period starts in 1954 when George Devol, Jr patented his multijoined robotic arm (the first modern robot). In 1956, together with Joseph Engelberger founded the world's first robot company called *Unimation* (from *Uni*versal Auto*mation*):

1961: The first industrial robot called *Unimate* joined a die-casting production line at *General Motors*.
1963: The *RanchoArm*, the first computer controlled robotic arm, was put in operation at the Rancho Los Amigos Hospital (Downey, CA). This was a prosthetic arm designed to aid the handicapped.
1969: The first truly flexible arm, known as the, *Stanford Arm*, was developed in the Stanford Artificial Intelligence Laboratory by Victor Scheinman. This arm soon became a standard, and is still influencing the design of today's robotic manipulators.
1970: This is the *starting year* of *mobile robotics*. The mobile robot *Shakey* was developed at the Stanford Research Institute (today known as *SRI Technology*), controlled by intelligent algorithms that observe via sensors, and react to their own actions (Figure 1.1).

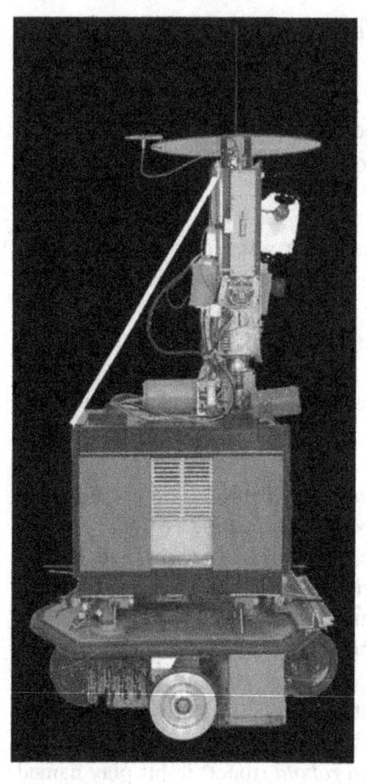

Figure 1.1 The wheeled mobile robot Shakey of SRI which is equipped with on-board logic, a camera, a range finder sensor, and a bump detector.
Source: http://www.thocp.net/reference/robotics/robotics2.htm

Shakey is referred to as "*the first electronic person.*" The name Shakey is due to its jerky motion.

1979: The *Stanford Cart*, originally designed in 1970 as a line follower, is rebuilt by Hans Moravec and equipped with a more robust 3-D vision that allows more autonomy (Figure 1.2). In an experiment the Stanford Cart crossed a chair-filled room autonomously using a TV camera that was taking pictures from several angles. These pictures were processed by a computer to analyze the distance of the cart from the obstacles.

1980–1989: This decade is dominated by the development of advanced Japanese robots, especially walking robots (humanoids). Among them the *WABOT-2* humanoid is mentioned (Figure 1.3B) which was developed in 1984, representing the first attempt to develop a personal robot with a "specialistic purpose" to play a keyboard musical instrument, rather than a versatile robot like WABOT-1 (Figure 1.3A).

Other robots developed in the 1980s are the 1983 British computer controlled micro robot vehicle *Prowler* (Figure 1.4), the 1985 Waseda-Hitachi Leg-11 (*WHL-11*), the 1989 *Aquarobot* (Figure 1.5), and the 1989 multilegged robot *Genghis* of MIT (Figure 1.6).

1990–1999: During this decade the emergence of "*explorer robots*" took place. These robots went where the human did not visit before or it was considered too risky or inconvenient. Examples of such robots are *Dante* (1993) and *Dante II* (1994) that explored Mt. Erebrus in Antarctica and Mt. Spurr in Alaska (Figure 1.7).

An example of NASA planetary missions aiming at studying the climate and geology of the Red Planet (Mars) is Path Finder. The Mars Observer was to touch down close to the south pole latitudes, carrying two scientific instruments and a lander (robotic rover).

Figure 1.2 The Stanford cart.
Source: http://www.thocp.net/reference/robotics/robotics2.htm.

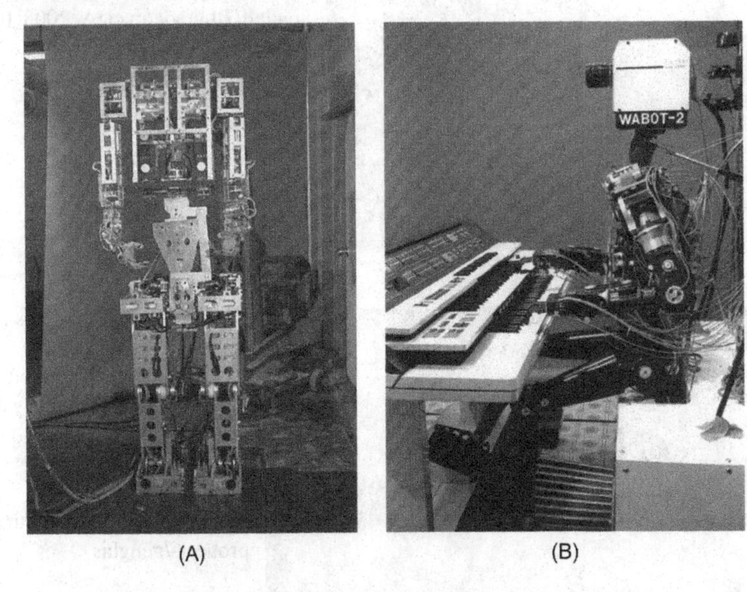

Figure 1.3 The Waseda University robots (A) WABOT-1, (B) WABOT-2.
Source: http://www.humanoid.waseda.ac..jp/booklet/kato_2.html.

The Pathfinder spacecraft landed successfully on Mars' Ares Vallis region on July 4, 1997. Its robotic rover was named Sojourner and is a 10.6 kg WMR (Figure 1.8). Sojourner conducted a number of experiments on the surface of Mars and continued to broadcast data until September 1997.
2000–Present: In the 2000s the development of numerous new intelligent mobile robots capable of almost fully interacting with humans recognizing voices, faces and gestures,

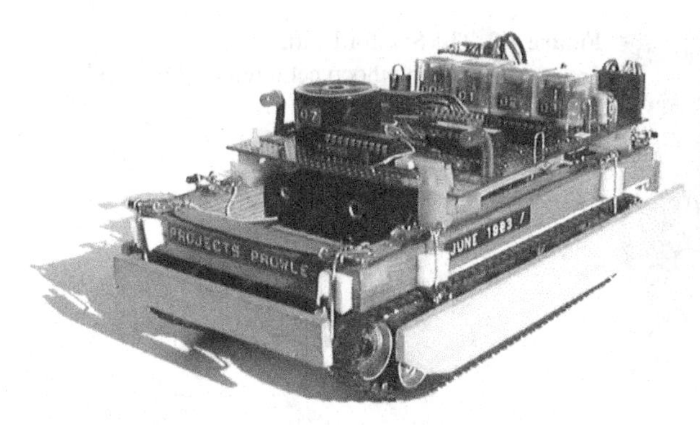

Figure 1.4 The expandable microrobot Prowler published in Sinclair projects, August 1983. *Source*: http://www.davidbuckley.net/DB/Prowler.htm.

Figure 1.5 The AQUA robot can take pictures of coral reefs and other aquatic organisms returning home after completion of its tasks. *Source*: http://www.rutgersprep.org/kendall/7thgrade/cycleA_2008_09/zi/robo_AQUA.html.

Figure 1.6 The Genghis robot. Genghis has a special multilegged walking mode known as the "Genghis gait." It is now retired at the Smithsonian Air Space Museum. *Source*: http://www.ai.mit.edu/prohects/genghis.

and expressing emotions through speech, dexterous walking or performing household chores, hospital chores, microsurgeries and the like, is continuing with growing rates. Notable examples are:

- HONDA humanoid ASIMO (2000) (Figure 1.9)
- LEGO Robotics Invention System-2 (2000)
- FDA Cyberknife for treating tumors anywhere in the human body (2001)

Figure 1.7 The Dante II explorer robot.
Source: http://www.frc.ri.cmu.edu/robots/robs/photos/1994_DanteII.jpg.

Figure 1.8 The NASA Sojourner robotic rover.
Source: http://haberlesmeplatformu.blogspot.com/2010_04_01_archive.html.

- SONY AIBO ERS-7: Third generation robotic pet (2003)
- iROBOT Roomba, a robotic vacuum cleaner (2003)
- TOMY i-SOBOT entertainment robot, a humanoid robot capable of walking like a human and performing entertainment actions, such as kicks and punches (2007)
- SHADOW dextrous hand robot (2008)
- ROLLIN JUSTIN robot of the German Air Space Agency (2009), a humanoid preparing and serving drinks (Figure 1.10)
- FLAME: the Toyota 130 cm running humanoid robot
- WowWee Roborover, Joebot, and Robosapien robots.

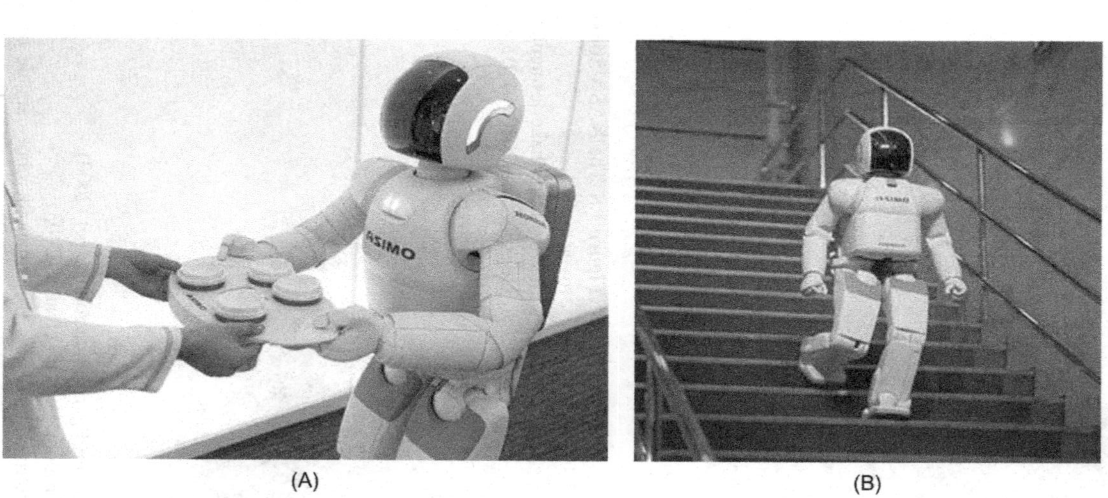

Figure 1.9 The Honda's humanoid ASIMO.
Source: http://www.gizmag.com/go/1765picture/2029; http://razorrobotics.com/safety.

Figure 1.10 The Rollin Justin robot mixing instant tea.
Source: http://inventors.about.com/od/robotart/ig/Robots-and-Robotics/Rollin-Justin-Robot.htm.

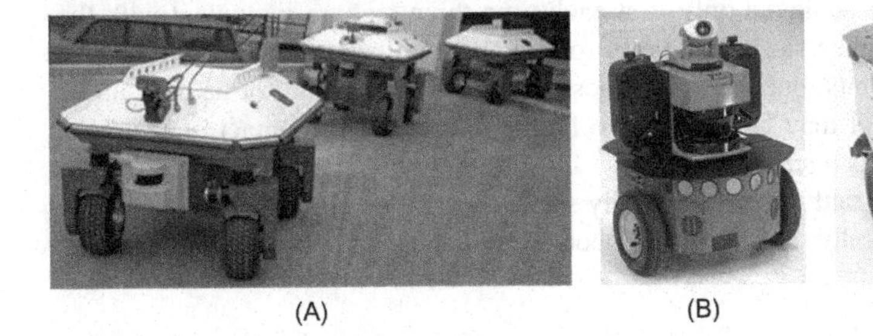

(A) (B) (C)

Figure 1.11 (A) Seekur (350 kg, dimensions 1.4 m × 1.3 m × 1.1 m), (B) Pioneer 3D-X, and (C) PowerBot.
Source: http://mobilerobots.com/ResearchRobots/ResearchRobots.aspx; http://www.conscious-robots.com/en/reviews/robots/mobilerobots-pioneer-3p3-dx-8.html.

A comprehensive presentation of the development and current maturity of robosapiens and socialized robots is provided in *Menzel* and *D'Aluisio* book (Robo Sapiens: Evolution and New Species, MIT Press, MA, 2000). Three currently commercially available mobile robotic platforms for research purposes are the following (Figure 1.11):

Seekur: An all weather large holonomic robot platform for security, inspection, and research

Pioneer 3-DX: A fully programmable platform equipped with motors-encoders, and 16 ultrasonic (front-facing and rear-facing sonars). It is used for research and rapid development (localization, monitoring, navigation, control, etc.)

PowerBot: A high payload (up to 100 kg) differential drive robot for research and rapid prototyping in universities and research institutes.

1.3 Ground Robot Locomotion

Locomotion of ground mobile robots is distinguished in [4–15][1] :

- Legged locomotion
- Wheeled locomotion

1.3.1 Legged Locomotion

The *wheel* is a human invention, but the *leg* is a biological element. Locomotion of most of the highly developed animals is through legs. Biological multilegged organisms can move in diverse and difficult environments with obstacles, rough grounds, etc. Insects have very small size and weight and possess strong robustness that cannot be achieved by artificial creatures. To be useful in real-life tasks, a legged robot must be statically stable. This condition is satisfied if the center of gravity lies always within the polygon defined by the actual contact points with the ground. This can be achieved only if at each time three feet are in contact with the floor. Thus to guarantee statical stability four legs are at least needed. If the feet do not have contact points, but lines or planes of contact, this might not be true, and a statically stable robot may have only two legs. In practice, the contact of the robot body with the floor is a small region.

A legged robot is said to be dynamically stable if it is does not fall over, despite the fact that it is not statically stable. Legged robots are distinguished in two main categories:

- Two-leg (bipedal) robots
- Many-leg robots

Bipedal locomotion is standing on two legs, walking and running.

Humanoid robots are bipedal robots with an overall appearance based on that of the human body, that is, head, torso, legs, arms, and hands. Some humanoids may model only part of the body, for example, from waist up, such as in the NASA's *Robonaut*, while others have also a "face" with "eyes" and "mouth."

Robot bipedal locomotion needs complex interaction of mechanical and control system features. Humans are able to perform bipedal locomotion because their spines are s-curved and their heels are round. During locomotion the legs need to be lifted from and return to the ground. The sequence and way of placing and lifting each foot (in time and space), synchronized with the body motion so as to move from one place to another, is called *gait.*

The human gait involves the following distinct phases:

- Rocking back and forth between feet

[1] UAVs and AUVs will not be studied in this book.

- Pushing with the toe to sustain speed
- Combined interruption in rocking and ankle twist to turn
- Shortening and extending the knees to prolong the "forward fall"

The basic cycle of a gait is called a *stride*, which describes the complete cycle from one occurrence of a leg motion to its repetition. The fraction of a stride during which the foot is in on-state is called the *duty factor*. For statically stable walking the minimum duty factor needed is "3: (number of legs)," where 3 is the minimum number of feet in on-state to assure static stability. A walking gait is one where at least one foot is on the ground at any time. A running gait is occurring if for some time periods all feet are in off-state. Robots without static stability need higher energy to achieve dynamic stability, and extra stabilizing motions to prevent the body from falling over (in addition to useful/productive motions). The initial research on multilegged walking robots was focused on robot locomotion design for smooth or easy rough terrain, by passing simple obstacles, motion on soft ground, body maneuvering, and so on. These requirements can be realized via periodic gaits and binary (yes/no) contact information with the ground. Newer studies are concerned with multilegged robots that can move over an impassable road or

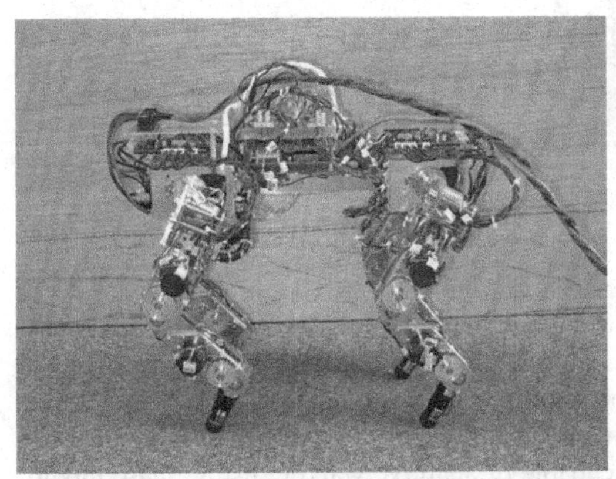

Figure 1.12 The quadruped robot "Kotetsu" (leg length at standing 18–22 cm).
Source: http://robotics.mech.kit.ac.jp/kimura/research/Quadruped/photo-movie-kotetsu-e.html.

Figure 1.13 DARPA quadruped robot (LC3).
Source: http://www.gizmag.com/darpa-lc3-robot-quadruped/14256/picture/111087.

Figure 1.14 Six-legged robot "SLAIR" capable of operating at "action level."
Source: http://www.uni-magdeburg.de/ieat/robotslab/images/Slair/CIMG1059.jpg.

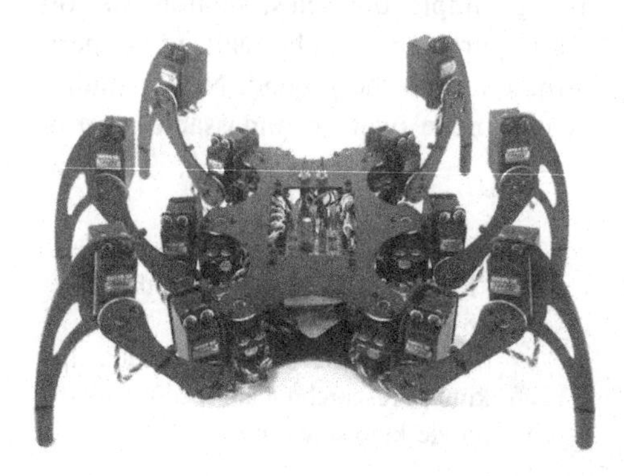

Figure 1.15 A typical example of hexabot robo-spider (Gadget Lab).
Source: http://www.wired.com/gadgetlab/2010/04/gallery-spider-robot/2/.

an extremely complex terrain, such as mountain areas, ditches, trenches, earthquake damaged areas, etc. In these cases, additional capabilities are needed, as well as detailed support reactions and robot stability prediction. Figures 1.12–1.15 show three advanced multilegged robots with capabilities of the above type. The quadruped robot Kotetsu of Figure 1.12 is capable of adaptive walking using phase modulations based on leg loading/unloading.

1.3.2 Wheeled Locomotion

The maneuverability of a WMR depends on the wheels and drives used. The WMRs that have three DOF are characterized by maximal maneuverability which is needed for planar motions, such as operating on a warehouse floor, a road, a hospital, a museum, etc. Nonholonomic WMRs have less than three DOF in the plane, but they are simpler in construction and cheaper because less than three motors are used. A holonomic vehicle can travel in every direction and function in tight areas. This capability is called *omnidirectionality*. Balance is inherently assured in WMRs with three or more wheels. But in the case of m-wheel WMRs ($m \geq 3$) a suspension system must be used to assure that all wheels can have ground contact in rough terrains. The main problems in WMR design are the traction, maneuverability, stability, and control that depend on the wheel types and configurations (drives).

1.3.2.1 Wheel Types

The types of wheels used in WMRs are:

- Conventional wheels
- Special wheels

Conventional wheels: These wheels are distinguished in powered fixed wheels, castor wheels, and powered steering wheels. *Powered fixed wheels* (Figure 1.16A) are driven by motors mounted on fixed positions of the vehicle. Their axis of rotation has a fixed direction with respect to the platform's coordinate frame. *Castor wheels* (Figure 1.16B) are not powered but they can also rotate freely about an axis perpendicular to their axis of rotation.

Powered steering wheels have a driving motor for their rotation and can be steered about an axis perpendicular to their axis of rotation. They can be without offset (Figure 1.16C) or with offset (Figure 1.16D) in which case the axes of rotation and steering do not intersect. To achieve omnidirectionality with conventional castor and powered steering wheels some kind of motion redundancy should be used, for example, n-wheel drives ($n>2$) with all wheels driven and steered. Conventional wheels have higher load capacities and higher tolerance for ground irregularities compared to special wheel configurations. But due to their nonholonomic constraints are not truly omnidirectional wheels.

Special wheels: These wheels are designed such that to have activated traction in one direction and passive motion in another, thus allowing greater maneuverability in congested environments. We have three main types of special wheels:

1. Universal wheel
2. Mecanum wheel
3. Ball wheel

The *universal wheel* provides a combination of constrained and unconstrained motion during turning. It contains small rollers around its outer diameter which are

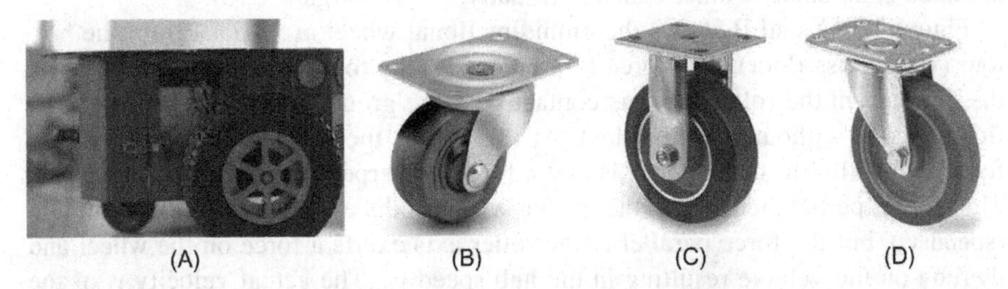

Figure 1.16 Conventional wheels (A) fixed wheel, (B) castor wheel, (C) powered steering wheel without any offset, and (D) power steering wheel with longitudinal offset.

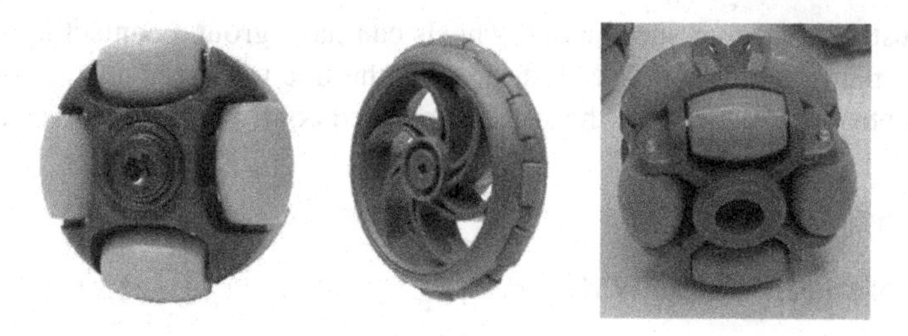

Figure 1.17 Three designs of universal wheel.
Source: http://www.generationrobots.com/2-omni-directional-wheel-robot-v.ex-robotics,us,4,2165-Omni-Directional-Wheel-kit.cfm; http://www.rotacaster.com.au/robot-wheels.html; http://www-scf.usc.edu/~csci445_final_contest/SearchAndRescue/OtherContests/2004_contest_Fall_RobotSoccer/Locomotion/omni_4wheel_encoders/omni_drive.pdf.

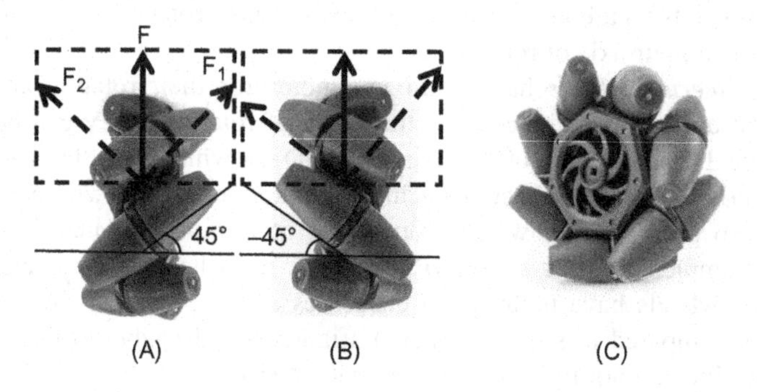

Figure 1.18 (A) Mecanum wheel with $\alpha = 45°$ (left wheel), (B) mecanum wheel with $\alpha = -45°$ (right wheel), and (C) an actual mecanum wheel.
Source: http://www.aceize.com/node/562.

mounted perpendicular to the wheel's rotation axis. This way the wheel can roll in the direction parallel to the wheel axis in addition to the normal wheel rotation (Figure 1.17).

The *mecanum wheel* is similar to the universal wheel except that the rollers are mounted at an angle α other than 90° (usually $\pm 45°$) (Figure 1.18).

Figure 1.18A and B shows the omnidirectional wheel as it looks from the bottom (via a glass floor). The force F produced by the rotation of the wheel acts on the ground via the roller that has contact with the ground (which is assumed sufficiently flat without irregularities). At this roller, the force is decomposed in a force F_1 parallel to the roller axis and a force F_2 perpendicular to the roller axis. The force perpendicular to the roller axis produces a small roller rotation (speed v_r), but the force parallel to the roller axis exerts a force on the wheel and thereby on the vehicle resulting in the hub speed v_h. The actual velocity v_t of the vehicle is the combination of v_h and v_r. Figure 1.18C shows a practical mecanum wheel.

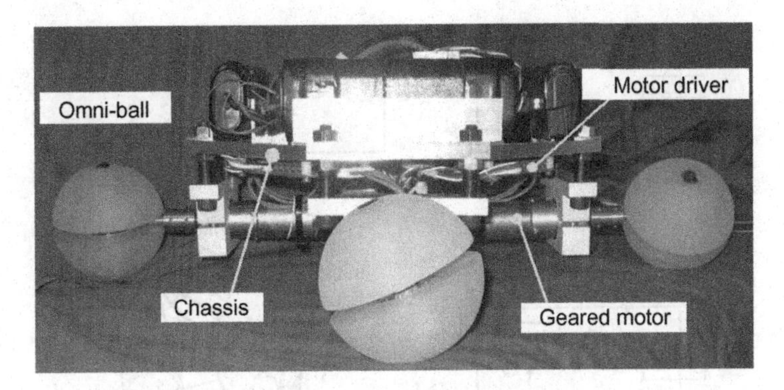

Figure 1.19 A practical implementation of a ball wheel.
Source: http://www-hh.mech.eng.osaka-u.ac.jp/robotics/Omni-Ball_e.html.

The *ball (or spherical) wheel* places no direct constraints on the motion, that is, it is omnidirectional like castor or special universal and mecanum wheels. In other words, the rotational axis of the wheel can have any arbitrary direction. One way to achieve this is by using an active ring driven by a motor and gearbox to transmit power to the ball via rollers and friction, which is free to rotate in any direction instantaneously. Because of its difficult construction the ball wheel is very rarely used in practice. A type of ball wheel is shown in Figure 1.19.

1.3.2.2 Drive Types

The drives of WMRs are distinguished in:

- Differential drive
- Tricycle
- Omnidirectional
- Synchro drive
- Ackerman steering
- Skid steering

Differential drive: This drive consists of two fixed powered wheels mounted on the left and right side of the robot platform. The two wheels are independently driven. One or two passive castor wheels are used for balance and stability. Differential drive is the simplest mechanical drive since it does not need rotation of a driven axis. If the wheels rotate at the same speed, the robot moves straight forward or backward. If one wheel is running faster than the other, the robot follows a curved path along the arc of an instantaneous circle. If both wheels are rotating at the same velocity in opposite directions, the robot turns about the midpoint of the two driving wheels. The above locomotion modes are illustrated in Figure 1.20. Clearly, this type of WMR cannot turn on the spot.

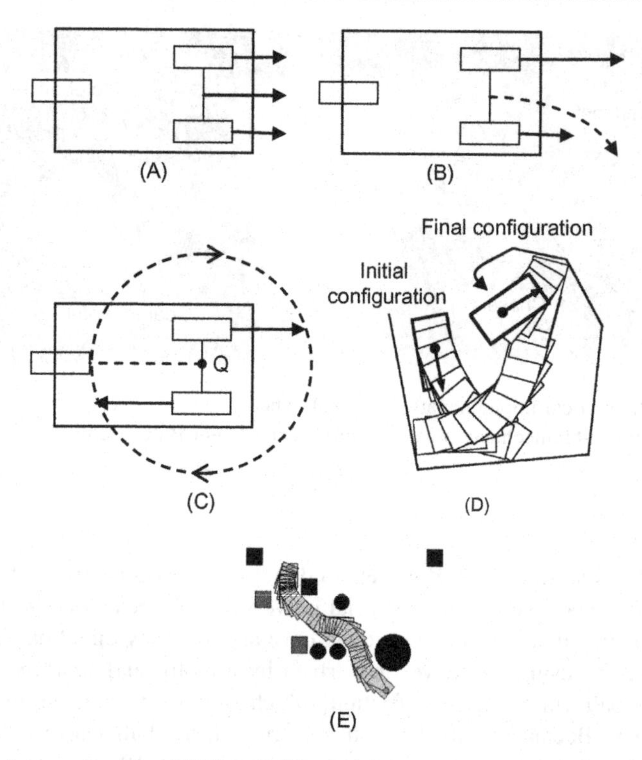

Figure 1.20 Locomotion possibilities of differential drive. (A) Straight path, (B) curved path, (C) circular path, (D) obstacle-free maneuvering to go from an initial to a final pose, and (E) maneuvering to go from an initial to a final pose while avoiding obstacles.
Source: Colored picture, courtesy of N. Katevas.

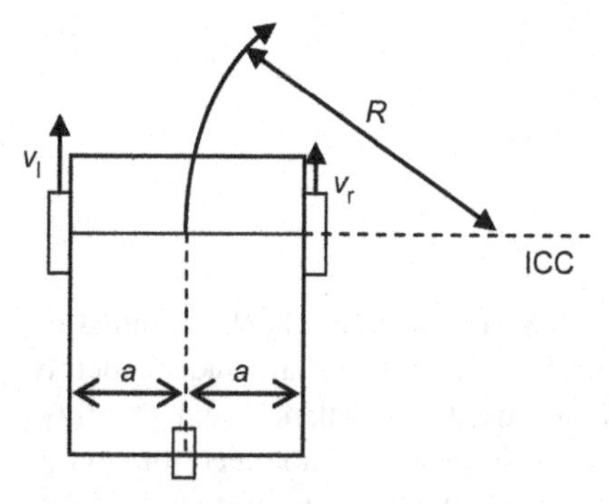

Figure 1.21 Calculation of the instantaneous radius R of WMR rotation.

The *instantaneous center of curvature* (ICC) of the WMR lies at the cross point of all axes of the wheels. ICC is the center of the circle with radius R depending on the speeds of the two wheels (Figure 1.21).

The radius R is determined by the relation: $(v_l - v_r)/2a = v_r/(R - a)$.

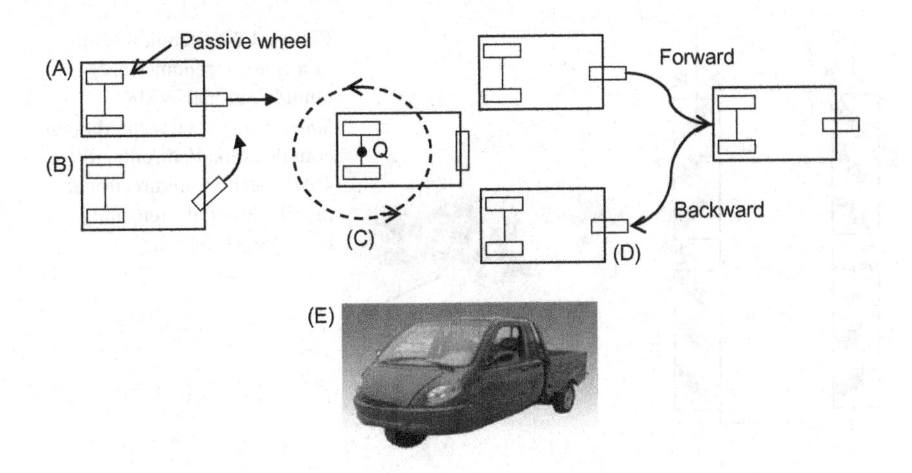

Figure 1.22 Tricycle WMR locomotion modes (A–D), (E) A tricycle example.
Source: http://www.asianproducts.com/product/A12391789884559774_p1240094409205655/cargo-tricycle(250cc).html.

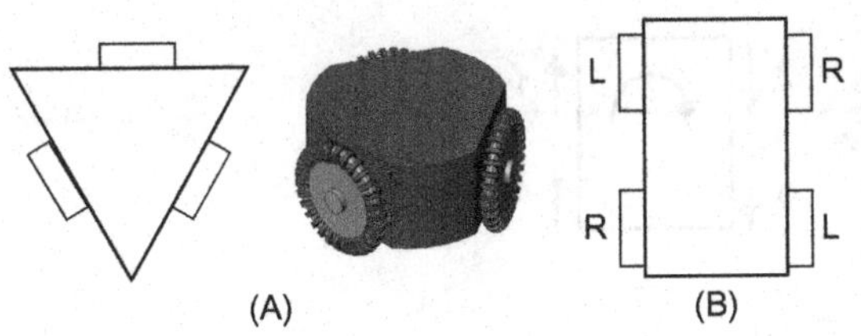

Figure 1.23 Omnidirectional WMRs. (A) Three-wheel case, (B) four-wheel case with roller angle different than 90° (typically $\alpha = \pm 45°$).
Source: Colored picture, courtesy of Jahobr.

Therefore:

$$R = a(v_l + v_r)/(v_l - v_r), \quad v_l \geq v_r \tag{1.1}$$

When $v_l = v_r$ we have $R = \infty$ (i.e., straight motion, and when $v_r = -v_l$ we have $R = 0$ (i.e., rotational motion).

Tricycle: This drive has a single wheel which is both driven (powered) and steered. For stability, two free-running (unpowered) fixed wheels in the back are used, in order to have always the three point contact required. The linear and angular velocities of the wheel are fully decoupled. For driving straight, the wheel is positioned in the middle position and driven at the desired speed (Figure 1.22A). When the front wheel is at an angle the vehicle follows a curved path (Figure 1.22B). If the front wheel is positioned at 90°, the robot will rotate following a circular path the center of which is in the middle point of the rear wheels and not in the robot's geometric center (Figure 1.22C). This means that this WMR cannot turn on the spot. Nonholonomic WMRs (like differential drive or tricycle

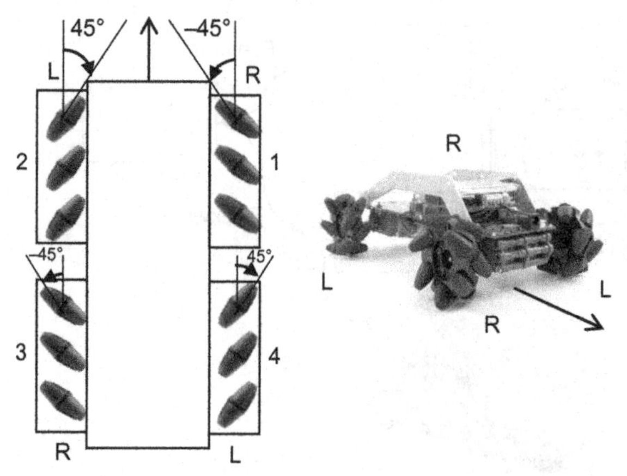

Figure 1.24 Standard setup of a four-mecanum-wheel omnidirectional WMR.
Source: http://www.interhopen.com/download/pdf/pdfs_id/465/InTech-Omnidirectional_mobile_robot_design_and_implementation.pdf.

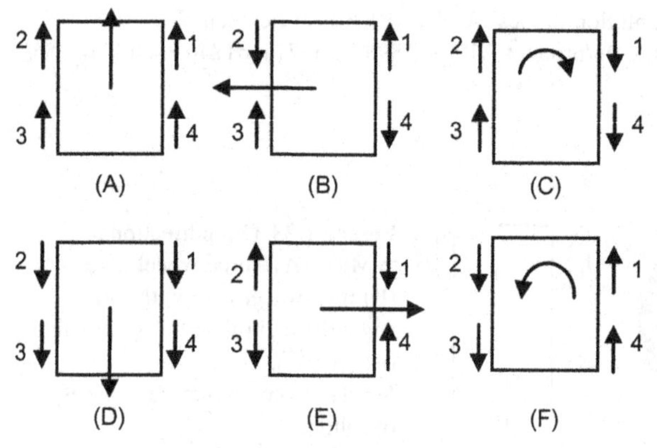

Figure 1.25 Six basic locomotion modes of a WMR with four mecanum wheels.

vehicles) cannot perform parallel parking directly but by a number of maneuvers with forward and backward movements as shown in Figure 1.22D.

Omnidirectional: This drive can be obtained using three, four, or more omnidirectional wheels as shown in Figure 1.23. WMRs with three wheels use universal wheels that have a 90° roller angle (Figure 1.17) as shown in Figure 1.23A. Omnidirectional WMRs with four wheels use mecanum wheels (Figure 1.18) in the configuration shown in Figure 1.23B.

From the four wheels in Figure 1.24, two are called *left-handed* (L) wheels and the other two *right-handed* (R) wheels. The left-handed wheels have a roller angle $\alpha = 45°$ and the right-handed ones an angle $\alpha = -45°$. Therefore, the four-wheel omnidirectional WMRs have the typical structure shown in Figure 1.24.

Figure 1.25 shows six basic motions of a four-wheel omnidirectional robot, namely (A) forward motion, (B) left sliding, (C) clockwise turning (on the spot), (D) backward motion, (E) right sliding, and (F) anticlockwise turning. The arrows on the left and the right side of the vehicle show the motion direction of the corresponding wheels.

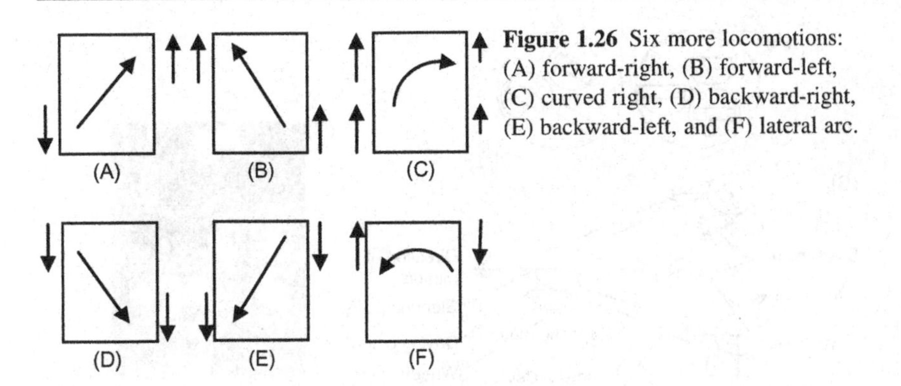

Figure 1.26 Six more locomotions: (A) forward-right, (B) forward-left, (C) curved right, (D) backward-right, (E) backward-left, and (F) lateral arc.

The arrows on the vehicle platform show the respective directions of the WMR motion, that is, for forward vehicle direction all wheels must move forward (Figure 1.25A), for left sliding wheels 1, 3 should move forward, and wheels 2, 4 backward, and so on. The locomotions shown in Figure 1.25 are obtained if all wheels move at the same speed. By varying the speed magnitude of the wheels one can realize WMR motion in any direction on the 2-D plane. A few cases are shown in Figure 1.26.

All locomotions of Figures 1.25 and 1.26 can be explained using the force or velocity diagrams of Figure 1.18A and B). For example, because of the symmetricity of left and right wheels (Figure 1.24), if all wheels are driven forward, there are four vectors pointing forward that are added up and four vectors pointing sideways, two to the left and two to the right, that cancel each other out. Thus, in overall the WMR moves forward. The L and R wheels may be interchanged (i.e., front wheels RL, back wheels LR). Again, with proper motion of the wheels all omnidirectional locomotion modes can be obtained with this setup too.

Synchro drive: This drive has three or more wheels that are mechanically coupled such that all of them rotate in the same direction at the same speed and pivot in unision about their own steering axes when perform a turn. This mechanical *steering synchronization* can be realized in several ways, for example, using a chain, a belt or gear drive. Actually, synchro drive is an extension of a single driven and steered wheel and so it still has only two DOF. But a synchro drive WMR is nearly a holonomous vehicle because it can move in any desired direction. However, it cannot drive and rotate at the same time. To change its driving from forward to sideways this WMR must stop and realign its wheels. Figure 1.27A shows pictorially how a three-wheel WMR with synchro drive is moving and rotated.

Chain- or belt-based synchro drive presents lower steering accuracy and alignment. This problem does not occur if a gear drive is used. Actually, two independent motor-drive subsystems (operating with chain, belt, and gear) must be used; one for the steering and one for the driving shaft (Figure 1.27B).

Ackerman steering: This is the standard steering used in automobiles. It consists of two combined driven rear wheels and two combined steered front wheels. An Ackerman-steered vehicle can move straight (because the rear wheels are driven by

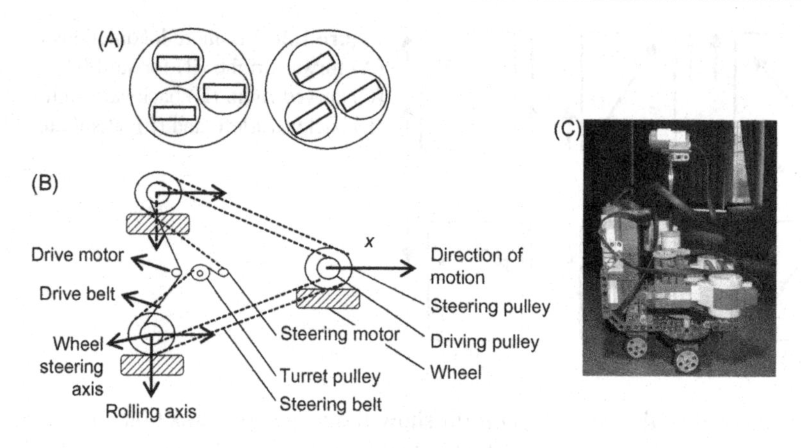

Figure 1.27 (A) Illustration of synchro drive WMR motion, (B) illustration of the two independent belt subsystems, and (C) a synchro drive WMR example.
Source: http://members.efn.org/~kirbyf/05syncro/1.jpg.

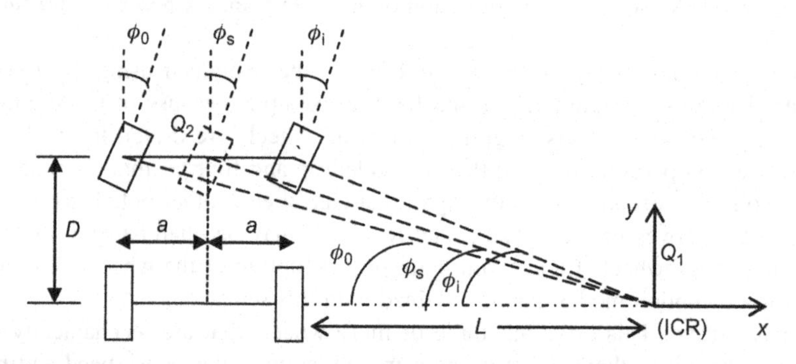

Figure 1.28 The rotation axes of all wheels intersect at the same point ICR.

a common axis), but cannot turn on the spot (it requires a certain minimum radius). The rear driving wheels experience slippage (when moving in curves).

Ackerman steering is designed so as to ensure that at turns the wheels of all axes have a common cross point (instantaneous center of rotation (ICR)), in order to avoid geometrically caused wheel slippage.

From Figure 1.28, we find the following relations:

$$ctg\phi_s = (a + L)/D, \quad ctg\phi_o = (2a + L)/D, \quad ctg\phi_i = L/D$$

which by elimination of L give:

$$ctg\phi_s = \frac{a}{D} + ctg\phi_i$$

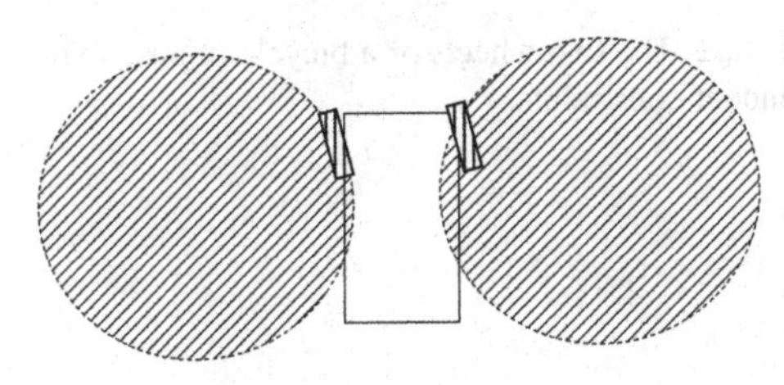

Figure 1.29 The shaded areas are inaccessible to the Ackerman-steered robot.

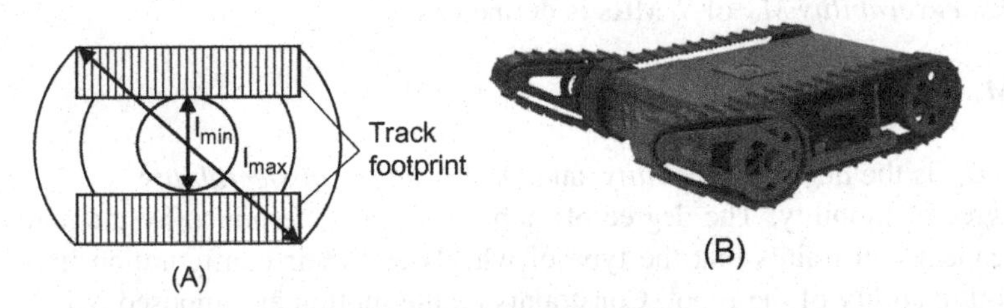

Figure 1.30 (A) Illustration of the effective contact point and (B) a typical tracked platform. *Source*: http://www.robotshop.com/Dr-robot-jaguar-tracked-mobile-platform-chassis-motors-3.html.

or

$$ctg\phi_s = ctg\phi_o - \frac{a}{D} \tag{1.2}$$

where ϕ_s is the vehicle's actual steering angle and ϕ_o, ϕ_i are the steering angles of the outer and inner wheel, respectively.

Figure 1.29 illustrates the constraint in the motion of an Ackerman-steered vehicle. There is a circular area on the left and on the right of its current position which is inaccessible to the vehicle. This is due to that the robot cannot turn (right or left) following a path with radius smaller than a minimum. Therefore, for parallel parking a considerable maneuvering is required.

Skid steering: This is a special implementation of differential drive and realized in track form on bulldozers and harmored vehicles. Its difference from a differential drive vehicle is the increased maneuverability in uneven terrains, and the higher friction which is due to its tracks and the multiple contact points with the terrain (rough or even). Figure 1.30A illustrates that the effective point contact for such a skid-steer robot is roughly constrained on either side by a rectangular uncertainty area which corresponds to the track footprint.

One can see from the concentric circles that for the vehicle to turn, a considerable slippage is needed. Figure 1.30B shows a typical tracked robotic platform that can carry a manipulator or special exploration equipment.

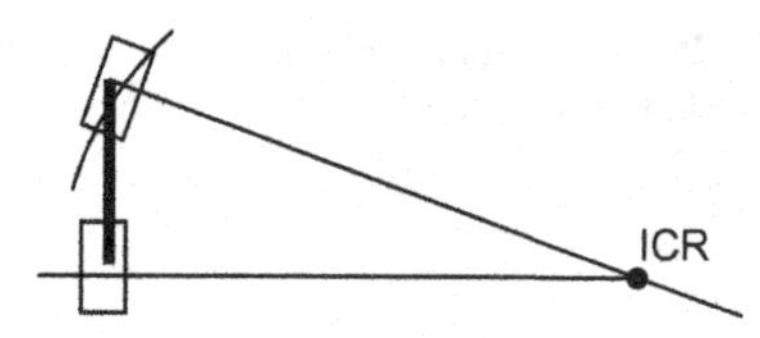

Figure 1.31 The two wheels of a bicycle impose two independent constraints.

1.3.2.3 WMR Maneuverability

The *maneuverability* M_w of WMRs is defined as

$$M_w = D_m + D_s \quad (1.3)$$

where D_m is the *degree of mobility*, and D_s the *degree of steerability*.

Degree of mobility: The degree of mobility D_m is determined by the number of independent constraints that the type of wheels and their configuration impose on the motion ability of the robot. Constraints on the motion are imposed only by conventional wheels (fixed or steered). Omnidirectional wheels do not impose any constraint on the robot's mobility. The best way to see the independent kinematic constraints of a WMR is by studying the geometric properties of the robot through the ICC or ICR. For example, a single conventional wheel cannot move laterally, that is, along the line determined by its axis of rotation. This line is called the *zero motion line* of the wheel. This means that the wheel can only move on an instantaneous circle of radius R with its center lying on the zero motion line. A bicycle has two wheels: the steered front wheel and the fixed wheel on the rear (Figure 1.31).

Each wheel introduces a separate (independent) zero motion line. The two lines intersect at the ICR. In the case of a differential drive WMR (Figure 1.21) the zero motion lines of the two (common axis) wheels coincide and so they are not independent. This means that there is only one independent kinematic constraint. Any point on the common zero motion line can be an ICR. In the Ackerman steering, the WMR has four conventional wheels but two independent kinematic constraints (Figure 1.28). The two rear wheels impose a single constraint (as in the differential drive), and also the two front steered wheels impose a second single kinematic constraint, because they cross on an ICR lying on the zero motion line determined by the common axis rear wheels. The maximum degree of mobility D_m is 3, which is true when no kinematic constraints are imposed. This is the case when all wheels of the WMR are omnidirectional. In general, the degree of mobility is equal to:

$$D_m = 3 - N_c \quad (1.4)$$

where N_c is the number of independent constraints.

Degree of steerability: The degree of steerability D_s depends on the number of independently controllable steering parameters and lies in the interval $0 \le D_s \le 2$. If no steerable wheels exist we have $D_s = 0$. The case $D_s = 2$ holds only if the robot has no fixed standard wheels. In this case we can have a platform with two separate

Table 1.1 Degree of Mobility and Steerability (D_m, D_s) of Typical WMRs

Configuration	D_m	D_s	M_w	Notation
Bicycle	1	1	2	(1,1)
Differential drive	2	0	2	(2,0)
Synchro drive	1	1	2	(1,1)
Tricycle	1	1	2	(1,1)
Ackerman steer	1	1	2	(1,1)
Two-steer	1	2	3	(1,2)
Omni-steer	2	1	3	(2,1)
Omnidirectional	3	0	3	(3,0)

steerable conventional wheels (as e.g., in a 2-steer bicycle or 3-wheeled 2-steer WMR). Actually, $D_s = 2$ means that the WMR can place its ICR at any point of the plane. The most common case is $D_s = 1$ which is obtained when the robot configuration has one or more steerable conventional wheels. A steered conventional wheel can decrease the robot's mobility, but also can increase the steerability. In fact, although an instantaneous orientation of the wheel imposes a kinematic constraint its capability to change this orientation may allow additional trajectories. The maneuverability (M_w), the degree of mobility (D_m) and the degree of steerability (D_s) of some typical WMR configurations are shown in Table 1.1.

Two other characteristic parameters of WMRs are the "*degrees of freedom*" and the "*differential degrees of freedom*" (DDOF), which satisfy the relation:

$$\text{DDOF} \leq M_w \leq \text{DOF}$$

The DDOF is equal to D_m and represents the number of independent velocities that can be achieved. DOF represents the ability of a WMR to achieve various poses (x, y, ϕ) in its environment (work space).

A bicycle can achieve any pose (x, y, ϕ) on the plane by some maneuver and so it has DOF = 3, but its DDOF is DDOF = D_m = 1. An omnirobot, with three omnidirectional wheels, has $D_m = 3$, that is, DDOF = 3, and also DOF = 3. Similarly, a tricycle has DDOF = D_m = 1 and DOF = 3, because it can reach any (x, y, ϕ) by appropriate maneuvering.

A list of possible wheel drive configurations is as follows [7]:

- One traction wheel in the back, one steering wheel in the front (bicycle, motor cycle)
- Two-wheel differential drive with the center of mass, below the wheels' axis (a balance controller is needed)
- Two-wheel differential drive centered with an omni wheel for stability (Nomad Scout robot)
- Three-wheel differential drive with an unpowered omni wheel, rear or front driven (typical indoor WMRs)
- Two connected powered wheels (differential) in the back, one free turning wheel in front
- One steered and driven wheel in front two free wheels in the back (e.g., Neptune)

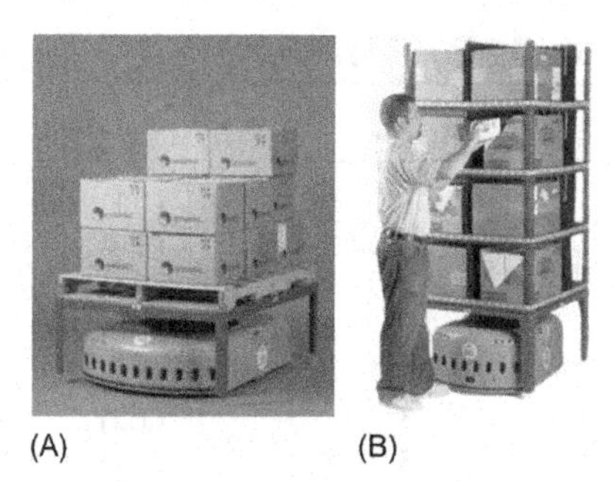

(A) (B)

Figure 1.32 KIVA autonomous mobile robots. (A) Pallet and case handling, (B) Order fulfillment.
Source: (A) www.kivasystems.com/solutions/picking/pick-from-pallets.
(B) www.kivasystems.com/about-us-the-kiva-approach.

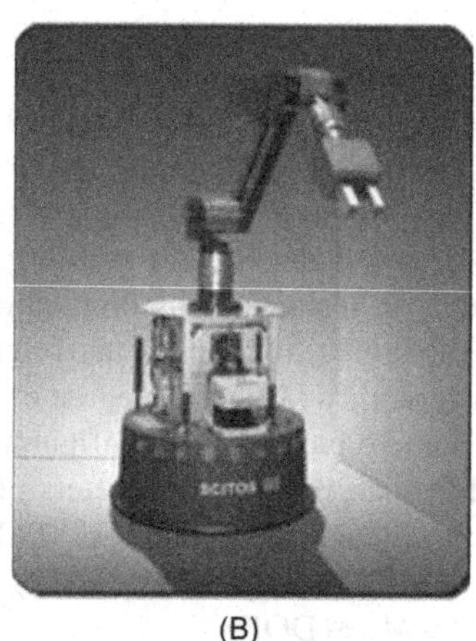

(A) (B)

Figure 1.33 (A) SCITOS mobile general platform (SCITOS G5). (B) Robotic manipulator mounted on SCITOS G5.

- Three omnidirectional wheels (universal)
- Three synchronous powered and steered wheels (synchronous drive)
- Car-like WMR (rear-wheel driven)
- Car-like WMR (front-wheel driven)
- Four-wheel drive, four-wheel steering (Hyperion)
- Differential wheel drive in the back two omni wheels in front
- Four Swedish omnidirectional wheels ($\alpha \neq 90°$) (Uranus)
- Four motorized and steered castor wheels (Nomad XR4000)
- Multi wheel walking drive (rovers, climbing robots).

A small set of modern WMRs falling in the above categories are shown in Figures 1.32–1.42.

Figure 1.42 shows the components of the AMiR swarm robot involving the main board, communication module, kinematic design, power module, and sensory system.

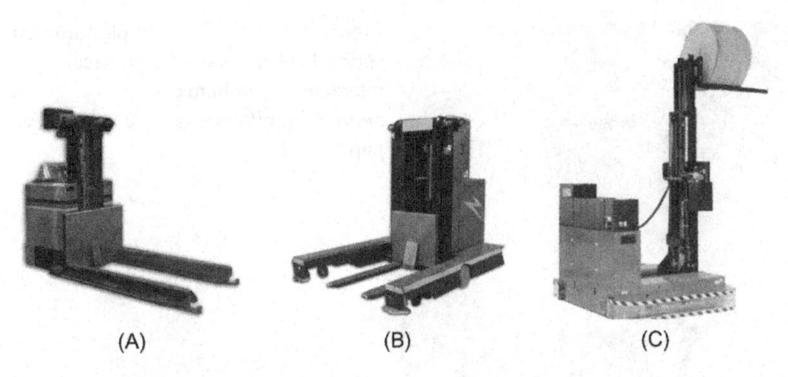

Figure 1.34 CORECON automated guided mobile robots (AGVs). (A) Horizontal roll handling, (B) low lift rear loader, and (C) high lift side loader.
Source: www.coreconagvs.com/products.

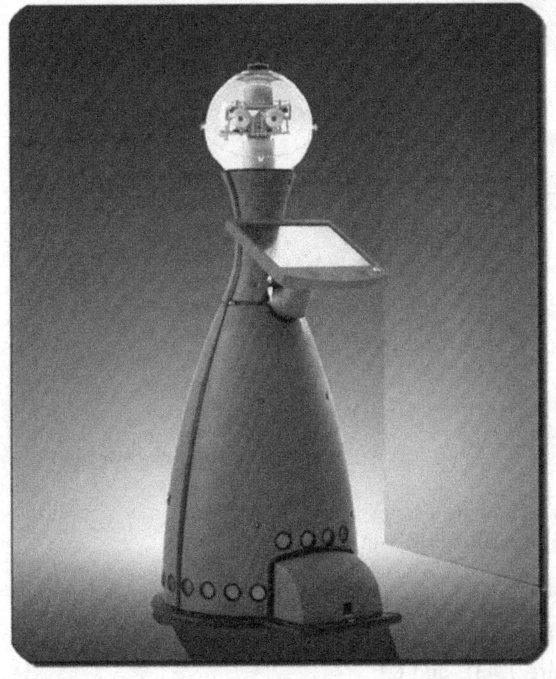

Figure 1.35 SCITOS mobile robot guide (it can provide users valuable information via speech or touch screen at any location).
Source: http://www.expo21xx.com/automation21xx/13582_st3_mobile-robots/default.htm.

Figure 1.43 shows the miniature WMR "Khepera" used for research for more than 15 years. It was developed at the LAM laboratory in EPFL (Lausanne, Switzerland). The initial version of Khepera (Figure 1.43A) is 55 mm diameter and 30 mm high robot. It has appropriate sensors and actuators to ensure that it can be

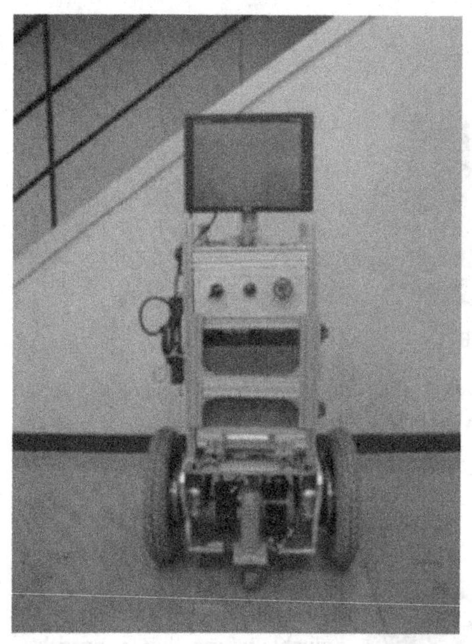

Figure 1.36 Mobile robot platform with spring-loaded castors for physical interaction with humans.
Source: http://robot.kaist.ac.kr/paper/view.php?n = 318.

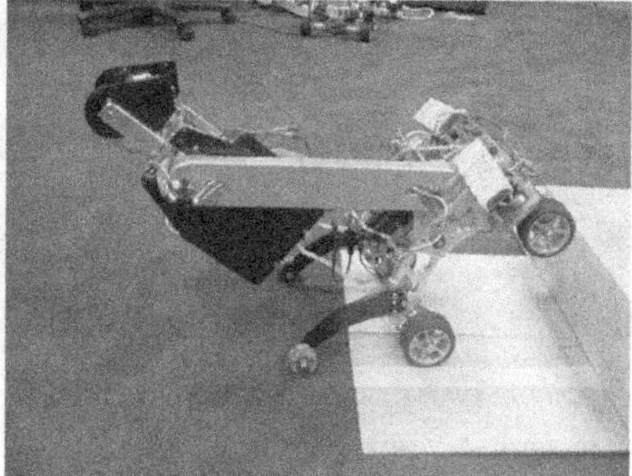

Figure 1.37 The CMU Rover 1 robot climbing a stair.
Source: http://www.cs.cmu.edu/~myrover/Rover1/robot.htm.

programmed to perform a large repertory of tasks. Khepera can run autonomously or tethered to a host computer. The newer versions of Khepera (K-II Version and K-III Version) are shown in Figure 1.43B and C).

Khepera is able to move on a table top as well as on a room floor for performing real-world swarm robotics. To enable fast development of portable applications, Khepera III supports a standard Linux operating system.

Finally, Figure 1.44 shows the famous mecanum-wheeled omnidirectional WMR "Uranus."

Figure 1.38 The Nomad robot (CACS Louisiana University).
Source: http://www.cacs.louisiana.edu/~sxg3148/nomad_pics/Nomad_robot_jpg.

Figure 1.39 The NASA nBot (two-wheel balancing robot).
Source: http://www.geology.smu.edu/~dpa-www/robot/nbot/nobot2/nb12.jpg.

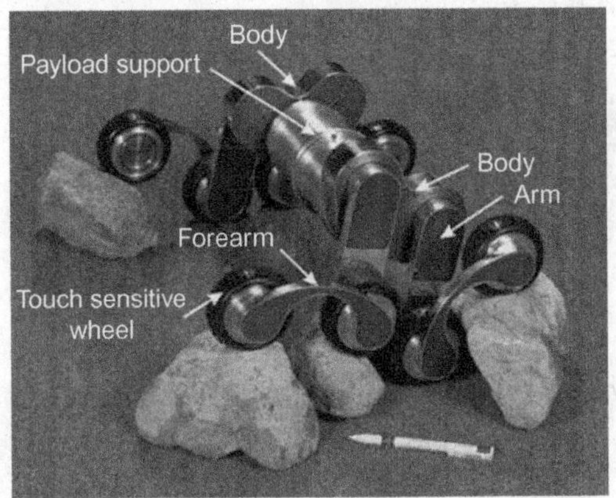

Figure 1.40 The EPFL wheeled climbing "Octopus Robot" (43 cm × 42 cm × 23 cm).
Source: http://www-robot.mes.titech.ac.jp/robot/walking/rollerwalker/roller_e.html.

Figure 1.41 Robotic swarms showing a "collective behavior."
Source: http://www.humansinvent.com/#!/8878/swarm-robots-the-droid-workforce-of-the-future/.

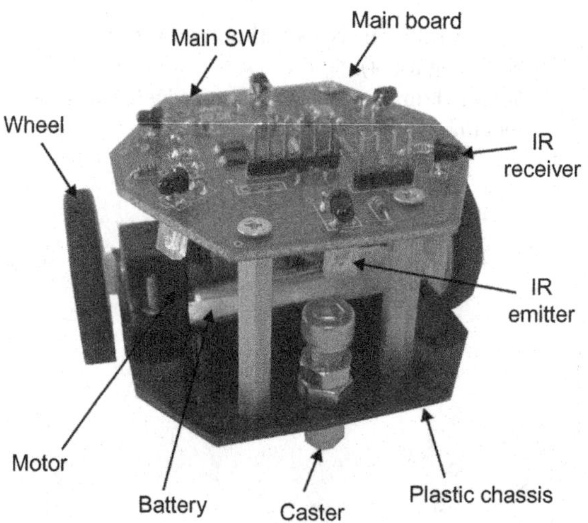

Figure 1.42 The AMiR swarm robot.
Source: http://www.swarmrobotic.com/Robot.htm.

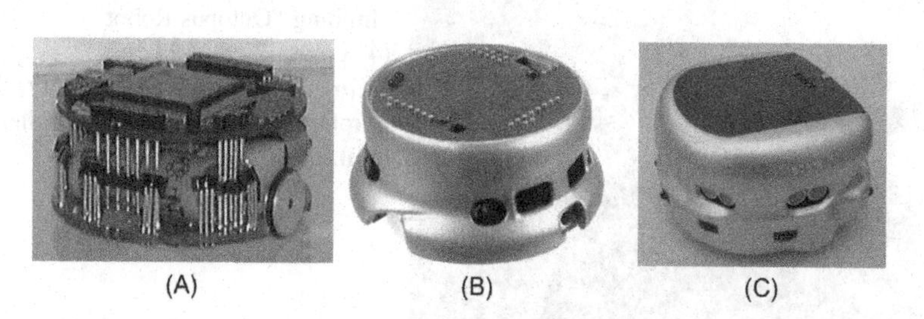

Figure 1.43 The evolution of Khepera WMR. (A) Original version, (B) Version K-II, (C) Version K-III.
Source: http://mobotica.blogspot.com/2011/08khepera.html; www.k-team.com/mobile-robotics-products/KheperaII.

Figure 1.44 "Uranus" four-wheel omnidirectional robot.
Source: http://www.cs.cmu.edu/afs/cs/user/gwp/www/robots/Uranus.jpg.

References

[1] Freedman J. Robots through history: robotics. New York, NY: Rosen Central; 2011.

[2] Mayr O. The origins of feedback control. Cambridge, MA: MIT Press; 1970.

[3] Lazos C. Engineering and technology in ancient Greece. Athens: Aeolos Editions; 1993.

[4] Campion G, Bastin G, D'Andréa-Novel B. Structural properties and classification of kinematic and dynamic models of wheeled mobile robots. IEEE Trans Rob Autom 1996;12(1):47–62.

[5] Floreano D, Zufferey J-C. Robots mobiles. EPFL Course-Mobile Robots. <www.cs.cmu.edu/~gwp/robots/Uranus.html>.

[6] Bekey G. Autonomous robots. Cambridge, MA: MIT Press; 2005.

[7] Siegwart R, Nourbakhsh I. Autonomous mobile robots. Cambridge, MA: MIT Press; 2005.

[8] Bräunl T. Embedded robotics: mobile robot design and applications with embedded systems. Berlin: Springer; 2006. <http://newplans.net/RDB>.

[9] Salih J, Rizon M, Yacacob S, Adom A, Mamat M. Designing omni-directional mobile robot with mecanum wheel. Am J Appl Sci 2006;3(5):1831–5.

[10] West M, Asada H. Design of ball wheel mechanisms for omnidirectional vehicles with full mobility and invariant kinematics. J Mech Des 1997;119:153–7.

[11] Holland J-M. Rethinking robot mobility. Rob Age 1988;7(1):26–30.

[12] Duro JR, Santos J, Grana M. Biologically inspired robot behavior engineering. Berlin/Heidelberg: Springer; 2002.

[13] Katevas N, editor. Mobile robotics in healthcare. Amsterdam: IOS Press; 2001.

[14] Tzafestas SG, editor. Autonomous mobile robots in health care services. J Intell Rob Syst 1998;22(3–4):177–350 [special issue].

[15] Fong T, Nourbakhsh IR, Dautenhahn K. A survey of socially interactive robots. Rob Auton Syst 2003;42(3–4):143–66.

References

2 Mobile Robot Kinematics

2.1 Introduction

Robot kinematics deals with the configuration of robots in their workspace, the relations between their geometric parameters, and the constraints imposed in their trajectories. The kinematic equations depend on the geometrical structure of the robot. For example, a fixed robot can have a Cartesian, cylindrical, spherical, or articulated structure, and a mobile robot may have one two, three, or more wheels with or without constraints in their motion [1−20]. The study of kinematics is a fundamental prerequisite for the study of dynamics, the stability features, and the control of the robot. The development of new and specialized robotic kinematic structures is still a topic of ongoing research, toward the end of constructing robots that can perform more sophisticated and complex tasks in industrial and societal applications [1−20].

The objectives of this chapter are as follows:

- To present the fundamental analytical concepts required for the study of mobile robot kinematics
- To present the kinematic models of nonholonomic mobile robots (unicycle, differential drive, tricycle, and car-like wheeled mobile robots (WMRs))
- To present the kinematic models of 3-wheel, 4-wheel, and multiwheel omnidirectional WMRs.

2.2 Background Concepts

As a preparation for the study of mobile robot kinematics the following background concepts are presented:

- Direct and inverse robot kinematics
- Homogeneous transformations
- Nonholonomic constraints

2.2.1 Direct and Inverse Robot Kinematics

Consider a fixed or mobile robot with generalized coordinates $q_1, q_2, \ldots, q_n$ in the joint (or actuation) space and $x_1, x_2, \ldots, x_m$ in the task space. Define the vectors:

$$\mathbf{q} = \begin{bmatrix} q_1 \\ q_2 \\ \vdots \\ q_n \end{bmatrix}, \quad \mathbf{p} = \begin{bmatrix} x_1 \\ x_2 \\ \vdots \\ x_m \end{bmatrix} \tag{2.1}$$

Introduction to Mobile Robot Control. DOI: http://dx.doi.org/10.1016/B978-0-12-417049-0.00002-X

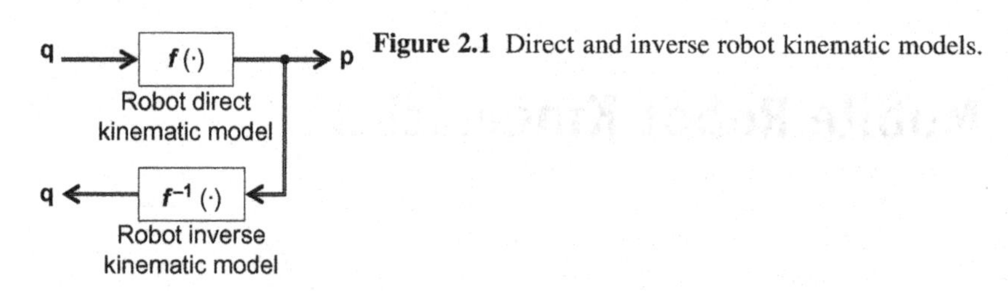

Figure 2.1 Direct and inverse robot kinematic models.

The problem of determining **p** knowing **q** is called the *direct kinematics* problem. In general $\mathbf{p} \in R^m$ and $\mathbf{q} \in R^n$ (R^n denotes the n-dimensional Euclidean space) are related by a nonlinear function (model) as:

$$\mathbf{p} = \mathbf{f}(\mathbf{q}), \quad \mathbf{f}(\mathbf{q}) = \begin{bmatrix} f_1(\mathbf{q}) \\ f_2(\mathbf{q}) \\ \vdots \\ f_m(\mathbf{q}) \end{bmatrix} \tag{2.2}$$

The problem of solving Eq. (2.2), that is of finding **q** from **p**, is called the *inverse kinematic* problem expressed by:

$$\mathbf{q} = \mathbf{f}^{-1}(\mathbf{p}) \tag{2.3}$$

The direct and inverse kinematic problems are pictorially shown in Figure 2.1.

In general, kinematics is the branch of mechanics that investigates the motion of material bodies without referring to their masses/moments of inertia and the forces/torques that produce the motion. Clearly, the kinematic equations depend on the fixed geometry of the robot in the fixed world coordinate frame.

To get these motions we must tune appropriately the motions of the joint variables, expressed by the velocities $\dot{\mathbf{q}} = [\dot{q}_1, \dot{q}_2, \ldots, \dot{q}_n]^T$. We therefore need to find the differential relation of **q** and **p**. This is called *direct differential kinematics* and is expressed by:

$$\mathbf{dp} = \mathbf{J}\mathbf{dq} \tag{2.4}$$

where

$$\mathbf{dq} = \begin{bmatrix} dq_1 \\ \vdots \\ dq_n \end{bmatrix}, \quad \mathbf{dp} = \begin{bmatrix} dx_1 \\ \vdots \\ dx_m \end{bmatrix}$$

and the $m \times n$ matrix:

$$\mathbf{J} = \begin{bmatrix} \frac{\partial x_1}{\partial q_1} & \frac{\partial x_1}{\partial q_2} & \cdots & \frac{\partial x_1}{\partial q_n} \\ \cdots & \cdots & \cdots & \cdots \\ \cdots & \cdots & \cdots & \cdots \\ \frac{\partial x_m}{\partial q_1} & \frac{\partial x_m}{\partial q_2} & \cdots & \frac{\partial x_m}{\partial q_n} \end{bmatrix} = [J_{ij}] \tag{2.5}$$

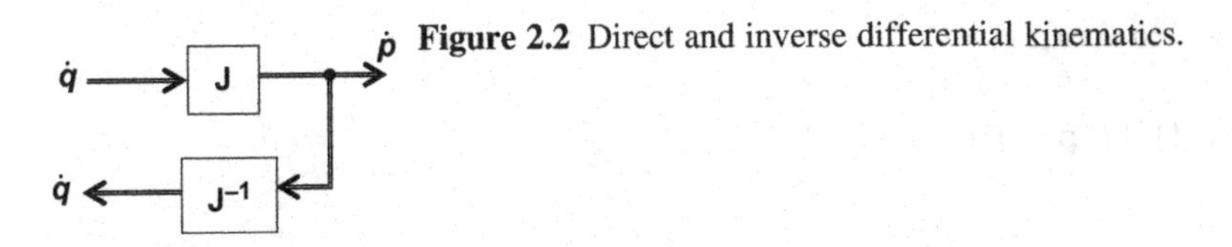

Figure 2.2 Direct and inverse differential kinematics.

with (i,j) element $J_{ij} = \partial x_i / \partial q_j$ is called the *Jacobian matrix* of the robot.[1]

For each configuration $q_1, q_2, \ldots, q_n$ of the robot, the Jacobian matrix represents the relation of the displacements of the joints with the displacement of the position of the robot in the task space.

Let $\dot{\mathbf{q}} = [\dot{q}_1, \ldots, \dot{q}_n]^T$ and $\dot{\mathbf{p}} = [\dot{x}_1, \dot{x}_2, \ldots, \dot{x}_m]^T$ be the velocities in the joint and task spaces.

Then, dividing Eq. (2.4) by dt we get formally:

$$\frac{\mathrm{d}\mathbf{p}}{\mathrm{d}t} = \mathbf{J}\frac{\mathrm{d}\mathbf{q}}{\mathrm{d}t} \text{ or } \dot{\mathbf{p}} = \mathbf{J}\dot{\mathbf{q}} \tag{2.6}$$

Under the assumption that $m = n$ ($\mathbf{J}$ square) and that the inverse Jacobian matrix $\mathbf{J}^{-1}$ exists (i.e., its determinant is not zero: det $\mathbf{J} \neq 0$), from Eq. (2.6) we get:

$$\dot{\mathbf{q}} = \mathbf{J}^{-1}\dot{\mathbf{p}} \tag{2.7}$$

This is the inverse differential kinematics equation, and is illustrated in Figure 2.2.

If $m \neq n$, then we have two cases:

Case 1 There are more equations than unknowns ($m > n$), that is, $\dot{\mathbf{q}}$ is *overspecified.* In this case $\mathbf{J}^{-1}$ in Eq. (2.7) is replaced by the generalized inverse $\mathbf{J}^{\dagger}$ given by:

$$\mathbf{J}^{\dagger} = (\mathbf{J}^T\mathbf{J})^{-1}\mathbf{J}^T \tag{2.8a}$$

under the condition that $\mathbf{J}$ is full rank (i.e., rank $\mathbf{J} = \min\ (m.n) = n$) so as $\mathbf{J}^T\mathbf{J}$ is invertible. The expression (2.8a) of $\mathbf{J}^{\dagger}$ follows by minimizing the squared norm of the difference $\dot{\mathbf{p}} - \mathbf{J}\dot{\mathbf{q}}$, that is of the function:

$$V = ||\dot{\mathbf{p}} - \mathbf{J}\dot{\mathbf{q}}||^2 = (\dot{\mathbf{p}} - \mathbf{J}\dot{\mathbf{q}})^T(\dot{\mathbf{p}} - \mathbf{J}\dot{\mathbf{q}})$$

with respect to $\dot{\mathbf{q}}$. The optimality condition is:

$$\partial V / \partial \dot{\mathbf{q}} = -2(\dot{\mathbf{p}} - \mathbf{J}\dot{\mathbf{q}})^T\mathbf{J} = \mathbf{0}$$

[1] It is remarked that in many works the Jacobian matrix is defined as the transpose of that defined in Eq. (2.5).

which, if solved for $\dot{\mathbf{q}}$, gives:

$$\dot{\mathbf{q}} = (\mathbf{J}^T\mathbf{J})^{-1}\mathbf{J}^T\dot{\mathbf{p}} = \mathbf{J}^\dagger\dot{\mathbf{p}}$$

Case 2 There are less equations than unknowns ($m < n$), that is, $\dot{\mathbf{q}}$ is *underspecified* and many choices of $\dot{\mathbf{q}}$ lead to the same $\dot{\mathbf{p}}$. In this case we select $\dot{\mathbf{q}}$ with the minimum norm, that is, we solve the constrained minimization problem:

$$\min||\dot{\mathbf{q}}||^2 \text{ subject to } \dot{\mathbf{p}} - \mathbf{J}\dot{\mathbf{q}} = \mathbf{0}, \quad \dot{\mathbf{q}} \in R^n$$

Introducing the Lagrange multiplier vector $\boldsymbol{\lambda}$, we get the augmented (unconstrained) Lagrangian minimization problem:

$$\min_{\dot{\mathbf{q}},\boldsymbol{\lambda}} L(\dot{\mathbf{q}}, \boldsymbol{\lambda}), \quad L(\dot{\mathbf{q}}, \boldsymbol{\lambda}) = \dot{\mathbf{q}}^T\dot{\mathbf{q}} + \boldsymbol{\lambda}^T(\dot{\mathbf{p}} - \mathbf{J}\dot{\mathbf{q}})$$

The optimality conditions are:

$$\partial L/\partial\dot{\mathbf{q}} = 2\dot{\mathbf{q}} - \mathbf{J}^T\boldsymbol{\lambda} = \mathbf{0}, \quad \partial L/\partial\boldsymbol{\lambda} = \dot{\mathbf{p}} - \mathbf{J}\dot{\mathbf{q}} = \mathbf{0}$$

Solving the first for $\dot{\mathbf{q}}$ we get $\dot{\mathbf{q}} = (1/2)\mathbf{J}^T\boldsymbol{\lambda}$.
Introducing this result into $\partial L/\partial\boldsymbol{\lambda} = \mathbf{0}$ yields:

$$\boldsymbol{\lambda} = 2(\mathbf{J}\mathbf{J}^T)^{-1}\dot{\mathbf{p}}$$

Therefore, finally we find:

$$\dot{\mathbf{q}} = \mathbf{J}^T(\mathbf{J}\mathbf{J}^T)^{-1}\dot{\mathbf{p}} = \mathbf{J}^\dagger\dot{\mathbf{p}}$$

where

$$\mathbf{J}^\dagger = \mathbf{J}^T(\mathbf{J}\mathbf{J}^T)^{-1} \tag{2.8b}$$

under the condition that rank $\mathbf{J} = m$ (i.e., $\mathbf{J}\mathbf{J}^T$ invertible). Therefore, when $m < n$ the generalized inverse Eq. (2.8b) should be used.

Formally, the generalized inverse $\mathbf{J}^\dagger$ of a $m \times n$ real matrix $\mathbf{J}$ is defined to be the unique $n \times m$ real matrix that satisfies the following four conditions:

$$\mathbf{J}\mathbf{J}^\dagger\mathbf{J} = \mathbf{J}, \ \mathbf{J}^\dagger\mathbf{J}\mathbf{J}^\dagger = \mathbf{J}^\dagger$$

$$(\mathbf{J}\mathbf{J}^\dagger)^T = \mathbf{J}\mathbf{J}^\dagger, (\mathbf{J}^\dagger\mathbf{J})^T = \mathbf{J}^\dagger\mathbf{J}$$

It follows that $\mathbf{J}^{\dagger}$ has the properties:

$$(\mathbf{J}^{\dagger})^{\dagger} = \mathbf{J}, (\mathbf{J}^{T})^{\dagger} = (\mathbf{J}^{\dagger})^{T}, (\mathbf{J}\mathbf{J}^{T})^{\dagger} = (\mathbf{J}^{\dagger})^{T}\mathbf{J}^{\dagger}$$

All the above relations are useful when dealing with overspecified or underspecified linear algebraic systems (encountered, e.g., in underactuated or overactuated mechanical systems).

2.2.2 Homogeneous Transformations

The position and orientation of a solid body (e.g., a robotic link) with respect to the fixed world coordinate frame *Oxyz* (Figure 2.3) are given by a 4 × 4 transformation matrix **A**, called *homogeneous transformation*, of the type:

$$\mathbf{A} = \left[\begin{array}{c|c} \mathbf{R} & \mathbf{p} \\ \hline \mathbf{0} & 1 \end{array}\right] \tag{2.9}$$

where **p** is the position vector of the center of gravity O'(or some other fixed point of the link) with respect to *Oxyz*, and **R** is a 3 × 3 matrix defined as:

$$\mathbf{R} = [\mathbf{n} \ \vdots \ \mathbf{o} \ \vdots \ \mathbf{a}] \tag{2.10}$$

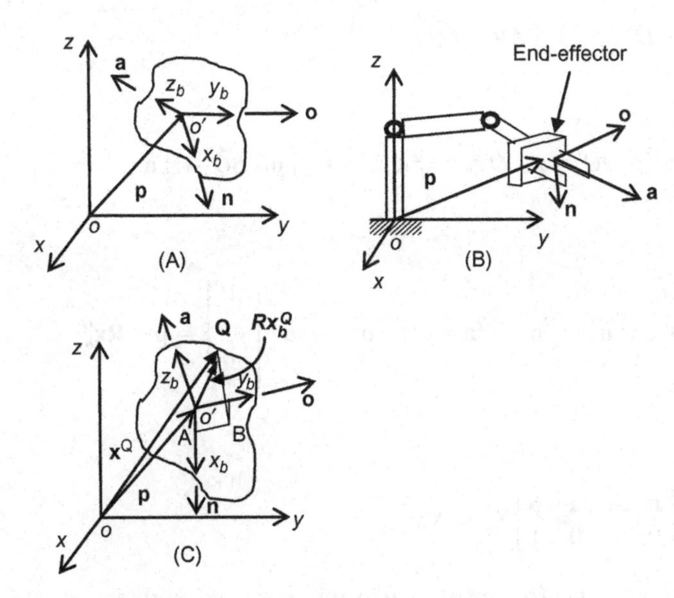

Figure 2.3 (A) Position and orientation of a solid body, (B) position and orientation of a robotic end-effector (**a** = approach vector, **n** = normal vector, **o** = orientation or sliding vector), and (C) position vectors of a point Q with respect to the frames $Oxyz$ and $O'x_by_bz_b$.

In Eq. (2.10), **n**, **o** and **a** are the unit vectors along the axes x_b, y_b, z_b of the local coordinate frame $O'x_by_bz_b$. The matrix **R** represents the rotation of $O'x_by_bz_b$ with respect to the reference (world) frame $Oxyz$. The columns **n**, **o**, and **a** of **R** are pairwise orthonormal, that is, $\mathbf{n}^T\mathbf{o}=0$, $\mathbf{o}^T\mathbf{a}=0$, $\mathbf{a}^T\mathbf{n}=0$, $|\mathbf{n}|=1$, $|\mathbf{o}|=1$, $|\mathbf{a}|=1$ where $\mathbf{b}^T$ denotes the transpose (row) vector of the column vector **b**, and $|\mathbf{b}|$ denotes the Euclidean norm of **b** ($|\mathbf{b}|=[b_x^2+b_y^2+b_z^2]^{1/2}$), with b_x, b_y, and b_z being the x, y, z components of **b**, respectively.

Thus the rotation matrix **R** is orthonormal, that is:

$$\mathbf{R}^{-1}=\mathbf{R}^T \tag{2.11}$$

To work with homogeneous matrices we use 4-dimensional vectors (called *homogeneous vectors*) of the type:

$$\mathbf{X}^Q=\begin{bmatrix} x^Q \\ y^Q \\ z^Q \\ \cdots \\ 1 \end{bmatrix}=\begin{bmatrix} \mathbf{x}^Q \\ \cdots \\ 1 \end{bmatrix}, \quad \mathbf{X}_b^Q=\begin{bmatrix} x_b^Q \\ y_b^Q \\ z_b^Q \\ \cdots \\ 1 \end{bmatrix}=\begin{bmatrix} \mathbf{x}_b^Q \\ \cdots \\ 1 \end{bmatrix} \tag{2.12}$$

Suppose that $\mathbf{X}_b^Q$ and $\mathbf{X}^Q$ are the homogeneous position vectors of a point Q in the coordinate frames $O'x_by_bz_b$ and $Oxyz$, respectively. Then, from Figure 2.3C we obtain the following vectorial equation:

$$\vec{OQ}=\vec{OO'}+\vec{O'A}+\vec{AB}+\vec{BQ}$$

where

$$\vec{OQ}=\mathbf{x}^Q, \quad \vec{OO'}=\mathbf{p}, \quad \vec{O'A}=x_b^Q\mathbf{n}, \quad \vec{AB}=y_b^Q\mathbf{o}, \quad \vec{BQ}=z_b^Q\mathbf{a}.$$

Thus:

$$\mathbf{x}^Q=\mathbf{p}+x_b^Q\mathbf{n}+y_b^Q\mathbf{o}+z_b^Q\mathbf{a}=\mathbf{p}+\begin{bmatrix}\mathbf{n} & \mathbf{o} & \mathbf{a}\end{bmatrix}\begin{bmatrix} x_b^Q \\ y_b^Q \\ z_b^Q \end{bmatrix}=\mathbf{p}+\mathbf{R}\mathbf{x}_b^Q \tag{2.13a}$$

or

$$\mathbf{X}^Q=\left[\begin{array}{ccc|c} \mathbf{n} & \mathbf{o} & \mathbf{a} & \mathbf{p} \\ \hline 0 & 0 & 0 & 1 \end{array}\right]\mathbf{X}_b^Q=\mathbf{A}\mathbf{X}_b^Q \tag{2.13b}$$

where **A** is given by Eqs. (2.9) and (2.10). Equation (2.13b) indicates that the homogeneous matrix **A** contains both the position and orientation of the local coordinate frame $O'x_by_bz_b$ with respect to the world coordinate frame $Oxyz$.

It is easy to verify that:

$$\mathbf{A}^{-1} = \begin{bmatrix} \mathbf{R}^T & -\mathbf{R}^T p \\ 0 & 1 \end{bmatrix} \tag{2.14}$$

Indeed, from Eq. (2.13a) we have: $\mathbf{x}_b^Q = -\mathbf{R}^{-1}\mathbf{p} + \mathbf{R}^{-1}\mathbf{x}^Q$, which by Eq. (2.11) gives:

$$\begin{bmatrix} \mathbf{x}_b^Q \\ 1 \end{bmatrix} = \begin{bmatrix} \mathbf{R}^T & -\mathbf{R}^T\mathbf{p} \\ 0 & 1 \end{bmatrix} \begin{bmatrix} \mathbf{x}^Q \\ 1 \end{bmatrix} = \mathbf{A}^{-1} \begin{bmatrix} \mathbf{x}^Q \\ 1 \end{bmatrix}$$

The columns **n**, **o**, and **a** of **R** consist of the direction cosines with respect to $Oxyz$. Thus the rotation matrices with respect to axes x, y, z which are represented as:

$$\mathbf{n}_x = \begin{bmatrix} 1 \\ 0 \\ 0 \end{bmatrix}, \quad \mathbf{o}_y = \begin{bmatrix} 0 \\ 1 \\ 0 \end{bmatrix}, \quad \mathbf{a}_z = \begin{bmatrix} 0 \\ 0 \\ 1 \end{bmatrix}$$

are given by:

$$\mathbf{R}_x(\phi_x) = \begin{bmatrix} 1 & 0 & 0 \\ 0 & \cos\phi_x & -\sin\phi_x \\ 0 & \sin\phi_x & \cos\phi_x \end{bmatrix} \tag{2.15a}$$

$$\mathbf{R}_y(\phi_y) = \begin{bmatrix} \cos\phi_y & 0 & -\sin\phi_y \\ 0 & 1 & 0 \\ \sin\phi_y & 0 & \cos\phi_y \end{bmatrix} \tag{2.15b}$$

$$\mathbf{R}_z(\phi_z) = \begin{bmatrix} \cos\phi_z & -\sin\phi_z & 0 \\ \sin\phi_z & \cos\phi_z & 0 \\ 0 & 0 & 1 \end{bmatrix} \tag{2.15c}$$

where ϕ_x, ϕ_y, and ϕ_z are the rotation angles with respect to x, y, and z, respectively.

In mobile robots moving on a horizontal plane, the robot is rotating only with respect to the vertical axis z, and so Eq. (2.15c) is used. Thus, for convenience, we drop the index z.

For better understanding, the upper left block of Eq. (2.15c) is obtained directly using the Oxy plane geometry shown in Figure 2.4.

Let a point $P(x_p, y_p)$ in the coordinate frame Oxy, which is rotated about the axis Oz by the angle ϕ. The coordinates of P in the frame $Ox'y'$ are x'_p and y'_p as shown in Figure 2.4. From this figure we see that:

$$\begin{aligned} x_P &= (OB) = (OD) - (BD) = x'_p \cos\phi - y'_p \sin\phi \\ y_P &= (OE) = (EZ) + (ZO) = x'_p \sin\phi + y'_p \cos\phi \end{aligned} \tag{2.16}$$

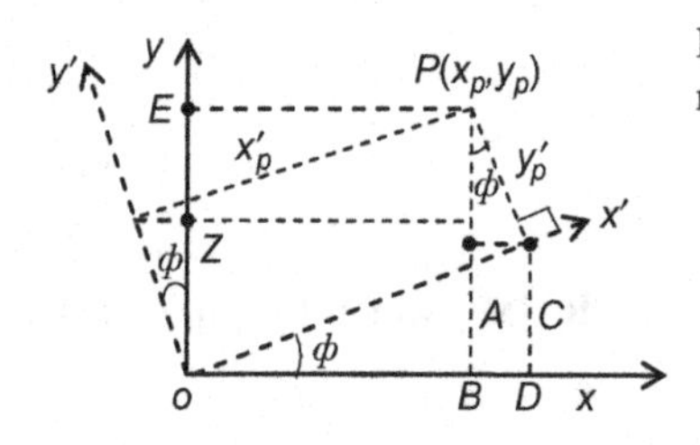

Figure 2.4 Direct trigonometric derivation of the rotation matrix with respect to axis z.

that is:

$$\begin{bmatrix} x_P \\ y_P \end{bmatrix} = \begin{bmatrix} \cos\phi & -\sin\phi \\ \sin\phi & \cos\phi \end{bmatrix} \begin{bmatrix} x'_P \\ y'_P \end{bmatrix} \tag{2.17}$$

Similarly, one can derive the respective 2×2 blocks $\mathbf{R}_x(\phi_x)$ and $\mathbf{R}_y(\phi_y)$ for the rotations about the x and y axes, respectively.

Given an open kinematic chain of n links, the homogeneous vector $\mathbf{X}^n$ of the local coordinate frame $O_n x_n y_n z_n$ of the nth link, expressed in the world coordinate frame $Oxyz$ can be found by successive application of Eq. (2.13b), that is, as:

$$\mathbf{X}^0 = \mathbf{A}_1^0 \mathbf{A}_2^1 \cdots \mathbf{A}_n^{n-1} \mathbf{X}^n \tag{2.18}$$

where $\mathbf{A}_i^{i-1}$ is the 4×4 homogeneous transformation matrix that leads from the coordinate frame of link i to that of link $i-1$. The matrices $\mathbf{A}_i^{i-1}$ can be computed by the so-called *Denavit–Hartenberg* (D–H) method (Section 10.2.1). The general relation (2.18) is of the form (2.2), and provides the robot Jacobian as indicated in Eq. (2.5).

2.2.3 Nonholonomic Constraints

A nonholonomic constraint (relation) is defined to be a constraint that contains time derivatives of generalized coordinates (variables) of a system and is not integrable. To understand what this means we first define a *holonomic constraint* as any constraint which can be expressed in the form:

$$F(\mathbf{q}, t) = 0 \tag{2.19}$$

where $\mathbf{q} = [q_1, q_2, \ldots, q_n]^T$ is the vector of generalized coordinates.

Now, suppose we have a constraint of the form:

$$f(\mathbf{q}, \dot{\mathbf{q}}, t) = 0 \tag{2.20}$$

If this constraint can be converted to the form:

$$F(\mathbf{q}, t) = 0 \tag{2.21}$$

we say that it is *integrable*. Therefore, although f in Eq. (2.20) contains the time derivatives $\dot{\mathbf{q}}$, it can be expressed in the holonomic form Eq. (2.21), and so it is actually a holonomic constraint. More specifically we have the following definition.

Definition 2.1 (*Nonholonomic constraint*)

A constraint of the form (2.20) is said to be *nonholonomic* if it cannot be rendered to the form (2.21) such that to involve only the generalized variables themselves.

Typical systems that are subject to nonholonomic constraints (and hence are called *nonholonomic systems*) are underactuated robots, WMRs, autonomous underwater vehicles (AUVs), and unmanned aerial vehicles (UAVs). It is emphasized that "holonomic" does not necessarily mean unconstrained. Surely, a mobile robot with no constraint is holonomic. But a mobile robot capable of only translations is also holonomic.

Nonholonomicity occurs in several ways. For example a robot has only a few motors, say $k<n$, where n is the number of degrees of freedom, or the robot has redundant degrees of freedom. The robot can produce at most k independent motions. The difference $n-k$ indicates the existence of nonholomicity. For example, a differential drive WMR has two controls (the torques of the two wheel motors), that is, $k=2$, and three degrees of freedom, that is, $n=3$. Therefore, it has one $(n-k=1)$ nonholonomic constraint.

Definition 2.2 (*Pfaffian constraints*)

A nonholonomic constraint is called a Pfaffian constraint if it is linear in $\dot{\mathbf{q}}$, that is, if it can be expressed in the form:

$$\boldsymbol{\mu}_i(\mathbf{q})\dot{\mathbf{q}} = 0, \quad i = 1, 2, \ldots, r$$

where $\boldsymbol{\mu}_i$ are linearly independent row vectors and $\mathbf{q} = [q_1, q_2, \ldots, q_n]^T$.

In compact matrix form the above r Pfaffian constraints can be written as:

$$\mathbf{M}(\mathbf{q})\dot{\mathbf{q}} = \mathbf{0}, \quad \mathbf{M}(\mathbf{q}) = \begin{bmatrix} \boldsymbol{\mu}_1(\mathbf{q}) \\ \boldsymbol{\mu}_2(\mathbf{q}) \\ \vdots \\ \boldsymbol{\mu}_r(\mathbf{q}) \end{bmatrix} \tag{2.22}$$

An example of integrable Pfaffian constraint is:

$$\mu(\mathbf{q})\dot{\mathbf{q}} = q_1\dot{q}_1 + q_2\dot{q}_2 + \cdots + q_n\dot{q}_n, \quad \mathbf{q} \in R^n \tag{2.23}$$

This is integrable because it can be derived via differentiation, with respect to time, of the equation of a sphere:

$$s(\mathbf{q}) = q_1^2 + q_2^2 + \cdots + q_n^2 - a^2 = 0$$

with constant radius a. The particular resulting sphere by integrating Eq. (2.23) depends on the initial state $\mathbf{q}(t)_{t=0} = \mathbf{q}_0$. The collection of all concentric spheres with center at the origin and radius "a" is called a *foliation* with spherical leaves. For example, if $n = 3$ the foliation produces a maximal integral manifold $(0, 0, a)$:

$$\mathcal{M} = \{\mathbf{q} \in R^3 : s(\mathbf{q}) = q_1^2 + q_2^2 + q_3^2 - a^2 = 0\}$$

The nonholonomic constraint encountered in mobile robotics is the motion constraint of a disk that rolls on a plane without slipping (Figure 2.5). The no-slipping condition does not allow the generalized velocities $\dot{x}, \dot{y}$, and $\dot{\phi}$ to take arbitrary values.

Let r be the disk radius. Due to the no-slipping condition the generalized coordinates are constrained by the following equations:

$$\dot{x} = r\dot{\theta} \cos \phi, \quad \dot{y} = r\dot{\theta} \sin \phi \tag{2.24}$$

which are not integrable. These constraints express the condition that the velocity vector of the disk center lies in the midplane of the disk. Eliminating the velocity $v = r\,\dot{\theta}$ in Eq. (2.24) gives:

$$v = r\dot{\theta} = \frac{\dot{x}}{\cos \phi} = \frac{\dot{y}}{\sin \phi}$$

or

$$\dot{x} \sin \phi - \dot{y} \cos \phi = 0 \tag{2.25}$$

This is the nonholonomic constraint of the motion of the disk. Because of the kinematic constraints (2.24), the disk can attain any final configuration $(x_2, y_2, \phi_2, \theta_2)$ starting from any initial configuration $(x_1, y_1, \phi_1, \theta_1)$. This can be done in two steps as follows:

Step 1: Move the contact point (x_1, y_1) to (x_2, y_2) by rolling the disk along a line of length $(2k\pi + \theta_2 - \theta_1)r, \quad k = 0, 1, 2, \ldots$.

Step 2: Rotate the disk about the vertical axis from ϕ_1 to ϕ_2.

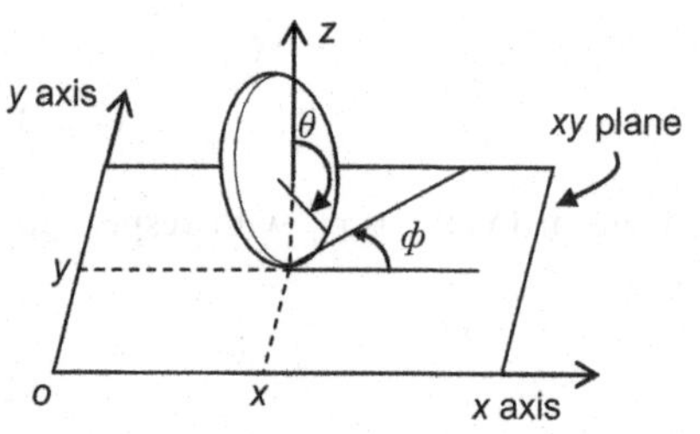

Figure 2.5 The generalized coordinates x, y, and ϕ.

Given a kinematic constraint one has to determine whether it is integrable or not. This can be done via the *Frobenius theorem* which uses the differential geometry concepts of *distributions* and *Lie Brackets*. We will come to this later (Section 6.2.1).

Two other systems that are subject to nonholonomic constraints are the *rolling ball* on a plane without spinning on place, and the *flying airplane* that cannot instantaneously stop in the air or move backward.

2.3 Nonholonomic Mobile Robots

The kinematic models of the following nonholonomic WMRs will be derived:

- Unicycle
- Differential drive WMR
- Tricycle WMR
- Car-like WMR

2.3.1 Unicycle

Unicycle has a kinematic model which is used as a basis for many types of nonholonomic WMRs. For this reason this model has attracted much theoretical attention by WMR controlists and nonlinear systems workers.

Unicycle is a conventional wheel rolling on a horizontal plane, while keeping its body vertical (Figure 2.5). The unicycle configuration (as seen from the bottom via a glass floor) is shown in Figure 2.6.

Its configuration is described by a vector of generalized coordinates: $\mathbf{p} = [x_Q, y_Q, \phi]^T$, that is, the position coordinates of the point of contact Q with the ground in the fixed coordinate frame Oxy, and its orientation angle ϕ with respect to the x axis. The linear velocity of the wheel is v_Q and its angular velocity about its instantaneous rotational axis is $v_\phi = \dot{\phi}$. From Figure 2.6, we find:

$$\dot{x}_Q = v_Q \cos\phi, \quad \dot{y}_Q = v_Q \sin\phi, \quad \dot{\phi} = v_\phi \tag{2.26}$$

Eliminating v_Q from the first two equations (2.26) we find the nonholonomic constraint (2.25):

$$-\dot{x}_Q \sin\phi + \dot{y}_Q \cos\phi = 0 \tag{2.27}$$

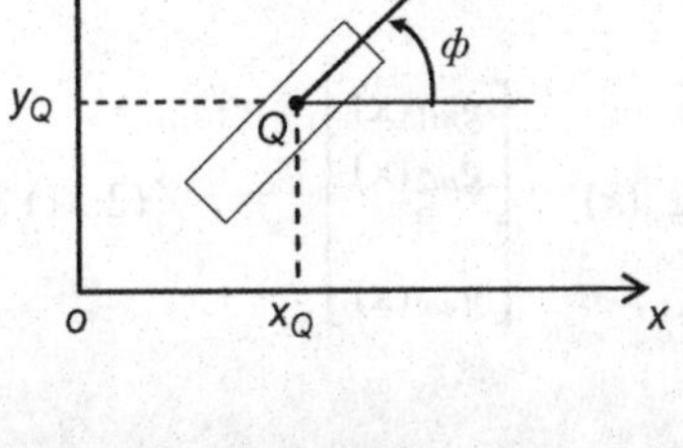

Figure 2.6 Kinematic structure of a unicycle.

Using the notation $v_1 = v_Q$ and $v_2 = v_\phi$, for simplicity, the kinematic model (2.26) of the unicycle can be written as:

$$\dot{\mathbf{p}} = \begin{bmatrix} \cos\phi \\ \sin\phi \\ 0 \end{bmatrix} v_1 + \begin{bmatrix} 0 \\ 0 \\ 1 \end{bmatrix} v_2, \quad \dot{\mathbf{p}} = \begin{bmatrix} \dot{x}_Q \\ \dot{y}_Q \\ \dot{\phi} \end{bmatrix} \tag{2.28a}$$

or

$$\dot{\mathbf{p}} = \mathbf{J}\dot{\mathbf{q}}, \quad \dot{\mathbf{q}} = [v_1, v_2]^T \tag{2.28b}$$

where $\mathbf{J}$ is the system Jacobian matrix:

$$\mathbf{J} = \begin{bmatrix} \cos\phi & 0 \\ \sin\phi & 0 \\ 0 & 1 \end{bmatrix} \tag{2.28c}$$

The linear velocity $v_1 = v_Q$ and the angular velocity $v_\phi = v_2$ are assumed to be the action (joint) variables of the system.

The model (2.28a) belongs to the special class of nonlinear systems, called *affine systems*, and described by a dynamic equation of the form (Chapter 6):

$$\dot{\mathbf{x}} = \mathbf{g}_0(\mathbf{x}) + \sum_{i=1}^{m} \mathbf{g}_i(\mathbf{x})u_i \tag{2.29a}$$

$$= \mathbf{g}_0(\mathbf{x}) + \mathbf{G}(\mathbf{x})\mathbf{u} \tag{2.29b}$$

where $u_i (i = 1, 2, \ldots, m)$ appear linearly, and:

$$\mathbf{x} = [x_1, x_2, \ldots, x_n]^T \in \mathcal{X}, \quad \mathbf{u} = [u_1, u_2, \ldots, u_m]^T \in \mathcal{U}$$

$$\mathbf{G}(\mathbf{x}) = [\mathbf{g}_1(\mathbf{x}) \vdots \mathbf{g}_2(\mathbf{x}) \vdots \cdots \vdots \mathbf{g}_m(\mathbf{x})] \tag{2.30}$$

If $m < n$ the system has a less number of actuation variables (controls) than the degrees of freedom under control and is known as *underactuated* system. If $m > n$ we have an *overactuated* system. In practice, usually $m < n$. The vector $\mathbf{x}$ is actually the state vector of the system and $\mathbf{u}$ the control vector. The term $\mathbf{g}_0(\mathbf{x})$ is called "*drift*," and the system with $\mathbf{g}_0(\mathbf{x}) = \mathbf{0}$ is called a "*driftless*" system. The column vector set:

$$\mathbf{g}_1(\mathbf{x}) = \begin{bmatrix} g_{11}(\mathbf{x}) \\ g_{12}(\mathbf{x}) \\ \vdots \\ g_{1n}(\mathbf{x}) \end{bmatrix}, \quad \mathbf{g}_2(x) = \begin{bmatrix} g_{21}(\mathbf{x}) \\ g_{22}(\mathbf{x}) \\ \vdots \\ g_{2n}(\mathbf{x}) \end{bmatrix}, \ldots, \quad \mathbf{g}_m(x) = \begin{bmatrix} g_{m1}(\mathbf{x}) \\ g_{m2}(\mathbf{x}) \\ \vdots \\ g_{mn}(\mathbf{x}) \end{bmatrix} \tag{2.31}$$

is referred to as the system's *vector field.* It is assumed that the set $\mathcal{U}$ contains at least an open set that involves the origin of R^m. If $\mathcal{U}$ does not contain the origin, then the system is not "driftless."

The unicycle model (2.28a) is a 2-input driftless affine system with two vector fields:

$$\mathbf{g}_1 = \begin{bmatrix} \cos\phi \\ \sin\phi \\ 0 \end{bmatrix}, \quad \mathbf{g}_2 = \begin{bmatrix} 0 \\ 0 \\ 1 \end{bmatrix} \tag{2.32}$$

The Jacobian formulation (2.28c) organizes the two column vector fields into a matrix $\mathbf{J} = \mathbf{G}$. Each action variable $u_i \in R$ in Eq. (2.29a) is actually a coefficient that determines how much of $\mathbf{g}_i(\mathbf{x})$ is contributing into the result $\mathbf{x}$. The vector field $\mathbf{g}_1(\phi)$ of the unicycle allows *pure translation*, and the field $\mathbf{g}_2$ allows *pure rotation.*

2.3.2 Differential Drive WMR

Indoor and other mobile robots use the differential drive locomotion type (Figure 1.20). The *Pioneer* WMR shown in Figure 1.11 is an example of differential drive WMR. The geometry and kinematic parameters of this robot are shown in Figure 2.7. The pose (position/orientation) vector of the WMR and its speed are respectively:

$$\mathbf{p} = \begin{bmatrix} x_Q \\ y_Q \\ \phi \end{bmatrix}, \quad \dot{\mathbf{p}} = \begin{bmatrix} \dot{x}_Q \\ \dot{y}_Q \\ \dot{\phi} \end{bmatrix} \tag{2.33}$$

The angular positions and speeds of the left and right wheels are $\{\theta_l, \dot{\theta}_l\}, \{\theta_r, \dot{\theta}_r\}$, respectively.

The following assumptions are made:

- Wheels are rolling without slippage
- The guidance (steering) axis is perpendicular to the plane Oxy
- The point Q coincides with the center of gravity G, that is, $||\overrightarrow{GQ}|| = 0$.[2]

Let v_l and v_r be the linear velocity of the left and right wheel respectively, and $v_Q$the velocity of the wheel midpoint Q of the WMR. Then, from Figure 2.7A we get:

$$v_r = v_Q + a\dot{\phi}, \quad v_l = v_Q - a\dot{\phi} \tag{2.34a}$$

Adding and subtracting v_r and v_l we get

$$v_Q = \frac{1}{2}(v_r + v_l), \quad 2a\dot{\phi} = v_r - v_l \tag{2.34b}$$

[2] In Figure 2.7, the points Q and G are shown distinct in order to use the same figure in all configurations with Q and G separated by distance b.

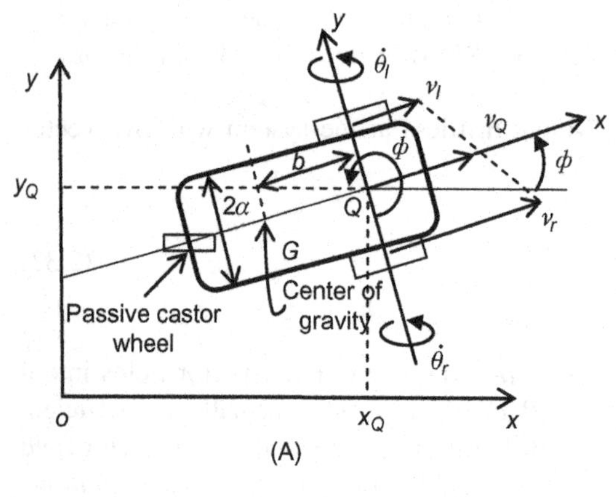

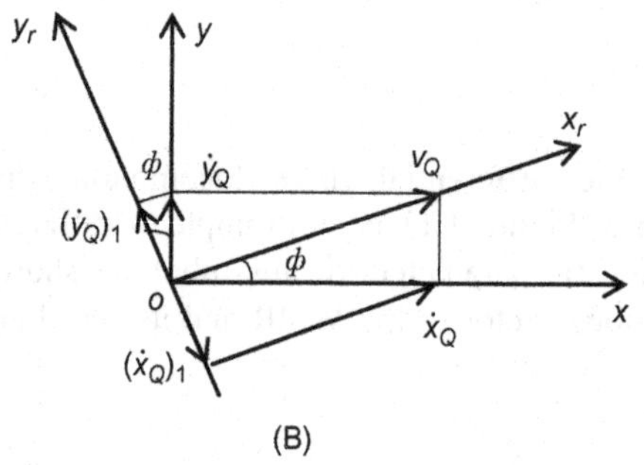

Figure 2.7 (A) Geometry of differential drive WMR, (B) Diagram illustrating the nonholonomic constraint.

where, due to the nonslippage assumption, we have $v_r = r\dot{\theta}_r$ and $v_l = r\dot{\theta}_l$. As in the unicycle case $\dot{x}_Q$ and $\dot{y}_Q$ are given by:

$$\dot{x}_Q = v_Q \cos\phi, \quad \dot{y}_Q = v_Q \sin\phi \tag{2.35}$$

and so the kinematic model of this WMR is described by the following relations:

$$\dot{x}_Q = \frac{r}{2}(\dot{\theta}_r \cos\phi + \dot{\theta}_l \cos\phi) \tag{2.36a}$$

$$\dot{y}_Q = \frac{r}{2}(\dot{\theta}_r \sin\phi + \dot{\theta}_l \sin\phi) \tag{2.36b}$$

$$\dot{\phi} = \frac{r}{2a}(\dot{\theta}_r - \dot{\theta}_l) \tag{2.36c}$$

Analogously to Eq. (2.28a,b) the kinematic model (2.36a–c) can be written in the *driftless affine* form:

$$\dot{\mathbf{p}} = \begin{bmatrix} (r/2)\cos\phi \\ (r/2)\sin\phi \\ r/2a \end{bmatrix} \dot{\theta}_r + \begin{bmatrix} (r/2)\cos\phi \\ (r/2)\sin\phi \\ -r/2a \end{bmatrix} \dot{\theta}_l \tag{2.37a}$$

or

$$\dot{\mathbf{p}} = \mathbf{J}\dot{\mathbf{q}} \tag{2.37b}$$

where

$$\dot{\mathbf{p}} = \begin{bmatrix} \dot{x}_Q \\ \dot{y}_Q \\ \dot{\phi} \end{bmatrix}, \quad \dot{\mathbf{q}} = \begin{bmatrix} \dot{\theta}_r \\ \dot{\theta}_1 \end{bmatrix} \tag{2.37c}$$

and $\mathbf{J}$ is the WMR's Jacobian:

$$\mathbf{J} = \begin{bmatrix} (r/2)\cos\phi & (r/2)\cos\phi \\ (r/2)\sin\phi & (r/2)\sin\phi \\ r/2a & -r/2a \end{bmatrix} \tag{2.37d}$$

Here, the two 3-dimensional vector fields are:

$$\mathbf{g}_1 = \begin{bmatrix} (r/2)\cos\phi \\ (r/2)\sin\phi \\ r/2a \end{bmatrix}, \quad \mathbf{g}_2 = \begin{bmatrix} (r/2)\cos\phi \\ (r/2)\sin\phi \\ -r/2a \end{bmatrix} \tag{2.38}$$

The field $\mathbf{g}_1$ allows the rotation of the right wheel, and $\mathbf{g}_2$ allows the rotation of the left wheel. Eliminating v_Q in Eq. (2.35) we get as usual the nonholonomic constraint (2.25) or (2.27).

$$-\dot{x}_Q\sin\phi + \dot{y}_Q\cos\phi = 0 \tag{2.39}$$

which expresses the fact that the point Q is moving along Qx_r, and its velocity along the axis Qy_r is zero (no lateral motion), that is (Figure 2.7B):

$$-(\dot{x}_Q)_1 + (\dot{y}_Q)_1 = 0$$

where $(\dot{x}_Q)_1 = \dot{x}_Q \sin\phi$ and $(\dot{y}_Q)_1 = \dot{y}_Q \cos\phi$.

The Jacobian matrix $\mathbf{J}$ in Eq. (2.37d) has three rows and two columns, and so it is not invertible. Therefore, the solution of Eq. (2.37b) for $\dot{\mathbf{q}}$ is given by:

$$\dot{\mathbf{q}} = \mathbf{J}^{\dagger}\dot{\mathbf{p}} \tag{2.40}$$

where $\mathbf{J}^{\dagger}$ is the generalized inverse of $\mathbf{J}$ given by Eq. (2.8a). However, here $\mathbf{J}^{\dagger}$ can be computed directly by using Eq. (2.34a), and observing from Figure 2.7B that:

$$v_Q = \dot{x}_Q \cos\phi + \dot{y}_Q \sin\phi$$

Thus, using this equation in Eq. (2.34a) we obtain:

$$\begin{aligned} r\dot{\theta}_{\mathrm{r}} &= \dot{x}_Q \cos\phi + \dot{y}_Q \sin\phi + a\dot{\phi} \\ r\dot{\theta}_{\mathrm{l}} &= \dot{x}_Q \cos\phi + \dot{y}_Q \sin\phi - a\dot{\phi} \end{aligned} \tag{2.41a}$$

that is:

$$\begin{bmatrix} \dot{\theta}_{\mathrm{r}} \\ \dot{\theta}_{\mathrm{l}} \end{bmatrix} = \frac{1}{r}\begin{bmatrix} \cos\phi & \sin\phi & a \\ \cos\phi & \sin\phi & -a \end{bmatrix}\begin{bmatrix} \dot{x}_Q \\ \dot{y}_Q \\ \dot{\phi} \end{bmatrix}$$

or

$$\dot{\mathbf{q}} = \mathbf{J}^{\dagger}\dot{\mathbf{p}} \tag{2.41b}$$

where[3] :

$$\mathbf{J}^{\dagger} = \frac{1}{r}\begin{bmatrix} \cos\phi & \sin\phi & a \\ \cos\phi & \sin\phi & -a \end{bmatrix} \tag{2.41c}$$

The nonholonomic constraint (2.39) can be written as:

$$\mathbf{M}\dot{\mathbf{p}} = 0, \quad \mathbf{M} = \begin{bmatrix} -\sin\phi & \cos\phi & 0 \end{bmatrix} \tag{2.42}$$

Clearly, if $\dot{\theta}_{\mathrm{r}} \neq \dot{\theta}_{\mathrm{l}}$, then the difference between $\dot{\theta}_{\mathrm{r}}$ and $\dot{\theta}_{\mathrm{l}}$ determines the robot's rotation speed $\dot{\phi}$ and its direction. The instantaneous curvature radius R is given by (Eq. 1.1):

$$R = \frac{v_Q}{\dot{\phi}} = a\left(\frac{v_{\mathrm{r}} + v_{\mathrm{l}}}{v_{\mathrm{r}} - v_{\mathrm{l}}}\right), \quad v_{\mathrm{r}} \geq v_{\mathrm{l}} \tag{2.43a}$$

and the instantaneous curvature coefficient is:

$$\kappa = 1/R \tag{2.43b}$$

Example 2.1

Derive the kinematic relations (2.35) using the rotation matrix concept (2.17).

Solution

Here, the point $P(x_p, y_p)$ of Figure 2.4 is the point $Q(x_Q, y_Q)$ in Figure 2.7. The WMR velocities along the local coordinate axes Qx_r and Qy_r are $\dot{x}_r$ and $\dot{y}_r$. The corresponding velocities in the world coordinate frame are $\dot{x}_Q$ and $\dot{y}_Q$. Therefore, for a given ϕ, (2.17) gives:

$$\begin{bmatrix} \dot{x}_Q \\ \dot{y}_Q \end{bmatrix} = \begin{bmatrix} \cos\phi & -\sin\phi \\ \sin\phi & \cos\phi \end{bmatrix}\begin{bmatrix} \dot{x}_r \\ \dot{y}_r \end{bmatrix} \tag{2.44}$$

[3] As an exercise, the reader is advised to derive Eq. (2.41c) using Eq. (2.8a).

Now, the condition of no lateral wheel movement implies that

$$\dot{y}_r = 0 \begin{matrix} \cos\phi & -\sin\phi \\ \sin\phi & \cos\phi \end{matrix}$$

and $\dot{x}_r = v_Q$. Therefore, the above relation gives:

$$\dot{x}_Q = v_Q \cos\phi \text{ and } \dot{y}_Q = v_Q \sin\phi$$

as desired.

Example 2.2

Derive the kinematic equations and constraints of a differential drive WMR by relaxing the no-slipping condition of the wheels' motion.

Solution

We will work with the WMR of Figure 2.7. Considering the rotation about the center of gravity G we get the following relations:

$$\dot{x}_G = \dot{x}_Q + b\dot{\phi}\sin\phi$$

$$\dot{y}_G = \dot{y}_Q - b\dot{\phi}\cos\phi$$

Therefore, the kinematic equations (2.41a) and the nonholonomic constraint (2.42) become:

$$r\dot{\theta}_r = \dot{x}_G\cos\phi + \dot{y}_G\sin\phi + a\dot{\phi}$$

$$r\dot{\theta}_l = \dot{x}_G\cos\phi + \dot{y}_G\sin\phi - a\dot{\phi}$$

$$-\dot{x}_G\sin\phi + \dot{y}_G\cos\phi + b\dot{\phi} = 0$$

Now, assume that the wheels are subject to longitudinal and lateral slip [10]. To include the slip into the kinematics of the robot, we introduce two variables w_r, w_l for the longitudinal slip displacements of the right wheel and left wheel, respectively, and two variables z_r, z_l for the corresponding lateral slip displacements. Thus, here:

$$\mathbf{p} = [x_G, y_G, \phi \vdots w_r, w_l, z_r, z_l]^T$$

The slipping wheels' velocities are now given by:

$$v_r = (r\dot{\theta}_r - \dot{w}_r)\cos\zeta_r, \quad v_l = (r\dot{\theta}_l - \dot{w}_l)\cos\zeta_l$$

where ζ_r and ζ_l are the steering angles of the wheels.

Using these relations for v_r and v_l the above kinematic equations are written as:

$$v_r = (r\dot{\theta}_r - \dot{w}_r)\cos\zeta_r = \dot{x}_G\cos\phi + \dot{y}_G\sin\phi + a\dot{\phi}$$

$$v_l = (r\dot{\theta}_l - \dot{w}_l)\cos\zeta_l = \dot{x}_G\cos\phi + \dot{y}_G\sin\phi - a\dot{\phi}$$

and the nonholonomic constraint becomes:

$$-\dot{x}_G\sin\phi + \dot{y}_G\cos\phi + b\dot{\phi} - \dot{z}_r\cos\zeta_r = 0$$

$$-\dot{x}_G\sin\phi + \dot{y}_G\cos\phi + b\dot{\phi} - \dot{z}_l\cos\zeta_l = 0$$

In our WMR the two wheels have a common axis and are unsteered. Therefore, $\zeta_r = \zeta_l = 0$. For WMRs with steered wheels we may have $\zeta_r \neq 0$, $\zeta_l \neq 0$. In our case $\cos\zeta_r = \cos\zeta_l = 1$, and so the two kinematic equations, solved for the angular wheel velocities $\dot{\boldsymbol{\theta}}_r$ and $\dot{\boldsymbol{\theta}}_l$, give:

$$\dot{\mathbf{q}} = \mathbf{J}^{\dagger}\dot{\mathbf{p}}, \quad \dot{\mathbf{q}} = \begin{bmatrix} \dot{\boldsymbol{\theta}}_r \\ \dot{\boldsymbol{\theta}}_l \end{bmatrix}$$

where the inverse Jacobian is:

$$\mathbf{J}^{\dagger} = \frac{1}{r}\begin{bmatrix} \cos\phi & \sin\phi & a & 1 & 0 & 0 & 0 \\ \cos\phi & \sin\phi & -a & 0 & 1 & 0 & 0 \end{bmatrix}$$

The nonholonomic constraints are written in Pfaffian form:

$$\mathbf{M(p)\dot{p} = 0}$$

where

$$\mathbf{M(p)} = \begin{bmatrix} -\sin\phi & \cos\phi & b & 0 & 0 & -1 & 0 \\ -\sin\phi & \cos\phi & b & 0 & 0 & 0 & -1 \end{bmatrix}$$

$$\dot{\mathbf{p}} = (\dot{x}_G, \;\; \dot{y}_G, \;\; \dot{\phi} \;\; \vdots \;\; \dot{w}_r, \;\; \dot{w}_l \;\; \dot{z}_r \;\; \dot{z}_l)^T$$

In the special case where only lateral slip takes place (i.e., $\dot{w}_r = 0$, $\dot{w}_l = 0$), the components $\dot{w}_r$ and $\dot{w}_l$ are dropped from $\dot{\mathbf{p}}$, and the matrices $\mathbf{J}^{\dagger}$ and $\mathbf{M(p)}$ are reduced appropriately, having only five columns. Note that here the wheels are fixed and so $\dot{z}_r = \dot{z}_l = \dot{y}_r$ where $\dot{y}_r$ is the lateral slipping velocity of the body of the WMR. Typically, the slipping variables, which are unknown and nonmeasurable are treated as disturbances via disturbance rejection and robust control techniques.

2.3.3 Tricycle

The motion of this WMR is controlled by the wheel steering angular velocity ω_ψ and its linear velocity v_w (or its angular velocity $\omega_w = \dot{\theta}_w = v_w/r$, where r is the radius of the wheel) (Figure 2.8).

The orientation angle and angular velocity are ϕ and $\dot{\phi}$, respectively. It is assumed that the vehicle has its guidance point Q in the back of the powered wheel (i.e., it has a central back axis). The state of the robot's motion is:

$$\mathbf{p} = [x_Q, \quad y_Q, \quad \phi, \quad \psi]^T$$

The kinematic variables are:

Steering wheel velocity: $v_w = r\dot{\theta}_w$.
Vehicle velocity: $v = v_w \cos\psi = r(\cos\psi)\dot{\theta}_w$
Vehicle orientation velocity: $\dot{\phi} = (1/D)v_w \sin\psi$
Steering angle velocity: $\dot{\psi} = \omega_\psi$

Using the above relations we find:

$$\dot{\mathbf{p}} = \begin{bmatrix} \dot{x}_Q \\ \dot{y}_Q \\ \dot{\phi} \\ \dot{\psi} \end{bmatrix} = \begin{bmatrix} v\cos\phi \\ v\sin\phi \\ \dot{\phi} \\ \dot{\psi} \end{bmatrix} = \begin{bmatrix} r\cos\psi\cos\phi \\ r\cos\psi\sin\phi \\ (r/D)\sin\psi \\ 0 \end{bmatrix} \dot{\theta}_w + \begin{bmatrix} 0 \\ 0 \\ 0 \\ 1 \end{bmatrix} \dot{\psi} = \mathbf{J}\dot{\boldsymbol{\theta}} \tag{2.45}$$

where $\dot{\boldsymbol{\theta}} = [\dot{\theta}_w, \dot{\psi}]^T$ is the vector of joint velocities (control variables), and

$$\mathbf{J} = \begin{bmatrix} r\cos\psi\cos\phi & 0 \\ r\cos\psi\sin\phi & 0 \\ (r/D)\sin\psi & 0 \\ 0 & 1 \end{bmatrix} \tag{2.46}$$

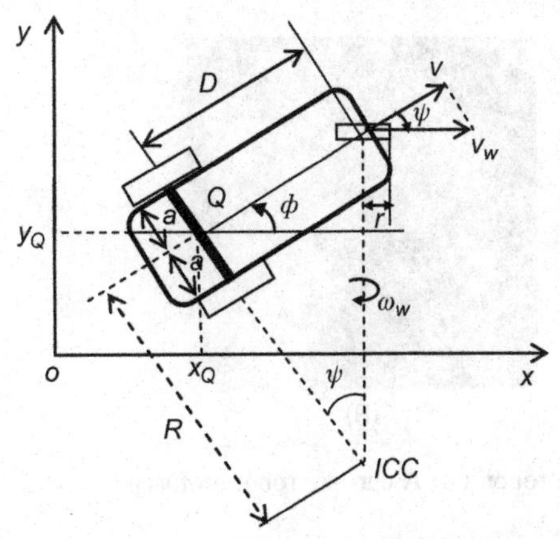

Figure 2.8 Geometry of the tricycle WMR (ψ is the steering angle).

is the Jacobian matrix. This Jacobian is again noninvertible, but we can find the inverse kinematic equations directly using the relations:

$$\dot{\phi}/v = (1/D)tg\psi \text{ or } \psi = arctg(D\dot{\phi}/v) \tag{2.47a}$$

and

$$\dot{\theta}_{\mathrm{w}} = \frac{v_{\mathrm{w}}}{r} = \frac{1}{r}\sqrt{v^2 + (D\,\dot{\phi})^2} \tag{2.47b}$$

The instantaneous curvature radius R is given by (Figure 2.8):

$$R = Dtg(\pi/2 - \psi(t)) \tag{2.48}$$

From Eq. (2.45) we see that the tricycle is again a 2-input driftless affine system with vector fields:

$$\mathbf{g}_1 = \begin{bmatrix} r\cos\psi\cos\phi \\ r\cos\psi\sin\phi \\ (r/D)\sin\psi \\ 0 \end{bmatrix}, \quad \mathbf{g}_2 = \begin{bmatrix} 0 \\ 0 \\ 0 \\ 1 \end{bmatrix}$$

that allow steering wheel motion $\dot{\theta}_{\mathrm{w}}$, and steering angle motion $\dot{\psi}$, respectively.

2.3.4 Car-Like WMR

The geometry of the car-like mobile robot is shown in Figure 2.9A and the A.W.E. S.O.M.-9000 line-tracking car-like robot prototype (Aalborg University) in Figure 2.9B.

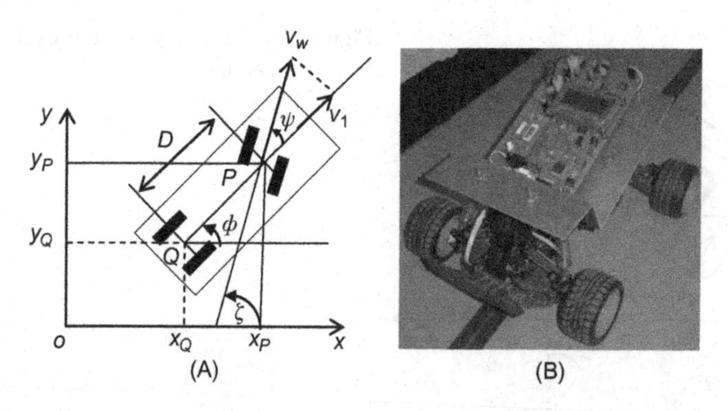

Figure 2.9 (A) Kinematic structure of a car-like robot, (B) A car-like robot prototype. *Source:* http://sqrt-1.dk/robot/robot.php

The state of the robot's motion is represented by the vector [20]:

$$\mathbf{p} = [x_Q, y_Q, \phi, \psi]^T \tag{2.49}$$

where x_Q, y_Q are the Cartesian coordinates of the wheel axis midpoint Q, ϕ is the orientation angle of the vehicle, and ψ is the steering angle. Here, we have two nonholonomic constraints, one for each wheel pair, that is:

$$-\dot{x}_Q \sin \phi + \dot{y}_Q \cos \phi = 0 \tag{2.50a}$$

$$-\dot{x}_p \sin(\phi + \psi) + \dot{y}_p \cos(\phi + \psi) = 0 \tag{2.50b}$$

where x_p and y_p are the position coordinates of the front wheels midpoint P. From Figure 2.9 we get:

$$x_p = x_Q + D \cos \phi, \quad y_p = y_Q + D \sin \phi$$

Using these relations the second kinematic constraint (2.50b) becomes:

$$-\dot{x}_Q \sin(\phi + \psi) + \dot{y}_Q \cos(\phi + \psi) + D(\cos \psi)\dot{\phi}$$

The two nonholonomic constraints are written in the matrix form:

$$\mathbf{M}(\mathbf{p})\dot{\mathbf{p}} = \mathbf{0} \tag{2.51a}$$

where

$$\mathbf{M}(\mathbf{p}) = \begin{bmatrix} -\sin \phi & \cos \phi & 0 & 0 \\ -\sin(\phi + \psi) & \cos(\phi + \psi) & D \cos \psi & 0 \end{bmatrix} \tag{2.51b}$$

The kinematic equations for a rear-wheel driving car are found to be (Figure 2.9):

$$\begin{aligned} \dot{x}_Q &= v_1 \cos \phi \\ \dot{y}_Q &= v_1 \sin \phi \\ \dot{\phi} &= \frac{1}{D} v_w \sin \psi \\ &= \frac{1}{D} v_1 tg\, \psi \\ \dot{\psi} &= v_2 \end{aligned} \tag{2.52}$$

These equations can be written in the affine form:

$$\begin{bmatrix} \dot{x}_Q \\ \dot{y}_Q \\ \dot{\phi} \\ \dot{\psi} \end{bmatrix} = \begin{bmatrix} \cos\phi \\ \sin\phi \\ (1/D)tg\psi \\ 0 \end{bmatrix} v_1 + \begin{bmatrix} 0 \\ 0 \\ 0 \\ 1 \end{bmatrix} v_2 \tag{2.53}$$

that has the vector fields:

$$\mathbf{g}_1 = \begin{bmatrix} \cos\phi \\ \sin\phi \\ (1/D)tg\psi \\ 0 \end{bmatrix}, \quad \mathbf{g}_2 = \begin{bmatrix} 0 \\ 0 \\ 0 \\ 1 \end{bmatrix}$$

allowing the driving motion v_1 and the steering motion $v_2 = \dot{\psi}$, respectively. The Jacobian form of Eq. (2.53) is:

$$\dot{\mathbf{p}} = \mathbf{J}\mathbf{v}, \quad \mathbf{v} = [v_1, v_2]^T \tag{2.54}$$

with Jacobian matrix:

$$\mathbf{J} = \begin{bmatrix} \cos\phi & 0 \\ \sin\phi & 0 \\ (tg\psi)/D & 0 \\ 0 & 1 \end{bmatrix} \tag{2.55}$$

Here, there is a singularity at $\psi = \pm\pi/2$, which corresponds to the "jamming" of the WMR when the front wheels are normal to the longitudinal axis of its body. Actually, this singularity does not occur in practice due to the restricted range of the steering angle $\psi\ (-\pi/2 < \psi < \pi/2)$.

The kinematic model for the front wheel driving vehicle is (Eqs. 2.45 and 2.46) [20]:

$$\dot{\mathbf{p}} = \mathbf{J}\mathbf{v}, \quad \mathbf{J} = \begin{bmatrix} \cos\phi\cos\psi & 0 \\ \sin\phi\cos\psi & 0 \\ (\sin\psi)/D & 0 \\ 0 & 1 \end{bmatrix} \tag{2.56a}$$

In this case the previous singularity does not occur, since at $\psi = \pm\pi/2$ the car can still (in principle) pivot about its rear wheels. Using the new inputs u_1 and u_2 defined as:

$$u_1 = v_1, u_2 = (1/D)\sin(\zeta - \phi)v_1 + v_2$$

the above model is transformed to:

$$\begin{bmatrix} \dot{x}_p \\ \dot{y}_p \\ \dot{\phi} \\ \dot{\zeta} \end{bmatrix} = \begin{bmatrix} \cos\zeta \\ \sin\zeta \\ (1/D)\sin(\zeta-\phi) \\ 0 \end{bmatrix} u_1 + \begin{bmatrix} 0 \\ 0 \\ 0 \\ 1 \end{bmatrix} u_2 \tag{2.56b}$$

where $\zeta = \phi + \psi$ is the total steering angle with respect to the axis Ox.

Indeed, from $x_p = x_Q + D\cos\phi$ and $y_p = y_Q + D\sin\phi$ (Figure 2.9), and Eq. (2.56a) we get:

$$\begin{aligned} \dot{x}_p &= \dot{x}_Q - D(\sin\phi)\dot{\phi} = (\cos\phi\cos\psi - \sin\phi\sin\psi)v_1 \\ &= [\cos(\phi+\psi)]v_1 = (\cos\zeta)u_1 \end{aligned}$$

$$\begin{aligned} \dot{y}_p &= \dot{y}_Q + D(\cos\phi)\dot{\phi} = (\sin\phi\cos\psi + \cos\phi\sin\psi)v_1 \\ &= [\sin(\phi+\psi)]v_1 = (\sin\zeta)u_1 \end{aligned}$$

$$\dot{\phi} = \frac{1}{D}[\sin(\zeta-\phi)]v_1 = \left[\frac{1}{D}\sin(\zeta-\phi)\right]u_1$$

$$\dot{\zeta} = \dot{\phi} + \dot{\psi} = \left[\frac{1}{D}\sin(\zeta-\phi)\right]v_1 + v_2 = u_2$$

We observe, from Eq. (2.56b), that the kinematic model for x_p, y_p, and ζ (i.e., the first, second, and fourth equation in Eq. (2.56b)) is actually a unicycle model (2.28a).

Two special cases of the above car-like model are known as:

- Reeds-Shepp car
- Dubins car

The *Reeds-Shepp car* is obtained by restricting the values of the velocity v_1 to three distinct values $+1$, 0, and -1. These values appear to correspond to three distinct "gears": "forward," "park," or "reverse." The *Dubins car* is obtained when the reverse motion is not allowed in the Reeds-Shepp car, that is, the value $v_1 = -1$ is excluded, in which case $v_1 = \{0, 1\}$.

Example 2.3

It is desired to find the steering angle ψ which is required for a rear-wheel driven car-like WMR to go from its present position $Q(x_Q, y_Q)$ to a given goal $F(x_f, y_f)$. The available data, which are obtained via proper sensors, are the distance L between (x_f, y_f) and (x_Q, y_Q) and the angle ε of the vector $\overrightarrow{QF}$ with respect to the current vehicle orientation.

Solution

We will work with the geometry of Figure 2.10 [19]. The kinematic equations of the WMR are given by Eq. (2.52). The WMR will go from the position Q to the goal F following a circular path with curvature:

$$\frac{1}{R_1} = \frac{1}{D} tg\psi$$

determined using the bicycle equivalent model, that combines the two front wheels and the two rear wheels (Figure 2.10, left).

On the other hand, the curvature $1/R_2$ of the circular path that passes through the goal, is obtained from the relation (Figure 2.10, right):

$$L/2 = R_2 \sin(\varepsilon)$$

that is:

$$\frac{1}{R_2} = \frac{2}{L}\sin(\varepsilon) \quad (2.57a)$$

To meet the goal tracking requirement the above two curvatures $1/R_1$ and $1/R_2$ must be the same, that is:

$$\frac{1}{D} tg\psi = \frac{2}{L}\sin(\varepsilon)$$

Therefore:

$$\psi = tg^{-1}\left(\frac{2D}{L}\sin(\varepsilon)\right) \quad (2.57b)$$

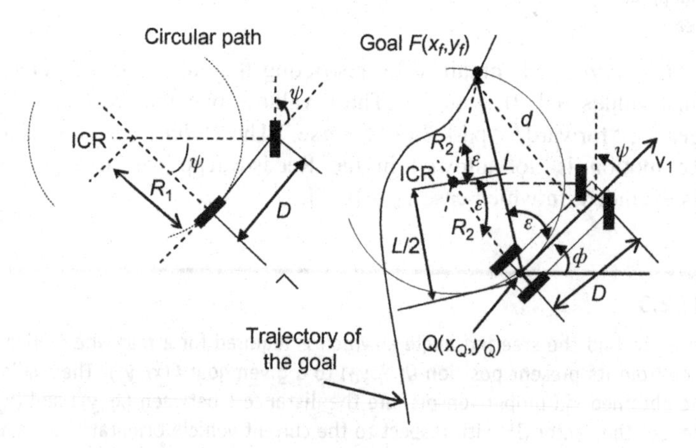

Figure 2.10 Geometry of the goal tracking problem.

Equation (2.57b) gives the steering angle ψ in terms of the data L and ε, and can be used to pursuit tracking of goals (targets) that are moving along given trajectories. In these cases the goal F lies at the intersection of the goal trajectory and the look-ahead circle. To get a better interpretation of (2.57a), we use the lateral distance d between the vehicles orientation (heading) vector and the goal point, which is given by (Figure 2.10):

$$d = L\sin(\varepsilon)$$

Then, the curvature $1/R_2$ in Eq. (2.57a) is given by:

$$\frac{1}{R_2} = \frac{2d}{L^2}$$

This indicates that the curvature $1/R_1$ of the path resulting from the steering angle ψ should be:

$$\frac{1}{R_1} = \left(\frac{2}{L^2}\right)d \tag{2.57c}$$

Equation (2.57c) is a *"proportional control law"* and shows that the curvature $1/R_1$ of the robot's path should be proportional to the cross track error d some look-ahead distance in front of the WMR with a gain $2/L^2$.

2.3.5 Chain and Brockett—Integrator Models

The general 2-input n-dimensional chain model (briefly $(2, n)$-chain model) is:

$$\begin{aligned}\dot{x}_1 &= u_1\\ \dot{x}_2 &= u_2\\ \dot{x}_3 &= x_2u_1\\ &\vdots\\ \dot{x}_n &= x_{n-1}u_1\end{aligned} \tag{2.58}$$

The *Brockett* (single) *integrator* model is:

$$\begin{aligned}\dot{x}_1 &= u_1\\ \dot{x}_2 &= u_2\\ \dot{x}_3 &= x_1u_2 - x_2u_1\end{aligned} \tag{2.59}$$

and the *double integrator* model is:

$$\begin{aligned}\dot{x}_1 &= u_1\\ \dot{x}_2 &= u_2\\ \dot{x}_3 &= x_1\dot{x}_2 - x_2\dot{x}_1\end{aligned} \tag{2.60}$$

The nonholonomic WMR kinematic models can be transformed to the above models. Here, the unicycle model (which also covers the differential drive model) and the car-like model will be considered.

2.3.5.1 Unicycle WMR

The unicycle kinematic model is given by Eq. (2.26):

$$\dot{x}_Q = v_Q \cos\phi, \quad \dot{y}_Q = v_Q \sin\phi, \quad \dot{\phi} = v_\phi \tag{2.61}$$

Using the transformation:

$$\begin{aligned} z_1 &= \phi \\ z_2 &= x_Q \cos\phi + y_Q \sin\phi \\ z_3 &= x_Q \sin\phi - y_Q \cos\phi \end{aligned} \tag{2.62}$$

the unicycle model is converted to the *(2,3)-chain* form:

$$\begin{aligned} \dot{z}_1 &= u_1 \\ \dot{z}_2 &= u_2 \\ \dot{z}_3 &= z_2 u_1 \end{aligned} \tag{2.63}$$

where $u_1 = v_\phi$ and $u_2 = v_Q - z_3 u_1$.

Defining new state variables:

$$x_1 = z_1, \quad x_2 = z_2, \quad x_3 = -2z_3 + z_1 z_2 \tag{2.64}$$

the (2,3)-chain model is converted to the *Brockett integrator*:

$$\begin{aligned} \dot{x}_1 &= u_1 \\ \dot{x}_2 &= u_2 \\ \dot{x}_3 &= x_1 u_2 - x_2 u_1 \end{aligned} \tag{2.65}$$

2.3.5.2 Rear-Wheel Driving Car

The rear-wheel driven car model is given by Eq. (2.52):

$$\dot{x}_Q = v_1 \cos\phi, \quad \dot{y}_Q = v_1 \sin\phi, \quad \dot{\phi} = \frac{v_1}{D} tg\psi, \quad \dot{\psi} = v_2 \tag{2.66}$$

Using the state transformation:

$$x_1 = x_Q, \quad x_2 = \frac{tg\psi}{D\cos^3\phi}, \quad x_3 = tg\phi, \quad x_4 = y_Q \tag{2.67}$$

and input transformation:

$$v_1 = u_1/\cos\phi \tag{2.68}$$

$$v_2 = -\frac{3\sin^2\psi\sin\phi}{D\cos^2\phi}u_1 + D\cos^2\psi(\cos^3\phi)u_2$$

for $\phi \neq \pi/2 \pm k\pi$ and $\psi \neq \pi/2 \pm k\pi$, the model (2.66) is converted to the (2,4)-chain form:

$$\begin{aligned}\dot{x}_1 &= u_1\\ \dot{x}_2 &= u_2\\ \dot{x}_3 &= x_2u_1\\ \dot{x}_4 &= x_3u_1\end{aligned} \tag{2.69}$$

2.3.6 Car-Pulling Trailer WMR

This is an extension of the car-like WMR, where N one-axis trailers are attached to a car-like robot with rear-wheel drive. This type of trailer is used, for example, at airports for transporting luggage. The form of equations depend crucially on the exact point at which the trailer is attached and on the choice of body frames. Here, for simplicity each trailer will be assumed to be connected to the axle midpoint of the previous trailer (*zero hooking*) as shown in Figure 2.11 [20].

The new parameter introduced here is the distance from the center of the back axle of trailer i to the point at which is hitched to the next body. This is called the *hitch* (or *hinge-to-hinge*) length denoted by L_i. The car length is D. Let ϕ_i be

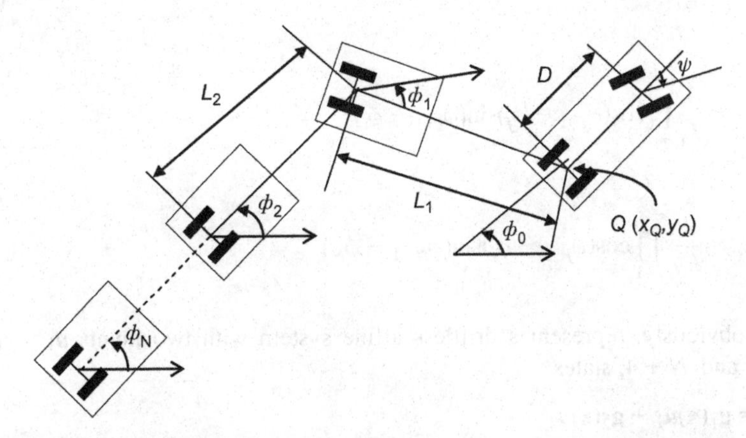

Figure 2.11 Geometrical structure of the N-trailer WMR.

the orientation of the ith trailer, expressed with respect to the world coordinate frame. Then from the geometry of Figure 2.11 we get the following equations:

$$x_i = x_Q - \sum_{j=1}^{i} L_j \cos \phi_j$$
$$i = 1, 2, \ldots, N$$
$$y_i = y_Q - \sum_{j=1}^{i} L_j \sin \phi_j$$

which give the following nonholonomic constraints:

$$\dot{x}_Q \sin \phi_0 - \dot{y}_Q \cos \phi_0 = 0$$

$$\dot{x}_Q \sin(\phi_0 + \psi) - \dot{y}_Q \cos(\phi_0 + \psi) - \dot{\phi}_0 D \cos \psi = 0$$

$$\dot{x}_Q \sin \phi_i - \dot{y}_Q \cos \phi_i + \sum_{j=1}^{i} \dot{\phi}_j L_j \cos(\phi_i - \phi_j) = 0$$

for $i = 1, 2, \ldots, N$.

In analogy to Eq. (2.52) the kinematic equations of the N-trailer are found to be:

$$\begin{aligned}
\dot{x}_Q &= v_1 \cos \phi_0 \\
\dot{y}_Q &= v_1 \sin \phi_0 \\
\dot{\phi}_0 &= (1/D) v_1 tg\psi \\
\dot{\psi} &= v_2 \\
\dot{\phi}_1 &= \frac{1}{L_1} \sin(\phi_0 - \phi_1) \\
\dot{\phi}_2 &= \frac{1}{L_2} \cos(\phi_0 - \phi_1) \sin(\phi_1 - \phi_2) \\
&\vdots \\
\dot{\phi}_i &= \frac{1}{L_i} \prod_{j=1}^{i-1} \cos(\phi_{j-1} - \phi_j) \sin(\phi_{i-1} - \phi_i) \\
&\vdots \\
\dot{\phi}_N &= \frac{1}{L_N} \prod_{j=1}^{N-1} \cos(\phi_{j-1} - \phi_j) \sin(\phi_{N-1} - \phi_N)
\end{aligned} \tag{2.70}$$

which, obviously, represent a driftless affine system with two inputs $u_1 = v_1$ and $u_2 = v_2$ and $N + 4$, states:

$$\dot{\mathbf{x}} = \mathbf{g}_1(\mathbf{x}) u_1 + \mathbf{g}_2(\mathbf{x}) u_2$$

We observe that the first four lines of the fields $\mathbf{g}_1$ and $\mathbf{g}_2$ represent the (powered) car-like WMR itself.

2.4 Omnidirectional WMR Kinematic Modeling

The following WMRs will be considered [2,4,11,12,16]:

- Multiwheel omnidirectional WMR with orthogonal (universal) wheels
- Four-wheel omnidirectional WMR with mecanum wheels that have a roller angle $\pm 45°$.

2.4.1 Universal Multiwheel Omnidirectional WMR

The geometric structure of a multiwheel omnirobot is shown in Figure 2.12A. Each wheel has three velocity components [16]:

- Its own velocity $v_i = r\dot{\theta}_i$, where r is the common wheel radius and $\dot{\theta}_i$ its own angular velocity
- An induced velocity $v_{i,\text{roller}}$ which is due to the free rollers (here assumed of the universal type; roller angle $\pm 90°$)
- A velocity component v_ϕ which is due to the rotation of the robotic platform about its center of gravity Q, that is, $v_\phi = D\dot{\phi}$, where $\dot{\phi}$ is the angular velocity of the platform and D is the distance of the wheel from Q.

Here, the roller angle is $\pm 90°$, and so:

$$v_h^2 = v_i^2 + v_{i,\text{roller}}^2 \tag{2.71a}$$

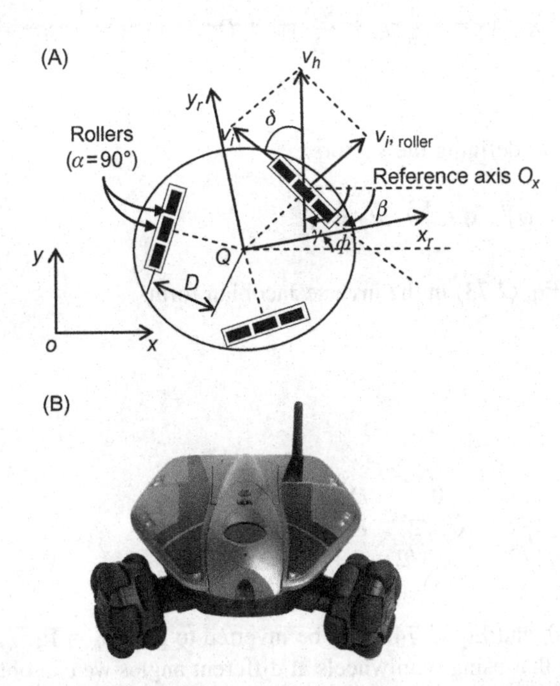

Figure 2.12 (A) Velocity vector of wheel i. The velocity v_h is the robot vehicle velocity due to the wheel motion, (B) An example of a 3-wheel setup.
Source: http://deviceguru.com/files/rovio-3.jpg.

where

$$\begin{aligned} v_i &= v_h \cos(\delta) \\ &= v_h \cos(\gamma - \beta) \\ &= v_h(\cos\gamma \cos\beta + \sin\gamma \sin\beta) \end{aligned} \tag{2.71b}$$

Thus the total velocity of the wheel i is:

$$\begin{aligned} v_i &= v_h(\cos\gamma \cos\beta + \sin\gamma \sin\beta) + D\dot{\phi} \\ &= v_{hx}\cos\beta + v_{hy}\sin\beta + D\dot{\phi} \end{aligned} \tag{2.72}$$

where v_{hx} and v_{hy} are the x, y components of v_h, that is:

$$v_{hx} = v_h\cos\gamma, \quad v_{hy} = v_h\sin\gamma$$

Equation (2.72) is general and can be used in WMRs with any number of wheels.

Thus, for example, in the case of a 3-wheel robot we may choose the angle β for the wheels 1, 2, and 3 as 0°, 120°, and 240°, respectively, and get the equations:

$$v_1 = v_{hx} + D\dot{\phi}, \quad v_2 = -\frac{1}{2}v_{hx} + \frac{\sqrt{3}}{2}v_{hy} + D\dot{\phi}, \quad v_3 = -\frac{1}{2}v_{hx} - \frac{\sqrt{3}}{2}v_{hy} + D\dot{\phi} \tag{2.73}$$

with $v_i = r\dot{\theta}_i$. Now, defining the vectors:

$$\dot{\mathbf{p}}_h = \left[v_{hx}, v_{hy}, \dot{\phi}\right]^T, \ \dot{\mathbf{q}} = \left[\dot{\theta}_1, \dot{\theta}_2, \dot{\theta}_3\right]^T$$

we can write Eq. (2.73) in the inverse Jacobian form:

$$\dot{\mathbf{q}} = \mathbf{J}^{-1}\dot{\mathbf{p}}_h \tag{2.74a}$$

where

$$\mathbf{J}^{-1} = \frac{1}{r}\begin{bmatrix} 1 & 0 & D \\ -1/2 & \sqrt{3}/2 & D \\ -1/2 & -\sqrt{3}/2 & D \end{bmatrix} \tag{2.74b}$$

Here $\det \mathbf{J} \neq 0$, and Eq. (2.74a) can be inverted to give $\dot{\mathbf{p}}_h = \mathbf{J}\dot{\mathbf{q}}$.

It is remarked that using omniwheels at different angles we can obtain an overall velocity of the WMR's platform which is greater than the maximum angular velocity of each wheel. For example, selecting in the above 3-wheel case $\beta = 60°$ and $\gamma = 90°$ we get from Eq. (2.71b):

$$v_i = \frac{\sqrt{3}}{2} v_h, \text{ i.e., } v_h = \frac{2}{\sqrt{3}} v_i > v_i$$

The ratio v_h/v_i is called the *velocity augmentation factor* (VAF) [16]:

$$\text{VAF} = v_h/v_i$$

and depends on the number of wheels used and their angular positions on the robot's body. As a further example, consider a 4-wheel robot with $\beta = 45°$ and $\gamma = 90°$. Then, Eq. (2.71b) gives:

$$\text{VAF} = \sqrt{2}$$

Example 2.4

We are given the 4-universal-wheel omnidirectional robot of Figure 2.13, where the angles of the wheels with respect to the axis Qx_r of the vehicle's coordinate frame are β_i $(i = 1, 2, 3, 4)$. Derive the kinematic equations of the robot in terms of the unit directional vectors $\mathbf{u}_i$ $(i = 1, 2, 3, 4)$ of the wheel velocities, with respect to the local coordinate frame Qx_ry_r.

Solution

Let $\dot{\phi}$ be the robot's angular velocity, and $\mathbf{v}_Q$ its linear velocity with world-frame coordinates $\dot{x}_Q$ and $\dot{y}_Q$.

The unit directional vectors of the wheel velocities are:

$$\mathbf{u}_1 = \begin{bmatrix} -\sin\beta_1 \\ \cos\beta_1 \end{bmatrix}, \quad \mathbf{u}_2 = \begin{bmatrix} -\sin\beta_2 \\ \cos\beta_2 \end{bmatrix}, \quad \mathbf{u}_3 = \begin{bmatrix} -\sin\beta_3 \\ \cos\beta_3 \end{bmatrix}, \quad \mathbf{u}_4 = \begin{bmatrix} 0 \\ 1 \end{bmatrix}$$

where it was assumed that the axis of wheel 4 coincides with axis Qx_r.

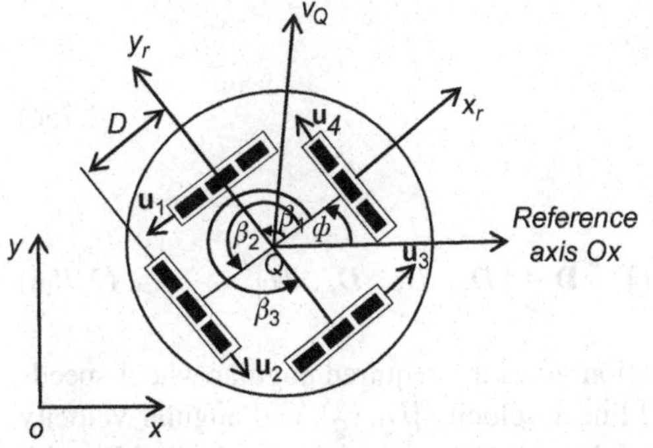

Figure 2.13 Four-wheel omnidirectional robot.

The relation between $\dot{x}_Q$, $\dot{y}_Q$ and $\dot{x}_r$, $\dot{y}_r$ is given by the rotational matrix $\mathbf{R}(\phi)$ (Eq. 2.17), that is:

$$\begin{bmatrix} \dot{x}_Q \\ \dot{y}_Q \end{bmatrix} = \begin{bmatrix} \cos\phi & -\sin\phi \\ \sin\phi & \cos\phi \end{bmatrix} \begin{bmatrix} \dot{x}_r \\ \dot{y}_r \end{bmatrix} = \mathbf{R}(\phi) \begin{bmatrix} \dot{x}_r \\ \dot{y}_r \end{bmatrix}$$

or

$$\begin{bmatrix} \dot{x}_r \\ \dot{y}_r \end{bmatrix} = \begin{bmatrix} \cos\phi & \sin\phi \\ -\sin\phi & \cos\phi \end{bmatrix} \begin{bmatrix} \dot{x}_Q \\ \dot{y}_Q \end{bmatrix} = \mathbf{R}^{-1}(\phi) \begin{bmatrix} \dot{x}_Q \\ \dot{y}_Q \end{bmatrix}$$

Now, we have:

$$v_1 = r\dot{\theta}_1 = \mathbf{u}_1^T \begin{bmatrix} \dot{x}_r \\ \dot{y}_r \end{bmatrix} + D\dot{\phi} = \mathbf{u}_1^T \mathbf{R}^{-1}(\phi) \begin{bmatrix} \dot{x}_Q \\ \dot{y}_Q \end{bmatrix} + D\dot{\phi}$$

$$v_2 = r\dot{\theta}_2 = \mathbf{u}_2^T \begin{bmatrix} \dot{x}_r \\ \dot{y}_r \end{bmatrix} + D\dot{\phi} = \mathbf{u}_2^T \mathbf{R}^{-1}(\phi) \begin{bmatrix} \dot{x}_Q \\ \dot{y}_Q \end{bmatrix} + D\dot{\phi}$$

$$v_3 = r\dot{\theta}_3 = \mathbf{u}_3^T \begin{bmatrix} \dot{x}_r \\ \dot{y}_r \end{bmatrix} + D\dot{\phi} = \mathbf{u}_3^T \mathbf{R}^{-1}(\phi) \begin{bmatrix} \dot{x}_Q \\ \dot{y}_Q \end{bmatrix} + D\dot{\phi}$$

$$v_4 = r\dot{\theta}_4 = \mathbf{u}_4^T \begin{bmatrix} \dot{x}_r \\ \dot{y}_r \end{bmatrix} + D\dot{\phi} = \mathbf{u}_4^T \mathbf{R}^{-1}(\phi) \begin{bmatrix} \dot{x}_Q \\ \dot{y}_Q \end{bmatrix} + D\dot{\phi}$$

or, in compact, form:

$$\dot{\mathbf{q}} = \mathbf{J}^{-1}\dot{\mathbf{p}}_Q \tag{2.75a}$$

where

$$\dot{\mathbf{q}} = \begin{bmatrix} \dot{\theta}_1 \\ \dot{\theta}_2 \\ \dot{\theta}_3 \\ \dot{\theta}_4 \end{bmatrix}, \quad \dot{\mathbf{p}}_Q = \begin{bmatrix} \dot{x}_Q \\ \dot{y}_Q \\ \dot{\phi} \end{bmatrix} \tag{2.75b}$$

$$\mathbf{J}^{-1} = \frac{1}{r}(\mathbf{U}^T\mathbf{R}^{-1}(\phi) + \overline{\mathbf{D}}) \tag{2.75c}$$

with:

$$\mathbf{U} = [\mathbf{u}_1 \;\vdots\; \mathbf{u}_2 \;\vdots\; \mathbf{u}_3 \;\vdots\; \mathbf{u}_4], \quad \overline{\mathbf{D}} = [D, \; D, \; D, \; D]^T \tag{2.75d}$$

As usual, this inverse Jacobian equation gives the required angular wheel speeds $\dot{\theta}_i (i = 1, 2, 3, 4)$ that lead to the desired linear velocity $[\dot{x}_Q, \dot{y}_Q]$, and angular velocity $\dot{\phi}$ of the robot. A discussion of the modeling and control problem of a WMR with this structure is provided in Ref. [17].

2.4.2 Four–Wheel Omnidirectional WMR with Mecanum Wheels

Consider the 4-wheel WMR of Figure 2.14, where the mecanum wheels have roller angle $\pm 45°$[2,4].

Here, we have four-wheel coordinate frames $O_{ci}(i = 1, 2, 3, 4)$. The angular velocity $\dot{q}_i$ of the wheel i has three components:

1. $\dot{\theta}_{ix}$: rotation speed around the hub
2. $\dot{\theta}_{ir}$: rotation speed of the roller i
3. $\dot{\theta}_{iz}$: rotation speed of the wheel around the contact point.

The wheel velocity vector $\mathbf{v}_{ci} = \left[\dot{x}_{ci}, \quad \dot{y}_{ci}, \quad \dot{\phi}_{ci}\right]^T$ in O_{ci} coordinates is given by:

$$\begin{bmatrix} \dot{x}_{ci} \\ \dot{y}_{ci} \\ \dot{\phi}_{ci} \end{bmatrix} = \begin{bmatrix} 0 & r_i \sin\alpha_i & 0 \\ R_i & -r_i \cos\alpha_i & 0 \\ 0 & 0 & 1 \end{bmatrix} \begin{bmatrix} \dot{\theta}_{ix} \\ \dot{\theta}_{ir} \\ \dot{\theta}_{iz} \end{bmatrix} \tag{2.76}$$

for $i = 1, 2, 3, 4$, where R_i is the wheel radius, r_i is the roller radius, and a_i the roller angle. The robot velocity vector $\dot{\mathbf{p}}_Q = [\dot{x}_Q, \dot{y}_Q, \dot{\phi}_Q]^T$ in the Ox_Qy_Q coordinate frame (Eqs. 2.9–2.13) is:

$$\dot{\mathbf{p}}_Q = \begin{bmatrix} \dot{x}_Q \\ \dot{y}_Q \\ \dot{\phi}_Q \end{bmatrix} = \begin{bmatrix} \cos\phi_{ci}^Q & -\sin\phi_{ci}^Q & d_{ciy}^Q \\ \sin\phi_{ci}^Q & \cos\phi_{ci}^Q & -d_{cix}^Q \\ 0 & 0 & 1 \end{bmatrix} \begin{bmatrix} \dot{x}_{ci} \\ \dot{y}_{ci} \\ \dot{\phi}_{ci} \end{bmatrix} \tag{2.77}$$

where ϕ_{ci}^Q denotes the rotation angle (orientation) of the frame O_{ci} with respect to Qx_Qy_Q, and d_{cix}^Q, d_{ciy}^Q are the translations of O_{ci} with respect to Qx_Qy_Q. Introducing Eq. (2.76) into Eq. (2.77) we get:

$$\dot{\mathbf{p}}_Q = \mathbf{J}_i\dot{\mathbf{q}}_i \quad (i = 1, 2, 3, 4) \tag{2.78}$$

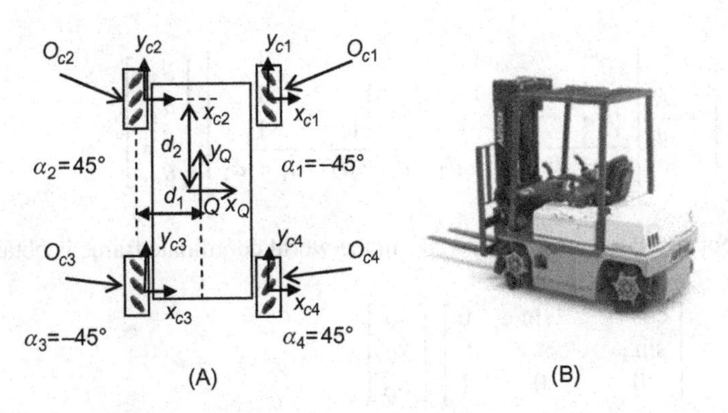

Figure 2.14 Four-mecanum-wheel WMR (A) Kinematic geometry (B) A real 4-mecanum-wheel WMR.
Source: http://www.automotto.com/entry/airtrax-wheels-go-in-any-direction.

where $\dot{\mathbf{q}}_i = \left[\dot{\theta}_{ix}, \dot{\theta}_{ir}, \dot{\theta}_{iz}\right]^T$, and

$$\mathbf{J}_i = \begin{bmatrix} -R_i \sin \phi_{ci}^Q & r_i \sin(\phi_{ci}^Q + \alpha_i) & d_{ciy}^Q \\ R_i \cos \phi_{ci}^Q & -r_i \cos(\phi_{ci}^Q + \alpha_i) & -d_{cix}^Q \\ 0 & 0 & 1 \end{bmatrix} \tag{2.79}$$

is the Jacobian matrix of wheel i, which is square and invertible. If all wheels are identical (except for the orientation of the rollers), the kinematic parameters of the robot in the configuration shown in Figure 2.14 are:

$$\begin{aligned} &R_i = R, \quad r_i = r, \quad \phi_{ci}^Q = 0 \\ &|d_{cix}^Q| = d_1, \quad |d_{ciy}^Q| = d_2 \\ &\alpha_1 = a_3 = -45^\circ, \quad \alpha_2 = \alpha_4 = 45^\circ \end{aligned} \tag{2.80}$$

Thus, the Jacobian matrices (2.79) are:

$$\begin{aligned} \mathbf{J}_1 &= \begin{bmatrix} 0 & -r\sqrt{2}/2 & d_2 \\ R & -r\sqrt{2}/2 & d_1 \\ 0 & 0 & 1 \end{bmatrix}, \quad \mathbf{J}_2 = \begin{bmatrix} 0 & r\sqrt{2}/2 & d_2 \\ R & -r\sqrt{2}/2 & -d_1 \\ 0 & 0 & 1 \end{bmatrix} \\ \mathbf{J}_3 &= \begin{bmatrix} 0 & -r\sqrt{2}/2 & -d_2 \\ R & -r\sqrt{2}/2 & -d_1 \\ 0 & 0 & 1 \end{bmatrix}, \quad \mathbf{J}_4 = \begin{bmatrix} 0 & r\sqrt{2}/2 & -d_2 \\ R & -r\sqrt{2}/2 & d_1 \\ 0 & 0 & 1 \end{bmatrix} \end{aligned} \tag{2.81}$$

The robot motion is produced by the simultaneous motion of all wheels.

In terms of $\dot{\theta}_{ix}$ (i.e., the wheels' angular velocities around their axles) the velocity vector $\dot{\mathbf{p}}_Q$ is given by:

$$\begin{bmatrix} \dot{x}_Q \\ \dot{y}_Q \\ \dot{\phi}_Q \end{bmatrix} = \frac{R}{4} \begin{bmatrix} -1 & 1 & -1 & 1 \\ 1 & 1 & 1 & 1 \\ \dfrac{1}{d_1 + d_2} & \dfrac{-1}{d_1 + d_2} & \dfrac{-1}{d_1 + d_2} & \dfrac{1}{d_1 + d_2} \end{bmatrix} \begin{bmatrix} \dot{\theta}_{1x} \\ \dot{\theta}_{2x} \\ \dot{\theta}_{3x} \\ \dot{\theta}_{4x} \end{bmatrix} \tag{2.82}$$

The robot speed vector $\dot{\mathbf{p}} = \left[\dot{x}, \dot{y}, \dot{\phi}\right]^T$ in the world coordinate frame is obtained as:

$$\begin{bmatrix} \dot{x} \\ \dot{y} \\ \dot{\phi} \end{bmatrix} = \begin{bmatrix} \cos \phi & -\sin \phi & 0 \\ \sin \phi & \cos \phi & 0 \\ 0 & 0 & 1 \end{bmatrix} \begin{bmatrix} \dot{x}_Q \\ \dot{y}_Q \\ \dot{\phi}_Q \end{bmatrix} \tag{2.83}$$

where ϕ is the rotation angle of the platform's coordinate frame Qx_Qy_Q around the z axis which is orthogonal to Oxy. Inverting Eqs. (2.82) and (2.83) we get the inverse kinematic model, which gives the angular speeds $\dot{\theta}_{ix}(i = 1, 2, 3, 4)$ of the wheels around their hubs required to get a desired speed $[\dot{x}, \dot{y}, \dot{\phi}]^T$ of the robot:

$$\begin{bmatrix} \dot{\theta}_{1x} \\ \dot{\theta}_{2x} \\ \dot{\theta}_{3x} \\ \dot{\theta}_{4x} \end{bmatrix} = \frac{1}{R} \begin{bmatrix} -1 & 1 & (d_1 + d_2) \\ 1 & 1 & -(d_1 + d_2) \\ -1 & 1 & -(d_1 + d_2) \\ 1 & 1 & (d_1 + d_2) \end{bmatrix} \begin{bmatrix} \dot{x}_Q \\ \dot{y}_Q \\ \dot{\phi}_Q \end{bmatrix} \tag{2.84}$$

$$\begin{bmatrix} \dot{x}_Q \\ \dot{y}_Q \\ \dot{\phi}_Q \end{bmatrix} = \begin{bmatrix} \cos\phi & \sin\phi & 0 \\ -\sin\phi & \cos\phi & 0 \\ 0 & 0 & 1 \end{bmatrix} \begin{bmatrix} \dot{x} \\ \dot{y} \\ \dot{\phi} \end{bmatrix} \tag{2.85}$$

For historical awareness, we mention here that the mecanum wheel was invented by the Swedish engineer *Bengt Ilon* in 1973 during his work at the Swedish company Mecanum AB. For this reason it is also known as *Ilon wheel* or *Swedish wheel.*

Example 2.5

It is desired to construct a mecanum wheel with n rollers of angle α. Determine the roller length D_r and the thickness d of the wheel.

Solution

We consider the wheel geometry shown in Figure 2.15, where R is the wheel radius [18].

From this figure we get the following relations:

$$n = 2\pi/\phi \tag{2.86a}$$

$$\sin(\phi/2) = b/R \tag{2.86b}$$

$$2b = D_r \sin\alpha \tag{2.86c}$$

$$d = D_r \cos\alpha \tag{2.86d}$$

From Eq. (2.86a–c) we have:

$$\sin\left(\frac{\pi}{n}\right) = \left(\frac{D_r}{2R}\right)\sin\alpha \tag{2.87}$$

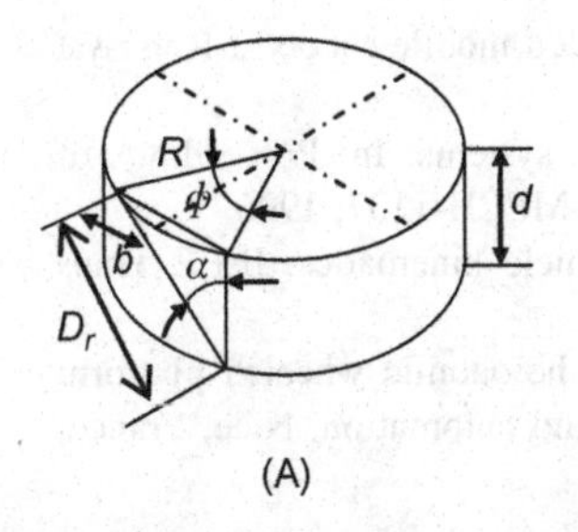

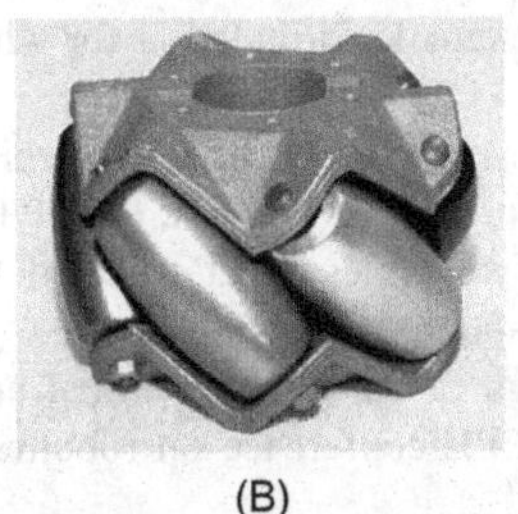

Figure 2.15 (A) Geometry of mecanum wheel where the rollers are assumed to be placed peripherally, (B) A 6-roller wheel example. *Source:* http://store.kornylak.com/SearchResults.asp?Cat=7.

whence:

$$D_r = 2R\frac{\sin(\pi/n)}{\sin\alpha} \tag{2.88}$$

Solving (2.87) for $\sin\alpha$ and noting that $\cos\alpha = \sin\alpha/tg\alpha$, (2.86d) gives:

$$d = 2R\frac{\sin(\pi/n)}{tg\alpha} \tag{2.89}$$

For a roller angle $\alpha = 45^0$, Eqs. (2.88) and (2.89) give:

$$D_r = 2\sqrt{2}R\sin(\pi/n) \tag{2.90a}$$

$$d = 2R\sin(\pi/n) \tag{2.90b}$$

For a roller angle $\alpha = 90^0$ (universal wheel) we get:

$$D_r = 2R\sin(\pi/n) \tag{2.91a}$$

$$d = 0 \quad \text{(ideally)} \tag{2.91b}$$

In this case, d can have any convenient value required by other design considerations.

References

[1] Angelo A. Robotics: a reference guide to new technology. Boston, MA: Greenwood Press; 2007.
[2] Muir PF, Neuman CP. Kinematic modeling of wheeled mobile robots. J Rob Syst 1987;4(2):281–329.
[3] Alexander JC, Maddocks JH. On the kinematics of wheeled mobile robots. Int J Rob Res 1981;8(5):15–27.
[4] Muir PF, Neuman C. Kinematic modeling for feedback control of an omnidirectional wheeled mobile robot. In: Proceedings of IEEE international conference on robotics and automation, Raleigh, NC; 1987, p. 1772–8.
[5] Kim DS, Hyun Kwon W, Park HS. Geometric kinematics and applications of a mobile robot. Int J Control Autom Syst 2003;1(3):376–84.
[6] Rajagopalan R. A generic kinematic formulation for wheeled mobile robots. J Rob Syst 1997;14:77–91.
[7] Sreenivasan SV. Kinematic geometry of wheeled vehicle systems. In: Proceedings of 24th ASME mechanism conference, Irvine, CA, 96-DETC-MECH-1137; 1996.
[8] Balakrishna R, Ghosal A. Two dimensional wheeled vehicle kinematics. IEEE Trans Rob Autom 1995;11(1):126–30.
[9] Killough SM, Pin FG. Design of an omnidirectional and holonomic wheeled platform design. In: Proceedings of IEEE conference on robotics and automation, Nice, France; 1992, p. 84–90.

[10] Sidek N, Sarkar N. Dynamic modeling and control of nonholonomic mobile robot with lateral slip. In: Proceedings of seventh WSEAS international conference on signal processing robotics and automation (ISPRA'08), Cambridge, UK; February 20–22, 2008, p. 66–74.
[11] Giovanni I. Swedish wheeled omnidirectional mobile robots: kinematics analysis and control. IEEE Trans Rob 2009;25(1):164–71.
[12] West M, Asada H. Design of a holonomic omnidirectional vehicle. In: Proceedings of IEEE conference on robotics and automation, Nice, France; May 1992, p. 97–103.
[13] Chakraborty N, Ghosal A. Kinematics of wheeled mobile robots on uneven terrain. Mech Mach Theory 2004;39:1273–87.
[14] Sordalen OJ, Egeland O. Exponential stabilization of nonholonomic chained systems. IEEE Trans Autom Control 1995;40(1):35–49.
[15] Khalil H. Nonlinear Systems. Upper Saddle River, NJ: Prentice Hall; 2001.
[16] Ashmore M, Barnes N. Omni-drive robot motion on curved paths: the fastest path between two points is not a straight line. In: Proceedings of 15th Australian joint conference on artificial intelligence: advances in artificial intelligence (AI'02). London: Springer; 2002. p. 225–36.
[17] Huang L, Lim YS, Li D, Teoh CEL. Design and analysis of a four-wheel omnidirectional mobile robot. In: Proceedings of second international conference on autonomous robots and agents, Palmerston North, New Zealand; December 2004. p. 425–8.
[18] Doroftei I, Grosu V, Spinu V. Omnidirectional mobile robot: design and implementation. In: Habib MK, editor. Bioinspiration and robotics: walking and climbing robots. Vienna, Austria: I-Tech; 2007. p. 512–27.
[19] Phairoh T, Williamson K. Autonomous mobile robots using real time kinematic signal correction and global positioning system control. In: Proceedings of 2008 IAJC-IJME international conference on engineering and technology, Sheraton, Nashville, TN; November 2008, Paper 087/IT304.
[20] De Luca A, Oriolo G, Samson C. Feedback control of a nonholonomic car-like robot. In: Laumond J-P, editor. Robot motion planning and control. Berlin, New York: Springer; 1998. p. 171–253.

3 Mobile Robot Dynamics

3.1 Introduction

The next problem in the study of all types of robots, after kinematics, is the *dynamic modeling* [1–4]. Dynamic modeling is performed using the laws of mechanics that are based on the three physical elements: inertia, elasticity, and friction that are present in any real mechanical system such as the robot. Mobile robot dynamics is a challenging field on its own and has attracted considerable attention by researchers and engineers over the years. Most mobile robots, employed in practice, use conventional wheels and are subject to nonholonomic constraints that need particular treatment. Delicate stability and control problems, that often have to be faced in the design of a mobile robot, are due to the existence of longitudinal and lateral slip in the movement of the wheeled mobile robot (WMR) wheels [5–7].

This chapter has the following objectives:

- To present the general dynamic modeling concepts and techniques of robots.
- To study the Newton–Euler and Lagrange dynamic models of differential-drive mobile robots.
- To study the dynamics of differential-drive mobile robots with longitudinal and lateral slip.
- To derive a dynamic model of car-like WMRs.
- To derive a dynamic model of three-wheel omnidirectional robots.
- To derive a dynamic model of four-wheel mecanum omnidirectional robots.

3.2 General Robot Dynamic Modeling

Robot dynamic modeling deals with the derivation of the dynamic equations of the robot motion. This can be done using the following two methodologies:

- Newton–Euler method
- Lagrange method

The complexity of the Newton–Euler method is $O(n)$, whereas the complexity of the Lagrange method can only be reduced up to $O(n^3)$, where n is the number of degrees of freedom.

Introduction to Mobile Robot Control. DOI: http://dx.doi.org/10.1016/B978-0-12-417049-0.00003-1

Like kinematics, dynamics is distinguished in:

- Direct dynamics
- Inverse dynamics

Direct dynamics provides the dynamic equations that describe the dynamic responses of the robot to given forces/torques $\tau_1, \tau_2, \ldots, \tau_m$ that are exerted by the motors.

Inverse dynamics provides the forces/torques that are needed to get desired trajectories of the robot links. Direct and inverse dynamic modeling is pictorially illustrated in Figure 3.1.

In the inverse dynamic model the inputs are the desired trajectories of the link variables, and outputs the motor torques.

3.2.1 Newton–Euler Dynamic Model

This model is derived by direct application of the Newton–Euler equations for translational and rotational motion. Consider the object B_i (robotic link, WMR, etc.) of Figure 3.2 to which a total force $\mathbf{f}_i$ is applied at its *center of gravity* (COG).

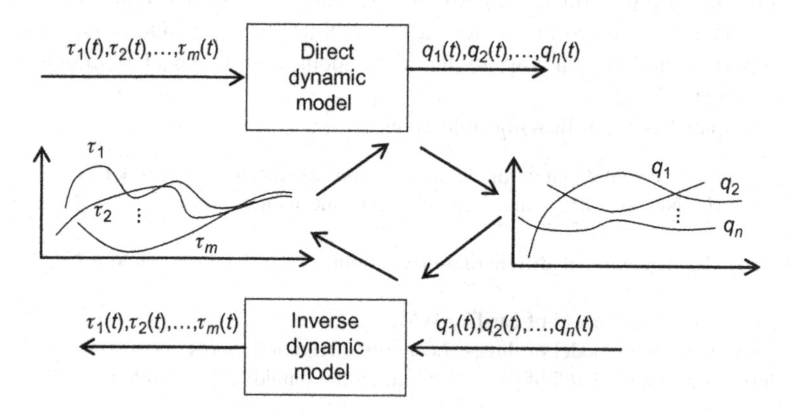

Figure 3.1 Direct and inverse dynamic modeling.

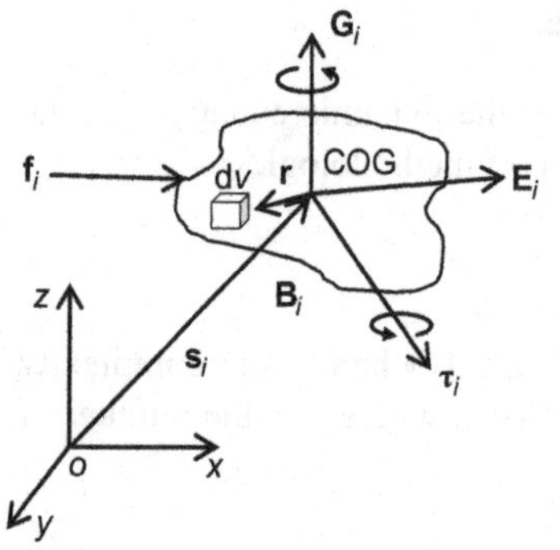

Figure 3.2 A solid body B_i and the inertial coordinate frame $Oxyz$.

Then, its translational motion is described by:

$$\frac{d\mathbf{E}_i}{dt} = \mathbf{f}_i \tag{3.1}$$

Here, $\mathbf{E}_i$ is the *linear momentum* given by:

$$\mathbf{E}_i = m_i \dot{\mathbf{s}}_i \tag{3.2}$$

where m_i is the mass of the body and $\mathbf{s}_i$ is the position of the COG with respect to the world (inertial) coordinate frame $Oxyz$. Assuming that m_i is constant, then Eqs. (3.1) and (3.2) give:

$$m_i \ddot{\mathbf{s}}_i = \mathbf{F}_i \tag{3.3}$$

which is the general *translational dynamic model.*

The rotational motion of B_i is described by:

$$\frac{d\mathbf{G}_i}{dt} = \boldsymbol{\tau}_i \tag{3.4}$$

where $\mathbf{G}_i$ is the total angular momentum of B_i with respect to COG, and $\boldsymbol{\tau}_i$ is the total external torque that produces the rotational motion of the body. The total momentum $\mathbf{G}_i$ is given by:

$$\mathbf{G}_i = \mathbf{I}_i \boldsymbol{\omega}_i \tag{3.5}$$

Here, $\mathbf{I}_i$ is the inertia tensor given by the volume integral:

$$\mathbf{I}_i = \int_{V_i} [\mathbf{r}^T \mathbf{r} \mathbf{I}_3 - \mathbf{r}\mathbf{r}^T] \rho_i dV \tag{3.6}$$

where ρ_i is the mass density of B_i, dV is the volume of an infinitesimal element of B_i lying at the position $\mathbf{r}$ with respect to COG, $\boldsymbol{\omega}_i$ is the angular velocity vector about the inertial axis passing from COG, $\mathbf{I}_3$ is the 3×3 unit matrix, and V_i is the volume of B_i.

3.2.2 Lagrange Dynamic Model

The general Lagrange dynamic model of a solid body (multilink robot, WMR, etc.) is described by[1]:

$$\frac{d}{dt}\left(\frac{\partial L}{\partial \dot{\mathbf{q}}}\right) - \frac{\partial L}{\partial \mathbf{q}} = \boldsymbol{\tau}, \quad \mathbf{q} = [q_1, q_2, \ldots, q_n]^T \tag{3.7}$$

[1] Derivations of Eq. (3.7) from first principles are provided in textbooks on mechanics.

where q_i is the ith-degree of freedom variable, $\boldsymbol{\tau}$ is the external generalized force vector applied to the body (i.e., force for translational motion, and torque for rotational motion), and L is the Lagrangian function defined by:

$$L = K - P \tag{3.8}$$

Here, K is the total kinetic energy, and P the total potential energy of the body given by:

$$\begin{aligned} K &= K_1 + K_2 + \cdots + K_n \\ P &= P_1 + P_2 + \cdots + P_n \end{aligned} \tag{3.9}$$

where K_i is the kinetic energy of link (degree of freedom) i, and P_i its potential energy. The kinetic energy K of the body is equal to:

$$K = \frac{1}{2} m \dot{\mathbf{s}}^{\mathrm{T}} \dot{\mathbf{s}} + \frac{1}{2} \boldsymbol{\omega}^{\mathrm{T}} \mathbf{I} \boldsymbol{\omega} \tag{3.10}$$

where $\dot{\mathbf{s}}$ is the linear velocity of the COG, $\boldsymbol{\omega}$ is the angular velocity of the rotation, m is the mass, and $\mathbf{I}$ the inertia tensor of the body.

3.2.3 Lagrange Model of a Multilink Robot

Given a general multilink robot, the application of Eq. (3.7) leads (always) to a dynamic model of the form:

$$\mathbf{D}(\mathbf{q})\ddot{\mathbf{q}} + \mathbf{h}(\mathbf{q},\dot{\mathbf{q}}) + \mathbf{g}(\mathbf{q}) = \boldsymbol{\tau} \tag{3.11a}$$

where, for any $\dot{\mathbf{q}} \neq \mathbf{0}, \mathbf{D}(\mathbf{q})$ is an $n \times n$ positive define matrix, and:

$$\mathbf{q} = [q_1, q_2, \ldots, q_n]^{\mathrm{T}} \tag{3.11b}$$

is the vector of generalized variables (linear, angular) $q_i, \mathbf{D}(\mathbf{q})\ddot{\mathbf{q}}$ represents the inertia force, $\mathbf{h}(\mathbf{q},\dot{\mathbf{q}})$ represents the centrifugal and Coriolis force, and $\mathbf{g}(\mathbf{q})$ stands for the gravitational force. Since $\boldsymbol{\tau}$ is the net force/torque applied to the robot, if there is also friction force/torque $\boldsymbol{\tau}_f$, then $\boldsymbol{\tau} = \boldsymbol{\tau}' - \boldsymbol{\tau}_f$ where $\boldsymbol{\tau}'$ is the force/torque exerted by the actuators at the joints.

It is useful to note that given the model (3.11a,b) one can express the kinetic energy of the robot as:

$$K = \frac{1}{2} \dot{\mathbf{q}}^T(t) \mathbf{D}(\mathbf{q}) \dot{\mathbf{q}}(t) \tag{3.12}$$

The expressions of $\mathbf{D}(\mathbf{q})$, $\mathbf{h}(\mathbf{q}, \dot{\mathbf{q}})$, and $\mathbf{g}(\mathbf{q})$ have to be derived in each particular case. General derivations of Eq. (3.11a) from Eq. (3.7) are given in standard industrial (fixed) robotics books.

Typically, the function $\mathbf{h}(\mathbf{q},\dot{\mathbf{q}})$ can be written in the form:

$$\mathbf{h}(\mathbf{q},\dot{\mathbf{q}}) = \mathbf{C}(\mathbf{q},\dot{\mathbf{q}})\dot{\mathbf{q}} \tag{3.13}$$

A useful universal property of the Lagrange model given by Eqs. (3.11a)–(3.13) is that the $n \times n$ matrix $\mathbf{A} = \dot{\mathbf{D}} - 2\mathbf{C}$ is antisymmetric, that is, $\mathbf{A}^{\mathrm{T}} = -\mathbf{A}$.

3.2.4 Dynamic Modeling of Nonholonomic Robots

The Lagrange dynamic model of a nonholonomic robot (fixed or mobile) has the form (3.7):

$$\frac{\mathrm{d}}{\mathrm{d}t}\left(\frac{\partial L}{\partial \dot{\mathbf{q}}}\right) - \frac{\partial L}{\partial \mathbf{q}} + \mathbf{M}^{\mathrm{T}}(\mathbf{q})\boldsymbol{\lambda} = \mathbf{E}\boldsymbol{\tau} \tag{3.14}$$

where $\mathbf{M}(\mathbf{q})$ is the $m \times n$ matrix of the m nonholonomic constraints:

$$\mathbf{M}(\mathbf{q})\dot{\mathbf{q}} = \mathbf{0} \tag{3.15}$$

and $\boldsymbol{\lambda}$ is the vector Lagrange multiplier. This model leads to:

$$\mathbf{D}(\mathbf{q})\ddot{\mathbf{q}} + \mathbf{C}(\mathbf{q},\dot{\mathbf{q}})\dot{\mathbf{q}} + \mathbf{g}(\mathbf{q}) + \mathbf{M}^{\mathrm{T}}(\mathbf{q})\boldsymbol{\lambda} = \mathbf{E}\boldsymbol{\tau} \tag{3.16}$$

where $\mathbf{E}$ is a nonsingular transformation matrix. To eliminate the constraint term $\mathbf{M}^{\mathrm{T}}(\mathbf{q})\boldsymbol{\lambda}$ in Eq. (3.16) and get a constraint-free model we use an $n \times (n-m)$ matrix $\mathbf{B}(\mathbf{q})$ which is defined such that:

$$\mathbf{B}^{\mathrm{T}}(\mathbf{q})\mathbf{M}^{\mathrm{T}}(\mathbf{q}) = \mathbf{0} \tag{3.17}$$

From Eqs. (3.15) and (3.17), one can verify that there exists an $(n-m)$-dimensional vector $\mathbf{v}(t)$ such that:

$$\dot{\mathbf{q}}(t) = \mathbf{B}(\mathbf{q})\mathbf{v}(t) \tag{3.18}$$

Now, premultiplying Eq. (3.16) by $\mathbf{B}^{\mathrm{T}}(\mathbf{q})$ and using Eqs. (3.15), (3.17), and (3.18) we get:

$$\overline{\mathbf{D}}(\mathbf{q})\dot{\mathbf{v}} + \overline{\mathbf{C}}(\mathbf{q},\dot{\mathbf{q}})\mathbf{v} + \overline{\mathbf{g}}(\mathbf{q}) = \overline{\mathbf{E}}\boldsymbol{\tau} \tag{3.19a}$$

where:

$$\begin{aligned}\overline{\mathbf{D}} &= \mathbf{B}^{\mathrm{T}}\mathbf{D}\mathbf{B}\\ \overline{\mathbf{C}} &= \mathbf{B}^{\mathrm{T}}\mathbf{D}\dot{\mathbf{B}} + \mathbf{B}^{\mathrm{T}}\mathbf{C}\mathbf{B}\\ \overline{\mathbf{g}} &= \mathbf{B}^{\mathrm{T}}\mathbf{g}\\ \overline{\mathbf{E}} &= \mathbf{B}^{\mathrm{T}}\mathbf{E}\end{aligned} \tag{3.19b}$$

The reduced (unconstrained) model (3.19a,b) describes the dynamic evolution of the n-dimensional vector $\mathbf{q}(t)$ in terms of the dynamic evolution of the $(n-m)$-dimensional vector $\mathbf{v}(t)$.

3.3 Differential-Drive WMR

The dynamic model of the differential-drive WMR will be derived here by both the Newton–Euler and Lagrange methods.

3.3.1 Newton–Euler Dynamic Model

In the present case use will be made of the Newton–Euler equations:

$$m\dot{\mathbf{v}} = \mathbf{F} \quad \text{(Translational motion)} \tag{3.20a}$$

$$\mathrm{I}\dot{\boldsymbol{\omega}} = \mathbf{N} \quad \text{(Rotational motion)} \tag{3.20b}$$

where $\mathbf{F}$ is the total force applied at the COG G, $\mathbf{N}$ is the total torque with respect to the COG m is the mass of the WMR, and I is the inertia of the WMR. Referring to Figure 2.7, and assuming that the COG G coincides with the midpoint Q (i.e., $b = 0$), we find:

$$F = F_{\mathrm{r}} + F_{\mathrm{l}}, \quad \tau_{\mathrm{r}} = rF_{\mathrm{r}}, \quad \tau_{\mathrm{l}} = rF_{\mathrm{l}} \tag{3.21a}$$

i.e:

$$F = \frac{1}{r}(\tau_{\mathrm{r}} + \tau_{\mathrm{l}}) \tag{3.21b}$$

where F_{r} and F_{l} are the forces that produce the torques τ_{r} and τ_{l}, respectively.

Also:

$$N = (F_{\mathrm{r}} - F_{\mathrm{l}})2a = \frac{2a}{r}(\tau_{\mathrm{r}} - \tau_{\mathrm{l}}) \tag{3.22}$$

Therefore, (3.20a,b) give the WMR's dynamic model:

$$\dot{v} = \frac{1}{mr}(\tau_{\mathrm{r}} + \tau_{\mathrm{l}}) \tag{3.23a}$$

$$\dot{\omega} = \frac{2a}{Ir}(\tau_{\mathrm{r}} - \tau_{\mathrm{l}}) \tag{3.23b}$$

3.3.2 Lagrange Dynamic Model

We refer to Figure 2.7 and assume again that the point Q is at the position of point G. Here, the nonholonomic constraint matrix is (Eq. (2.42)):

$$\mathbf{M}(\mathbf{q}) = \begin{bmatrix} -\sin\phi & \cos\phi & 0 \end{bmatrix} \tag{3.24}$$

Since the WMR moves on a horizontal planar terrain, the terms $\mathbf{C}(\mathbf{q},\dot{\mathbf{q}})$ and $\mathbf{g}(\mathbf{q})$ in Eq. (3.16) are zero. Therefore, the model (3.16) becomes:

$$\mathbf{D}(\mathbf{q})\ddot{\mathbf{q}} + \mathbf{M}^{\mathrm{T}}(\mathbf{q})\boldsymbol{\lambda} = \mathbf{E}\boldsymbol{\tau} \tag{3.25}$$

where:

$$\mathbf{q} = \begin{bmatrix} x_Q \\ y_Q \\ \phi \end{bmatrix}, \quad \boldsymbol{\tau} = \begin{bmatrix} \tau_{\mathrm{r}} \\ \tau_{\mathrm{l}} \end{bmatrix} \tag{3.26a}$$

$$\mathbf{D}(\mathbf{q}) = \begin{bmatrix} m & 0 & 0 \\ 0 & m & 0 \\ 0 & 0 & I \end{bmatrix}, \quad \mathbf{E} = \frac{1}{r}\begin{bmatrix} \cos\phi & \cos\phi \\ \sin\phi & \sin\phi \\ 2a & -2a \end{bmatrix} \tag{3.26b}$$

To convert the model (3.25) to the corresponding unconstrained model (3.19a,b), we need the matrix $\mathbf{B}(\mathbf{q})$ in (3.17). Here, this matrix is:

$$\mathbf{B}(\mathbf{q}) = \begin{bmatrix} \cos\phi & 0 \\ \sin\phi & 0 \\ 0 & 1 \end{bmatrix} \tag{3.27}$$

which satisfies Eq. (3.17). Therefore, Eq. (3.19b) gives:

$$\overline{\mathbf{D}} = \mathbf{B}^{\mathrm{T}}\mathbf{D}\mathbf{B} = \begin{bmatrix} m & 0 \\ 0 & I \end{bmatrix}, \quad \overline{\mathbf{E}} = \frac{1}{r}\begin{bmatrix} 1 & 1 \\ 2a & -2a \end{bmatrix} \tag{3.28}$$

and the model (3.19a) becomes:

$$\begin{bmatrix} m & 0 \\ 0 & I \end{bmatrix}\begin{bmatrix} \dot{v}_1 \\ \dot{v}_2 \end{bmatrix} = \frac{1}{r}\begin{bmatrix} 1 & 1 \\ 2a & -2a \end{bmatrix}\begin{bmatrix} \tau_{\mathrm{r}} \\ \tau_{\mathrm{l}} \end{bmatrix} \tag{3.29}$$

Noting that v_1 is the translation velocity v and v_2 the angular velocity ω of the robot, Eq. (3.29) gives the model:

$$\dot{v} = \frac{1}{mr}(\tau_{\mathrm{r}} + \tau_{\mathrm{l}}) \tag{3.30a}$$

$$\dot{\omega} = \frac{2a}{Ir}(\tau_r - \tau_l) \tag{3.30b}$$

which is identical to Eq. (3.23a,b), as expected. Finally, using Eq. (3.18) we get:

$$\begin{bmatrix} \dot{x}_Q \\ \dot{y}_Q \\ \dot{\phi} \end{bmatrix} = \begin{bmatrix} \cos\phi & 0 \\ \sin\phi & 0 \\ 0 & 1 \end{bmatrix} \begin{bmatrix} v \\ \omega \end{bmatrix} = \begin{bmatrix} v\cos\phi \\ v\sin\phi \\ \omega \end{bmatrix} \tag{3.31}$$

which is the WMR's kinematic model. The dynamic and kinematic Eqs. (3.30a,b) and (3.31) describe fully the motion of the differential-drive WMR.

Example 3.1

The problem is to derive the Lagrange dynamic model using directly the Lagrangian function L for a differential-drive WMR in which:

1. There is linear friction in the wheels with the same friction coefficient.
2. The wheel midpoint Q does not coincide with the COG G.
3. The wheel–motor assemblies have nonzero inertia.

Solution

We will work with the WMR of Figure 2.7.The kinetic energy K of the robot is given by:

$$K = K_1 + K_2 + K_3 \tag{3.32}$$

where:

$$\begin{aligned} K_1 &= \frac{1}{2}mv_G^2 = \frac{1}{2}m(\dot{x}_G^2 + \dot{y}_G^2) \\ K_2 &= \frac{1}{2}I_Q\,\dot{\phi}^2 \\ K_3 &= \frac{1}{2}I_o\dot{\theta}_r^2 + \frac{1}{2}I_o\dot{\theta}_l^2 \end{aligned} \tag{3.33}$$

and (Example 2.2):

$$\dot{x}_G = \dot{x}_Q + b\dot{\phi}\sin\phi$$
$$\dot{y}_G = \dot{y}_Q - b\dot{\phi}\cos\phi$$

with:

m = mass of the entire robot
v_G = linear velocity of the COG G
I_Q = moment of inertia of the robot with respect to Q
I_o = moment of inertia of each wheel plus the corresponding motor's rotor moment of inertia.

The velocities $\dot{x}_Q$, $\dot{y}_Q$, and $\dot{\phi}$ are given by Eq. (2.36a–c):

$$\dot{x}_Q = \frac{r}{2}(\dot{\theta}_r \cos\phi + \dot{\theta}_l \cos\phi) = \frac{r}{2}(\dot{\theta}_r + \dot{\theta}_l)\cos\phi$$

$$\dot{y}_Q = \frac{r}{2}(\dot{\theta}_r \sin\phi + \dot{\theta}_l \sin\phi) = \frac{r}{2}(\dot{\theta}_r + \dot{\theta}_l)\sin\phi \tag{3.34}$$

$$\dot{\phi} = \frac{r}{2a}(\dot{\theta}_r - \dot{\theta}_l)$$

Using Eqs. (3.33) and (3.34) in Eq. (3.32) the total kinetic energy K of the robot is found to be:

$$K(\dot{\theta}_r, \dot{\theta}_l) = \left[\frac{mr^2}{8} + \frac{(I_Q + mb^2)r^2}{8a^2} + \frac{I_o}{2}\right]\dot{\theta}_r^2 + \left[\frac{mr^2}{8} + \frac{(I_Q + mb^2)r^2}{8a^2} + \frac{I_o}{2}\right]\dot{\theta}_l^2 + \left[\frac{mr^2}{4} - \frac{(I_Q + mb^2)r^2}{4a^2}\right]\dot{\theta}_r\dot{\theta}_l \tag{3.35}$$

Here the kinetic energy is expressed directly in terms of the angular velocities $\dot{\theta}_r$ and $\dot{\theta}_l$ of the driving wheels. The Lagrangian L is equal to K, since the robot is moving on a horizontal plane and so the potential energy P is zero. Therefore, the Lagrange dynamic equations of this robot are:

$$\frac{d}{dt}\left(\frac{\partial K}{\partial \dot{\theta}_r}\right) - \frac{\partial K}{\partial \theta_r} = \tau_r - \beta\dot{\theta}_r$$
$$\frac{d}{dt}\left(\frac{\partial K}{\partial \dot{\theta}_l}\right) - \frac{\partial K}{\partial \theta_l} = \tau_l - \beta\dot{\theta}_l \tag{3.36}$$

where β is the wheels' common friction coefficient, and τ_r, τ_l are the right and left actuation torques. Using (3.35) in (3.36) we get:

$$D_{11}\ddot{\theta}_r + D_{12}\ddot{\theta}_l + \beta\dot{\theta}_r = \tau_r$$
$$D_{21}\ddot{\theta}_r + D_{22}\ddot{\theta}_l + \beta\dot{\theta}_l = \tau_l \tag{3.37}$$

where:

$$D_{11} = D_{22} = \left[\frac{mr^2}{4} + \frac{(I_Q + mb^2)r^2}{8a^2} + I_o\right]$$
$$D_{12} = D_{21} = \left[\frac{mr^2}{4} - \frac{(I_Q + mb^2)r^2}{8a^2}\right] \tag{3.38}$$

Using the known relations:

$v_r = r\dot{\theta}_r$, $v_l = r\dot{\theta}_l$, $v = (v_r + v_l)/2$, $\omega = (v_r - v_l)/2a$, one can easily verify that in the case where the above conditions 1, 2, and 3 are relaxed (i.e., $\beta = 0$, $b = 0$, $I_o = 0$), the above dynamic model reduces to the model (3.30a,b). This model was implemented and validated using MATLAB/SIMULINK in Ref. [8].

3.3.3 Dynamics of WMR with Slip

Here we will derive the Newton–Euler dynamic model of the slipping differential-drive WMR considered in Example 2.2. The case where both longitudinal slip (variables w_r, w_l) and lateral slip (variables z_r, z_l) are present will be considered [5].

For convenience we write again the kinematic equations of the robot (with steering angles $\zeta_r = \zeta_l = 0$):

$$\dot{\gamma}_r = \dot{x}_G \cos\phi + \dot{y}_G \sin\phi + a\dot{\phi}, \quad \gamma_r = r\theta_r - w_r$$

$$\dot{\gamma}_l = \dot{x}_G \cos\phi + \dot{y}_G \sin\phi - a\dot{\phi}, \quad \gamma_l = r\theta_l - w_l$$

$$\dot{z}_r = -\dot{x}_G \sin\phi + \dot{y}_G \cos\phi + b\dot{\phi}$$

$$\dot{z}_l = -\dot{x}_G \sin\phi + \dot{y}_G \cos\phi + b\dot{\phi}$$

Defining the generalized variables' vector $\mathbf{q}$ as:

$$\mathbf{q} = [x_G, y_G, \phi, z_r, z_l, \gamma_r, \gamma_l, \theta_r, \theta_l] \tag{3.39}$$

the above relations can be written in the following Pfaffian matrix form:

$$\mathbf{M(q)\dot{q} = 0}$$

where:

$$\mathbf{M(q)} = \begin{bmatrix} \cos\phi & \sin\phi & a & 0 & 0 & -1 & 0 & 0 & 0 \\ \cos\phi & \sin\phi & -a & 0 & 0 & 0 & -1 & 0 & 0 \\ -\sin\phi & \cos\phi & b & -1 & 0 & 0 & 0 & 0 & 0 \\ -\sin\phi & \cos\phi & b & 0 & -1 & 0 & 0 & 0 & 0 \end{bmatrix} \tag{3.40}$$

The matrix $\mathbf{B(q)}$ and the velocity vector $\mathbf{v}(t)$ that satisfy the relations (Eqs. (3.17)–(3.18)):

$$\mathbf{B}^{\mathrm{T}}(\mathbf{q})\mathbf{M}^{T}(\mathbf{q}) = \mathbf{0}, \quad \dot{\mathbf{q}} = \mathbf{B(q)v}(t) \tag{3.41}$$

are found to be:

$$\mathbf{B(q)} = \begin{bmatrix} -\sin\phi & \mathrm{A} & C & 0 & 0 \\ \cos\phi & \mathrm{B} & D & 0 & 0 \\ 0 & \dfrac{1}{2a} & -\dfrac{1}{2a} & 0 & 0 \\ 1 & 0 & 0 & 0 & 0 \\ 1 & 0 & 0 & 0 & 0 \\ 0 & 1 & 0 & 0 & 0 \\ 0 & 0 & 1 & 0 & 0 \\ 0 & 0 & 0 & 1 & 0 \\ 0 & 0 & 0 & 0 & 1 \end{bmatrix} \tag{3.42}$$

where:

$$\mathbf{v} = [\dot{z}_1, \dot{\gamma}_r, \dot{\gamma}_l, \dot{\theta}_r, \dot{\theta}_l]^{\mathrm{T}} \quad \text{(note that } z_r = z_l = y_r) \tag{3.43a}$$

$$A = \frac{a\cos\phi - b\sin\phi}{2a}, \quad B = \frac{b\cos\phi + a\sin\phi}{2a}$$
$$C = \frac{a\cos\phi + b\sin\phi}{2a}, \quad D = \frac{a\sin\phi - b\cos\phi}{2a} \tag{3.43b}$$

To derive the Newton–Euler dynamic model we draw the free-body diagram of the WMR body without the wheels (Figure 3.3), and the free-body diagram of the two wheels (Figure 3.4).

In Figures 3.3 and 3.4, we have the following dynamic parameters and variables:

τ_b: The torque given to the wheels by the body of the WMR
τ_r, τ_l: The driving torques exerted to the right and left wheel by their motors
m_b: The mass of the WMR body without the wheels
m_w: The mass of each driving wheel assembly (wheel plus DC motor of the robot body)
I_{bz}: The moment of inertia about a vertical axis passing via G (without the wheels)

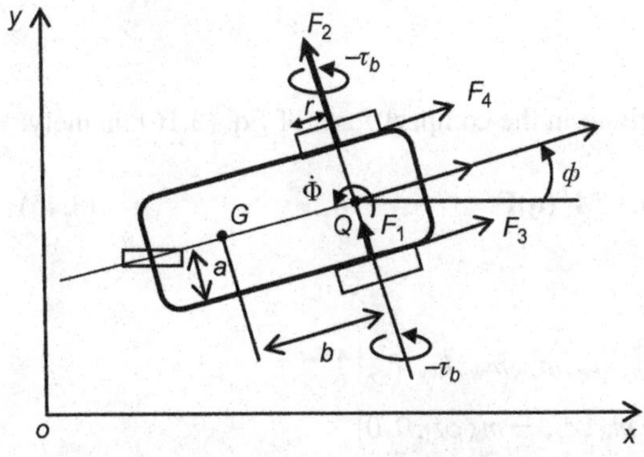

Figure 3.3 Force–torque diagram of the body of the WMR (without the wheels).

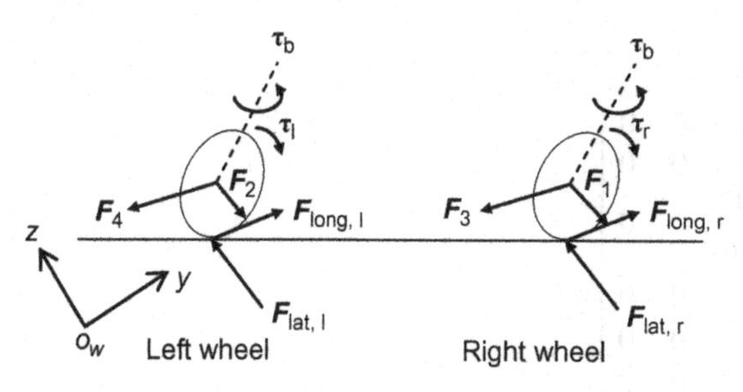

Figure 3.4 Force–torque diagram of the two wheels. The coordinate frame $O_w\gamma z$ is fixed at the driving wheels' midpoint.

I_{wy}: The wheel moment of inertia about its axis
I_{wz}: The wheel moment of inertia about its diameter
F_i: The reaction forces between the WMR body and the wheels
$F_{lat,r}, F_{lat,l}$: The lateral traction force at each wheel.
$F_{long,r}, F_{long,l}$: The longitudinal traction force at each wheel.

Using the above notation, the Newton–Euler dynamic equations of the robot, written down for each generalized variable in the vector (3.39), are:

$$\begin{aligned}
m_b\ddot{x}_G &= (F_1 + F_2)\sin\phi - (F_3 + F_4)\cos\phi \\
m_b\ddot{y}_G &= -(F_1 + F_2)\cos\phi - (F_3 + F_4)\sin\phi \\
(I_{bz} + 2I_{wz})\ddot{\phi} &= (F_1 + F_2)b - (F_3 + F_4)a \\
m_w\ddot{z}_r + m_w\dot{\phi}\dot{\gamma}_r &= F_{lat,r} - F_1 \\
m_w\ddot{z}_l + m_w\dot{\phi}\dot{\gamma}_l &= F_{lat,l} - F_2 \\
m_w\ddot{\gamma}_r - m_w\dot{\phi}\dot{z}_r &= F_{long,r} - F_3 \\
m_w\ddot{\gamma}_l - m_w\dot{\phi}\dot{z}_l &= F_{long,l} - F_4 \\
I_{wy}\ddot{\theta}_r &= \tau_r - F_{long,r}r \\
I_{wy}\ddot{\theta}_l &= \tau_l - F_{long,l}r
\end{aligned} \tag{3.44}$$

where $\tau_b = I_{wz}\ddot{\phi}$.

The detailed Eq. (3.44) can be written in the compact form of Eq. (3.16), namely:

$$\mathbf{D(q)\ddot{q} + h(q,\dot{q}) = E(q)\tau + f(\dot{q}) + M^T(q)F} \tag{3.45}$$

where $\mathbf{M(q)}$ is given by (3.40), and:

$$\mathbf{D} = \text{diag}\left[m_b, m_b, I_{bz} + 2I_{wz}, m_w, m_w, m_w, m_w, I_{wy}, I_{wy}\right]$$

$$\mathbf{h} = \left[0, 0, 0, m_w\dot{\phi}\dot{\gamma}_r, m_w\dot{\phi}\dot{\gamma}_l, -m_w\dot{\phi}\dot{z}_r, -m_w\dot{\phi}\dot{z}_l, 0, 0\right]$$

$$\mathbf{E} = \left[\frac{\mathbf{O}_{2\times 7}}{\mathbf{I}_{2\times 2}} \right], \boldsymbol{\tau} = \begin{bmatrix} \tau_r \\ \tau_l \end{bmatrix}$$

$$\mathbf{f} = \left[0, 0, 0, F_{\text{lat},r}, F_{\text{lat},l}, F_{\text{long},r}, F_{\text{long},l}, -rF_{\text{long},r}, -rF_{\text{long},l}\right]^{\text{T}}$$

$$\mathbf{F} = [F_1, F_2, F_3, F_4]^{\text{T}}$$

Applying to Eq. (3.45) the procedure of Section 3.2.4, we get the following reduced model (3.19a,b):

$$\overline{\mathbf{D}}(\mathbf{q})\dot{\mathbf{v}} + \overline{\mathbf{C}}_1(\mathbf{q},\dot{\mathbf{q}})\mathbf{v} + \overline{\mathbf{h}}(\mathbf{q},\dot{\mathbf{q}}) = \overline{\mathbf{E}}\boldsymbol{\tau} + \overline{\mathbf{f}}(\mathbf{q}) \tag{3.46}$$

where:

$$\overline{\mathbf{D}}(\mathbf{q}) = \mathbf{B}^{\text{T}}\mathbf{D}\mathbf{B}, \quad \overline{\mathbf{C}}_1(\mathbf{q},\dot{\mathbf{q}}) = \mathbf{B}^{\text{T}}\mathbf{D}\dot{\mathbf{B}} \quad \overline{\mathbf{h}}(\mathbf{q},\dot{\mathbf{q}}) = \mathbf{B}^{\text{T}}\mathbf{h}, \quad \overline{\mathbf{E}} = \mathbf{B}^{\text{T}}\mathbf{E}, \quad \overline{\mathbf{f}}(\mathbf{q}) = \mathbf{B}^{\text{T}}\mathbf{f}$$

The model (3.46) can be split as:

$$\overline{\hat{\mathbf{D}}}(\hat{\mathbf{q}})\dot{\hat{\mathbf{v}}} + \overline{\hat{\mathbf{C}}}_1(\hat{\mathbf{q}},\dot{\hat{\mathbf{q}}})\hat{\mathbf{v}} + \overline{\hat{\mathbf{h}}}(\hat{\mathbf{q}},\dot{\hat{\mathbf{q}}}) = \overline{\hat{\mathbf{f}}} \tag{3.47a}$$

$$I_{wy}\ddot{\theta}_{\text{r}} = \tau_{\text{r}} - rF_{\text{long},r} \tag{3.47b}$$

$$I_{wy}\ddot{\theta}_{\text{l}} = \tau_{\text{l}} - rF_{\text{long},l} \tag{3.47c}$$

where:

$$\overline{\hat{\mathbf{D}}} = \hat{\mathbf{B}}^{\text{T}}\hat{\mathbf{D}}\hat{\mathbf{B}}, \quad \overline{\hat{\mathbf{C}}}_1 = \hat{\mathbf{B}}^{\text{T}}\hat{\mathbf{D}}\dot{\hat{\mathbf{B}}}$$

$$\overline{\hat{\mathbf{h}}} = \hat{\mathbf{B}}^{\text{T}}\hat{\mathbf{h}}, \quad \overline{\hat{\mathbf{f}}} = \hat{\mathbf{B}}^{\text{T}}\hat{\mathbf{f}}$$

$$\hat{\mathbf{D}} = \text{diag}[m_{\text{b}}, m_{\text{b}}, I_{bz} + 2I_{wz}, m_{\text{w}}, m_{\text{w}}, m_{\text{w}}m_{\text{w}}]^{\text{T}}$$

$$\hat{\mathbf{v}} = [\dot{z}_1, \dot{\gamma}_{\text{r}}, \dot{\gamma}_1]^{\text{T}}$$

$$\hat{\mathbf{h}} = [0, 0, 0, m_{\text{w}}\dot{\phi}\dot{\gamma}_{\text{r}}, m_{\text{w}}\dot{\phi}\dot{\gamma}_{\text{l}}, -m_{\text{w}}\dot{\phi}\dot{z}_{\text{r}}, -m_{\text{w}}\dot{\phi}\dot{z}_{\text{l}}]^{\text{T}}$$

$$\hat{\mathbf{f}} = [0, 0, 0, F_{\text{lat,r}}, F_{\text{lat,l}}, F_{\text{long,r}}, F_{\text{long,l}}]^{\text{T}}$$

$$\hat{\mathbf{q}} = [x_G, y_G, \phi, z_{\text{r}}, z_{\text{l}}, \gamma_{\text{r}}, \gamma_{\text{l}}]^{\text{T}}$$

$$\mathbf{B}(\hat{\mathbf{q}}) = \begin{bmatrix} -\sin\phi & \text{A} & C \\ \cos\phi & \text{B} & D \\ 0 & \dfrac{1}{2a} & -\dfrac{1}{2a} \\ 1 & 0 & 0 \\ 1 & 0 & 0 \\ 0 & 1 & 0 \\ 0 & 0 & 1 \end{bmatrix} \tag{3.48}$$

with A, B, C, and D as in Eq. (3.43b). Writing $\hat{\mathbf{h}}(\hat{\mathbf{q}},\dot{\hat{\mathbf{q}}})$ in the form of Eq. (3.13), that is, $\hat{\mathbf{h}}(\hat{\mathbf{q}},\dot{\hat{\mathbf{q}}}) = \hat{\mathbf{C}}_2(\hat{\mathbf{q}},\dot{\hat{\mathbf{q}}})\dot{\hat{\mathbf{q}}}$, the model (3.47a) takes the form:

$$\bar{\hat{\mathbf{D}}}(\hat{\mathbf{q}})\dot{\hat{\mathbf{v}}} + \bar{\hat{\mathbf{C}}}(\hat{\mathbf{q}},\dot{\hat{\mathbf{q}}})\hat{\mathbf{v}} = \bar{\hat{\mathbf{f}}} \tag{3.49}$$

where:

$$\bar{\hat{\mathbf{C}}}(\hat{\mathbf{q}},\dot{\hat{\mathbf{q}}}) = \bar{\hat{\mathbf{C}}}_1(\hat{\mathbf{q}},\dot{\hat{\mathbf{q}}}) + \bar{\hat{\mathbf{C}}}_2(\hat{\mathbf{q}},\dot{\hat{\mathbf{q}}}), \quad \bar{\hat{\mathbf{C}}}_2(\hat{\mathbf{q}},\dot{\hat{\mathbf{q}}}) = \hat{\mathbf{B}}^{\mathrm{T}}\hat{\mathbf{C}}_2(\hat{\mathbf{q}},\dot{\hat{\mathbf{q}}})$$

Finally, Eq. (3.47b,c) can be written in the matrix form:

$$\mathbf{I}\dot{\boldsymbol{\theta}} = \boldsymbol{\tau} - r\mathbf{f} \tag{3.50}$$

where:

$$\mathbf{I} = \begin{bmatrix} I_{wy} & 0 \\ 0 & I_{wy} \end{bmatrix}, \quad \boldsymbol{\theta} = \begin{bmatrix} \dot{\theta}_{\mathrm{r}} \\ \dot{\theta}_{\mathrm{l}} \end{bmatrix}, \quad \boldsymbol{\tau} = \begin{bmatrix} \tau_{\mathrm{r}} \\ \tau_{\mathrm{l}} \end{bmatrix}, \quad \mathbf{f} = \begin{bmatrix} F_{\mathrm{long},r} \\ F_{\mathrm{long},l} \end{bmatrix} \tag{3.51}$$

To summarize, the dynamic model of the differential-drive WMR with slip is:

$$\bar{\hat{\mathbf{D}}}(\hat{\mathbf{q}})\dot{\hat{\mathbf{v}}} + \bar{\hat{\mathbf{C}}}(\hat{\mathbf{q}},\dot{\hat{\mathbf{q}}})\hat{\mathbf{v}} = \bar{\hat{\mathbf{f}}} \tag{3.52a}$$

$$\mathbf{I}\dot{\boldsymbol{\theta}} = \boldsymbol{\tau} - r\mathbf{f} \tag{3.52b}$$

with:

$$\begin{aligned} \hat{\mathbf{q}} &= [x_G, y_G, \phi, z_{\mathrm{r}}, z_{\mathrm{l}}, \gamma_{\mathrm{r}}, \gamma_{\mathrm{l}}]^{\mathrm{T}} \\ \hat{\mathbf{v}} &= [\dot{z}_{\mathrm{l}}, \dot{\gamma}_{\mathrm{r}}, \dot{\gamma}_{\mathrm{l}}]^{\mathrm{T}} \end{aligned} \tag{3.52c}$$

3.4 Car-Like WMR Dynamic Model

The kinematic equations of the car-like WMR were derived in Section 2.6. Here, we will derive the Newton–Euler dynamic model for a four-wheel rear-drive front-steer WMR [9]. The equivalent bicycle model shown in Figure 3.5 will be used.

The WMR is subject to the driving force F_{d} and two lateral slip forces F_{r} and F_{f} applied perpendicular to the corresponding wheels. Before presenting the dynamic equations we derive the nonholonomic constraints that apply to the COG G which lies at a distance b from Q, and a distance d from P. To this end, we start with the nonholonomic constraints (2.50a,b) that refer to the points Q and P:

$$\begin{aligned} \dot{x}_Q \sin\phi - \dot{y}_Q \cos\phi &= 0 \\ \dot{x}_P \sin(\phi + \psi) - \dot{y}_P \cos(\phi + \psi) &= 0 \end{aligned} \tag{3.53}$$

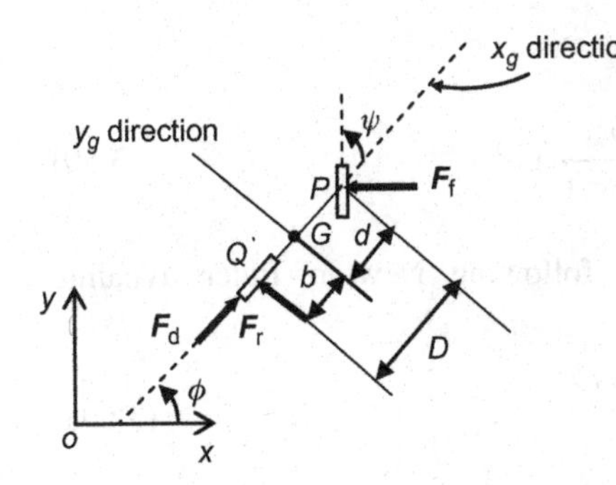

Figure 3.5 Free-body diagram of the equivalent bicycle.

Denoting by x_G and y_G the coordinates of the point G in the world coordinate frame we get:

$$\begin{aligned} x_Q &= x_G - b\cos\phi, \quad x_P = x_G + d\cos\phi \\ y_Q &= y_G - b\sin\phi, \quad y_P = y_G + d\sin\phi \\ \dot{x}_Q &= \dot{x}_G + b\dot{\phi}\sin\phi, \quad \dot{x}_P = \dot{x}_G - d\dot{\phi}\sin\phi \\ \dot{y}_Q &= \dot{y}_G - b\dot{\phi}\cos\phi, \quad \dot{y}_P = \dot{y}_G + d\dot{\phi}\cos\phi \end{aligned} \tag{3.54}$$

Introducing the relations (3.54) into (3.53) yields:

$$\begin{aligned} \dot{x}_G\sin\phi - \dot{y}_G\cos\phi + b\dot{\phi} = 0 \\ \dot{x}_G\sin(\phi+\psi) - \dot{y}_G\cos(\phi+\psi) - d\dot{\phi}\cos\psi = 0 \end{aligned} \tag{3.55}$$

Now, denoting by $[\dot{x}_g, \dot{y}_g]$ the velocity of the COG G in the local coordinate frame Gx_gy_g and applying the rotational transformation (2.17) (Example 2.1) we get:

$$\begin{bmatrix} \dot{x}_G \\ \dot{y}_G \end{bmatrix} = \begin{bmatrix} \cos\phi & -\sin\phi \\ \sin\phi & \cos\phi \end{bmatrix} \begin{bmatrix} \dot{x}_g \\ \dot{y}_g \end{bmatrix} \tag{3.56}$$

Introducing Eq. (3.56) into Eq. (3.55) yields:

$$\dot{y}_g = b\dot{\phi}, \quad \dot{\phi} = \frac{tg\psi}{D}\dot{x}_g \tag{3.57}$$

which, upon differentiation, give:

$$\ddot{y}_g = b\ddot{\phi} \tag{3.58a}$$

$$\ddot{\phi} = \frac{tg\psi}{D}\ddot{x}_g + \frac{1}{D\cos^2\psi}\dot{x}_g\dot{\psi} \tag{3.58b}$$

From Eqs. (3.57) and (3.58a,b) we get:

$$\dot{y}_g = \frac{b}{D}(tg\psi)\dot{x}_g, \quad \ddot{y}_g = \frac{b}{D}(tg\psi)\ddot{x}_g + \frac{b}{D\cos^2\psi}\dot{x}_g\dot{\psi} \tag{3.59}$$

Referring to Figure 3.5 we obtain the following Newton–Euler dynamic equations:

$$m(\ddot{x}_g - \dot{y}_g\dot{\phi}) = F_d - F_f \sin\psi \tag{3.60a}$$

$$m(\ddot{y}_g + \dot{x}_g\dot{\phi}) = F_r + F_f \cos\psi \tag{3.60b}$$

$$J\ddot{\phi} = dF_f \cos\psi - bF_r \tag{3.60c}$$

$$\dot{\psi} = -\frac{1}{T}\psi + \frac{K}{T}u_s \tag{3.60d}$$

where:

m = mass of the WMR
J = moment of inertia of the WMR about G
F_d = driving force
F_f, F_r = front and rear wheel lateral forces
T = time constant of the steering system
u_s = steering control input
K = a constant coefficient (gain)
$F_d = (1/r)\tau_d$, with r the rear wheel radius, and τ_d the applied motor torque.[2]

We will now put the above dynamic equations into state space form, where the state vector is:

$$\mathbf{x} = [x_G, y_G, \phi, \dot{x}_g, \psi]^T \tag{3.61}$$

To this end, we solve Eq. (3.60c) for F_r and introduce it, together with Eq. (3.58a) into (3.60b) to get:

$$m(b\ddot{\phi} + \dot{x}_g\dot{\phi}) = F_f \cos\psi + \frac{d}{b}F_f \cos\psi - \frac{J}{b}\ddot{\phi}$$

$$= \frac{DF_f \cos\psi - J\ddot{\phi}}{b} \quad (D = b + d)$$

[2] Here, the driving motor dynamics is not included. Derivation of DC motor dynamic models is provided in Example 5.1.

whence:

$$F_{\mathrm{f}} = \left(\frac{mb^2 + J}{D\cos\psi}\right)\ddot{\phi} + \left(\frac{mb\dot{x}_g}{D\cos\psi}\right)\dot{\phi} \tag{3.62}$$

Now introducing Eqs. (3.57), (3.58b), and (3.62) into Eq. (3.60a) we get:

$$\ddot{x}_g = \frac{\dot{x}_g(mb^2 + J)tg\psi}{a}\dot{\psi} + \frac{D^2\cos^2\psi}{a}F_{\mathrm{d}} \tag{3.63a}$$

where:

$$a = (\cos^2\psi)(mD^2 + (mb^2 + J)tg^2\psi) \tag{3.63b}$$

Finally, introducing Eq. (3.57) into Eq. (3.56) gives:

$$\begin{aligned}
\dot{x}_G &= \{\cos\phi - (b/D)(tg\psi)\sin\phi\}\dot{x}_g \\
\dot{y}_G &= \{\sin\phi + (b/D)(tg\psi)\cos\phi\}\dot{x}_g \\
\dot{\phi} &= [(1/D)tg\psi]\dot{x}_g \\
\ddot{x}_g &= (1/a)[(mb^2 + J)(tg\psi)\dot{\psi}\dot{x}_g] + (1/a)(D^2\cos^2\psi)F_{\mathrm{d}} \\
\dot{\psi} &= -(1/T)\psi + (\mathrm{K}/T)u_{\mathrm{s}}
\end{aligned} \tag{3.64}$$

with $F_{\mathrm{d}} = (1/r)\tau_{\mathrm{d}}$. The model (3.64) represents an affine system with two inputs $(\tau_{\mathrm{d}}, u_{\mathrm{s}})$, a five-dimensional state vector:

$$\mathbf{x} = [x_G, y_G, \phi, \dot{x}_g, \psi]^T \tag{3.65a}$$

a drift term:

$$\mathbf{g}_0(\mathbf{x}) = \begin{bmatrix} [\cos\phi - (b/D)(tg\psi)\sin\phi]\dot{x}_g \\ [\sin\phi + (b/D)(tg\psi)\cos\phi]\dot{x}_g \\ (1/D)(tg\psi)\dot{x}_g \\ (1/a)[(mb^2 + J)(tg\psi)\dot{\psi}\dot{x}_g] \\ -(1/T)\psi \end{bmatrix} \tag{3.65b}$$

and the two input fields:

$$\mathbf{g}_1(\mathbf{x}) = \left[0, 0, 0, \frac{1}{ra}D^2\cos^2\psi, 0\right]^{\mathrm{T}}, \quad \mathbf{g}_2(\mathbf{x}) = [0, 0, 0, 0, \mathrm{K/T}]^{\mathrm{T}} \tag{3.65c}$$

namely:

$$\dot{\mathbf{x}} = \mathbf{g}_0(\mathbf{x}) + \mathbf{g}_1(\mathbf{x})\tau_{\mathrm{d}} + \mathbf{g}_2(\mathbf{x})u_{\mathrm{s}} \tag{3.66}$$

3.5 Three-Wheel Omnidirectional Mobile Robot

Here, we will derive the dynamic model of a three-wheel omnidirectional robot using the Newton–Euler method [10]. This derivation is the same for any number of universal (orthogonal) omniwheels. Some further studies of omnidirectional WMRs are presented in Refs. [11–14].

Consider the WMR in the pose shown in Figure 3.6 in which the three wheels are placed at angles 30°, 150°, and 270°, respectively.

The rotation matrix of the robot's local coordinate frame Qx_ry_r with respect to the world coordinate frame Oxy is:

$$\mathbf{R}(\phi) = \begin{bmatrix} \cos\phi & -\sin\phi \\ \sin\phi & \cos\phi \end{bmatrix} \tag{3.67}$$

Let $\mathbf{s}_Q = \begin{bmatrix} x_Q & y_Q \end{bmatrix}^{\mathrm{T}}$ be the position vector of the COG Q. Then:

$$\mathbf{M}\ddot{\mathbf{s}}_Q = \mathbf{F}_Q, \quad \mathbf{F}_Q = \begin{bmatrix} F_{Qx} & F_{Qy} \end{bmatrix}^{\mathrm{T}} \tag{3.68}$$

where $\mathbf{M} = \mathrm{diag}(m, \mathrm{m})$, m is the robot's mass, and $\mathbf{F}_Q$ is the force exerted at the COG expressed in the world coordinate frame. Denoting by:

$$\mathbf{s}_{\mathrm{r}} = \begin{bmatrix} x_{\mathrm{r}} & y_{\mathrm{r}} \end{bmatrix}^{\mathrm{T}}, \quad \mathbf{F}_{\mathrm{r}} = [F_{x_{\mathrm{r}}}, F_{y_{\mathrm{r}}}]^{\mathrm{T}} \tag{3.69}$$

the position vector of the COG and the force vector expressed in the local (moving) coordinate frame, then Eq. (3.67) implies that:

$$\dot{\mathbf{s}}_Q = \mathbf{R}(\phi)\dot{\mathbf{s}}_{\mathrm{r}}, \quad \mathbf{F}_Q = \mathbf{R}(\phi)\mathbf{F}_{\mathrm{r}} \tag{3.70}$$

Thus, using Eq. (3.70) in Eq. (3.68) we get:

$$\mathbf{M}(\ddot{\mathbf{s}}_{\mathrm{r}} + \mathbf{R}^T(\phi)\dot{\mathbf{R}}(\phi)\dot{\mathbf{s}}_{\mathrm{r}}) = \mathbf{F}_{\mathrm{r}} \tag{3.71}$$

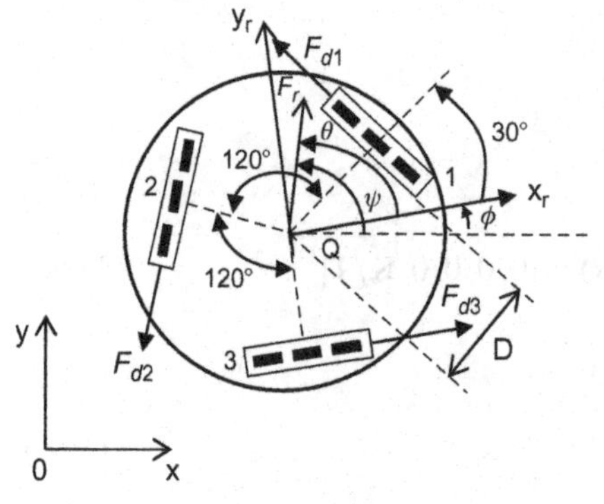

Figure 3.6 Geometry of the three-wheel omnidirectional WMR.

Now, the dynamic equation of the rotation about the COG Q is:

$$I_Q\ddot{\phi} = \tau_Q \tag{3.72}$$

where I_Q is the moment of inertia of the robot about Q, and τ_Q is the applied torque at Q.

From the geometry of Figure 3.6 we obtain:

$$\begin{aligned} F_{xr} &= -\frac{1}{2}F_{d1} - \frac{1}{2}F_{d2} + F_{d3} \\ F_{yr} &= \frac{\sqrt{3}}{2}F_{d1} - \frac{\sqrt{3}}{2}F_{d2} \\ \tau_Q &= (F_{d1} + F_{d2} + F_{d3})D \end{aligned} \tag{3.73}$$

where D is the distance of the wheels from the rotation point Q, and $F_{di}(i = 1,2,3)$ are the driving forces of the wheels.

The rotation of each wheel is described by the dynamic equation:

$$I_o\ddot{\theta}_i + \beta\dot{\theta}_i = K\tau_i - rF_{di} \quad (i = 1,2,3) \tag{3.74}$$

where I_o is the common moment of inertia of the wheels, θ_i is the angular position of wheel i, β is the linear friction coefficient, r is the common radius of the wheels, τ_i is the driving input torque of wheel i, and K is the driving torque gain.

Now, from the geometry of Figure 3.6 we see that the angles of the wheel velocities in the local coordinate frame are: $30° + 90° = 120°$, $120° + 120° = 240°$, and $240° + 120° = 360°$

Thus, we obtain the inverse kinematics equations from $\dot{\mathbf{p}}_r = [\dot{x}_r, \dot{y}_r, \dot{\phi}]^T$ to $\dot{\mathbf{q}} = [\dot{\theta}_1, \dot{\theta}_2, \dot{\theta}_3]^T = [\omega_1, \omega_2, \omega_3]^T$ as:

$$r\omega_1 = -\frac{1}{2}\dot{x}_r + \frac{\sqrt{3}}{2}\dot{y}_r + D\dot{\phi} \tag{3.75a}$$

$$r\omega_2 = -\frac{1}{2}\dot{x}_r - \frac{\sqrt{3}}{2}\dot{y}_r + D\dot{\phi} \tag{3.75b}$$

$$r\omega_3 = \dot{x}_r + D\dot{\phi} \tag{3.75c}$$

Using Eqs. (3.69)–(3.75c) we get after some algebraic manipulation:

$$\ddot{x}_r = a_1\dot{x}_r + a_2^*\dot{y}_r\dot{\phi} - b_1(\tau_1 + \tau_2 - 2\tau_3) \tag{3.76a}$$

$$\ddot{y}_r = a_1\dot{y}_r - a_2^*\dot{x}_r\dot{\phi} + \sqrt{3}b_1(\tau_1 - \tau_2) \tag{3.76b}$$

$$\ddot{\phi} = a_3\dot{\phi} + b_2(\tau_1 + \tau_2 + \tau_3) \tag{3.76c}$$

where:

$$a_1 = -3\beta(3I_o + 2mr^2), \quad a_2^* = 2mr^2/(3I_o + 2mr^2)$$

$$a_3 = -3\beta D^2/(3I_o D^2 + I_Q r^2)$$

$$b_1 = \mathrm{K}r/(3I_o + 2mr^2), \quad b_2 = \mathrm{K}r/(3I_o D^2 + I_Q r^2)$$

Finally, combining Eqs. (3.67), (3.68), and (3.76a–c) we get the following state–space dynamic model of the WMR's motion:

$$\dot{\mathbf{x}} = \mathbf{A}(\mathbf{x})\mathbf{x} + \mathbf{B}(\mathbf{x})\mathbf{u} \tag{3.77a}$$

$$\mathbf{y} = \mathbf{C}\mathbf{x} \tag{3.77b}$$

where:

$$\mathbf{x} = [x_Q, y_Q, \phi, \dot{x}_Q, \dot{y}_Q, \dot{\phi}]^{\mathrm{T}}$$

$$\mathbf{y} = [\dot{x}_Q, \dot{y}_Q, \phi]^{\mathrm{T}}$$

$$\mathbf{u} = [\tau_1, \tau_2, \tau_3]^{\mathrm{T}}$$

$$\mathrm{A}(\mathbf{x}) = \begin{bmatrix} 0 & 0 & 0 & 1 & 0 & 0 \\ 0 & 0 & 0 & 0 & 1 & 0 \\ 0 & 0 & 0 & 0 & 0 & 1 \\ 0 & 0 & 0 & a_1 & -a_2\dot{\phi} & 0 \\ 0 & 0 & 0 & a_2\dot{\phi} & a_1 & 0 \\ 0 & 0 & 0 & 0 & 0 & a_3 \end{bmatrix}$$

$$\mathbf{B}(\mathbf{x}) = \begin{bmatrix} 0 & 0 & 0 \\ 0 & 0 & 0 \\ 0 & 0 & 0 \\ b_1\beta_1 & b_1\beta_2 & 2b_1\cos\phi \\ b_1\beta_3 & b_1\beta_4 & 2b_1\sin\phi \\ b_2 & b_2 & b_2 \end{bmatrix} = [\,\bar{\mathbf{b}}_1(\mathbf{x}) \quad \bar{\mathbf{b}}_2(\mathbf{x}) \quad \bar{\mathbf{b}}_3(\mathbf{x})\,]$$

$$\mathbf{C} = \begin{bmatrix} 0 & 0 & 0 & 1 & 0 & 0 \\ 0 & 0 & 0 & 0 & 1 & 0 \\ 0 & 0 & 1 & 0 & 0 & 0 \end{bmatrix}$$

$$a_2 = 1 - a_2^* = 3I_o/(3I_o + 2mr^2)$$

$$\beta_1 = -\sqrt{3}\sin\phi - \cos\phi, \quad \beta_2 = \sqrt{3}\sin\phi - \cos\phi$$

$$\beta_3 = \sqrt{3}\cos\phi - \sin\phi, \quad \beta_4 = -\sqrt{3}\cos\phi - \sin\phi$$

The azimuth of the robot in the world coordinate frame is denoted by ψ, where $\psi=\phi+\theta$ (θ denotes the angle between Qx_r and F_r, that is, the azimuth of the robot in the moving coordinate frame). Then:

$$\dot{x}_Q=v\cos\psi,\quad \dot{y}_Q=v\sin\psi,\quad v=\sqrt{\dot{x}_Q^2+\dot{y}_Q^2}$$

Whence:

$$\psi=tg^{-1}(\dot{y}_Q/\dot{x}_Q) \tag{3.78}$$

where the positive direction is the counterclockwise rotation direction. It is noted that the motions along x_Q and y_Q are coupled because the dynamic equations are derived in the world coordinate frame. But since the rotational angle is always equal to $\phi=\psi-\theta$, despite the fact that θ may be changing arbitrarily, the WMR can realize a translational motion without changing the pose (i.e., the WMR is holonomic). The model (3.77a) can be written in a three-input affine form with drift as follows:

$$\dot{\mathbf{x}}=\mathbf{g}_o(\mathbf{x})+\sum_{i=1}^{3}\mathbf{g}_i(\mathbf{x})u_i \tag{3.79}$$

where:

$$\mathbf{g}_o(\mathbf{x})=\mathrm{A}(\mathbf{x})\mathbf{x},\quad \mathbf{g}_i(\mathbf{x})=\overline{\mathbf{b}}_i(\mathbf{x})\quad (i=1,2,3)$$

The dynamic model (3.77a) of a three-wheel omnidirectional robot (with universal wheels) is one of the many different models available. Actually, many other equivalent models were derived in the literature.

Example 3.2

Derive the dynamic equations of the three-wheel omnidirectional robot using the unit directional vectors $\varepsilon_1,\varepsilon_2,\varepsilon_3$ of the wheel velocities.

Solution

To simplify the derivation we select the pose of the WMR in which the wheel 1 orientation is perpendicular to the local coordinate axis Qx_r as shown in Figure 3.7 [15]. Thus the unit directional vectors $\varepsilon_1,\varepsilon_2$, and ε_3 are:

$$\varepsilon_1=\begin{bmatrix}0\\1\end{bmatrix},\quad \varepsilon_2=-\begin{bmatrix}\sqrt{3}/2\\1/2\end{bmatrix},\quad \varepsilon_3=\begin{bmatrix}\sqrt{3}/2\\-1/2\end{bmatrix} \tag{3.80}$$

The rotational matrix of Qx_ry_r with respect to Oxy is given by (3.67).

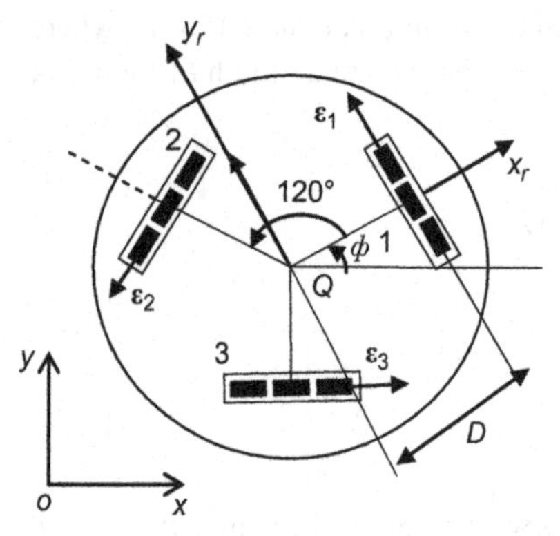

Figure 3.7 The three-wheel omnidirectional mobile robot (ε_1 perpendicular to Qx_r).

Therefore, the driving velocities $v_i (i = 1, 2, 3)$ of the wheels are:

$$v_1 = r\dot{\theta}_1 = -\dot{x}_Q \sin\phi + \dot{y}_Q \cos\phi + D\dot{\phi}$$
$$v_2 = r\dot{\theta}_2 = -\dot{x}_Q \sin(\pi/3 - \phi) - \dot{y}_Q \cos(\pi/3 - \phi) + D\dot{\phi}$$
$$v_3 = r\dot{\theta}_3 = \dot{x}_Q \sin(\pi/3 + \phi) - \dot{y}_Q \cos(\pi/3 + \phi) + D\dot{\phi}$$

or

$$\dot{\mathbf{q}} = \mathbf{J}^{-1}(\phi)\dot{\mathbf{p}}_Q \tag{3.81}$$

where:

$$\mathbf{J}^{-1}(\phi) = \frac{1}{r}\begin{bmatrix} -\sin\phi & \cos\phi & D \\ -\sin(\pi/3 - \phi) & -\cos(\pi/3 - \phi) & D \\ \sin(\pi/3 + \phi) & -\cos(\pi/3 + \phi) & D \end{bmatrix}$$
$$\dot{\mathbf{q}} = [\dot{\theta}_1, \dot{\theta}_2, \dot{\theta}_3]^{\mathrm{T}}, \ \dot{\mathbf{p}}_Q = [\dot{x}_Q, \dot{y}_Q, \dot{\phi}]^{\mathrm{T}}$$

This is the inverse kinematic model of the robot. Now, applying the Newton–Euler method to the robot we get:

$$m\begin{bmatrix} \ddot{x}_Q \\ \ddot{y}_Q \end{bmatrix} = \mathbf{s}_1(\phi)F_{d1} + \mathbf{s}_2(\phi)F_{d2} + \mathbf{s}_3(\phi)F_{d3} \tag{3.82a}$$

$$I_Q\ddot{\phi} = D(F_{d1} + F_{d2} + F_{d3}) \tag{3.82b}$$

where $F_{di} (i = 1, 2, 3)$ is the magnitude of the driving force of the ith wheel, m is the robot mass, I_Q is the robot moment of inertia about Q, and:

$$\mathbf{s}_i(\phi) = \mathbf{R}(\phi)\varepsilon_i \quad (i = 1, 2, 3) \tag{3.83}$$

are two-dimensional vectors found using Eqs. (3.67) and (3.80). The driving forces $F_{di} (i = 1, 2, 3)$ are given by the relation:

$$F_{di} = aV_i - \beta r\dot{\theta}_i \quad (i = 1, 2, 3) \tag{3.84}$$

where V_i is the voltage applied to the motor of the ith wheel, "a" is the voltage–force constant, and β is the friction coefficient.

Combining Eqs. (3.82a,b), (3.83), and (3.84) we obtain the model:

$$\mathbf{D}\ddot{\mathbf{p}}_Q + \mathbf{C}(\phi)\dot{\mathbf{p}}_Q = \mathbf{E}\mathbf{v} \tag{3.85a}$$

where:

$$\mathbf{D} = \begin{bmatrix} m & 0 & 0 \\ 0 & m & 0 \\ 0 & 0 & I_Q \end{bmatrix}, \quad \mathbf{C}(\phi) = \left(\frac{\beta r}{a}\right)\mathbf{E}(\phi)\mathbf{J}^{-1}(\phi) \tag{3.85b}$$

$$\mathbf{E}(\phi) = a\begin{bmatrix} \frac{\mathbf{s}_1(\phi)}{D} & \frac{\mathbf{s}_2(\phi)}{D} & \frac{\mathbf{s}_3(\phi)}{D} \end{bmatrix}, \quad \mathbf{v} = \begin{bmatrix} V_1 \\ V_2 \\ V_3 \end{bmatrix} \tag{3.85c}$$

The model (3.85a–c) has the standard form of the robot model described by Eqs. (3.11) and (3.13).

Example 3.3

We are given a car-like robot where there are lateral slip forces and longitudinal friction forces on all wheels. It is desired to write down the Newton–Euler dynamic equations in the form (3.60a–d).

Solution

The free-body diagram of the robot is as shown in Figure 3.8.

We use the following definitions:

$F_d = F_d^l + F_d^r$ (Total driving force)

$F_{x_g,r} = F_{x_g,r}^l + F_{x_g,r}^r$ (Rear total longitudinal friction)

$F_{x_g,f} = F_{x_g,f}^l + F_{x_g,f}^r$ (Front total longituditial friction)

$F_{y_g,r} = F_{y_g,r}^l + F_{y_g,r}^r$ (Rear total lateral force)

$F_{y_g,f} = F_{y_g,f}^l + F_{y_g,f}^r$ (Front total lateral force)

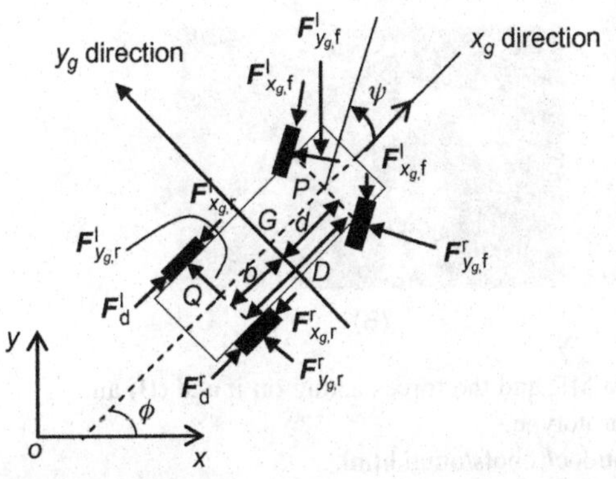

Figure 3.8 Free-body diagram of the car-like WMR.

where the upper index r refers to the right wheel and the upper index l to the left wheel. Therefore, considering the bicycle model of the WMR we get the following Newton–Euler dynamic equations in the local coordinate frame:

$$m(\ddot{x}_g - \dot{y}_g\dot{\phi}) = F_d - F_{x_g,r} - F_{x_g,f}\cos\psi - F_{y_g,f}\sin\psi$$
$$m(\ddot{y}_g + \dot{x}_g\dot{\phi}) = F_{y_g,r} - F_{x_g,f}\sin\psi + F_{y_g,f}\cos\psi$$
$$J\ddot{\phi} = dF_{y_g,f}\cos\psi - bF_{y_g,r}$$
$$\dot{\psi} = -(1/T)\psi + (K/T)u_s$$

where all variables have the meaning presented in Section 3.4. From this point the development of the full model can be done as in Section 3.4.

3.6 Four Mecanum-Wheel Omnidirectional Robot

We consider the four-wheel omnidirectional robot of Figure 3.9A [3].

The total forces F_x and F_y acting on the robot in the x and y directions are:

$$F_x = (F_{x1} + F_{x2} + F_{x3} + F_{x4}) \quad (3.86a)$$
$$F_y = (F_{y1} + F_{y2} + F_{y3} + F_{y4}) \quad (3.86b)$$

where F_{xi}, F_{yi} $(i = 1, 2, 3, 4)$ are the forces acting on the wheels along the x and y axes. In the absence of separate rotation motion, the direction of motion is defined by an angle δ where:

$$\delta = tg^{-1}(F_y/F_x) \quad (3.87)$$

The torque τ that produces pure rotation is:

$$\tau = (F_{x1} - F_{x2} - F_{x3} + F_{x4})d_1 + (F_{y3} + F_{y4} - F_{y1} - F_{y2})d_2 \quad (3.88)$$

where positive rotation is in the counterclockwise direction.

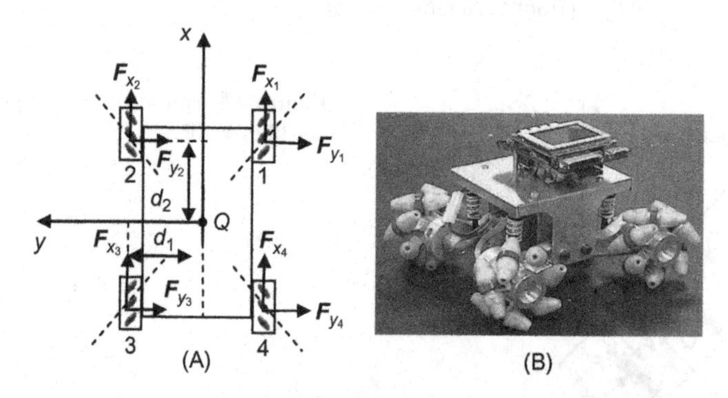

Figure 3.9 (A) The four-wheel mecanum WMR and the forces acting on it and (B) an experimental four-wheel mecanum WMR prototype.
Source: www.robotics.ee.uwa.edu.au/eyebot/doc/robots/omni.html.

The Newton–Euler motion equations are:

$$m\ddot{x} = F_x - \beta_x\dot{x} \tag{3.89a}$$

$$m\ddot{y} = F_y - \beta_y\dot{y} \tag{3.89b}$$

$$I_Q\ddot{\phi} = \tau - \beta_z\dot{\phi} \tag{3.89c}$$

where β_x, β_y, and β_z are the linear friction coefficients in the x, y, and ϕ motion, and m, I_Q are the mass and moment of inertia of the robot. Equations (3.89a–c) show that the robot can achieve steady-state velocities ($\ddot{x} = 0, \ddot{y} = 0, \ddot{\phi} = 0$) $\dot{x}_{ss}, \dot{y}_{ss}$ and $\dot{\phi}_{ss}$ equal to:

$$\dot{x}_{ss} = \frac{F_x}{\beta_x}, \quad \dot{y}_{ss} = \frac{F_y}{\beta_y}, \quad \dot{\phi}_{ss} = \frac{\tau}{\beta_z} \tag{3.90}$$

Example 3.4

The problem is to calculate the wheel angular velocities which are required to achieve a desired translational and rotational motion (WMR velocities v and $\dot{\phi}$) of the mecanum mobile robot of Figure 3.9A.

Solution

A solution to this problem was provided by the relations (2.84) and (2.85). Here we will provide an alternative method [3]. We draw the displacement and velocity vectors of a single wheel which are as shown in Figure 3.10A and B, where "a" is the roller angle ($a = \pm 45°$).

The vector $\mathbf{s}_p$ represents the displacement due to the wheel rotation (in the positive direction), $\mathbf{s}_r$ represents the displacement vector due to rolling which is orthogonal to the roller axis, and $\mathbf{s}$ represents the total displacement vector. The dotted horizontal lines represent the discontinuities where the roller contact point transfers from one roller to the next. In Figure 3.10, the point A was selected in the middle of this discontinuity line to facilitate the calculation.

From Figure 3.10B we get $(\omega r)\cos a = v\cos(a - \gamma)$ because the components of ωr and $v = ds/dt$ along the roller axis are equal. Therefore:

$$\omega r = v\cos(a - \gamma)/\cos a \tag{3.91}$$

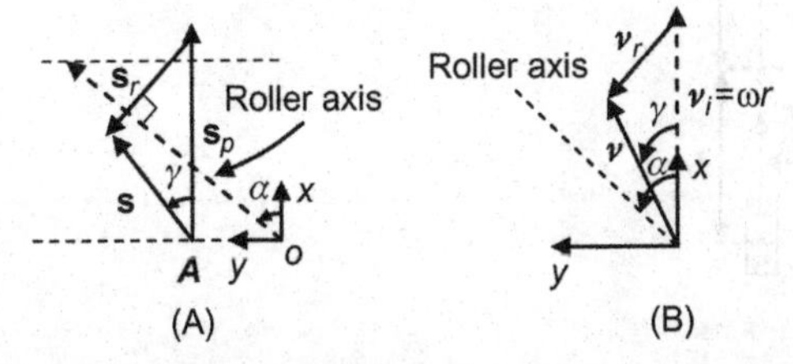

Figure 3.10 (A) Displacement vectors of a wheel and roller and (B) velocity vectors (a is the rollers angle).

for $a \neq \pi/2 + k\pi$ $(k = 0, 1, 2, \ldots)$. If $a = \pi/2 + k\pi$, the rotation of the wheel does not produce translation motion of the rollers. Solving Eq. (3.91) for v we get:

$$v = \omega r \frac{\cos a}{\cos(\alpha - \gamma)}, \quad a - \gamma \neq \frac{\pi}{2} + k\pi \ (k = 0, 1, 2, \ldots) \tag{3.92}$$

When $a - \gamma = \pi/2$, the rotational speed ω of the wheel must be zero, but the wheel can have any value of translation velocity because of the motion of the other wheels.

We now calculate the wheel angular velocity ω_i for a desired translational motion velocity v. From Eq. (3.90), we see that the wheel velocity is proportional to the forces applied by each wheel, that is:

"v proportional to $(F_1 + F_2 + F_3 + F_4)$," and because the WMR is a rigid body, all wheels should have the same translational speed, that is:

$$v_i = v \quad (i = 1, 2, 3, 4)$$

Therefore, from Eq. (3.92) we have:

$$\omega_i = \frac{v \cos(a_i - \gamma)}{r_i \cos a_i}, \quad a_i \neq \frac{\pi}{2} + k\pi \ (k = 0, 1, 2, \ldots) \tag{3.93}$$

Equation (3.93) gives the velocities $\omega_i (i = 1, 2, 3, 4)$ of the four wheels needed to get a desired translational velocity v of the robot.

Finally, we will calculate the wheel angular velocities required to get a desired rotational speed $\dot{\phi}$. Consider a robot with velocity v. Then the instantaneous curvature radius (ICR) of its path is:

$$R = v/\dot{\phi} \text{ with } \varepsilon = ctg^{-1}(v_x/v_y) \tag{3.94}$$

These relations give the following world frame coordinates x_{ICR}, y_{ICR} of the ICR (Figure 3.11):

$$x_{ICR} = -R \sin \varepsilon$$
$$y_{ICR} = R \cos \varepsilon$$

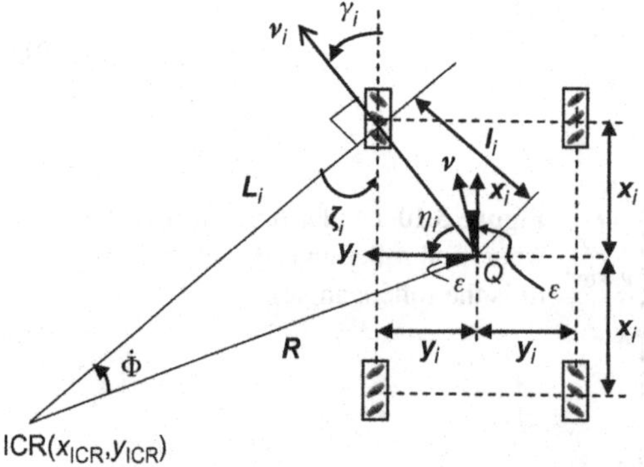

Figure 3.11 Geometry of wheel i with reference to the coordinate frame $Q_{x_i y_i}$. Each wheel has its own total velocity v_i and angle γ_i.

The geometry of each wheel is defined by its position (x_i, y_i) and the orientation a_i of its rollers. Let Σ_i be the contact point of wheel i then:

$$\eta_i = tg^{-1}(x_i/y_i), \quad l_i = \sqrt{x_i^2 + y_i^2} \tag{3.95}$$

where l_i is the same for all wheels due to the symmetry of the wheels with respect to the point Q. The distance L_i of the ICR and the contact point $\Sigma_i(x_i, y_i)$ is given by the triangle formula, as:

$$L_i = \sqrt{l_i^2 + R^2 - 2l_i R \cos(\eta_i + \varepsilon)} \tag{3.96}$$

The velocity v_i of the wheel should be perpendicular to the line L_i. Therefore, the angle ζ_i of the line L_i with the x-axis is determined by the relation:

$$tg\zeta_i = (y_{\rm ICR} - y_i)/(x_{\rm ICR} - x_i) \tag{3.97}$$

whence:

$$\gamma_i = \pi/2 + \zeta_i \tag{3.98}$$

because, in Figure 3.11, ζ_i is actually negative $(x_{\rm ICR} - x_i < 0)$. Now, in view of Eq. (3.94) we have:

$$|v_i| = L_i|\dot{\phi}| \tag{3.99}$$

Finally, using Eq. (3.99), Eq. (3.93) gives:

$$\omega_i = L_i\dot{\phi}\cos(a_i - \gamma_i)/r\cos a_i \quad (i = 1, 2, 3, 4) \tag{3.100}$$

for $a_i \neq \pi/2 + k\pi$ $(k = 0, 1, 2, \ldots)$, where r is the common radius of the wheels. Equation (3.100) gives the required rotational speeds of the wheels in terms of the position $(x_{\rm ICR}, y_{\rm ICR})$ of ICR and the desired rotational speed $\dot{\phi}$ of the robot. Actually, in view of Eqs. (3.94) and (3.96), Eq. (3.100) gives ω_i in terms of the known (desired) values of v, $\dot{\phi}$, x_i, and y_i.

Example 3.5

It is desired to describe a way for identifying the dynamic parameters of a differential-drive WMR by converting its dynamic model to a linear form which uses the robot linear displacement in place of x_Q, y_Q, and ϕ.

The nonlinear dynamic model of this WMR is given by Eqs. (3.30a,b) and (3.31) which is written as:

$$\begin{aligned}
\dot{v} &= \gamma_1 u_1, \quad u_1 = \tau_r + \tau_l \\
\dot{\omega} &= \gamma_2 u_2, \quad u_2 = \tau_r - \tau_l \\
\dot{x}_Q &= v\cos\phi \\
\dot{y}_Q &= v\sin\phi \\
\dot{\phi} &= \omega
\end{aligned}$$

where $\gamma_1 = 1/mr$ and $\gamma_2 = 2a/Ir$ are the dynamic parameters to be identified. Using the robot's linear displacement:

$$l = x_Q \cos\phi + y_Q \sin\phi$$

in place of the components x_Q and y_Q (Figure 2.7) we get the linear model:

$$\dot{v} = \gamma_1 u_1, \quad \dot{\omega} = \gamma_2 u_2, \quad \dot{l} = v, \quad \dot{\phi} = \omega$$

which reduces to:

$$\ddot{l} = \gamma_1 u_1, \quad \ddot{\phi} = \gamma_2 u_2 \tag{3.101}$$

This is a linear model with two outputs l and ϕ. The identification will be made rendering the model (3.101) to the standard linear regression model:

$$\mathbf{y} = \mathbf{M}(\mathbf{u})\boldsymbol{\xi} + \mathbf{e} \tag{3.102}$$

where:

$$\mathbf{y} = [y_1, y_2, \ldots, y_m]^{\mathrm{T}} \quad \text{(the vector of measurable signals)}$$

$$\boldsymbol{\xi} = \left[\xi_1, \xi_2, \ldots, \xi_n\right]^{\mathrm{T}} \quad \text{(the vector of unknown parameters)}$$

$$\mathbf{e} = [e_1, e_2, \ldots, e_m]^{\mathrm{T}} \quad \text{(the vector of measurement errors)}$$

$$\mathbf{M}(\mathbf{u}) = \begin{bmatrix} \mu_{11}(\mathbf{u}) & \ldots & \mu_{1n}(\mathbf{u}) \\ \mu_{m1}(\mathbf{u}) & \ldots & \mu_{mn}(\mathbf{u}) \end{bmatrix} \cdots \quad \text{(an } m \times n \text{ known matrix)}$$

The matrix $\mathbf{M}(\mathbf{u})$ is known as "*regressor* matrix." Under the assumption that $\mathbf{e}$ is independent of $\mathbf{M}(\mathbf{u})$, and $m > n$ the solution for $\boldsymbol{\xi}$ is given by[3]:

$$\hat{\boldsymbol{\xi}} = (\mathbf{M}^{\mathrm{T}}(\mathbf{u})\mathbf{M}(\mathbf{u}))^{-1}\mathbf{M}^{\mathrm{T}}(\mathbf{u})\mathbf{y} \tag{3.103}$$

To convert Eq. (3.101) into the regression form (3.102) we discretize it in time using the first order approximation: $dx/dt \simeq (x_{k+1} - x_k)/\mathrm{T}$, where $x_k = x(t)_{t=k\mathrm{T}}$ $(k = 0, 1, 2, \ldots)$, and T is the sampling period. Then, Eq. (3.101) becomes:

$$\begin{aligned} \Delta l_k &= \Delta l_{k-1} + \xi_1 u_{1,k} \\ \Delta\phi_k &= \Delta\phi_{k-1} + \xi_2 u_{2,k} \end{aligned} \tag{3.104}$$

[3] Equation (3.103) can be found by minimizing the function $J(\boldsymbol{\xi}) = \mathbf{e}^T\mathbf{e} = (\mathbf{y} - \mathbf{M}\boldsymbol{\xi})^T(\mathbf{y} - \mathbf{M}\boldsymbol{\xi})$ with respect to $\boldsymbol{\xi}$. An easy way to get Eq. (3.103) is by multiplying Eq. (3.102) by $\mathbf{M}^T(\mathbf{u})$and solving for $\boldsymbol{\xi}$, under the assumption that $\mathbf{M}^T(\mathbf{u})\mathbf{e} = 0$ ($\mathbf{e}$ independent of $\mathbf{M}(\mathbf{u})$,and rank $\mathbf{M}(\mathbf{u}) = \min(m, n) = n$. Actually, Eq. (3.103) is $\hat{\boldsymbol{\xi}} = \mathbf{M}^{\dagger}\mathbf{y}$ where $\mathbf{M}^{\dagger}$ is the generalized inverse of $\mathbf{M}$ (Eq. (2.8a)).

where $\xi_i = T\gamma_i, (i = 1, 2)$. Clearly, the parameters ξ_1 and ξ_2 are identifiable, but the difficulty is that l cannot be measured. To overcome this difficulty we use a second-order parametric representation of $x_Q(t)$, $y_Q(t)$, and $\phi(t)$, namely [16]:

$$x_Q(\mu) = a_2\mu^2 + a_1\mu + a_0, \quad y_Q(\mu) = b_2\mu^2 + b_1\mu + b_0$$

$$tg\phi(\mu) = f(\mu) = \frac{dy_Q/d\mu}{dx_Q/d\mu} = \frac{2b_2\mu + b_1}{2a_2\mu + a_1}$$

with boundary conditions:

$$x_Q(0) = x_Q^0 = a_0, \quad x_Q(1) = x_Q^1 = a_2 + a_1 + a_0$$

$$y_Q(0) = y_Q^0 = b_0, \quad y_Q(1) = y_Q^1 = b_2 + b_1 + b_0$$

$$f(0) = tg\phi(0) = b_1/a_1$$

$$f(1) = tg\phi(1) = (2b_2 + b_1)/(2a_2 + a_1)$$

From the above conditions, the parameters a_i and b_i $(i = 0, 1, 2, \ldots)$ are computed as:

$$a_0 = x_Q^0, \quad a_1 = \frac{2(tg\phi(1))(x_Q^1 - x_Q^0) - y_Q^1 + y_Q^0}{tg\phi(1) - tg\phi(0)} \tag{3.105a}$$

$$b_0 = y_Q^0, \quad b_1 = a_1 tg\phi(0), \quad b_2 = y_Q^1 - y_Q^0 - a_1 \tag{3.105b}$$

The approximate length increment $\Delta\hat{l}$ of Δl is given by:

$$|\Delta\hat{l}| = \int_0^1 \sqrt{\left(\frac{dx_Q}{d\mu}\right)^2 + \left(\frac{dy_Q}{d\mu}\right)^2} d\mu = \int_0^1 \sqrt{k_2\mu^2 + k_1\mu + k_0} d\mu \tag{3.106a}$$

where:

$$k_0 = a_1^2 + b_1^2, \quad k_1 = 4a_1a_2 + 4b_1b_2, \quad k_2 = 4a_2^2 + 4b_2^2 \tag{3.106b}$$

Equation (3.106a) is an integral equation, the close solution of which is available in integral tables. The sign of Δl, which determines whether the robot moved forward or backward, can be determined by the following relation:

$$\Delta l = l_1 - l_0 = (x_Q^1 \cos\phi(0) + y_Q^1 \sin\phi(0)) - (x_Q^0 \cos\phi(0) + y_Q^0 \sin\phi(0))$$

where x_Q^1, y_Q^1 is the current position and x_Q^0, y_Q^0 is the past position of the robot. Several numerical experiments performed using the above WMR identification method gave very satisfactory results for both Δl and $\Delta\phi$ [16].

Example 3.6

Outline a method for applying the least squares identification model (3.102)–(3.103) to the general nonholonomic WMR model (3.19a–b)

Solution

The model (3.19a,b):

$$\bar{\mathbf{D}}(\mathbf{q})\dot{\mathbf{v}} + \bar{\mathbf{C}}(\mathbf{q},\dot{\mathbf{q}})\mathbf{v} + \bar{\mathbf{g}}(\mathbf{q}) = \bar{\mathbf{E}}\boldsymbol{\tau}$$

is put in the form:

$$\dot{\mathbf{v}} = \mathbf{M}(\mathbf{v},\boldsymbol{\tau})\boldsymbol{\xi}$$

where $\boldsymbol{\xi}$ is the vector of unknown parameters The above model can be written in the form (3.102) by computing $\mathbf{y}_k$ as:

$$\mathbf{y}_k = \mathbf{v}_k - \mathbf{v}_{k-1} = \left(\int_{(k-1)\mathrm{T}}^{k\mathrm{T}} [\mathbf{M}(\mathbf{v},t), \boldsymbol{\tau}(t)]\mathrm{d}t \right)\boldsymbol{\xi} \tag{3.107}$$

where $\mathbf{v}_k = \mathbf{v}(t)]_{t=k\mathrm{T}}$ and T is the sampling period of the measurements.

Now, defining $\bar{\mathbf{y}}_N$ and $\bar{\mathbf{M}}_N$ as:

$$\bar{\mathbf{y}}_N = [\mathbf{y}_1, \mathbf{y}_2, \ldots, \mathbf{y}_N]^{\mathrm{T}}$$
$$\bar{\mathbf{M}}_N = \left[\int_0^{\mathrm{T}} \mathbf{M}\mathrm{d}t, \int_0^{2\mathrm{T}} \mathbf{M}\mathrm{d}t, \ldots, \int_{(N-1)T}^{N\mathrm{T}} \mathbf{M}\mathrm{d}t \right]^{\mathrm{T}}$$

The model (3.107), after N measurements, becomes:

$$\bar{\mathbf{y}}_N = \bar{\mathbf{M}}_N\boldsymbol{\xi} \tag{3.108}$$

This method overcomes the noise problem posed by the calculation of the acceleration. The optimal estimate $\hat{\boldsymbol{\xi}}$ of $\boldsymbol{\xi}$ is given by Eq. (3.103):

$$\hat{\boldsymbol{\xi}} = (\bar{\mathbf{M}}_N^{\mathrm{T}}\bar{\mathbf{M}}_N)^{-1}\bar{\mathbf{M}}_N^{\mathrm{T}}\bar{\mathbf{y}}_N \tag{3.109}$$

The identification process can be simplified if it is split in two parts, namely: (i) identification when the robot is moving on a line without rotation, and (ii) identification when the robot has pure rotation movement. In the pure linear motion we keep the angle ϕ constant (i.e., $\dot{\phi} = 0$), and in the pure rotation motion we have $v_Q = 0$ (i.e., $\dot{x}_Q = 0$ and $\dot{y}_Q = 0$) [17].

References

[1] McKerrow PK. Introduction to robotics. Reading, MA: Addison-Wesley; 1999.

[2] Dudek G, Jenkin M. Computational principles of mobile robotics. Cambridge: Cambridge University Press; 2010.

[3] De Villiers M, Bright G. Development of a control model for a four-wheel mecanum vehicle. In: Proceedings of twenty fifth international conference of CAD/CAM robotics and factories of the future conference. Pretoria, South Africa; July 2010.

[4] Song JB, Byun KS. Design and control of a four-wheeled omnidirectional mobile robot with steerable omnidirectional wheels. J Rob Syst 2004;21:193–208.

[5] Sidek SN. Dynamic modeling and control of nonholonomic wheeled mobile robot subjected to wheel slip. PhD Thesis, Vanderbilt University, Nashville, TN, December 2008.

[6] Williams II RL, Carter BE, Gallina P, Rosati G. Dynamic model with slip for wheeled omni-directional robots. IEEE Trans Rob Autom 2002;18(3):285–93.

[7] Stonier D, Se-Hyoung C, Sung–Lok C, Kuppuswamy NS, Jong-Hwan K. Nonlinear slip dynamics for an omniwheel mobile robot platform. In: Proceedings of IEEE international conference on robotics and automation. Rome, Italy; April 10–14, 2007. p. 2367–72.

[8] Ivanjko E, Petrinic T, Petrovic I. Modeling of mobile robot dynamics. In: Proceedings of seventh EUROSIM congress on modeling and simulation. Prague, Czech Republic; September 6–9, 2010. p. 479–86.

[9] Moret EN. Dynamic modeling and control of a car-like robot. MSc Thesis, Virginia Polytechnic Institute and State University, Blacksburg, VA, February 2003.

[10] Watanabe K, Shiraishi Y, Tzafestas SG, Tang J, Fukuda T. Feedback control of an omnidirectional autonomous platform for mobile service robots. J Intell Rob Syst 1998;22:315–30.

[11] Pin FG, Killough SM. A new family of omnidirectional and holonomic wheeled platforms for mobile robots. IEEE Trans Rob Autom 1994;10(4):480–9.

[12] Rojas R. Omnidirectional control. Freie University, Berlin; May 2005. <http://robocup.mi.fu-berlin.de/buch/omnidrive.pdf>.

[13] Connette CP, Pott A, Hagele M, Verl A. Control of a pseudo-omnidirectional, nonholonomic, mobile robot based on an ICM representation in spherical coordinates. In: Proceedings of 47th IEEE conference on decision and Control. Canum, Mexico; December 9–11, 2008. p. 4976–83.

[14] Moore KL, Flann NS. A six-wheeled omnidirectional autonomous mobile robot. IEEE Control Syst Mag 2000;20(6):53–66.

[15] Kalmar-Nagy T, D'Andrea R, Ganguly P. Near-optimal dynamic trajectory generation and control of an omnidirectional vehicle. Rob Auton Syst 2007;46:47–64.

[16] Cuerra PN, Alsina PJ, Medeiros AAD, Araujo A. Linear modeling and identification of a mobile robot with differential drive. In: Proceedings of ICINCO international conference on informatics in control automation and robotics. Setubal, Portugal; 2004. p. 263–9.

[17] Handy A, Badreddin E. Dynamic modeling of a wheeled mobile robot for identification, navigation and control. In: Proceedings of IMACS conference on modeling and control of technological systems. Lille, France; 1992. p. 119–28.

4 Mobile Robot Sensors

4.1 Introduction

The sensors designed for robots resemble the sensors of the human sensory system (e.g., vision, hearing, kinesthetic) that provide input signals to the brain for processing, utilization, and action. The use of sensors in robotics is of paramount importance for closing the feedback control loops that secure efficient and automated/autonomous operation of robots in real-life applications. Sensing methods provide the robots (fixed, mobile, and hybrid) with higher level and intelligence capabilities that go far beyond the "preprogrammed" style of operation through repetitive execution of a set of programmed tasks.

The aim of this chapter is to provide a conceptual introduction to a number of important sensors used in both fixed and mobile robots (wheeled mobile robots (WMRs), mobile manipulators, humanoid robots). The particular objectives of the chapter are the following:

- To provide a popular classification of sensors, along with their operational features
- To discuss sonar, laser, and infrared sensors
- To present an outline of robotic vision and its principal functions (including omnidirectional vision)
- To list the operation principles of gyroscope, compass, and force/tactile sensors
- To give a brief introduction to the global positioning system

In compatibility with the book's scope, the material of the chapter is presented at an introductory descriptive level. Physical, design, and operational details are provided in textbooks dedicated to sensors [1–9]. The use of sensors in some representative mobile robot applications can be found in Refs. [10–22].

4.2 Sensor Classification and Characteristics

4.2.1 Sensor Classification

In general, robotic sensors are distinguished into

- analog sensors
- digital sensors

Analog sensors provide analog output signals which need analog-to-digital (A/D) conversion. Examples of analog sensors are analog infrared distance sensor, microphone, and analog compass.

Introduction to Mobile Robot Control. DOI: http://dx.doi.org/10.1016/B978-0-12-417049-0.00004-3

Digital sensors are more accurate than analog sensors, and their outputs may be of different form. For example, they may have a "synchronous serial" form (i.e., bit by bit data reading), or a "parallel form" (e.g., 8 or 16 digital output lines).

In all cases, the desired sensor "features" are *high resolution*, *wide operation range*, *fast response*, *easy calibration*, *high reliability*, and *low cost* (for purchase, support, maintenance). The sensors include also the *visual cameras* with all their auxiliary equipments needed in computational or artificial vision.

The transfer of data from the sensor to the computer (CPU) can be initiated by the computer (called *pollin*) or by the sensor itself (via interrupt). In the CPU-initiated case, the CPU must properly check whether the sensor is ready, by reading a status line in the loop. In the sensor initiating case, the availability of an interrupt line is needed. The second way is much faster since once an interrupt (indicating that data is ready) is made, the CPU reacts immediately to this request.

The sensory systems for robot positioning can be grouped in the following categories:

- Mechanical
- Acoustic
- Electromagnetic
- Magnetic
- Optical

Some of them are suitable for a position measurement of a stationary robot, whereas others are suitable for a position measurement during a motion of the robot. *Mechanical systems* require a physical contact between the robot and the sensor. Frequently, they are integrated in the robot body. *Acoustic* and *electromagnetic sensors* use the directionality and the time-of-flight measurement of sent and received signals in order to compute the angular position and the linear position of the object of interest.

Acoustic systems employ ultrasound frequencies, and electromagnetic sensor systems include optical, laser, and radar equipment. In both cases, a free "line of sight" between the transmitter and the receiver is required. *Magnetic sensors* employ the spatial configuration of static magnetic fields of the Earth and solenoids for the calculation of the position. Finally, optical sensors use appropriate vision cameras (monocular, binocular, omnidirectional).

Some further classifications of robot sensors are the following:

From a robot viewpoint:

- On-board (local) sensors, that is, sensors mounted on the robot
- Global sensors, that is, sensors mounted outside the robot in the environment and sending the sensor data back to the robot.

From a passive/active viewpoint:

- Passive sensors, that is, sensors monitoring the environment without affecting it (e.g., gyroscope, vision camera)
- Active sensors, that is, sensors that stimulate the environment for monitoring it (e.g., infrared sensor, laser scanner, sonar sensor).

Mobile robots' sensors:

- Internal (proprioceptive) sensors, that is, sensors that monitor the robot internal state
- External (exteroceptive) sensors, that is, sensors that monitor the robots environment.

Internal sensors include the sensors that measure motor speed, wheel load, robot arm joint angles, and battery voltage. External sensors include sensors that measure distance, sound amplitude, and light intensity. Examples of passive sensors include temperature probes, microphones, and *charge-coupled device* (CCD) or CMOS cameras. Examples of active sensors, which emit energy into the environment, are wheel quadrature encoders and laser range finders.

4.2.2 Sensor Characteristics

The sensors used in stationary or mobile robotics have a variety of performance characteristics. These characteristic features are different in controlled environments (indoor, laboratory environments) and noncontrolled environments (outdoor, real-world environments).

The basic features of sensors are as follows:

- Dynamic range (i.e., the spread between the lower and upper limits of input values for which the sensor is working normally). Typically, in order to cover both very small and very large signal ranges use is made of the logarithm of the ratio of the maximum and minimum values, that is,

$$\text{Range} = 20\log\left(\frac{\text{Maximum input}}{\text{Minimum input}}\right)(\text{dB})$$

For example, if the interval of voltage values measured by a voltmeter is between $V_{\min} = 1$ mV and $V_{\max} = 20$ V, then,

$$\text{Range} = 20\log\frac{V_{\max}}{V_{\min}} = 20\log\left[\frac{20}{0.001}\right] = 86\text{ dB}$$

- Resolution (i.e., the minimum difference of the measured variable that can be recognized by the sensor). Typically, in analog sensors, the resolution coincides with the lower limit of the sensor's operational range. However, this is not true in digital sensors, where the analog input is converted into binary form.
- Linearity (i.e., the property that the output value $f(x+y)$ of a sensor to a sum of inputs x and y is equal to the sum $f(x)+f(y)$ of the output values of the sensor obtained separately by each input). In a more general formulation, linearity implies that $f(k_1x+k_2y) = k_1f(x)+k_2f(y)$, where k_1 and k_2 are constant parameters.
- Bandwidth (i.e., the maximum rate or frequency of readings/data the sensor can provide). The number of readings (measurements) per second provides the sensor's frequency measured in *hertz*.

Other features that are important for nonlaboratory environments include the following:

- Sensitivity (i.e., the degree to which the changes of the input signal affect the output signal).

- Accuracy (i.e., in what degree the sensors reading coincides with the true value of the input). If $e = y - x =$ sensor reading − true value, then accuracy is given by

  $$\text{Accuracy} = 1 - |\text{error}|/x$$

- Precision (which applies in case the errors are random). Precision is defined as

  $$\text{Precision} = \text{Range}/\sigma$$

  where σ is the standard deviation of the error from a mean value m.

4.3 Position and Velocity Sensors

4.3.1 Position Sensors

The *position sensors* (or *position transducers)* are used to determine whether the joints (linear/rotational axes) of robotic links or mobile platforms have moved in the correct position, which drives the end effectors or mobile platform in the desired position/orientation of the Cartesian space. Similarly, the velocity sensors measure the speed of the motion (linear, angular) of the robot's joints or platform.

The three basic position sensors are the following:

1. Potentiometers
2. Resolvers
3. Encoders

The *potentiometer* (linear or angular) is a device that gives an output $V_0(t)$ proportional to the position of the pointer, that is, $V_0 = K_P\Theta(t)$, where K_p is the potentiometers coefficient.

The *resolver* is an analog sensor that gives an output proportional to the rotation angle of another object with respect to a fixed element. In the simplest case, a resolver has a simple winding in the rotor and a couple of windings in the stator that have a relative angle of 90°. If the rotor receives a signal $A\sin(\omega t)$, the voltages at the two terminals of the stator are

$$V_{s1}(t) = A\sin(\omega t)\sin\theta$$
$$V_{s2}(t) = A\sin(\omega t)\cos\theta$$

where θ is the rotation angle of the rotor with respect to the stator.

The *encoders* are distinguished into *differential* (or incremental) encoders and *absolute* encoders. Encoders constitute a basic feedback sensor for motor control. The two typical techniques for building an encoder is the *Hall effect sensor* (magnetic sensor) and the *optical encoder* (a sector disk with black and white segments together with a *light-emitting diode* (LED) and a photo-diode; Figure 4.1A). The photodiode detects reflected light during a white segment, but not during a black

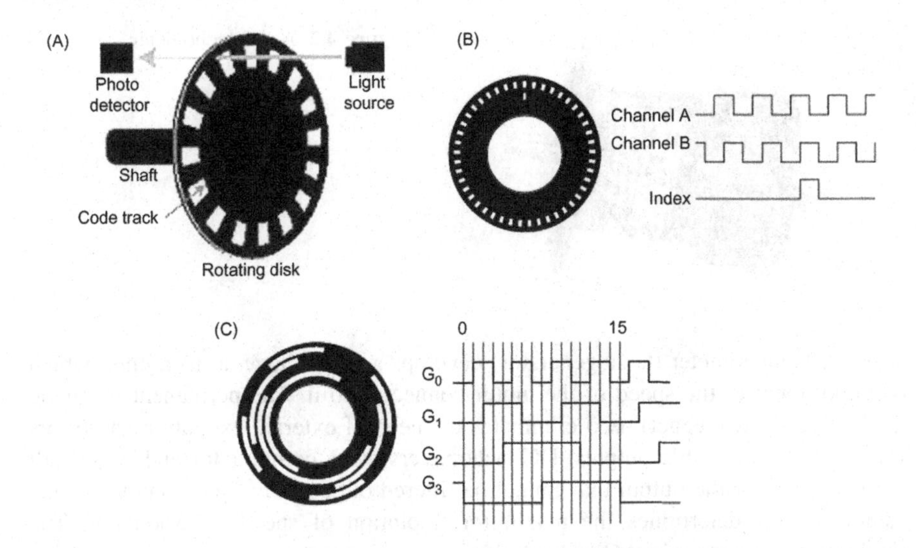

Figure 4.1 (A) An encoder disk with 16 white and 16 black segments. (B) A differential encoder with an index track. (C) A 4-bit absolute encoder.

segment. Therefore, if this disk has 16 white and 16 black segments, the sensor will receive 16 pulses during a revolution.

The *differential* (or *quadrature*) *encoder* is the most common feedback device of robots. An encoder is mounted on each joint (motor) axis.

A differential encoder has a second track added to generate a pulse that occurs once per revolution (index signal) which is used to indicate an absolute position (Figure 4.1B). To obtain information about the direction of rotation, the lines on the disk are read out by two different photodiode elements that look at the disk pattern with mechanical shift of one-fourth the pitch of line pair between them. As the disk rotates, the two photodiodes generate signals that have a 90° phase difference. These two signals are usually called the *quadrature* A and B signals (Figure 4.1B). The clockwise direction is typically defined as positive when the channel A moves before the channel B.

The disk of an *absolute encoder* is patterned with a number of discrete tracks which correspond to the word length (Figure 4.1C). The pattern shown in Figure 4.1C for this 4-bit absolute encoder is the so-called *Gray* (or *reflected binary*) code.

4.3.2 Velocity Sensors

Typically, the velocity of robotic joints is measured directly with the aid of tachometers. The indirect way of velocity measurement through numerical differentiation of the position signal is not preferred because of the occurring differentiation noise. Tachometers are distinguished in *direct current* (DC) and *alternate current* (AC) tachometers. In robotics, the DC tachometer is mostly used (Figure 4.2).

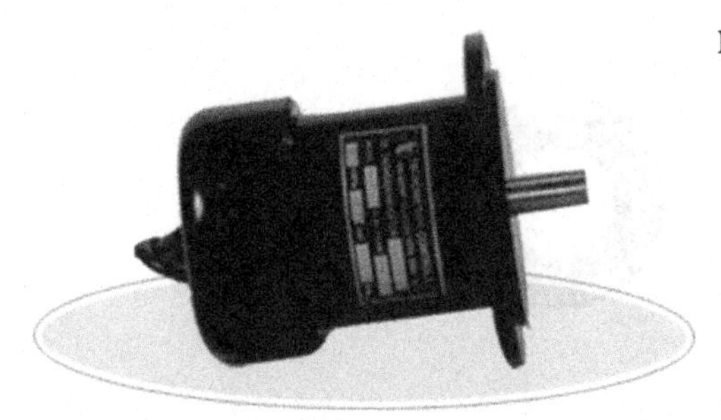

Figure 4.2 A DC tachometer.

The DC tachometer (tachogenerator) develops a DC voltage, at its output, which is proportional to the speed of the motor connected to it. The permanent magnetic field (permanent magnet) used eliminates the need of external excitation and offers very reliable and stable outputs. DC tachometers use a commutator and so a small ripple appears in the output that cannot be filtered out entirely. The accuracy of the tachogenerator determines the maximum resolution of speed measurement. The ripple does not appear in AC tachometers.

4.4 Distance Sensors

Sensors which measure the distance of a robot with the obstacles around it include the following:

- Sonar sensors
- Laser sensors
- Infrared sensors

4.4.1 Sonar Sensors

Sonar sensors (or, simply, *sonars* from *sound and navigation and ranging sensors*) have a relatively narrow cone (Figure 4.3) and so for a 360° coverage, a typical mobile robot sensor configuration is to use 24 sensors, each one mapping a cone of about 15° each. Actually, there are available for selected commercial sonars with a wide range of cone angles, such as to fit all possible practical applications.

The operation principle of sonar sensors includes the emission of a short acoustic signal (of duration about 1 ms) at an ultrasonic frequency of 50–250 kHz and the measurement of the time from signal emission until the echo returns to the sensor. The measured *time-of-flight* is proportional to twice the distance of the nearest obstacle in the sensor cone. If no signal is received within a maximum time period, then no obstacle is detected within the corresponding distance. Measurements are repeated about 20 times per second (which correspond to its typical clicking sound).

As shown in Figure 4.3D, the sonar sensors beam is not fully confined to a narrow cone because of the presence of *side lobes* which, if reflected first, might

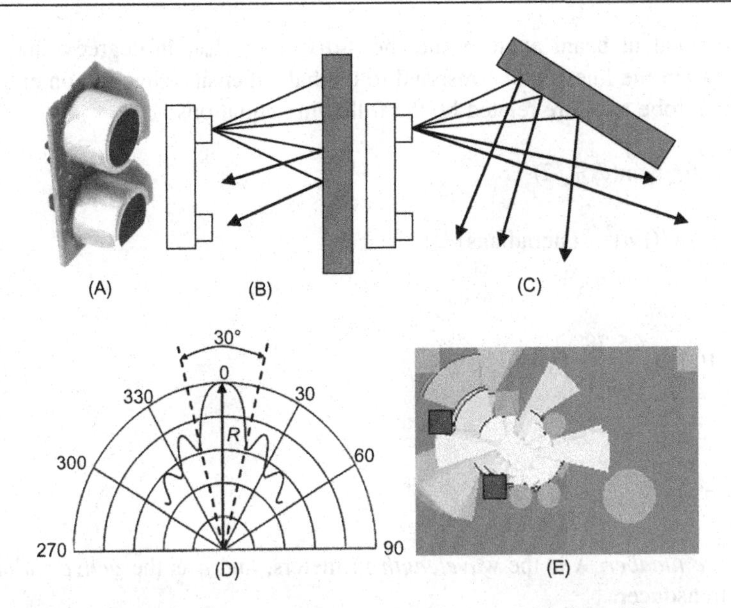

Figure 4.3 (A) A commercial sonar sensor (MPC). (B) A cone of about 30° is mapped by each sonar sensor. (C) If the obstacle in front of the sensor is far from perpendicular, then no signal will be reflected toward the sensor and so the object will not be detected. (D) Typical beam intensity pattern. (E) 360° scan in an environment with several obstacles.
Source: (D) Reprinted from Ref. [22], with permission of Elsevier Science Ltd.; (E) Courtesy of N. Katevas.

confuse the interpretation of the time-of-flight information. Very often, the actual scenes appear differently on the observer's perspective. For example, in case a sonar sensor is mounted on a mobile robot, due to the relatively wide beam width, important characteristics of the environment (obstacles, doors, etc.) only show up when the robot is close enough to observe them. The distance L of an object that causes the reflection of the wave is given by

$$L = \frac{1}{2} v_s t_0$$

where v_s is the speed of sound (about 330–350 m/s in air) and t_0 is the time-of-flight.

From physics we know that the speed v_s of sound in air is given by

$$v_s = \sqrt{\gamma R T}$$

where R is the gas constant, γ is the ratio of specific heat, and T is the absolute temperature (K). The *effective range* of most sonars used in mobile robotics varies approximately between 12 cm and 5 m, with 98–99.1% accuracy.

The equivalent beam angle δ and the 3-dB angle $\theta_{3\text{dB}}$ in degrees, that is, the angle between the lines that correspond to the half-intensity direction on either side of the main lobe axis, are related by the following equations:

$$a = 1.6/\mu \sin(\theta_{3\text{dB}}/2)$$

$$\delta = 5.78/(\mu a)^2 \quad \text{(steradians)}$$

or

$$\delta = 10 \log\left(\frac{5.78}{(\mu a)^2}\right) \text{ (dB)}$$

where

$$\mu = 2\pi/\lambda$$

is the *wave number*, λ is the *wavelength* in meters, and a is the *active radius* of a circular transducer.

4.4.2 Laser Sensors

Laser proximity sensors constitute a special case of optical sensors that have a range from centimeters to meters. They are commonly called *laser radars* (or *lidars* = light direction and ranging sensors). Energy is emitted at impulses. The distance is computed from the time-of-flight. They can also be employed as *laser altimeters* for obstacle avoidance or for vehicle detection in highways. A typical laser range finder is shown in Figure 4.4.

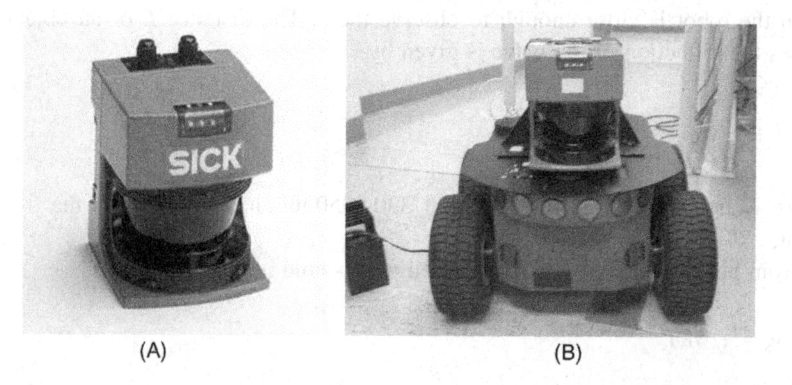

Figure 4.4 (A) A laser range finder (SICK LMS 210). (B) SICK LRF mounted on a P2AT WMR.
Source: http://www.ai.sri/centibots/tech_design/robot.html.

Laser sensors supply *speed* and *height*. *Rotary lasers* are based on rotation of the wave 360° at more than 1 or 2 rpm, and a mirror at 45°. This overcomes the problem of hidden areas. Very often, they work in conjunction with reflecting beacons. Rotary lasers measure angular position. *Scanning laser range finders* combine the lidar with the rotary lasers by providing both range and angular position of the detected object. They do not need reflecting beacons and so they are very useful because they can work in unstructured environments. Unfortunately, laser sensors are very large and heavy (and also too expensive) for small mobile robots. For this reason, infrared distance sensors are very popular in mobile robots. The time-of-flight measurement in a laser range finder is done using a pulsed laser and measuring the elapsed time directly as in the sonar sensor. The easiest way is to measure the phase shift of the reflected light—as explained in Section 4.4.3.

4.4.3 Infrared Sensors

Near-infrared light can be produced by an LED or a laser. The typical wavelength of the infrared light emitted is somewhere between 820 and 880 nm. Therefore, most surfaces have a roughness greater than the wavelength of the incident light and so diffuse reflection takes place, that is, the light is reflected almost isotropically. The component of the infrared light that falls within the aperture of the sensor returns almost parallel to the transmitted beam for distant objects (Figure 4.5A).

The sensor sends a 100% amplitude modulated light at a given frequency f and measures the phase shift between the transmitted and the reflected signals. If ϕ is the electronically measured phase shift and λ the wavelength, then the distance $2D$ is equal to $(\phi/2\pi)\lambda$, that is, (Figure 4.5B)

$$D = \left(\frac{\lambda}{4\pi}\right)\phi$$

A real infrared sensor in front of a wall obstacle is shown in Figure 4.6.

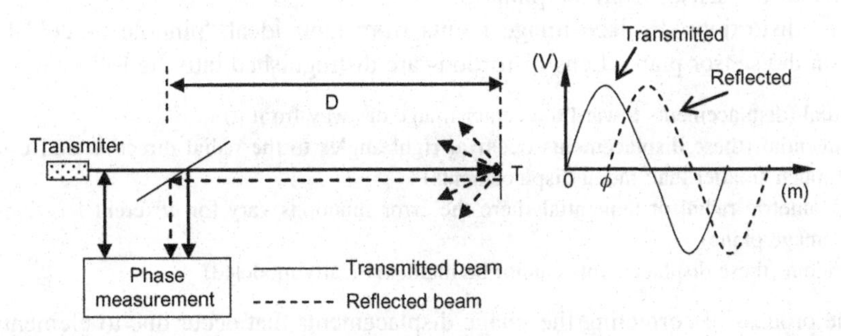

Figure 4.5 (A) Infrared distance measurement use the phase-shift method. (B) Phase shift between transmitted and received signals.

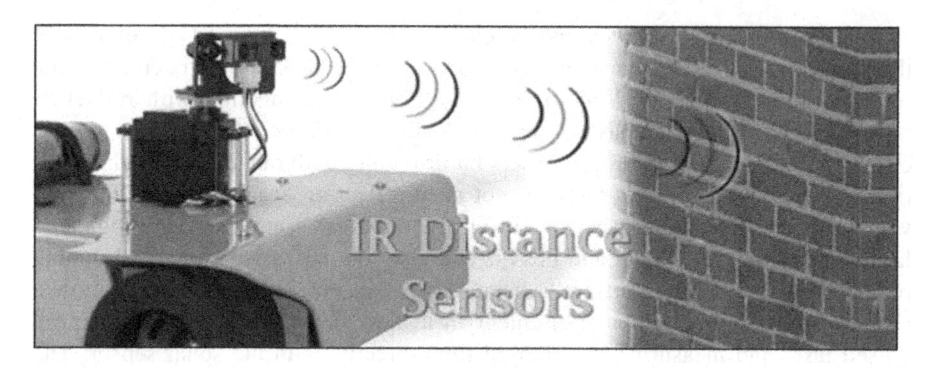

Figure 4.6 Infrared sensor mounted on a mobile robotic platform (infrared sensors can be used for obstacle avoidance, line following, and map building).
Source: http://www.trossenrobotics.com/c/robot-IR-sensors.aspx.

4.5 Robot Vision

4.5.1 General Issues

Robot vision is the ability of a robot to see and recognize objects via the collection into an image of the light reflected by these objects, and then interpreting and processing this image. Robot vision employs *optical* or *visual sensors* (cameras) and proper electronic equipment to process/analyze visual images and recognize objects of concern in each robotic application.

An ideal *pinhole* camera model is used to represent an ideal lens, and assumes that rays of light travel in straight lines from the object through the pinhole to the image (sensor) plane. Pinhole camera is the simplest device that captures accurately the geometry of perspective projection. The pinhole is an infinitesimally small aperture. The image of the object is formed by the intersection of the light rays with the image plane (Figure 4.7). This mapping from the three dimensions onto two dimensions is called *perspective projection.* The best radius r of the pinhole is approximately equal to $r \simeq \sqrt{\lambda d}$, where λ is the wavelength of the light and d the distance of the sensor from the pinhole.

Lens distortions displace image points from their ideal "pinhole model" locations on the sensor plane. Lens distortions are distinguished into the following:

- Radial (displacements toward the center image or away from it)
- Tangential (these displacements occur at right angles to the radial direction and usually are much smaller than radial displacements)
- Asymmetric radial or tangential (here the error functions vary for different locations on the image plane)
- Random (these displacements cannot be mathematically modeled)

The process of correcting the image displacements that occur due to elements of the camera's interior orientation is called *camera calibration.* The ideal pinhole camera model can be achieved by very accurate and expensive lenses. A *stereo*

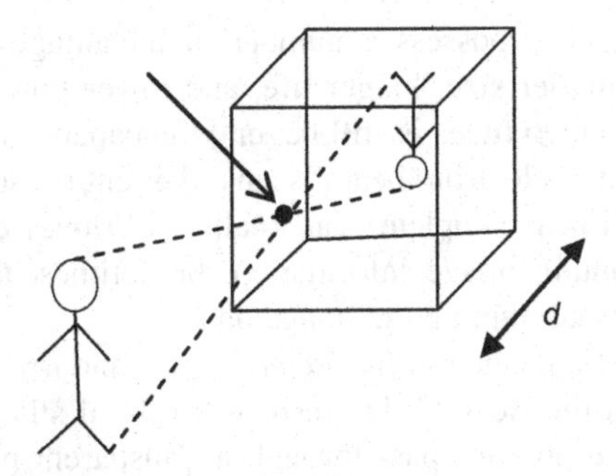

Figure 4.7 Schematic of pinhole camera geometry (for $d = 50$ mm and $\lambda = 550$ nm, we get $r = 0.165$ mm).

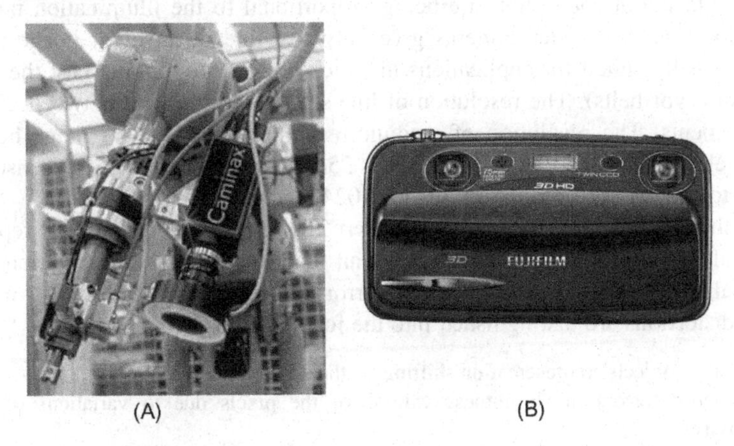

Figure 4.8 (A) A commercial pinhole camera mounted on a robot. (B) A typical stereo camera. *Source*: http://www.caminax.com, http://www.benezin.com/3d/.

camera has two or more lenses with a separate image sensor or film frame for each lens and can simulate human binocular vision. Thus, it has the ability to capture 3-D images, a process called *stereophotography*. The distance between the lenses is typically near-equal to the distance between the humans eyes. A pinhole camera and a stereocamera are shown in Figure 4.8A and B.

Image (*electronic image*) is an array of *pixels* (picture elements) that has been digitized into the memory of a computer. A binary number is stored in each pixel to represent the intensity and the wavelength of the light falling on that part of the image.

A pixel, which is a vector of binary numbers, represents a particular color. A digital image which is a matrix of pixels (vectors) filled with corresponding colors creates a picture of the scene from the point of view of the camera. Cameras used in robotic vision include primarily television cameras that consist either of a tube or a solid-state imaging sensor, and the associated electronics. A common representative of tube family of television cameras is the *vidicon tube*. The principal representative of solid-state imaging sensors is the so-called *charge-coupled device*.

Solid-state imaging devices possess a number of advantages over tube cameras, e.g., smaller weight, smaller size, longer life, and lower power consumption. But, the resolution of some video tubes is still beyond the capabilities of CCDs.

In vidicon tubes, the electron beam scans the entire surface of the object 30 times per second. Each complete scan (called a *frame*) consists of 525 lines of which 480 lines contain image information. Sometimes, the frame consists of 559 lines with 512 lines containing the image data.

CCD devices are distinguished in *line-scan sensors* and *area sensors*. The fundamental component of a line-scan CCD sensor is a row of silicon imaging elements called *photosites*. Image photons pass through a transparent polycrystalline silicon gate structure and are absorbed in the silicon crystal, creating electron–hole pairs. The resulting photoelectrons are collected in the photosites, with the amount of charge collected at each photosite being proportional to the illumination intensity at that location. Line-scan cameras give only one line of an input image and so they are ideally suited for applications in which objects are moving past the sensor (as in conveyor belts). The resolution of line-scan sensors ranges between 256 and 2048 elements. The resolution of medium-resolution area sensors range between 32×32 elements at the low end and 256×256 elements. Current CCD sensors are capable to achieve a resolution of 1024×1024 elements or higher.

It is obvious that cameras actually convert 2-D or 3-D reality into 2-D representations, although via postprocessing they can reconstruct 3-D representations from the initial 2-D images. Cameras are not error-free devices. Broadly speaking, the camera distortions are distinguished into the following:

- Geometric (objects' representation shifting on the sensor plane)
- Radiometric (errors in "brightness values" of the pixels due to variations of pixels sensitivity)
- Spectral (errors in pixel "brightness values" due to the varying response of the sensor to different wavelengths of light)

Robot (computer) vision can be grouped in the following principal subareas:

- Sensing
- Preprocessing
- Segmentation
- Description
- Recognition
- Interpretation

4.5.2 Sensing

The sensing subarea of robotic vision includes the following (besides the design of cameras in their own):

- Camera calibration
- Image acquisition
- Illumination
- Imaging geometry

4.5.2.1 Camera Calibration

Camera calibration, the first requirement in sensing, is concerned with the correction of image displacements that occur due to the characteristics of the camera's interior orientation. Lens distortion is one of the reasons of the displacement of image points from their ideal "pinhole model" positions on the sensor plane. The methods of camera calibration are classified as follows:

- Model-based methods (where specific dominant features of the error are modeled and corrected)
- Mapping-based methods (where a proper reality-to-image or image-to-reality mapping is produced, without the need to understand the underlying causes)

A basic contributing factor which is modeled is the radial lens distortion. The model typically used is a second-, third-, or fourth-order order polynomial. Then, least squares or interpolating techniques are used to determine the values of the coefficients that lead to a best model of the observed error.

In the mapping-based approach, no attempt is made to understand the particular causes of the error. All that matters is to have an explicit image-to-reality mapping preserved.

In practice, both approaches to camera calibration can remove most of the displacement errors caused by elements of interior camera orientation, prior to image preprocessing.

4.5.2.2 Image Acquisition

As described in Section 4.5.1, visual information is converted to electrical signals by visual sensors and associated electronic equipment. When sampled spatially and quantized in amplitude, these visual signals yield a digital image.

Three basic topics that are included in image acquisition are as follows:

1. Imaging techniques
2. Effects of sampling on spatial resolution
3. Effects of amplitude quantization on intensity resolution

Let $f(x, y)$ be a 2-D image, which for computer processing is digitized both spatially and in amplitude (intensity). Digitization of the spatial coordinates x and y is known as *image sampling*, while the amplitude digitization is called *intensity* or *gray-level quantification* (for monochrome images, i.e., black–white variations in shades of gray).

The digitized form of $f(x, y)$ (called *digital image*) is represented by a matrix of the type

$$f(x,y) \simeq \begin{bmatrix} f(0,0) & f(0,1) & \cdots & f(0,M-1) \\ f(1,0) & f(1,1) & \cdots & f(1,M-1) \\ \vdots & \vdots & \ddots & \vdots \\ f(N-1,0) & f(N-1,1) & \cdots & f(N-1,M-1) \end{bmatrix}$$

where now x and y have discrete values at $x = 0, 1, 2, \ldots, N-1$ and $y = 0, 1, 2, \ldots, M-1$. Each element in the array is what we call a *pixel* or *image element*. Clearly, $f(0, 0)$ represents the pixel at the origin of the image, $f(0, 1)$ the next pixel to the right, and so on. The value $f(x, y)$ at any point (x, y) represents the intensity of the image $f(x, y)$ at that point. The greater N and M are the better (of higher resolution) image representation obtained (Figure 4.9).

4.5.2.3 Illumination

Illumination of a scene is a critical factor that affects the complexity of the vision algorithms. A good lighting system illuminates a scene so as to minimize the complexity of the produced image and improve (enhance) the information needed for object detection and extraction. The four typical illumination techniques used in robotics are as follows:

1. Diffuse lighting (for objects with smooth and regular surfaces)
2. Back lighting (for applications where silhouettes of objects are sufficient for recognition or other purposes)
3. Structured lighting (i.e., projection of points, stripes, or grids onto the work surface, e.g., when a block is illuminated by parallel light planes that become *light stripes* when intersecting a flat surface)
4. Directional lighting (where a highly directed light (or laser) beam is used to inspect object surfaces and detect defects on the surface such as pits and scratches)

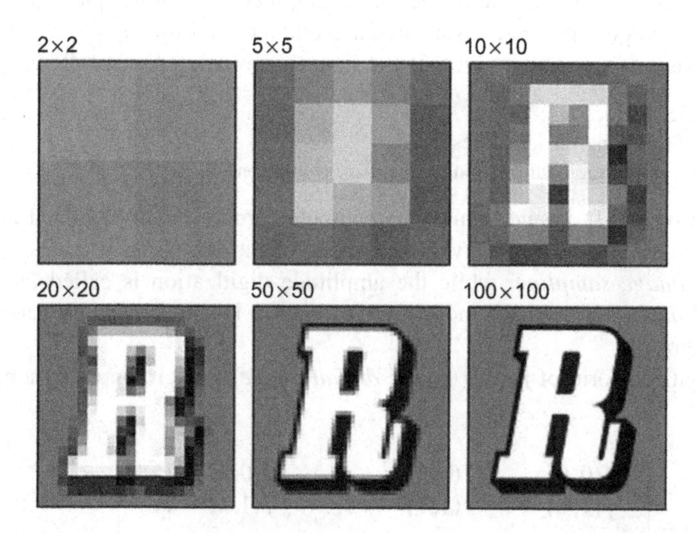

Figure 4.9 Sequence of increasing number of pixels and the corresponding image resolution.
Source: http://patriottruckleasing.com/resolution.html.

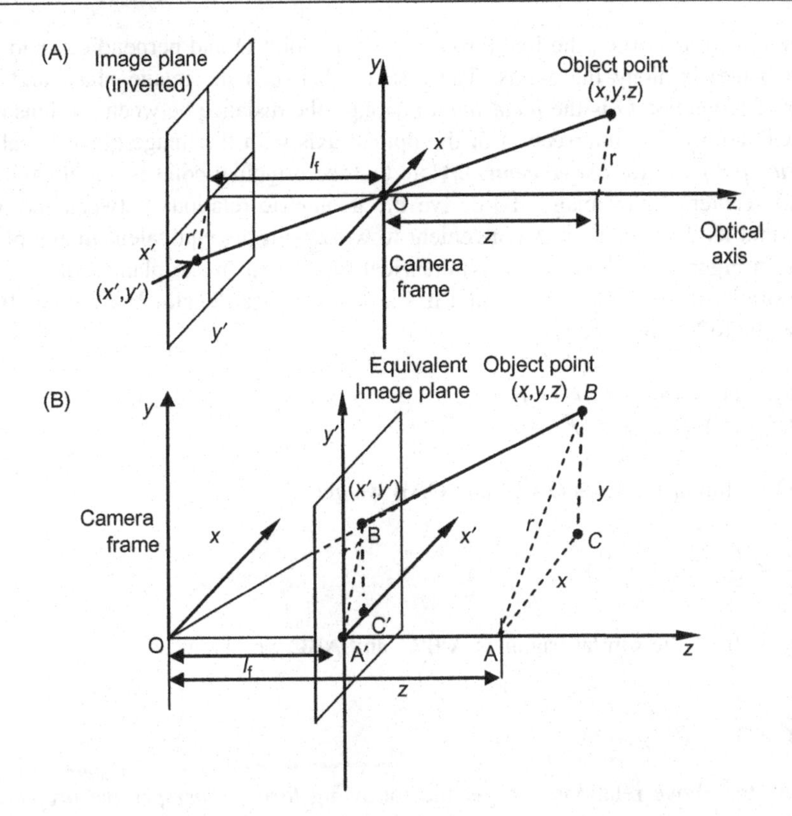

Figure 4.10 (A) Perspective imaging projection (called *reverse perspective* projection). The camera coordinate frame (x, y, z) is aligned with the world frame. In the camera frame the focal length l_f has negative value. (B) Equivalent forward image plane, symmetrical to real image plane with respect to the origin of the camera frame.

4.5.2.4 Imaging Geometry

The relation of the position of the point $f(x', y')$ in the 2-D image plane and the corresponding point $F(x_0, y_0, z_0)$ in the real 3-D space is determined by the laws of optics. Because the aperture of camera lenses via which the light falls onto the image (sensor) plane is very much smaller than the size of the objects under study, one can replace the lenses by the *pinhole* model. Thus, points from space are projected onto the image plane by lines intersecting in a common point which is known as *center of projection*. In the pinhole camera model, the center of projection is located at the center of lenses[1]. The camera itself represents a rigid body to which a coordinate frame is assigned. This coordinate frame describes the poses of the camera.

Figure 4.10A shows the basic model of the *perspective* (or *imaging*) transformation.

[1] It is recalled that the equation of a convex lens is $1/l_o + 1/l_{im} = 1/l_f$ where l_f is the focal length (i.e., the distance of the focus from the lens) and l_o, l_{im} are the distances of the object and the image from the lens, respectively (see Section 4.8).

The *optical axis* (i.e., the line through the 3-D point O and perpendicular to the image plane) is along the z-axis. The distance between the image plane and the center of projection O is the *focal length* l_f (e.g., the distance between the lens and the CCD array). The intersection of the optical axis with the image plane is called the *principal point* (or *image center*). Note that the principal point is not always the "actual" center of the image. To derive the geometric relations between the 3-D object and its 2-D image, it is convenient to work with the equivalent image plane shown in Figure 4.10B, which is symmetrical to the real image plane with respect to the origin of the camera frame and has a positive focal length. In Figure 4.10B, we use the following notations:

$$(x, y, z) \to (X, Y, Z), r \to R$$
$$(x', y', z') \to (x, y, z), r' \to r$$

Using the similar triangles $OA'B'$ and OAB we get

$$\frac{l_f}{Z} = \frac{r}{R}$$

Similarly, from the similar triangles A′B′C′ and ABC we obtain

$$\frac{x}{X} = \frac{y}{Y} = \frac{r}{R}$$

From the above relations, we get the following *forward perspective projection equations*:[2]

$$x = \frac{l_f X}{Z}, \quad y = \frac{l_f Y}{Z}, \quad z = l_f$$

which are nonlinear (because they involve division by Z). However, using homogeneous representations, we can obtain a *forward linear perspective transformation* from $\lambda[X, Y, Z, 1]^T$ to $[x, y, z, 1]^T$, where λ is a scaling factor, namely,

$$\begin{bmatrix} x \\ y \\ z \\ 1 \end{bmatrix} = \begin{bmatrix} l_f & 0 & 0 & 0 \\ 0 & l_f & 0 & 0 \\ 0 & 0 & l_f & 0 \\ 0 & 0 & 1 & 0 \end{bmatrix} \lambda \begin{bmatrix} X \\ Y \\ Z \\ 1 \end{bmatrix}$$

[2] The reverse perspective equations can be found from Figure 4.10A, and are: $x = -l_f X/(Z - l_f) = l_f X/(l_f - Z)$, $y = -l_f Y/(Z - l_f) = l_f Y/(l_f - Z), z = -l_f$

with $\lambda = 1/Z$. Thus, our forward linear perspective transformation matrix is

$$\mathbf{P} = \begin{bmatrix} l_f & 0 & 0 & 0 \\ 0 & l_f & 0 & 0 \\ 0 & 0 & l_f & 0 \\ 0 & 0 & 1 & 0 \end{bmatrix}$$

It is now easy to verify that by simply using $\mathbf{P}^{-1}$ (the inverse) we cannot recover a 3-D point from its image. This can be done only if at least one of the coordinates of the 3-D point is known.

4.5.3 Preprocessing

Image preprocessing can be performed by two general methodologies:

1. Spatial domain methodology
2. Frequency domain methodology

Spatial domain preprocessing methods work directly with image pixel arrays. In general, spatial domain preprocessing functions have the form

$$p(x, y) = H(f(x, y))$$

where $f(x, y)$ is the input image, $H(\cdot)$ is an operator on $f(x, y)$, and $p(x, y)$ is the outcome of preprocessing (i.e., the preprocessed image). In the simplest case, H has the form of an *intensity mapping* I, that is,

$$u = I(v)$$

where v and u denote the intensities of $f(x, y)$ and $p(x, y)$, respectively. Another typical image preprocessing technique is the so-called *window* (or *template* or *filter*) technique which uses convolution masks.

Frequency domain preprocessing methods use the Fourier transform of an image which converts the image to an aggregate of complex-valued pixels. To reduce the noise and other spurious effects resulting from the operations of sampling, quantization transmission, and any other disturbances of the environment, appropriate smoothing operations are employed. For example,

- Neighborhood averaging (where the smoothed image is obtained by averaging the intensity values of the pixels contained in a predefined region around (x, y))
- Median filtering (where the smoothed image is obtained by using the median instead of the average of the pixels in the desired surrounding region)

Other preprocessing operations include the following:

- Image enhancement (automatic adaptation to illumination variations)
- Edge detection (this is the preliminary step for numerous detection algorithms)
- Image thresholding (i.e., the selection of a threshold T that separates the intensity modes, e.g., in images that have intensities that are grouped into two dominant models, viz., light objects on a dark background)

4.5.4 Image Segmentation

Image segmentation is the process that splits a source image into its constituent parts or objects. The regions of interest are selected on the basis of several criteria. For example, it may be necessary to find a single part out of a bin. For navigation purposes, it may be useful to extract only *floor lines* from an image. In general, by segmentation, objects are extracted from a scene for subsequent recognition and analysis.

The two basic principles used in segmentation algorithms are as follows:

1. Discontinuity (e.g., edge detection)
2. Similarity (e.g., thresholding, region growing)

These processes can be applied to both static and time-varying scenes. Edge detection is based on intensity discontinuities and yield pixels that are on the boundary between the objects and the background. However, the boundaries detected in this way are not sharp in many cases due to noise and other spurious effects. This is avoided by using linking and other boundary detection methods by which edge pixels are assembled into a proper and meaningful set of objects.

4.5.5 Image Description

Image description is the process of extracting features from an object for recognition purposes. Descriptors must be independent of the size, location, and orientation, and provide sufficient discriminatory information. Typical boundary descriptors are as follows:

- Chain codes (these represent a boundary as a set of straight line segments of specified length and direction)
- Polygonal approximations (here a digital boundary can be approximated to any desired degree by a polygon)
- Fourier descriptors (here, a 2-D boundary can be represented by 1-D transform, that is, a point (x, y) is reduced to the complex number $x + jy$)

Typical regional descriptors include the following:

- Texture (this provides quantitative measures of properties such as smoothness, regularity, and coarseness)
- Region skeleton (here a thinning algorithm (or skeletonizing algorithm) is used to obtain the skeleton of the region)
- Moment invariants (here normalized central moments of order $p + q$ are used as descriptors because they are invariant to translation, rotation, and scaling)

The segmentation and description methods mentioned in Section 4.5.4 and the present section are applicable to 2-D scene data. The corresponding processes for 3-D scenes are more complex and include the following:

- Constructing planar patches

- Gradient technique (to obtain patch representations)
- Generalized cones or cylinders (moving along a cross section of the object, e.g., a ring, along a straight line spine)

4.5.6 Image Recognition

Image recognition is called the labeling process applied to a segmented object of a scene. That is, the image recognition presumes that objects in a scene have been segmented as individual elements (e.g., a bolt, a seal, a wrench). The typical constraint here is that images are acquired in a known viewing geometry (often perpendicular to the workspace).

Image recognition methodologies are distinguished into the following:

- Decision theoretic methods (these methods use proper decision or discriminant functions for matching the objects to one of several prototypes)
- Structural methods (here, an object is decomposed in a set of primitive element pattern of predefined length and direction). A simple example is shown in Figure 4.11. Three other structural methods are (i) shape number matching, (ii) string matching, and (iii) syntactic methods (string grammars, semantics, etc.).

4.5.7 Image Interpretation

Image interpretation is a higher level process which uses combinations of the methods discussed earlier, namely, sensing, preprocessing, segmentation, description, and recognition. A machine vision system is ranked according to its general ability to extract useful information from a scene under a wide repertory of viewing conditions, needing minimal knowledge about the objects at hand. Factors that make image interpretation a difficult task include variations in illumination conditions, viewing geometry, and occluding bodies.

Occlusion problems occur when we have a multiplicity of objects in an unconstrained work terrain.

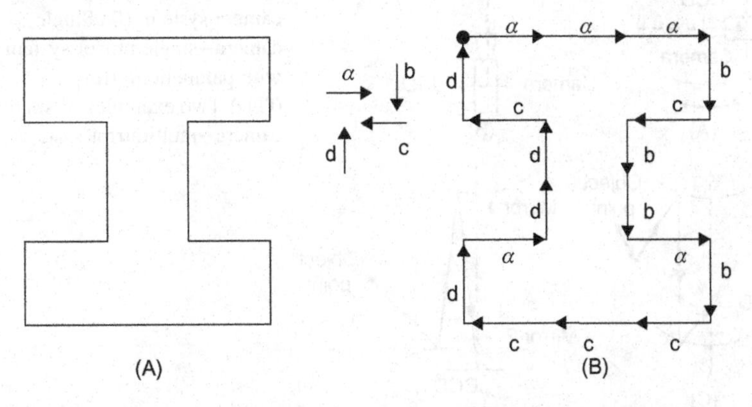

Figure 4.11 Representation of an object's boundary (A), by primitives (B).

Other subsidiary processes that help in better image interpretation are as follows:

- Multiscaling (this is useful whenever multiple scales of an image can be obtained). Many camera image systems can emit a thumbnail along the main image. This can be used as a basis for searching the main image, that is, any regions in the low-resolution image lacking in pixels probably represent empty or very sparse areas in the full image.
- Sequential search (here, each and every pixel is examined once and only once). New pixels are compared against previously found groups of pixels and inserted into the matching group. If a pixel is found to belong to two groups, the groups must be combined.

4.5.8 Omnidirectional Vision

Omnidirectional vision deals with the capture and interpretation of images that depict full 360° view (i.e., horizontal panoramic view) of the environment. Combining separate horizontal panoramic views one can get full spherical projection. The ability of omnidirectional vision to capture 360° view leads to improved results for optical flow, feature selection, and feature matching.

The vision systems that can generate omnidirectional/panoramic images fall into the following categories:

- *Camera-only systems*: Here, a standard camera is rotated about its vertical axis, and the perspective images are produced by sticking them so as to obtain 360° panoramic views (Figure 4.12A). The overall resolution does not depend on the camera resolution but on the resolution of the angular rotation.
- *Single camera-single mirror systems*: These systems are very popular in vision-based mobile robot navigation. In general, a CCD camera is pointed vertically up as shown in Figure 4.12B.

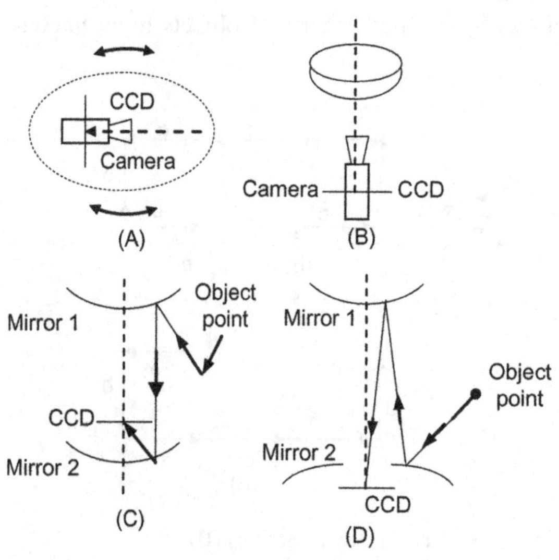

Figure 4.12 Schematic representation of omnidirectional vision systems: (A) Rotating camera system. (B) Single camera–single mirror system with parabolic mirror. (C,D) Two examples of single camera–multimirror systems.

- *Single camera-multimirror systems*: These are also called catadioptric cameras, and have the important feature of compactness. Two single camera-double mirror systems are shown in Figure 4.12C and D.
- *Multicamera-multimirror systems*: A group of cameras is arranged in some way together with an equal number of mirrors, for example, placing a camera under each mirror, and merging the images obtained from all cameras to produce a 360° panoramic view of the environment.

An example of single camera-single mirror system is shown in Figure 4.13A, and a multicamera-multimirror system in Figure 4.13B.

In the system of Figure 4.13B, four triangular planar mirrors are placed side by side in a pyramidal shape, and a camera is placed under each mirror. The images of all cameras obtained are merged to provide a 360° panoramic view of the environment. This configuration provides high resolution, and the possibility of a single view, but it is not isotropic. The camera-only systems present some problems with the registration and the delay of the camera motion. In single camera-single mirror systems, there is no need to wait for a moving camera or to synchronize multiple cameras.

Actually, there are several types of mirrors:

- Planar mirrors, which possess no radial distortion and no radial loss of resolution.
- Radially curved mirrors, which include three popular quadric surface mirrors (elliptic, hyperbolic, parabolic) possessing a single viewpoint at their focal points.
- Cone mirrors, which combine the desired features of planar mirrors and rotational symmetricity of radially curved mirrors. These sensors are simple to set up and function, and isotropic (i.e., the projection is the same for all directions). They capture the entire environment in a single instant and so they are suited for dynamic environments analysis and moving cameras. The image resolution and distortion of the cone mirror are better than other catadioptric system.

In many applications, true fish-eye lenses are used which are hemispherical (i.e., they have a wide field of view of at least 180°). True fish-eye systems are easy to set up and use, but the lens need correction and calibration. They are isotropic and

Figure 4.13 (A) A single camera–single mirror omnidirectional vision system. (B) A pyramidal multicamera–multimirror system.
Source: (B) (Full View, Inc.). http://www.lkl.ac.uk/niall/nwdis/, http://www.english.pan.pl/images/stories/pliki/publikacje/academia/2005/06/28-29_siemiatkowska.pdf.

dynamic, and have very low image resolution at the periphery, and medium to high in the center.

The three primary problems that can be faced by omnidirectional vision are as follows:

1. Determine the distance of objects in a purely passive visual way (i.e., by strereopsis). This is very useful in mobile robots because it provides higher precision in range estimation than that obtained with active ranging.
2. Find the position and motion of the camera, very useful in mobile robot navigation.
3. Determine the position and motion of the robot and the objects.

Tasks that have to be performed for solving the above problems include the following:

- Registration, that is, identification and estimation of the camera and mirror intrinsic and extrinsic parameters.
- Unwarping, which is done by transformation of the original circular image from polar (ρ, θ) coordinates to rectangular Cartesian (x, y) coordinates that can be more easily interpreted by humans. The unwarped image (which is a distortion-free image) is known as "panoramic image."
- Image mapping, that is, matching pairs of images by which the rotation of the robot in between two positions can be found more easily than using a compass.
- Estimation of the distance of objects, which can be done by matching individual image features along the radial epipolar lines.
- Estimation of the instantaneous velocities of object edges from pairs of images of a moving object.

Figure 4.14A shows an omnidirectional camera sensor mounted on the Pioneer 1 WMR. The sensor is made by a camera pointed upward at the vertex of a spherical

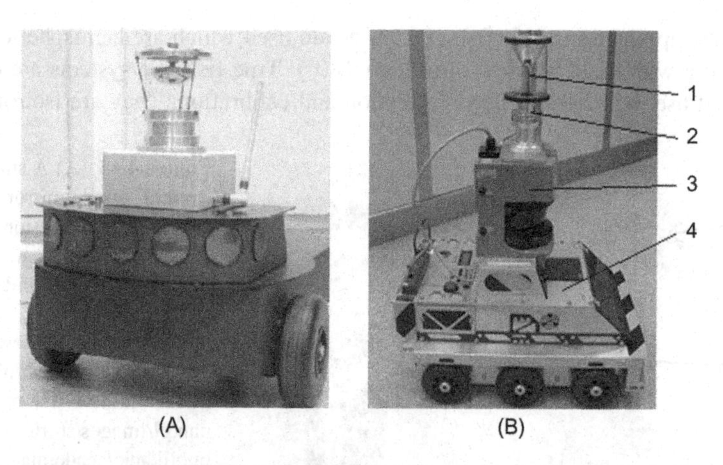

Figure 4.14 (A) The Pioneer 1 WMR equipped with an omnicamera. (B) A mobile robot (4) with a catadioptric sensor (1), a camera (2), and a laser scanner (3).
Source: http://www.ippt.gov.pl/~bsiem/ecmr_last.pdf, http://home.elka.pw.edu.pl/~mmajchro/romansy2006/romansy06.pdf.

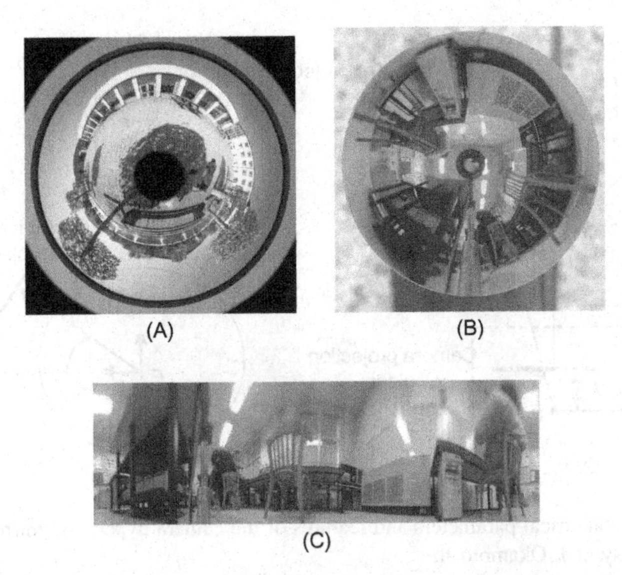

Figure 4.15 (A) Omnidirectional image acquired by a Canon EOS35OD combined with a 0–360 omnidirectional optic. (B) An omnidirectional image obtained by hypercatadioptric system. (C) Reprojection of (B) on cylindrical surface.
Source: http://www.oru.se/PageFiles/15214/Valgren_Licentiate_Thesis_Highres.pdf, http://cmp.felk.cvut.cz/demos/Omnivis/Hyp2Img/.

mirror, with the camera optical axis and the mirror optical axis aligned. Due to the spherical shape of the mirror, the sensor resolution depends on the distance between the camera and the observed region. The resolution of an image is maximum near the robot and so it is possible to localize very precisely objects that are nearer to the robot.

Figure 4.15A shows an omnidirectional image obtained by a Canon camera equipped with the 0–360 panoramic optic. Finally, Figure 4.15B shows an image obtained by a standard perspective camera observing a hyperbolic mirror, with its reprojection on cylindrical surface shown in Figure 4.15C.

Example 4.1

We are given a single camera-single mirror omnidirectional vision system with hyperbolic mirror.

(a) Investigate the hyperbolic mirror geometry.
(b) Using the property of single center of projection, show how a perspective image from the image sensed by the omnidirectional system can be created.

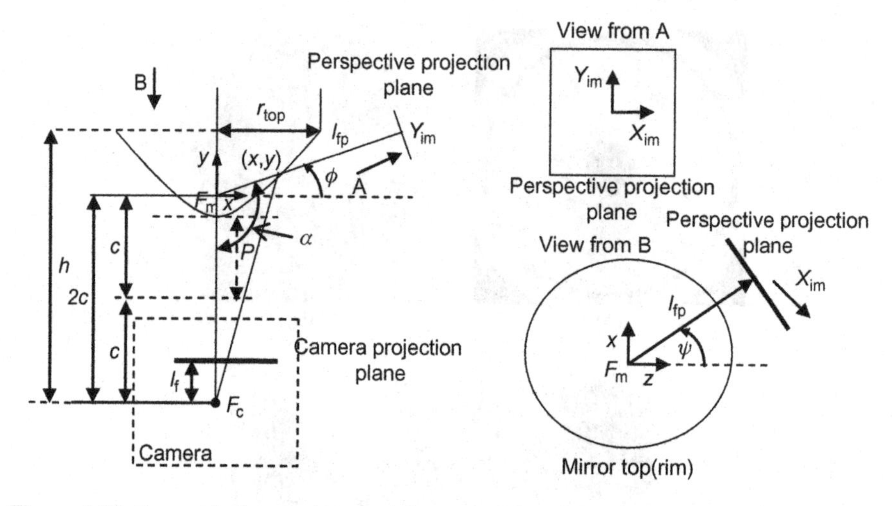

Figure 4.16 Geometrical parameters and features of the camera hyperbolic mirror system. *Source*: Courtesy of J. Okamoto Jr.

Solution

(a) We will work with the omnidirectional vision system of Figure 4.16[18]. The hyperbolic mirror is described by:[3]

$$y = \sqrt{p^2(1 + x^2/q^2)} - \sqrt{p^2 + q^2} \tag{4.1}$$

where the origin of the coordinate frame (x, y) is at the mirror focal point F_m, and p, q are the mirror parameters. The other symbols in Figure 4.16 are as follows:

F_c, focal point of the camera;
h, distance between the mirror top (rim) and the camera focal point;
r_{top}, the x coordinate of the mirror top;
c, half of the distance (F_m, F_c), the mirror eccentricity $(c = \sqrt{p^2 + q^2})$
y_{top}, the y coordinate of the mirror top $(y_{top} = h - 2c)$;
α, the vertical angle of view.

The maximum angle of view α_{max} of the mirror is determined by

$$tg(\alpha_{max} - \pi/2) = (h - 2\sqrt{p^2 + q^2})/r_{top} \tag{4.2}$$

Applying the mirror equation at $(x, y) = (r_{top}, y_{top})$, we obtain $h - \sqrt{p^2 + q^2} = p^2(1 + r_{top}^2/q^2)$, which if squared gives $h^2 + q^2 - 2h\sqrt{p^2 + q^2} = p^2 r_{top}^2/q^2$. Completing the squares gives $(h\sqrt{1+p^2/q^2} - q)^2 = (p^2/q^2)(h^2 + r_{top}^2)$. So, finally we have

[3] This equation is derived from the standard hyperbola equation opening along the axis y: $(y - y_0)^2/p^2 - (x - x_0)^2/q^2 = 1$(see Eq. (9.107d)), where the origin is at the mirror focal point $F_m(x_0 = 0,\ \ y_0 = -c,\ \ c = \sqrt{p^2 + q^2})$.

$$q + (p/q)\sqrt{h^2 + r_{\text{top}}^2} = h\sqrt{1 + p^2/q^2} \quad (4.3)$$

We can easily verify that the relation of h with the focal length l_f of the camera, the mirror top radius in pixels (r_{pixel}), and the size of each pixel in the camera CCD in millimeters (t_{pixel}) is the following:

$$h = l_f r_{\text{top}} / r_{\text{pixel}} \cdot t_{\text{pixel}} \quad (4.4)$$

The parameters l_f, t_{pixel}, and r_{pixel} depend on the camera and mirror used and on the image acquired by the system. The above relation is valid when the horizontal and vertical scale factors are the same in which case the image of the mirror top is a circle and not an ellipse.

Equations (4.1)–(4.4) are used to design a proper omnidirectional camera–mirror pair that can be placed on a WMR. Specifically, to get a desired value of h we need Eq. (4.4) to select r_{top}. Having available the value of h we choose a value for the ratio p/q, and using Eq. (4.3), we find the value of q, and then the value of p. Finally, using h, p, and q we can find the mirror equations, and the maximum angle of view using Eqs. (4.1) and (4.2).

The perspective images are created by mapping the pixels of the omnidirectional image into a plane perpendicular to a ray passing via the center of projection F_m. The distortion-free image that is obtained is equivalent to an image acquired by a perspective camera with its focus located at F_m. The transformation of omnidirectional images into perspective or panoramic images is called *unwrapping*. Obviously, if the creation of equivalent perspective images is possible, all available vision techniques for perspective projection images (including those of visual servoing) can be applied.

Figure 4.16 (right-hand side) shows also the perspective images obtained in the directions A and B. A pixel in the perspective image is described by the coordinates $(x_{\text{im}}, y_{\text{im}})$. A perspective projection plane is defined by $(l_{\text{fp}}, \phi_a, \phi_e)$, where l_{fp} is the distance in pixels from the focal point of the hyperbolic mirror to the plane concerned, ϕ_a is the azimuthal angle of the plane direction, and ϕ_e is the elevation angle of the plane (Figure 4.16).

(b) As shown in Figure 4.16, the hyperbolic mirror has the property of single projection center, that is, the light rays that contribute to the image formation, after their reflection by the mirror surface, intersect at the (virtual) focal point F_m, which is the center of projection of the mirror. The two most popular ways to construct an omnidirectional vision system with this property are the following (see Section 9.9.2):

1. Use a parabolic mirror with an orthographic projection lens camera.
2. Use a hyperbolic mirror with a perspective projection lens camera.

Figure 4.17 is referred to the second way and shows how perspective and panoramic images can be created by projecting the acquired image into properly defined projection planes [18].

By assuming that the above perspective plane parameters define the pose of a virtual perspective camera placed on a pan/tilt mechanism, l_{fp} would represent the focal distance of the virtual camera, and ϕ_a, ϕ_e would be the *pan* and *tilt* angles (i.e., azimuth and zenith angles), respectively.

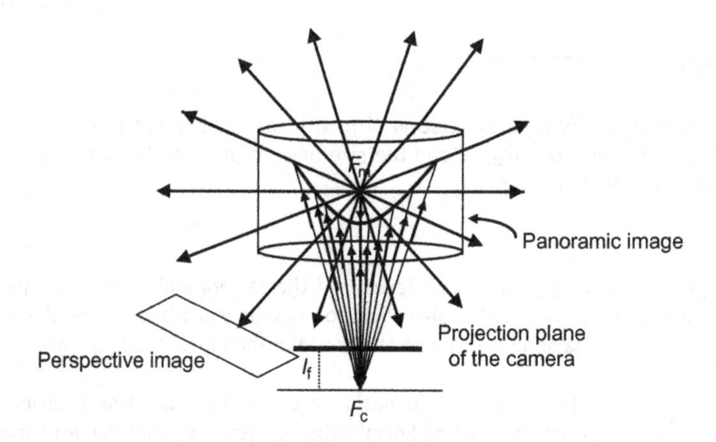

Figure 4.17 Creation of perspective (planar) image and panoramic (on cylindrical surface) image.
Source: Courtesy of J. Okamoto Jr.

It can be verified that the relation of the pixel (x,y) on the mirror surface and the pixel (x_{im}, y_{im}) on the perspective projection plane is the following [18]:

$$x_{im} = \frac{x(2c + y_{top})r_{pixel}}{(xtg\phi + 2c)}\cos\psi \quad (4.5a)$$

$$y_{im} = \frac{x(2c + y_{top})r_{pixels}}{(xtg\phi + 2c)}\sin\psi \quad (4.5b)$$

The relations of (ϕ, ψ) and (l_{fp}, ϕ_a, ϕ_e) are

$$tg\phi = \frac{l_{fp}\sin\phi_e + y_{im}\cos\phi_e}{l_{fp}\cos\phi_e} \quad (4.6a)$$

$$tg\psi = \frac{(l_{fp}\cos\phi_e - y_{im}\sin\phi_e)\sin\phi_a - x_{im}\cos\phi_a}{(l_{fp}\cos\phi_e - y_{im}\sin\phi_e)\cos\phi_a + x_{im}\cos\phi_a} \quad (4.6b)$$

4.6 Some Other Robotic Sensors

In this section, some further important robotic sensors are briefly described:

- Gyroscope
- Compass
- Force and tactile sensors

4.6.1 Gyroscope

Gyroscopes are devices measuring angular position by preserving their orientation with respect to a world coordinate frame. Thus, they are used in aeronautics, navigation, and mobile robot navigation. The gyroscope is distinguished into free gyroscope and two *degrees of freedom* (DOF) gyroscope (Figure 4.18).

The disc is rotating with very large angular speed (in the range 10,000 rad/min up to 25,000 rad/min). The *Cardano mounting system* via the two of gimbals allows the angular momentum axis of the rotating disk to have three DOF of motion about the axes x, y, and z. The free gyroscope can permanently indicate either the south-north direction (*horizontal direction gyroscope*) or the vertical direction if its angular momentum axis is aligned with the gravity direction (*vertical direction gyroscope*), that is, if the gyroscope is tipped, the gimbals will reorient to keep the spin axis of the disk in the same direction. This is the implication of the angular momentum conservation principle. The gyroscope with two DOF is obtained from the free gyroscope by placing a torsional spring along the axis x between the gimbals I and II. In this way, the rotating disc can rotate only about the axis $y$and the axis x. If we apply a torque T about the axis x, then the angular momentum axis of the disc will rotate around the axis z with a velocity

$$\Omega = T/\omega J$$

where ω is the speed of the disc and J its moment of inertia. The 2-DOF gyroscope is used to measure angular speeds instead of absolute position (*rate gyro*). The rotation of the spin axis around the axis z (velocity Ω) is known as *twist rotation* or *recession*.

4.6.2 Compass

The magnetic compass was first used in China in 2634 BC, when a piece of naturally magnetic iron ore (magnetite) was suspended by a silk thread to allow its alignment with the lines of the Earth's magnetic field, which are horizontal at the equator and vertical at the magnetic poles.

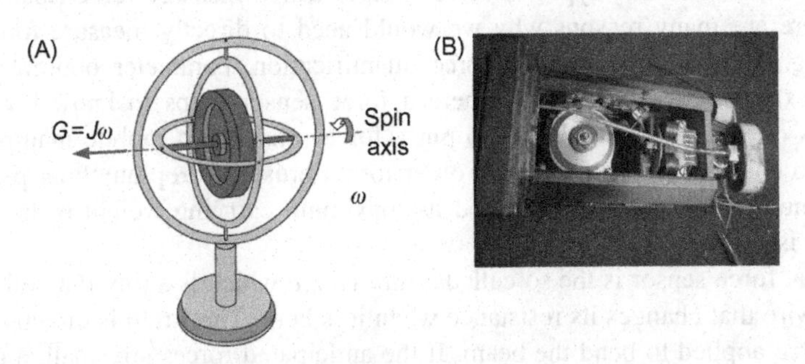

Figure 4.18 (A) Schematic gyroscope. (B) An aircraft gyroscope.
Source: http://hyperphysics.phy-astr.gsu.edu/hbase/gyr.html.

Figure 4.19 Pioneer II mobile robot integrated with gyroscope and compass.
Source: http://www.intechopen.com/source/pdfs/15967/InTech-Mobile_robot_integrated_with_gyroscope_by_using_ikf.pdf.

Today, many electronic compass modules are available for on-board use of mobile robots. Analog compass modules can be easily integrated with robot controllers. A simple analog compass can distinguish only eight directions which are represented by respective voltage levels. This is used in many four-wheel-drive systems. Digital compass modules are much more complex than analog ones and provide a considerably higher directional resolution (e.g., one resolution in indoor applications).

Some commercially available compasses for mobile robots include the Dinsmore Starguide Magnetic Compass, fluxgate compasses (Zemco compass, Watson gyro compass, KVA compasses, Philips magnetoresistive compass, etc.). A well-known mobile robot with a gyroscope and a compass integrated is Pioneer II (Figure 4.19).

4.6.3 Force and Tactile Sensors

4.6.3.1 Force Sensors

Actually, there are many types of force sensors which measure force (load) or torque. There are many reasons why we would need to directly measure forces in a robot, e.g., weight measurement, force quantification, parameter optimization. A notable example is a humanoid, where a force sensor helps to know how much weight is on each leg. Another is to put a force sensor in a mobile manipulator's gripper to control gripper friction in order to not crush or drop anything picked up and to know if the robot has reached its maximum carrying weight or how much weight it is carrying.

A usual force sensor is the so-called *strain gauge* which is a tiny flat coil of conductive wire that changes its resistance when it is bent. The strain is directly related to the force applied to bend the beam. If the anticipated forces are small, a conductive elastomer of foam can be used as a strain gauge a conductive foam. Compressing the foam lowers the electrical resistance. To face the problem of a

(A)

(B)

Figure 4.20 Force sensor uses (FANUC): (A) gear assembly; (B) clutch assembly.
Source: http://lrmate.com/forcesensor.htm.

typical strain gauge, due to which it has a very low change in resistance when bent, the strain gauges are connected inside the force sensor, usually a multiple of four, in a Wheatstone bridge configuration. In this way, the very small change in resistance is converted into a usable electrical signal. Passive components such as resistors and temperature-dependent wires are employed to compensate and calibrate the bridge output signal. The primary reason for transducer failure is the force overload which should be properly selected. The range of values of mass for which the result is not affected by other limit error is called *measuring range*. Two other parameters that secure proper operation of the force sensor are the *safe load/torque* limit (the maximum load/torque that can be applied without producing a permanent shift) and the *safe side load* (i.e., the maximum load that cannot act 90° to the axis along the sensor and cannot produce a permanent shift in the performance beyond specified). Two industrial applications of force sensor use are shown in Figure 4.20.

4.6.3.2 Tactile Sensors

Touch and *tactile sensors* are devices that measure the parameters of a contact between the sensor and an object, which is confined to a small well-defined region. This is in contrast with a force sensor which measures the total forces that are applied to an object.

Touch sensing is the detection and measurement of a contact force at a defined point (which can also be binary, i.e., touch, no touch).

Tactile sensing is the detection and measurement of the spatial distribution of forces perpendicular to a defined sensory area, followed by an interpretation of this force distribution. A tactile sensing array is obtained by the coordination of a group of touch sensors.

Slip is the detection and measurement of the movement of an object relative to a sensor. This can be done either by a specially designed slip sensor or via interpretation of the data obtained by a touch sensor or tactile array. Some desired properties of tactile sensors are as follows:

- A proper sensory area (ideally a single-point) contact

- A proper sensor sensitivity depending on the application concerned (typical sensitivity range being 0.4–10 N)
- A minimum sensor bandwidth of 100 Hz
- Stable and repeatable sensor features with low hysteresis
- Robustness and protection against large environmental changes

The principal physical principles used for the design of touch/tactile sensors capable of working with rigid objects are as follows:

- Conventional mechanical switch (binary touch sensor).
- Resistive-based sensors (based on the measurement of the resistance of a conductive elastomer of foam between two points; Figure 4.21).
- Force-sensing resistor (a piezoresistive conductive polymer that changes resistance in a predictable way following the application of force to its surface). As with resistive-based sensors, a relative simple interface is needed.
- Capacitance-based sensors (i.e., sensors using the capacitance between two parallel plates, $C = \varepsilon A/l$, where A is the plate area, l the distance between plates, and ε the permittivity of the dielectric medium).
- Magnetic-based sensors (utilizing the movement of a small magnet by an applied force that causes a change to the flux density, or using a magnetoelastic material whose magnetic properties change when external forces are applied to it).

Other ways of constructing touch/tactile sensors include *optical fiber-based sensors* (using the internal state microbending of optical fibers), *piezoelectric sensors* (using polymeric materials such as polyvinylidene fluoride (PVDF)), *silicon-based sensors* (silicon possesses a tensile strength comparable to steel, and elastic to breaking point with very small mechanical hysteresis), and *strain gauge-based tactile sensors.*Figure 4.22 shows a tactile sensor pad mounted on RIKEN'S RI-MAN

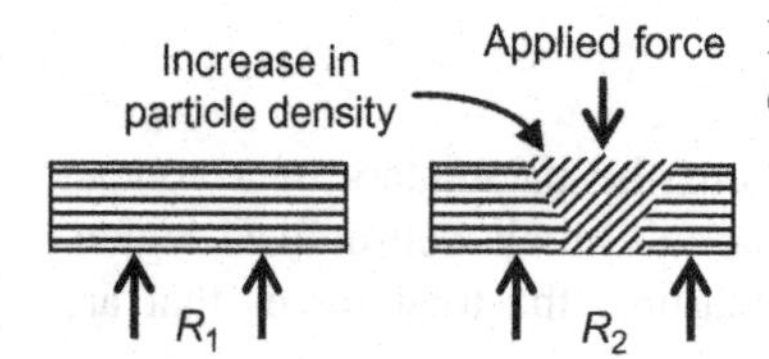

Figure 4.21 The resistance R of the elastomer changes according to the force applied.

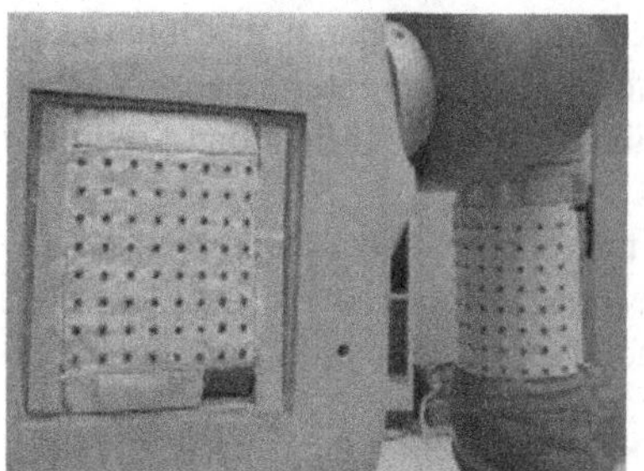

Figure 4.22 Tactile sensor array mounted on the torso and arm of humanoid.
Source: http://lh6.ggpht.com/touchuiresourcecenter/SNHVwL98kPI/AAAAAAAAAR8/uPv6Qy_e0s0/s1600-h/image%5B3%5D.png.

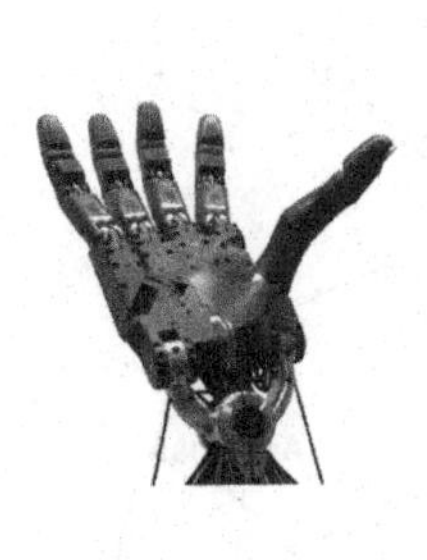

Figure 4.23 A tactile sensor mounted on robotic hand fingers and thumb.
Source: https://www.shadowrobot.com/products/dexterous-hand.

humanoid robot, and Figure 4.23 shows Shadow's tactile sensor mounted on the fingers of a robotic hand (sensitivity of sensor tactile elements from 0.1 to 25 N) suitable for both delicate handling and power grasping.

4.7 Global Positioning System

The *global positioning system* (GPS) is a space-based radio-positioning and time-transfer system. GPS satellites transmit signals to proper equipment on the ground. These signals provide accurate *position*, *velocity*, and *time* (PVT) information to an unlimited number of users on ground, sea, air, and space. GPS receivers need an unobstructed view of the sky, so they can only be used outdoors, and they usually do not operate well in near tall buildings or forestry areas. Passive PVT fixes are offered worldwide in all weathers in a worldwide common grid system. The three primary parts of GPS are *space, control*, and *user* segments (Figure 4.24A). Each satellite transmits data continuously indicating its location and current time, and all satellite signals are sent at the same time (synchronous transition). The GPS receiver has a quartz clock, but although three satellites can give the 3-D position, four satellites are used (the fourth satellite for time correction).

By knowing its distance from a satellite, each receiver also "knows" it is located somewhere on the surface of an imaginary sphere centered at the satellite. By determining the sizes of several spheres, one for each satellite, the receiver finds its location at the intersection point of these spheres (Figure 4.24B).

The *space segment* involves a minimum constellation of 24 satellites. The *control segment* consists of a network of monitoring and control facilities (master stations, monitoring stations, uploading stations) that are used to manage the satellite constellation and to update the satellite navigation data messages. The *users segment* involves all the radio navigation receivers that can receive, decode, and process the GPS satellite ranging codes and navigation data messages. Because GPS satellites

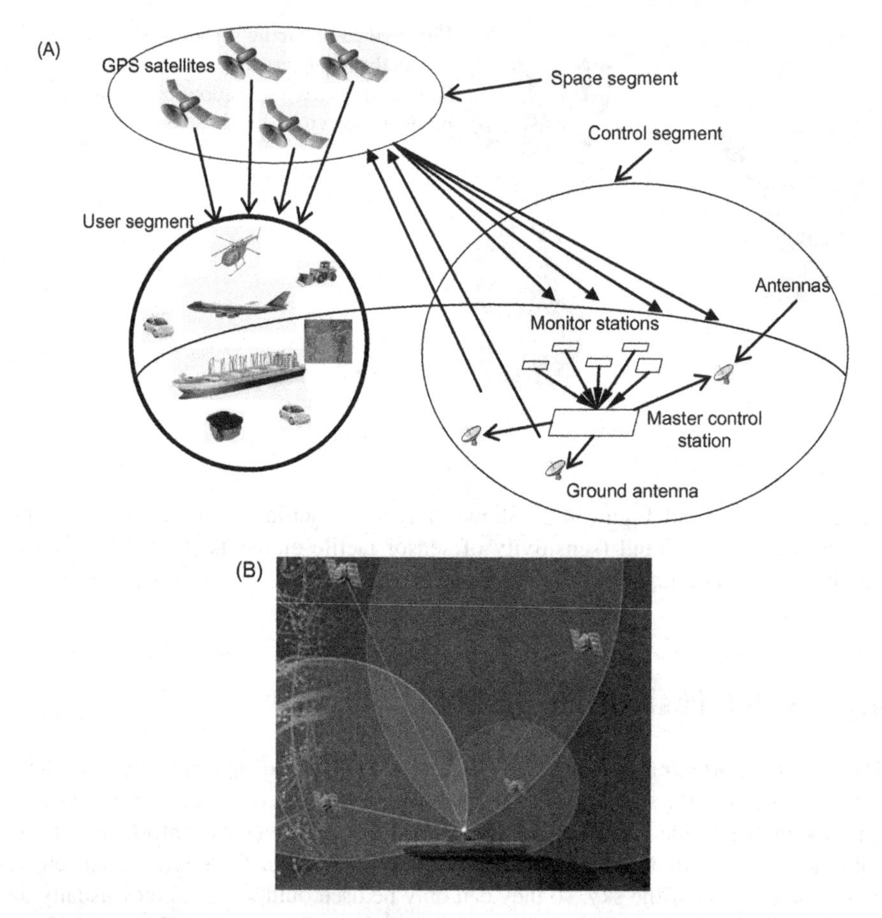

Figure 4.24 (A) The three segments of GPS. (B) The receiver lies at the intersection of the spheres centered at the four satellites.
Source: http://www.nasm.si.edu/gps/work.html.

are merely an information source, the localization resolution they provide depends strongly on the strategies employed. The basic strategy (called *pseudorange strategy*) gives a resolution of 15 m and can be improved to 1 m if a second receiver, which is static and at a known exact position, is employed. This technique is called *differential GPS* or *dual frequency GPS*. In general, the basic performance parameters that are used to compare different GPS receivers are as follows:

- Position accuracy
- Velocity accuracy
- Time accuracy
- Time to first failure

The real-time satellite tracking (GPS Operational Satellites) can be found in the site www.n2yo.com/satellites/?c=20.

4.8 Appendix: Lens and Camera Optics

We start by deriving the formula of a convex lens (Figure 4.25).

It is assumed that the distance (OA) = l_o of the object AB, which is perpendicular to the principal axis Ox of the lens, is greater than the focal length $(F_1\text{O}) = (F_2\text{O}) = l_f$ of the lens. From the similarity of the triangles OAB and OA′B′, we get

$$\frac{(\text{A}'\text{B}')}{(\text{AB})} = \frac{(\text{OA}')}{(\text{OA})} = \frac{l_{im}}{l_o}$$

where (OA′) = l_{im} is the distance of the image from the lens. Also, from the similarity of the triangles OF_2C and F_2A′B′, we have

$$\frac{(\text{A}'\text{B}')}{(\text{OC})} = \frac{(F_2\text{A}')}{(\text{O}F_2)} = \frac{(\text{OA}') - (\text{O}F_2)}{l_f} = \frac{l_{im} - l_f}{l_f}$$

From the above equations, taking into account that (OC) = (AB), we get

$$\frac{l_{im}}{l_o} = \frac{l_{im} - l_f}{l_f}$$

or

$$l_{im}l_f = l_{im}l_o - l_f l_o$$

Dividing this equation throughout by $l_{im}l_o l_f$ gives the desired lens equation:

$$\frac{1}{l_o} + \frac{1}{l_{im}} = \frac{1}{l_f}$$

This formula is used in *camera optics* for determining *depth* from focus, based on the fact that image properties change both as a function of the scene and as a function of the intrinsic camera parameters (focal length l_f, distance l_e from the lens to the focal point, and the parameter ε) (Figure 4.26).

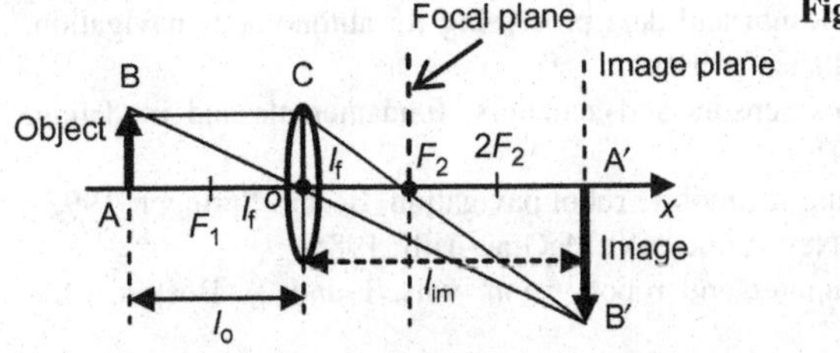

Figure 4.25 Geometry of convex lens.

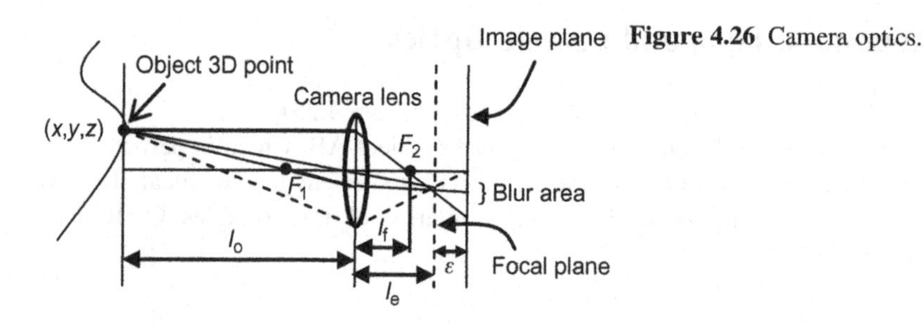

Figure 4.26 Camera optics.

From Figure 4.26, we have $1/l_f = 1/l_o + 1/l_e$. In order to obtain a sharp image of the point (x, y, z), the *image plane* of the camera must coincide with the *focal plane*. Otherwise, the image of the point (x, y, z) will be *blurred* as can be seen from this figure. This is because if the image plane is located at distance l_e from the lens, then for the specific object voxel depicted, all light will be focused at a single point on the image plane, and the object voxel will be *focused*. But, when the image plane is not at distance l_e (as shown in Figure 4.26), the light from the object voxel will be mapped on the image plane as a *blur circle*. It is easy to show (using similar angles calculation) that the radius R of the blur circle is equal to

$$R = D\varepsilon/2l_e$$

where D is the diameter (aperture) of the lens. Clearly, $R = 0$ if $\varepsilon = 0$, or $D = 0$ (as in the pinhole camera, where the lens reduces to a point). This is in agreement with the fact that decreasing the *iris aperture* opening causes the *depth of field* to increase until all objects are in focus (of course with less light entering to form the image on the image plane).

References

[1] Borenstein J, Everett HR, Feng L. Navigating mobile robots: sensors and techniques. Wellesley, MA: Peters A K Ltd; 1999.

[2] Everett HR. Sensors for mobile robots: theory and applications. New York, NY: Peters A K Ltd; 1995.

[3] DeSilva CW. Control sensors and actuators. Upper Saddle River, NJ: Prentice Hall; 1989.

[4] Adams MD. Sensors modeling, design and data processing for autonomous navigation. Singapore: World Scientific Publishers; 1999.

[5] Bishop RH. Mechatronic systems sensors and actuators: fundamentals and modeling. Boca Raton, FL: CRC Press; 2007.

[6] Leonard JL. Directed sonar sensing for mobile robot navigation. Berlin: Springer; 1992.

[7] Gonzalez RG. Computer vision. New York, NY: McGraw-Hill; 1985.

[8] Haralick RM, Shapiro LG. Computer and robot vision, vols. 1 and 2). Boston, MA: Addison Wesley; 1993.

[9] Davies ER. Machine vision: theory algorithms, practicalities. Amsterdam, The Netherlands: Morgan Kaufmann/Elsevier; 2005.

[10] Tzafestas SG, editor. Intelligent robotic systems. New York/Basel: Marcel Dekker; 1991.

[11] Tzafestas SG, editor. Advances in intelligent autonomous systems. Dordrecht/Boston: Kluwer; 1999.

[12] Panich S, Afzulpurkar N. Mobile robot integrated with gyroscope by using IKF. Int J Adv Robot Syst (INTECH Open Access) 2011;8(2):122–36.

[13] Kleeman L, Kuc R. Mobile robot sensor for target localization and classification. Int J Robot Res 1995;14(4):295–318.

[14] Phairoh T, Williamson K. Autonomous mobile robots using real time kinematic signal correction and global positioning system control. Proceedings of 2008 IAJC-IJME international conference paper 087. IT304, Nashville, TN; November 17–19, 2008.

[15] Tzafestas SG. Sensor integration and fusion techniques in robotic applications. J Int Robot. Syst. 2005;43(1):1–110 [special issue]

[16] Taha Z, Chew JY, Yap HJ. Omnidirectional vision for mobile robot navigation. J Adv Comput Intell Intell Inform 2010;14(1):55–62.

[17] Goh M, Lee S. Indoor robot localization using adaptive omnidirectional vision system. Int J Comput Sci Netw Secur 2010;10(4):66–70.

[18] Grassi Jr. V, Okamoto Jr J. Development of an omnidirectional vision system. J Braz Soc Mech Sci Eng 2006;28(1):1–18.

[19] Menegatti E. Omnidirectional vision for mobile robots. PhD thesis. Italy: Department of Information Engineering, University of Padova; December, 2002.

[20] Svoboda T, Pajda T, Hlavac V. Central panoramic cameras geometry and design. Research report no. K355/97/147. December 5, 1997. <ftp://cmp.felk.cvut.cz/pub/cmp/articles/svoboda/TR-K355-97-147.ps.gz>

[21] Benosman R, Kang R. Panoramic vision. Berlin: Springer; 2000.

[22] Velagic J, Lacevic B, Perunicic BA. 3-level autonomous mobile robot navigation system designed by using reasoning/search approaches. Robot Auton Syst 2006;5 (12):999–1004.

5 Mobile Robot Control I: The Lyapunov-Based Method

5.1 Introduction

Robot control deals with the problem of determining the forces and torques that must be developed by the robotic actuators in order for the robot to go at a desired position, track a desired trajectory, and, in general, to perform some task with desired performance requirements. The solution to control problems in robotics (fixed and mobile) is more complicated than usual due to the inertial forces, coupling reaction forces, and gravity effects. The performance requirements concern both the transient period and the steady-state period. In well-structured and fixed environments, such as the factory, the environment can be arranged to match the capabilities of the robot. In these cases, it can be assured that the robot knows certainly the configuration of the environment, and people are protected from the robot's operation. In such controlled environments, it is sufficient to employ some type of model-based control, but in uncertain and varying (uncontrolled) environments, the control algorithms must be more sophisticated involving some kind of intelligence. The techniques to be presented in this chapter assume that the goal of the control and the robot kinematic and dynamic parameters are precisely known, and if this goal (posture or path) is changing, the change is compatible with the environment, and risk free.

Specifically, the objectives of the chapter are as follows:

- To provide a minimal set of general control concepts and methods that are used in the control of robots
- To study the basic general robot controllers that are applicable to all types of robots
- To present a number of feedback controllers, designed using the Lyapunov-based control theory, for the differential drive, car-like, and omnidirectional mobile robots.

These controllers refer to the problems of position (posture) tracking, trajectory tracking, parking, and leader following. In all cases, the control design involves two stages, viz., *kinematic control* (where only the kinematic models are used), and *dynamic control* (where the robot dynamics and actuators are also taken into account).

Introduction to Mobile Robot Control. DOI: http://dx.doi.org/10.1016/B978-0-12-417049-0.00005-5

5.2 Background Concepts

In this section, the following fundamental control concepts and techniques are briefly discussed:

- State-space model
- Lyapunov stability
- State feedback control
- Second-order systems

Knowledge of these concepts is a basic prerequisite for the understanding of the material presented in the chapter. Full accounts are given in standard control textbooks [1].

5.2.1 State-Space Model

The state-space model of a control system is based on the concept of *state vector* $\mathbf{x}(t) \in R^n$, which is the minimum dimensionality Euclidean vector, with components called state variables, the knowledge of which at an initial time $t = t_0$, together with the input vector $u(t)$, for $t \geq t_0$, determines completely the behavior of the system for any time $t \geq t_0$. The dimension n of the state vector specifies the system's dimensionality.

The above definition of the state means that the state of the system is determined by its initial value $x(t_0)$ at $t = t_0$ and the input for $t \geq t_0$, and is independent of the state and the inputs for times previous to $t = t_0$.

It is noted that the state variables $x_1(t), x_2(t), \ldots, x_n(t)$ of an n-dimensional system may not necessarily be measurable physical quantities, although in practice, an effort is made to use as more as possible measurable variables, because the state feedback control laws need all of them.

The expression of $\mathbf{x}(t)$ as a function of $t, t_0, \mathbf{x}(t_0) = \mathbf{x}_0$, and $\mathbf{u}(\tau) = [u_1(\tau), \ldots, u_m(\tau)]^T$, $\tau \geq t_0$, that is, $\mathbf{x}(t) = \varphi(t; t_0, \mathbf{x}_0, \mathbf{u}(\tau))$, is called the *system's trajectory*.

The output $\mathbf{y}(t)$ of the system is a similar function of $\mathbf{x}(t)$, $\mathbf{u}(t)$, and t, that is:

$$\mathbf{y}(t) = \eta(t; \varphi(t, t_0, \mathbf{x}_0, \mathbf{u}(\tau)), \mathbf{u}(t)) \quad \text{for all } t \geq t_0$$

The trajectories satisfy the transition property:

$$\varphi(t; t_0, \mathbf{x}(t_0), u(\tau)) = \varphi(t; t_1, \mathbf{x}(t_1), u(\tau))$$

for all $t_0 < t_1 < t$, where $\mathbf{x}(t_1) = \varphi(t_1; t_0, \mathbf{x}_0, \mathbf{u}(\tau))$.

In state-space model, the dynamic model of a nonlinear system (in continuous time) has the form:

$$\dot{\mathbf{x}}(t) = \mathbf{f}(\mathbf{x}, \mathbf{u}, t) \quad (t \geq t_0) \tag{5.1a}$$

$$\mathbf{y}(t) = \mathbf{g}(\mathbf{x}, \mathbf{u}, t) \quad (t \geq t_0) \tag{5.1b}$$

where $\mathbf{f}(\cdot)$ and $\mathbf{g}(\cdot)$ are nonlinear vector functions of their arguments with proper dimensionality and the continuity and smoothness properties required in each case. The state vector $\mathbf{x}$ belongs to the state space $\mathcal{X}$, the input (control) $\mathbf{u}$ belongs to the input space $\mathcal{U}$, and the output $\mathbf{y}$ to the output space $\mathcal{Y}$, where $\mathcal{X} \subset \mathcal{R}^n$, $\mathcal{U} \in \mathcal{R}^m$, $Y \subset R^p$, and $\mathcal{R}^n$ is the n-dimensional Euclidean space.

If the vector functions $\mathbf{f}$ and $\mathbf{g}$ are linear, then the system is linear and is described by the model:

$$\dot{\mathbf{x}} = \mathbf{A}\mathbf{x} + \mathbf{B}\mathbf{u} \quad (\mathbf{x}(t_0) \text{ known}) \tag{5.2a}$$

$$\mathbf{y} = \mathbf{C}\mathbf{x} + \mathbf{D}\mathbf{u} \quad (\mathbf{x}(t_0) \text{ known}) \tag{5.2b}$$

where $\mathbf{A}$, $\mathbf{B}$, $\mathbf{C}$, $\mathbf{D}$ may be time-invariant or time-varying matrices of proper dimensionality (in many cases, $\mathbf{D} = \mathbf{0}$). A linear state-space model with the following matrices, $\mathbf{A}$, $\mathbf{B}$, and $u \in R$ (scalar), is called the *controllable canonical model* of the system that represents:

$$\mathbf{x} = \begin{bmatrix} x_1 \\ x_2 \\ \vdots \\ x_n \end{bmatrix}, \quad \mathbf{A} = \begin{bmatrix} 0 & 1 & \cdots & 0 \\ 0 & 0 & \cdots & 0 \\ \vdots & \vdots & \ddots & \vdots \\ 0 & 0 & \cdots & 1 \\ -a_n & -a_{n-1} & \cdots & -a_1 \end{bmatrix}, \quad \mathbf{B} = \begin{bmatrix} 0 \\ 0 \\ \vdots \\ 0 \\ 1 \end{bmatrix}, \quad \mathbf{u} = u \tag{5.3}$$

For a scalar output $y \in R$, a linear time-invariant system described by the nth-order differential equation:

$$(D^n + a_1 \mathrm{D}^{n-1} + \cdots + a_{n-1}\mathrm{D} + a_n) y(t) = (b_0 \mathrm{D}^n + b_1 \mathrm{D}^{n-1} + \cdots + b_n) u(t), \quad \mathrm{D} = d/dt$$

or transfer function:

$$\frac{\overline{y}(s)}{\overline{u}(s)} = \frac{b_0 s^n + b_1 s^{n-1} + \cdots + b_{n-1} s + b_n}{s^n + a_1 s^{n-1} + \cdots + a_{n-1} s + a_n} \tag{5.4}$$

where $s = a + j\omega$ is the complex frequency variable, can be modeled as in Eqs. (5.2a), (5.2b), and (5.3), if we define the state variables $x_1, x_2, \ldots, x_n$ as:

$$\mathrm{D}x_1 = x_2, \ \mathrm{D}x_2 = x_3, \ldots, \mathrm{D}x_{n-1} = x_n \tag{5.5}$$

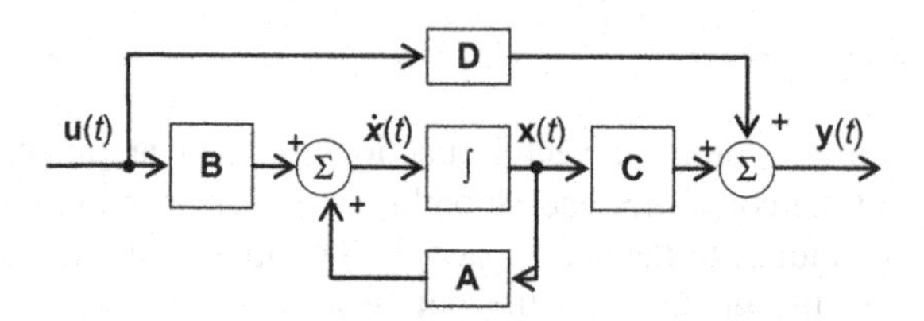

Figure 5.1 Block diagram of a general linear state-space model.

Indeed, using Eq. (5.5) we get:

$$Dx_n = -a_1x_n - a_2x_{n-1} - \cdots - a_{n-1}x_2 - a_nx_1 + u$$

$$y = b_nx_1 + b_{n-1}x_2 + \cdots + b_1x_n + b_0(u - a_1x_n - \cdots - a_{n-1}x_2 - a_nx_1)$$

which gives the state-space model ((5.2a), (5.2b), and (5.3)) with:

$$\mathbf{D} = [b_0], \quad \mathbf{C} = [b_n - a_nb_0, b_{n-1} - a_{n-1}b_0, \ldots, b_1 - a_1b_0] \tag{5.6}$$

This model, also called *phase variables canonical model*, is very convenient for the pole-placement (or assignment) state feedback controller design.

The block diagram representation of the general model ((5.2a) and (5.2b)) has the form shown in Figure 5.1.

Other state-space canonical models of the system ((5.2a) and (5.2b)) are the *observable canonical form* and the *Jordan canonical form* fully described in control textbooks. To convert a given model ((5.2a) and (5.2b)) to some canonical form, use is made of a proper nonsingular linear (*similarity*) transformation $\mathbf{x} = \mathbf{Tz}$, where $\mathbf{z}$ is the new state vector.

Example 5.1

In this example, we derive the dynamic models (transfer function, state-space model) of the direct current (DC) electrical motor which is used in mobile robots to provide the torques that lead to the desired acceleration and velocity of them. DC motors are distinguished into motors controlled by the rotor (*armature controlled*), and motors controlled by the stator (*field controlled*). Both the motors will be considered.

Armature Controlled DC Motor

The rotor involves the armature and the commutator. A schematic of this motor is shown in Figure 5.2, where R_a and L_a are the resistance and inductance of the rotor, I_L is the load moment of inertia, and β is the linear friction coefficient.

The mechanical torque is given by:

$$T_m(t) = K_a i_a(t) \tag{5.7}$$

where K_a is the motor's torque constant. The back electromotive force (emf) e_b which is subtracted from the input voltage v_a is proportional to ω_m, that is:

$$e_b = K_b\omega_m(t), \quad \omega_m(t) = d\theta_m(t)/dt \tag{5.8}$$

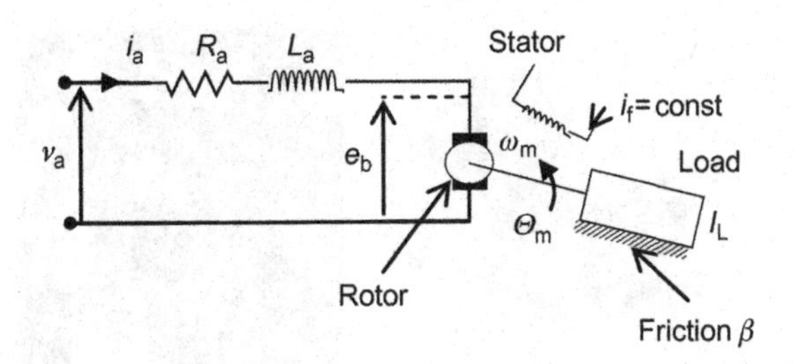

Figure 5.2 Schematic of the armature controlled DC motor (θ_m = rotation angle, ω_m = motor angular speed, I_L = load moment of inertia).

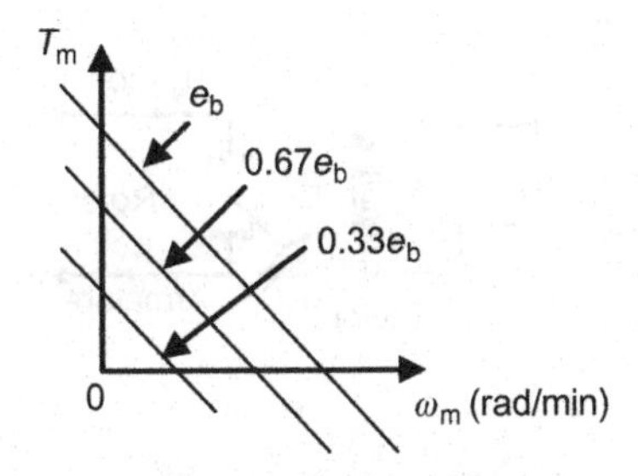

Figure 5.3 Characteristic curves of armature controlled system.

The characteristic curves of "T_m versus ω_m" are obtained by plotting the following function:

$$T_m = K_a \left[\frac{v_a - K_b \omega_m}{R_a} \right]$$

and have the linear form shown in Figure 5.3.

At the point where $e_b = v_a$, the motor maintains a constant angular speed (assuming of course that no external disturbances affect the motor). The differential equation of the motor can be found using the following relations:

$$T_m = K_a i_a = I_L \, D^2 \theta_m + \beta \, D\theta_m$$

$$v_a = R_a i_a + L_a \, Di_a + K_b \, D\theta_m$$

where $D = d/dt$, I_L is the moment of inertia of the load (plus the moment of inertia of the motor), and β is the linear friction coefficient, and eliminating i_a. The result is:

$$(\tau_a \, D^2 + \tau_b \, D + 1)D\theta_m(t) = Kv_a(t)$$

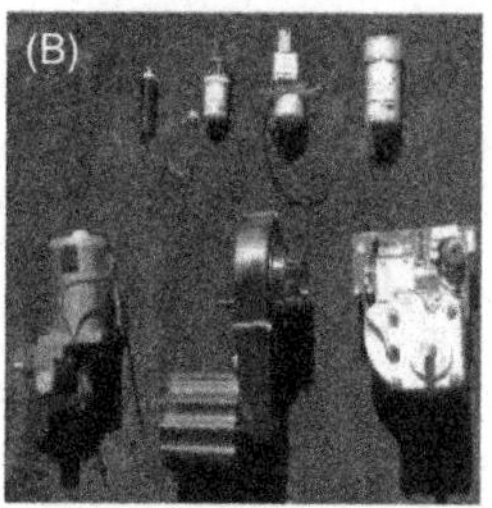

Figure 5.4 (A) A heavy-duty DC robot motor. (B) Several DC motors with embedded gear boxes.
Source: http://www.goldmine-elec-products.com/prodinfo.asp?number=0; http://www.robojrr.tripod.com/motortech.htm.

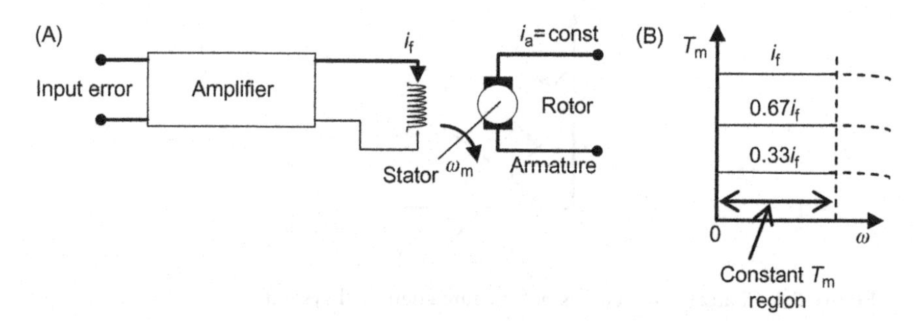

Figure 5.5 (A) Field controlled DC motor. (B) Characteristic curves for different i_f (for very large ω_m, we have a fall of T_m).

where $K = K_a/(\beta R_a + K_a K_b)$, $\tau_a = I_L L_a/(\beta R_a + K_a K_b)$, $\tau_b = (\beta L_a + I_L R_a)(\beta R_a + K_a K_b)$, $\tau_a = I_L L_a/(\beta R_a + K_a K_b)$, $\tau_b = (\beta L_a + I_L R_a)(\beta R_a + K_a K_b)$. If L_a is negligible, then $\tau_a \simeq 0$, and the above differential equation reduces to:

$$(\tau_b D^2 + D)\theta_m(t) = K v_a(t) \tag{5.9}$$

where K is as above, and $\tau_b = I_L R_a/(\beta R_a + K_a K_b)$. In practice, the model in Eq. (5.9) represents an adequate approximation of the motor dynamics. The controllable state-space canonical model of the motor is found from Eq. (5.9) using Eqs. (5.3) and (5.6), that is:

$$\mathbf{x} = \begin{bmatrix} x_1 \\ x_2 \end{bmatrix}, \quad \mathbf{A} = \begin{bmatrix} 0 & 1 \\ 0 & -1/\tau_b \end{bmatrix}, \quad \mathbf{B} = \begin{bmatrix} 0 \\ 1 \end{bmatrix}, \quad \mathbf{C} = \left[\frac{K}{\tau_b}, 0\right] \tag{5.10}$$

with $u = v_a$ and $x_1 = \theta_m$. A DC motor is shown in Figure 5.4A. Figure 5.4B shows DC motors of several sizes.

Field Controlled DC motor

In this motor, the rotor's current is kept constant (coming from a constant-current source), whereas the error is fed to the magnetic field of the stator via a phase-sensitive amplifier (Figure 5.5). The characteristic curves $T_m = T_m(\omega_m, i_f)$ show that when ω_m is not

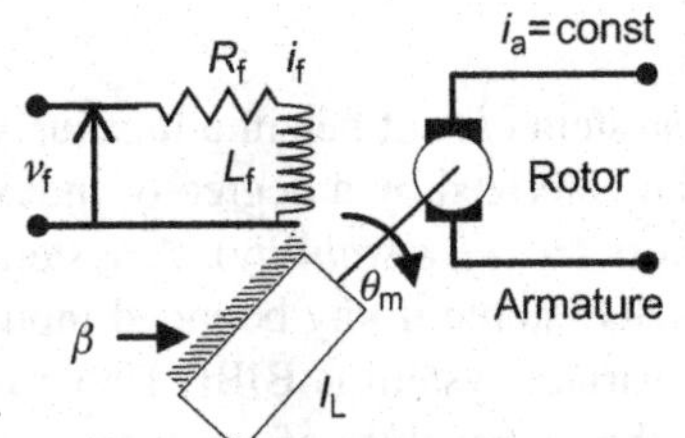

Figure 5.6 Schematic of the field controlled DC motor.

excessively large, the mechanical torque T_m is independent of ω_m, depending only on the field current i_f (Figure 5.5B).

A schematic of this motor is shown in Figure 5.6.

Here, we have the following relations:

$$\begin{aligned} T_m &= K_f i_f \\ v_f &= R_f i_f + L_f \, di_f/dt \\ T_m &= I_L \, d^2\theta_m/dt^2 + \beta \, d\theta_m/dt \end{aligned} \tag{5.11}$$

where K_f is the field current/torque constant and L_f is the inductance of the stator, and all other symbols have the same meaning as in the armature controlled motor.

Combining the above equations we get:

$$(1 + D\tau_f)(1 + D\tau_m)D\theta_m(t) = Kv_f(t) \tag{5.12}$$

where $K = K_f/\beta R_f$ (DC gain), and τ_f, τ_m are the time constants of the magnetic field and the mechanical time constant of the motor, respectively, given by:

$$\tau_f = L_f/R_f, \quad \tau_m = I_L/\beta$$

In practice, τ_f is much smaller than τ_m, and so Eq. (5.12) reduces to:

$$(\tau_m D^2 + D)\theta_m(t) = Kv_f(t) \tag{5.13}$$

which is of the same form as Eq. (5.9). Therefore, the motor has the state-space model (similar to Eq. (5.10)):

$$\mathbf{x} = \begin{bmatrix} x_1 \\ x_2 \end{bmatrix}, \quad \mathbf{A} = \begin{bmatrix} 0 & 1 \\ 0 & -1/\tau_m \end{bmatrix}, \quad \mathbf{B} = \begin{bmatrix} 0 \\ 1 \end{bmatrix}, \quad \mathbf{C} = \begin{bmatrix} \dfrac{K}{\tau_m}, 0 \end{bmatrix}$$

with $x_1 = \theta_m$ and $u = v_f$. If the motor dynamics (Eq. (5.9) or (5.13)) is expressed in terms of the angular velocity $\omega_m = D\theta_m = \dot{\theta}_m$, then (whenever the motor dynamics is not neglected) we use the form:

$$\dot{\omega}_m(t) = -(1/\tau)\omega_m(t) + (K/\tau)u(t)$$

where $\tau = \tau_b$ or $\tau = \tau_f$. This is analogous to a first-order RC circuit with time constant $\tau = RC$.

5.2.2 Lyapunov Stability

Stability is a binary property of a system, that is, a system cannot be simultaneously stable or not stable. However, a stable system is characterized by a degree or index that shows how much near to instability the system is (*relative stability*). A system is defined to be *bounded-input bounded-output* (BIBO) stable if any bounded input leads always to a bounded output. A linear time-invariant system is BIBO stable if and only if all the poles of its transfer function or the eigenvalues of the matrix **A** of its state-space model ((5.2a) and (5.2b)) lie strictly on the left-hand complex semiplane $s = a + j\omega$. The matrix **A** with the above property is called a *Hurwitz matrix*. The Routh and Hurwitz algebraic criteria specify the conditions that the coefficients of the system's characteristic polynomial must satisfy in order for the system to be stable. A first-order system $\dot{x} + ax = bu$ (with a real pole $-a$) is stable if $a > 0$, and has the impulse response:

$$x(t) = b\,\mathrm{e}^{-at}$$

Since $x(t)$ tends to zero as $t \to \infty$, asymptotically, the system is said to be *asymptotically stable*. Furthermore, since the convergence is exponential, that is, according to $b\,\mathrm{e}^{-at}$, this system is called *exponentially stable*. For a second-order system where the matrix **A** has the eigenvalues $-a_1 \pm j\omega$, $a_1 > 0$, the impulse response has the form:

$$x(t) = (K/\omega)e^{-a_1 t}\sin(\omega t), \quad a_1 > 0$$

Since $x(t) \to 0$ as $t \to \infty$, the system is asymptotically stable, and because $|x(t)| \leq (K/\omega)e^{-a_1 t}$, the system is exponentially stable. The above results hold for any combination of first-order and second-order systems.

The study of stability of a system and its stabilization via state or output feedback are two of the central problems in control theory. But the Routh and Hurwitz stability criteria can only be used for time-invariant linear single-input single-output (SISO) systems.

Lyapunov's stability method can also be applied to time-varying systems and to nonlinear systems. Lyapunov has introduced a generalized notion of *energy* (called *Lyapunov function*) and studied dynamic systems without external input. Combining Lyapunov's theory with the concept of BIBO stability we can derive stability conditions for *input-to-state* stability (ISS).

Lyapunov has introduced two stability methods. The first method requires the availability of the system's time response (i.e., the solution of the differential equations). The second method, also called *direct Lyapunov method*, does not require the knowledge of the system's time response.

Definition 5.1 The equilibrium state $\mathbf{x} = \mathbf{0}$ of the free system $\dot{\mathbf{x}} = \mathbf{A}(t)\mathbf{x}$ is stable in the Lyapunov sense (*L-stable*) if for every initial time t_0 and every real number $\varepsilon > 0$, there exists some number $\delta > 0$ as small as desired, that depends on t_0 and ε, such that: if $||\mathbf{x}_0|| < \delta$, then $||\mathbf{x}(t)|| < \varepsilon$ for all $t \geq t_0$, where $||\cdot||$ denotes the norm of the vector **x**, that is, $||\mathbf{x}|| = (x_1^2 + x_2^2 + \cdots + x_n^2)^{1/2}$. ■

Theorem 5.1 The transition matrix $\mathbf{\Phi}(t, t_0)$ of a linear system is bounded by $||\mathbf{\Phi}(t, t_0)|| < k(t_0)$ for all $t \geq t_0$ if and only if the equilibrium state $\mathbf{x} = 0$ of $\dot{\mathbf{x}} = \mathbf{A}(t)\mathbf{x}$ is L-stable. ■

The bound of $||\mathbf{x}(t)||$ of a linear system does not depend on $\mathbf{x}_0$. In general, if the system stability (of any kind) does not depend on $\mathbf{x}_0$, we say that we have *global* (*total*) *stability or stability in the large*. If the stability depends on $\mathbf{x}_0$, then it is called *local stability*. Clearly, total stability of a linear system implies also local stability.

Definition 5.2 The equilibrium state $\mathbf{x} = \mathbf{0}$ is asymptotically stable if:

(i) It is L-stable.
(ii) For every t_0 and $\mathbf{x}_0$ sufficiently near to $\mathbf{x} = \mathbf{0}$, the condition $\mathbf{x}(t) \to \mathbf{0}$, for $t \to \infty$ holds. ■

Definition 5.3 If the parameter δ in Definition 5.1 does not depend on t_0, then we have uniform L-stability. ■

Definition 5.4 If the system $\dot{\mathbf{x}}(t) = \mathbf{A}(t)\mathbf{x}$ is uniformly L-stable, and for all t_0 and for arbitrarily large ρ, the relation $||\mathbf{x}_0|| < \rho$ implies $\mathbf{x}(t) \to \mathbf{0}$ for $t \to \infty$, then the system is called *uniformly asymptotically stable*. ■

Theorem 5.2 The linear system $\dot{\mathbf{x}} = \mathbf{A}(t)\mathbf{x}$ is uniformly asymptotically stable if and only if there exist two constant parameters k_1 and k_2 such that: $||\mathbf{\Phi}(t, t_0)|| \leq k_1 \, \mathrm{e}^{-k_2(t-t_0)}$ for all t_0 and all $t \geq t_0$. ■

Definition 5.5 The equilibrium state $\mathbf{x} = \mathbf{0}$ of $\dot{\mathbf{x}} = \mathbf{A}(t)\mathbf{x}$ is said to be *unstable* if for some real number $\varepsilon > 0$, some $t_1 > t_0$ and any real number δ arbitrarily small, there always exists an initial state $||\mathbf{x}_0|| < \delta$ such that $||\mathbf{x}(t)|| > \varepsilon$ for $t \geq t_1$. ■

Figure 5.7 illustrates geometrically the concepts of L-stability, L-asymptotic stability, and instability.

Direct Lyapunov method: Let $d(\mathbf{x}(t), \mathbf{0})$ be the distance of the state $\mathbf{x}(t)$ from the origin $\mathbf{x} = \mathbf{0}$ (defined using any valid norm). If we find some distance $d(\mathbf{x}(t), \mathbf{0})$ which tends to zero for $t \to \infty$, then we conclude that the system is asymptotically stable. To show that a system is asymptotically stable using Lyapunov's direct

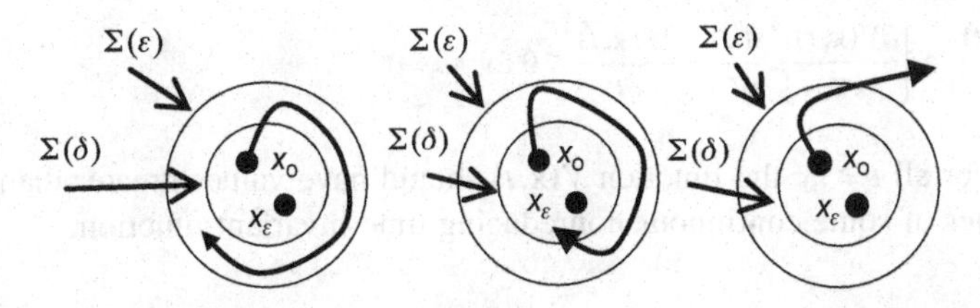

Figure 5.7 Illustration of L-stability (A), L-asymptotic stability (B), and instability (C). $\Sigma(\varepsilon)$ and $\Sigma(\delta)$ symbolize n-dimensional balls (spheres) with radii ε and δ, respectively.

method, we do not need to find such a *distance* (norm), but a *Lyapunov function* which is actually a generalized energy function.

Definition 5.6 *Time-invariant Lyapunov function* is called any scalar function $V(\mathbf{x})$ of $\mathbf{x}$ which for all $t \geq t_0$ and $\mathbf{x}$ in the vicinity of the origin satisfies the following four conditions:[1]

(i) $V(\mathbf{x})$ is continuous and has continuous derivatives
(ii) $V(\mathbf{0}) = 0$
(iii) $V(\mathbf{x}) > 0$ for all $\mathbf{x} \neq 0$
(iv) $\dfrac{dV(\mathbf{x})}{dt} = \left[\dfrac{\partial V(\mathbf{x})}{\partial \mathbf{x}}\right]^T \dfrac{d\mathbf{x}}{dt} < 0$ for $\mathbf{x} \neq \mathbf{0}$

Theorem 5.3 If a Lyapunov function $V(\mathbf{x})$ can be found for the state of a nonlinear or linear system $\dot{\mathbf{x}}(t) = \mathbf{f}(\mathbf{x}(t), t)$, where $\mathbf{f}(\mathbf{0}, t) = \mathbf{0}$ ($\mathbf{f}$ is a general function), then the state $\mathbf{x} = \mathbf{0}$ is asymptotically stable. ■

Remarks

(i) If Definition 5.6 holds for all t_0, then we have "uniformly asymptotic stability."
(ii) If the system is linear, or we replace in Definition 5.6 the condition "$\mathbf{x}$ in the vicinity of the origin," by the condition "$\mathbf{x}$ everywhere," then we have "total asymptotic stability."
(iii) If the condition (iv) of Definition 5.6 becomes $dV(\mathbf{x})/dt \leq 0$, then we have (simple) L-stability.

Clearly, to establish L-stability of a system, we must find a Lyapunov function. Unfortunately, there does not exist a general methodology for this.

Analogous results hold for the case of time-varying Lyapunov functions $V(\mathbf{x}(t), t)$, namely:

Definition 5.7 Time-varying Lyapunov function $V(\mathbf{x}, t)$ for the system's state is any scalar function of $\mathbf{x}$ and t, which, for all $t \geq t_0$ and $\mathbf{x}$ near the origin $\mathbf{x} = \mathbf{0}$, has the following properties:

(i) $V(\mathbf{x}, t)$ and its partial derivatives exist and are continuous
(ii) $V(\mathbf{0}, t) = 0$
(iii) $V(\mathbf{x}, t) \geq a(||\mathbf{x}||)$ for $\mathbf{x} \neq \mathbf{0}$ and $t \geq t_0$, where $a(\mathbf{0}) = 0$ and $a(\xi)$ is a scalar continuous non decreasing function of ξ
(iv) $\dfrac{dV(\mathbf{x}, t)}{dt} = \left[\dfrac{\partial V(\mathbf{x}, t)}{\partial \mathbf{x}}\right]^T \dfrac{d\mathbf{x}}{dt} + \dfrac{\partial V(\mathbf{x}, t)}{\partial t} < 0$ for $\mathbf{x} \neq 0$

Remark For all $t \geq t_0$, the function $V(\mathbf{x}, t)$ should have values greater than or equal to the values of some continuous nonreducing time-invariant function.

[1] Note that here $\partial V(\mathbf{x})/\partial \mathbf{x}$ is considered to be a column vector, that is, $\partial V(\mathbf{x})/\partial \mathbf{x} = [\partial V/\partial x_1, \partial V/\partial x_2, \ldots, \partial V/\partial x_n]^T$.

Theorem 5.4 If a time-varying Lyapunov function $V(\mathbf{x}, t)$ can be found for the state $\mathbf{x}(t)$ of the system $\dot{\mathbf{x}}(t) = \mathbf{f}(\mathbf{x}(t), t)$, then the state $\mathbf{x} = 0$ is asymptotically stable. ■

Definition 5.8

(i) If the conditions of Definition 5.7 hold for all t_0 and $V(\mathbf{x}, t) \leq \beta(||\mathbf{x}||)$, where $\beta(\xi)$ is a continuous nondecreasing scalar function of ξ with $\beta(\mathbf{0}) = 0$, then we have *uniformly asymptotic stability*.

(ii) If the system is linear or the conditions of Definition 5.7 hold everywhere (not only in a region of the origin $\mathbf{x} = \mathbf{0}$), and if we have $a(||\mathbf{x}||) \to \infty$, when $||\mathbf{x}|| \to \infty$, then we say that the system is *uniformly totally asymptotically stable*. ■

In the case of a linear time-varying system $\dot{\mathbf{x}} = \mathbf{A}(t)\mathbf{x}$ the Lyapunov (time varying) function $V(\mathbf{x}, t)$ is given by the quadratic (energy) function:

$$V(\mathbf{x}, t) = \mathbf{x}^T\mathbf{P}(t)\mathbf{x} \tag{5.14a}$$

where $\mathbf{P}(t)$ satisfies the following matrix differential equation:

$$d\mathbf{P}(t)/dt + \mathbf{A}^T(t)\mathbf{P}(t) + \mathbf{P}(t)\mathbf{A}(t) = -\mathbf{Q}(t), \quad \mathbf{Q}(t) > 0 \tag{5.14b}$$

If the system is time-invariant $\mathbf{A}(t) = \mathbf{A} = \text{const.}$, then $\mathbf{P}(t) = \mathbf{P} = \text{const.}$ and the above differential equation for $\mathbf{P}(t)$ reduces to the following algebraic equation for $\mathbf{P}$:

$$\mathbf{A}^T\mathbf{P} + \mathbf{P}\mathbf{A} = -\mathbf{Q} \tag{5.15}$$

In this case, we can select a positive definite matrix $\mathbf{Q} > \mathbf{0}$ and solve the $n(n+1)/2$ equations (**P** is symmetric) for the elements of **P**. Then, if $\mathbf{P} > \mathbf{0}$ (i.e., if **P** is positive definite) the system is asymptotically stable.

Remark Using Eq. (5.15), we can show that the Lyapunov stability criterion is equivalent to the Hurwitz stability criterion.

5.2.3 State Feedback Control

State feedback control is more powerful than classical control because the design of a total controller for a *multiple-input multiple-output* (MIMO) system is performed in a unified way for all control loops simultaneously, and not serially one loop after the other which does not guarantee the overall system stability and robustness. In this section we will briefly review the *eigenvalue placement* controller for SISO systems.

Let a SISO system:

$$\dot{\mathbf{x}}(t) = \mathbf{A}\mathbf{x}(t) + \mathbf{B}u(t), \quad y(t) = \mathbf{C}\mathbf{x}(t) + Du(t), \quad u \in R, \quad y \in R, \quad \mathbf{x} \in R^n$$

where $\mathbf{A}$ is a $n \times n$ constant matrix, $\mathbf{B}$ is an $n \times 1$ constant matrix (column vector), $\mathbf{C}$ is an $1 \times n$ matrix (row vector), u is a scalar input, and D is a scalar constant. In this case, a state feedback controller has the form:

$$u(t) = \mathbf{F}\mathbf{x}(t) + \upsilon(t) \tag{5.16}$$

where $\upsilon(t)$ is a new control input and $\mathbf{F}$ is an n-dimensional constant row vector: $\mathbf{F} = [f_1, f_2, \ldots, f_n]$. Introducing this control law into the system, we get the state equations of the closed-loop (feedback) system:

$$\dot{\mathbf{x}}(t) = (\mathbf{A} + \mathbf{B}\mathbf{F})\mathbf{x}(t) + \mathbf{B}\upsilon(t), \quad y(t) = (\mathbf{C} + D\mathbf{F})\mathbf{x}(t) + D\upsilon(t) \tag{5.17}$$

The eigenvalue placement design problem is to select the controller gain matrix $\mathbf{F}$ such that the eigenvalues of the closed-loop matrix $\mathbf{A} + \mathbf{B}\mathbf{F}$ are placed at desired positions $\lambda_1, \lambda_2, \ldots, \lambda_n$. It can be shown that this can be done (i.e., the system eigenvalues are controllable by state feedback) if and only if the system $(\mathbf{A}, \mathbf{B})$ is totally controllable. The concept of *controllability* has been developed to study the ability of a controller to alter the performance of the system in an arbitrary desired way. As it is known, the positions of the eigenvalues specify the performance characteristics of a system.

Definition 5.9 A state $\mathbf{x}_0$ of a system is called *totally controllable* if it can be driven to a final state $\mathbf{x}_f$ as quickly as desired independently of the initial time t_0. A system is said to be *totally controllable* if all of its states are totally controllable. ■

Intuitively, we can see that if some state variables do not depend on the control input $\mathbf{u}(t)$, no way exists that can drive it to some other desired state. Thus, this state is called a *noncontrollable state*. If a system has at least one noncontrollable state, it is said to be nontotally controllable or, simply, noncontrollable. The above controllability concept refers to the states of a system and so it is characterized as *state controllability*. If the controllability is referred to the outputs of a system then we have the so-called *output controllability*. In general, state controllability and output controllability are not the same.

Theorem 5.5 The necessary and sufficient condition for a linear system $(\mathbf{A}, \mathbf{B})$ to be totally state controllable is that the controllability matrix:

$$\mathbf{Q}_c = [\mathbf{B} \,\vdots\, \mathbf{A}\mathbf{B} \,\vdots\, \mathbf{A}^2\mathbf{B} \,\vdots\, \cdots \,\vdots\, \mathbf{A}^{n-1}\mathbf{B}] \tag{5.18a}$$

has

$$\text{rank}\, \mathbf{Q}_c = n \tag{5.18b}$$

where n is the dimensionality of the state vector $\mathbf{x}$. ■

The most straightforward technique for selecting the feedback matrix $\mathbf{F}$ is through the use of the controllable canonical form. This technique involves the following steps:

Step 1: We write down the characteristic polynomial $\chi_{\mathbf{A}}(s)$ of the matrix $\mathbf{A}$:

$$\chi_{\mathbf{A}}(s) = |s\mathbf{I} - \mathbf{A}| = s^n + a_1 s^{n-1} + \cdots + a_{n-1}s + a_n$$

Step 2: Then, we find a similarity transformation $\mathbf{T}$ that converts the given system to its controllable canonical form $\hat{\mathbf{A}} = \mathbf{T}^{-1}\mathbf{A}\mathbf{T}$.

Step 3: From the desired eigenvalues of the closed-loop system, we determine the desired characteristic polynomial:

$$\chi_{\text{desired}}(s) = s^n + \tilde{a}_1 s^{n-1} + \cdots + \tilde{a}_{n-1}s + \tilde{a}_n$$

The feedback gain matrix $\hat{\mathbf{F}}$ of the controllable canonical model is given by:

$$\hat{\mathbf{F}} = \mathbf{F}\mathbf{T} = [\hat{f}_n, \hat{f}_{n-1}, \ldots, \hat{f}_1]$$

Step 4: Equating the last rows of $\mathbf{A} + \mathbf{B}\mathbf{F}$ and $\hat{\mathbf{A}} + \hat{\mathbf{B}}\hat{\mathbf{F}}$ we find:

$$a_1 - \hat{f}_1 = \tilde{a}_1, \;\; a_2 - \hat{f}_2 = \tilde{a}_2, \ldots, a_n - \hat{f}_n = \tilde{a}_n$$

and so solving for $\mathbf{F} = \hat{\mathbf{F}}\mathbf{T}^{-}1$ we find:

$$\mathbf{F} = [\hat{f}_n, \hat{f}_{n-1}, \ldots, \hat{f}_1]\mathbf{T}^{-1} = [a_n - \tilde{a}_n, a_{n-1} - \tilde{a}_{n-1}, \ldots, a_1 - \tilde{a}_1]\mathbf{T}^{-1} \tag{5.19}$$

5.2.4 Second-Order Systems

The state vector $\mathbf{x}$ of a second-order system contains the position and velocity of the variable (physical quantity) of interest, namely:

$$\mathbf{x}(t) = \begin{bmatrix} x(t) \\ \dot{x}(t) \end{bmatrix}$$

Suppose that it is desired to design the state feedback controller such that the system's state follows a desired trajectory $\mathbf{x}_d(t)$. In this case, the feedback must use the measured error:[2]

$$\tilde{\mathbf{x}}(t) = \mathbf{x}_d(t) - \mathbf{x}(t) \tag{5.20}$$

and the controller should reduce the sensitivity of the system to the inaccuracy and the uncertainty in the parameter values used in the dynamic model.

[2] It is remarked that the error $\tilde{\mathbf{x}}(t)$ can also be defined as $\tilde{\mathbf{x}}(t) = \mathbf{x}(t) - \mathbf{x}_d(t)$. In this case, the feedback gains have opposite signs and actually lead to the same negative feedback controller (Section 5.3.1).

A fundamental characteristic of control systems is the bandwidth Ω which determines the operation speed of a system and capability of fast trajectory tracking. The greater the bandwidth is the better, but Ω should not be very high because it may excite possible high-frequency components that have not been included in the system model.

As an example, consider the following simple second-order system:

$$\ddot{x}(t) = u(t) \tag{5.21}$$

which by Eq. (5.20) gives the error equation:

$$\ddot{\tilde{x}}(t) = \ddot{x}_{\mathrm{d}}(t) - \ddot{x}(t) = \ddot{x}_{\mathrm{d}} - u(t) \tag{5.22}$$

Defining the state vector:

$$\tilde{\mathbf{x}} = \begin{bmatrix} \tilde{x} \\ \dot{\tilde{x}} \end{bmatrix} = \begin{bmatrix} \tilde{x}_1 \\ \tilde{x}_2 \end{bmatrix}$$

we get the following controllable canonical form:

$$\dot{\tilde{\mathbf{x}}} = \mathbf{A}\tilde{\mathbf{x}} + \mathbf{b}(\ddot{x}_{\mathrm{d}} - u) \tag{5.23a}$$

where:

$$\mathbf{A} = \begin{bmatrix} 0 & 1 \\ -a_2 & -a_1 \end{bmatrix} = \begin{bmatrix} 0 & 1 \\ 0 & 0 \end{bmatrix}, \quad \mathbf{b} = \begin{bmatrix} 0 \\ 1 \end{bmatrix} \tag{5.23b}$$

To get a desired bandwidth Ω, we select the desired closed-loop characteristic polynomial as:

$$\chi_{\mathrm{desired}}(s) = s^2 + \tilde{a}_1 s + \tilde{a}_2 = s^2 + 2\zeta\omega_{\mathrm{n}} s + \omega_{\mathrm{n}}^2 \tag{5.24}$$

where ω_{n} is the undamped natural frequency (taken equal to the desired bandwidth Ω) and ζ is the damping coefficient (usually selected $\zeta \geq 0.7$).

Then, the feedback controller is selected as in Eqs. (5.16) and (5.19) with $\mathbf{T} = \mathbf{I}$ (unit matrix), namely:

$$\ddot{x}_{\mathrm{d}} - u = [f_2, f_1] \begin{bmatrix} \tilde{x} \\ \dot{\tilde{x}} \end{bmatrix} + v$$

or

$$\begin{aligned} u &= \ddot{x}_{\mathrm{d}} - [f_2, f_1] \begin{bmatrix} \tilde{x} \\ \dot{\tilde{x}} \end{bmatrix} - v \\ &= \ddot{x}_{\mathrm{d}} + [\tilde{a}_2, \tilde{a}_1] \begin{bmatrix} \tilde{x} \\ \dot{\tilde{x}} \end{bmatrix} - v \\ &= \ddot{x}_{\mathrm{d}} + 2\zeta\omega_{\mathrm{n}}\dot{\tilde{x}} + \omega_{\mathrm{n}}^2\tilde{x} - v \end{aligned} \tag{5.25}$$

The closed loop error system is obtained using Eqs. (5.22) and (5.25), that is:

$$\ddot{\tilde{x}}(t) + 2\zeta\omega_n\dot{\tilde{x}}(t) + \omega_n^2\tilde{x}(t) = v(t) \quad (5.26)$$

which has the desired damping and bandwidth specifications.

The control law (5.25) contains the proportional term $\omega_n^2\tilde{x}$ and the derivative term $2\zeta\omega_n\dot{\tilde{x}}$, that is, it is a PD (*proportional plus derivative controller*) which is one of the most popular and efficient controllers. For second-order systems, the PD controller gives exact results.

Example 5.2

Consider the case where the system (5.21) is corrupted by a disturbance $\xi(t)$ as:

$$\ddot{x}(t) = u(t) + \xi(t)$$

(a) It is desired to determine the steady-state position error $\tilde{x}_{ss} = \lim_{t\to\infty}\tilde{x}(t)$ of the closed-loop PD controlled system when $\xi(t)$ is a step disturbance of amplitude ξ_0:

$$\xi(t) = \begin{cases} \xi_0, & t \geq 0 \\ 0, & t < 0 \end{cases} \quad (\xi_0 \neq 0)$$

(b) Show that this steady-state error vanishes if, instead of the PD controller, a PID (proportional plus integral plus derivative) controller is used.

Solution

(a) With the PD controller, the closed-loop disturbed system (5.26) becomes:

$$\ddot{\tilde{x}}(t) + 2\zeta\omega_n\dot{\tilde{x}} + \omega_n^2\tilde{x} = v(t) + \xi(t)$$

With $v(t) = 0$ and $\xi(t)$, the above step disturbance, the steady-state position error, obtained by setting $\lim_{t\to\infty}\dot{\tilde{x}}(t) = 0$ and $\lim_{t\to\infty}\ddot{\tilde{x}}(t) = 0$, satisfies the relation:

$$\omega_n^2 \lim_{t\to\infty}\tilde{x}(t) = \omega_n^2 x_{ss} = \xi_0$$

Thus:

$$\tilde{x}_{ss} = \xi_0/\omega_n^2$$

We see that $\tilde{x}_{ss}$ has a nonzero finite value which is proportional to ξ_0 and inversely proportional to the square of the bandwidth $\Omega = \omega_n$.

(b) If we use a PID controller:

$$u = \ddot{x}_d + 2\zeta\Omega\dot{\tilde{x}} + \Omega^2\tilde{x} + \Omega^3\int_0^t \tilde{x}(\tau)d\tau$$

the closed-loop error system becomes:

$$\ddot{\tilde{x}}(t) + 2\zeta\Omega\dot{\tilde{x}}(t) + \Omega^2\tilde{x}(t) + \Omega^3\int_0^t \tilde{x}(t)d\tau = v(t) + \xi(t)$$

Then, in the limit as $t \to \infty$, for $\lim_{t\to\infty}\dot{\tilde{x}}(t) = 0$, $\lim_{t\to\infty}\ddot{\tilde{x}}(t) = 0$ and $v(t) = 0$, $t \geq 0$ we get:

$$\Omega^2\tilde{x}_{ss} + \Omega^3\lim_{t\to\infty}\int_0^t \tilde{x}(\tau)d\tau = \xi_0$$

which implies $\tilde{x}_{ss} = \lim_{t\to\infty}\tilde{x}(t) = 0$. Otherwise, $\lim_{t\to\infty}\int_0^t \tilde{x}(t)d\tau$ cannot have a finite value.

Remark The above results can also be obtained using the well-known Laplace transform final-value property $\tilde{x}_{ss} = \lim_{t\to\infty}\tilde{x}(t) = \lim_{s\to 0} s\tilde{x}(s)$ and computing $\tilde{x}(s)$ via the error system's transfer function.

Example 5.3

It is desired to check the stability of the following systems using the Lyapunov method:

(a) $\dot{x} = -x$

(b) $\dot{x}_1 = x_2 - x_1(x_1^2 + x_2^2)$

$\dot{x}_2 = -x_1 - x_2(x_1^2 + x_2^2)$

Solution

System (a): This system has the equilibrium state $x = 0$. We examine if the function $V(x) = x^2$ is a Lyapunov function. This is done by checking if all conditions for $V(x)$ to be a Lyapunov function are satisfied. We have:

- $V(x) = x^2$ and $dV/dx = 2x$ are continuous
- $V(0) = 0$
- $V(x) = x^2 > 0$ for $x \neq 0$
- $\dot{V}(x) = (2x)\dot{x} = (2x)(-x) = -2x^2 < 0$ for all $x \neq 0$

We see that all properties of the candidate function required to be a Lyapunov function hold, and so the system $\dot{x} = -x$ is uniformly asymptotically stable.

System (b): We try the following candidate Lyapunov function:

$$V(\mathbf{x}) = x_1^2 + x_2^2$$

This function has the following properties:

- $V(\mathbf{x})$ and $dV/d\mathbf{x}$ are continuous
- $V(\mathbf{0}) = 0$
- $V(\mathbf{x}) > 0$ for $\mathbf{x} \neq 0$

- $\dot{V}(\mathbf{x}) = 2x_1\dot{x}_1 + 2x_2\dot{x}_2 = -2(x_1^2 + x_2^2)^2 < 0$ for $\mathbf{x} \neq \mathbf{0}$

that is, it possesses all properties of a Lyapunov function. Therefore, the only equilibrium state $\mathbf{x}(t) = [x_1(t), x_2(t)]^T = \mathbf{0}$ (i.e., the origin) is totally asymptotically stable.

5.3 General Robot Controllers

The following general controllers, which are standard in robotics [2–4], will be examined:

- Proportional plus derivative control
- Lyapunov function based control
- Computed torque control
- Resolved motion rate control
- Resolved motion acceleration control

5.3.1 Proportional Plus Derivative Position Control

Here, it will be shown that PD control leads to satisfactory results in the control of the position of a general robot described in Eqs. (3.11a) and (3.11b):

$$\mathbf{D}(\mathbf{q})\ddot{\mathbf{q}} + \mathbf{h}(\mathbf{q}, \dot{\mathbf{q}}) + \mathbf{g}(\mathbf{q}) = \boldsymbol{\tau}$$

$$\mathbf{q} = [q_1, q_2, \ldots, q_n]^T$$

where for any $\dot{\mathbf{q}} \neq 0$, $\mathbf{D}$ is a known positive definite matrix. Assuming that the friction is negligible and omitting the gravity term $\mathbf{g}(\mathbf{q})$, which anyway is zero in mobile robots moving on an horizontal terrain, we get:

$$\mathbf{D}(\mathbf{q})\ddot{\mathbf{q}} + \mathbf{C}(\mathbf{q}, \dot{\mathbf{q}})\dot{\mathbf{q}} = \boldsymbol{\tau} \tag{5.27}$$

where $\mathbf{C}(\mathbf{q}, \dot{\mathbf{q}})$ is defined as in Eq. (3.13), and the matrix $\dot{\mathbf{D}} - 2\mathbf{C}$ is antisymmetric.

Let $\tilde{\mathbf{q}} = \mathbf{q} - \mathbf{q}_\mathrm{d}$ be the error between $\mathbf{q}$ and $\mathbf{q}_\mathrm{d}$. Then, the PD controller has the form

$$\begin{aligned} \boldsymbol{\tau} &= \mathbf{K}_\mathrm{p}(\mathbf{q}_\mathrm{d} - \mathbf{q}) + \mathbf{K}_\mathrm{d}(\dot{\mathbf{q}}_\mathrm{d} - \dot{\mathbf{q}}) \\ &= -\mathbf{K}_\mathrm{p}\tilde{\mathbf{q}} - \mathbf{K}_\mathrm{d}\dot{\tilde{\mathbf{q}}} = -\mathbf{K}_\mathrm{p}\tilde{\mathbf{q}} - \mathbf{K}_\mathrm{d}\dot{\mathbf{q}} \quad (\text{since } \mathbf{q}_\mathrm{d} = \text{const.}) \end{aligned} \tag{5.28}$$

where $\mathbf{K}_\mathrm{p}$ and $\mathbf{K}_\mathrm{d}$ are positive definite symmetric matrices. The resulting feedback control scheme has the form of Figure 5.8.

Let us try the following candidate Lyapunov function:

$$V(\tilde{\mathbf{q}}) = \frac{1}{2}(\tilde{\mathbf{q}}^T\mathbf{K}_\mathrm{p}\tilde{\mathbf{q}} + \dot{\mathbf{q}}^T\mathbf{D}\dot{\mathbf{q}}) \tag{5.29}$$

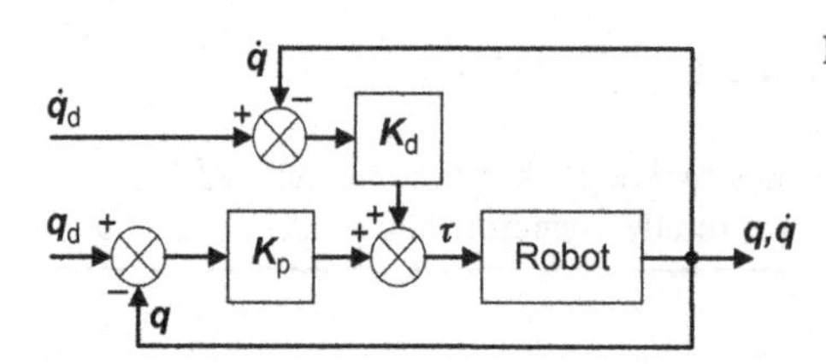

Figure 5.8 PD robot control.

where the term $(1/2)\dot{\mathbf{q}}^T\mathbf{D}\dot{\mathbf{q}}$ is the robot's kinetic energy, and the term $(1/2)\tilde{\mathbf{q}}^T\mathbf{K}_p\tilde{\mathbf{q}}$ represents the proportional control term. Thus, the function V can be considered as representing the total energy of the closed-loop system. Since $\mathbf{K}_p$ and $\mathbf{D}$ are symmetric positive definite matrices, we have $V(\mathbf{0})=0$ and $V(\tilde{\mathbf{q}})>0$ for $\tilde{\mathbf{q}}\neq\mathbf{0}$. Therefore, we have to check the validity of property (iv) of Definition 5.6.

Here, Eq. (5.29) gives:

$$\begin{aligned}\dot{V} &= \tilde{\mathbf{q}}^T\mathbf{K}_p\dot{\mathbf{q}} + \dot{\mathbf{q}}^T\mathbf{D}\ddot{\mathbf{q}} + (1/2)\dot{\mathbf{q}}^T\dot{\mathbf{D}}\dot{\mathbf{q}} \\ &= \tilde{\mathbf{q}}^T\mathbf{K}_d\dot{\mathbf{q}} + \dot{\mathbf{q}}^T(\boldsymbol{\tau} - \mathbf{C}\dot{\mathbf{q}}) + (1/2)\dot{\mathbf{q}}^T\dot{\mathbf{D}}\dot{\mathbf{q}}\end{aligned} \tag{5.30}$$

Now, introducing the control law (5.28) into Eq. (5.30) we get:

$$\begin{aligned}\dot{V} &= \tilde{\mathbf{q}}^T\mathbf{K}_p\dot{\mathbf{q}} + \dot{\mathbf{q}}^T(\mathbf{C}+\mathbf{K}_d)\dot{\mathbf{q}} - \dot{\mathbf{q}}^T\mathbf{K}_p\tilde{\mathbf{q}} + (1/2)\dot{\mathbf{q}}^T\dot{\mathbf{D}}\dot{\mathbf{q}} \\ &= -\dot{\mathbf{q}}^T(\mathbf{C}+\mathbf{K}_d)\dot{\mathbf{q}} + (1/2)\dot{\mathbf{q}}^T\dot{\mathbf{D}}\dot{\mathbf{q}}\end{aligned} \tag{5.31}$$

Therefore, since the matrix $\dot{\mathbf{D}} - 2\mathbf{C}$ is antisymmetric, Eq. (5.31) finally gives:

$$\dot{V} = -\dot{\mathbf{q}}^T\mathbf{K}_d\dot{\mathbf{q}} \leq 0 \tag{5.32}$$

We observe that while the Lyapunov function V in Eq. (5.29) depends on $\mathbf{K}_p$, its derivative $\dot{V}$ depends on $\mathbf{K}_d$, which is analogous of the known property of classical SISO PD control. The condition (5.32) ensures that the feedback error control system (Figure 5.8) is L-stable. It is also useful to remark that the PD control (5.28) is particularly robust with respect to mass variations because it does not require knowledge of the parameters that depend on mass.

A special case of the controller (5.28) is:

$$\tau_j = -K_{jp}\tilde{q}_j - K_{jd}\dot{\tilde{q}}_j \quad (j=1,2,\ldots,n)$$

which is applied to each joint separately. If the motion is subject to friction (assumed linear) the robot model (5.27) must simply be replaced by:

$$\mathbf{D}(\mathbf{q})\ddot{\mathbf{q}} + \mathbf{C}(\mathbf{q},\dot{\mathbf{q}})\dot{\mathbf{q}} + \mathbf{B}_f\dot{\mathbf{q}} = \boldsymbol{\tau}$$

where $\mathbf{B}_f$ is the diagonal matrix of friction coefficients. The present PD controller can be enhanced with an integral term as given in Example 5.2.

5.3.2 Lyapunov Stability-Based Control Design

The control design method applied to the above problem is known as *Lyapunov-based controller design* and constitutes a widely used method for both linear and nonlinear systems. The steps of this method are the following:

Step 1: Select a trial (candidate) Lyapunov function, which is typically some kind of energy-like function for the system, and possesses the first three properties of Lyapunov functions (Definition 5.6, Section 5.2.2).
Step 2: Derive the equation for the derivative $\dot{V}(\mathbf{x})$ along the system trajectory:

$$\dot{\mathbf{x}} = \mathbf{f}(\mathbf{x}, \mathbf{u}, t)$$

and select a feedback control law:

$$\mathbf{u} = \mathbf{u}(\mathbf{x})$$

which ensures that:

$$\frac{dV(\mathbf{x})}{dt} < 0 \quad \text{for } \mathbf{x} \neq \mathbf{0}$$

Typically, $\mathbf{u}(\mathbf{x})$ is a nonlinear function of $\mathbf{x}$ that contains some parameters and gains which can be selected to make $dV/dt < 0$, and thus ensures that the closed-loop system is asymptotically stable.

Remark For a given system, one may find several Lyapunov functions and corresponding stabilization controllers. If the system is linear time-invariant $\dot{\mathbf{x}} = \mathbf{A}\mathbf{x} + \mathbf{B}\mathbf{u}$, then it is not necessary to work using the Lyapunov method. In this case, the controller is a static linear state feedback controller $\mathbf{u}(t) = \mathbf{F}\mathbf{x}(t) + \upsilon(t)$ (Eq. (5.16)), which can be selected to make the closed-loop matrix a *Hurwitz matrix* (with all its eigenvalues on the strict left-hand s-plane). This assures that the closed-loop system is *asymptotically* and *exponentially stable*. The Lyapunov-based controller design method will be applied as a rule in most cases of the discussions that follow.

5.3.3 Computed Torque Control

The *computed torque control* technique reduces the effects of the uncertainty in all the terms of the Lagrange model. The controller $\boldsymbol{\tau}$ is selected to have the same form as the dynamic model (Eq. (3.11a)), that is:

$$\boldsymbol{\tau} = \mathbf{D}(\mathbf{q})\mathbf{u} + \mathbf{h}(\mathbf{q}, \dot{\mathbf{q}}) + \mathbf{g}(\mathbf{q}) \tag{5.33}$$

Thus, since the inertia matrix is positive definite (and so invertible), introducing the control law (5.33) in the system (3.11a), we get:

$$\ddot{\mathbf{q}}(t) = \mathbf{u}(t) \tag{5.34}$$

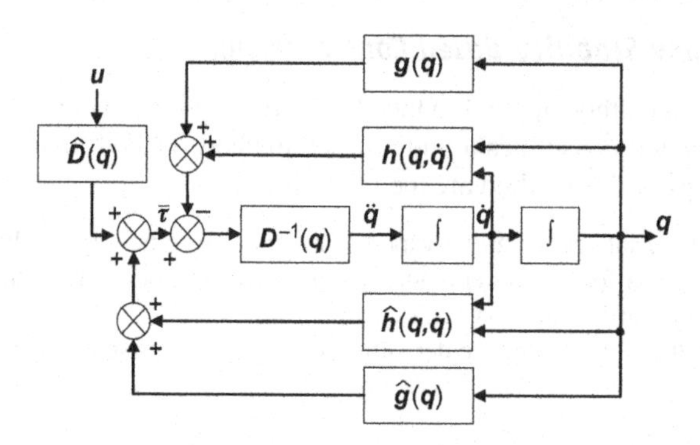

Figure 5.9 Closed-loop computed torque control system.

This implies that $\mathbf{u}(t)$ can be a decoupled controller (PD, PID) that can control each joint (motor axis) independently. The basic problem of the computed torque method is that we do not have available the exact values of $\mathbf{D}(\mathbf{q})$, $\mathbf{h}(\mathbf{q}, \dot{\mathbf{q}})$, and $\mathbf{g}(\mathbf{q})$, but only approximate values $\hat{\mathbf{D}}(\mathbf{q})$, $\hat{\mathbf{h}}(\mathbf{q}, \dot{\mathbf{q}})$, and $\hat{\mathbf{g}}(\mathbf{q})$. Then, instead of Eq. (5.33) we get:

$$\boldsymbol{\tau} = \hat{\mathbf{D}}(\mathbf{q})\mathbf{u} + \hat{\mathbf{h}}(\mathbf{q}, \dot{\mathbf{q}}) + \hat{\mathbf{g}}(\mathbf{q}) \tag{5.35}$$

and so Eq. (5.34) is replaced by:

$$\ddot{\mathbf{q}}(t) = (\mathbf{D}^{-1}\hat{\mathbf{D}})\mathbf{u} + \mathbf{D}^{-1}(\hat{\mathbf{h}} - \mathbf{h}) + \mathbf{D}^{-1}(\hat{\mathbf{g}} - \mathbf{g}) \tag{5.36}$$

A problem with the model in Eq. (5.35) is the investigation of its robustness to modeling uncertainties which include uncertainties in the parameter values and nonmodeled high-frequency components (e.g., structural resonance, sampling rate, or omitted time delays). The computed torque control method belongs to the general class of linearization techniques via nonlinear state feedback (see Section 6.3). Solving the model (Eq. (3.11a)) for $\ddot{\mathbf{q}}$ we get:

$$\ddot{\mathbf{q}} = \mathbf{D}^{-1}(\mathbf{q})[\boldsymbol{\tau} - \mathbf{h}(\mathbf{q}, \dot{\mathbf{q}}) - \mathbf{g}(\mathbf{q})] \tag{5.37}$$

Introducing the controller (5.35) into (5.37) we obtain the block diagram of the overall closed-loop system shown in Figure 5.9.

5.3.4 Robot Control in Cartesian Space

The controllers presented thus far work in the joints' (motors') space and are based on the error $\tilde{\mathbf{q}} = \mathbf{q} - \mathbf{q}_d$ between the actual and desired generalized joint variables $q_1, q_2, \ldots q_n$ (called internal variables). The motion of the robot in the Cartesian (or task, or working) space is obtained indirectly from the motion of the joints. However, in many cases, it is required to design the controller so as to work directly with the Cartesian variables, called external variables.

In Cartesian space, we have three types of controllers which are known as resolved motion controllers:

- Resolved motion rate control
- Resolved motion acceleration control
- Resolved force control

Here, we will study the resolved motion rate control which is mostly used in mobile robots. The resolved acceleration controller is actually an extension of the resolved motion rate control that includes the acceleration, a fact that will also be briefly considered.

5.3.4.1 Resolved Motion Rate Control

The resolved motion rate control is the control where the joints are moved simultaneously in different velocities, such that a desired motion in Cartesian (or task) space is obtained.

In general, the relation of the linear and angular velocity (motion rate) vector:

$$\dot{\mathbf{p}}(t) = \begin{bmatrix} \mathbf{v}(t) \\ \omega(t) \end{bmatrix}$$

in Cartesian space and the velocities $\dot{\mathbf{q}}(t)$ of the robotic joints is given by the Jacobian relation (Eq. (2.6)):

$$\dot{\mathbf{p}} = \mathbf{J}(\mathbf{q})\dot{\mathbf{q}} \tag{5.38}$$

where the Jacobian matrix $\mathbf{J}$ is given by Eq. (2.5), which has the inverse (see Eqs. (2.7b) and (2.8)):

$$\dot{\mathbf{q}} = \mathbf{J}^{-1}(\mathbf{q})\dot{\mathbf{p}} \quad (\text{if } \mathbf{J} \text{ is invertible}) \tag{5.39a}$$

or the generalized inverse:

$$\dot{\mathbf{q}} = \mathbf{J}^{\dagger}(\mathbf{q})\dot{\mathbf{p}} \quad (\text{if } \mathbf{J} \text{ is not square}) \tag{5.39b}$$

Differentiating Eq. (5.38) we get:

$$\begin{bmatrix} \dot{\mathbf{v}}(t) \\ \dot{\omega}(t) \end{bmatrix} = \dot{\mathbf{J}}(\mathbf{q})\dot{\mathbf{q}} + J(\mathbf{q})\ddot{\mathbf{q}} \tag{5.40}$$

Thus, introducing into Eq. (5.40) the expression of $\dot{\mathbf{q}}(t)$ given by Eq. (5.39a) yields:

$$\ddot{\mathbf{q}}(t) = \mathbf{J}^{-1}(\mathbf{q})\begin{bmatrix} \dot{\mathbf{v}}(t) \\ \dot{\omega}(t) \end{bmatrix} - \mathbf{J}^{-1}(\mathbf{q})\dot{\mathbf{J}}(\mathbf{q})\mathbf{J}^{-1}(q)\begin{bmatrix} \mathbf{v}(t) \\ \omega(t) \end{bmatrix} \tag{5.41}$$

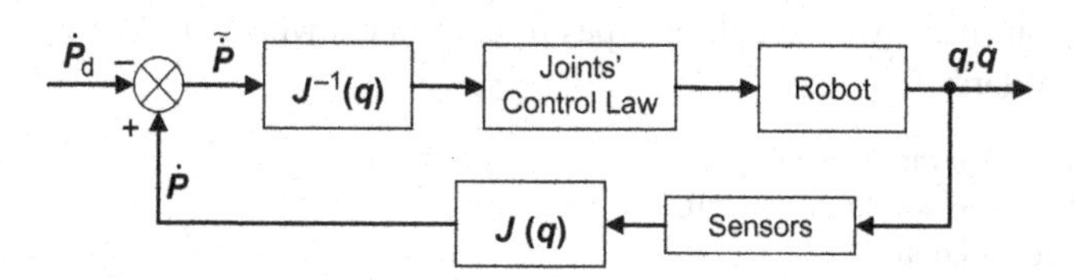

Figure 5.10 Block diagram of the robot resolved rate control based on Eq. (5.39a).

This relation gives the accelerations of the joints for given linear/angular velocity and acceleration of the robot end-effector in Cartesian space. If $\mathbf{J}(\mathbf{q})$ is not square, we use the generalized inverse $\mathbf{J}^{\dagger}$ in place of $\mathbf{J}^{-1}$.

The block diagram of the resolved velocity control is shown in Figure 5.10.

In the simplest case, the joints control can be a proportional control law with gain K_a. In many cases, the task space control is required to be in a coordinate frame attached to the robot (and not in the world coordinate frame). The velocity $\dot{\mathbf{p}}(t)$ is then given by:

$$\dot{\mathbf{p}}(t) = \mathbf{R}_{\mathrm{m}}^{0}\dot{\mathbf{r}}(t) \tag{5.42}$$

where $\dot{\mathbf{r}}(t)$ is the desired velocity of the robot and $\mathbf{R}_{\mathrm{m}}^{0}$ is a matrix that relates $\dot{\mathbf{r}}(t)$ to $\dot{\mathbf{p}}(t)$. Then, from Eqs. (5.39a) and (5.39b) we get:

$$\dot{\mathbf{q}}(t) = \mathbf{J}^{\dagger}(\mathbf{q})\dot{\mathbf{p}}(t) = \mathbf{J}^{\dagger}(\mathbf{q})\mathbf{R}_{\mathrm{m}}^{0}\dot{\mathbf{r}}(t) \tag{5.43}$$

The relation (5.43) is typically used in vision-based robot control (visual servoing).

5.3.4.2 Resolved Motion Acceleration Control

The resolved motion acceleration control method is based on the equation:

$$\ddot{\mathbf{p}}(t) = \mathbf{J}(\mathbf{q})\ddot{\mathbf{q}}(t) + \dot{\mathbf{J}}(\mathbf{q})\dot{\mathbf{q}}(t) \tag{5.44}$$

which is found by differentiating Eq. (5.38). The desired position, velocity, and acceleration of the robot in Cartesian space are assumed to be known from the trajectory planner. Thus, to reduce the position error, we must apply appropriate forces/torques at the robot joints, such that the acceleration in Cartesian space satisfies the relation:

$$\dot{\mathbf{v}}(t) = \dot{\mathbf{v}}_{\mathrm{d}}(t) + \mathbf{K}_{\mathrm{dv}}[\mathbf{v}_{\mathrm{d}}(t) - \mathbf{v}(t)] + K_{\mathrm{pv}}[\mathbf{s}_{\mathrm{d}}(t) - \mathbf{s}(t)] \tag{5.45}$$

where $\mathbf{s}_{\mathrm{d}}(t)$, $\mathbf{v}_{\mathrm{d}}(t)$, and $\dot{\mathbf{v}}_{\mathrm{d}}(t)$ are the desired translation position, velocity, and acceleration, respectively.

Here, the position error is:

$$\mathbf{e}_{\mathrm{p}}(t) = \mathbf{s}_{\mathrm{d}}(t) - \mathbf{s}(t)$$

Therefore, in terms of $\mathbf{e}_{\mathrm{p}}(t)$, Eq. (5.45) is written as:

$$\ddot{\mathbf{e}}_{\mathrm{p}}(t) + \mathbf{K}_{\mathrm{ds}}\dot{\mathbf{e}}_{\mathrm{p}}(t) + \mathbf{K}_{\mathrm{ps}}\mathbf{e}_{\mathrm{p}}(t) = \mathbf{0} \tag{5.46}$$

and the gains $\mathbf{K}_{\mathrm{ps}}$, $\mathbf{K}_{\mathrm{ds}}$ should be selected such that $\mathbf{e}_{\mathrm{p}}(t)$ tends asymptotically to zero. A similar error equation can be derived for the angular acceleration $\dot{\omega}(t)$, using the control law:

$$\dot{\omega}(t) = \dot{\omega}_{\mathrm{d}}(t) + \mathbf{K}_{\mathrm{d}\phi}[\omega_{\mathrm{d}}(t) - \omega(t)] + \mathbf{K}_{\mathrm{p}\phi}[\varphi_{\mathrm{d}}(t) - \varphi(t)] \tag{5.47}$$

where $\varphi(t)$ is the orientation angle in the Cartesian space. Combining Eqs. (5.45) and (5.47), and inserting into Eq. (5.44), we get:

$$\ddot{\mathbf{q}}(t) = \mathbf{J}^{-1}(\mathbf{q})\{\ddot{\mathbf{p}}_{\mathrm{d}}(t) + \mathbf{K}_{\mathrm{d}}[\dot{\mathbf{p}}_{\mathrm{d}}(t) - \dot{\mathbf{p}}(t)] + \mathbf{K}_{\mathrm{p}}\mathbf{e}(t) - \dot{\mathbf{J}}(\mathbf{q})\dot{\mathbf{q}}\} \tag{5.48}$$

where:

$$\dot{\mathbf{p}}_{\mathrm{d}}(t) = \begin{bmatrix} \mathbf{v}_{\mathrm{d}} \\ \omega_{\mathrm{d}} \end{bmatrix}, \quad \mathbf{e}(t) = \begin{bmatrix} \mathbf{s}_{\mathrm{d}} - \mathbf{s} \\ \varphi_{\mathrm{d}} - \varphi \end{bmatrix} = \begin{bmatrix} \mathbf{e}_{\mathrm{p}} \\ \mathbf{e}_{\phi} \end{bmatrix}, \quad \mathbf{K}_{\mathrm{p}} = \begin{bmatrix} \mathbf{K}_{\mathrm{ps}} & \mathbf{0} \\ \mathbf{0} & \mathbf{K}_{\mathrm{p}\phi} \end{bmatrix}, \quad \mathbf{K}_{d} = \begin{bmatrix} \mathbf{K}_{\mathrm{ds}} & \mathbf{0} \\ \mathbf{0} & \mathbf{K}_{\mathrm{d}\phi} \end{bmatrix}$$

Equation (5.48) constitutes the basis for the resolved motion acceleration control of robots. The position $\mathbf{q}(t)$ and velocity $\dot{\mathbf{q}}(t)$ are measured by potentiometers or optical encoders.

5.4 Control of Differential Drive Mobile Robot

The control procedure will involve two stages:

1. Kinematic stabilizing control
2. Dynamic stabilizing control

The resulting linear and angular velocities in the kinematic stage will be used as reference inputs for the dynamic stage. Thus, this procedure belongs to the general class of *backstepping control* [5–13].

5.4.1 Nonlinear Kinematic Tracking Control

The robot motion is governed by the dynamic model (Eqs. (3.23a) and (3.23b), the kinematic model (Eq. (2.26)), and the nonholonomic constraint (Eq. (2.27)), namely:

$$\dot{v} = \frac{1}{mr}(\tau_{\mathrm{r}} + \tau_{\mathrm{l}}) = \frac{1}{m}\tau_{\mathrm{a}}, \quad \tau_{\mathrm{a}} = \frac{1}{r}(\tau_{\mathrm{r}} + \tau_{\mathrm{l}}) \tag{5.49a}$$

$$\dot{\omega} = \frac{2a}{Ir}(\tau_r - \tau_l) = \frac{1}{I}\tau_b, \quad \tau_b = \frac{2a}{r}(\tau_r - \tau_l) \tag{5.49b}$$

$$\begin{aligned} \dot{x} &= v \cos\phi \\ \dot{y} &= v \sin\phi \end{aligned} \tag{5.49c}$$

where, for notational simplicity, the index Q was dropped from x_Q, y_Q and $\omega = \dot{\phi}$, τ_a, τ_b are the control inputs, with:

$$\mathbf{p} = [x, y, \phi]^T \tag{5.49d}$$

being the state vector.

The problem is to track a desired state trajectory:

$$\mathbf{p}_d(t) = [x_d(t), y_d(t), \phi_d(t)]^T \tag{5.50}$$

with error that goes asymptotically to zero.

To this end, the Lyapunov stabilizing method will be used. For realizability, the desired trajectory must satisfy both the kinematic equations and the nonholonomic constraint, that is:[3]

$$\begin{bmatrix} \dot{x}_d \\ \dot{y}_d \\ \dot{\phi}_d \end{bmatrix} = \begin{bmatrix} v_d \cos\phi_d \\ v_d \sin\phi_d \\ \omega_d \end{bmatrix}, \quad \dot{x}_d \sin\phi_d = \dot{y}_d \cos\phi_d \tag{5.51}$$

The errors $\tilde{x} = (x_d - x)$, $\tilde{y} = (y_d - y)$, and $\tilde{\phi} = \phi_d - \phi$, expressed in the wheeled mobile robots' (WMR's) local (moving) coordinate frame Qx_ry_r, are given by (see Eq. (2.17)):

$$\begin{bmatrix} \tilde{x}_r \\ \tilde{y}_r \end{bmatrix} = \begin{bmatrix} \cos\phi & -\sin\phi \\ \sin\phi & \cos\phi \end{bmatrix}^{-1} \begin{bmatrix} \tilde{x} \\ \tilde{y} \end{bmatrix} = \begin{bmatrix} \cos\phi & \sin\phi \\ -\sin\phi & \cos\phi \end{bmatrix} \begin{bmatrix} \tilde{x} \\ \tilde{y} \end{bmatrix} \tag{5.52a}$$

and

$$\tilde{\phi}_r = \tilde{\phi} \tag{5.52b}$$

Differentiating Eqs. (5.52a) and (5.52b) and taking into account Eqs. (5.49c) and (5.51), we get the following kinematic model for the error $[\tilde{x}_r, \tilde{y}_r, \tilde{\phi}_r]^T$:

$$\begin{aligned} \dot{\tilde{x}}_r &= v_d \cos\tilde{\phi}_r - v + \tilde{y}_r\omega \\ \dot{\tilde{y}}_r &= v_d \sin\tilde{\phi}_r - \tilde{x}_r\omega \\ \dot{\tilde{\phi}}_r &= \omega_d - \omega \end{aligned} \qquad \dot{\tilde{\mathbf{p}}}_r = \begin{bmatrix} \dot{\tilde{x}}_r \\ \dot{\tilde{y}}_r \\ \dot{\tilde{\phi}}_r \end{bmatrix} \tag{5.53}$$

[3] This is equivalent to considering that our WMR has to track a similar (virtual) differential-drive WMR which is moving with linear velocity v_d and angular velocity ω_d.

where the linear and angular velocities v and ω are the kinematic control variables. Clearly, Eq. (5.53) satisfies the kinematic and nonholonomic equations of the WMR.

Therefore, the kinematic feedback controller will be based on Eq. (5.53). The Lyapunov stabilizing method of Section 5.3.2 will be applied. Since here the controller should be nonlinear, we cannot select its structure beforehand. Its structure will be determined by the choice of the candidate Lyapunov function. Here, the following candidate function is selected [14]:

$$V(\tilde{\mathbf{p}}_{\mathrm{r}}) = \frac{1}{2}(\tilde{x}_{\mathrm{r}}^2 + \tilde{y}_{\mathrm{r}}^2) + (1 - \cos\tilde{\phi}_{\mathrm{r}}) \tag{5.54}$$

This function satisfies the first three properties of Lyapunov functions, namely:

(i) $V(\tilde{\mathbf{p}}_{\mathrm{r}})$ is continuous and has continuous derivatives
(ii) $V(\mathbf{0}) = 0$
(iii) $V(\tilde{\mathbf{p}}_{\mathrm{r}}) > 0$ for all $\tilde{\mathbf{p}}_{\mathrm{r}} \neq 0$

We therefore have to check under what conditions the fourth property can be satisfied.

Differentiating Eq. (5.54) with respect to time we get:

$$\dot{V}(\tilde{\mathbf{p}}_{\mathrm{r}}) = (-v + v_{\mathrm{d}}\cos\tilde{\phi}_{\mathrm{r}})\tilde{x}_{\mathrm{r}} + (-\omega + v_{\mathrm{d}}\tilde{y}_{\mathrm{r}} + \omega_{\mathrm{d}})\sin\tilde{\phi}_{\mathrm{r}} \tag{5.55}$$

To make $\dot{V}(\tilde{\mathbf{p}}_{\mathrm{r}}) \leq 0$, the control inputs v and ω are selected such that:

$$\dot{V}(\tilde{\mathbf{p}}_{\mathrm{r}}) = -(K_{\mathrm{x}}\tilde{x}_{\mathrm{r}}^2 + K_{\mathrm{j}}\sin^2\tilde{\phi}_{\mathrm{r}}) \tag{5.56}$$

which leads to:

$$v = v_{\mathrm{c}} = K_x\tilde{x}_{\mathrm{r}} + v_{\mathrm{d}}\cos\tilde{\phi}_{\mathrm{r}} \tag{5.57a}$$

$$\omega = \omega_{\mathrm{c}} = K_\phi \sin\tilde{\phi}_{\mathrm{r}} + v_{\mathrm{d}}\tilde{y}_{\mathrm{r}} + \omega_{\mathrm{d}} \tag{5.57b}$$

Clearly, for $K_x > 0$ and $K_\phi > 0$, we have $\dot{V}(\tilde{\mathbf{p}}_{\mathrm{r}}) \leq 0$ with the equality obtained only when $\tilde{x}_{\mathrm{r}} \equiv 0$ and $\tilde{\phi}_{\mathrm{r}} \equiv 0$. Thus, the controller ((5.57a) and (5.57b) guarantees total asymptotic tracking to the desired trajectory.

Remark 1 We can also add a gain K_y in the second term of Eq. (5.56b), that is, choose ω as

$$\omega = \omega_{\mathrm{c}} = \omega_{\mathrm{d}} + K_y v_{\mathrm{d}}\tilde{y}_{\mathrm{r}} + K_\phi \sin\tilde{\phi}_{\mathrm{r}} \tag{5.58a}$$

In this case, total asymptotic tracking to the desired trajectory $[x_{\mathrm{d}}(t), y_{\mathrm{d}}(t), \phi_{\mathrm{d}}(t)]^T$ can be proved using the following Lyapunov function:

$$V = \frac{1}{2}(\tilde{x}_{\mathrm{r}}^2 + \tilde{y}_{\mathrm{r}}^2) + \left(\frac{1}{K_y}\right)(1 - \cos\tilde{\phi}_{\mathrm{r}}), \quad K_y > 0 \tag{5.58b}$$

Remark 2 A more general kinematic controller can be obtained by using the Lyapunov function:

$$V(\tilde{\mathbf{p}}_r) = K_p(\tilde{x}_r^2 + \tilde{y}_r^2)^{\mu} + K_q(1 - \cos\tilde{\phi}_r) \tag{5.58c}$$

with $K_p > 0$, $K_q > 0$, and $\mu > 1$. In the following, we will derive this controller. To this end, we define new control variables $u_1 = v_d \cos\tilde{\phi}_r - v$, $u_2 = \omega - \omega_d$ and write the model (5.53) as:

$$\dot{\tilde{x}}_r = \omega\tilde{y}_r + u_1, \quad \dot{\tilde{y}}_r = v_d \sin\tilde{\phi}_r - \tilde{x}_r\omega, \quad \dot{\tilde{\phi}}_r = u_2 \tag{5.58d}$$

Differentiating Eq. (5.58c) and using the model (5.58d) we get:

$$\begin{aligned}\dot{V} &= 2K_p\mu(\tilde{x}_r^2 + \tilde{y}_r^2)^{\mu-1}\tilde{x}_r u_1 + K_q(\sin\tilde{\phi}_r)u_2 \\ &\quad + 2K_p\mu(\tilde{x}_r^2 + \tilde{y}_r^2)^{\mu-1}\tilde{y}_r v_d \sin\tilde{\phi}_r\end{aligned}$$

To assure that $\dot{V} \leq 0$, we use two functions $F(\mathbf{x}) \geq M > 0$ and $G(\mathbf{x}) \geq M > 0$ for all $\mathbf{x} = \tilde{\mathbf{p}}_r \in R^3$, and select u_1 and u_2 such that:

$$\dot{V} = -2K_p^2\mu^2\tilde{x}_r^2(\tilde{x}_r^2 + \tilde{y}_r^2)^{\mu-1}F(\mathbf{x}) - K_q^2(\sin^2\tilde{\phi}_r)G(\mathbf{x}) < 0 \tag{5.58e}$$

Then, it follows that:

$$\begin{aligned}u_1 &= -\mu K_p\tilde{x}_r F(\tilde{\mathbf{p}}_r) \\ u_2 &= -(2K_p/K_q)\mu(\tilde{x}_r^2 + \tilde{y}_r^2)^{\mu-1}v_d y_r - K_q(\sin\phi_r)G(\tilde{\mathbf{p}}_r)\end{aligned}$$

The above controller assures that $\tilde{x}_r^2 + \tilde{y}_r^2 \to 0$ and $\tilde{\phi}_r \to k\pi$ $(k = 0, 1, 2, \ldots)$. Since $\dot{V}(\tilde{\mathbf{p}}_r)$ is uniformly continuous, Barbalat's Lemma (Sec. 6.2.3) implies that $\dot{V}(\tilde{\mathbf{p}}_r) \to 0$. This by Eq. (5.58e) implies that $x_r \to 0$ and $\tilde{\phi}_r \to k\pi$ $(k = 0, 1, 2, \ldots)$. Now, obviously, $\dot{\tilde{\phi}}_r \to 0$ (i.e., $\omega \to \omega_d$). One way to overcome the fact that $\dot{\tilde{\phi}}_r$ does not converge only to 0 but also to $k\pi$ $(k = 1, 2, \ldots)$ is to take care that the WMR, before trying to track immediately the desired trajectory (i.e., the virtual WMR), is rotating about its own axis with an increasing angular velocity ω^* until it sees the virtual robot. The reader can verify that this can be done by the controller [11]:

$$u_1^* = \gamma(t)u_1(t), \quad u_2^* = \gamma(t)u_2(t) + [1 - \gamma(t)]\omega^*(t)$$

where $\gamma(t)$ is given by the dynamic model:

$$a_2\ddot{\gamma}(t) + a_1\dot{\gamma}(t) + \gamma(t) = \sigma(t)$$

with $\sigma(t)$ being a step input function defined by:

$$\sigma(t) = \begin{cases} 1 & \text{for } t_1 \in [0, t] \text{ with } \phi(t_1) = \operatorname{atan2}(\tilde{y}_r(t_1), \tilde{x}_r(t_1)) \\ 0 & \text{otherwise} \end{cases}$$

5.4.2 Dynamic Tracking Control

Having selected v and ω as in Eqs. (5.57a) and (5.57b) (or Eq. (5.58a)), we select the control inputs (torques) τ_a and τ_b in Eq. (5.49a) as:

$$\tau_a = m\dot{v}_c + K_a\tilde{v}_c \tag{5.59a}$$

$$\tau_b = I\dot{\omega}_c + K_b\tilde{\omega}_c \tag{5.59b}$$

where:

$$\tilde{v}_c = v_c - v \quad \tilde{\omega}_c = \omega_c - \omega \tag{5.59c}$$

Introducing Eqs. (5.59a)–(5.59c) into Eqs. (5.49a) and (5.49b) we get the velocities' error equations:

$$\dot{\tilde{v}}_c + (K_a/m)\tilde{v}_c = 0, \quad \dot{\tilde{\omega}}_c + (K_b/I)\tilde{\omega}_c = 0$$

which for $K_a > 0$ and $K_b > 0$ are stable and $\tilde{v}_c$, $\tilde{\omega}_c$ converge to zero asymptotically. Therefore, in selecting the feedback control inputs (torques) as in Eqs. (5.59a) and (5.59b) with v_c and ω_c given by Eqs. (5.57a) and (5.57b), the tracking of the desired trajectory $[x_d(t), y_d(t), \phi_d(t)]^T$ is achieved asymptotically, as required. The block diagram of the feedback tracking controller is depicted in Figure 5.11.

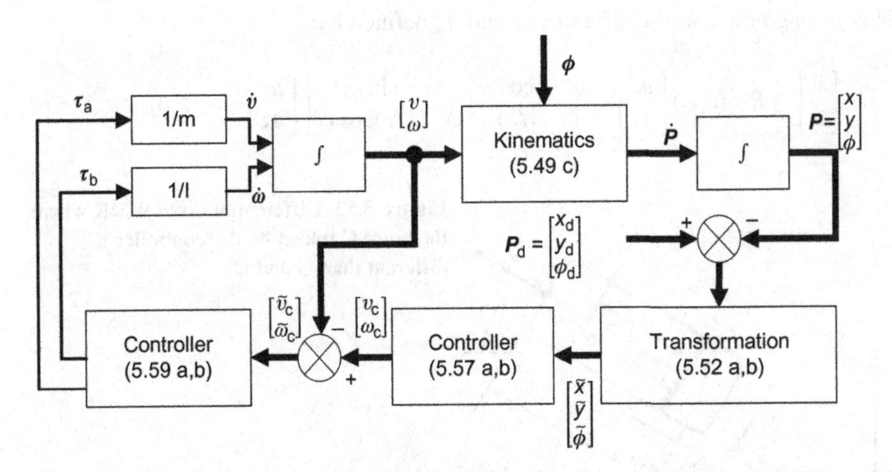

Figure 5.11 The complete trajectory tracking feedback control system of the WMR.

5.5 Computed Torque Control of Differential Drive Mobile Robot

The control design procedure involves again two stages: *kinematic control* followed by *dynamic control*. We will work on the WMR shown in Figure 5.12, where the motor dynamics includes a gear box (of ratio N) [13].

Here, Q is the midpoint of wheel baseline, G the center of gravity, and C the point traced by the controller (different than the point Q).

The meaning of the remaining symbols are self-evident (the same as in Figure 2.7).

5.5.1 Kinematic Tracking Control

The kinematic equations of the robot are:

$$\begin{aligned}\dot{x} &= v\cos\phi - [\dot{y}_r + (c-b)\omega]\sin\phi \\ &= v\cos\phi - c\omega\sin\phi\end{aligned} \tag{5.60a}$$

$$\begin{aligned}\dot{y} &= v\sin\phi + [\dot{y}_r + (c-b)\omega]\sin\phi \\ &= v\sin\phi + c\omega\cos\phi\end{aligned} \tag{5.60b}$$

$$\dot{\phi} = \omega \tag{5.60c}$$

where $\dot{y}_r$ is the lateral velocity in the local coordinate frame Gx_ry_r.

Equations (5.60a) and (5.60b) are written in the matrix form:

$$\begin{bmatrix}\dot{x}\\ \dot{y}\end{bmatrix} = \begin{bmatrix}\cos\phi & -c\sin\phi\\ \sin\phi & c\cos\phi\end{bmatrix}\begin{bmatrix}v\\ \omega\end{bmatrix} = R(c,\phi)\begin{bmatrix}v\\ \omega\end{bmatrix}$$

Now, using new control variables u_v and u_ϕ defined by:

$$\begin{bmatrix}v\\ \omega\end{bmatrix} = R^{-1}(c,\phi)\begin{bmatrix}u_v\\ u_\phi\end{bmatrix} = \begin{bmatrix}\cos\phi & \sin\phi\\ -(1/c)\sin\phi & (1/c)\cos\phi\end{bmatrix}\begin{bmatrix}u_v\\ u_\phi\end{bmatrix} \quad (c \neq 0) \tag{5.61}$$

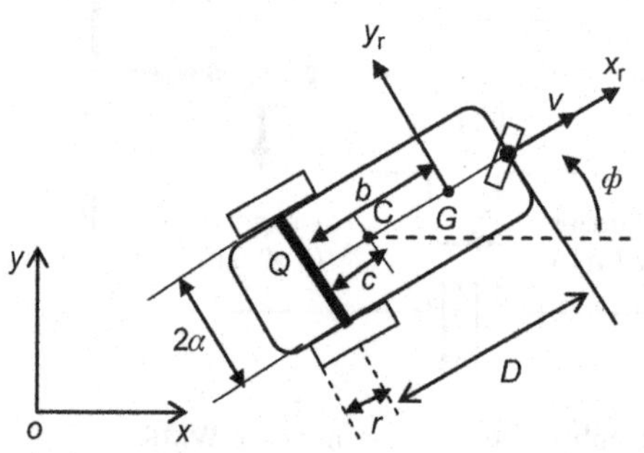

Figure 5.12 Differential drive WMR where the point C traced by the controller is different than Q and G.

we get:

$$\dot{x} = u_v, \quad \dot{y} = u_\phi \tag{5.62}$$

The dynamic system (5.62) is linear and decoupled, and so the state-feedback law:

$$u_v = \dot{x}_\mathrm{d} + K_x\tilde{x}, \quad u_\phi = \dot{y}_\mathrm{d} + K_y\tilde{y} \tag{5.63}$$

yields the error dynamics:

$$\dot{\tilde{x}} + K_x\tilde{x} = 0, \quad \dot{\tilde{y}} + K_y\tilde{y} = 0 \tag{5.64}$$

with $\tilde{x} = x_\mathrm{d} - x$ and $\tilde{y} = y_\mathrm{d} - x$.

Therefore, from Eq. (5.64), it follows that for any positive gains:

$$K_x > 0, \quad K_y > 0$$

the tracking error tends exponentially to zero.

Combining Eqs. (5.61) and (5.63) we get the overall nonlinear kinematic control law:

$$\begin{bmatrix} v \\ \omega \end{bmatrix} = \begin{bmatrix} \cos\phi & \sin\phi \\ -(1/c)\sin\phi & (1/c)\cos\phi \end{bmatrix} \left(\begin{bmatrix} \dot{x}_\mathrm{d} \\ \dot{y}_\mathrm{d} \end{bmatrix} + \begin{bmatrix} K_x & 0 \\ 0 & K_y \end{bmatrix} \begin{bmatrix} \tilde{x} \\ \tilde{y} \end{bmatrix} \right) \tag{5.65}$$

5.5.2 Dynamic Tracking Control

The feedback kinematic controller (5.65) incorporates the WMR kinematic equations and so one can now use the reduced (unconstrained) dynamic model ((3.19a) and (3.19b)) of the robot for the selection of the control inputs (motor torques or motor voltages), as described in Section 5.4.2, where actually the computed torque method was applied.

For the robot of Figure 5.12, with the motor dynamics included, the reduced model has the following form:

$$\overline{\mathbf{D}}\dot{\mathbf{v}} + \overline{\mathbf{C}}\mathbf{v} = \overline{\mathbf{E}}\mathbf{V} \tag{5.66}$$

where:

$$\mathbf{v} = \begin{bmatrix} v \\ \omega \end{bmatrix}, \quad \overline{\mathbf{D}} = \begin{bmatrix} \overline{D}_{11} & 0 \\ 0 & \overline{D}_{22} \end{bmatrix}, \quad \overline{\mathbf{C}} = \begin{bmatrix} \overline{C}_{11} & \overline{C}_{12} \\ \overline{C}_{21} & \overline{C}_{22} \end{bmatrix}, \quad \overline{\mathbf{E}} = \begin{bmatrix} \overline{E}_1 & 0 \\ 0 & \overline{E}_2 \end{bmatrix}, \quad \mathbf{V} = \begin{bmatrix} V_\mathrm{a} \\ V_\mathrm{b} \end{bmatrix} \tag{5.67}$$

with:

$$\overline{D}_{11} = (1 + 2I_{\mathrm{m}}/mr^2)$$

$$\overline{D}_{22} = \left(I_z + \frac{2a^2}{r^2}I_{\mathrm{m}} + b^2 m\right)$$

$$\overline{C}_{11} = \frac{2}{m}\left(\frac{\beta_{\mathrm{m}}}{r^2} + \frac{N^2K_1K_2}{Rr^2}\right), \quad \overline{C}_{12} = -b\omega \tag{5.68}$$

$$\overline{C}_{22} = 2a^2\left(\frac{\beta_{\mathrm{m}}}{r^2} + \frac{N^2K_1K_2}{Rr^2}\right), \quad \overline{C}_{21} = bm\omega$$

$$\overline{E}_1 = (NK_1/Rrm)V_{\mathrm{a}}, \quad \overline{E}_2 = (NK_2\, a/Rr)V_{\mathrm{b}}$$

$$V_{\mathrm{a}} = V_{\mathrm{r}} + V_{\mathrm{l}}, \quad V_{\mathrm{b}} = V_{\mathrm{r}} - V_{\mathrm{l}}$$

Here:

I_{m} = combined wheel, motor rotor, and gearbox inertia
β_{m} = combined wheel, motor, and gearbox friction coefficient
$V_{\mathrm{r}}, V_{\mathrm{l}}$ = right and left wheel motor voltage
R = electrical resistance
K_1, K_2 = motor voltage/torque constants

Now, applying the computed torque (linearization) technique to Eq. (5.66), we choose the voltage control vector $\mathbf{V}$ as:

$$\mathbf{V} = \overline{\mathbf{E}}^{-1}(\overline{\mathbf{D}}\mathbf{u} + \overline{\mathbf{C}}\mathbf{v}) \tag{5.69}$$

where $\mathbf{u}$ is the new control vector. Introducing Eq. (5.69) into Eq. (5.66) we get:

$$\dot{\mathbf{v}} = \mathbf{u}$$

with:

$$\mathbf{v} = \begin{bmatrix} v \\ \omega \end{bmatrix}, \quad \mathbf{u} = \begin{bmatrix} u_1 \\ u_2 \end{bmatrix}$$

Therefore, selecting the linear state feedback control law:

$$\mathbf{u} = \mathbf{K}\tilde{\mathbf{v}} + \dot{\mathbf{v}}_{\mathrm{d}} \tag{5.70}$$

yields the error system:

$$\dot{\tilde{\mathbf{v}}} + \mathbf{K}\tilde{\mathbf{v}} = \mathbf{0}, \quad \mathbf{K} = \mathrm{diag}[K_1, K_2]$$

which for $\mathbf{K} > 0$ is asymptotically stable with equilibrium state $\tilde{\mathbf{v}} = \mathbf{v}_\mathrm{d} - \mathbf{v} = \mathbf{0}$.

Combining Eq. (5.69) with Eq. (5.70), we get the full dynamic controller:

$$\mathbf{V} = \mathbf{E}^{-1}(\overline{\mathbf{D}}\mathbf{K}\tilde{\mathbf{v}} + \dot{\mathbf{v}}_\mathrm{d} + \overline{\mathbf{C}}\mathbf{v}) \tag{5.71}$$

The overall tracking controller of the robot is given by Eqs. (5.65) and (5.71).

Example 5.4

It is desired to formulate and solve the kinematic problem of a unicycle-like WMR to go asymptotically from an initial state (position and orientation) to a goal state (position and orientation) using polar coordinates.

Solution

The unicycle-like robots are described by the kinematic Eq. (2.26):

$$\dot{x} = v \cos \phi, \quad \dot{y} = v \sin \phi, \quad \dot{\phi} = \omega \tag{5.72}$$

where v is the velocity and ϕ is the angle of the WMR x_r axis with the goal (world) coordinate frame x-axis. All WMRs with the above kinematic equations, for example, the differential drive WMR, are said to belong to the unicycle-like class of mobile robots. The geometry of the WMR that will be used for formulating the polar coordinates is shown in Figure 5.13 [15].

The kinematic control variables of the robot are v and ω. The polar coordinates (position and orientation) of the WMR are its distance l from the goal, and its orientation ψ with respect to the goal's coordinate frame Gxy. The steering angle is $\zeta = \psi - \phi$. Then, in polar coordinates, the kinematic model in Eq. (5.72) is replaced by:

$$\dot{l} = -v \cos \zeta \tag{5.73a}$$

$$\dot{\zeta} = -\omega + (v/l)\sin \zeta \tag{5.73b}$$

$$\dot{\psi} = (v/l)\sin \zeta \tag{5.73c}$$

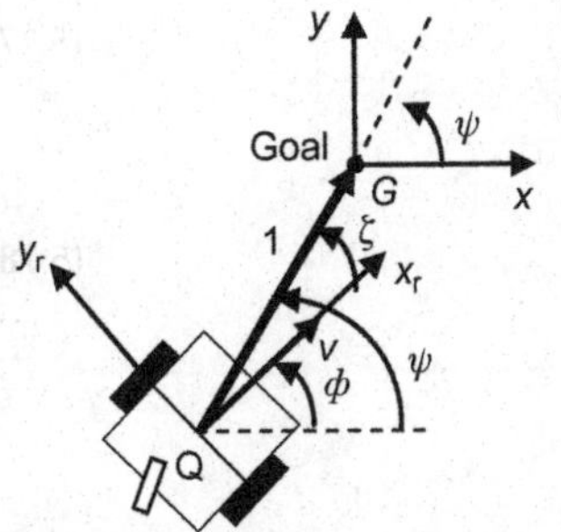

Figure 5.13 Polar coordinates of the unicycle. Here, the goal coordinate frame Gxy is considered to be the world coordinate frame.

These relations hold for $l > 0$, a condition which will be always satisfied by an asymptotic reduction of l to zero (since for any finite time there always be $l > 0$).

Our goal tracking control problem is to find a state-feedback law:

$$\begin{bmatrix} v \\ \omega \end{bmatrix} = \mathbf{u}(l, \zeta, \psi) \tag{5.74}$$

which guarantees that $l \to 0$, $\zeta \to 0$, and $\psi \to 0$, asymptotically.

To this end, we will apply the Lyapunov-based control method. Let us choose the following candidate Lyapunov function [15]:

$$V(\mathbf{x}) = \frac{1}{2}\mathbf{x}^T\mathbf{Q}\mathbf{x}, \quad \mathbf{x} = [l, \zeta, \psi]^T$$

$$\mathbf{Q} = \begin{bmatrix} q_1 & & \mathbf{0} \\ & 1 & \\ \mathbf{0} & & q_2 \end{bmatrix}, \quad q_1 > 0, \quad q_2 > 0 \tag{5.75}$$

Clearly, the function $V(\mathbf{x})$ possesses the first three properties of Lyapunov functions. We will determine the controller (5.74) which will assure that the fourth property $\dot{V} \leq 0$ is also possessed by V, along the system trajectory.

The time derivative of $V(\mathbf{x})$, along the trajectory determined by Eqs. (5.73a)–(5.73c), is found to be:

$$\begin{aligned} \dot{V}(\mathbf{x}) &= \mathbf{x}^T Q\dot{\mathbf{x}} \\ &= q_1 l\dot{l} + \zeta\dot{\zeta} + q_2\psi\dot{\psi} = \dot{V}_1 + \dot{V}_2 \end{aligned} \tag{5.76a}$$

where:

$$\dot{V}_1 = q_1 l\dot{l} = -q_1 lv \cos\zeta \tag{5.76b}$$

$$\begin{aligned} \dot{V}_2 &= \zeta\dot{\zeta} + q_2\psi\dot{\psi} \\ &= \zeta[-\omega + (v/l)\sin\zeta] + (q_2/l)(\sin\zeta)\psi v \end{aligned} \tag{5.76c}$$

Choosing v as:

$$v = K_1(\cos\zeta)l, \quad K_1 > 0 \tag{5.77}$$

yields:

$$\dot{V}_1 = -K_1 q_1(\cos^2\zeta)l^2 \leq 0 \tag{5.78}$$

Now, introducing Eq. (5.77) into Eq. (5.76c) we get:

$$\dot{V}_2 = \zeta[-\omega + K_1(\cos\zeta)(\sin\zeta)(\zeta + q_2\psi)/\zeta]$$

Thus, selecting ω as:

$$\omega = K_2\zeta + K_1(\cos\zeta)(\sin\zeta)(\zeta + q_2\psi)/\zeta,\ K_2 > 0 \tag{5.79}$$

yields:

$$\dot{V}_2 = -K_2\zeta^2 \leq 0 \tag{5.80}$$

The inequalities (5.78) and (5.80) give:

$$\dot{V} = -K_1 q_1(\cos^2\zeta)l^2 - K_2\zeta^2 \leq 0 \tag{5.81}$$

This implies, by the Lyapunov stability theorem, that ζ and l go asymptotically to zero for any ψ. Thus, we have to see what happens to ψ. To this end, we get the closed-loop system kinematics by introducing the control laws ((5.77) and (5.79)) into Eqs. (5.73a)–(5.73c), namely:

$$\dot{l} = -K_1 l\cos^2\zeta,\quad l(0) > 0 \tag{5.82a}$$

$$\dot{\zeta} = -K_2\zeta - K_1 q_2(\cos\zeta)\left(\frac{\sin\zeta}{\zeta}\right)\psi \tag{5.82b}$$

$$\dot{\psi} = K_1(\cos\zeta)(\sin\zeta) \tag{5.82c}$$

These equations show that the asymptotic convergence of l and ζ to zero implies the asymptotic convergence of ψ to its only equilibrium state $\psi_s = 0$. In fact, from Eq. (5.82c) it follows that $\zeta \to 0$ implies $\dot{\psi} \to 0$, that is, ψ tends to some finite value ψ_s. Then, we see from Eq. (5.82b) that the uniformly continuous function $\dot{\zeta}$ tends necessarily to $-K_1 q_2\psi_s$.[4] On the other hand, by Barbalat's lemma, $\dot{\zeta}$ tends to zero, which in turn implies that $\psi_s = 0$ (see Section 6.2.3). Therefore, the smooth kinematic control laws ((5.77) and (5.79)) assure the asymptotic tracking of the goal position and orientation by the WMR, as desired. The fact that the control laws ((5.77) and (5.79)) are continuously differentiable does not contradict Brockett's Theorem 6.6 and its corollary (c) because this theorem is valid for the Cartesian state-space representation (Eq. (5.72)) of the unicycle-like WMR [15].

5.6 Car-Like Mobile Robot Control

For the car-like WMR, we will study the following two representative problems:

1. Parking (or posture) control
2. Leader–follower (formation) control

[4] Recall that $(\sin\zeta)/\zeta \to 1$ as $\zeta \to 0$.

5.6.1 Parking Control

Consider a car-like WMR (Figure 2.9) which controls the steering angle ψ and the rear-wheels' velocity v_1. The orientation of the car body (i.e., of v_1) is ϕ. The kinematic equations of the robot are given by Eq. (2.52):

$$\begin{aligned} \dot{x} &= v_1 \cos\phi \\ \dot{y} &= v_1 \sin\phi \\ \dot{\phi} &= (v_1/D) tg\,\psi \\ \dot{\psi} &= v_2 \end{aligned} \tag{5.83}$$

The problem is to control the WMR (using v_1 and ψ) so as to move it to a desired parking position and orientation, which here is assumed to be $x=0$, $y=0$, and $\phi=0$ (as it was actually done in Example 5.4 for the unicycle WMR). Here, this problem will be solved by a two-step maneuver to overcome the turning radius limitation of the car-like mobile robot, namely [16]:

Step 1: Controller stabilizing y and ϕ
Step 2: Controller stabilizing x and ϕ

Use of the Lyapunov-based control method will again be made as usual.

Step 1: (y,φ) control
We select the following candidate Lyapunov function:

$$V(\mathbf{x}) = \frac{1}{2}\mathbf{x}^T\mathbf{Q}\mathbf{x}, \quad \mathbf{x} = \begin{bmatrix} y \\ \phi \end{bmatrix}, \quad \mathbf{Q} = \begin{bmatrix} q_1 & 0 \\ 0 & 1 \end{bmatrix}$$

which satisfies the first three conditions of Lyapunov functions. We will check if the third condition can be satisfied, along the trajectory of the system in Eq. (5.83). We have:

$$\begin{aligned} \dot{V} &= \mathbf{x}^T\mathbf{Q}\dot{\mathbf{x}} = q_1 y\dot{y} + \phi\dot{\phi} \\ &= \left[q_1 y \sin\phi + (\phi/D) tg\,\psi\right] v_1 \end{aligned} \tag{5.84}$$

Choosing:

$$\begin{aligned} &v_1 = +|V_1| = \text{const.} \quad \text{or} \quad v_1 = -|V_1| = \text{const.} \\ &tg\,\psi = -\frac{D}{v_1}\left(q_1 v_1 \frac{\sin\phi}{\phi} y + K_1\phi\right) \end{aligned} \tag{5.85}$$

and introducing into Eq. (5.84) yields:

$$\dot{V} = -K_1\phi^2 \leq 0 \quad \text{for } K_1 > 0 \tag{5.86}$$

which, by Lyapunov theorem, implies that ϕ tends asymptotically to zero.

The closed loop kinematics equation is:

$$\dot{\phi} = -K_1\phi + q_1 v_1\left(\frac{\sin\phi}{\phi}\right)y \tag{5.87}$$

Let $\phi = 0$. For ϕ to stay zero, $\dot{\phi}$ should be zero. Then, Eq. (5.87) implies that y should also tend to zero, for $v_1 = \text{const.} \neq 0$.

The change of the sign of v_1 is needed when the WMR cannot move with its current velocity due to the presence of obstacles or when the measured states exceed some predetermined bounds. Of course, the initial selection of $|V_1|$ affects the efficiency of the path. Actually, the controller (5.85) is a nonlinear bang-bang controller. Therefore, a switching rule is needed to determine when the change from $v_1 = +|V_1|$ to $v_1 = -|V_1|$ must occur.

Defining $\varepsilon = tg(y/x)$, the switching rule is:

If $\cos(\varepsilon - \phi) > 0$
Then $v < 0$; otherwise $v > 0$

This rule means that if the front part of the WMR is closer to the origin, then it will go forward or backward.

Step 2: (x,φ) control

We use the same form of the candidate Lyapunov function:

$$V(\mathbf{x}) = \frac{1}{2}\mathbf{x}^T\mathbf{Q}\mathbf{x}$$

with $\mathbf{x} = [x, \phi]^T$ and $\mathbf{Q} = \text{diag}\,[q_1, 1]$.

The time derivative $\dot{V}$ along the trajectory of Eq. (5.83) is:

$$\dot{V} = q_1 x\dot{x} + \phi\dot{\phi} = q_1 x v_1 \cos\phi + \phi(v_1/D)tg\,\psi$$

Choosing:

$$v_1 = -K_2 x, \; tg\,\psi = -v_1\phi \tag{5.88}$$

gives:

$$\begin{aligned}\dot{V} &= -K_2 q_1 x^2 \cos\phi - (1/D)(v_1^2\phi^2) \\ &= -K_2 x^2[q_1 \cos\phi + (K_2/D)\phi^2] \geq -K_2 x^2\end{aligned} \tag{5.89}$$

which is negative semidefinite for $K_2 > Dq_1$.

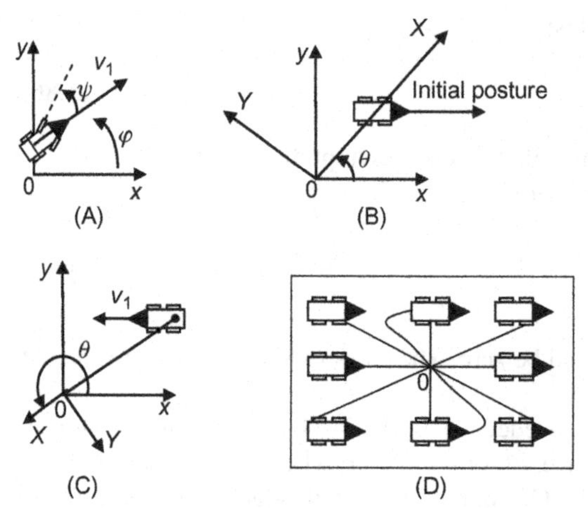

Figure 5.14 (A) For ϕ to converge, $|x|$ must be increased. (B) Case $|\text{atan2}(x,y)-\phi|<90°$. (C) Case $|\text{atan2}(x,y)-\phi|>90°$. (D) Convergence to the origin $(x,y)=(0,0)$ using Eq. (5.90).

For $\phi=0$, in which case, by Eq. (5.88), $\psi=0$, we have

$$\dot{V}=-K_2q_1x^2\leq 0$$

Therefore, the mobile robot is uniformly L-stable at $x=0$. However, ϕ cannot converge without increasing $|x|$, a fact which is due to the low bound of the WMR turning radius (Figure 5.14A) [16,17]. Actually, ϕ can be made arbitrarily small at the first step. Therefore, when we achieve a very small ϕ (i.e., $|\phi|<\varepsilon$), we use $\psi=0$, and $v_1=-K_2x$.

This situation is overcome if we use the transformation [16]:

$$\begin{aligned}
\theta &= \text{atan2}(y,x) && \text{for} \quad |\text{atan2}(y,x)-\phi|<90° \\
\theta &= \text{atan2}(y,x)-\text{sgn}(\text{atan2}(y,x)-\phi)\times 180° && \text{for} \quad |\text{atan2}(y,x)-\phi|>90°
\end{aligned} \tag{5.90}$$

In this way, the distance from the origin to the WMR is equal to the error of x, and the difference angle of the WMR with the x-axis becomes the error of ϕ. As shown in Figure 5.14D, the controller (5.88) with the above switching type transformation assures that the WMR can go to $(x,y)=(0,0)$ starting from any initial posture.

5.6.2 Leader–Follower Control

Consider two car-like robots that follow a path with the first car acting as the leader and the second being a follower (Figure 5.15). For more WMRs, one following the other in front of it, this problem is known as *formation control* [7].

The leader–follower control problem under consideration is to find a velocity control input for the follower that assures convergence of the relative distance L_{lf} and relative bearing angle θ_{lf} of the WMRs to their desired values, under the

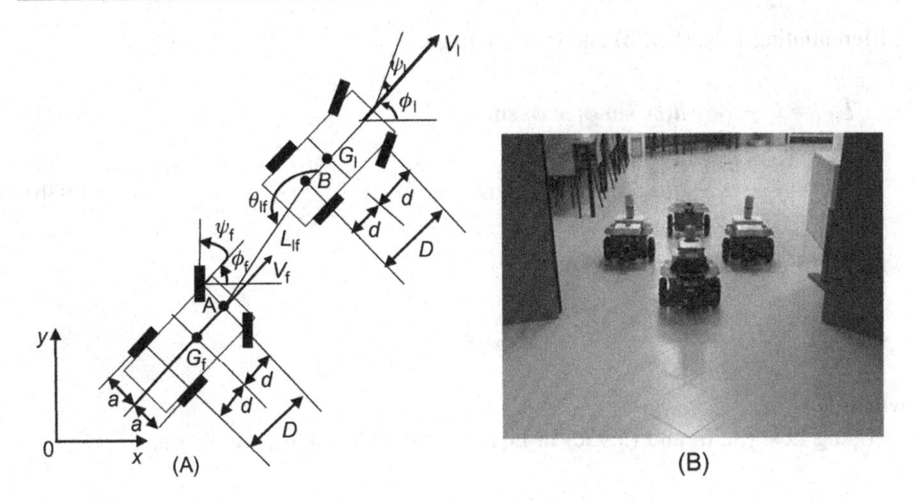

Figure 5.15 (A) Two car-like WMRs (leader–follower structure). (B) Four WMRs in a typical formation (diamond structure).
Source: www.robot.uji.es/lab/plone/research/pnebot/index_html2.

assumption that the leader motion is known and is the result of an independent control law [7]. To solve the problem, we will apply the Lyapunov-based control design method using the kinematic and dynamic equations of the bicycle equivalent presented in Section 3.4 (see Eqs. (3.56), (3.57), (3.60a)–(3.60d)).

In Figure 5.15, v_l, ϕ_l, and ψ_l are the linear velocity, orientation angle, and steering angle of the leader, and v_f, ϕ_f, ψ_f are the respective variables of the follower. The coordinates of points G_l and G_f are denoted by (x_l, y_l) and (x_f, y_f).

We first derive the dynamic equations for the errors:

$$\varepsilon_1 = x_{fd} - x_f, \quad \varepsilon_2 = y_{fd} - y_f, \quad \varepsilon_3 = \phi_{fd} - \phi_f \tag{5.91}$$

where x_{fd}, y_{fd}, and ϕ_{fd} represent the desired trajectory of the follower in world coordinates, which are transformed to ε_{f1}, ε_{f2}, and ε_{f3} in the local coordinate frame of the follower. From Figure 5.15 we get:

$$L_{lf}^2 = L_{lf,x}^2 + L_{lf,y}^2 \tag{5.92a}$$

$$L_{lf,x} = x_l - x_f - d(\cos\phi_l + \cos\phi_f) \tag{5.92b}$$

$$L_{lf,y} = y_l - y_f - d(\sin\phi_l + \sin\phi_f) \tag{5.92c}$$

$$tg(\theta_{lf} + \phi_l - \pi) = L_{lf,y}/L_{lf,x} \tag{5.92d}$$

Differentiating Eqs. (5.92b) and (5.92c) gives:

$$\dot{L}_{\mathrm{lf},x} = \dot{x}_{\mathrm{l}} - \dot{x}_{\mathrm{f}} + d(\dot{\phi}_{\mathrm{l}} \sin \phi_{\mathrm{l}} + \dot{\phi}_{\mathrm{f}} \sin \phi_{\mathrm{f}}) \tag{5.93a}$$

$$\dot{L}_{\mathrm{lf},y} = \dot{y}_{\mathrm{l}} - \dot{y}_{\mathrm{f}} - d(\dot{\phi}_{\mathrm{l}} \cos \phi_{\mathrm{l}} + \dot{\phi}_{\mathrm{f}} \cos \phi_{\mathrm{f}}) \tag{5.93b}$$

with (see Eqs. (3.56) and (3.57)):

$$\dot{y}_{g\mathrm{l}} = (d/D)\dot{x}_{g\mathrm{l}} tg\psi_{\mathrm{l}}, \quad \dot{y}_{g\mathrm{f}} = (d/D)\dot{x}_{g\mathrm{f}} tg\psi_{\mathrm{f}} \tag{5.93c}$$

where $D = 2d$.

Using Eqs. (3.56) and (5.93c) in Eqs. (5.93a) and (5.93b), we obtain:

$$\dot{L}_{\mathrm{lf},x} = \dot{x}_{g\mathrm{l}} \cos \phi_{\mathrm{l}} - \dot{x}_{g\mathrm{f}} \cos \phi_{\mathrm{f}} + \dot{x}_{g\mathrm{f}}(tg\ \psi_{\mathrm{f}}) \sin \phi_{\mathrm{f}} \tag{5.94a}$$

$$\dot{L}_{\mathrm{lf},y} = \dot{x}_{g\mathrm{l}} \sin \phi_{\mathrm{l}} - \dot{x}_{g\mathrm{f}} \sin \phi_{\mathrm{l}} - \dot{x}_{g\mathrm{f}}(tg\ \psi_{\mathrm{f}}) \cos \phi_{\mathrm{f}} \tag{5.94b}$$

while from Figure 5.13 we have:

$$\frac{L_{\mathrm{lf},x}}{L_{\mathrm{lf}}} = \cos(\theta_{\mathrm{lf}} + \phi_{\mathrm{l}} - \pi), \quad \frac{L_{\mathrm{lf},y}}{L_{\mathrm{lf}}} = \sin(\theta_{\mathrm{lf}} + \phi_{\mathrm{l}} - \pi) \tag{5.95}$$

Then, differentiating Eqs. (5.92a) and (5.92d), introducing Eqs. (5.94a) and (5.94b), and using the auxiliary variable:

$$\zeta_{\mathrm{f}} = \theta_{\mathrm{lf}} + \phi_{\mathrm{l}} - \phi_{\mathrm{f}}$$

we get, after some algebraic manipulation:

$$\dot{L}_{\mathrm{lf}} = -\dot{x}_{g\mathrm{l}} \cos(\theta_{\mathrm{lf}}) + \dot{x}_{g\mathrm{f}} tg(\psi_{\mathrm{f}}) \sin \zeta_{\mathrm{f}} + \dot{x}_{g\mathrm{f}} \cos \zeta_{\mathrm{f}} \tag{5.96a}$$

$$\dot{\theta}_{\mathrm{lf}} = (1/L_{\mathrm{lf}})[(\dot{x}_{g\mathrm{l}} \sin \theta_{\mathrm{lf}} - \dot{x}_{g\mathrm{f}} \sin \zeta_{\mathrm{f}}) + \dot{x}_{g\mathrm{f}} tg\ \psi_{\mathrm{f}} \cos \zeta_{\mathrm{f}}] - (1/D)\dot{x}_{g\mathrm{l}} tg\ \psi_{\mathrm{l}}] \tag{5.96b}$$

Using Figure 5.15, the actual and desired coordinates of the follower's point A can be expressed in terms of the coordinates of the leader's point B. Therefore, we use the variables $\{L_{\mathrm{lf}}, \theta_{\mathrm{lf}}\}$ and $\{L_{\mathrm{lfd}}, \theta_{\mathrm{lfd}}\}$ and get the error equations (in the world coordinate frame):

$$\varepsilon_{\mathrm{f1}} = L_{\mathrm{lfd}} \cos(\theta_{\mathrm{lfd}} + \varepsilon_{\mathrm{f3}}) - L_{\mathrm{lf}} \cos(\theta_{\mathrm{lf}} + \varepsilon_{\mathrm{f3}}) - d \cos(\varepsilon_{\mathrm{f3}}) + d \tag{5.97a}$$

$$\varepsilon_{\mathrm{f2}} = L_{\mathrm{lfd}} \sin(\theta_{\mathrm{lfd}} + \varepsilon_{\mathrm{f3}}) - L_{\mathrm{lf}} \sin(\theta_{\mathrm{lf}} + \varepsilon_{\mathrm{f3}}) - d \sin(\varepsilon_{\mathrm{f3}}) \tag{5.97b}$$

$$\varepsilon_{\mathrm{f3}} = \phi_{\mathrm{l}} - \phi_{\mathrm{f}} \tag{5.97c}$$

Finally, differentiating Eqs. (5.97a)–(5.97c), we obtain the dynamic equations for ε_{f1}, ε_{f2}, and ε_{f3}:

$$\dot{\varepsilon}_{f1} = \dot{x}_{gl}\cos(\varepsilon_{f3}) - \dot{x}_{gf} + \dot{y}_{gl}[-L_{lf}\sin\zeta_f - \varepsilon_{f2}] + \dot{y}_{gf}\varepsilon_{f2} \tag{5.98a}$$

$$\dot{\varepsilon}_{f2} = \dot{x}_{gl}\sin(\varepsilon_{f3}) - D\dot{y}_{gf} + \dot{y}_{gl}(\varepsilon_{f1} - d) - \dot{y}_{gf}(\varepsilon_{f1} - d) + \dot{y}_{gl}L_{lf}\cos\zeta_f \tag{5.98b}$$

$$\dot{\varepsilon}_{f3} = (1/D)[\dot{x}_{gl}\, tg\, \psi_l - \dot{x}_{gf}\, tg\, \psi_f] \tag{5.98c}$$

We are now ready to apply the usual two-stage (kinematic, dynamic) backstep controller design.

5.6.2.1 Kinematic Controller

We select the candidate Lyapunov function [7]:

$$V = \frac{1}{2}(q_1\varepsilon_{f1}^2 + q_2\varepsilon_{f2}^2) + q_3(1 - \cos(\varepsilon_{f3})) \tag{5.99}$$

which is similar to Eq. (5.54), which possesses the first three properties of Lyapunov functions. The feedback control inputs $\dot{x}_{gf}$ and $\dot{y}_{gf}$ will be selected such that to make $\dot{V} < 0$. Differentiating Eq. (5.99) gives:

$$\dot{V} = q_1\varepsilon_{f1}\dot{\varepsilon}_{f1} + q_2\varepsilon_{f2}\dot{\varepsilon}_{f2} + q_3(\sin\varepsilon_{f3})\dot{\varepsilon}_{f3} \tag{5.100}$$

Introducing Eqs. (5.98a)–(5.98c) into Eq. (5.100), we find that the selection of $\dot{x}_{gf}$ and $\dot{y}_{gf}$ as:

$$\dot{x}_{gf} = K_{xf}\varepsilon_{f1} + \dot{x}_{gl}\cos(\varepsilon_{f3}) - \dot{y}_{gl}L_{lf}\sin(\zeta_f) \tag{5.101a}$$

$$\begin{aligned}\dot{y}_{gf} = &-\dot{y}_{gl} + \left(\frac{1}{d}\right)(K_x + \dot{x}_{gl})\sin(\varepsilon_{f3}) + \left(\frac{1}{d}\right)\dot{y}_{gl}L_{lf}\cos(\zeta_f) + \varepsilon_{f2}\\ &-\frac{1}{qd|\varepsilon_{f2}| + q_3}\{2q_3\dot{y}_{gl} + (1/d)q_3\dot{y}_{gl}L_{lf} + q_3|\varepsilon_{f2}| + qK_x|\varepsilon_{f2}|\}\end{aligned} \tag{5.101b}$$

with $q_1 = q_2 = q$ makes $\dot{V} < 0$. Indeed, introducing Eqs. (5.98a)–(5.98c), (5.101a), and (5.101b) into Eq. (5.100) yields:

$$\dot{V} < -\{qK_{xf}\varepsilon_{f1}^2 + qd\varepsilon_{f2}^2 + (1/d)[q_3(K_x + \dot{x}_{gl})\sin^2\varepsilon_{f3}]\}$$

Since $\dot{x}_{gl} > 0$, choosing $q > 0$, $q_3 > 0$, and $K_x > 0$ makes $\dot{V} < 0$.

5.6.2.2 Dynamic Controller

Use will be made of the dynamic model (Eqs. (3.60a)–(3.60d)). The desired velocity and steering angle v_{fd}, ψ_{fd} of the follower are given by the results of the kinematic controller. We define the error:

$$\tilde{\mathbf{z}}_{\mathrm{f}} = \mathbf{z}_{\mathrm{fd}} - \mathbf{z}_{\mathrm{f}}, \quad \mathbf{z}_{\mathrm{f}} = \begin{bmatrix} v_{\mathrm{f}} \\ \psi_{\mathrm{f}} \end{bmatrix}, \quad \mathbf{z}_{\mathrm{fd}} = \begin{bmatrix} v_{\mathrm{fd}} \\ \psi_{\mathrm{fd}} \end{bmatrix} \tag{5.102}$$

From Eqs. (3.60a)–(3.60c), applied to the follower WMR, we get:

$$\dot{v}_{\mathrm{f}} = \frac{1}{m}\left(-F_{\mathrm{f}} \sin \psi_{\mathrm{f}} + \frac{d}{D^2} v_{\mathrm{f}}^2 \, tg^2 \, \psi_{\mathrm{f}} + \frac{\tau_{\mathrm{f}}}{r}\right) \tag{5.103}$$

where $\tau_{\mathrm{f}} = rF_{\mathrm{d}}$ is the driving torque of the steering wheel of radius r, and the relation $\dot{x}_{g\mathrm{f}} = v_{\mathrm{f}}$ was used. Combining Eqs. (5.103) and (3.60d) gives:

$$\dot{\mathbf{z}}_{\mathrm{f}} = -\mathbf{A}(\mathbf{z}_{\mathrm{f}})\mathbf{z}_{\mathrm{f}} - \mathbf{G} + \mathbf{E}\boldsymbol{\tau} \tag{5.104}$$

where:

$$\mathbf{A}(\mathbf{z}_{\mathrm{f}}) = \begin{bmatrix} a_{11} & a_{12} \\ a_{21} & a_{22} \end{bmatrix}, \quad \mathbf{G} = \begin{bmatrix} g_1 \\ g_2 \end{bmatrix}, \quad \mathbf{E} = \begin{bmatrix} e_{11} & 0 \\ 0 & e_{22} \end{bmatrix}, \quad \boldsymbol{\tau} = \begin{bmatrix} \tau_{\mathrm{f}} \\ u_{\mathrm{s}} \end{bmatrix}$$

$$a_{11} = -(d/D^2)v_{\mathrm{f}} \, tg^2 \, \psi_{\mathrm{f}}, \quad a_{12} = a_{21} = 0, \quad a_{22} = 1/T$$

$$g_1 = (1/m)F_{\mathrm{f}} \sin \psi_{\mathrm{f}}, \quad g_2 = 0, \quad e_{11} = 1/rm, \quad e_{22} = K/T$$

Subtracting both sides of Eq. (5.104) from $\dot{\mathbf{z}}_{\mathrm{fd}}$ we get:

$$\dot{\tilde{\mathbf{z}}}_{\mathrm{f}} = \dot{\mathbf{z}}_{\mathrm{fd}} + \mathbf{A}(\mathbf{z}_{\mathrm{f}})\mathbf{z}_{\mathrm{f}} + \mathbf{G} - \mathbf{E}\boldsymbol{\tau}$$

Now, adding and subtracting $\mathbf{A}(\mathbf{z}_{\mathrm{f}})\mathbf{z}_{\mathrm{fd}}$ to the right-hand side of this equation yields:

$$\begin{aligned} \dot{\tilde{\mathbf{z}}}_{\mathrm{f}} &= -\mathbf{A}(\mathbf{z}_{\mathrm{f}})\tilde{\mathbf{z}}_{\mathrm{f}} + \dot{\mathbf{z}}_{\mathrm{fd}} + \mathbf{A}(\mathbf{z}_{\mathrm{f}})\mathbf{z}_{\mathrm{fd}} + \mathbf{G} - \mathbf{E}\boldsymbol{\tau} \\ &= -\mathbf{A}(\mathbf{z}_{\mathrm{f}})\tilde{\mathbf{z}}_{\mathrm{f}} + \mathbf{F}(\mathbf{x}^0) - \mathbf{E}\boldsymbol{\tau} \end{aligned} \tag{5.105a}$$

where:

$$\mathbf{F}(\mathbf{x}^0) = \mathbf{A}(\mathbf{z}_{\mathrm{f}})\mathbf{z}_{\mathrm{fd}} + \dot{\mathbf{z}}_{\mathrm{fd}} + \mathbf{G} \tag{5.105b}$$

and $\mathbf{x}^0 = [\varepsilon_{\mathrm{f}1}, \varepsilon_{\mathrm{f}2}, \varepsilon_{\mathrm{f}3}]^T$:

The torque $\boldsymbol{\tau}$ in Eq. (5.105a) can be found using the computed torque technique as:

$$\boldsymbol{\tau} = \mathbf{E}^{-1}[\mathbf{K}\tilde{\mathbf{z}}_{\mathrm{f}} + \mathbf{F}(\mathbf{x}^0)] \tag{5.106}$$

which is introduced into Eq. (5.105a) to give the closed-loop error equation:

$$\dot{\tilde{\mathbf{z}}}_{\mathrm{f}} = -(\mathbf{A} + \mathbf{K})\tilde{\mathbf{z}}_{\mathrm{f}} \tag{5.107}$$

It only remains to select $\mathbf{K}$ such that $\tilde{\mathbf{z}}_{\mathrm{f}}$ tends to zero asymptotically. We select the candidate Lyapunov function:

$$V_0 = V + \frac{1}{2}\tilde{\mathbf{z}}_{\mathrm{f}}^T\tilde{\mathbf{z}}_{\mathrm{f}} \tag{5.108}$$

where V is given by Eq. (5.99).

Differentiating V_0 and introducing the result into Eq. (5.107) gives:

$$\dot{V}_0 = \dot{V} - \tilde{\mathbf{z}}_{\mathrm{f}}^T(\mathbf{A} + \mathbf{K})\tilde{\mathbf{z}}_{\mathrm{f}}$$

Since $\dot{V} < 0$ by the kinematic controller design, we can assure that $\dot{V}_0 < 0$ by selecting the gain matrix $\mathbf{K}$ such that the matrix $\mathbf{A} + \mathbf{K}$ is positive definite, that is:

$$\mathbf{K} = \begin{bmatrix} K_1 + (d/D^2)v_{\mathrm{f}}\, tg^2\, \psi_{\mathrm{f}} & 0 \\ 0 & K_2 - 1/T \end{bmatrix} \quad \text{or} \quad \mathbf{A} + \mathbf{K} = \begin{bmatrix} K_1 & 0 \\ 0 & K_2 \end{bmatrix}$$

with $K_1 > 0$ and $K_2 > 0$. Then, we find:

$$\dot{V}_0 = \dot{V} - K_1\tilde{v}_{\mathrm{f}}^2 - K_2\tilde{\psi}_{\mathrm{f}}^2$$

This implies that the error $\tilde{\mathbf{z}}_{\mathrm{f}}$ in Eq. (5.105a) tends asymptotically to zero, that is, $v_{\mathrm{f}} \to v_{\mathrm{fd}}$ and $\psi_{\mathrm{f}} \to \psi_{\mathrm{fd}}$, as required. The function $\mathbf{F}(\mathbf{x}^0)$ in the control law Eq. (5.106) can be approximated by a neural network using a weight updating rule as described in Section 8.5 [7].

5.7 Omnidirectional Mobile Robot Control

We will consider the three-wheel omnidirectional dynamic robot model (Eqs. (3.77a) and (3.77b)) derived in Section 3.5:

$$\dot{\mathbf{x}} = \mathbf{A}(\mathbf{x})\mathbf{x} + \mathbf{B}(\mathbf{x})\mathbf{u}, \quad \mathbf{y} = \mathbf{C}\mathbf{x} \tag{5.109}$$

and will apply the resolved motion acceleration technique, combined with PI or PD control [18]. Solving Eq. (5.109) for $u_i (i = 1, 2, 3)$, we get the inverse dynamic resolved acceleration equations:

$$\begin{aligned} u_1 = &(\beta_1/6b_1)[\ddot{x}_0 - a_1\dot{x} + a_2\dot{\phi}\dot{y}] \\ &+ (\beta_3/6b_1)[\ddot{y}_0 - a_1\dot{y} - a_2\dot{\phi}\dot{x}] + (1/3b_2)(\ddot{\phi}_0 - a_3\dot{\phi}) \end{aligned} \tag{5.110a}$$

$$\begin{aligned} u_2 = &(\beta_2/6b_1)[\ddot{x}_0 - a_1\dot{x} + a_2\dot{\phi}\dot{y}] \\ &+ (\beta_4/6b_1)[\ddot{y}_0 - a_1\dot{y} - a_2\dot{\phi}\dot{x}] + (1/3b_2)(\ddot{\phi}_0 - a_3\dot{\phi}) \end{aligned} \tag{5.110b}$$

$$\begin{aligned} u_3 = &(\cos\phi/3b_1)[\ddot{x}_0 - a_1\dot{x} + a_2\dot{\phi}\dot{y}] \\ &+ (\sin\phi/3b_1)[\ddot{y}_0 - a_1\dot{y} - a_2\dot{\phi}\dot{x}] + (1/3b_2)[\ddot{\phi}_0 - a_3\dot{\phi}] \end{aligned} \tag{5.110c}$$

where:

$$\ddot{x}_0 = \ddot{x}_\mathrm{d} + K_{\mathrm{p}\dot{x}}\dot{\tilde{x}} + K_{\mathrm{i}\dot{x}}\int_0^t \dot{\tilde{x}}\,\mathrm{d}\tau \tag{5.111a}$$

$$\ddot{y}_0 = \ddot{y}_\mathrm{d} + K_{\mathrm{p}\dot{y}}\dot{\tilde{y}} + K_{\mathrm{i}\dot{y}}\int_0^t \dot{\tilde{y}}\,\mathrm{d}t \tag{5.111b}$$

$$\ddot{\phi}_0 = \ddot{\phi}_\mathrm{d} + K_{v\phi}\dot{\tilde{\phi}} + K_{\mathrm{p}\phi}\tilde{\phi} \tag{5.111c}$$

with:

$$\dot{\tilde{x}} = \dot{x}_\mathrm{d} - \dot{x}, \quad \dot{\tilde{y}} = \dot{y}_\mathrm{d} - \dot{y}, \quad \tilde{\phi} = \phi_\mathrm{d} - \phi$$

being the errors between the desired and actual trajectories, and $K_{\mathrm{p}\dot{x}}$, $K_{\mathrm{p}\dot{y}}$, $K_{\mathrm{p}\phi}$ being proportional gains, $K_{\mathrm{i}\dot{x}}$, $K_{\mathrm{i}\dot{y}}$ integral gains, and $K_{v\phi}$ the derivative (velocity) gain of ϕ. Note that the factors b_1 and b_2 in Eqs. (5.110a)–(5.110c) are always nonzero (see Eqs. (3.76a)–(3.76c)) and so the above resolved acceleration controllers u_1, u_2, and u_3 exist for all t.[5]

The block diagram of the overall feedback control WMR system is shown in Figure 5.16.

The above controller was applied to a real robot with physical parameters:

$$I_\mathrm{Q} = 11.25\,\mathrm{kg\,m^2},\ I_0 = 0.02108\,\mathrm{kg\,m^2},\ \beta = 5.983\times 10^{-6}\mathrm{kg\,m^2/s},\ M = 9.4\,\mathrm{kg},$$
$$L = 0.178\,\mathrm{m},\ r = 0.0245\,\mathrm{m},\ \text{and}\ K = 1.0$$

[5] All the coordinates are those of the center of gravity (and symmetry) Q. The index Q was dropped for notational simplicity.

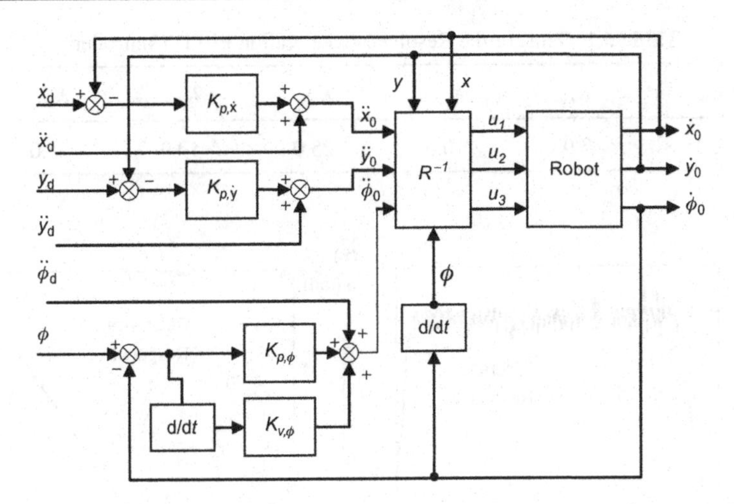

Figure 5.16 Resolved acceleration control system for the three-wheel omnidirectional robot (R^{-1} represents the robot inverse dynamics Eqs. (5.110a)–(5.110c)).

The initial state used is:

$$\mathbf{x}(0) = [x_Q(0), y_Q(0), \phi(0), \dot{x}_Q(0), \dot{y}_Q(0), \dot{\phi}(0)]^T = \mathbf{0}$$

A basic experiment was to check the WMRs holonomic property, that is, the ability of the robot to independently achieve translational and rotational motion around the center of gravity in the x-y plane.

To check this property, it was assumed that the robot must travel with a single azimuth $\psi_d = \pi/4$ rad (see Figure 3.6) for 20 s, but with zero rotational angle for 0–10 s, and a uniformly varying rotational angle from $\phi_d = 0$ rad to $\phi_d = \pi/2$ rad for 10–20 s. Although the moving velocity was to be $v_d = 0.05$ m/s in the steady state, a sinusoidal reference velocity was set for each 2 s at the starting and ending period. Using these v_d and ψ_d values, the corresponding $\dot{x}_d$ and $\dot{y}_d$ were derived from:

$$\dot{x}_d = v \cos \psi, \quad \dot{y}_d = v \sin \psi$$

where the positive rotational direction of motion is the counterclockwise direction. The desired accelerations $\ddot{x}_d$ and $\ddot{y}_d$ were derived by differentiation of $\dot{x}_d$ and $\dot{y}_d$. A selection of the gains used is given in Table 5.1.

The responses obtained for $\dot{x}$, $\dot{y}$, and ϕ, and the (x,y) trajectory are shown in Figure 5.17.

We observe that despite the occurrence of some oscillations in the velocities $\dot{x}$ and $\dot{y}$, the (x, y, ϕ) trajectory matches perfectly the desired trajectory (x_d, y_d, ϕ_d). Another experiment was carried out to follow a circular path with a radius less than half the distance between the wheels (say 0.1 m with distance between the wheels 0.356 m). Using the same gains as before, the results were very good, with perfect tracking of the circular path [18].

Table 5.1 Gains of the Resolved Acceleration PI/PD Controller

$K_{p\phi}$	$K_{v\phi}$	$K_{p\dot{x}}$	$K_{i\dot{x}}$	$K_{p\dot{y}}$	$K_{i\dot{y}}$
2.25	3.0	10.0	25.0	10.0	25.0

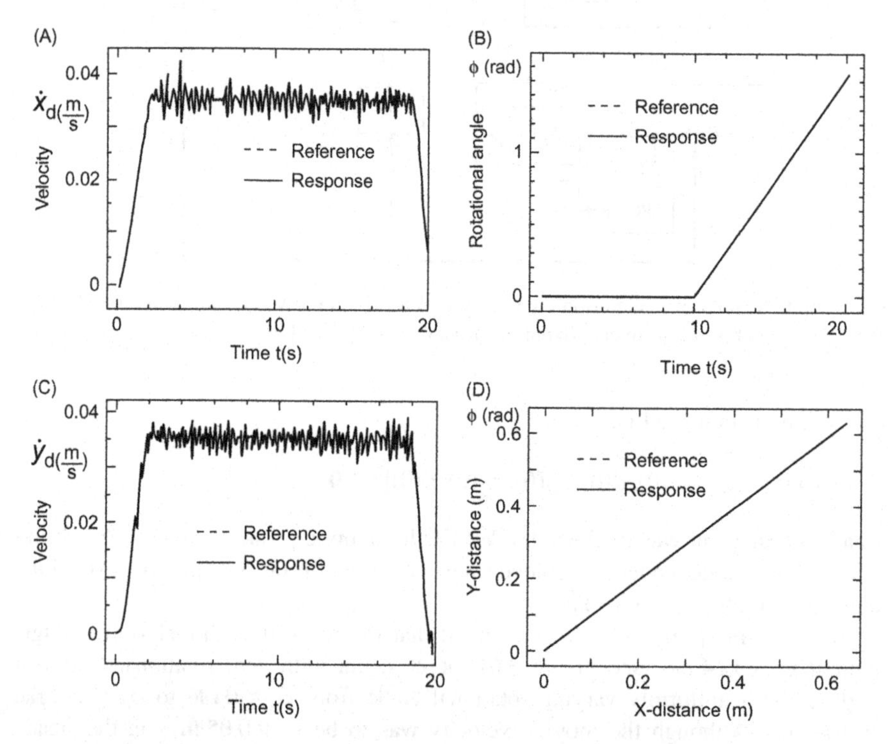

Figure 5.17 (A,C) Actual velocity responses compared to the desired velocity responses $\dot{x}_d$ and $\dot{y}_d$. (B) Trajectory of ϕ, (D) (x,y) trajectory.
Source: Reprinted from Ref. [18], with permission from Springer Science+Business Media BV.

Example 5.5

Apply the computed torque technique to derive a PD path tracking controller for a three-wheel omnidirectional robot.

Solution

We will work with the dynamic model derived for the omnidirectional robot of Figure 3.7 in Example 3.2, namely:

$$\mathbf{D}(\mathbf{p}_Q)\ddot{\mathbf{p}}_Q + \mathbf{C}(\phi)\dot{\mathbf{p}}_Q = \mathbf{E}\mathbf{v} \tag{5.112}$$

The computed torque control law has the form (Eq. (5.33)):

$$\mathbf{E}\mathbf{v} = \mathbf{D}(\mathbf{p}_Q)\mathbf{u} + \mathbf{C}(\phi)\dot{\mathbf{p}}_Q \tag{5.113}$$

and the PD control has the form:

$$\mathbf{u} = -\mathbf{K}_p\tilde{\mathbf{p}}_Q - \mathbf{K}_d\dot{\tilde{\mathbf{p}}}_Q + \ddot{\mathbf{p}}_{Q,d} \tag{5.114}$$

where:

$$\tilde{\mathbf{p}}_Q(t) = \mathbf{p}_Q(t) - \mathbf{p}_{Q,d}(t) \tag{5.115}$$

is the tracking error of the resulting path from the desired path $\mathbf{p}_{Q,d}(t)$. Combining Eqs. (5.113) and (5.114) and introducing into Eq. (5.112), we get the closed-loop system:

$$\mathbf{D}\ddot{\mathbf{p}}_Q = -\mathbf{D}(\mathbf{K}_p\tilde{\mathbf{p}}_Q + \mathbf{K}_d\dot{\tilde{\mathbf{p}}}_Q) + \mathbf{D}\ddot{\mathbf{p}}_{Q,d}$$

or the tracking error dynamic equation:

$$\ddot{\tilde{\mathbf{p}}}_Q + \mathbf{K}_d\dot{\tilde{\mathbf{p}}}_Q + \mathbf{K}_p\tilde{\mathbf{p}}_Q = \mathbf{0} \tag{5.116}$$

since the matrix $\mathbf{D}$ is nonsingular. We see that the origin $\tilde{\mathbf{p}}_Q = \mathbf{0}$ is an equilibrium point of this system. Therefore, selecting proper values of $\mathbf{K}_d$ and $\mathbf{K}_p$, we can assure that $\tilde{\mathbf{p}}_Q(t)$ tends to zero, that is, $\mathbf{p}_Q(t) \to \mathbf{p}_{Q,d}(t)$, asymptotically. To this end, we use the following candidate Lyapunov function:

$$V(\tilde{\mathbf{p}}_Q) = \frac{1}{2}\tilde{\mathbf{p}}_Q^T(\mathbf{K}_p + \gamma\mathbf{K}_d)\tilde{\mathbf{p}} + \frac{1}{2}\dot{\tilde{\mathbf{p}}}_Q^T\dot{\tilde{\mathbf{p}}}_Q + \gamma\tilde{\mathbf{p}}_Q^T\dot{\tilde{\mathbf{p}}}_Q \tag{5.117}$$

which, for sufficiently small constant $\gamma > 0$, satisfies the first two properties of Lyapunov functions.

We will now examine under what conditions $\dot{V}(\tilde{\mathbf{p}}_Q)$ is negative. The time derivative of $V(\tilde{\mathbf{p}}_Q)$ along the trajectory of $\ddot{\mathbf{p}}_Q(t) = -\mathbf{K}_d\dot{\tilde{\mathbf{p}}}_Q(t) - \mathbf{K}_p\tilde{\mathbf{p}}_Q(t)$ is found to be:

$$\begin{aligned}
\dot{V} &= \tilde{\mathbf{p}}_Q^T(\mathbf{K}_p + \gamma\mathbf{K}_d)\dot{\tilde{\mathbf{p}}}_Q + \dot{\tilde{\mathbf{p}}}_Q{}^T\ddot{\tilde{\mathbf{p}}}_Q + \gamma\dot{\tilde{\mathbf{p}}}_Q^T\dot{\tilde{\mathbf{p}}}_Q + \gamma\tilde{\mathbf{p}}_Q{}^T\ddot{\tilde{\mathbf{p}}}_Q \\
&= \tilde{\mathbf{p}}_Q^T(\mathbf{K}_p + \gamma\mathbf{K}_d)\dot{\tilde{\mathbf{p}}}_Q + \dot{\tilde{\mathbf{p}}}^T(-\mathbf{K}_d\dot{\tilde{\mathbf{p}}}_Q - \mathbf{K}_p\tilde{\mathbf{p}}_Q) \\
&\quad + \gamma\tilde{\mathbf{p}}_Q^T(-\mathbf{K}_d\dot{\tilde{\mathbf{p}}}_Q - \mathbf{K}_p\tilde{\mathbf{p}}_Q) + \gamma\dot{\tilde{\mathbf{p}}}_Q{}^T\dot{\tilde{\mathbf{p}}}_Q \\
&= -\dot{\tilde{\mathbf{p}}}_Q{}^T(\mathbf{K}_d - \gamma\mathbf{I})\dot{\tilde{\mathbf{p}}}_Q - \gamma\tilde{\mathbf{p}}_Q^T\mathbf{K}_p\tilde{\mathbf{p}}_Q
\end{aligned}$$

Choosing $\mathbf{K}_d - \gamma\mathbf{I} > \mathbf{0}$ and $\mathbf{K}_p > \mathbf{0}$ (positive definite), we get $\dot{V} < 0$ and so $\tilde{\mathbf{p}}_Q \to \mathbf{0}$ asymptotically.

Table 5.2 Gains of the Computed Torque PD Controller

K_{px}	K_{py}	$K_{p\phi}$	K_{dx}	K_{dy}	$K_{d\phi}$
40	30	20	70	55	10

The performance of the controller was tested on the WMR of Section 5.7 with the same values of physical parameters and the same desired path using the PD gains given in Table 5.2.

The responses obtained for the (x, y, ϕ) and $(\dot{x}, \dot{y})$ trajectories are similar to those shown in Figure 5.17. As an exercise, the reader may test the performance of the controller for a circular desired path with a given radius R (e.g., $R = 10$ or 40 cm). A parameterized form of the circle equation, namely, $x_d(t) = R \cos a_d(t)$, $y_d(t) = R \sin a_d(t)$ can be used where $a_d(t)$ is the angle of the point (x_d, y_d) with reference to the x-axis of the world coordinate frame. In this case, a suitable polynomial or other series representation of the angle $a(t)$ can be used. A more complex desired path that might be examined is an 8 shaped path of a given size.

References

[1] Ogata K. State space analysis of control systems. Upper Saddle River, NJ: Prentice Hall; 1997.

[2] Asada H, Slotine JJ. Robot analysis and control. New York, NY: Wiley; 1986.

[3] Spong MW, Vidyasagar M. Robot dynamics and control. New York, NY: Wiley; 1989.

[4] Wolovich W. Robotics: Basic Analysis and Design. Birmingham, UK: Holt Rinehart and Winston, Dreyden Press;1987.

[5] Kanayama Y, Kimura Y, Noguchi T. A stable tracking control method for a nonholonomic mobile robot. IEEE Trans Robot Autom 1991;7:1236–41.

[6] Yiaoping Y, Yamamoto Y. Dynamic feedback control of vehicles with two steerable wheels. Proceedings of IEEE international conference on robotics and automation. Minneapolis, MN; 1996. 12(1), p. 1006–1010.

[7] Panimadai Ramaswamy SA, Balakrishnan SN. Formation control of car-like mobile robots: a Lyapunov function based approach. Proceedings of 2008 American Control Conference. Seattle, Washington; June 11–13, 2008.

[8] Tian Y, Sidek N, Sarkar N. Modeling and Control of a nonholonomic wheeled mobile robot with wheel slip dynamics. Proceedings of IEEE symposium on computational intelligence in control and automation. Nashville, TN; March 30–April 2, 2009. p. 7–14.

[9] Chang CF, Huang CI, Fu LC. Nonlinear control of a wheeled mobile robot with nonholonomic constraints. Proceedings of 2004 IEEE International conference on systems, man, and cybernetics. The Hague, The Netherlands; October 10–13, 2004. p. 5404–9.

[10] Samson C, Ait Abderrahim K. Feedback control of nonholonomic wheeled cart in Cartesian space. Proceedings of IEEE Conference on robotics and automation. Sacramento, CA; 1990. p. 1136–41.

[11] Velagic J, Lacevic B, Osmic N. Nonlinear motion control of mobile robot dynamic model [Chapter 27] In: Jing X-J, editor. Motion Planning. In Tech, Open Books; 2008. p. 534–56.

[12] Zhang Y, Hong D, Chung JH, Velinsky SA. Dynamic model based robust tracking control of a differentially steered wheeled mobile robot. Proceedings of american control conference (ACC '88). Philadelphia, PA; June 1988. p. 850–5.

[13] Ashoorizad M, Barzamini R, Afshar A, Zouzdani J. Model reference adaptive path following for wheeled mobile robots. Proceedings of international conference on information and automation (IEEE/ICIA'06). Colombo, Sri Lanka; 2006. p. 289–94.

[14] Gholipour A, Dehgham SM, Ahmadabadi MN. Lyapunov based tracking control of nonholonomic mobile robot. Proceedings of 10th Iranian conference on electrical engineering. Tabeiz, Iran; 2002. 3, p. 262–69.

[15] Aicardi M, Casalino G, Bicchi A, Balestrino A. Closed-loop steering of unicycle-like vehicles via Lyapunov techniques. IEEE Robot Autom Mag 1995; March:27–35.

[16] Lee S, Kim M, Youm Y, Chung W. Control of a car-like mobile robot for parking problem. Proceedings of 1999 IEEE international conference on robotics and automation. Detroit, MI; May 1999. p. 1–5.

[17] Lee S, Youm Y, Chung. Control of car-like mobile robots for posture stabilization. Proceedings IEEE/RSJ international conference on intelligent robots and systems (IROS'99). Kyongju, Korea; October 1999. p. 1745–50.

[18] Watanabe K, Shiraishi Y, Tang J, Fukuda T, Tzafestas SG. Autonomous control for an omnidirectional mobile robot with feedback control system [Chapter 13] In: Tzafestas SG, editor. Advances in intelligent autonomous systems. Boston / Dordrecht: Kluwer; 1999. p. 289–308.

6 Mobile Robot Control II: Affine Systems and Invariant Manifold Methods

6.1 Introduction

The control of wheeled mobile robots (WMRs) is a challenging subject for both its theoretical and practical value. Mobile robots, omnidirectional and nonholonomic, are highly nonlinear, and especially nonholonomic constraints have motivated the development of highly nonlinear control techniques. Most of the control results available in the literature were developed for the two major categories of nonholonomic WMRs, namely:

- *Unicycle-type WMRs* that involve two independently driven wheels on a common axis and one or more passive/castor wheels.
- *Rear-drive car-like WMRs* that involve a motorized real-axis at the rear of the chassis, and one (or two) orientable front steering wheel(s).

As discussed in Chapter 5, the three control problems of mobile robots are as follows:

1. *Path tracking*: Given a curved planar path, the WMR has to follow this path with a prespecified longitudinal velocity.
2. *Trajectory tracking*: Here, the velocity is not prespecified. The WMR has, in addition to path following, the goal to control the distance gone along the curve.
3. *Posture stabilization*: The goal is to stabilize at zero the posture (position and orientation) of the WMR with respect to a fixed coordinate frame (e.g., in car parking).

In Chapter 5, we have presented the basic computed torque/motion and Lyapunov-based control design techniques drawn from standard robotic manipulator control theory. The goal of this present chapter is to provide further advanced WMR control methods either utilizing the state feedback linearization of affine systems (before a linear controller design), or following the invariant manifolds methodology (which leads directly to overall nonlinear controllers).

In particular, the chapter:

- provides an introduction to the fundamental concepts of affine control systems, invariant/attractive manifolds, and related extended Lyapunov stability theory;

Introduction to Mobile Robot Control. DOI: http://dx.doi.org/10.1016/B978-0-12-417049-0.00006-7

- presents a number of feedback linearization (and linear trajectory tracking) controllers for differential-drive and car-like WMRs;
- derives kinematic and dynamic controllers of Brockett integrator and (2,n)-chained models of differential-drive and car-like WMRs.

6.2 Background Concepts

6.2.1 Affine Dynamic Systems

As we saw in Chapters 2 and 3, mobile robots belong to the *affine systems* class of nonlinear systems described by the general state-space equation [1–4]:

$$\begin{aligned}\dot{\mathbf{x}} &= \mathbf{g}_0(\mathbf{x}) + \sum_{i=1}^{m}\mathbf{g}_i(\mathbf{x})u_i, \quad \mathbf{x}\in R^n, \quad \mathbf{g}_i(\mathbf{x})\in R^n \\ &= \mathbf{g}_0(\mathbf{x}) + \mathbf{G}(\mathbf{x})\mathbf{u}\end{aligned} \tag{6.1}$$

with:

$$\begin{aligned}\mathbf{x} &= [x_1, x_2, \dots, x_n]^{\mathrm{T}} \\ \mathbf{u} &= [u_1, u_2, \dots, u_m]^{\mathrm{T}}, \quad m \le n \\ \mathbf{G}(x) &= [\mathbf{g}_1(\mathbf{x}) \vdots \mathbf{g}_2(\mathbf{x}) \vdots \dots \vdots \mathbf{g}_m(\mathbf{x})] \\ \mathbf{g}_i(\mathbf{x}) &= [g_{i1}(\mathbf{x}), g_{i2}(\mathbf{x}), \dots, g_{in}(\mathbf{x})]^{\mathrm{T}}\end{aligned}$$

where the drift term $\mathbf{g}_0(\mathbf{x})$ represents the general kinematic constraints of the system.

Consider a nonlinear mapping from the n-dimensional Euclidean space R^n to itself:

$$\mathbf{z} = \varphi(\mathbf{x}), \quad \mathbf{x}\in R^n, \quad \mathbf{z}\in R^n \tag{6.2}$$

where $\mathbf{z}$ is a new vector, and $\varphi(\mathbf{x})\in R^n$ is a vector function (field):

$$\varphi(\mathbf{x}) = \begin{bmatrix}\phi_1(\mathbf{x}) \\ \phi_2(\mathbf{x}) \\ \vdots \\ \phi_n(\mathbf{x})\end{bmatrix}, \quad \mathbf{x} = \begin{bmatrix}x_1 \\ x_2 \\ \vdots \\ x_n\end{bmatrix}, \quad \mathbf{z} = \begin{bmatrix}z_1 \\ z_2 \\ \vdots \\ z_n\end{bmatrix}$$

with the following properties:

1. The function $\varphi(\mathbf{x})$ is invertible, that is, there exists a function $\varphi^{-1}(\mathbf{z})$ such that:

$$\varphi^{-1}(\varphi(\mathbf{x})) = \mathbf{x}, \quad \varphi(\varphi^{-1}(\mathbf{z})) = \mathbf{z} \tag{6.3}$$

   for all $\mathbf{x}\in R^n$ and $\mathbf{z}\in R^n$.

2. Both functions $\varphi(\mathbf{x})$ and $\varphi^{-1}(\mathbf{z})$ have continuous partial derivatives of any order (i.e., they are *smooth* functions).

Definition 6.1 (*Diffeomorphism*)

A function of the type (6.2) with the above two properties is called *diffeomorphism.* Sometimes, it is not possible to find a diffeomorphism valid for all $\mathbf{x}$ in a domain of interest. In these cases, we can define a diffeomorphism in the vicinity of a given point $\mathbf{x}^0$ in this domain. A transformation of this type is called a *local diffeomorphism.*

The condition for the existence of a local diffeomorphism in the vicinity of a point $\mathbf{x}=\mathbf{x}^0$ is the nonsingularity of the Jacobian matrix $\partial\varphi(\mathbf{x})/\partial\mathbf{x}$ at the point $\mathbf{x}^0$. ■

Definition 6.2 (*Lie derivative*)

Given a smooth real-valued scalar function:

$$s(\mathbf{x}) = s(x_1, x_2, ..., x_n) \in R \tag{6.4}$$

and a vector-valued function (vector field):

$$\mathbf{f}(\mathbf{x}) = \begin{bmatrix} f_1(x_1, ..., x_n) \\ \vdots \\ f_n(x_1, ..., x_n) \end{bmatrix} \in R^n \tag{6.5}$$

of the vector variable $\mathbf{x} = [x_1, x_2, \ldots, x_n]^T$, the scalar function $L_{\mathbf{f}}\mathbf{s}(x)$ is defined as:

$$\begin{aligned} L_{\mathbf{f}}s(\mathbf{x}) &= L_{\mathbf{f}}s(x_1, x_2, ..., x_n) \\ &= \left[\frac{\partial s(\mathbf{x})}{\partial \mathbf{x}}\right]\mathbf{f}(\mathbf{x}) = (\nabla s(\mathbf{x}))\mathbf{f}(\mathbf{x}) \\ &= \sum_{i=1}^{n} \frac{\partial s}{\partial x_i} f_i(x_1, x_2, ..., x_n) \end{aligned} \tag{6.6}$$

where:

$$\frac{\partial s(\mathbf{x})}{\partial \mathbf{x}} = \nabla s(\mathbf{x}) = \left[\frac{\partial s(\mathbf{x})}{\partial x_1}, \frac{\partial s(\mathbf{x})}{\partial x_2}, \ldots, \frac{\partial s(\mathbf{x})}{\partial x_n}\right]$$

is called the *Lie derivative* of $s(\mathbf{x})$ along the field $\mathbf{f}(\mathbf{x})$. ■

Clearly, the function $L_{\mathbf{f}}s(\mathbf{x})$ represents the projection of the gradient vector $\nabla s(\mathbf{x})$ along the vector $\mathbf{f}(\mathbf{x})$, that is, the directional derivative of $s(\mathbf{x})$ along the direction of the vector $\mathbf{f}(\mathbf{x})$.

Successive application of Eq. (6.6) gives, for example, the Lie derivative of $s(\mathbf{x})$ first along $\mathbf{f}(\mathbf{x})$ and then along another function $\mathbf{g}(\mathbf{x})$, that is:

$$L_{\mathbf{g}}L_{\mathbf{f}}s(\mathbf{x}) = \left[\frac{\partial L_{\mathbf{f}}s(\mathbf{x})}{\partial \mathbf{x}}\right]\mathbf{g}(\mathbf{x})$$

In general:

$$L_{\mathbf{f}}^{k}s(\mathbf{x})=\left[\frac{\partial L_{\mathbf{f}}^{k-1}s(\mathbf{x})}{\partial \mathbf{x}^{k-1}}\right]\mathbf{f}(\mathbf{x}) \tag{6.7}$$

For example, given $\dot{\mathbf{x}}=\mathbf{f}(\mathbf{x})$, $y=h(\mathbf{x})$ we find:

$$\dot{y}=\left[\frac{\partial h}{\partial \mathbf{x}}\right]\dot{\mathbf{x}}=L_{\mathbf{f}}h, \quad \ddot{y}=\left[\frac{\partial (L_{\mathbf{f}}h)}{\partial \mathbf{x}}\right]\dot{\mathbf{x}}=L_{\mathbf{f}}^{2}h$$

and so on.

Using the Lie derivative, we can introduce the concept of *relative degree* of a system, as follows:

Definition 6.3 (*Relative degree*)

We say that the affine system:

$$\dot{\mathbf{x}}=\mathbf{f}(\mathbf{x})+\mathbf{g}(\mathbf{x})u, \quad y=h(\mathbf{x}), \quad u\in R \tag{6.8}$$

has *relative degree* r at the point $\mathbf{x}^0$ if:

(a) $L_{\mathbf{g}}L_{\mathbf{f}}^{k}h(\mathbf{x})=0$ for all $\mathbf{x}$ in the vicinity of $\mathbf{x}^0$ and all $k<r-1$
(b) $L_{\mathbf{g}}L_{\mathbf{f}}^{r-1}h(\mathbf{x}^0)\neq 0$ ■

It is easy to verify that the relative degree r of a system is equal to the number of times the output y must be differentiated such that the input $u(t)$ appears in the derivative equation. This shows the importance of the functions $h(\mathbf{x}),\ L_{\mathbf{f}}h(\mathbf{x}),\ldots,L_{\mathbf{f}}^{r-1}h(\mathbf{x})$, because they can be used for finding a local transformation in the vicinity of $\mathbf{x}^0$, where $\mathbf{x}^0$ is a point for which the relation:

$$L_{\mathbf{g}}L_{\mathbf{f}}^{r-1}h(\mathbf{x}^0)\neq \mathbf{0} \tag{6.9}$$

holds.

As an example, consider the system:

$$\dot{\mathbf{x}}=\mathbf{f}(\mathbf{x})+\mathbf{g}(\mathbf{x})u, \quad y=h(\mathbf{x})=x_3$$

where:

$$\mathbf{x}=\begin{bmatrix}x_1\\x_2\\x_3\end{bmatrix}, \quad \mathbf{f}(\mathbf{x})=\begin{bmatrix}0\\x_1^2+\sin x_2\\-x_2\end{bmatrix}, \quad \mathbf{g}(\mathbf{x})=\begin{bmatrix}e^{x_2}\\1\\0\end{bmatrix}$$

For this system, we have:

$$\frac{\partial h}{\partial \mathbf{x}} = [0 \quad 0 \quad 1], \quad L_{\mathbf{g}}h(\mathbf{x}) = 0, \quad L_{\mathbf{f}}h(\mathbf{x}) = -x_2$$

$$\frac{\partial L_{\mathbf{f}}h}{\partial \mathbf{x}} = [0 \quad -1 \quad 0], \quad L_{\mathbf{g}}L_{\mathbf{f}}h(\mathbf{x}) = -1$$

Thus, the system has relative degree $r = 2$ at any point $\mathbf{x}^0$. If we choose $y = h(\mathbf{x}) = x_2$, we get $L_{\mathbf{g}}h(\mathbf{x}) = 1$, and so the relative degree of the system is $r = 1$ at any point $\mathbf{x}^0$.

Definition 6.4 (*Flow induced by a vector field*)
Consider the system:

$$\dot{\mathbf{x}} = \mathbf{f}(\mathbf{x}), \quad \mathbf{x} \in X \subset R^n \tag{6.10}$$

and assume that there is a unique solution $\mathbf{x}(\mathbf{x}_0, t)$ for each initial state $\mathbf{x}(\mathbf{x}_0, 0) = \mathbf{x}_0$ (which is not necessary to be found analytically). Then, the mapping:

$$(\mathbf{x}_0, t) \to \mathbf{x}(\mathbf{x}_0, t) \tag{6.11}$$

is called the *flow* (or *dynamic*) *system* induced by the vector field $\mathbf{f}$. ■

Essential properties of the flow are the following:

1. $\mathbf{x}(\mathbf{x}_0, 0) = \mathbf{x}_0$ for each $\mathbf{x}_0 \in X$
2. $\mathbf{x}(\mathbf{x}_0, t + s) = \mathbf{x}(\mathbf{x}(\mathbf{x}_0, t), s) = \mathbf{x}(\mathbf{x}(\mathbf{x}_0, s), t)$ for all $\mathbf{x}_0 \in X$ and s, t on the real line R
3. $\frac{\partial}{\partial t}\mathbf{x}(\mathbf{x}_0, t) = \mathbf{f}(\mathbf{x}(\mathbf{x}_0, t))$

Denote by $\exp t\mathbf{f}$ the transformation on X induced by $\mathbf{f}$. Clearly, $\exp \mathbf{f}$ maps each initial point onto $\mathbf{x}(\mathbf{x}_0, t)$. From properties 1 and 2 it follows that $\exp 0\mathbf{f}$ = identity, and $\exp(t + s)\mathbf{f} = (\exp t\mathbf{f})(\exp s\mathbf{f})$, which denotes the *mapping composition.*

Since $\exp 0\mathbf{f} = \mathbf{I}$, it follows that:

$$(\exp t\mathbf{f})^{-1} = \exp - t\mathbf{f}$$

and so, the mapping$\{\exp t\mathbf{f}: t \in R\}$ constitutes a commutative group of (local) diffeomorphisms. This group is called the *1-parameter group of diffeomorphisms induced by* $\mathbf{f}$.

If $\mathbf{f}$ is linear vector field, that is, $\mathbf{f} = \mathbf{A}\mathbf{x}$ on X, then $\mathbf{x}(0, t) = e^{t\mathbf{A}}\mathbf{x}_0$ with $e^{t\mathbf{A}} = \sum_{k=0}^{\infty}(t^k/k!)\mathbf{A}^k$. Therefore, in this case, $\exp t\mathbf{f}$ is a linear transformation on X equal to the exponential of a linear mapping (matrix) $\mathbf{A}$.

Definition 6.5 (*Lie bracket*)

Let $\mathbf{f}(\mathbf{x})$ and $\mathbf{g}(\mathbf{x})$ be two vector functions (fields) of the vector variable $\mathbf{x} = [x_1, x_2, ..., x_n]^{\mathrm{T}}$. Then, we define a new vector function of $\mathbf{x}$ denoted by $[\mathbf{f}, \mathbf{g}](\mathbf{x})$ and called the *Lie bracket* (or *Lie product*) as:

$$[\mathbf{f}, \mathbf{g}](\mathbf{x}) = \left(\frac{\partial \mathbf{g}}{\partial \mathbf{x}}\right)\mathbf{f}(\mathbf{x}) - \left(\frac{\partial \mathbf{f}}{\partial \mathbf{x}}\right)\mathbf{g}(\mathbf{x}) \tag{6.12}$$

where $\partial \mathbf{f}/\partial \mathbf{x}$ and $\partial \mathbf{g}/\partial \mathbf{x}$ are the Jacobian matrices of $\mathbf{f}(\mathbf{x})$ and $\mathbf{g}(\mathbf{x})$, with elements $(\partial \mathbf{f}/\partial \mathbf{x})_{ij} = \partial f_i/\partial x_j$ and $(\partial \mathbf{g}/\partial \mathbf{x})_{ij} = \partial g_i/\partial x_j$, respectively.■

For example, the Lie bracket of the system $\dot{x}_1 = -2x_1 + bx_2 + \sin x_1$, $\dot{x}_2 = -x_2 \cos x_1 + u \cos 2x_1$, which can be written as:

$$\dot{\mathbf{x}} = \mathbf{f}(\mathbf{x}) + \mathbf{g}(\mathbf{x})u, \quad \mathbf{f}(\mathbf{x}) = \begin{bmatrix} -2x_1 + bx_2 + \sin x_1 \\ -x_2 \cos x_1 \end{bmatrix}, \quad \mathbf{g}(\mathbf{x}) = \begin{bmatrix} 0 \\ \cos 2x_1 \end{bmatrix}$$

is:

$$\begin{aligned}
[\mathbf{f}, \mathbf{g}](\mathbf{x}) &= \begin{bmatrix} 0 & 0 \\ -2 \sin x_1 & 0 \end{bmatrix} \begin{bmatrix} -2x_1 + bx_2 + \sin x_1 \\ -x_2 \cos x_1 \end{bmatrix} \\
&\quad - \begin{bmatrix} -2 + \cos x_1 & b \\ x_2 \sin x_1 & -\cos x_1 \end{bmatrix} \begin{bmatrix} 0 \\ \cos 2x_1 \end{bmatrix} \\
&= \begin{bmatrix} b \cos 2x_1 \\ \cos x_1 \cos 2x_1 - 2(\sin 2x_1)(-2x_1 + bx_2 + \sin x_1) \end{bmatrix}
\end{aligned}$$

The Lie bracket can be applied successively, using the following notation:

$$\begin{aligned}
&ad_{\mathbf{f}}\mathbf{g}(\mathbf{x}) = [\mathbf{f}, \mathbf{g}](\mathbf{x}) \\
&ad_{\mathbf{f}}^2\mathbf{g}(\mathbf{x}) = [\mathbf{f}, [\mathbf{f}, \mathbf{g}]](\mathbf{x}) \\
&\quad \vdots \\
&ad_{\mathbf{f}}^k\mathbf{g}(\mathbf{x}) = [\mathbf{f}, ad_{\mathbf{f}}^{k-1}\mathbf{g}(\mathbf{x})](\mathbf{x})
\end{aligned} \tag{6.13}$$

with initial condition $ad_{\mathbf{f}}^0\mathbf{g}(\mathbf{x}) = \mathbf{g}(\mathbf{x})$. Clearly, the Lie bracket has the properties:

1. $[\mathbf{f}, \mathbf{g}](\mathbf{x}) = -[\mathbf{g}, \mathbf{f}](\mathbf{x})$
2. If $[\mathbf{f}, \mathbf{g}](\mathbf{x}) = \mathbf{0}$, then $(\exp t\mathbf{f})(\exp s\mathbf{g}) = (\exp s\mathbf{g})(\exp t\mathbf{f})$ and conversely (*communication of flows*)

For example, if $\mathbf{f}(\mathbf{x}) = a$ and $\mathbf{g}(\mathbf{x}) = b$ are two arbitrary constant fields, and so $[\mathbf{f}, \mathbf{g}](\mathbf{x}) = \mathbf{0}$, then $(\exp t\mathbf{f})(\mathbf{x}) = x + ta$ and $(\exp s\mathbf{g}) = x + sb$. Therefore:

$$\begin{aligned}(\exp t\mathbf{f})(\exp s\mathbf{g})(\mathbf{x}) &= (x + sb) + ta = (x + ta) + sb \\ &= (\exp s\mathbf{g})(\exp t\mathbf{f})(\mathbf{x})\end{aligned}$$

which confirms that the flows commute.

Definition 6.6 (*Involutive set*)

A set of n-dimensional column vector functions (fields) $\boldsymbol{\Delta}(\mathbf{x}) = \{\mathbf{X}_1(\mathbf{x}), \mathbf{X}_2(\mathbf{x}), \ldots, \mathbf{X}_m(\mathbf{x})\}$ for which the matrix $[\mathbf{X}_1(\mathbf{x}), \mathbf{X}_2(\mathbf{x}), \ldots, \mathbf{X}_m(\mathbf{x})]$ has rank m at the point $\mathbf{x} = \mathbf{x}^0$ is called *involutive* in the vicinity of $\mathbf{x}^0$, if for every pair (i,j), $(i,j = 1, 2, ..., m)$, the matrix:

$$\overline{\boldsymbol{\Delta}}(\mathbf{x}) = [\mathbf{X}_1(\mathbf{x}), \mathbf{X}_2(\mathbf{x}), \ldots, \mathbf{X}_m(\mathbf{x}), [\mathbf{X}_i, \mathbf{X}_j](\mathbf{x})] \tag{6.14}$$

has also rank m for all $\mathbf{x}$ in a vicinity of $\mathbf{x}^0$. ■

The *field* distribution $\boldsymbol{\Delta}(\mathbf{x}) = \{\mathbf{X}_1(\mathbf{x}), \mathbf{X}_2(\mathbf{x}), \ldots, \mathbf{X}_m(\mathbf{x})\}$ is said to be *nonsingular* if $\dim \boldsymbol{\Delta}(\mathbf{x}) = r$ is constant for all $\mathbf{x}$, in which case r is called the *dimension of the distribution.* The distribution $\overline{\boldsymbol{\Delta}}$ in Eq. (6.14) that has the same rank as $\boldsymbol{\Delta}$ for all $\mathbf{x}$ in a vicinity of $\mathbf{x}^0$ is called the *involutive closure* of Δ *under the Lie bracket operation.*

The concepts of Lie derivative, Lie bracket, and relative degree play a key role in the analysis and design of affine control systems, namely, in the solution of state feedback linearization, total stabilization via state feedback, adaptive control, robust control, and controllability/observability problems.

A central result used in the above studies is the *Frobenius theorem* which gives the necessary and sufficient condition for a set (distribution) of m vector fields to be completely integrable. We will formulate this theorem.

Let a driftless system be of the form:

$$\dot{\mathbf{x}} = \sum_{i=1}^{m} \mathbf{g}_i(\mathbf{x}) u_i, \quad \mathbf{x} \in X \subseteq R^n \tag{6.15}$$

From Definition 6.6, the distribution:

$$\boldsymbol{\Delta}(\mathbf{x}) = \{\mathbf{g}_1(\mathbf{x}), \mathbf{g}_2(\mathbf{x}), \ldots, \mathbf{g}_m(\mathbf{x})\} \tag{6.16a}$$

is involutive if for each Lie bracket $[\mathbf{g}_i, \mathbf{g}_j](\mathbf{x})$, there exist m real coefficients β_k $(k = 1, 2, ..., m)$ such that:

$$[\mathbf{g}_i, \mathbf{g}_j](\mathbf{x}) = \sum_{k=1}^{m} \beta_k \mathbf{g}_k(\mathbf{x}) \tag{6.16b}$$

which means that every Lie bracket can be expressed as a linear combination of the system vector fields, and so it already belongs to Δ. In other words, the Lie brackets cannot escape Δ and produce new directions of motion. It is noted that because of the property $[\mathbf{g}_i, \mathbf{g}_j](\mathbf{x}) = -[\mathbf{g}_j, \mathbf{g}_i](\mathbf{x})$ $(i,j = 1,2,\ldots,m)$, it is not necessary to consider all m^2 possible brackets, but it is sufficient to consider $\binom{m}{2} = \frac{m(m-1)}{2}$ Lie brackets. Using the above concepts, the Frobenius theorem is stated as follows.

Theorem 6.1 (*Frobenius theorem*)
A smooth affine system with a nonsingular distribution is completely integrable if and only if it is involutive. ■

Example 6.1

Consider a two-input driftless affine system with:

$$\mathbf{g}_1(\mathbf{x}) = \begin{bmatrix} x_2 \\ 0 \\ 1 \end{bmatrix}, \quad \mathbf{g}_2(\mathbf{x}) = \begin{bmatrix} x_3 \\ 1 \\ 0 \end{bmatrix}$$

The dimension of $\Delta = \{\mathbf{g}_1, \mathbf{g}_2\}$ is 2 at any point $\mathbf{x} \in R^3$. The Lie product is found to be:

$$[\mathbf{g}_1, \mathbf{g}_2](\mathbf{x}) = \mathbf{0}$$

Therefore, Δ is involutive, and hence Δ and the system is integrable. Indeed, the system gives:

$$\dot{x}_1 = x_2 u_1 + x_3 u_2, \quad \dot{x}_2 = u_2, \quad \dot{x}_3 = u_1$$

that is, $\dot{x}_1 = x_2 \dot{x}_3 + x_3 \dot{x}_2 = d(x_2 x_3)/dt$, whence $x_1 - x_2 x_3 = k$, where k is a real constant.

Example 6.2

The unicycle WMR (2.28a) has two fields:

$$\mathbf{g}_1 = \begin{bmatrix} \cos\phi \\ \sin\phi \\ 0 \end{bmatrix}, \quad \mathbf{g}_2 = \begin{bmatrix} 0 \\ 0 \\ 1 \end{bmatrix}$$

and so it yields a two-dimensional distribution:

$$\Delta = \{\mathbf{g}_1, \mathbf{g}_2\} = \left\{ \begin{bmatrix} \cos\phi \\ \sin\phi \\ 0 \end{bmatrix}, \begin{bmatrix} 0 \\ 0 \\ 1 \end{bmatrix} \right\}$$

This distribution is nonsingular because for any $(x \quad y \quad \phi)$ in the coordinate neighborhood, the resulting vector space $\Delta(x,y,\phi)$ is two-dimensional. We form the distribution $\overline{\Delta}$ by adding to Δ a third column equal to $[\mathbf{g}_1,\mathbf{g}_2](\mathbf{x})$. Noting that $\mathbf{x}=[x,y,\phi]^{\mathrm{T}}$, this is found to be:

$$[\mathbf{g}_1,\mathbf{g}_2](\mathbf{x})=\frac{\partial \mathbf{g}_2}{\partial \mathbf{x}}\mathbf{g}_1-\frac{\partial \mathbf{g}_1}{\partial \mathbf{x}}\mathbf{g}_2=\begin{bmatrix}\sin\phi\\ -\cos\phi\\ 0\end{bmatrix}$$

Hence, we get the matrix:

$$\overline{\Delta}=\begin{bmatrix}\cos\phi & 0 & \sin\phi\\ \sin\phi & 0 & -\cos\phi\\ 0 & 1 & 0\end{bmatrix}$$

which has det $\overline{\Delta}=1\neq 0$ and rank $=3>2$. Thus, $[\mathbf{g}_1,\mathbf{g}_2](\mathbf{x})$ is linearly independent of $\mathbf{g}_1$ and $\mathbf{g}_2$, and the distribution $\Delta=\{\mathbf{g}_1,\mathbf{g}_2\}$ is not involutive. Therefore, by Frobenius theorem, the unicycle system is nonintegrable (nonholonomic). The same can be verified to hold for all unicycle-like WMRs, the car-like WMRs, and the Brockett integrator (2.59), $\dot{x}_1=u_1,\ \dot{x}_2=u_2,\ \dot{x}_3=x_1u_2-x_2u_1$, which has the two fields:

$$\mathbf{g}_1=\begin{bmatrix}1 & 0 & -x_2\end{bmatrix}^{\mathrm{T}},\quad \mathbf{g}_2=\begin{bmatrix}0 & 1 & x_1\end{bmatrix}^{\mathrm{T}}$$

Example 6.3

Let two linear vector fields $\mathbf{f}(\mathbf{x})=\mathbf{Bx}$ and $\mathbf{g}(\mathbf{x})=\mathbf{Ax}$ on X. Then $[\mathbf{f},\mathbf{g}](\mathbf{x})$ is a linear vector field given by $[\mathbf{f},\mathbf{g}](\mathbf{x})=(\mathbf{AB}-\mathbf{BA})\mathbf{x}$. For example, if:

$$\mathbf{A}=\begin{bmatrix}1 & 0\\ 0 & -1\end{bmatrix}\quad\text{and}\quad \mathbf{B}=\begin{bmatrix}0 & 1\\ 1 & 0\end{bmatrix}$$

are the matrices corresponding to $\mathbf{f}$ and $\mathbf{g}$ relative to a linear system of coordinates, then the matrix $\mathbf{C}$ that corresponds to $[\mathbf{f},\mathbf{g}](\mathbf{x})$ is:

$$\mathbf{C}=2\begin{bmatrix}0 & 1\\ -1 & 0\end{bmatrix}$$

The flows of these fields have the form shown in Figure 6.1.

6.2.2 Manifolds

A manifold is a topological space which is locally Euclidean, that is, around every point, there is a neighborhood which is topologically the same as the *open unit ball* in R^n.

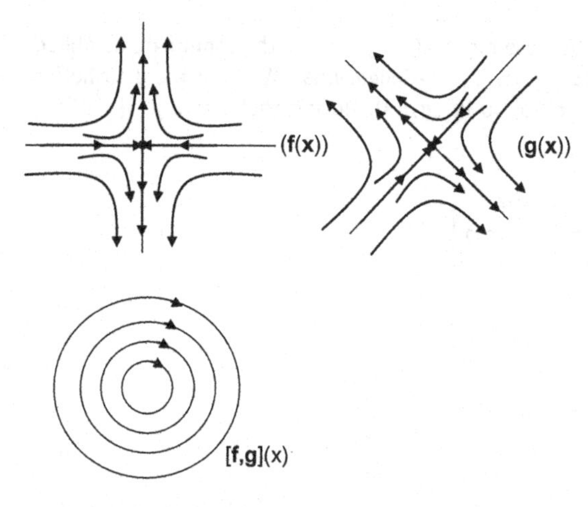

Figure 6.1 Flows of the fields $\mathbf{f}(\mathbf{x})$,$\mathbf{g}(\mathbf{x})$, and $[\mathbf{f},\mathbf{g}](\mathbf{x})$.

Practically, any geometric object that can be charted (drawn, parameterized, etc.) is a manifold. One of the aims of manifold theory is to find ways of distinguishing manifolds. For example, a circle is topologically the same as any closed loop despite the differences possibly observed. In the same way, the surface of a coffee mug with a handle is topologically the same as the surface of a ring, called a *one-handled torus*. A manifold may be compact or noncompact in the topological sense, connected or disconnected, with or without boundary. The closed ball in R^n is a manifold with boundary the unit hypersphere.

Definition 6.7 (*Differentiable manifold*)

A *differentiable manifold* is a continuously and smoothly parameterizable geometric object for which it is possible to establish a system under which every point within the object can be labeled with a unique identifier (e.g., coordinates), and for which the labels vary continuously and smoothly as one moves across the object. ■

The word *smooth* means that the parameterization of a differentiable manifold has at least as many continuous derivatives as required. Since differentiable manifolds are continuously parameterizable, then in any specified smooth coordinate frame (e.g., a state/phase space), it is possible to express the manifold locally (i.e., in some neighborhood of any given point) as the graph of a function.

Formally, a differentiable manifold M is defined as:

$$M = \{\mathbf{x} \in R^n : \mathbf{s}(\mathbf{x}) = 0\} \tag{6.17}$$

where $\mathbf{s}: R^n \rightarrow R^m$ is a smooth map.

Definition 6.8 (*Invariant set*)

An *invariant* set Σ_i is defined any set of points (states) in a dynamic system which are mapped into other points in the same set by the dynamic evolution operator. ■

Similarly, a trajectory is an invariant set because each point in the trajectory evolves into another point in the same trajectory under the action of the evolution operator.

Definition 6.9 (*Invariant manifold*)
An *invariant manifold* is an invariant set that happens to be a differentiable manifold. More specifically, let $\mathbf{s}: R^n \to R^m$ be a smooth map. A manifold $M = \{\mathbf{x} \in R^n: \mathbf{s}(\mathbf{x}) = 0\}$ is invariant for the dynamic system $\dot{\mathbf{x}} = \mathbf{f}(\mathbf{x}, \mathbf{u})$ if all system trajectories starting in M at $t = t_0$ remain in this manifold for all $t \geq t_0$. ■

This implies that the Lie derivative of **s** along the vector field **f** is zero, that is:

$$L_{\mathbf{f}}\mathbf{s}(\mathbf{x}) = \mathbf{0} \quad \text{for all } \mathbf{x} \in M \tag{6.18}$$

It is remarked that a single equilibrium point is an invariant manifold (actually a trivial, zero-dimensional, manifold). But a set of equilibrium points is not an invariant manifold because it lacks continuity.

Definition 6.10 (*Attractive manifold*)
An invariant manifold $M = \{\mathbf{x} \in R^n: \mathbf{s}(\mathbf{x}) = \mathbf{0}\}$ is said to be an *attractive manifold* in an open domain X of R^n, where $X \notin M$, if for all $t_0 \geq 0$ such that $\mathbf{x}(t_0) \in X$, then $\lim_{t \to \infty} \mathbf{x}(t) \in M$ which implies that any $\mathbf{x}(t_0)$ outside M is always attracted toward M. ■

A sufficient condition for $M \in R$ to be attractive is:

$$s(\mathbf{x})\dot{s}(\mathbf{x}) < 0 \quad \text{for all } \mathbf{x} \in X \tag{6.19}$$

that is:

$$\begin{aligned} &\dot{s}(\mathbf{x}) < 0 \text{ for } s(\mathbf{x}) > 0, \quad \mathbf{x} \in X \\ &\dot{s}(\mathbf{x}) > 0 \text{ for } s(\mathbf{x}) < 0, \quad \mathbf{x} \in X \end{aligned}$$

The condition (6.19) is derived using the Lyapunov function $V = (1/2)s^2(\mathbf{x})$ and requiring $\dot{V}(\mathbf{x}) = s(\mathbf{x})\dot{s}(\mathbf{x}) < 0$ in accordance with the Lyapunov theorem.

6.2.3 Lyapunov Stability Using Invariant Sets

The concepts of invariant set (Definition 6.8) and invariant manifold (Definition 6.9) extend the concept of equilibrium point which is an invariant monoset. Using this concept, we can find ways of determining (constructing) Lyapunov functions for nonlinear systems. The Lyapunov functions V defined for invariant sets have the physical property that their reduction rate should be gradually reduced (i.e., $\dot{V}$ must be zero), since V is bounded from below. This is expressed by the Barbalat lemma which states:

Barbalat Lemma If the Lyapunov function $V(\mathbf{x})$ of Definition 5.6 has the additional property of $\dot{V}(\mathbf{x})$ to be uniformly continuous (i.e., to have finite second derivative $\ddot{V}(\mathbf{x})$), then $\dot{V}(\mathbf{x}) \to 0$. ■

The invariant set-based Lyapunov stability is based on the following two theorems [2].

Theorem 6.2 (*LaSalle local invariant sets theorem*)

Let an autonomous system $\dot{\mathbf{x}} = \mathbf{f}(\mathbf{x})$ with $\mathbf{f}$ continuous, and $V(\mathbf{x})$ a scalar function with continuous first derivatives. We assume that:

1. For some $\gamma > 0$, the region Ω_γ which is defined by $V(\mathbf{x}) < \gamma$ is bounded
2. $\dot{V}(\mathbf{x}) \leq 0$ for all $\mathbf{x} \in \Omega_\gamma$

Let G be the set of points inside Ω_γ, for which $\dot{V}(\mathbf{x}) = 0$, and S be the larger invariant set in $G(S \subset G)$. Then, every solution $\mathbf{x}(t)$ of the system that starts in Ω_γ tends to S for $t \rightarrow \infty$. ■

Here, the term "larger" is used in the set theory sense, that is, S is the union of all the invariant sets (e.g., equilibrium points or limit cycles). If the entire set G is invariant, then $S = G$. The geometric interpretation of this theorem is shown in Figure 6.2, where a trajectory that departs in a bounded Ω_γ converges to the larger invariant set S.

Theorem 6.3 (*LaSalle global invariant sets theorem*)

Consider the nonlinear autonomous system $\dot{\mathbf{x}} = \mathbf{f}(\mathbf{x})$, where $\mathbf{f}$ is a continuous function, and a nonlinear function $V(\mathbf{x})$ with continuous first-order partial derivatives.

We assume that:

1. $\dot{V}(\mathbf{x}) \leq 0$ over the entire state space
2. $V(\mathbf{x}) \rightarrow \infty$ for $||x|| \rightarrow \infty$

Let G be the set of points for which $\dot{V} = 0$, and S be the larger invariant set within G. Then, all solutions of the system possess total (global) asymptotic convergence. ■

This theorem shows, for example, that the convergence to a limit cycle is global, and all trajectories of the system converge to the limit cycle. The construction of Lyapunov functions for linear systems is made as described in Section 5.2.2 (Eqs. (5.14a), (5.14b), and (5.15)). A general method for constructing Lyapunov

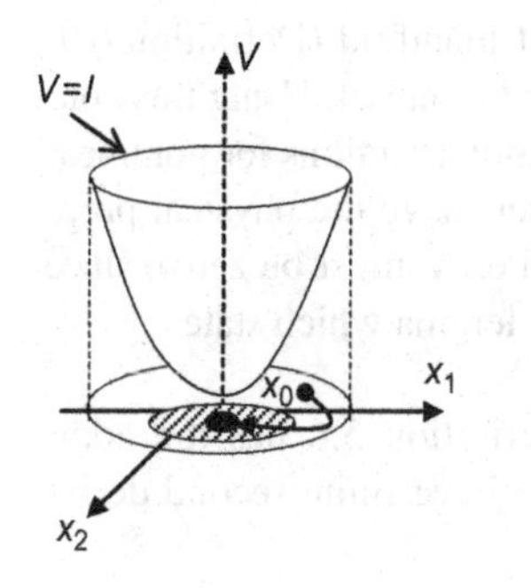

Figure 6.2 Geometric representation of the local invariant sets theorem (convergence to the larger invariant set S).

functions for nonlinear systems was developed by Krasovskii and is formulated by the following two theorems.

Theorem 6.4 (*Krasovskii theorem*)

Let the autonomous system $\dot{\mathbf{x}} = \mathbf{f}(\mathbf{x})$, where the equilibrium point of concern is at the origin $\mathbf{x} = \mathbf{0}$. Let $\mathbf{A}(\mathbf{x}) = \partial\mathbf{f}/\partial\mathbf{x}$ be the Jacobian matrix of $\mathbf{f}$. If the matrix $\mathbf{F} = \mathbf{A} + \mathbf{A}^{\mathrm{T}}$ is negative definite in a region Ω, then the equilibrium point $\mathbf{x} = \mathbf{0}$ is asymptotically stable. A Lyapunov function for this system is:

$$V(\mathbf{x}) = \mathbf{f}^{\mathrm{T}}(\mathbf{x})\mathbf{f}(\mathbf{x}) \tag{6.20}$$

If Ω is the entire state space, and $V(\mathbf{x}) \to \infty$ for $||\mathbf{x}|| \to \infty$, then the equilibrium point is totally asymptotically stable. ■

Theorem 6.5 (*Generalized Krasovskii theorem*)

Consider the system $\dot{\mathbf{x}} = \mathbf{f}(\mathbf{x})$, where the equilibrium point of concern is the origin $\mathbf{x} = \mathbf{0}$. If $\mathbf{A}(\mathbf{x})$ is the Jacobian matrix of the system, then a sufficient condition for $\mathbf{x} = \mathbf{0}$ to be asymptotically stable is the existence of symmetric positive definite matrices $\mathbf{P}$ and $\mathbf{Q}$ such that, for every $\mathbf{x} \neq \mathbf{0}$, the matrix:

$$\mathbf{F}(\mathbf{x}) = \mathbf{A}^{\mathrm{T}}\mathbf{P} + \mathbf{P}\mathbf{A} + \mathbf{Q} \tag{6.21}$$

is negative semidefinite in a region Ω of the origin. Then, the function $V(\mathbf{x}) = \mathbf{f}^{\mathrm{T}}\mathbf{P}\mathbf{f}$ is a Lyapunov function of the system. If the region Ω is the entire state space and $V(\mathbf{x}) \to \infty$ for $\|\mathbf{x}\| \to \infty$, then the system is globally asymptotically stable. ■

Clearly, $V(\mathbf{x})$ possesses the first three properties of Lyapunov functions. The derivative $\dot{V}$ is computed as:

$$\begin{aligned}\dot{V} &= \frac{\partial V}{\partial \mathbf{x}}\mathbf{f}(\mathbf{x}) = \mathbf{f}^{\mathrm{T}}\mathbf{P}\mathbf{A}(\mathbf{x})\mathbf{f} + \mathbf{f}^{\mathrm{T}}\mathbf{A}^{\mathrm{T}}(\mathbf{x})\mathbf{P}\mathbf{f} \\ &= \mathbf{f}^{\mathrm{T}}\mathbf{F}\mathbf{f} - \mathbf{f}^{\mathrm{T}}\mathbf{Q}\mathbf{f}\end{aligned} \tag{6.22}$$

Since $\mathbf{F}$ is negative semidefinite, and $\mathbf{Q}$ is positive definite, then $\dot{V}$ is negative, and the system is asymptotically stable. If $V(\mathbf{x}) \to \infty$ for $||\mathbf{x}|| \to \infty$, the global asymptotic stability of the system follows from the conventional Lyapunov stability theory (Section 5.2.2).

Example 6.4 (*Application of Krasovskii theorem*)

We will examine the stability of the nonlinear system:

$$\dot{x}_1 = -6x_1 + 2x_2, \quad \dot{x}_2 = 2x_1 - 6x_2 - 2x_2^3$$

using Theorem 6.4.

Here, we have:

$$\mathbf{A} = \frac{\partial \mathbf{f}}{\partial \mathbf{x}} = \begin{bmatrix} -6 & 2 \\ 2 & -6-6x_2^2 \end{bmatrix}, \quad \mathbf{F} = \mathbf{A} + \mathbf{A}^{\mathrm{T}} = \begin{bmatrix} -12 & 4 \\ 4 & -12-12x_2^2 \end{bmatrix}$$

It is easy to show that $\mathbf{F}$ is negative definite, and so the origin $\mathbf{x} = \mathbf{0}$ is asymptotically stable, and a Lyapunov function is $V = \mathbf{f}^{\mathrm{T}}\mathbf{f}$, that is:

$$V(\mathbf{x}) = (-6x_1 + 2x_2)^2 + (2x_1 - 6x_2 - 2x_2^3)^2$$

Since $V(\mathbf{x}) \to \infty$ for $||\mathbf{x}|| \to \infty$, the equilibrium point $\mathbf{x} = \mathbf{0}$ is totally asymptotically stable.

In many cases, it is difficult to check if the matrix $\mathbf{F} = \mathbf{A} + \mathbf{A}^{\mathrm{T}}$ is negative definite for all $\mathbf{x}$, and also, the Jacobian matrices of many systems do not satisfy the condition of the theorem. In these cases, we must work with the generalized Krasovskii theorem and $\mathbf{F}(\mathbf{x})$ given by Eq. (6.21).

The theorem of Brockett provides a necessary condition for the existence of a stabilizing control law of a nonlinear system. This theorem is stated as follows [5]:

Theorem 6.6 (*Brockett theorem*)

Consider the nonlinear system:

$$\dot{\mathbf{x}} = \mathbf{f}(\mathbf{x}, \mathbf{u}), \quad \mathbf{f}(\mathbf{x}_0, \mathbf{0}) = \mathbf{0}$$

where $\mathbf{f}$ is a continuously differentiable function in a neighborhood of $(\mathbf{x}_0, \mathbf{0})$. Then, a necessary condition for the existence of a *continuously differentiable control* law that makes $(\mathbf{x}_0, \mathbf{0})$ asymptotically stable is that:

1. The linearized system $\dot{\mathbf{x}} = \mathbf{A}\mathbf{x} + \mathbf{B}\mathbf{u}$, $\mathbf{A} = (\partial \mathbf{f}/\partial \mathbf{x})_{\mathbf{x}=\mathbf{x}_0}$, $\mathbf{B} = (\partial \mathbf{f}/\partial \mathbf{u})_{\mathbf{u}=\mathbf{0}}$ has no uncontrollable modes associated with eigenvalues whose real part is positive.
2. There exists a neighborhood Ω of $(\mathbf{x}_0, \mathbf{0})$ such that for each $\xi \in \Omega$, there exists a control $\mathbf{u}(\xi, t)$ defined for $t \geq 0$ such that this control steers the state response (solution) of the system from $\mathbf{x} = \xi$ at $t = 0$ to $\mathbf{x} = \mathbf{x}_0$ at $t = \infty$.
3. The mapping $\Gamma:(\mathbf{x}, \mathbf{u}) \to \mathbf{f}(\mathbf{x}, \mathbf{u})$ is onto an open set containing $\mathbf{0}$. ■

Corollary

(a) For an affine system with drift: $\dot{\mathbf{x}} = \mathbf{g}_0(\mathbf{x}) + \sum_{i=1}^{m} \mathbf{g}_i(\mathbf{x})u_i$, $\mathbf{x}(t) \in \Omega \subset R^n$ the condition (1) implies that the stabilization problem cannot have a solution if there is a smooth distribution $\boldsymbol{\Delta}$ which contains $\mathbf{g}_0(\cdot)$ and $\mathbf{g}_1(\cdot), \ldots, \mathbf{g}_m(\cdot)$ with $\dim\boldsymbol{\Delta} < n$.

(b) If the affine system is driftless (i.e., $\mathbf{g}_0(\mathbf{x}) = \mathbf{0}$), with the vectors $\mathbf{g}_i(\mathbf{x})$ being linearly independent at $\mathbf{x}_0$, then there exists a solution to the stabilization problem only if $m = n$.

(c) There is no continuously differentiable stabilizing control law for $\dot{x} = u_1$, $\dot{y} = u_2$, $\dot{z} = xu_2 - yu_1$ because this system satisfies only the conditions (1) and (2) of the theorem, failing to satisfy condition (3). ■

Remark If $\mathbf{A} = (\partial \mathbf{f}/\partial \mathbf{x})_{\mathbf{x}=\mathbf{x}_0}$ has an eigenvalue with zero real part, then the equilibrium state $\mathbf{x}_0$ is said to be an asymptotically stable critical point. If there exists a control law that makes $\mathbf{x} = \mathbf{0}$ an asymptotically stable equilibrium point for the system $\dot{\mathbf{x}} = \mathbf{A}\mathbf{x} + \mathbf{B}\mathbf{u}$, then there exists a control law which makes $\mathbf{x}_0$ an asymptotically stable critical point, provided that the algebraic system $\mathbf{A}\mathbf{x}_0 + \mathbf{B}\mathbf{u}_0 = \mathbf{0}$ can be solved for $\mathbf{u}_0$. In fact, if $\mathbf{u} = \mathbf{K}\mathbf{x}$ makes $\mathbf{x} = \mathbf{0}$ an asymptotically stable equilibrium point, then $\mathbf{u} = \mathbf{K}\mathbf{x} + \mathbf{u}_0$ makes $\mathbf{x}_0$ an asymptotically stable equilibrium point. Therefore, if $\mathbf{x} = \mathbf{0}$ can be made asymptotically stable, then there is an entire subspace $U = \{\mathbf{x}\colon \mathbf{A}\mathbf{x} \in \text{range } \mathbf{B}\}$ of points that can be made asymptotically stable.

6.3 Feedback Linearization of Mobile Robots

6.3.1 General Issues

The methodology of linearization via state feedback consists in the algebraic transformation of the nonlinear system dynamics to an equivalent linear form so that linear control laws to be applicable. We distinguish two cases of feedback linearization:

1. *Input-state linearization*: In this case, we seek to find a state transformation $\mathbf{z} = \mathbf{z}(\mathbf{x})$ and then an input/transformation $\mathbf{u} = \mathbf{u}(\mathbf{x}, \boldsymbol{\upsilon})$, where $\boldsymbol{\upsilon}$ is the new manipulable input. The purpose of the above transformation is to bring the system $\dot{\mathbf{x}} = \mathbf{f}(\mathbf{x}, \mathbf{u})$ to the linear form $\dot{\mathbf{z}} = \mathbf{A}\mathbf{z} + \mathbf{B}\boldsymbol{\upsilon}$.
2. *Input–output linearization*: In this case, we have the system $\dot{\mathbf{x}} = \mathbf{f}(\mathbf{x}, \mathbf{u})$ with output $\mathbf{y} = \mathbf{h}(\mathbf{x})$. The basic feature of this system is that the output $\mathbf{y}$ is connected to $\mathbf{u}$ only indirectly through $\mathbf{x}$. Therefore, to achieve input–output linearization, we must find a direct relation between the input and the output of the system. This may be done by successive differentiation of the output $\mathbf{y} = \mathbf{h}(\mathbf{x})$ until all inputs appear in the resulting derivative equations.

The nonholonomic mobile robots can be modeled by the affine system:

$$\dot{\mathbf{x}} = \mathbf{f}(\mathbf{x}) + \mathbf{g}(\mathbf{x})\mathbf{u}, \quad \mathbf{x} \in R^n$$

where $\mathbf{x}$ is the n-dimensional state and $\mathbf{u}$ the m-dimensional control input. For driftless systems (which typically represent first-order kinematic models), we have $\mathbf{f}(\mathbf{x}) = \mathbf{0}$, and the state involves the robot generalized coordinates. The *input-state* linearization of nonholonomic mobile robots is not/possible by means of smooth state feedback because the fields $\mathbf{f}(\mathbf{x})$ and $\mathbf{g}(\mathbf{x})$ are not involutive as discussed in Example 6.2. But these robots may be *input–output linearizable* (and decoupled). Here, we will study the case of a differential-drive WMR with two outputs: the coordinates x_Q and y_Q of the center point Q. In this case, the input–output linearization is not possible by static feedback, but it can be achieved using dynamic feedback.

For single-input systems, we have the following results which are directly applicable to many practical cases [1–4].

Theorem 6.7 (*Feedback linearization conditions*)
Let a single-input affine system be:

$$\dot{\mathbf{x}} = \mathbf{f}(\mathbf{x}) + \mathbf{g}(\mathbf{x})u, \quad \mathbf{x} \in R^n \tag{6.23a}$$

Then, there exists an output function $y = h(\mathbf{x})$ for which the system has relative degree n at the point $\mathbf{x} = \mathbf{x}^0$, if and only if the following conditions are satisfied:

1. The matrix:

$$[\mathbf{g}(\mathbf{x}^0) \vdots ad_{\mathbf{f}}\mathbf{g}(\mathbf{x}^0) \vdots \cdots \vdots ad_{\mathbf{f}}^{n-2}\mathbf{g}(\mathbf{x}^0) \vdots ad_{\mathbf{f}}^{n-1}\mathbf{g}(\mathbf{x}^0)] \tag{6.23b}$$

has rank n.

2. The distribution:

$$\Delta = \{\mathbf{g}(\mathbf{x}), ad_{\mathbf{f}}\mathbf{g}(\mathbf{x}), \ldots, ad_{\mathbf{f}}^{n-2}\mathbf{g}(\mathbf{x})\} \tag{6.23c}$$

is involutive in the vicinity of $\mathbf{x}^0$.

Theorem 6.8 (*Controllable canonical form*)
Defining new state variables as:

$$z_i = \phi_i(\mathbf{x}) = L_f^{i-1}h(\mathbf{x}), \quad i = 1, 2, \ldots, n \tag{6.24a}$$

we can transform the system (6.23a) to the controllable canonical form:

$$\begin{aligned} \dot{z}_1 &= z_2 \\ \dot{z}_2 &= z_3 \\ &\vdots \\ \dot{z}_{n-1} &= z_n \\ \dot{z}_n &= b(\mathbf{z}) + a(\mathbf{z})u \end{aligned} \tag{6.24b}$$

where $\mathbf{z} = [z_1, z_2, \ldots, z_n]^{\mathrm{T}}$, and the function $a(\mathbf{z})$ is nonzero in a vicinity of $\mathbf{z}^0$ (since $z_i^0 = \phi_i(\mathbf{x}^0)$). ■

As described in Definition 5.9, a system state $\mathbf{x}_0$ is called controllable if it is possible to find a control $\mathbf{u}(t)$ that drives $\mathbf{x}_0$ to a desired final state $\mathbf{x}_\mathrm{f}$. In this case, $\mathbf{x}_\mathrm{f}$ is said to be *reachable from* $\mathbf{x}_0$. The system is said to be *totally controllable* if all of its states are controllable. To understand better the controllability concept, consider a driftless two-input system:

$$\dot{\mathbf{x}} = \mathbf{g}_1(\mathbf{x})u_1 + \mathbf{g}_2(\mathbf{x})u_2, \quad \mathbf{x} \in R^3$$

The question is to find which states can be reached from a given initial state $\mathbf{x}_0$. If the input fields $\mathbf{g}_1(\mathbf{x})$ and $\mathbf{g}_2(\mathbf{x})$ are linearly independent, then the system can move in two independent directions, namely, along $\mathbf{g}_1(\mathbf{x})$ when $u_2 = 0$ and $u_1 = \pm 1$, or along $\mathbf{g}_2(\mathbf{x})$ when $u_1 = 0$ and $u_2 = \pm 1$. Of course, the system can move in any direction which is a linear combination of the directions $\mathbf{g}_1$ and $\mathbf{g}_2$. The question is whether the system can move in any direction independent of $\mathbf{g}_1$ and $\mathbf{g}_2$. This can be done if $\mathbf{g}_1(\mathbf{x}_0), \mathbf{g}_2(\mathbf{x}_0)$, and $ad_{\mathbf{g}_1}\mathbf{g}_2(\mathbf{x}_0)$ are linearly independent, where $ad_{\mathbf{g}_1}\mathbf{g}_2(\mathbf{x}_0) = [\mathbf{g}_1, \mathbf{g}_2](\mathbf{x}_0)$. The motion in the $[\mathbf{g}_1, \mathbf{g}_2](\mathbf{x}_0)$ direction can be done by appropriate switching between $\mathbf{g}_1$ and $\mathbf{g}_2$, namely:

$$\mathbf{u}(t) = \begin{bmatrix} u_1(t) \\ u_2(t) \end{bmatrix} = \begin{cases} \begin{bmatrix} 1 \\ 0 \end{bmatrix} & \text{if} \quad 0 \leq t \leq \Delta t \\ \begin{bmatrix} 0 \\ 1 \end{bmatrix} & \text{if} \quad \Delta t \leq t \leq 2\Delta t \\ \begin{bmatrix} -1 \\ 0 \end{bmatrix} & \text{if} \quad 2\Delta t \leq t \leq 3\Delta t \\ \begin{bmatrix} 0 \\ -1 \end{bmatrix} & \text{if} \quad 3\Delta t \leq t \leq 4\Delta t \end{cases}$$

Then, we get:

$$\mathbf{x}(4\Delta t) = \mathbf{x}_0 + (1/2)(\Delta t)^2 [\mathbf{g}_1, \mathbf{g}_2](\mathbf{x}_0) + \mathbf{0}(\Delta t^3)$$

which is illustrated in Figure 6.3.

The above results can be extended to the general case, where there are $m > 2$ inputs, and we have the following reachability and controllability theorem.

Theorem 6.9 (*Reachability–Controllability*)

The system

$$\dot{\mathbf{x}} = \mathbf{f}(\mathbf{x}) + \sum_{i=1}^{m} \mathbf{g}_i(\mathbf{x})u_i, \quad \mathbf{x} \in R^n$$

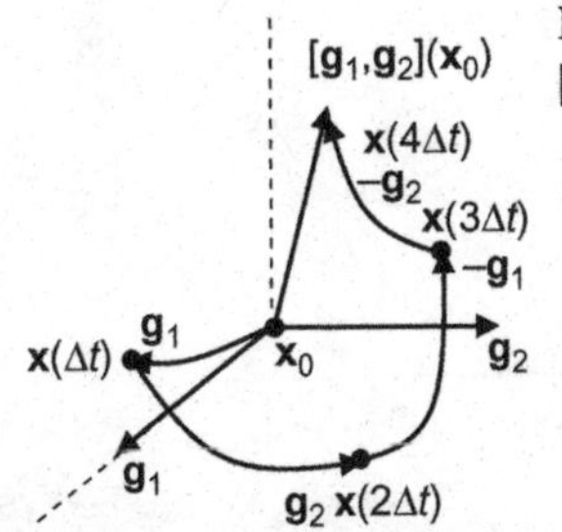

Figure 6.3 Illustration of approximating the Lie bracket $[\mathbf{g}_1, \mathbf{g}_2](\mathbf{x}_0)$ by successive switching between $\mathbf{g}_1$ and $\mathbf{g}_2$.

is locally reachable from $\mathbf{x}_0$ if the following reachability field distribution $\mathbf{Q}_c$ spans the n-dimensional space, where:

$$\mathbf{Q}_c = [\mathbf{g}_1, \mathbf{g}_2, \ldots, \mathbf{g}_m, [\mathbf{g}_i, \mathbf{g}_j], \ldots, ad_{\mathbf{g}_i}^k \mathbf{g}_j, \ldots, [\mathbf{f}, \mathbf{g}_i], \ldots, ad_{\mathbf{f}}^k \mathbf{g}_i, \ldots]$$

if it has rank n. If $\mathbf{f}(\mathbf{x}) = \mathbf{0}$ and $\mathbf{Q}_c$ has rank n, then the system is controllable. ■

This theorem extends to nonlinear systems, the linear controllability condition ((5.18a) and (5.18b)). The $\mathbf{g}_i$ terms correspond to the $\mathbf{B}$ terms, the terms $[\mathbf{g}_i, \mathbf{g}_j]$ are new coming from the nonlinearity of $\mathbf{g}_i$, and the $[\mathbf{f}, \mathbf{g}_i]$ terms correspond to the $\mathbf{AB}$ terms, and so on. Denoting $\mathbf{B} = [\mathbf{b}_1, \mathbf{b}_2, \ldots, \mathbf{b}_m]$ and writing the linear system $\dot{\mathbf{x}}(t) = \mathbf{Ax} + \mathbf{Bu}(t)$ as $\dot{\mathbf{x}} = \mathbf{Ax} + \mathbf{Bu} = \mathbf{Ax} + \mathbf{b}_1 u_1 + \ldots + \mathbf{b}_m u_m$, we have $\mathbf{f}(\mathbf{x}) = \mathbf{Ax}$, $\mathbf{g}_i(\mathbf{x}) = \mathbf{b}_i$, $\partial \mathbf{f}/\partial \mathbf{x} = \mathbf{A}$, $\partial \mathbf{g}/\partial \mathbf{x} = \mathbf{0}$, and so on. Then, the controllability matrix ((5.18a) and (5.18b)) takes the form:

$$\mathbf{Q}_c = [\mathbf{b}_1, \mathbf{b}_2, \ldots, \mathbf{b}_m, ad_{\mathbf{f}} \mathbf{b}_1, \ldots, ad_{\mathbf{f}} \mathbf{b}_m, \ldots, ad_{\mathbf{f}}^{n-1} \mathbf{b}_1, \ldots, ad_{\mathbf{f}}^{n-1} \mathbf{b}_m]$$

and so the controllability condition remains the same.

Example 6.5

Let us check the reachability and controllability of the unicycle (2.26) which has the affine representation (see Eq. (2.32)):

$$\dot{\mathbf{x}} = \mathbf{f}(\mathbf{x}) + \mathbf{g}_1 u_1 + \mathbf{g}_2 u_2, \quad \mathbf{x} = [x \quad y \quad \phi]^{\mathrm{T}} \in R^3$$

where $u_1 = v_Q$, $u_2 = \dot{\phi}$, and:

$$\mathbf{f}(\mathbf{x}) = \mathbf{0}, \quad \mathbf{g}_1 = \begin{bmatrix} \cos\phi \\ \sin\phi \\ 0 \end{bmatrix}, \quad \mathbf{g}_2 = \begin{bmatrix} 0 \\ 0 \\ 1 \end{bmatrix}$$

Here, we have:

$$\mathbf{Q}_c = [\mathbf{g}_1, \mathbf{g}_2, [\mathbf{g}_1, \mathbf{g}_2]]$$

where:

$$\begin{aligned} [\mathbf{g}_1, \mathbf{g}_2] &= \frac{\partial \mathbf{g}_2}{\partial \mathbf{x}} \mathbf{g}_1 - \frac{\partial \mathbf{g}_1}{\partial \mathbf{x}} \mathbf{g}_2 \\ &= -\begin{bmatrix} 0 & 0 & -\sin\phi \\ 0 & 0 & \cos\phi \\ 0 & 0 & 0 \end{bmatrix} \begin{bmatrix} 0 \\ 0 \\ 1 \end{bmatrix} = \begin{bmatrix} \sin\phi \\ -\cos\phi \\ 0 \end{bmatrix} \end{aligned}$$

Therefore:

$$\mathbf{Q}_c = \begin{bmatrix} \cos\phi & 0 & \sin\phi \\ \sin\phi & 0 & -\cos\phi \\ 0 & 1 & 0 \end{bmatrix}$$

Since $\det \mathbf{Q}_c = -1$, that is, $\text{rank}\,\mathbf{Q}_c = 3$ everywhere, the unicycle is *locally reachable* from *everywhere*, and since $\mathbf{f}(\mathbf{x}) = \mathbf{0}$, the system is *controllable*.

Theorem 6.10 (*Generalized controllable canonical form*)
Consider the system:

$$\dot{\mathbf{x}} = \mathbf{f}(\mathbf{x}) + \mathbf{g}(\mathbf{x})u, \quad y = h(\mathbf{x}) \tag{6.25a}$$

with relative degree $r \leq n$ at $\mathbf{x} = \mathbf{x}^0$. We set,

$$\phi_1(\mathbf{x}) = h(\mathbf{x}), \quad \phi_2(\mathbf{x}) = L_\mathbf{f} h(\mathbf{x}), \ldots, \phi_r(\mathbf{x}) = L_\mathbf{f}^{r-1} h(\mathbf{x}).$$

If r is strictly less than n, then we can find $n - r$ additional functions $\phi_{r+1}(\mathbf{x}), \ldots, \phi_n(\mathbf{x})$ such that the mapping:

$$\boldsymbol{\Phi}(\mathbf{x}) = [\phi_1(\mathbf{x}), \phi_2(\mathbf{x}), \ldots, \phi_n(\mathbf{x})] \tag{6.25b}$$

has invertible Jacobian matrix at $\mathbf{x} = \mathbf{x}^0$. It can therefore be used as local transformation in a region around $\mathbf{x}^0$. It is always possible to select the functions $\phi_{r+1}(\mathbf{x}), \ldots, \phi_n(\mathbf{x})$ such that:

$$L_g \phi_i(\mathbf{x}) = 0 \tag{6.25c}$$

for all i with $r + 1 \leq i \leq n$ and all $\mathbf{x}$ in the vicinity of $\mathbf{x}^0$. Then, using the new state variables $z_i = \phi_i(\mathbf{x})$, $i = 1, 2, \ldots, n$, the system can be written in the canonical form:

$$\begin{aligned}
\dot{z}_1 &= z_2 \\
\dot{z}_2 &= z_3 \\
&\vdots \\
\dot{z}_{r-1} &= z_r \\
\dot{z}_r &= b(\mathbf{z}) + a(\mathbf{z})u, \quad a(\mathbf{z}^0) = a(\boldsymbol{\Phi}(\mathbf{x}^0)) \neq 0 \\
\dot{z}_{r+1} &= c_{r+1}(\mathbf{z}) \\
&\vdots \\
\dot{z}_n &= c_n(\mathbf{z}) \\
y &= z_1 = \phi_1(\mathbf{x})
\end{aligned} \tag{6.25d}$$

Proof To derive the canonical mode ((2.25c) and (2.25d)) we differentiate $z_i(i = 1, 2, \ldots, r)$ and get:

$$\dot{z}_1 = \left(\frac{\partial \phi_1}{\partial \mathbf{x}}\right)\dot{\mathbf{x}} = L_{\mathbf{f}}h(\mathbf{x}(t)) = \phi_2(\mathbf{x}(t)) = z_2(t)$$

$$\vdots$$

$$\dot{z}_{r-1} = \left(\frac{\partial \phi_{r-1}}{\partial \mathbf{x}}\right)\dot{\mathbf{x}} = \left(\frac{\partial L_{\mathbf{f}}^{r-2}h(\mathbf{x})}{\partial \mathbf{x}}\right)\dot{\mathbf{x}} = L_{\mathbf{f}}^{r-1}h(\mathbf{x}) = \phi_r(\mathbf{x}) = z_r(t)$$

$$\dot{z}_r = L_{\mathbf{f}}^r h(\mathbf{x}) + L_g L_{\mathbf{f}}^{r-1}h(\mathbf{x})u$$

where $\mathbf{x}(t)$ is replaced by $\mathbf{x}(t) = \mathbf{\Phi}^{-1}(\mathbf{z}(t))$ to give:

$$\dot{z}_r = b(\mathbf{z}(t)) + a(\mathbf{z}(t))u$$

where $a(\mathbf{z}) = L_g L_{\mathbf{f}}^{r-1}h(\mathbf{\Phi}^{-1}(\mathbf{z})),\ \ b(\mathbf{z}) = L_{\mathbf{f}}^r h(\mathbf{\Phi}^{-1}(\mathbf{z})) \neq 0$.

For the remaining variables $z_i(i = r+1, \ldots, n)$ we may not have a special form, but choosing them such that $L_g \mathbf{\Phi}_i(\mathbf{x}) = 0,\ \ i = r+1, \ldots, n$, we get:

$$\dot{z}_i = (\partial \phi_i/\partial \mathbf{x})[\mathbf{f}(\mathbf{x}) + \mathbf{g}(\mathbf{x})u] = L_{\mathbf{f}}\phi_i(\mathbf{x}) + L_g\phi_i(\mathbf{x})u = L_{\mathbf{f}}\phi_i(\mathbf{x})$$

Therefore:

$$\dot{z}_i = L_{\mathbf{f}}(\phi_i(\mathbf{\Phi}^{-1}(\mathbf{z}))) = c_i(\mathbf{z}), \quad i = r+1, \ldots, n$$

The model (6.24b) is a special case of (6.25d) obtained when $r = n$.

Theorem 6.11 (*Linearizing feedback law*)

Selecting the following state feedback law:

$$u = \frac{1}{a(\mathbf{z})}[-b(\mathbf{z}) + v] \tag{6.26}$$

in the canonical model (6.24b), the following linear and controllable system is obtained:

$$\dot{z}_1 = z_2,\ \ \dot{z}_2 = z_3, \ldots, \dot{z}_{n-1} = z_n, \dot{z}_n = v \tag{6.27}$$

which is valid in the vicinity of $\mathbf{z}^0$ where $a(\mathbf{z}) \neq 0$. ■

Theorem 6.7 provides the necessary and sufficient conditions under which Theorems 6.8 and 6.10 are valid. Theorems 6.7 and 6.11 suggest the following: three-step procedure for linearizing a system of the form (6.23a) namely:

Step 1: Find the output $y = h(\mathbf{x})$ for which the system (6.23a) satisfies the conditions of Theorem 6.7.

Step 2: Compute the state transformation:

$$\mathbf{z} = \boldsymbol{\Phi}(\mathbf{x}) = \begin{bmatrix} \phi_1(\mathbf{x}) \\ \phi_2(\mathbf{x}) \\ \vdots \\ \phi_n(\mathbf{x}) \end{bmatrix} = \begin{bmatrix} h(\mathbf{x}) \\ L_\mathbf{f} h(\mathbf{x}) \\ \vdots \\ L_\mathbf{f}^{n-1} h(\mathbf{x}) \end{bmatrix} \tag{6.28}$$

for $\mathbf{x}$ in the vicinity of $\mathbf{x}^0$.

Step 3: Compute the local state feedback law:

$$\begin{aligned} u &= \frac{1}{a(\boldsymbol{\Phi}(\mathbf{x}))}[-b(\boldsymbol{\Phi}(\mathbf{x})) + v] \\ &= \frac{1}{L_\mathbf{g} L_\mathbf{f}^{n-1} h(\mathbf{x})}\left[-L_\mathbf{f}^n h(\mathbf{x}) + v\right] \end{aligned} \tag{6.29}$$

One can see that Eq. (6.29) provides the expression of the state feedback linearizing controller in terms of the functions $\mathbf{f}(\mathbf{x})$, $\mathbf{g}(\mathbf{x})$, and $h(\mathbf{x})$ of the original system:

$$\dot{\mathbf{x}} = \mathbf{f}(\mathbf{x}) + \mathbf{g}(\mathbf{x})u, \quad y = h(\mathbf{x}), \quad \mathbf{x} \in R^n \tag{6.30}$$

Now, using a new linear feedback controller:

$$v = \mathbf{Kz}, \quad \mathbf{K} = [k_1, k_2, \ldots, k_n] \tag{6.31a}$$

we can choose the gains $k_i (i = 1, 2, \ldots, n)$ such that to achieve desired specification as usual.

Introducing Eq. (6.28) into Eq. (6.31a) we get the overall controller:

$$v = k_1 h(\mathbf{x}) + k_2 L_\mathbf{f} h(\mathbf{x}) + \cdots + k_n L_\mathbf{f}^{n-1} h(\mathbf{x}) \tag{6.31b}$$

which is actually a nonlinear state feedback controller for the system (6.30).

Remarks:

1. The steps 2 and 3 of the algorithm can be interchanged, if it is more convenient.
2. Although the algorithm was presented, for simplicity, for single-input systems, it is applicable to mobile robots that have two or more inputs (of course with the proper selection of the output, the diffeomorphic state transformation, and the state feedback controller).
3. If $n = 2$ (second-order nonlinear system), then the conditions of Theorem 6.7 hold always since $[\mathbf{g}, \mathbf{g}](\mathbf{x}) = \mathbf{0}$.

4. If the system (6.25a) has relative degree $r < n$, but satisfies the conditions (1) and (2) of Theorem 6.7, then there exists another output function $c(\mathbf{x})$ for which the relative degree of the system is n, and the theorem is applicable. However, the real output, expressed in terms of the state variables z, that is $y = h(\Phi^{-1}(\mathbf{z}))$ remains nonlinear. Therefore, we naturally need to determine whether there exists a transformation and a feedback controller that can make the overall system (state and output) linear and controllable. This can be done if some more conditions are satisfied as it is described in the following theorem.

Theorem 6.12 (*Generalized linearizing feedback law*)
Suppose that the system:

$$\dot{\mathbf{x}} = \mathbf{f}(\mathbf{x}) + \mathbf{g}(\mathbf{x})u, \quad y = h(\mathbf{x}), \quad \mathbf{x} \in R^n$$

has at the point $\mathbf{x} = \mathbf{x}^0$ relative degree $r < n$. Then, there exists a static state feedback controller and a state transformation, defined in a vicinity of $\mathbf{x}^0$, such that the system is transformed into linear and controllable form:

$$\dot{\mathbf{x}} = \mathbf{A}\mathbf{x} + \mathbf{B}u, \quad y = \mathbf{C}\mathbf{x}$$

if and only if the following conditions are satisfied:

1. The matrix $[\mathbf{g}(\mathbf{x}^0) \vdots ad_{\mathbf{f}}\mathbf{g}(\mathbf{x}^0) \vdots \cdots \vdots ad_{\mathbf{f}}^{n-2}\mathbf{g}(\mathbf{x}^0) \vdots ad_{\mathbf{f}}^{n-1}\mathbf{g}(\mathbf{x}^0)]$ has rank n.
2. The n-dimensional vector fields $\tilde{\mathbf{f}}(\mathbf{x})$ and $\tilde{\mathbf{g}}(\mathbf{x})$ that are defined by:

$$\tilde{\mathbf{f}}(\mathbf{x}) = \mathbf{f}(\mathbf{x}) - \frac{L_{\mathbf{f}}^r h(\mathbf{x})}{L_{\mathbf{g}} L_{\mathbf{f}}^{r-1} h(\mathbf{x})}, \quad \tilde{\mathbf{g}}(\mathbf{x}) = \frac{\mathbf{g}(\mathbf{x})}{L_{\mathbf{g}} L_{\mathbf{f}}^{r-1} h(\mathbf{x})}$$

are such that:

$$[ad_{\tilde{\mathbf{f}}}^i \tilde{\mathbf{g}}(\mathbf{x}) \vdots ad_{\tilde{\mathbf{f}}}^j \tilde{\mathbf{g}}(\mathbf{x})] = \mathbf{0}$$

for all pairs (i, j) with $i, j = 1, 2, \ldots, n$.

Example 6.6

It is desired to check: (a) if the system:

$$\dot{\mathbf{x}} = \mathbf{f}(\mathbf{x}) + \mathbf{g}(\mathbf{x})u, \quad y = x_2, \quad \mathbf{f}(\mathbf{x}) = \begin{bmatrix} x_3 - x_2 \\ 0 \\ x_3 + x_1^2 \end{bmatrix}, \quad \mathbf{g}(\mathbf{x}) = \begin{bmatrix} 0 \\ \exp(x_1) \\ \exp(x_1) \end{bmatrix}$$

can be put in controllable form according to Theorem 6.8, and (b) if the entire system (state and output) can be transformed to a linear and controllable form according to Theorem 6.12.

Solution

(a) This system has relative degree $r = 1$ at all points $\mathbf{x}$ because $L_{\mathbf{g}}h(\mathbf{x}) = \exp(x_1)$. We can easily verify that conditions (1) and (2) of Theorem 6.7 (and 6.8) are satisfied. Thus, there exists an output function $h(\mathbf{x})$ for which the system has relative degree $r = n = 3$. The output function $h(\mathbf{x})$ must satisfy the condition:

$$\left(\frac{\partial h}{\partial \mathbf{x}}\right)[\mathbf{g}(\mathbf{x}) \quad ad_{\mathbf{f}}\mathbf{g}(\mathbf{x})] = \mathbf{0}$$

The function $h(\mathbf{x}) = x_1$ satisfies this condition. Therefore, the state feedback controller and the state transformation that give a linear controllable closed-loop system are:

$$u = \frac{-L_{\mathbf{f}}^3 h(\mathbf{x}) + v}{L_{\mathbf{g}}L_{\mathbf{f}}^2 h(\mathbf{x})} = \frac{2x_1x_2 - 2x_1x_3 - x_3 - x_1^2 + v}{\exp(x_1)}$$

$$z_1 = h(\mathbf{x}) = x_1, \quad z_2 = L_{\mathbf{f}}h(\mathbf{x}) = x_3 - x_2, \quad z_3 = L_{\mathbf{f}}^2 h(\mathbf{x}) = x_3 + x_1^2$$

(b) The original output $y = x_2$ expressed in terms of the new state variables z_1, z_2, and z_3 is $y = -z_2 + z_3 - z_1^2$. To check whether the entire (state and output) system can be transformed to linear controllable form, we must check if condition (2) of theorem is satisfied for the functions $\tilde{\mathbf{f}}(\mathbf{x})$ and $\tilde{\mathbf{g}}(\mathbf{x})$. Because here $L_{\mathbf{f}}h(\mathbf{x}) = 0$, we have:

$$\tilde{\mathbf{f}}(\mathbf{x}) = \mathbf{f}(\mathbf{x}) \text{ and } \tilde{\mathbf{g}}(\mathbf{x}) = \begin{bmatrix} 0 \\ 1 \\ 1 \end{bmatrix}$$

Now, we easily find that:

$$ad_{\tilde{\mathbf{f}}}\tilde{\mathbf{g}} = \begin{bmatrix} 0 \\ 0 \\ -1 \end{bmatrix}, \quad ad_{\tilde{\mathbf{f}}}^2\tilde{\mathbf{g}} = \begin{bmatrix} 1 \\ 0 \\ 1 \end{bmatrix}, \quad ad_{\tilde{\mathbf{f}}}^3\tilde{\mathbf{g}} = \begin{bmatrix} -1 \\ 0 \\ -2x_1 - 1 \end{bmatrix}$$

Thus,

$$\left[ad_{\tilde{\mathbf{f}}}^2\tilde{\mathbf{g}} \quad ad_{\tilde{\mathbf{f}}}^3\tilde{\mathbf{g}}\right] \neq \mathbf{0}$$

that is, the condition (2) of Theorem 6.11 is not satisfied. Therefore, the entire system (with its output) cannot be put in linear controllable form.

6.3.2 Differential-Drive Robot Input–Output Feedback Linearization and Trajectory Tracking

6.3.2.1 Kinematic Constraint Revisited

We consider the differential-drive WMR of Figure 6.4, where the (x_Q, y_Q) coordinates of the wheel axis midpoint Q are denoted by x and y, and work following the

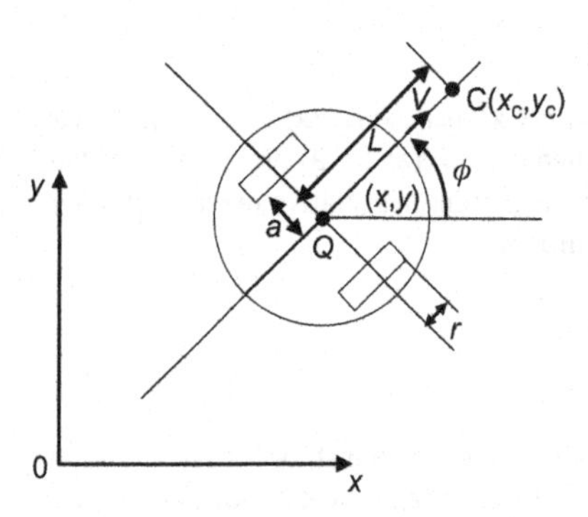

Figure 6.4 The differential-drive robot.

path of Yun and Yamamoto [6]. An analogous derivation for car-like robots is provided in Ref. [7].

The generalized coordinates of the WMR are:

$$\mathbf{x} = [x, y, \phi, \theta_{\mathrm{r}}, \theta_{\mathrm{l}}]^{\mathrm{T}}$$

The robot has the three constraints defined by Eqs. (2.39) and (2.41a) which can be written as:

$$\mathbf{M}(\mathbf{x})\dot{\mathbf{x}} = \mathbf{0} \tag{6.32a}$$

with:

$$\mathrm{M}(\mathbf{x}) = \begin{bmatrix} -\sin\phi & \cos\phi & 0 & 0 & 0 \\ -\cos\phi & -\sin\phi & -a & r & 0 \\ -\cos\phi & -\sin\phi & a & 0 & r \end{bmatrix} \tag{6.32b}$$

We define, the 5×2 dimensional matrix $\mathbf{B}(\mathbf{x})$:

$$\mathbf{B}(\mathbf{x}) = [\mathbf{b}_1(\mathbf{x})\vdots\mathbf{b}_2(\mathbf{x})] = \begin{bmatrix} \rho a\cos\phi & \rho a\cos\phi \\ \rho a\sin\phi & \rho a\sin\phi \\ \rho & -\rho \\ 1 & 0 \\ 0 & 1 \end{bmatrix} \tag{6.33}$$

(where $\rho = r/2a$) which has two independent columns and the property:

$$\mathbf{M}(\mathbf{x})\mathbf{B}(\mathbf{x}) = \mathbf{0} \tag{6.34}$$

Consider the field distribution $\mathbf{\Delta}(\mathbf{x})$ formed by the two columns of $\mathbf{B}(\mathbf{x})$. From Frobenius theorem we know that if $\mathbf{\Delta}(\mathbf{x})$ is involutive, all the constraints are integrable (holonomic). If the smallest involutive distribution that involves $\mathbf{\Delta}(\mathbf{x})$, denoted by $\overline{\boldsymbol{\Delta}}^*$ (see Eq. (6.14)) spans[1] the entire 5-dimensional space (where 5 is the dimensionality of $\mathbf{x}$), then all the constraints are holonomic. If $\dim \overline{\boldsymbol{\Delta}}^* = 5 - s$, then s constraints are holonomic and the remaining nonholonomic.

To determine the involutivity of $\boldsymbol{\Delta}(\mathbf{x})$, we find the Lie bracket of $\mathbf{b}_1(\mathbf{x})$ and $\mathbf{b}_2(\mathbf{x})$:

$$\mathbf{b}_3(x) = \begin{bmatrix}\mathbf{b}_1 & \mathbf{b}_2\end{bmatrix}(\mathbf{x}) = \frac{\partial \mathbf{b}_2}{\partial \mathbf{x}}\mathbf{b}_1 - \frac{\partial \mathbf{b}_1}{\partial \mathbf{x}}\mathbf{b}_2 = \begin{bmatrix} -r\rho \sin\phi \\ r\rho\cos\phi \\ 0 \\ 0 \\ 0 \end{bmatrix}$$

which is linearly independent of $\mathbf{b}_1(\mathbf{x})$ and $\mathbf{b}_2(\mathbf{x})$. Thus, one of the constraints is surely nonholonomic. We continue by computing the Lie bracket of $\mathbf{b}_1(\mathbf{x})$ and $\mathbf{b}_3(\mathbf{x})$, that is:

$$\mathbf{b}_4(\mathbf{x}) = \begin{bmatrix}\mathbf{b}_1 & \mathbf{b}_3\end{bmatrix}(\mathbf{x}) = \frac{\partial \mathbf{b}_3}{\partial \mathbf{x}}\mathbf{b}_1 - \frac{\partial \mathbf{b}_1}{\partial \mathbf{x}}\mathbf{b}_3 = \begin{bmatrix} -r\rho^2 \cos\phi \\ -r\rho^2\cos\phi \\ 0 \\ 0 \\ 0 \end{bmatrix}$$

which is linearly independent of $\mathbf{b}_1(\mathbf{x}), \mathbf{b}_2(\mathbf{x})$, and $\mathbf{b}_3(\mathbf{x})$. Here, the distribution spanned by $\mathbf{b}_1(\mathbf{x}), \mathbf{b}_2(\mathbf{x}), \mathbf{b}_3(\mathbf{x})$, and $\mathbf{b}_4(\mathbf{x})$ is involutive, that is:

$$\overline{\boldsymbol{\Delta}}^* = \text{span}[\mathbf{b}_1(\mathbf{x}), \mathbf{b}_2(\mathbf{x}), \mathbf{b}_3(\mathbf{x}), \mathbf{b}_4(\mathbf{x})]$$

Therefore, we conclude that two of the three constraints in Eqs. (6.32a) and (6.32b) are nonholonomic. To find the holonomic constraint, we subtract the second Eq. (2.41a) from the first Eq. (2.41a), that is, the third line of Eq. (6.32a) with $\mathbf{M}$ given by Eq.(6.32b) from the second line, and get:

$$2a\dot{\phi} = r(\dot{\theta}_r - \dot{\theta}_1)$$

which can be integrated to give:

$$\phi = \rho(\theta_r - \theta_1), \quad \rho = r/2a \tag{6.35}$$

[1] If every vector $\mathbf{v} \in V$ (where V is a vector space) can be written as a linear combination of the vectors of some set, then we say that this set *spans* the vector space V (as in Eq. (6.16b)).

This is obviously an holonomic constraint. We can therefore eliminate ϕ from the generalized coordinate vector, and obtain a four-dimensional vector:

$$\mathbf{x} = \begin{bmatrix} x_1 \\ x_2 \\ x_3 \\ x_4 \end{bmatrix} = \begin{bmatrix} x \\ y \\ \theta_r \\ \theta_l \end{bmatrix} \tag{6.36}$$

It follows that the two nonholonomic constraints are:

$$\begin{aligned} &\dot{x} \sin \phi - \dot{y} \cos \phi = 0 \\ &\dot{x} \cos \phi + \dot{y} \sin \phi = \rho a(\dot{\theta}_r + \dot{\theta}_l) \end{aligned} \tag{6.37a}$$

The above two constraints are written in matrix form as:

$$\mathrm{M}(\mathbf{x})\dot{\mathbf{x}} = \mathbf{0}, \quad \mathrm{M}(\mathbf{x}) = \begin{bmatrix} -\sin \phi & \cos \phi & 0 & 0 \\ -\cos \phi & -\sin \phi & \rho a & \rho a \end{bmatrix} \tag{6.37b}$$

6.3.2.2 Input–Output Linearization

Since the WMR has two inputs, it is natural to select an output equation with its two independent components, the coordinates of the point Q, that is:

$$\mathbf{y} = \begin{bmatrix} y_1 \\ y_2 \end{bmatrix} = \mathbf{h}(\mathbf{x}) = \begin{bmatrix} x \\ y \end{bmatrix} \tag{6.38}$$

The Lagrangian of this WMR which is equal to the kinetic energy (due to the absence of gravitational effects) is given by:

$$\begin{aligned} L = K = &\frac{1}{2}m(\dot{x}_1^2 + \dot{x}_2^2) + m_p \rho b(\dot{\theta}_r - \dot{\theta}_l)(\dot{x}_2 \cos \phi - \dot{x}_1 \sin \phi) \\ &+ \frac{1}{2}\mathrm{I_w}(\dot{\theta}_r^2 + \dot{\theta}_l^2) + \frac{1}{2}\mathrm{I}\rho^2(\dot{\theta}_r - \dot{\theta}_l)^2 \end{aligned} \tag{6.39}$$

where b is the displacement of Q from the center of gravity, and:

$$m = m_p + 2m_w, \quad I = I_p + 2m_w a^2 + 2I_m$$

Here, m is the total mass of the robot, m_p the mass of the platform, m_w the mass of each wheel, I_p the moment of inertia of the robot without the driving wheels and the motors' rotors, and I_m the moment of inertia of each driving wheel and its rotor about a wheel diameter. Working as usual, using Eqs. (3.14) and (3.15), we get the dynamic model (3.16) of the robot:

$$\mathbf{D}(\mathbf{q})\ddot{\mathbf{q}}+\mathbf{C}(\mathbf{q},\dot{\mathbf{q}})\dot{\mathbf{q}}+\mathbf{M}^{\mathrm{T}}(\mathbf{q})\boldsymbol{\lambda}=\mathbf{E}\boldsymbol{\tau} \tag{6.40}$$

where:

$$\mathbf{D}(\mathbf{q})=\begin{bmatrix} m & 0 & -m_{\mathrm{p}}\rho b\sin\phi & m_{\mathrm{p}}\rho b\sin\phi \\ 0 & m & m_{\mathrm{p}}\rho b\cos\phi & -m_{\mathrm{p}}\rho b\cos\phi \\ -m_{\mathrm{p}}\rho b\sin\phi & m_{\mathrm{p}}\rho b\cos\phi & I\rho^2+I_{\mathrm{w}} & -I\rho^2 \\ m_{\mathrm{p}}\rho b\sin\phi & -m_{\mathrm{p}}\rho b\cos\phi & -I\rho^2 & I\rho^2+I_{\mathrm{w}} \end{bmatrix}$$

$$\mathbf{C}(\mathbf{q},\dot{\mathbf{q}})\dot{\mathbf{q}}=\begin{bmatrix} -m_{\mathrm{p}}b\dot{\phi}^2\cos\phi \\ -m_{\mathrm{p}}b\dot{\phi}^2\sin\phi \\ 0 \\ 0 \end{bmatrix},\quad \mathbf{E}=\begin{bmatrix} 0 & 0 \\ 0 & 0 \\ 1 & 0 \\ 0 & 1 \end{bmatrix},\quad \boldsymbol{\tau}=\begin{bmatrix} \tau_{\mathrm{r}} \\ \tau_{\mathrm{l}} \end{bmatrix},\quad \boldsymbol{\lambda}=\begin{bmatrix} \lambda_1 \\ \lambda_2 \end{bmatrix}$$

The nonholonomic constraint matrix $\mathbf{M}(\mathbf{q})$ is given by Eq. (6.37b) with $\mathbf{q}$ in place of $\mathbf{x}$. Eliminating the constraint $\mathbf{M}^{\mathrm{T}}(\mathbf{q})\boldsymbol{\lambda}$ from Eq. (6.40), as usual, we get the unconstrained Lagrange model ((3.19a) and (3.19b)):

$$\overline{\mathbf{D}}(\mathbf{q})\dot{\mathbf{v}}+\overline{\mathbf{C}}(\mathbf{q},\dot{\mathbf{q}})\mathbf{v}=\overline{\mathbf{E}}\boldsymbol{\tau},\quad \dot{\mathbf{q}}=\mathbf{B}(\mathbf{q})\mathbf{v} \tag{6.41a}$$

where $\overline{\mathbf{D}}=\mathbf{B}^{\mathrm{T}}\mathbf{D}\mathbf{B}$, $\overline{\mathbf{C}}=\mathbf{B}^{\mathrm{T}}\mathbf{D}\dot{\mathbf{B}}+\mathbf{B}^{\mathrm{T}}\mathbf{C}\mathbf{B}$, and $\overline{\mathbf{E}}=\mathbf{B}^{\mathrm{T}}\mathbf{E}=\mathbf{I}_{2\times 2}$, and:

$$\mathbf{B}(\mathbf{q})=\begin{bmatrix} \rho a\cos\phi & \rho a\cos\phi \\ \rho a\sin\phi & \rho a\sin\phi \\ 1 & 0 \\ 0 & 1 \end{bmatrix} \tag{6.41b}$$

Choosing the following state vector:

$$\begin{aligned}\mathbf{x}&=[x_1,x_2,x_3,x_4,x_5,x_6]^{\mathrm{T}}\\ &=[x,y,\theta_{\mathrm{r}},\theta_{\mathrm{l}}\vdots v_1,v_2]^{\mathrm{T}}\\ &=[\mathbf{q}^{\mathrm{T}}\vdots\mathbf{v}^{\mathrm{T}}]^{\mathrm{T}}\end{aligned} \tag{6.42}$$

the model (6.41a) can be written in the following affine state-space form:

$$\dot{\mathbf{x}}=\mathbf{f}(\mathbf{x})+\mathbf{g}(\mathbf{x})\mathbf{u}(t),\quad \mathbf{u}=\boldsymbol{\tau} \tag{6.43a}$$

where:

$$\mathbf{f}(\mathbf{x})=\begin{bmatrix} \mathbf{B}\mathbf{v} \\ -\overline{\mathbf{D}}^{-1}\overline{\mathbf{C}}\mathbf{v} \end{bmatrix},\quad \mathbf{g}(\mathbf{x})=\begin{bmatrix} \mathbf{0} \\ \overline{\mathbf{D}}^{-1}\overline{\mathbf{E}} \end{bmatrix}=\begin{bmatrix} \mathbf{0} \\ \overline{\mathbf{D}}^{-1} \end{bmatrix} \tag{6.43b}$$

We will now examine the input–output linearization of Eqs. (6.43a) and (6.43b) with output (6.38) using the state feedback control law:

$$\mathbf{u}(t) = \mathbf{F}(\mathbf{x}) + \mathbf{G}(\mathbf{x})\upsilon(t) \tag{6.44a}$$

where $\upsilon(t)$ is the new control variable, and:

$$\mathbf{F}(\mathbf{x}) = \overline{\mathbf{E}}^{-1}\mathbf{C}\mathbf{v} = \overline{\mathbf{C}}\mathbf{v}, \quad \mathbf{G}(\mathbf{x}) = \mathbf{E}^{-1}\overline{\mathbf{D}} = \overline{\mathbf{D}} \tag{6.44b}$$

according to the *static computed torque* method.

To this end, we introduce Eqs. (6.44a) and (6.44b) into Eqs. (6.43a) and (6.43b) and get the closed-loop system:

$$\dot{\mathbf{x}}(t) = \mathbf{f}_c(\mathbf{x}) + \mathbf{g}_c(\mathbf{x})\upsilon(t) \tag{6.45a}$$

where:

$$\mathbf{f}_c(\mathbf{x}) = \mathbf{f}(\mathbf{x}) + \mathbf{g}(\mathbf{x})\mathbf{F}(\mathbf{x}) = \begin{bmatrix} \mathbf{B}\mathbf{v} \\ \vdots \\ \mathbf{0} \end{bmatrix} \tag{6.45b}$$

$$\mathbf{g}_c(\mathbf{x}) = \mathbf{g}(\mathbf{x})\mathbf{G}(\mathbf{x}) = \begin{bmatrix} \mathbf{0} \\ \vdots \\ \mathbf{I}_{2\times 2} \end{bmatrix} \tag{6.45c}$$

To see if the control law ((6.44a) and (6.44b)) input–output linearizes the system ((6.45a)–(6.45c) and (6.38)), we differentiate the output $\mathbf{y}(t)$, and obtain:

$$\begin{aligned} \dot{\mathbf{y}} &= \left(\frac{\partial \mathbf{h}}{\partial \mathbf{x}}\right)\dot{\mathbf{x}} = \left(\frac{\partial \mathbf{h}}{\partial \mathbf{x}}\right)[\mathbf{f}_c(\mathbf{x}) + \mathbf{g}_c(\mathbf{x})\upsilon] \\ &= \mathbf{B}_1(\mathbf{x})\mathbf{v} \end{aligned}$$

where:

$$\frac{\partial \mathbf{h}}{\partial \mathbf{x}} = [\mathbf{I}_{2\times 2} \vdots \mathbf{0}], \quad \mathbf{B}_1(\mathbf{x}) = \begin{bmatrix} \rho a \cos\phi & \rho a \cos\phi \\ \rho a \sin\phi & \rho a \sin\phi \end{bmatrix} \tag{6.46}$$

We see that $\dot{\mathbf{y}}$ does not involve the input υ, and so we compute the second derivative $\ddot{\mathbf{y}}$, namely:

$$\begin{aligned} \ddot{\mathbf{y}} &= \mathbf{B}_1(\mathbf{x})\dot{\mathbf{v}} + \dot{\mathbf{B}}_1(\mathbf{x})\mathbf{v} \\ &= \mathbf{B}_1(\mathbf{x})\upsilon + \dot{\mathbf{B}}_1(\mathbf{x})\mathbf{v} \end{aligned} \tag{6.47a}$$

where:

$$\dot{\mathbf{B}}_1(\mathbf{x})\mathbf{v} = \rho^2 a(v_1^2 - v_2^2)\begin{bmatrix} -\sin\phi \\ \cos\phi \end{bmatrix} \tag{6.47b}$$

We see that now the input υ appears explicitly in $\ddot{\mathbf{y}}$, and so $\mathbf{B}_1(\mathbf{x})$ is the decoupling matrix. However, $\mathbf{B}_1(\mathbf{x})$ is not invertible, which means that the system cannot be decoupled by a static state feedback of the form ((6.44a) and (6.44b)). Actually, no static state feedback exists that input–output linearizes the differential-drive WMR when the outputs are the coordinates of the point Q.

However, as it is given below, the system can be input–output linearized by dynamic state feedback law of the form [6]:

$$\dot{\mathbf{z}} = \mathbf{f}_1(\mathbf{x}, \mathbf{z}) + \mathbf{g}_1(\mathbf{x}, \mathbf{z})\mathbf{w} \tag{6.48a}$$

$$\upsilon = \mathbf{F}(\mathbf{x}, \mathbf{z}) + \mathbf{G}(\mathbf{x}, \mathbf{z})\mathbf{w} \tag{6.48b}$$

To this end, we start by using the static decoupling law ((6.44a) and (6.44b)) to linearize and decouple one of the outputs, which can be done since rank $\mathbf{B}_1(\mathbf{x}) = 1$ for all $\mathbf{x}$. Choosing to linearize the output y_1, and introducing the following feedback law into Eq. (6.47a):

$$\upsilon = \mathbf{F}_1(\mathbf{x}) + \mathbf{G}_1(\mathbf{x})\mathbf{w}, \quad \mathbf{w} = [w_1, w_2]^{\mathrm{T}} \tag{6.49a}$$

where:

$$\mathbf{F}_1(\mathbf{x}) = \begin{bmatrix} \rho(v_1^2 - v_2^2)tg\phi \\ 0 \end{bmatrix}, \quad \mathbf{G}_1(\mathbf{x}) = \begin{bmatrix} 1/\rho a \cos\phi & 1 \\ 0 & -1 \end{bmatrix} \tag{6.49b}$$

we get:

$$\ddot{\mathbf{y}} = \mathbf{A}_2(\mathbf{x}) + \mathbf{B}_2(\mathbf{x})\begin{bmatrix} w_1 \\ w_2 \end{bmatrix}$$

where:

$$\mathbf{A}_2(\mathbf{x}) = \begin{bmatrix} 0 \\ \rho^2 a(v_1^2 - v_2^2)/\cos\phi \end{bmatrix}, \quad \mathbf{B}_2(\mathbf{x}) = \begin{bmatrix} 1 & 0 \\ tg\,\phi & 0 \end{bmatrix}$$

We see that:

$$\ddot{y}_1 = w_1$$

that is, y_1 is linearized and decoupled, controlled only by w_1. But y_2 is still nonlinear and controlled by w_1.

Then, we continue by introducing the static feedback law ((6.49a) and (6.49b)) into Eqs. (6.45a)–(6.45c), which yields the new closed-loop state-space equation:

$$\begin{aligned}\dot{\mathbf{x}} &= \mathbf{f}_c(\mathbf{x}) + \mathbf{g}_c(\mathbf{x})\boldsymbol{\upsilon} \\ &= \mathbf{f}_c(\mathbf{x}) + \mathbf{g}_c(\mathbf{x})[\mathbf{F}_1(\mathbf{x}) + \mathbf{G}_1(\mathbf{x})\mathbf{w}] \\ &= \mathbf{f}_c^1(\mathbf{x}) + \mathbf{g}_c^1(\mathbf{x})\mathbf{w}\end{aligned} \tag{6.50a}$$

where:

$$\mathbf{f}_c^1(\mathbf{x}) = \begin{bmatrix} \mathbf{Bv} \\ \rho(v_1^2 - v_2^2)tg\ \phi \\ 0 \end{bmatrix}, \quad \mathbf{g}_c^1(\mathbf{x}) = \begin{bmatrix} 0 & 0 \\ 1/\rho a \cos\phi & 1 \\ 0 & -1 \end{bmatrix} \tag{6.50b}$$

Now, differentiating y_2 with the system ((6.50a) and (6.50b)), and considering w_1 as a time varying parameter, we get:

$$\begin{aligned}\dot{y}_2 &= \rho a(v_1 + v_2)\sin\phi \\ \ddot{y}_2 &= \rho^2 a(v_1^2 - v_2^2)/\cos\phi + (tg\ \phi)w_1\end{aligned}$$

$$\begin{aligned}\dddot{y}_2 &= \rho^3 a(v_1^2 - v_2^2)(v_1 - v_2)(\sin\phi)/\cos^2\phi + [\rho(v_1 - v_2)/\cos^2\phi]w_1 \\ &\quad + (2\rho^2 a v_1/\cos\phi)[\rho(v_1^2 - v_2^2)tg\ \phi + (1/\rho a \cos\phi)w_1] \\ &\quad + (tg\ \phi)\dot{w}_1 + [2\rho^2 a(v_1 + v_2)/\cos\phi]w_2\end{aligned}$$

We see that now w_2 appears explicitly in $\dddot{y}_2$, which can be written as:

$$\dddot{y}_2 = P_1(\mathbf{x}) + P_2(\mathbf{x})w_1 + P_3(\mathbf{x})\dot{w}_1 + P_4(\mathbf{x})w_2 \tag{6.51}$$

where $P_i(\mathbf{x})$ have obvious definition. Finally, in view of Eq. (6.51), y_2 can be linearized by using the feedback law:

$$w_2 = P_4^{-1}(\mathbf{x})(\upsilon_2^* - P_1(\mathbf{x}) - P_2(\mathbf{x})w_1 - P_3(\mathbf{x})\dot{w}_1) \tag{6.52a}$$

The derivative $\dot{w}_1$ appearing in Eq. (6.52a) can be eliminated by using an integrator to the first input channel, that is:

$$\dot{z} = \upsilon_1^*, \quad w_1 = z \tag{6.52b}$$

From the above, it follows that the dynamic state feedback law has the form:

$$\begin{aligned}\dot{z} &= f_2(\mathbf{x}, z) + \mathbf{g}_2(\mathbf{x}, z)\boldsymbol{\upsilon}^* \\ \mathbf{w} &= \begin{bmatrix} w_1 \\ w_2 \end{bmatrix} = \mathbf{F}_2(\mathbf{x}, z) + \mathbf{G}_2(\mathbf{x}, z)\boldsymbol{\upsilon}^*, \quad \boldsymbol{\upsilon}^* = [\upsilon_1^*, \upsilon_2^*]^T\end{aligned} \tag{6.53}$$

where:

$$f_2(\mathbf{x}, z) = 0, \quad \mathbf{g}_2(\mathbf{x}, z) = [1, \ 0]$$

$$\mathbf{F}_2(\mathbf{x}, z) = \begin{bmatrix} z \\ -P_4^{-1}(\mathbf{x})[P_1(\mathbf{x}) + P_2(\mathbf{x})z] \end{bmatrix}, \quad \mathbf{G}_2(\mathbf{x}, z) = \begin{bmatrix} 0 & 0 \\ -P_4^{-1}(\mathbf{x})P_3(\mathbf{x}) & P_4^{-1}(\mathbf{x}) \end{bmatrix} \tag{6.54}$$

Using this dynamic feedback law, we get the overall linearized and decoupled system:

$$\dddot{y}_1 = v_1^*, \quad \dddot{y}_2 = v_2^* \tag{6.55}$$

which is of third order.

6.3.2.3 Trajectory Tracking Control

Having available the model (6.55):

$$\dddot{y}_1 = w_1, \quad \dddot{y}_2 = w_2$$

and the desired trajectory$[y_{1\mathrm{d}}(t), y_{2\mathrm{d}}(t)]^{\mathrm{T}}$, we select as usual the linear controllers:

$$w_1 = \dddot{y}_{1\mathrm{d}} + K_{21}(\ddot{y}_{1\mathrm{d}} - \ddot{y}_1) + K_{11}(\dot{y}_{1\mathrm{d}} - \dot{y}_1) + K_{01}(y_{1\mathrm{d}} - y_1) \tag{6.56a}$$

$$w_2 = \dddot{y}_{2\mathrm{d}} + K_{22}(\ddot{y}_{2\mathrm{d}} - \ddot{y}_2) + K_{12}(\dot{y}_{2\mathrm{d}} - \dot{y}_2) + K_{02}(y_{2\mathrm{d}} - y_2) \tag{6.56b}$$

The error dynamics for y_1 and y_2 are:

$$\dddot{\tilde{y}}_1 + K_{21}\ddot{\tilde{y}}_1 + K_{11}\dot{\tilde{y}}_1 + K_{01}\tilde{y}_1 = 0$$
$$\dddot{\tilde{y}}_2 + K_{22}\ddot{\tilde{y}}_2 + K_{12}\dot{\tilde{y}}_2 + K_{02}\tilde{y}_2 = 0$$

The corresponding characteristic polynomials are:

$$\chi_i(\lambda) = \lambda_i^3 + K_{2i}\lambda_i^2 + K_{1i}\lambda_i + K_{0i} \quad (i = 1, 2) \tag{6.57}$$

Therefore, selecting appropriate values for the position, velocity, and acceleration gains K_{0i}, K_{1i}, and K_{2i} $(i = 1, 2)$ we assure, as usual, convergence to the desired trajectory, with a desired rate of convergence, from any initial state. Especially, if the initial point belongs to the desired trajectory, that is:

$$\tilde{y}_i(0) = y_{id}(0) - y_i(0) = 0, \quad \dot{\tilde{y}}_i(0) = \dot{y}_{id}(0) - \dot{y}_i(0) = 0, \quad \ddot{\tilde{y}}_i(0) = \ddot{y}_{id}(0) - \ddot{y}_i(0) = 0$$

the state will always remain on this trajectory.

Example 6.7

It is desired to derive a kinematic feedback input–output linearizing controller for the car-like WMR (2.52):

$$\begin{bmatrix} \dot{x} \\ \dot{y} \\ \dot{\phi} \\ \dot{\psi} \end{bmatrix} = \begin{bmatrix} \cos\phi \\ \sin\phi \\ (tg\,\psi)/D \\ 0 \end{bmatrix} v_1 + \begin{bmatrix} 0 \\ 0 \\ 0 \\ 1 \end{bmatrix} v_2 \tag{6.58}$$

employing the (2,4) chained model (2.69):

$$\begin{aligned} \dot{x}_1 &= u_1 \\ \dot{x}_2 &= u_2 \\ \dot{x}_3 &= x_2 u_1 \\ \dot{x}_4 &= x_3 u_1 \end{aligned} \tag{6.59}$$

Solution

We select as output vector **y** the following [8]:

$$\mathbf{y}(t) = \mathbf{h}(\mathbf{x}) = \begin{bmatrix} x(t) \\ y(t) \end{bmatrix} = \begin{bmatrix} y_1(t) \\ y_2(t) \end{bmatrix} \tag{6.60}$$

Knowledge of this output determines fully the entire trajectory $[x(t), y(t), \phi(t), \psi(t)]^{\mathrm{T}}$, and the corresponding input $[v_1(t), v_2(t)]^{\mathrm{T}}$. Indeed, from the first two rows of Eq. (6.58) we get the first input:

$$v_1(t) = \pm\sqrt{\dot{x}^2(t) + \dot{y}^2(t)} \tag{6.61}$$

where the sign "+" corresponds to the forward motion, and the sign "−" to the back motion. Division of y by x gives:

$$\phi(t) = tg^{-1}[y(t)/x(t)] \tag{6.62}$$

Differentiating the first two rows of Eq. (6.58) gives:

$$\dot{\phi}(t) = [\ddot{y}(t)\dot{x}(t) - \ddot{x}(t)\dot{y}(t)]/v_1^2(t) \tag{6.63}$$

Introducing Eq. (6.63) into the third row of Eq. (6.58) we obtain:

$$\psi(t) = tg^{-1}D[\ddot{y}(t)\dot{x}(t) - \ddot{x}(t)\dot{y}(t)]/v_1^3(t) \tag{6.64}$$

which is valid in the interval $(-\pi/2, \pi/2)$. Finally, the second input $v_2(t)$ can be obtained by introducing the derivative of Eq. (6.64) into $\dot{\psi}(t) = v_2(t)$, namely:

$$v_2(t) = v_{\mathrm{num}}(t)/[v_1^6 + D^2(\ddot{y}\dot{x} - \ddot{x}\dot{y})^2] \tag{6.65}$$

where:

$$v_{\text{num}}(t) = Dv_1\{(\dddot{y}\dot{x} - \dddot{x}\dot{y})v_1^2 - 3(\dot{y}\ddot{x} - \ddot{x}\dot{y})(\dot{x}\ddot{x} + \dot{y}\ddot{y})\}$$

Equations (6.61)–(6.65) determine all the state and input variables in terms of the output variables $x(t)$ and $y(t)$. This means that if we are given desired outputs $x_d(t)$ and $y_d(t)$, they specify via Eqs. (6.61)–(6.65) the desired trajectories for all the other state variables of the car-like WMR.

On the basis of the above, we can use the following input–output model for the car-like robot:

$$\dot{\mathbf{y}} = \mathbf{H}_1(\phi)\mathbf{v}, \quad \mathbf{y} = \begin{bmatrix} x \\ y \end{bmatrix}, \quad \mathbf{H}_1(\phi) = \begin{bmatrix} \cos\phi & 0 \\ \sin\phi & 0 \end{bmatrix}, \quad \mathbf{v} = \begin{bmatrix} v_1 \\ v_2 \end{bmatrix} \tag{6.66}$$

We see that at least one of the inputs v_1 and v_2 appears in $\dot{y}_1$ and $\dot{y}_2$. Therefore, $\mathbf{H}_1(\phi)$ is the decoupling matrix, which however is not invertible and cannot be diagonalized by any static similarity transformation. To overcome this difficulty, we must use a dynamic controller:

$$\begin{aligned} \mathbf{v} &= \mathbf{F}(\mathbf{y}, \mathbf{z}) + \mathbf{G}(\mathbf{y}, \mathbf{z})\mathbf{w} \\ \dot{\mathbf{z}} &= \mathbf{f}_1(\mathbf{y}, \mathbf{z}) + \mathbf{g}_1(\mathbf{y}, \mathbf{z})\mathbf{w} \end{aligned} \tag{6.67}$$

where $\mathbf{z}(t)$ is the dynamic controller state and $\mathbf{w}(t)$ is the new input.

The controller Eq. (6.67) will be applied to the chain representation Eq. (6.59) of the WMR, where (see Eq. (2.67)):

$$x_1 = x, \quad x_2 = (tg\,\psi)/D\cos^3\phi, \quad x_3 = tg\,\phi, \quad x_4 = y$$

and so (due to (2.68)):

$$\mathbf{y} = \begin{bmatrix} x \\ y \end{bmatrix} = \begin{bmatrix} x_1 \\ x_4 \end{bmatrix}, \quad \dot{\mathbf{y}} = \begin{bmatrix} \dot{x}_1 \\ \dot{x}_4 \end{bmatrix} = \begin{bmatrix} 1 & 0 \\ x_3 & 0 \end{bmatrix}\begin{bmatrix} u_1 \\ u_2 \end{bmatrix} \tag{6.68}$$

In Eq. (6.68), u_2 does not appear, and the decoupling matrix:

$$\mathbf{H}_2 = \begin{bmatrix} 1 & 0 \\ x_3 & 0 \end{bmatrix}$$

is again noninvertible. Thus, a dynamic feedback controller of the form (6.67) will be used. We start by adding an integrator with state z_1 to the first input u_1, that is:

$$\dot{z}_1 = u_1^*, \quad u_1 = z_1 \tag{6.69}$$

where u_1^* is an auxiliary control input. Now, the model (6.68) becomes:

$$\dot{\mathbf{y}} = \begin{bmatrix} z_1 \\ x_3 z_1 \end{bmatrix} \tag{6.70}$$

Differentiating Eq. (6.70) we get, from the third line of Eq. (6.59):

$$\begin{aligned}\mathbf{y} &= \begin{bmatrix} \dot{z}_1 \\ \dot{x}_3 z_1 + x_3 \dot{z}_1 \end{bmatrix} = \begin{bmatrix} 0 \\ \dot{x}_3 z_1 \end{bmatrix} + \begin{bmatrix} 1 & 0 \\ x_3 & 0 \end{bmatrix} \begin{bmatrix} u_1^* \\ u_2 \end{bmatrix} \\ &= \begin{bmatrix} 0 \\ x_2 z_1^2 \end{bmatrix} + \begin{bmatrix} 1 & 0 \\ x_3 & 0 \end{bmatrix} \begin{bmatrix} u_1^* \\ u_2 \end{bmatrix}\end{aligned} \tag{6.71}$$

Again, u_2 does not appear in $\ddot{\mathbf{y}}$, and so we add one more integrator:

$$\dot{z}_2 = u_1^{**}, \quad u_1^* = z_2 \tag{6.72}$$

to obtain:

$$\ddot{y} = \mathbf{H}_0 + \mathbf{H}_3 \begin{bmatrix} u_1^{**} \\ u_2 \end{bmatrix}, \quad \mathbf{H}_0 = \begin{bmatrix} 0 \\ 3x_2 z_1 z_2 \end{bmatrix}, \quad \mathbf{H}_3 = \begin{bmatrix} 1 & 0 \\ x_3 & z_1^2 \end{bmatrix} \tag{6.73}$$

Now, the decoupling matrix $\mathbf{H}_3$ is invertible for $z_1 \neq 0$, and so selecting the feedback control law:

$$\begin{aligned}\begin{bmatrix} u_1^{**} \\ u_2 \end{bmatrix} &= \mathbf{H}_3^{-1} \left\{ \begin{bmatrix} w_1 \\ w_2 \end{bmatrix} - \mathbf{H}_0 \right\} \\ &= \begin{bmatrix} w_1 \\ (w_2 - x_3 w_1 - 3x_2 z_1 z_2)/z_1^2 \end{bmatrix}\end{aligned} \tag{6.74}$$

we get the linear and fully decoupled system:

$$\ddot{y}_1 = w_1, \quad \ddot{y}_2 = w_2 \tag{6.75}$$

as required. This is of the same form as the linearized-decoupled model of the differential-drive robot.

Example 6.8

Show that by using output as $\mathbf{y}$ the coordinates of a point $C(x_c, y_c)$ in front of the car-like robot that lies on its steering line (at an angle $\phi + \psi$ in the O_{xy} coordinate frame), it is possible to decouple the inputs and outputs using a static feedback linearizing controller.

Solution

Consider the bicycle model of the car-like robot shown in Figure 6.5, where $L \neq 0$.

From Figure 6.5 we see that [8]:

$$\mathbf{y} = \begin{bmatrix} y_1 \\ y_2 \end{bmatrix} = \begin{bmatrix} x_c \\ y_c \end{bmatrix} = \begin{bmatrix} x + D\cos\phi + L\cos(\phi + \psi) \\ y + D\sin\phi + L\sin(\phi + \psi) \end{bmatrix} \quad (L \neq 0)$$

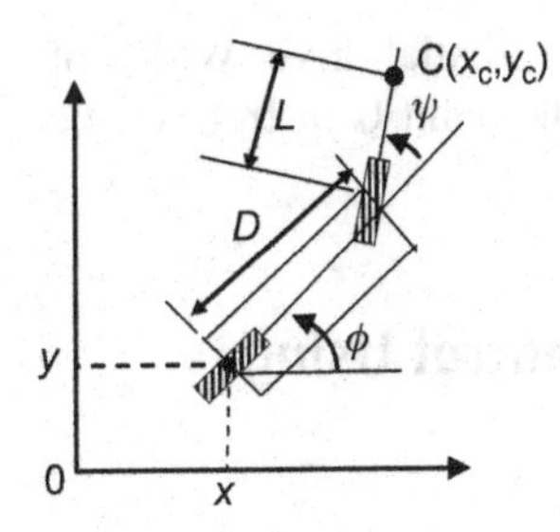

Figure 6.5 Using the coordinates x_c, y_c of the point C as output components y_1 and y_2.

Differentiating **y** we get:

$$\dot{\mathbf{y}} = \mathbf{H}(\phi, \psi)\mathbf{v}, \quad \mathbf{v} = [v_1, v_2]^T$$

where:

$$\mathbf{H}(\phi, \psi) = \begin{bmatrix} \cos\phi - [\sin\phi + L\sin(\phi+\psi)/D]tg\,\psi & -L\sin(\phi+\psi) \\ \sin\phi + [\cos\phi + L\cos(\phi+\psi)/D]\,tg\,\psi & L\cos(\phi+\psi) \end{bmatrix}$$

Here $\det \mathbf{H}(\phi, \psi) = L/\cos\psi \neq 0$ for $0° \leq \psi < 90°$, and so the system can be linearized and decoupled by a state feedback law:

$$\mathbf{v} = \mathbf{F}(\phi, \psi)\boldsymbol{\upsilon}$$

where $\mathbf{F}(\phi, \psi) = \mathbf{H}^{-1}(\phi, \psi)$. The resulting closed-loop system is:

$$\begin{aligned} \dot{y}_1 &= \upsilon_1 \\ \dot{y}_2 &= \upsilon_2 \\ \dot{\phi} &= (1/D)[\cos(\phi+\psi)\upsilon_1 + \sin(\phi+\psi)\upsilon_2]\sin\psi \\ \dot{\psi} &= -[(1/D)\cos(\phi+\psi)\sin\psi + (1/L)\sin(\phi+\psi)]\upsilon_1 \\ &\quad -[(1/D)\sin(\phi+\psi)\sin\psi - (1/L)\cos(\phi+\psi)]\upsilon_2 \end{aligned}$$

Obviously, this system is a first-order input—output linearized and decoupled system, but not input to state-decoupled. This means that the outputs $y_1(t)$ and $y_2(t)$ can track any desired trajectories $y_{1d}(t)$ and $y_{2d}(t)$ by a standard linear state feedback law:

$$\upsilon_i = \dot{y}_{di} + k_{pi}(y_{di} - y_i), \quad k_{pi} > 0 \quad (i = 1, 2)$$

as usual, but the trajectories of $\phi(t)$ and $\psi(t)$ cannot follow specified (desired) trajectories. This problem can be alleviated by applying the Lyapunov-based control technique to the closed-loop system as a whole.

Remark 1 Another selection of the output vector $y = [y_1, y_2]^T$ which can linearize and decouple the car-like robot is to use the first two variables $x_1 = x$ and $x_2 = (tg\,\psi/D\cos^3\phi)$ of the (2,4) chained model (6.59) studied in Example 6.7 (see Eq. (2.67)).

Remark 2 Similar results can be obtained for the differential-drive WMR of Figure 6.4 if we use as outputs the coordinates x_c, y_c of the point C in front of the vehicle, that is, $\mathbf{y} = [x_c, y_c]^T$.

6.4 Mobile Robot Feedback Stabilizing Control Using Invariant Manifolds

The mobile robot stabilization approach which is based on invariant manifolds leads to nonlinear controllers directly without prior feedback linearization. Its mathematical power and beauty have been applied by many scientists, and this approach is further enhanced in many ways. The two general model categories used for mobile robots and other nonholonomic systems are the nonholonomic (Brockett) integrators (simple, double, extended) and the (2-n)-chain models. This approach treats in an elegant way the nonholonomic constraints, and an extended literature exists with a large repertory of different controllers (that are valid under different conditions) [5,9–15]. On the theoretical side, the key result is Brockett's stabilization condition (theorem), which establishes the fact that nonholonomic systems cannot be asymptotically stabilized to a single equilibrium state using a smooth (or even continuous) static feedback controller (see Theorem 6.6).

6.4.1 Stabilizing Control of Unicycle in Chained Model Form

We have seen in Section 2.3.5.1 that the unicycle kinematic model $\dot{x} = v_Q \cos \phi$, $\dot{y} = v_Q \sin \phi$, $\dot{\phi} = v_\phi$ can be transformed to the (2,3) chain form (2.63):

$$\begin{aligned} \dot{z}_1 &= u_1 \\ \dot{z}_2 &= u_2 \\ \dot{z}_3 &= z_2 u_1 \end{aligned} \tag{6.76}$$

where $u_1 = v_\phi$ and $u_2 = v_Q - z_3 u_1$. The problem to be considered here is to determine a static quasi-continuous state feedback control law $\mathbf{u} = \mathbf{u}(\mathbf{z})$ which asymptotically stabilizes the system (6.76) to the origin [10].

One can directly verify that the control law:

$$\mathbf{u} = [-k_1 z_1, -k_1 z_2]^T, \quad \mathbf{u} = [u_1, u_2]^T, \quad k_1 > 0 \tag{6.77}$$

makes the origin $\mathbf{z} = [z_1, z_2, z_3]^T = \begin{bmatrix} 0 & 0 & 0 \end{bmatrix}^T$ globally asymptotically stable. The resulting closed-loop system is:

$$\dot{\mathbf{z}} = \mathbf{f}(\mathbf{z}), \quad \mathbf{f}(\mathbf{z}) = [-k_1 z_1, -k_1 z_2, -k_1 z_1 z_2]^T \tag{6.78}$$

We can easily verify that the manifold:

$$M = \{\mathbf{z} \in R^3 : s(\mathbf{z}) = z_1 z_2 - 2z_3 = 0\} \tag{6.79}$$

is an *invariant manifold* of the system (6.78). Indeed, we get (see Definition 6.9):

$$\begin{aligned} L_f s(\mathbf{z}) &= \sum_{i=1}^{3} \frac{\partial s}{\partial z_i} f_i(z_1, z_2, z_3) \\ &= z_2(-k_1 z_1) + z_1(-k_1 z_2) + (-2)(-k_1 z_1 z_2) \\ &= 0 \end{aligned}$$

The time derivative of $s(\mathbf{z})$ along the trajectories of (6.78) is found to be:

$$\dot{s}(\mathbf{z}) = z_1 u_2 - z_2 u_1 = 0$$

which means that the trajectories once on the surface (manifold) Mremain there. Furthermore, since $z_1(t) \to 0$ and $z_2(t) \to 0$, as $t \to \infty$, for any trajectory on M we have $z_3(t) \to 0$ as $t \to \infty$, and so:

$$[z_1(t), z_2(t), z_3(t)]^{\mathrm{T}} \to [0 \quad 0 \quad 0]^{\mathrm{T}} \text{ as } t \to \infty$$

One can observe that M does not depend on k_1.

We will now construct a stabilizing control law, which makes M an attractive manifold. To this end, the feedback controller (6.77) must be enhanced such that to satisfy the *attractivity condition*(6.19), namely:

$$\begin{aligned} &\text{If } s(\mathbf{z}) < 0, \text{ then } \dot{s}(\mathbf{z}) > 0 \quad \mathbf{z} \in R^3 \\ &\text{If } s(\mathbf{z}) > 0, \text{ then } \dot{s}(\mathbf{z}) < 0 \quad \mathbf{z} \in R^3 \end{aligned} \tag{6.80}$$

A possible enhancement of the controller (6.77) is the following:

$$\mathbf{u} = \begin{bmatrix} -k_1 z_1 - \dfrac{z_2 \mathrm{H}(s)}{z_1^2 + z_2^2} \\ -k_1 z_2 + \dfrac{z_1 \mathrm{H}(s)}{z_1^2 + z_2^2} \end{bmatrix} \quad \text{for } z_1^2 + z_2^2 \neq 0 \tag{6.81a}$$

where the scalar mapping $\mathrm{H}(s)$ satisfies the condition:

$$s\mathrm{H}(s) < 0 \tag{6.81b}$$

to assure the satisfaction of Eq. (6.80). A function $\mathrm{H}(s)$ with this property is:

$$\mathrm{H}(s) = -k_2 s \tag{6.82}$$

The closed-loop system obtained using the controller ((6.81a), (6.81b), and (6.82)) is:

$$\begin{bmatrix} \dot{z}_1 \\ \dot{z}_2 \\ \dot{s} \end{bmatrix} = \mathbf{f}\left(\begin{bmatrix} z_1 \\ z_2 \\ s \end{bmatrix}\right), \quad \mathbf{f} = \begin{bmatrix} -k_1 z_1 + \dfrac{k_2 z_2 s}{z_1^2 + z_2^2} \\ -k_1 z_2 - \dfrac{k_2 z_1 s}{z_1^2 + z_2^2} \\ -k_2 s(\mathbf{z}) \end{bmatrix} \tag{6.83}$$

where $s(\mathbf{z})$ is given by Eq. (6.79). It is remarked that the transformation:

$$\mathbf{z} \rightarrow \tilde{\mathbf{z}}, \quad \mathbf{z} = \begin{bmatrix} z_1 \\ z_2 \\ z_3 \end{bmatrix}, \quad \tilde{\mathbf{z}} = \begin{bmatrix} z_1 \\ z_2 \\ s \end{bmatrix} \tag{6.84}$$

is a *diffeomorphism*.

To summarize, with the controller ((6.81a), (6.81b), and (6.82)) and $k_2 > 2k_1$, the trajectory $\tilde{\mathbf{z}}(t) = [z_1(t), z_2(t), s(t)]^T$ is bounded for all $t \geq 0$, and converges exponentially to $[0,0,0]^T$ with a decay rate greater than or equal to k_1. Likewise, the control $\mathbf{u}(t)$ is bounded for all $t \geq 0$ and tends exponentially to $[0,0]^T$ with a decay rate at least k_1. In other words, for any initial condition with $z_1(0)^2 + z_2^2(0) \neq 0$, the feedback control law ((6.81a), (6.81b), and (6.82)) is well defined for all $t \geq 0$ and drives the unicycle to the origin, while avoiding the manifold:

$$M^* = \{\mathbf{z} \in R^3 : \; z_1^2 + z_2^2 = 0, \quad z_1 z_2 - 2z_3 \neq 0\}$$

6.4.2 Dynamic Control of Differential-Drive Robots Modeled by the Double Brockett Integrator

The nonlinear controller designed in Section 6.4.1 is a stabilizing controller for the kinematic performance of the unicycle. Here, we will derive a quasi-continuous dynamic controller, using invariant manifolds, for the combined kinematic and dynamic performance of a differential-drive WMR using the double Brockett integrator model (2.60) [15]. The full dynamic model of the WMR is (see Eqs. (3.22), (3.23a), (3.23b), and (3.31)):

$$\dot{x} = v \cos \phi \tag{6.85a}$$

$$\dot{y} = v \sin \phi \tag{6.85b}$$

$$\dot{\phi} = \omega \tag{6.85c}$$

$$m\dot{v} = F \tag{6.85d}$$

$$I\dot{\omega} = N \tag{6.85e}$$

where the symbols have the standard meaning, and the index Q was deleted from x_Q, y_Q for notational convenience.

The chained form of the kinematic equations (6.85a)–(6.85c) is given by Eq. (2.63) (see also Eq. (6.76)). Defining new variables:

$$x_1 = z_1, \quad x_2 = z_2, \quad x_3 = -2z_3 + z_1 z_2 \tag{6.86}$$

the chained model is transformed to the Brockett (nonholonomic) integrator model (2.65):

$$\dot{x}_1 = u_1, \quad \dot{x}_2 = u_2, \quad \dot{x}_3 = x_1 u_2 - x_2 u_1 \tag{6.87}$$

In the previous section, we have derived a stabilizing kinematic controller for Eqs. (6.85a)–(6.85c) using the chained form. A similar controller can also be derived using the Brockett integrator model (6.87). This controller should make the manifold:

$$M = \{\mathbf{x} \in R^3 : x_3 = 0\}$$

an invariant manifold. Working as in Section 6.4.1 we can prove that the nonlinear feedback controller:

$$\mathbf{u} = \begin{bmatrix} u_1 \\ u_2 \end{bmatrix} = \begin{bmatrix} -k_1 x_1 + \dfrac{k_2 x_3 x_2}{(x_1^2 + x_2^2)} \\ -k_1 x_2 - \dfrac{k_2 x_3 x_1}{x_1^2 + x_2^2} \end{bmatrix} \quad x_1^2 + x_2^2 \neq 0$$

with $k_1 > 0$ and $k_2 > 0$, brings the state $\mathbf{x} = [x_1, x_2, x_3]^{\mathrm{T}}$ to the origin asymptotically. We now consider the full model (6.85a)–(6.85e). By using the state transformation:

$$z_1 = \phi, \quad z_2 = x\cos\phi + y\sin\phi, \quad z_3 = x\sin\phi - y\cos\phi \tag{6.88a}$$

and input transformation:

$$u_1 = N/I, \quad u_2 = F/m - (N/I)z_3 - \omega^2 z_2 \tag{6.88b}$$

the full model can be transformed to the extended chained form:

$$\ddot{z}_1 = u_1, \quad \ddot{z}_2 = u_2, \quad \dot{z}_3 = z_2 \dot{z}_1$$

which in terms of the new variables in Eq. (6.86) takes the form of the extended (double) Brockett integrator (2.60):

$$\begin{aligned} \ddot{x}_1 &= u_1 \\ \ddot{x}_2 &= u_2 \\ \dot{x}_3 &= x_1 \dot{x}_2 - x_2 \dot{x}_1 \end{aligned} \tag{6.89}$$

We will work with the model (6.89). The control problem is to derive a nonlinear state feedback controller that asymptotically stabilizes the system at the origin [11]. We define the manifold:

$$M = \{\mathbf{x} \in R^5 : x_3 = 0\} \tag{6.90}$$

where:

$$\mathbf{x} = [x_1, x_2, x_3, \dot{x}_1, \dot{x}_2]^T \tag{6.91}$$

As usual, we will select a control law that makes M invariant and assures that $\mathbf{x} \to \mathbf{0}$ asymptotically. If $\mathbf{x} \notin M$, the control law drives x_3 to M asymptotically. The controller that we will examine is [11]:

$$\mathbf{u} = \begin{bmatrix} u_1 \\ u_2 \end{bmatrix} = \begin{bmatrix} -k_1 x_1 - k_2 \dot{x}_1 + \dfrac{k_3 x_3 x_2}{x_1^2 + x_2^2} \\ -k_1 x_2 - k_2 \dot{x}_2 - \dfrac{k_3 x_3 x_1}{x_1^2 + x_2^2} \end{bmatrix} \quad x_1^2 + x_2^2 \neq 0 \tag{6.92a}$$

with:

$$k_2 > 0, \quad k_2^2/4 > k_1 > 0, \quad k_2^2/4 > k_3 > 0 \tag{6.92b}$$

It will be shown that for any $\mathbf{x} \notin M^*$ where:

$$M^* = \{\mathbf{x} \in R^5 : x_1^2 + x_2^2 = 0, \quad x_3 \neq 0\}$$

the controller ((6.92a) and (6.92b)) stabilizes the system at the origin. Indeed, introducing Eq. (6.92a) into Eq. (6.89) we get the closed-loop system:

$$\ddot{x}_1 + k_1 x_1 + k_2 \dot{x}_1 - k_3 x_2 x_3 / (x_1^2 + x_2^2) = 0 \tag{6.93a}$$

$$\ddot{x}_2 + k_1 x_2 + k_2 \dot{x}_2 + k_3 x_1 x_3 / (x_1^2 + x_2^2) = 0 \tag{6.93b}$$

$$\ddot{x}_3 + k_2 \dot{x}_3 + k_3 x_3 = 0 \tag{6.93c}$$

Equation (6.93c) represents a second-order linear system with characteristic polynomial:

$$\chi(\lambda) = \lambda^2 + k_2 \lambda + k_3$$

which is exponentially stable when the eigenvalues $\lambda_{1,2} = -k_2/2 \mp (\sqrt{k_2^2 - 4k_3})/2$ are negative. The conditions for $\lambda_i < 0$ $(i = 1, 2)$ are $k_2 > 0$ and $k_2^2/4 > k_3 > 0$ which give $\lambda_1 < \lambda_2 < 0$. Clearly, for any initial condition $x_3(0) \neq 0$, $x_3(t) \to 0$ asymptotically, that is, $x_3(t)$ is attracted to go in the invariant manifold M given by Eq. (6.90). Once x_3 is in this manifold, Eqs. (6.93a) and (6.93b) become:

$$\ddot{x}_1 + k_2\dot{x}_1 + k_1x_1 = 0, \quad \ddot{x}_2 + k_2\dot{x}_2 + k_1x_2 = 0$$

which are exponentially stable when:

$$k_2^2/4 > k_1 > 0, \quad k_2 > 0$$

The above results show that using the controller ((6.92a) and (6.92b)), the origin $\mathbf{x} = 0$ becomes exponentially stable for all $\mathbf{x} \in M^*$. It is remarked that the controller ((6.92a) and (6.92b)) is valid when $x_1^2 + x_2^2 \neq 0$.

6.4.3 Stabilizing Control of Car-Like Robot in Chained Model Form

We consider the (2,4) chained form (2.69) of the rear-wheel driven car:

$$\begin{aligned} \dot{x}_1 &= u_1 \\ \dot{x}_2 &= u_2 \\ \dot{x}_3 &= x_2u_1 \\ \dot{x}_4 &= x_3u_1 \end{aligned} \qquad \mathbf{x} = \begin{bmatrix} x_1 \\ x_2 \\ x_3 \\ x_4 \end{bmatrix} \in X_1 \subset R^n, \quad \mathbf{u} = \begin{bmatrix} u_1 \\ u_2 \end{bmatrix} \in U \subset R^2 \tag{6.94}$$

The problem is to derive a static discontinuous nonlinear controller $\mathbf{u} = \mathbf{u}(\mathbf{x})$ which stabilizes the system (6.94).

We will use the invariant manifold approach, starting by constructing an invariant manifold of the system via the linear state feedback controller:

$$u_1(x) = -k_1x_1, \quad u_2(x) = -k_2x_1 - k_3x_2 \tag{6.95}$$

where k_1, k_2, k_3 are real constant gains with $k_1 > 0$, $k_2 \in R$, $k_3 > 0$, and $k_1 \neq k_3$. Then, we will enhance this controller to make the constructed manifold an attractive manifold. Introducing Eq. (6.95) into Eq. (6.94), we get the closed-loop system, which has the following solution [12]:

$$\begin{aligned} x_1(t) &= \mathrm{X}_1^1\, \mathrm{e}^{-k_1t} \\ x_2(t) &= \mathrm{X}_2^1\, \mathrm{e}^{-k_1t} + \mathrm{X}_2^2\, \mathrm{e}^{-k_3t} \\ x_3(t) &= \mathrm{X}_3^1\, \mathrm{e}^{-2k_1t} + \mathrm{X}_3^2\, \mathrm{e}^{-\mathrm{K}_bt} + s_3(\mathbf{x}_0) \\ x_4(t) &= \mathrm{X}_4^1\, \mathrm{e}^{-k_1t} + \mathrm{X}_4^2\, \mathrm{e}^{-3k_1t} + \mathrm{X}_4^3\, \mathrm{e}^{-(k_1+K_b)t} + s_4(\mathbf{x}_0) \end{aligned} \tag{6.96}$$

where:

$$X_1^1 = x_{10}, \quad X_2^1 = (k_2/K_a)x_{10}, \quad X_2^2 = [x_{20} - (k_2/K_a)x_{10}]$$
$$X_3^1 = (k_2/2K_a)x_{10}^2, \quad X_3^2 = (k_1/K_b)[x_{20} - (k_2/K_a)x_{10}]x_{10}$$
$$X_4^1 = x_{10}s_3(\mathbf{x}_0), \quad X_4^2 = \left(\frac{K_2}{6K_a}\right)x_{10}^3, \quad X_4^3 = \frac{k_1^2 x_{10}^2}{K_b(k_1 + K_b)}\left[x_{20} - \frac{k_2}{K_a}x_{10}\right]$$
$$K_a = k_1 - k_3, \; K_b = k_1 + k_3 \tag{6.97}$$

Here, $s_3(\mathbf{x}_0)$ and $s_4(\mathbf{x}_0)$ are integration constants determined by the initial conditions $x_i(0) = x_{i0}$ at $t = 0$. Clearly:

$$x_1(t) \to 0, \quad x_2(t) \to 0, \quad x_3(t) \to s_3(\mathbf{x}_0), \quad x_4(t) \to s_4(\mathbf{x}_0)$$

Thus, if we select the initial conditions such that $s_3(\mathbf{x}_0) = 0$ and $s_4(\mathbf{x}_0) = 0$, then the entire state $\mathbf{x} = [x_1, x_2, x_3, x_4]^T$ tends to the origin. Now, we set $t = 0$ in Eq. (6.96), solve for $s_3(\mathbf{x}_0)$ and $s_4(\mathbf{x}_0)$, and then replace $\mathbf{x}_0$ by $\mathbf{x}(t)$. In this way, we construct the functions:

$$\begin{aligned} s_3(\mathbf{x}) &= x_3 - \left(\frac{k_1}{K_b}\right)x_1x_2 + \left(\frac{k_2}{2K_b}\right)x_1^2 \\ s_4(\mathbf{x}) &= x_4 - x_1x_3 + \frac{k_1}{k_1 + K_b}x_1^2x_2 - \frac{k_2}{3(2k_1 + K_b)}x_1^3 \end{aligned} \tag{6.98}$$

These functions define the manifold:

$$M = \{\mathbf{x} \in R^4 : s_i(\mathbf{x}) = 0, \quad i = 3, 4\} \tag{6.99}$$

Cleary, since $x_1(t)$ and $x_2(t)$ tend exponentially to zero, if the state $\mathbf{x}(t)$ belongs to M, then it tends to zero. It is easy to verify that the manifold M is an invariant manifold for the closed-loop system ((6.94) and (6.95)), that is, M is invariant for the system (6.94) under the linear control law (6.95). Indeed, evaluating the Lie derivative of $s_j(\mathbf{x})$ along the vector fields of (6.94) with the controller (6.95) we get:

$$\dot{s}_j(\mathbf{x}) = L_f s_j = 0, \quad j = 3, 4 \text{ for all } \mathbf{x} \in M$$

This is the condition (6.18) that assures the invariance of M for the closed-loop system. This means that once the trajectories of the closed-loop system are in M, they remain there for all future times. Now, since the mapping:

$$(x_1, x_2, x_3, x_4) \to (x_1, x_2, s_3, s_4)$$

is a diffeomorphism, the stabilization of (x_1, x_2, x_3, x_4) is equivalent to the stabilization of (x_1, x_2, s_3, s_4). Therefore, to stabilize the system (6.94), it is sufficient to lead (x_1, x_2, x_3, x_4) into M by an additional state feedback, that is, to make M an attractive manifold. The closed-loop system for $[x_1, x_2, s_3, s_4]^{\mathrm{T}}$ with the controller (6.95) is found to be:

$$\begin{aligned} \dot{x}_1 &= u_1 \\ \dot{x}_2 &= u_2 \\ \dot{s}_3 &= (1/\mathrm{K}_b)[(k_3x_2 + k_2x_1)u_1 - k_1x_1u_2] \\ \dot{s}_4 &= -[1/(k_1 + \mathrm{K}_b)][(k_3x_2 + k_2x_1)u_1 - k_1x_1u_2]x_1 \end{aligned} \tag{6.100}$$

The enhanced controller to be used is:

$$u_1 = -k_1x_1, \quad u_2 = -k_2x_1 - k_3x_2 + v \tag{6.101}$$

where v is a control term added to the u_2 channel. Introducing Eq. (6.101) into Eq. (6.100) we get:

$$\begin{aligned} \dot{x}_1 &= -k_1x_1 \\ \dot{x}_2 &= -k_2x_1 - k_3x_2 + v \\ \dot{s}_3 &= -(k_1/K_b)x_1v \\ \dot{s}_4 &= [k_1/(k_1 + K_b)]x_1^2v \end{aligned} \tag{6.102}$$

The system of the last two equations (6.102) can be written as:

$$\dot{\mathbf{s}} = \mathbf{E}(x_1)\mathbf{b}v, \quad \mathbf{s} = [s_3, s_4]^{\mathrm{T}} \tag{6.103a}$$

where:

$$\mathbf{E}(x_1) = \begin{bmatrix} x_1 & 0 \\ 0 & x_1^2 \end{bmatrix}, \quad \mathbf{b} = \begin{bmatrix} -k_1/\mathrm{K}_b \\ k_1/(k_1 + K_b) \end{bmatrix} \tag{6.103b}$$

Now, introducing a transformed variable $\mathbf{z}$:

$$\mathbf{z} = \mathbf{E}^{-1}(x_1)\mathbf{s} \tag{6.104}$$

which is valid for $x_1 \neq 0$, we get:

$$\begin{aligned} \dot{\mathbf{z}} &= \frac{\mathrm{d}}{\mathrm{d}t}\mathbf{E}^{-1}(x_1)\mathbf{s} + \mathbf{E}^{-1}(\mathbf{x})\dot{\mathbf{s}} \\ &= \begin{bmatrix} -\dot{x}_1/x_1^2 & 0 \\ 0 & -2\dot{x}_1/x_1^3 \end{bmatrix}\mathbf{E}(x_1)\mathbf{z} + \mathbf{b}v \\ &= \begin{bmatrix} k_1/x_1 & 0 \\ 0 & 2k_1/x_1^2 \end{bmatrix}\mathbf{E}(x_1)\mathbf{z} + \mathbf{b}v \end{aligned}$$

that is:

$$\dot{\mathbf{z}} = \mathbf{A}\mathbf{z} + \mathbf{b}\upsilon \tag{6.105a}$$

where:

$$\mathbf{A} = \begin{bmatrix} k_1 & 0 \\ 0 & 2k_1 \end{bmatrix} \tag{6.105b}$$

The system ((6.105a) and (6.105b)) will be stabilized by using a feedback controller of the form:

$$\upsilon = \mathbf{g}^{\mathrm{T}}\mathbf{z}, \quad \mathbf{g} = [g_1, g_2]^{\mathrm{T}} \tag{6.106}$$

Using Eq. (6.106) the closed-loop system is:

$$\dot{\mathbf{z}} = \mathbf{A}_{\mathrm{c}}\mathbf{z} \tag{6.107}$$

where:

$$\mathbf{A}_{\mathrm{c}} = \mathbf{A} + \mathbf{b}\mathbf{g}^{\mathrm{T}} \tag{6.108}$$

is the closed-loop matrix. Therefore, to obtain exponential convergence to $\mathbf{z} = \mathbf{0}$, the gain vector $\mathbf{g}$ must be selected such as all eigenvalues of $\mathbf{A}_{\mathrm{c}}$ are negative real numbers. When $\mathbf{z} \to \mathbf{0}$ asymptotically, we have:

$$\mathbf{E}^{-1}(x_1)\mathbf{s} = \begin{bmatrix} 1/x_1 & 0 \\ 0 & 1/x_1^2 \end{bmatrix}\mathbf{s} \to 0 \tag{6.109}$$

which assures that $\mathbf{s}$ tends to zero faster than x_1. Therefore, the manifold M is reached before x_1 becomes zero (which assures the boundedness of the control law (6.106)).

For example, selecting $k_1 = 2,\ k_2 = 0,\ k_3 = 4,\ \mathrm{K}_b = k_1 + k_3 = 6$, we get:

$$\mathbf{A} = \begin{bmatrix} 2 & 0 \\ 0 & 4 \end{bmatrix}, \quad \mathbf{b} = \begin{bmatrix} -1/3 \\ 1/4 \end{bmatrix}, \quad \mathbf{A} + \mathbf{b}\mathbf{g}^{\mathrm{T}} = \begin{bmatrix} 2 - \dfrac{g_1}{3} & -\dfrac{g_2}{3} \\ \dfrac{g_1}{4} & 4 + \dfrac{g_2}{4} \end{bmatrix}$$

If we wish the eigenvalues of $\mathbf{A} + \mathbf{b}\mathbf{g}^{T}$ to be $\lambda_1 = -2$ and $\lambda_2 = -3$, the gain vector $\mathbf{g}$ must be:

$$\mathbf{g} = \begin{bmatrix} -30 \\ -84 \end{bmatrix}$$

From the above values of $k_1, \lambda_2, \lambda_3$, and Eq. (6.109), it follows that x_1, s_3, and s_4 converge to zero according to:

$$x_1 = C_1\,\mathrm{e}^{-2t},\ s_3 = C_3\,\mathrm{e}^{-4t} \text{ and } s_4 = C_4\,\mathrm{e}^{-7t}$$

Example 6.9

It is desired to apply the method of constructing invariant manifolds presented in Section 6.43 for the feedback stabilization to $\mathbf{x} = \mathbf{0}$ of the following Brockett type integrator models:

(a) Double integrator:

$$\dot{x}_1 = u_1,\quad \dot{x}_2 = u_2,\quad \dot{x}_3 = x_1u_2 - x_2u_1 \tag{6.110}$$

(b) Extended double integrator:

$$\dot{x}_1 = y_1,\quad \dot{x}_2 = y_2,\quad \dot{x}_3 = x_1y_2 - x_2y_1,\quad \dot{y}_1 = u_1,\quad \dot{y}_2 = u_2 \tag{6.111}$$

Justify the type of WMRs that can be modeled by the above integrators.

Solution

(a) Double integrator:

The double integrator model describes the kinematic performance of a differential-drive WMR and is derived as described in Section 6.4.2 (Eq. (6.87)). To derive an invariant manifold, we introduce the linear control law [13]:

$$u_1 = -k_1x_1,\quad u_2 = -k_1x_2;\quad k_1 > 0 \tag{6.112}$$

into Eq. (6.110), and get the closed-loop system:

$$\dot{x}_1 + k_1x_1 = 0,\quad \dot{x}_2 + k_1x_2 = 0 \tag{6.113a}$$

$$\dot{x}_3 + k_1x_1x_2 - k_1x_1x_2 = 0 \tag{6.113b}$$

The response of Eqs. (6.113a) and (6.113b) when the initial conditions are $x_1(0) = x_{10}$, $x_2(0) = x_{20}$ and $x_3(0) = x_{30}$, is:

$$x_1(t) = x_{10}\,\mathrm{e}^{-k_1t},\quad x_2(t) = x_{20}\,\mathrm{e}^{-k_1t},\quad x_3(t) = x_{30} \tag{6.114a}$$

The candidate manifold $M = \{x \in R^3, s(\mathbf{x}) = 0\}$ is selected here as:

$$s(\mathbf{x}) = x_3(t) \tag{6.114b}$$

for which we obtain:

$$\dot{s}(\mathbf{x}) = \dot{x}_3(t) = 0 \tag{6.114c}$$

that is:

$$s(\mathbf{x}) = \text{const.} \tag{6.114d}$$

Clearly, once the system state reaches M (i.e., $s(\mathbf{x}) = 0$), then by Eq. (6.114b) the entire state $\mathbf{x} = [x_1, x_2, x_3]^T$ tends to zero exponentially. Therefore, M is an invariant manifold for the double integrator (6.110) under the feedback control law (6.112). To make M an attractive manifold, we work with the candidate Lyapunov function:

$$V(\mathbf{x}) = (1/2)s^2(\mathbf{x}) \tag{6.115}$$

for which, in view of Eq. (6.110), we obtain:

$$\dot{V}(\mathbf{x}) = s(\mathbf{x})\dot{s}(\mathbf{x}) = s(\mathbf{x})(x_1u_2 - x_2u_1)$$

To make $\dot{V}(\mathbf{x}) \leq 0$, we select u_1 and u_2 as:

$$u_1 = k_2 s(\mathbf{x})x_2(t), \quad u_2 = -k_2 s(\mathbf{x})x_1(t), \quad k_2 > 0$$

which give:

$$\dot{V}(x) = -k_2 P(t)s^2(\mathbf{x}) < 0 \tag{6.116a}$$

for

$$P(t) = x_1^2(t) + x_2^2(t) \tag{6.116b}$$

Here, we have:

$$\begin{aligned} \dot{P}(t) &= 2(x_1\dot{x}_1 + x_2\dot{x}_2) \\ &= 2(x_1u_1 + x_2u_2) = k_2 s(\mathbf{x})[x_1x_2 - x_2x_1] \equiv 0 \end{aligned} \tag{6.116c}$$

Therefore, for:

$$P(t) = P(0) = x_1^2(0) + x_2^2(0) \neq 0$$

the manifold $s(\mathbf{x})$ tends to zero, and assures that $x_1(t) \to 0$, $x_2(t) \to 0$, and $x_3(t) \to 0$, provided that $x_1(0) \neq 0$ and/or $x_2(0) \neq 0$.

The total quasi-continuous stabilizing controller is selected as:

$$\mathbf{u} = \begin{bmatrix} u_1 \\ u_2 \end{bmatrix} = -k_1 \begin{bmatrix} x_1 \\ x_2 \end{bmatrix} + k_2 \left(\frac{s}{P}\right) \begin{bmatrix} x_2 \\ -x_1 \end{bmatrix} \tag{6.117a}$$

which leads to (see Eq. (6.116c)):

$$\dot{P}(t) = 2(x_1u_1 + x_2u_2) = -2k_1P(t)$$

and

$$\dot{s}(t) = \dot{x}_3(t) = x_1u_2 - x_2u_1 = -k_2s(t)$$

It follows that:

$$P(t) = P(0)e^{-2k_1t}, \quad s(t) = s(0)e^{-k_2t} \tag{6.117b}$$

To ensure that Eq. (6.117a) is bounded, $s(t)$ must converge to zero faster than $P(t)$. From Eq. (6.117b), we see that this holds if:

$$k_2 > 2k_1$$

(b) Extended double integrator:

The extended double integrator describes the full (kinematic and dynamic) performance of the differential-drive WMR as explained in Section 6.4.2(see Eqs. (6.85a)–(6.85e), (6.88a), (6.88b), and (6.89)). The solution to be given here is different than the one provided in Section 6.4.2. Here, to construct an invariant manifold, we start with the control law [13,15]:

$$u_1 = -2k_1y_1 - k_1^2x_1, \quad u_2 = -2k_1y_2 - k_1^2x_2 \tag{6.118}$$

The state vector of the system is:

$$x = [x_1, x_2, y_1, y_2]^{\mathrm{T}} \tag{6.119}$$

Introducing the control (6.118) into Eq. (6.111), we get the closed-loop system:

$$\begin{aligned}
\dot{x}_1 &= y_1 \\
\dot{x}_2 &= y_2 \\
\dot{y}_1 &= -2k_1y_1 - k_1^2x_1 \\
\dot{y}_2 &= -2k_1y_2 - k_1^2x_2
\end{aligned} \tag{6.120}$$

which gives the response:

$$\begin{aligned}
x_1(t) &= x_{10}[e^{-k_1t} + k_1t\,e^{-k_1t}] + y_{10}t\,e^{-k_1t} \\
x_2(t) &= x_{20}[e^{-k_1t} + k_1t\,e^{-k_1t}] + y_{20}t\,e^{-k_1t} \\
y_1(t) &= x_{10}[-k_1^2t\,e^{-k_1t}] + y_{10}[e^{-k_1t} - k_1t\,e^{-k_1t}] \\
y_2(t) &= x_{20}[-k_1^2t\,e^{-k_1t}] + y_{20}[e^{-k_1t} - k_1t\,e^{-k_1t}]
\end{aligned} \tag{6.121}$$

where x_{i0}, $y_{i0}(i = 1, 2)$ are the initial conditions. The expression for $x_3(t)$ is derived by integrating the third equation of (6.111): $\dot{x}_3 = x_1y_2 - x_2y_1$. The result is:

$$x_3(t) = s_3(\mathbf{x}_0) - (x_{10}y_{20}/2k_1)[e^{-2k_1t} - 1] + (y_{10}x_{20}/2k_1)[e^{-2k_1t} - 1] \tag{6.122}$$

where $s_3(\mathbf{x}_0)$ is the integration constant determined using the initial conditions. Clearly, $x_1(t) \to 0$, $x_2(t) \to 0$, $y_1(t) \to 0$, $y_2(t) \to 0$, and $x_3(t) \to s_3(\mathbf{x}_0)$. Therefore, selecting the initial conditions such that $s_3(\mathbf{x}_0) = 0$, we get $x_3(t) \to 0$. Now, we set $t = 0$ in Eq. (6.122), solve for $s_3(\mathbf{x}_0)$, and replace $\mathbf{x}(0)$ by $\mathbf{x}(t)$ to get the function:

$$s_3(\mathbf{x}) = (1/2k_1)x_1(t)y_2(t) - (1/2k_1)y_1(t)x_2(t) \tag{6.123}$$

which constitutes an invariant manifold, because:

$$\dot{s}_3(\mathbf{x}) = L_{\mathbf{f}} s_3(\mathbf{x}) = 0, \quad \mathbf{x} \in R^4 \tag{6.124}$$

where $L_{\mathbf{f}}$ is the derivative of s_3 along the system field $\mathbf{f}$ of the closed-loop system (6.120):

$$\mathbf{f} = [y_1, y_2, -2k_1y_1 - k_1^2x_1, -2k_1y_2 - k_1^2x_2]^T \tag{6.125}$$

This means that once $x_1(t), x_2(t), y_1(t)$, and $y_2(t)$ enter the manifold at some time $t = T$:

$$M = \{\mathbf{x} \in R^4 : s_3(\mathbf{x}) = 0\} \tag{6.126}$$

they remain there for all subsequent times $t \geq T$. Overall, the controller (6.118) assures that $x_1(t) \to 0$, $x_2(t) \to 0$, $y_1(t) \to 0$, $y_2(t) \to 0$, and $x_3(t) \to 0$ because $s_3(\mathbf{x}) = 0$ has already been satisfied. It remains to enhance the controller (6.118), as usual, in order to assure that M is an attractive manifold. To this end, we use the Lyapunov function:

$$V(\mathbf{x}) = \frac{1}{2}s_3^2(\mathbf{x})$$

and check if the control law:

$$u_1 = -2k_1y_1, \quad u_2 = -k_2s_3(\mathbf{x})/x_1(t) - 2k_1y_2 \quad (k_2 > 0) \tag{6.127}$$

makes $\dot{V}(\mathbf{x}) \leq 0$. Indeed from Eq. (6.123) we have:

$$\begin{aligned}\dot{V}(\mathbf{x}) &= s_3(\mathbf{x})\dot{s}_3(\mathbf{x}) = s_3(\mathbf{x})\left[x_1y_2 - x_2y_1 + \frac{1}{2k_1}(x_1u_2 - x_2u_1)\right] \\ &= s_3(\mathbf{x})[x_1y_2 - x_2y_1 + (1/2k_1)[-k_2s_3(\mathbf{x})]/x_1(t) - x_1y_2 + x_2y_1] \\ &= -(k_2/2k_1)s^2 \leq 0 \quad \text{for } k_2 > 0, k_1 > 0\end{aligned}$$

Now, using the controller (6.127), it is easy to verify that:

$$\dot{s}_3(t) = -(k_2/2k_1)s_3(t) \tag{6.128}$$

which has the response:

$$s_3(t) = s_{30}\, e^{-(k_2/2k_1)t} \tag{6.129}$$

To assure that the controller u_2 in Eq. (6.127) is bounded, we must select k_1 and k_2 such that:

$$\left|\frac{s_3(\mathbf{x})}{x_1(t)}\right| < \infty \quad \text{as } t \to \infty \tag{6.130a}$$

This can be done easily using Eqs. (6.121) and (6.129).

The overall controller is formed by joining Eqs. (6.118) and (6.127) as:

$$\begin{aligned} u_1 &= -2k_1 y_1 - k_1^2 x_1 \\ u_2 &= -2k_1 y_2 - k_1^2 x_2 - k_2 s_3(\mathbf{x})/x_1(t) \end{aligned} \tag{6.130b}$$

Now, it follows again that $\dot{s}_3(t) = -(k_2/2k_1)s_3(t)$, and so for keeping u_2 bounded, the same condition (6.130a) should be assured with proper choice of k_1 and k_2.

Example 6.10 (Treatment of controller singularity)

It is desired to find ways of avoiding the controller singularity of:

(a) the controller ((6.81a) and (6.81b)) when $z_1^2 + z_2^2 = 0$

(b) the controller (6.127) when $x_1(t) = 0$

(a) Singularity at $z_1^2 + z_2^2 = 0$

One way to avoid the controller singularity at $z_1^2 + z_2^2 = 0$, that is, at $z_1 = z_2 = 0$, is to create a region around the z_3-axis where this controller is not used, and replaced by a new controller [16]. Defining the new variable ζ as:

$$\zeta = s/\sqrt{z_1^2 + z_2^2} = s/l \tag{6.131a}$$

It can be shown that such a region is:

$$U_{\zeta^*} = \{(z_1, z_2, z_3) \in R^3 : |\zeta| \geq \zeta^*\} \tag{6.131b}$$

where ζ^* is a chosen large positive bound. In the region U_{ζ^*}, we may use, instead of Eqs. (6.81a) and (6.81b), the control law:

$$u_2 = b\,\mathrm{sgn}(s), \quad u_1 = 0 \tag{6.132}$$

where b is a constant gain that specifies how near to the singularity the switching between the two controllers takes place. It can be easily verified that using the control law (6.132), the system leaves the region U_{ζ^*} in finite time. It can also be verified that:

$$\dot{\zeta} = -(k_2/2 - k_1)\zeta, \quad k_2 > 2k_1 \tag{6.133}$$

and so the region $\hat{U}_{\varsigma^*}$, outside U_{ς^*}, in R^3 is invariant. Therefore, once the system goes outside U_{ς^*}, it stays there for all future times, and the controller ((6.81a) and (6.81b)) can be used without any singularity problem.

(b) Singularity at $x_1(t) = 0$

In this case, k_1 and k_2 must be chosen such that $|s_3(\mathbf{x}(t))/x_1(t)|$ is bounded at all times. Using Eqs. (6.111), (6.127), and (6.129), we get [13,15]:

$$\lim_{t\to\infty} \frac{s_3(\mathbf{x}(t))}{x_1(t)} = \lim_{t\to\infty} \frac{s_{30}\exp[-(k_2/2k_1)t]}{x_{10} - (1/2k_1)y_{10}[\exp(-2k_1 t) - 1]} = 0$$

for $x_{10} + (1/2k_1)y_{10} \neq 0$, and $x_{10} \neq 0$. But, if $x_{10} \neq 0$ and $x_{10} + (1/2k_1)y_{10} = 0$, we get:

$$\left|\frac{s_3(\mathbf{x})}{x_1(t)}\right| = \left|\frac{s_{30}\exp\left[-(k_2/k_1)t\right]}{-(1/2k_1)y_{10}\exp(-2k_1 t)}\right|$$
$$\leq |(2k_1 s_{30}/y_{10})\exp[-(k_2/2k_1 - 2k_1)t]|$$

Clearly, since $2k_1 s_{30}/y_{10}$ is bounded, $|s_3(\mathbf{x})/x_1(t)|$ decreases exponentially if k_1 and k_2 are such that $k_2/2k_1 - 2k_1 > 0$, that is, if:

$$k_2 > 4k_1^2$$

Now, we have to assure that $y_2(t) \to 0$ as $t \to \infty$. From Eqs. (6.111) and (6.127), we get:

$$\dot{y}_2 = -2k_1 y_2 + \sigma(\mathbf{x}, t)$$

where $\sigma(\mathbf{x}, t) = -k_2 s_3(\mathbf{x})/x_1(t)$. It can be easily verified that if $\sigma(\mathbf{x}, t)$ decreases faster than $\exp(-2k_1)$, then $y_2(t)$ converges exponentially to zero with rate $2k_1$. Therefore k_1 and k_2 must be such that:

$$k_2/2k_1 - 2k_1 > 2k_1, \quad \text{i.e.,} \quad k_2 \geq 8k_1^2 \tag{6.134}$$

From the above, it follows that the controller (6.127), with k_1 and k_2 satisfying (6.134), leads to a closed-loop system free of singularities.

References

[1] Isidori A. Nonlinear control systems: an introduction. Berlin/New York: Springer; 1985.

[2] Slotine JJ, Li W. Applied nonlinear control. Englewood Cliffs: Prentice Hall; 1991.

[3] Nijmeijer H, Van der Schaft HR. Nonlinear dynamical control systems. Berlin/New York: Springer; 1990.

[4] Sastry S. Nonlinear systems: analysis stability and control. Berlin/New York: Springer; 1999.

[5] Brockett RW. Asymptotic stability and feedback stabilization: differential geometric control theory. Boston, MA: Birkhauser; 1983.

[6] Yun X, Yamamoto Y. On feedback linearization of mobile robots technical report (CIS). University of Pennsylvania: Department of Computer and Information Science; 1992.
[7] Yang E, Gu D, Mita T, Hu H. Nonlinear tracking control of a car-like mobile robot via dynamic feedback linearization. Proceedings of control 2004. University of Bath: UK; September 2004 [paper 1D-218].
[8] DeLuca A, Oriolo G, Samson C. Feedback control of a nonholonomic car-like robot. In: Laumont JP, editor. Robot motion planning and control. Berlin/New York: Springer; 1998. p. 171–253.
[9] Astolfi A. Exponential stabilization of a wheeled mobile robot via discontinuous control. J Dyn Syst Meas Control 1999;121:121–6.
[10] Reyhanoglu M. On the stabilization of a class of nonholonomic systems using invariant manifold technique. Proceedings of the 34th IEEE conference on decision and control. New Orlean, LA; December 1995. p. 2125–26.
[11] DeVon D, Bretl T. Kinematic and dynamic control of a wheeled mobile robot. Proceedings of IEEE/RSJ international conference on intelligent robots and systems. San Diego, CA; October 29–November 2, 2007. p. 4065–70.
[12] Tayebi A, Tadijne M, Rachid A. Invariant manifold approach for the stabilization of nonholonomic chained systems: application to a mobile robot. Nonlinear Dynamics 2001;24:167–81.
[13] Watanabe K, Yamamoto K, Izumi K, Maeyama S. Underactuated control for nonholonomic mobile robots by using double integrator model and invariant manifold theory. Proceedings of IEEE/RSJ international conference on intelligent robots and systems. Taipei, Taiwan; October 18–22, 2010. p. 2862–67.
[14] Peng Y, Liu M, Tang Z, Xie S, Luo J. Geometry stabilizing control of the extended nonholonomic double integrator. Proceedings of IEEE international conference on robotics and biomimetics. Tianijn, China; December 14–18, 2010. p. 926–31.
[15] Izumi K, Watanabe K. Switching manifold control for an extended nonholonomic double integrator. Proceedings of international conference on control and automation systems. Kintex, Gyeonggi-do, Korea; October 27–30, 2010. p. 896–99.
[16] Kim BM, Tsiotras P. Controllers for unicycle-type wheeled robots: theoretical results and experimental validation. IEEE Trans Robot Autom 2002;18(3):294–307.

7 Mobile Robot Control III: Adaptive and Robust Methods

7.1 Introduction

Most advanced control systems, such as robots, aircrafts, and missiles, have slowly varying unknown parameters and contain important uncertainties or disturbances due to load variation, fuel consumption, and other effects. All controllers presented in Chapters 5 and 6 were based on the assumption that the wheeled mobile robots (WMRs) do not involve such unknown parameters or disturbances. One of the basic methodologies for treating such uncertain systems is the *adaptive control* methodology which always employs an algorithm for identifying (estimating) the varying parameters in real time [1–9]. The alternative methodology is the *robust control* methodology which requires a priori knowledge of the bounds of the parameter variations. The more precise the knowledge of these bounds is, the better is the robustness achieved by these controllers [10–17].

The adaptive controllers (control laws or algorithms) improve their performance as the adaptation evolves with time. On the other hand, robust controllers are trying to keep an acceptable performance right from the beginning. The adaptive controllers require little or no a priori knowledge for the parameters under estimation. But the robust controllers can face large disturbances, fast variations, and nonmodeled characteristics. Almost always, the adaptive control techniques require some linear parameterization of the dynamic of the nonlinear under control.

The two widely used adaptive control methods are as follows:

1. The model reference adaptive control (MRAC) method
2. The self-tuning control (STC) method

In WMRs, the MRAC method is the typical adaptive control method used, and will be studied in this chapter.

Specifically, the objectives of the chapter are as follows:

- To provide the necessary background concepts for understanding the material of the chapter with minimum prior control knowledge
- To present a number of implementations of model reference adaptive control applied to mobile robots
- To study the application of sliding mode and Lyapunov-based robust control to mobile robots.

Introduction to Mobile Robot Control. DOI: http://dx.doi.org/10.1016/B978-0-12-417049-0.00007-9

7.2 Background Concepts

7.2.1 Model Reference Adaptive Control

The general architecture of a model reference adaptive control system is shown in Figure 7.1, and contains four basic components (units) [1,2]:

1. The system to be controlled that involves unknown parameters
2. A reference model for the overall and compact determination of the desired system output
3. A feedback controller with adaptive (adjusted) parameters
4. An adaptation mechanism for updating the controller parameters

It is assumed that the structure of the controlled system is known, and only its parameters are unknown. The reference model provides the ideal response of the system that must be achieved through the parameters adaptation. The control law is parameterized with a number of adaptable parameters. The control law must have the ability to follow perfectly or asymptotically the reference response (trajectory). This means that when the system parameters are exactly known, the respective controller parameters must make the system output identical with the output of the reference model. The adaptation law searches to find the parameter values which assure that the system response under MRAC is ultimately the same with the reference model response, that is, assure convergence of the error between the two responses to zero. Essentially, the basic difference between conventional and adaptive controllers is the use of such parameter adaptation law. The two most popular methods of designing the controller's adaptation law are as follows:

1. The steepest descent method
2. The Lyapunov stability method

7.2.1.1 Steepest Descent Parameter Adaptation Law

This scheme is known as MIT rule since it was developed at the Massachusetts Institute of Technology. Let $\boldsymbol{\beta}$ be the parameter vector, and e the error between the actual and reference outputs. We use the following criterion:

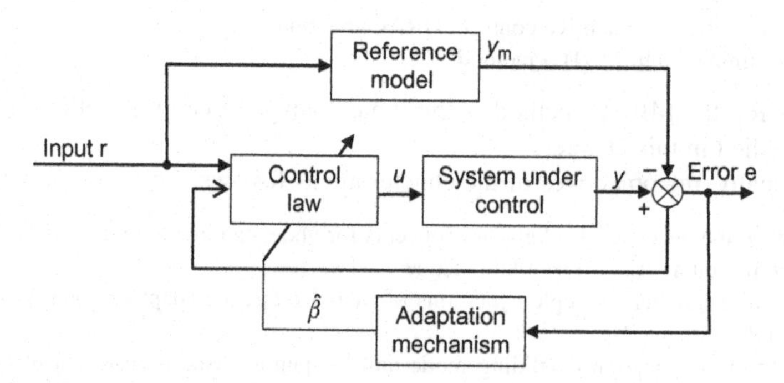

Figure 7.1 MRAC system architecture ($\hat{\beta}$ represents the estimated parameter vector).

$$I(\beta) = \frac{1}{2}e^2$$

To reduce $I(\beta)$, it is logical to vary the parameters in the opposite direction shown by $\mathrm{d}I/d\beta$, that is:

$$\frac{\mathrm{d}\beta}{\mathrm{d}t} = -\gamma\frac{\mathrm{d}I}{\mathrm{d}\beta} = -\gamma e\frac{\vartheta e}{\vartheta\beta} \tag{7.1}$$

For slowly varying parameters (much slower than the other variables of the system), the derivative $\vartheta e/\vartheta\beta$ can be computed for β constant. This derivative is called *sensitivity derivative*. If we use the error criterion:

$$I(\beta) = |e|,$$

then the adaptation law is:

$$\frac{\mathrm{d}\beta}{\mathrm{d}t} = -\gamma\frac{\vartheta e}{\vartheta\beta}\mathrm{sgn}(e) \tag{7.2}$$

where $\mathrm{sgn}(e)$ is the known *signum function*. The adaptation laws are applicable to both linear and nonlinear system. In all cases, the error dynamics must first be determined.

7.2.1.2 Lyapunov-Based Adaptation Law

This adaptation law assumes right from the beginning that the error $e(t)$ will really converge to zero. For clarity, we will illustrate this method for the simple scalar system:

$$\dot{y}(t) = -ay(t) + bu(t) \tag{7.3}$$

where $u(t)$ is the control variable and $y(t)$ is the measured output. The stable reference model is assumed to be:

$$\dot{y}_\mathrm{m}(t) = -a_\mathrm{m}y_\mathrm{m}(t) + b_\mathrm{m}v(t), \quad a_\mathrm{m} > 0 \tag{7.4}$$

The control law is selected as:

$$u(t) = -k_1y(t) + k_0v(t) \tag{7.5}$$

Defining the output error $e = y - y_\mathrm{m}$, we get the dynamic equation for the closed-loop error, obtained using the control law (7.5), as:

$$\dot{e}(t) = -a_\mathrm{m}e + (a_\mathrm{m} - a - bk_1)y + (bk_0 - b_\mathrm{m})v \tag{7.6}$$

Clearly, if $a+bk_1=a_m$, that is, $k_1=(a_m-a)/b$, and $bk_0-b_m=0$ (i.e., $k_0=b_m/b$), the closed loop system is identical with the reference model and so $e(t)\to 0$ for $t\to\infty$. To construct an adaptation law that leads the controller parameters k_0 and k_1 to the above ideal values $k_0=b_m/b$ and $k_1=(a_m-a)/b$, we use the following candidate Lyapunov function:

$$V(e,k_0,k_1)=\frac{1}{2}\left[e^2+\frac{1}{b\gamma}(bk_1+a-a_m)^2+\frac{1}{b\gamma}(bk_0-b_m)^2\right] \tag{7.7}$$

This function satisfies the first three properties of Lyapunov functions, and has zero value when k_0 and k_1 have the ideal values. Differentiating V, we get:

$$\begin{aligned}\dot{V}&=e\dot{e}+\frac{1}{\gamma}(bk_1+a-a_m)\dot{k}_1+\frac{1}{\gamma}(bk_0+b_m)\dot{k}_0\\&=-a_me^2+\frac{1}{\gamma}(bk_1+a-a_m)(\dot{k}_1-\gamma ye)\\&\quad+\frac{1}{\gamma}(bk_0+b_m)(\dot{k}_0+\gamma ve)\end{aligned} \tag{7.8}$$

Thus, we choose the parameter adaptation (updating) laws as:

$$\begin{aligned}\dot{k}_0&=-\gamma ve\\\dot{k}_1&=\gamma ye\end{aligned} \tag{7.9}$$

Then, Eq. (7.6) gives:

$$\dot{V}=-a_me^2<0 \quad (a_m>0) \tag{7.10}$$

Therefore, by the Lyapunov stability criterion, $e(t)$ tends asymptotically to zero as $t\to\infty$. Of course, the convergence of k_0 and k_1 to the ideal values is not assured unless some other appropriate conditions are posed. We observe that the adaptation laws (7.9) are of the general form:

$$\dot{\boldsymbol{\beta}}=\gamma\boldsymbol{\psi}\mathbf{e} \tag{7.11}$$

where $\boldsymbol{\beta}$ is the parameter vector, $\mathbf{e}$ is the error between the closed loop output and the reference output, and $\boldsymbol{\psi}$ is a known function depending on v and y. The above Lyapunov adaptation method is applicable to MIMO and nonlinear systems, as it will be shown in the case of WMRs.

7.2.2 Robust Nonlinear Sliding Mode Control

The *sliding mode control* methodology is the robust control methodology most commonly used in robotics, including mobile robots. For convenience, we will describe this method for the single-input single-output (SISO) canonical nonlinear model (see Eq. (6.24b)) [16]:

$$\begin{aligned} \dot{x}_1 &= x_2 \\ \dot{x}_2 &= x_3 \\ &\vdots \\ \dot{x}_{n-1} &= x_n \\ \dot{x}_n &= b(\mathbf{x}) + a(\mathbf{x})u + d(t) \\ y &= x_1 \end{aligned} \tag{7.12}$$

where $u(t)$ is the scalar input, $y(t)$ the scalar output, and $d(t)$ a scalar disturbance input. The state vector is:

$$\mathbf{x} = [x_1, x_2, ..., x_n]^T = [y, \mathrm{d}y/\mathrm{d}t, \mathrm{d}^2y/\mathrm{d}t^2, \ldots, \mathrm{d}^{n-1}y/\mathrm{d}t^{n-1}]^T$$

The nonlinear function $b(\mathbf{x})$ is not exactly known but with some error (or imprecision) $|\Delta b(\mathbf{x})|$ which is bounded from above by a known continuous function of $\mathbf{x}$. Similarly, the control input gain $a(\mathbf{x})$ is not exactly known. We know its sign and an upper bounding function. The problem under consideration is the following: *It is desired to find that* $u(t)$ *which drives the state on a desired trajectory* $\mathbf{x}_\mathrm{d} = [y_\mathrm{d}, \mathrm{d}y_\mathrm{d}/\mathrm{d}t, \ldots, \mathrm{d}^{n-1}y_\mathrm{d}/\mathrm{d}t^{n-1}]$ *in spite of the presence of the disturbance* $d(t)$ *and the fact that* $b(\mathbf{x})$ *and* $a(\mathbf{x})$ *are known with uncertainty.* This problem will be first treated under the assumption that $\mathbf{x}_\mathrm{d}(t=0) = \mathbf{x}(0) = \mathbf{x}_0$. The tracking error $\tilde{\mathbf{x}}(t)$ is $\tilde{\mathbf{x}}(t) = \mathbf{x}(t) - \mathbf{x}_\mathrm{d}(t) = [\tilde{y}, \mathrm{d}\tilde{y}/\mathrm{d}t, \ldots, \mathrm{d}^{n-1}\tilde{y}/\mathrm{d}t^{n-1}]$. We define a time-varying sliding surface $S(t)$ within the state space R^n as:

$$s(\mathbf{x}, t) = 0, \quad s(\mathbf{x}, t) = (\mathrm{d}/\mathrm{d}t + \Lambda)^{n-1}\tilde{\mathbf{x}}(t) \tag{7.13}$$

where Λ is a positive constant that represents the control signal bandwidth. Under the above condition $\mathbf{x}_\mathrm{d}(0) = \mathbf{x}_0$, the trajectory tracking problem $\mathbf{x}(t) = \mathbf{x}_\mathrm{d}(t)$ is equivalent to the problem of remaining on the sliding surface $S(t)$ for all t. This follows from the fact that $s(\mathbf{x}, t) = 0$ is a differential equation which, with initial condition $\tilde{\mathbf{x}}(0) = \mathbf{0}$, has the unique solution $\tilde{\mathbf{x}}(t) = \mathbf{0}$ for all t. Thus, to assure the trajectory tracking $\mathbf{x}(t) \to \mathbf{x}_\mathrm{d}(t)$, we must maintain $s(\mathbf{x}, t) = 0$ which can be done if $u(t)$ is selected such that outside the surface $S(t)$, the following *sliding condition* holds:

$$\frac{1}{2}\frac{\mathrm{d}}{\mathrm{d}t}s^2(\mathbf{x}, t) \leq -\gamma|s| \tag{7.14}$$

where γ is a positive constant. This condition forces all trajectories to slide toward the surface $S(t)$, and for this reason, the technique was named *sliding mode control* technique (Figure 7.2).

The basic idea behind Eqs. (7.13) and (7.14) is to find a suitable function s of the error and then choose a control law such that the function s^2 to be (and remain) a Lyapunov function despite the presence of the disturbance and the

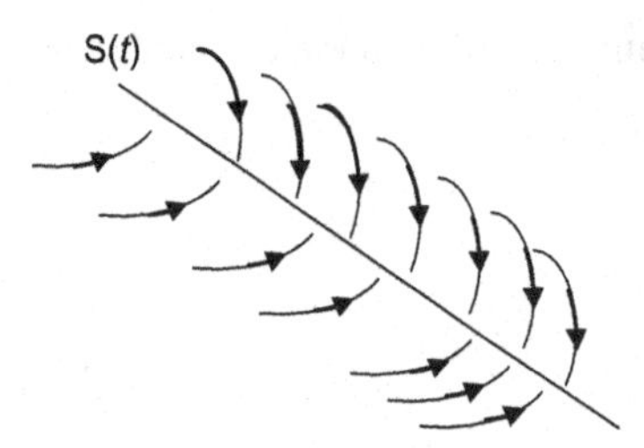

Figure 7.2 . The sliding condition (7.14) forces all trajectories to be directed toward the sliding surface $S(t)$.

model uncertainty. Moreover, the condition (7.14) secures that if $x_d(t=0)=x(0)=x_0$ is not valid, then again the trajectory will arrive at the surface $S(t)$ after the elapse of some time less than or equal to $|s(t=0)|/\gamma$. Indeed, integrating Eq. (7.14) with $s(t=0)>0$, from $t=0$ up to $t=t_a$ we get $0-s(t=0)=s(t-t_a)-\quad s(t=0)\leq-\gamma(t_a-0)$ from which it follows that $t_a\leq s(t=0)/\gamma$. The same result is obtained when $s(t=0)<0$. Moreover, Eq. (7.13) implies that once the trajectory arrives at the surface $S(t)$, the tracking error converges asymptotically to zero with time constant $(n-1)/\Lambda$ or decay rate $\Lambda/(n-1)$. To face robustly the disturbances and model uncertainties, the sliding mode controller that satisfies the condition (7.14) must be discontinuous when crossing $S(t)$. This is not desired in practice, since it may excite high frequency unmodeled dynamics (chattering effect).

On the basis of the above, the design of the sliding mode robust controllers involves three steps:

Step 1: Select a control law that satisfies the sliding condition. This controller follows to be of the switching (discontinuous) type involving the signum function:

$$\mathrm{sgn}(s)=\begin{cases}+1, & s>0\\ -1, & s<0\end{cases} \tag{7.15}$$

Step 2: To avoid the chattering effect, the switching type controller is smoothened to get a compromise between trajectory tracking accuracy and control signal bandwidth. This can be typically achieved by approximating the sharply varying "sgn" function by a saturation function and using a sliding boundary layer $B(t)$, instead of the sliding surface, as shown in Figure 7.3, where $B(t)$ is defined as:

$$B(t)=\{\mathbf{x}:|s(\mathbf{x},t)\leq U,\ \mathbf{x}\in R^n\},\quad U>0 \tag{7.16}$$

The boundary layer $B(t)$ is an invariant region of the state space (all trajectories that depart from a state inside $B(t)$ for $t=0$, remain always inside it for all $t>0$). Within the boundary layer, the function sgn(s) is replaced by the smooth linear function $z=s/U$, where U is the width of the boundary layer. Specifically, the sliding mode controller that uses the "saturation" function, is:

$$\mathrm{sat}(z)=\begin{cases}z & \text{if } |z|\leq 1\\ \mathrm{sgn}(z) & \text{if } |z|>1\end{cases} \tag{7.17}$$

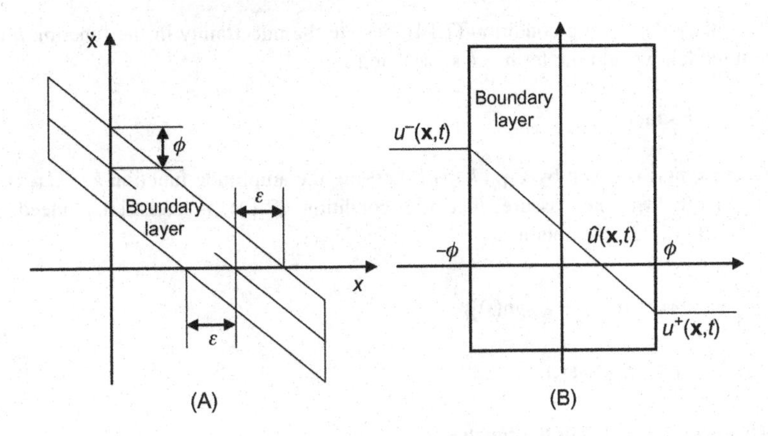

Figure 7.3 (A) Boundary layer for $n = 2$. (B) Control signal smoothing inside the boundary layer.

Of course, in this case the trajectory tracking is achieved with a certain maximum error ε, that is, for all trajectories originating inside $B(t = 0)$, the following condition holds:

$$|\mathrm{d}^i\tilde{y}(t)/\mathrm{d}t^i| \leq 2(\Lambda)^i\varepsilon, \quad i = 0, 1, 2, ..., n-1$$

Step 3: Outside the boundary layer $B(t)$, the control law is defined as before, that is, to satisfy the standard sliding condition (7.14). To illustrate how a sliding mode controller is designed, we consider the following simple, but representative, system:

$$\ddot{x} = b + u$$

where x is the scalar output, u the scalar control input, and the function $b(x)$(possibly nonlinear or time varying) is not precisely known but only approximately with uncertainty bound $\rho_{\max}$, that is:

$$|\hat{b} - b| \leq \rho_{\max} \tag{7.18}$$

The sliding surface $s = 0$ that assures $x(t) = x_\mathrm{d}(t)$ is given by Eq. (7.13), that is:

$$s(t) = (\mathrm{d}/\mathrm{d}t + \Lambda)\tilde{x} = \dot{\tilde{x}} + \Lambda\tilde{x} \tag{7.19}$$

where $\tilde{x}(t) = x(t) - x_\mathrm{d}(t)$. Differentiating Eq. (7.19) we get:

$$\dot{s} = \ddot{x} - \ddot{x}_\mathrm{d} + \Lambda\dot{\tilde{x}} = b + u - \ddot{x}_\mathrm{d} + \Lambda\dot{\tilde{x}} \tag{7.20}$$

Thus, the best approximation of the continuous control law that gives $\dot{s} = 0$, is:

$$\hat{u} = -\hat{b} + \ddot{x}_\mathrm{d} - \Lambda\dot{\tilde{x}} \tag{7.21}$$

To satisfy the sliding condition (7.14), despite the uncertainty in the function b of the system model, we add the term $-k\,\mathrm{sgn}(s)$, and so:

$$u = \hat{u} - k\,\mathrm{sgn}(s) \tag{7.22}$$

where $\mathrm{sgn}(s)$ is given by Eq. (7.15). Choosing the amplitude function $k = k(x, \dot{x})$ sufficiently large, we can assure that the condition (7.14) is satisfied. Indeed, from Eqs. (7.20)–(7.22), we obtain:

$$\begin{aligned}\frac{1}{2}\frac{\mathrm{d}}{\mathrm{d}t}s^2 &= \dot{s}s = [b - \hat{b} - k\,\mathrm{sgn}(s)]s \\ &= (b - \hat{b})s - k|s|\end{aligned} \tag{7.23}$$

Therefore, if we select the function $k(x, \dot{x})$ as:

$$k = \rho_{\max} + \gamma \tag{7.24a}$$

and take into account the condition (7.18), the relation (7.23) gives:

$$\frac{1}{2}\frac{\mathrm{d}}{\mathrm{d}t}s^2 \leq -\gamma|s| \tag{7.24b}$$

which is the desired condition (7.14). In a similar way, we can treat a system of the type $\ddot{x} = b + au$, where the gain function a (possibly nonlinear) is not exactly known but by an estimated value $\hat{a}$ satisfying the inequality:

$$1/\eta \leq \hat{a}/a \leq \eta \tag{7.24c}$$

where $\eta = \eta(x)$ is a given bounding function known as *gain margin function*. In this case, one can easily verify that the controller:

$$u = \hat{a}^{-1}\{\hat{u} - k\,\mathrm{sgn}(s)\} \tag{7.24d}$$

with $k = \eta(\rho_{\max} + \gamma) + (\eta - 1)|\hat{u}|$, satisfies the sliding condition (7.24b). If the uncertainty in $a(x)$ is specified by:

$$\eta_{\min} \leq \hat{a}/a \leq \eta_{\max} \tag{7.24e}$$

then we can put it in the form (7.24c) by setting $\eta = (\eta_{\max}/\eta_{\min})^{1/2}$ and replacing the estimate $\hat{a}$ by $(\eta_{\min}\eta_{\max})^{-1/2}\hat{a}$. Finally, if the uncertainty in $a(x)$ is determined by $\eta_{\min} \leq a \leq \eta_{\max}$, then we can express it in the form (7.24c) by setting $\hat{a} = (\eta_{\min}\eta_{\max})^{1/2}$ and $\eta = (\eta_{\max}/\eta_{\min})^{1/2}$.

7.2.3 Robust Control Using the Lyapunov Stabilization Method

This is an alternative robust control method which is based on Lyapunov stabilization. In this method, we construct a Lyapunov function V for the nominal closed loop system, and then, we use it for the design of the controller that assures the robustness against the system uncertainties. Consider the nonlinear system [16]:

$$\dot{\mathbf{x}} = \mathbf{f}(\mathbf{x}) + \mathbf{g}(\mathbf{x})u + \mathbf{d}(\mathbf{x}, t), \quad \mathbf{x} \in R^n \tag{7.25}$$

where u is the scalar control input, $\mathbf{f}(\mathbf{x})$, $\mathbf{g}(\mathbf{x})$ have the standard meaning, and $\mathbf{d}(x, t)$ is an uncertain function which is bounded by a known function $\rho(\mathbf{x})$ as:

$$||\mathbf{d}(\mathbf{x}, t)|| \leq \rho(\mathbf{x}) \tag{7.26}$$

We assume that the nominal system is stabilizable, that is, there exists a state feedback controller $\hat{u}(\mathbf{x})$ which gives an asymptotically closed loop system:

$$\dot{\mathbf{x}} = \mathbf{f}(\mathbf{x}) + \mathbf{g}(\mathbf{x})\hat{u}(\mathbf{x}) \tag{7.27}$$

at the equilibrium point $\mathbf{x} = 0$. We assume that we know a Lyapunov function V such that:

$$\left[\frac{\vartheta V(\mathbf{x})}{\vartheta \mathbf{x}}\right]^T [\mathbf{f}(\mathbf{x}) + \mathbf{g}(\mathbf{x})\hat{u}(\mathbf{x})] < 0 \quad (\text{for all } \mathbf{x} \neq \mathbf{0}) \tag{7.28}$$

The problem is to design an additional stabilizing controller $u_{\text{robust}}(\mathbf{x})$ such that the total controller:

$$u(\mathbf{x}) = \hat{u}(\mathbf{x}) + u_{\text{robust}}(\mathbf{x}) \tag{7.29}$$

robustly stabilizes the uncertain system (7.25). The requirement of robust stability is satisfied if $\dot{V}$ is negative along the trajectories of the system for all allowable uncertainties. Here:

$$\dot{V} = \left[\frac{\vartheta V(\mathbf{x})}{\vartheta \mathbf{x}}\right]^{\mathrm{T}} [\mathbf{f}(\mathbf{x}) + \mathbf{g}(\mathbf{x})\hat{u}(\mathbf{x})] + \left[\frac{\vartheta V(\mathbf{x})}{\vartheta \mathbf{x}}\right]^{\mathrm{T}} [\mathbf{g}(\mathbf{x})u_{\text{robust}}(\mathbf{x}) + \mathbf{d}(\mathbf{x}, t)] \tag{7.30}$$

Thus, $u_{\text{robust}}(\mathbf{x})$ should be selected such that $\dot{V} < 0$. We observe that the first term of Eq. (7.30) is negative due to the selection of $\hat{u}(\mathbf{x})$ (see the condition (7.28)). A solution for such a $u_{\text{robust}}(\mathbf{x})$ can be found if the disturbance $d(\mathbf{x}, t)$ has the form:

$$\mathbf{d}(\mathbf{x}, t) = \mathbf{g}(\mathbf{x})\overline{d}(\mathbf{x}, t) \tag{7.31}$$

for some uncertain function $\overline{d}(\mathbf{x}, t)$. Then, obviously:

$$[\vartheta V(\mathbf{x})/\vartheta \mathbf{x}]^T \mathbf{d}(\mathbf{x}, t) = [\vartheta V(\mathbf{x})/\vartheta \mathbf{x}]^{\mathrm{T}} \mathbf{g}(\mathbf{x})\overline{d}(\mathbf{x}, t) = 0$$

for all $\mathbf{x}$ for which $[\vartheta V(\mathbf{x})/\vartheta \mathbf{x}]^{\mathrm{T}}\mathbf{g}(\mathbf{x})=0$. The structural condition (7.31) is known as *matching condition,* because it allows to write the system (7.25) as:

$$\dot{\mathbf{x}}=\mathbf{f}(\mathbf{x})+\mathbf{g}(\mathbf{x})[u+\overline{d}(\mathbf{x},t)] \tag{7.32}$$

which means that the uncertainty $\overline{d}(\mathbf{x},t)$ enters the system from the same input channel. If the matching condition (7.31) holds, the robustifying controller $u_{\mathrm{robust}}(\mathbf{x})$ can be determined in several ways. For example, if the uncertainty $\overline{d}(\mathbf{x},t)$ is bounded as:

$$||\overline{d}(\mathbf{x},t)||\leq\overline{\rho}(\mathbf{x}) \tag{7.33}$$

for some known function $\overline{\rho}(\mathbf{x})$, then the controller:

$$u_{\mathrm{robust}}(\mathbf{x})=\begin{cases}-\overline{\rho}(\mathbf{x})\dfrac{[(\vartheta V/\vartheta\mathbf{x})^{\mathrm{T}}\mathbf{g}(\mathbf{x})]^{\mathrm{T}}}{||(\vartheta V/\vartheta\mathbf{x})^{\mathrm{T}}\mathbf{g}(\mathbf{x})||}, & ||(\vartheta V/\vartheta\mathbf{x})^{\mathrm{T}}\mathbf{g}(\mathbf{x})||\neq 0\\ 0, & ||(\vartheta V/\vartheta\mathbf{x})^{\mathrm{T}}\mathbf{g}(\mathbf{x})||=0\end{cases} \tag{7.34}$$

gives:

$$\dot{V}\leq\left[\frac{\vartheta V(\mathbf{x})}{\vartheta\mathbf{x}}\right]^{T}[\mathbf{f}(\mathbf{x})+\mathbf{g}(\mathbf{x})\hat{u}(\mathbf{x})]+\left\|\left[\frac{\vartheta V(\mathbf{x})}{\vartheta\mathbf{x}}\right]^{\mathrm{T}}\mathbf{g}(\mathbf{x})\right\|\{-\overline{\rho}(\mathbf{x})+||\overline{d}(\mathbf{x},t)||\}<0$$

because the first term is negative by the condition (7.28), and the second term is negative by Eq. (7.33). Thus, the controller:

$$u(\mathbf{x})=\hat{u}(\mathbf{x})+u_{\mathrm{robust}}(\mathbf{x})$$

where $\hat{u}$ satisfies Eq. (7.28) and u_{robust} given by Eq. (7.34) assures the robust stabilization of the system for all uncertainties $\overline{d}(\mathbf{x},t)$ that are bounded as in Eq. (7.33).

In the scalar control input case, the controller $u_{\mathrm{robust}}(\mathbf{x})$ in Eq. (7.34) reduces to:

$$u_{\mathrm{robust}}(\mathbf{x})=-\overline{\rho}(\mathbf{x})\,\mathrm{sgn}[(\vartheta V/\vartheta\mathbf{x})^{\mathrm{T}}\mathbf{g}(\mathbf{x})]$$

We see that, as in the sliding mode case, at the points $\mathbf{x}$ where $(\vartheta V/\vartheta\mathbf{x})^{\mathrm{T}}\mathbf{g}(\mathbf{x})=0$, the controller $u_{\mathrm{robust}}(\mathbf{x})$ is discontinuous. For this reason, some smooth approximation must be used, which will assure convergence not at $\mathbf{x}=\mathbf{0}$, but in an arbitrarily small region around $\mathbf{x}=\mathbf{0}$. A simple way to do this is to replace the function $\mathbf{z}^{\mathrm{T}}(\mathbf{x})/||\mathbf{z}(\mathbf{x})||$ in Eq. (7.34), by the function:

$$\sigma(\mathbf{x})=\frac{\mathbf{z}^{T}(\mathbf{x})}{||\mathbf{z}(\mathbf{x})||+\delta(\mathbf{x})},\quad \mathbf{z}(\mathbf{x})=\left(\frac{\vartheta V(\mathbf{x})}{\vartheta\mathbf{x}}\right)^{\mathrm{T}}\mathbf{g}(\mathbf{x})$$

where $\delta(\mathbf{x})$ is a smooth strictly positive function which is once differentiable and reduces to $\mathbf{z}^{\mathrm{T}}(\mathbf{x})/||\mathbf{z}(\mathbf{x})||$ when $\delta(\mathbf{x}) \equiv 0$. It is easy to verify that using $\boldsymbol{\sigma}(\mathbf{x})$, the smooth control law $\hat{u}(\mathbf{x})$ gives $\dot{V} \leq 0$ for all $\mathbf{x}$ except for $\mathbf{x} = \mathbf{0}$, provided that $\delta(\mathbf{x})$ is sufficiently small.

7.3 Model Reference Adaptive Control of Mobile Robots

7.3.1 Differential Drive WMR

We consider the feedback tracking controller of the differential drive WMR shown in Figure 5.11, and assume that the inertial parameters m and I are unknown [8,9]. Therefore, if $\hat{m}$ and $\hat{I}$ are the estimates of m and I, the control laws ((5.59a) and (5.59b)) are replaced by:

$$\tau_{\mathrm{a}} = \hat{m}\dot{v}_{\mathrm{d}} + K_{\mathrm{a}}\tilde{v}_{\mathrm{c}}, \quad \tilde{v}_{\mathrm{c}} = v_{\mathrm{d}} - v_{\mathrm{c}} \tag{7.35a}$$

$$\tau_{\mathrm{b}} = \hat{I}\dot{\omega}_{\mathrm{d}} + K_{\mathrm{b}}\tilde{\omega}_{\mathrm{c}}, \quad \tilde{\omega}_{\mathrm{c}} = \omega_{\mathrm{d}} - \omega_{\mathrm{c}} \tag{7.35b}$$

Now, introducing Eqs. (5.49a) and (5.49b) into Eqs. (7.35a) and (7.35b) we find:

$$\dot{v}_{\mathrm{c}} = \beta_1\dot{v}_{\mathrm{d}} + \beta_2(v_{\mathrm{d}} - v_{\mathrm{c}}) \tag{7.36a}$$

$$\dot{\omega}_{\mathrm{c}} = \beta_3\dot{\omega}_{\mathrm{d}} + \beta_4(\omega_{\mathrm{d}} - \omega_{\mathrm{c}}) \tag{7.36b}$$

where:

$$\beta_1 = \hat{m}/m, \quad \beta_2 = K_{\mathrm{a}}/m, \quad \beta_3 = \hat{I}/I, \quad \beta_4 = K_{\mathrm{b}}/I \tag{7.36c}$$

are linearly appearing parameters that will be estimated (updated) by the adaptation law. The reference models for v_{c} and ω_{c} are taken to be:

$$\dot{v}_{\mathrm{r}} + \beta_{\mathrm{rv}}v_{\mathrm{r}} = 0, \quad \beta_{\mathrm{rv}} > 0 \tag{7.37a}$$

$$\dot{\omega}_{\mathrm{r}} + \beta_{\mathrm{r}\omega}\omega_{\mathrm{r}} = 0, \quad \beta_{\mathrm{r}\omega} > 0 \tag{7.37b}$$

where β_{rv} and $\beta_{\mathrm{r}\omega}$ are linear and angular damping coefficients, respectively. We will now work with the linear velocity system (7.36a) and the corresponding reference model (7.37a) which is written in the form:

$$\dot{v}_{\mathrm{m}} = \dot{v}_{\mathrm{d}} + \beta_{\mathrm{rv}}v_{\mathrm{d}} - \beta_{\mathrm{rv}}v_{\mathrm{m}} \tag{7.38}$$

where $v_m = v_d - v_r$. Now, using Eqs. (7.36a) and (7.38) we find that the error $e = v_c - v_m$ between the system and reference model velocities is described by the dynamic equation:

$$\dot{e}(t) = -\beta_{rv}e + (\beta_1 - 1)\dot{v}_d + (\beta_2 - \beta_{rv})(v_d - v_c) \tag{7.39}$$

To construct an adaptation law for β_1 and β_2, we use the following candidate function, in analogy to Eq. (7.7):

$$V(e, \beta_1, \beta_2) = \frac{1}{2}\left[e^2 + \frac{1}{\gamma_1}(\beta_1 - 1)^2 + \frac{1}{\gamma_2}(\beta_2 - \beta_{rv})^2\right]$$

Differentiating V and using Eq. (7.39) we get:

$$\begin{aligned}\dot{V} &= e\dot{e} + (1/\gamma_1)(\beta_1 - 1)\dot{\beta}_1 + (1/\gamma_2)(\beta_2 - \beta_{rv})\dot{\beta}_2 \\ &= e[-\beta_{rv}e + (\beta_1 - 1)\dot{v}_d + (\beta_2 - \beta_{rv})(v_d - v_c)] + (1/\gamma_1)(\beta_1 - 1)\dot{\beta}_1 \\ &\quad + (1/\gamma_2)(\beta_2 - \beta_{rv})\dot{\beta}_2 \\ &= -\beta_{rv}e^2 + (\beta_1 - 1)[e\dot{v}_d + (1/\gamma_1)\dot{\beta}_1] + (\beta_2 - \beta_{rv})[e(v_d - v_c) + (1/\gamma_2)\dot{\beta}_2]\end{aligned}$$

Thus, selecting the adaptation laws for β_1 and β_2 as:

$$\dot{\beta}_1 = -\gamma_1\dot{v}_d e = \gamma_1\psi_1 e, \quad \psi_1 = -\dot{v}_d \tag{7.40a}$$

$$\dot{\beta}_2 = -\gamma_2(v_d - v_c)e = \gamma_2\psi_2 e, \quad \psi_2 = v_c - v_d \tag{7.40b}$$

we get:

$$\dot{V} = -\beta_{rv}e^2 \leq 0 \quad (\text{since } \beta_{rv} > 0)$$

which, by the Lyapunov stability criterion, implies that $e(t)$ converges asymptotically to zero. The corresponding adaptation laws for β_3 and β_4, which have the form (7.40a) and (7.40b), are:

$$\dot{\beta}_3 = \gamma_3\psi_3 e', \quad \dot{\beta}_4 = \gamma_4\psi_4 e' \tag{7.41}$$

where e', ψ_3, and ψ_4 have obvious corresponding definitions.

On the basis of the above, the tracking control system of Figure 5.11 can be upgraded to an adaptive control system by embedding in the boxes for 1/m and $1/I$ the adaptation laws (7.40a), (7.40b), and (7.41).

7.3.2 Adaptive Control Via Input–Output Linearization

7.3.2.1 Tracking Control for Known Parameters

In general, nonholonomic WMRs (unicycle-type and car-like type) have two inputs and two outputs. It was described, in Example 6.8, that using properly the outputs, the system can be input–output decoupled using a static feedback linearizing/decoupling controller. For car-like robots, these outputs may be the coordinates x_c, y_c of a point C in front of the robot that lies on the steering line. For the differential drive robots, the proper outputs are selected to be the coordinates x_c, y_c of a point C in front of the robot that lies on the axis of the linear velocity v of the robot (see Figure 6.4) which has an angle ϕ with respect to the x-axis of the world coordinate frame. Therefore, the output vector is:

$$y = \begin{bmatrix} y_1 \\ y_2 \end{bmatrix} = \begin{bmatrix} x + L\cos\phi \\ y + L\sin\phi \end{bmatrix} = \begin{bmatrix} x_c \\ y_c \end{bmatrix} \tag{7.42}$$

The proof that the use of the output (7.42) allows static input–output decoupling is analogous to the proof given in Example 6.8 for the car-like robot. Now, assuming that the outputs were selected such that the system can be decoupled by a static state feedback controller, we will develop the general feedback linearization process for an m-input m-output affine system of the form [7]:

$$\dot{\mathbf{x}} = \mathbf{f}(\mathbf{x}) + \mathbf{g}_1(\mathbf{x})u_1 + \cdots + \mathbf{g}_m(\mathbf{x})u_m, \quad \mathbf{x} \in R^n \tag{7.43a}$$

$$\mathbf{y} = \begin{bmatrix} y_1 \\ \vdots \\ y_m \end{bmatrix} = \begin{bmatrix} h_1(\mathbf{x}) \\ \vdots \\ h_m(\mathbf{x}) \end{bmatrix} \in R^m, \quad \mathbf{u} = \begin{bmatrix} u_1 \\ \vdots \\ u_m \end{bmatrix} \in R^m \tag{7.43b}$$

Differentiating the output y_i we get:

$$\dot{y}_i = L_{\mathbf{f}} h_i(\mathbf{x}) + \sum_{k=1}^{m} L_{\mathbf{g}_k} h_i(\mathbf{x}) u_k \tag{7.44a}$$

where:

$$L_{\mathbf{f}} h_i(\mathbf{x}) = \sum_{j=1}^{n} \frac{\vartheta h_i(\mathbf{x})}{\vartheta x_j} \dot{x}_{jf} \tag{7.44b}$$

$$L_{\mathbf{g}_k} h_i(\mathbf{x}) = \sum_{j=1}^{n} \frac{\vartheta h_i(\mathbf{x})}{\vartheta x_j} \dot{x}_{jg} \tag{7.44c}$$

where $\dot{x}_{jf}$ and $\dot{x}_{jg}$ are the parts of the jth state equation (7.43a) that are due to $\mathbf{f}(\mathbf{x})$ and $\mathbf{g}(\mathbf{x})$, respectively.

Clearly, if all $L_{\mathbf{g}_k} h_i(\mathbf{x})$ in Eq. (7.44a) are zero, then no input appears in $\dot{y}_i$. Suppose that r_i is the *relative degree* corresponding to y_i, that is, the lowest integer for which at least one of the inputs appears explicitly in $\mathrm{d}^{r_i} y_i / \mathrm{d}t^{r_i}$ (see Definition 6.3). Then:

$$y_i^{(r_i)} = \mathrm{d}^{r_i} y_i / \mathrm{d}t^{r_i} = L_{\mathbf{f}}^{r_i} h_i(\mathbf{x}) + \sum_{k=1}^{m} L_{\mathbf{g}_k} (L_{\mathbf{f}}^{r_i - 1} h_i(\mathbf{x})) u_k \tag{7.45a}$$

for $i = 1, 2, \ldots, m$, where:

$$L_{\mathbf{f}}^{r_i} h_i(\mathbf{x}) = \sum_{j=1}^{m} \frac{\vartheta L_{\mathbf{f}}^{r_i - 1} h_i(\mathbf{x})}{\vartheta x_j} \dot{x}_{jf} \tag{7.45b}$$

$$L_{\mathbf{g}_k} L_{\mathbf{f}}^{r_i - 1} h_i(\mathbf{x}) = \sum_{j=1}^{n} \frac{\vartheta L_{g_k} L_{\mathbf{f}}^{r_i - 2} h_i(\mathbf{x})}{\vartheta x_j} \dot{x}_{jg} \tag{7.45c}$$

and

$$L_{\mathbf{g}_k} L_{\mathbf{f}}^{r_i - 1} h_i(\mathbf{x}) \neq \mathbf{0}$$

for at least one k and all $\mathbf{x}$ in the linearization region of interest.

Now defining $\mathbf{\Theta}(\mathbf{x})$ as:

$$\mathbf{\Theta}(\mathbf{x}) = \begin{bmatrix} L_{\mathbf{g}_1} L_{\mathbf{f}}^{r_i - 1} h_1 \ldots L_{\mathbf{g}_m} L_{\mathbf{f}}^{r_i - 1} h_1 \\ \vdots \\ L_{\mathbf{g}_1} L_{\mathbf{f}}^{r_m - 1} h_m \ldots L_{\mathbf{g}_m} L_{\mathbf{f}}^{r_m - 1} h_m \end{bmatrix} \tag{7.46}$$

the relation (7.45a), $i = 1, 2, \ldots, m$, is written in the compact form:

$$\begin{bmatrix} y_1^{(r_1)} \\ \vdots \\ y_m^{(r_m)} \end{bmatrix} = \begin{bmatrix} L_{\mathbf{f}}^{r_1} h_1 \\ \vdots \\ L_{\mathbf{f}}^{r_m} h_m \end{bmatrix} + \mathbf{\Theta}(\mathbf{x}) \begin{bmatrix} u_1 \\ \vdots \\ u_m \end{bmatrix} \tag{7.47}$$

Therefore, if the inverse $\mathbf{\Theta}^{-1}(\mathbf{x})$ of the decoupling matrix $\mathbf{\Theta}(\mathbf{x})$ exists for $\mathbf{x}$ in the region of interest, we can use the state feedback law [7]:

$$\mathbf{u}(\mathbf{x}) = \mathbf{F}(\mathbf{x}) + \mathbf{G}(\mathbf{x})\boldsymbol{\upsilon}, \quad \boldsymbol{\upsilon} = [\upsilon_1, \upsilon_2, \ldots, \upsilon_m]^{\mathrm{T}} \tag{7.48a}$$

where $\boldsymbol{\upsilon}$ is the new input vector, and:

$$\mathbf{F}(\mathbf{x}) = -\mathbf{\Theta}^{-1}(\mathbf{x}) \begin{bmatrix} L_f^{r_1} h_1 \\ \vdots \\ L_f^{r_m} h_m \end{bmatrix}, \quad \mathbf{G}(\mathbf{x}) = \mathbf{\Theta}^{-1}(\mathbf{x}) \tag{7.48b}$$

Introducing the controller ((7.48a) and (7.48b)) in Eq. (7.47) we get the closed loop system:

$$\begin{bmatrix} y_1^{(r_1)} \\ \vdots \\ y_m^{(r_m)} \end{bmatrix} = \begin{bmatrix} v_1 \\ \vdots \\ v_m \end{bmatrix} \tag{7.48c}$$

which is linear and input–output decoupled for the selected outputs that allow input–output decoupling via static state feedback.

Now, the above decoupling m-input m-output method will be applied to the dynamic model ((6.43a) and (6.43b)) of the differential drive WMR:

$$\dot{\mathbf{x}} = \mathbf{f}(\mathbf{x}) + \mathbf{g}(\mathbf{x})\mathbf{u}, \quad \mathbf{x} \in R^6 \tag{7.49a}$$

$$\mathbf{f}(\mathbf{x}) = \begin{bmatrix} \mathbf{B}\mathbf{v} \\ \vdots \\ -\overline{\mathbf{D}}^{-1}\overline{\mathbf{C}}\mathbf{v} \end{bmatrix} = \begin{bmatrix} \mathbf{B}\mathbf{v} \\ \vdots \\ \mathbf{f}_2 \end{bmatrix} \in R^6 \tag{7.49b}$$

$$\mathbf{g}(\mathbf{x}) = \begin{bmatrix} \mathbf{0} \\ \vdots \\ \overline{\mathbf{D}}^{-1} \end{bmatrix} \in R^6, \quad \mathbf{u} = \begin{bmatrix} u_1 \\ u_2 \end{bmatrix} = \begin{bmatrix} \tau_r \\ \tau_l \end{bmatrix} \in R^2 \tag{7.49c}$$

with the output (7.42), which expressed in the world coordinate frame (see Eq. (2.17)), is:

$$\mathbf{y} = \begin{bmatrix} y_1 \\ y_2 \end{bmatrix} = \begin{bmatrix} x \\ y \end{bmatrix} + \begin{bmatrix} \cos\phi & -\sin\phi \\ \sin\phi & \cos\phi \end{bmatrix} \begin{bmatrix} x_c \\ y_c \end{bmatrix} = \begin{bmatrix} h_1(\mathbf{x}) \\ h_2(\mathbf{x}) \end{bmatrix} \tag{7.50}$$

Differentiating $\mathbf{y}$ in Eq. (7.50) twice with respect to time yields:

$$\ddot{y} = L_{\mathbf{f}}^2\mathbf{h}(\mathbf{x}) + \boldsymbol{\Theta}(\mathbf{x})\mathbf{u}(\mathbf{x}) \tag{7.51a}$$

where:

$$L_{\mathbf{f}}^2\mathbf{h}(\mathbf{x}) = \mathbf{p} + \mathbf{q}\mathbf{f}_2, \quad \boldsymbol{\Theta}(\mathbf{x}) = \mathbf{q}\overline{\mathbf{D}}^{-1} \tag{7.51b}$$

with $\mathbf{p}$ and $\mathbf{q}$ being computed from $\mathbf{f}(\mathbf{x})$. The model ((7.51a) and (7.51b)) has the form ((7.45a), (7.45b)), or (7.47)), and so using the state feedback control law:

$$\mathbf{u}(\mathbf{x}) = \mathbf{F}(\mathbf{x}) + \mathbf{G}(\mathbf{x})\boldsymbol{v}, \quad \boldsymbol{v} = [v_1, v_2]^{\mathrm{T}} \tag{7.52a}$$

with:

$$\mathbf{F}(\mathbf{x}) = \boldsymbol{\Theta}^{-1}(\mathbf{x})L_{\mathbf{f}}^2\mathbf{h}(\mathbf{x}), \quad \mathbf{G}(\mathbf{x}) = \boldsymbol{\Theta}^{-1}(\mathbf{x}) \tag{7.52b}$$

we get:

$$\begin{aligned} \ddot{y}_1(t) &= v_1(t) \\ \ddot{y}_2(t) &= v_2(t) \end{aligned} \tag{7.53}$$

Now, having available the desired output trajectory $\mathbf{y}_d = [y_{1d}, y_{2d}]^T$, the linear tracking controllers are selected as:

$$\begin{aligned} v_1 &= \ddot{y}_{1d} + k_{11}(\dot{y}_{1d} - \dot{y}_1) + k_{01}(y_{1d} - y_1) \\ v_2 &= \ddot{y}_{2d} + k_{12}(\dot{y}_{2d} - \dot{y}_2) + k_{02}(y_{2d} - y_2) \end{aligned}$$

which lead to the closed loop error dynamics:

$$\ddot{\tilde{y}} + k_{11}\dot{\tilde{y}} + k_{01}\tilde{y}_1 = 0, \quad \ddot{\tilde{y}}_2 + k_{12}\dot{\tilde{y}} + k_{02}\tilde{y}_2 = 0$$

Selecting the parameters $k_{ij}(i,j = 1,2)$ such that to obtain a desired damping ratio ζ and undamped natural frequency (bandwith) $\Omega = \omega_n^2$, we get the desired asymptotic tracking performance.

7.3.2.2 Adaptive Tracking Controller

If the system ((7.43a) and (7.43b)) has unknown parameters, then the linearization obtained by Eqs. (7.48a) and (7.48b) is not perfect. In this case, we use estimates of these parameters, which lead to estimates $\hat{\mathbf{f}}(\mathbf{x})$, $\hat{\mathbf{g}}(\mathbf{x})$, and $\hat{\mathbf{h}}(\mathbf{x})$ of $\mathbf{f}(\mathbf{x})$, $\mathbf{g}(\mathbf{x})$, and $\mathbf{h}(\mathbf{x})$, and the controller ((7.52a) and (7.52b)) is replaced by:

$$\hat{\mathbf{u}}(\mathbf{x}) = \hat{\mathbf{F}}(\mathbf{x}) + \hat{\mathbf{G}}(\mathbf{x})\upsilon \tag{7.54a}$$

where:

$$\hat{\mathbf{F}}(\mathbf{x}) = \hat{\mathbf{\Theta}}^{-1}(\mathbf{x})\hat{L}_{\mathbf{f}}^2\hat{\mathbf{h}}(\mathbf{x}), \quad \hat{\mathbf{G}}(\mathbf{x}) = \hat{\mathbf{\Theta}}^{-1}(\mathbf{x}) \tag{7.54b}$$

involves the estimates $\hat{\boldsymbol{\beta}}$ of the unknown parameters. Working as usual, we can get the adaptation law:

$$\dot{\hat{\boldsymbol{\beta}}} = \gamma\boldsymbol{\psi}\mathbf{e} \tag{7.55}$$

where $\mathbf{e}$ is the error between the closed loop output and the reference model output, $\boldsymbol{\psi}$ depends on the system structure, and γ is a constant specifying the rate of convergence.

7.3.3 Omnidirectional Robot

We will work with the model (3.77a):

$$\dot{\mathbf{x}} = \mathbf{A}(\mathbf{x})\mathbf{x} + \mathbf{B}(\mathbf{u})\mathbf{u}, \quad \mathbf{x} \in R^6, \quad \mathbf{u} \in R^3 \tag{7.56}$$

which involves unknown parameters in $\mathbf{A}(\mathbf{x})$ and $\mathbf{B}(\mathbf{x})$. Let the reference model be:

$$\dot{\mathbf{x}}_{\mathrm{m}} = \mathbf{A}_{\mathrm{m}}\mathbf{x}_{\mathrm{m}} + \mathbf{B}_{\mathrm{m}}\mathbf{u}, \quad \mathbf{x}_{\mathrm{m}} \in R^6, \quad \mathbf{u} \in R^3 \tag{7.57}$$

We define the generalized state vector error as:

$$\mathbf{e} = \mathbf{x}_{\mathrm{m}} - \mathbf{x}$$

in which case Eq. (7.56) can be written as:

$$\dot{\mathbf{x}} = \mathbf{A}(\mathbf{e}, \mathbf{t})\mathbf{x} + \mathbf{B}(\mathbf{e}, t)\mathbf{u}, \quad \mathbf{x} \in R^6, \quad \mathbf{u} \in R^3 \tag{7.58}$$

Therefore, the error equation is:

$$\dot{\mathbf{e}} = \mathbf{A}_{\mathrm{m}}\mathbf{e} + [\mathbf{A}_{\mathrm{m}} - \mathbf{A}(e, t)]\mathbf{x} + [\mathbf{B}_{\mathrm{m}} - \mathbf{B}(\mathbf{e}, t)]\mathbf{u} \tag{7.59}$$

To find the adaptation laws for $\mathbf{A}(\mathbf{e}, t)$ and $\mathbf{B}(\mathbf{e}, t)$, we define the following candidate Lyapunov function V in the increased state space $R^6 \times R^{6\times 6} \times R^{6\times 3}$:[1]

$$\begin{aligned} V = {} & \frac{1}{2}\mathbf{e}^{\mathrm{T}}\mathbf{P}\mathbf{e} + \mathrm{trace}\{[\mathbf{A}_{\mathrm{m}} - \mathbf{A}(\mathbf{e}, t)]^{\mathrm{T}}\mathbf{F}_{\mathbf{A}}^{-1}[\mathbf{A}_{\mathrm{m}} - \mathbf{A}(\mathbf{e}, t)]\} \\ & + \mathrm{trace}\{[\mathbf{B}_{\mathrm{m}} - \mathbf{B}(\mathbf{e}, t)]^{T}\mathbf{F}_{\mathbf{B}}^{-1}[\mathbf{B}_{\mathrm{m}} - \mathbf{B}(\mathbf{e}, t)]\} \end{aligned} \tag{7.60}$$

where $\mathbf{P}$, $\mathbf{F}_{\mathbf{A}}^{-1}$, and $\mathbf{F}_{\mathbf{B}}^{-1}$ are positive define matrices. The matrix $\mathbf{P}$ will be determined below, but the matrices $\mathbf{F}_{\mathrm{A}}$ and $\mathbf{F}_{\mathrm{B}}$ can be arbitrary. Differentiating V along the error trajectory (7.59) we find:

$$\begin{aligned} \dot{V} = {} & \mathbf{e}^{\mathrm{T}}(\mathbf{A}_{\mathrm{m}}^{\mathrm{T}}\mathbf{P} + \mathbf{P}\mathbf{A}_{\mathrm{m}})\mathbf{e} \\ & + \mathrm{trace}\{[\mathbf{A}_{\mathrm{m}} - \mathbf{A}(\mathbf{e}, t)]^{\mathrm{T}}[\mathbf{P}\mathbf{e}\mathbf{x}^{\mathrm{T}} - \mathbf{F}_{\mathbf{A}}^{-1}\dot{\mathbf{A}}(\mathbf{e}, t)]\} \\ & + \mathrm{trace}\{[\mathbf{B}_{\mathrm{m}} - \mathbf{B}(\mathbf{e}, t)]^{\mathrm{T}}[\mathbf{P}\mathbf{e}\mathbf{u}^{\mathrm{T}} - \mathbf{F}_{\mathbf{B}}^{-1}\dot{\mathbf{B}}(\mathbf{e}, t)]\} \end{aligned} \tag{7.61}$$

Therefore, if $\mathbf{A}_{\mathrm{m}}$ is a Hurwitz matrix, then:

$$\mathbf{A}_{\mathrm{m}}^{\mathrm{T}}\mathbf{P} + \mathbf{P}\mathbf{A}_{\mathrm{m}} = -\mathbf{Q} \tag{7.62}$$

where $\mathbf{Q}$ is a positive definite matrix that allows the computation of the proper matrix $\mathbf{P}$. Therefore, the first term in $\dot{V}$ is negative for all $\mathbf{e} \neq \mathbf{0}$ and the other two terms become zero if we select the adaptation laws for $\mathbf{A}(\mathbf{e}, t)$ and $\mathbf{B}(\mathbf{e}, t)$ as:

$$\dot{\mathbf{A}}(\mathbf{e}, t) = \mathbf{F}_{\mathbf{A}}\mathbf{P}\mathbf{e}\mathbf{x}^{\mathrm{T}} \tag{7.63a}$$

$$\dot{\mathbf{B}}(\mathbf{e}, t) = \mathbf{F}_{\mathrm{B}}\mathbf{P}\mathbf{e}\mathbf{u}^{\mathrm{T}} \tag{7.63b}$$

[1] It is recalled that the trace of a $n \times n$ matrix $\mathbf{A} = [a_{ij}]$ is defined as the sum of its diagonal elements, that is, trace $\mathbf{A} = \sum_{i=}^{n} a_{ii}$.

The result is that the laws ((7.63a) and (7.63b)) assure the asymptotic convergence of the MRAC scheme for any $\mathbf{F_A} > \mathbf{0}$, $\mathbf{F_B} > \mathbf{0}$, and any input vector $\mathbf{u}$. We will now see under what conditions the zero error $\mathbf{e}(t) \equiv \mathbf{0}$ implies that $\mathbf{A}(\mathbf{e}, t) = \mathbf{A}_\mathrm{m}$ and $\mathbf{B}(\mathbf{e}, t) = \mathbf{B}_\mathrm{m}$. From Eqs. (7.63a) and (7.63b), it follows (after integration) that if $\lim_{t\to\infty} \mathbf{e}(t) = 0$, then:

$$\lim_{t\to\infty} [\mathbf{A}_\mathrm{m} - \mathbf{A}(\mathbf{e}, t)] = \tilde{\mathbf{A}} \text{ and } \lim_{t\to\infty} [\mathbf{B}_\mathrm{m} - \mathbf{B}(\mathbf{e}, t)] = \tilde{\mathbf{B}}$$

where $\tilde{\mathbf{A}}$ and $\tilde{\mathbf{B}}$ represent the asymptotic difference in the parameters. Now, from Eq. (7.59) it follows that if $\mathbf{e}(t) \equiv \mathbf{0}$, then:

$$\tilde{\mathbf{A}}\mathbf{x} + \tilde{\mathbf{B}}\mathbf{u} \equiv \mathbf{0} \tag{7.64}$$

The identity (7.64) can hold for all t if:

1. The vectors $\mathbf{x}$ and $\mathbf{u}$ are linearly dependent and $\tilde{\mathbf{A}} \neq \mathbf{0}, \tilde{\mathbf{B}} \neq \mathbf{0}$
2. The vectors $\mathbf{x}$ and $\mathbf{u}$ are identically equal to zero
3. The vectors $\mathbf{x}$ and $\mathbf{u}$ are linearly independent and $\tilde{\mathbf{A}} = \mathbf{0}, \tilde{\mathbf{B}} = \mathbf{0}$

From the above it follows that only in the third case the parameters converge with certainty. Any controller PID or other (such as the one presented in Section 5.7) can be used in conjunction with the above parameter adaptation laws ((7.63a) and (7.63b)). Typically, the parameters that vary and are unknown while the robot is at work (e.g., while it carries and delivers objects) are the robot mass m, and moment of inertia I about the rotation axis.

Example 7.1

Consider a differential drive WMR where the mass m and moment of inertia I are constant (or very slowly variant) but unknown. Derive an adaptive tracking controller of your choice, different than that presented in Section 7.3.1.

Solution

The solution will be derived using the dynamic equation of the WMR expressed in terms of x, y, and ϕ. For tracking feasibility, we assume (as usual) that the desired trajectory to be tracked obeys the same kinematic equations as the robot at hand, namely:

$$\dot{x}_\mathrm{d} = v_\mathrm{d}\cos\phi_\mathrm{d}, \quad \dot{y}_\mathrm{d} = v_\mathrm{d}\sin\phi_\mathrm{d}, \quad \dot{\phi}_\mathrm{d} = \omega_\mathrm{d} \tag{7.65}$$

The kinematic and dynamic equations of the robot are (see Eqs. (3.30a), (3.30b), and (3.31)):

$$\dot{x} = v\cos\phi, \quad \dot{y} = v\sin\phi, \quad \dot{\phi} = \omega \tag{7.66a}$$

$$\dot{v} = (1/mr)u_1, \quad \dot{\omega} = (2a/Ir)u_2 \tag{7.66b}$$

where $u_1 = \tau_\mathrm{r} + \tau_\mathrm{l}$, $u_2 = \tau_\mathrm{r} - \tau_\mathrm{l}$, and τ_r and τ_l are the torques exerted by the right and left wheel, respectively. Multiplying $\dot{x}$ by $\cos\phi$, and $\dot{y}$ by $\sin\phi$ and adding we get

$$v = \dot{x}\cos\phi + \dot{y}\sin\phi \tag{7.67}$$

Differentiating $\dot{x}, \dot{y}$, and $\dot{\phi}$ in Eq. (7.66a), and introducing Eqs. (7.66b) and (7.67) we obtain the following dynamic model for x, y, and ϕ:

$$\begin{aligned}\ddot{x} &= -(\dot{x}\cos\phi + \dot{y}\sin\phi)\dot{\phi}\sin\phi + \beta_1(\cos\phi)u_1 \\ \ddot{y} &= (\dot{x}\cos\phi + \dot{y}\sin\phi)\dot{\phi}\cos\phi + \beta_1(\sin\phi)u_1 \\ \ddot{\phi} &= \beta_2 u_2\end{aligned} \tag{7.68}$$

where $\beta_1 = 1/mr$ and $\beta_2 = 2a/Ir$ are the parameters to be adaptively estimated. Note that the kinematic parameters r and $2a$ are precisely known (or assumed to be known). We define $\mathbf{x} = \begin{bmatrix} x & y & \phi \end{bmatrix}^{\mathrm{T}}$ and $\mathbf{x}_{\mathrm{d}} = [x_{\mathrm{d}}, y_{\mathrm{d}}, \phi_{\mathrm{d}}]^{\mathrm{T}}$; the problem is to design a feedback controller that drives the error $\tilde{\mathbf{x}}(t) = \mathbf{x}_{\mathrm{d}}(t) - \mathbf{x}(t)$ asymptotically to zero, while estimating adaptively the unknown parameters β_1 and β_2. The first step is to find the dynamics of the tracking error $\tilde{\mathbf{x}}$. For convenience, we use the equivalent error $\boldsymbol{\varepsilon}$ in the local coordinate frame, that is [8]:

$$\boldsymbol{\varepsilon} = \mathbf{E}\tilde{\mathbf{x}}, \quad \mathbf{E} = \begin{bmatrix} \cos\phi & \sin\phi & 0 \\ -\sin\phi & \cos\phi & 0 \\ 0 & 0 & 1 \end{bmatrix} \tag{7.69}$$

where $\mathbf{E}$ is invertible and so $\boldsymbol{\varepsilon} \to 0$ if and only if $\tilde{\mathbf{x}} \to \mathbf{0}$. Taking into account the nonholonomic constraint $-\dot{x}\sin\phi + \dot{y}\cos\phi = 0$, we get the following error dynamics:

$$\begin{aligned}\dot{\varepsilon}_1 &= \omega\varepsilon_2 - v + v_{\mathrm{d}}\cos\varepsilon_3 \\ \dot{\varepsilon}_2 &= -\omega\varepsilon_1 + v_{\mathrm{d}}\sin\varepsilon_3 \\ \dot{\varepsilon}_3 &= \omega_{\mathrm{d}} - \omega\end{aligned} \tag{7.70}$$

where $\omega = \dot{\phi}$ and v is given by Eq. (7.67). We observe that the control inputs do not appear explicitly in Eq. (7.70), but instead, we have the variables v and ω given by Eq. (7.66b). Therefore, we will first select v and ω such that $\varepsilon_1 \to 0$, $\varepsilon_2 \to 0$, and $\varepsilon_3 \to 0$ asymptotically, and then use them as inputs for the next step of choosing u_1 and u_2. If v_{m} and ω_{m} are the desired values of the conceptual intermediate controls v and ω, then we have the errors $\tilde{v} = v - v_{\mathrm{m}}$ and $\tilde{\omega} = \omega - \omega_{\mathrm{m}}$. Therefore:

$$v = v_{\mathrm{m}} + \tilde{v}, \quad \omega = \omega_{\mathrm{m}} + \tilde{\omega} \tag{7.71}$$

Now, in analogy to Eq. (5.58b) we use the following Lyapunov function:

$$V = \frac{1}{2}(\varepsilon_1^2 + \varepsilon_2^2) + (1/K_2)(1 - \cos\varepsilon_3) \tag{7.72}$$

$$\begin{aligned}\dot{V} &= \varepsilon_1\dot{\varepsilon}_1 + \varepsilon_2\dot{\varepsilon}_2 + (1/K_2)(\sin\varepsilon_3)\dot{\varepsilon}_3 \\ &= \varepsilon_1(\omega\varepsilon_2 - v_{\mathrm{m}} - \tilde{v} + v_{\mathrm{d}}\cos\varepsilon_3) \\ &\quad + \varepsilon_2(-\omega\varepsilon_1 + v_{\mathrm{d}}\sin\varepsilon_3) + (1/K_2)(\sin\varepsilon_3)(\omega_{\mathrm{d}} - \omega_{\mathrm{m}} - \tilde{\omega}) \\ &= \varepsilon_1(-v_{\mathrm{m}} + v_{\mathrm{d}}\cos\varepsilon_3) + \left(\varepsilon_2 v_{\mathrm{d}} + \frac{1}{K_2}\omega_{\mathrm{d}} - \frac{1}{K_2}\omega_{\mathrm{m}}\right)\sin\varepsilon_3 - \varepsilon_1\tilde{v} - (1/K_2)\tilde{\omega}\sin\varepsilon_3\end{aligned}$$

Selecting v_m and ω_m such that:

$$-v_m + v_d \cos \varepsilon_3 = -K_1 \varepsilon_1, \quad K_1 > 0$$
$$-\frac{1}{K_2}\omega_m + \frac{1}{K_2}\omega_d + \varepsilon_2 v_d = -\frac{K_3}{K_2}\sin \varepsilon_3, \quad K_2 > 0$$

that is:

$$v_m = v_d \cos \varepsilon_3 + K_1 \varepsilon_1 \tag{7.73a}$$

$$\omega_m = \omega_d + K_2 v_d \varepsilon_2 + K_3 \sin \varepsilon_3 \tag{7.73b}$$

the derivative of V becomes:

$$\dot{V} = -K_1 \varepsilon_1^2 - (K_3/K_2)\sin^2 \varepsilon_3 - \varepsilon_1 \tilde{v} - (1/K_2)(\sin \varepsilon_3)\tilde{\omega} \tag{7.74}$$

Now, from Eqs. (7.66b) and (7.71) we obtain:

$$\dot{\tilde{v}} = \beta_1 u_1 - \dot{v}_m, \quad \dot{\tilde{\omega}} = \beta_2 u_2 - \dot{\omega}_m \tag{7.75}$$

The first two terms of $\dot{V}$ are negative, and so for asymptotic stability ($\varepsilon_1 \to 0$, $\varepsilon_3 \to 0$) and parameter convergence, we have to make $\tilde{v} \to 0$ and $\tilde{\omega} \to 0$. To this end, we add to the Lyapunov function V a second term V' defined as [8]:

$$V' = \frac{1}{2}(\tilde{v}^2 + \tilde{\omega}^2) + \frac{1}{2}\left[\frac{|\beta_1|}{\gamma_1}\tilde{\theta}_1^2 + \frac{|\beta_2|}{\gamma_2}\tilde{\theta}_2^2\right] \tag{7.76}$$

where $\tilde{\theta}_1 = \theta_1 - \hat{\theta}_1$, $\hat{\theta}_2 = \theta_2 - \hat{\theta}_2$, $\theta_1 = 1/\beta_1$, and $\theta_2 = 1/\beta_2$. The total Lyapunov function is $V_0 = V + V'$, which if differentiated in time along the trajectories of the error systems described by Eqs. (7.70) and (7.75), and choosing:

$$\begin{aligned}
u_1 &= \hat{\theta}_1(-K_4\tilde{v} + \varepsilon_1 + \dot{v}_d), \quad K_4 > 0 \\
u_2 &= \hat{\theta}_2(-K_5\tilde{\omega} + (1/K_2)\sin \varepsilon_3 + \dot{\omega}_d), \quad K_5 > 0 \\
\dot{\hat{\theta}}_2 &= \gamma_1 \psi_1 \tilde{v}, \quad \psi_1 = -(-K_4\tilde{v} + \varepsilon_1 + \dot{v}_d)\mathrm{sgn}(\beta_1) \\
\dot{\hat{\theta}}_2 &= \gamma_2 \psi_2 \tilde{\omega}, \quad \psi_2 = -[-K_5\tilde{\omega} + (1/K_2)\sin \varepsilon_3 + \dot{\omega}_d]\mathrm{sgn}(\beta_2)
\end{aligned} \tag{7.77}$$

gives:

$$\dot{V}_0 = -(K_1\varepsilon_1^2 + (K_3/K_2)\sin^2 \varepsilon_3 + K_4\tilde{v}^2 + K_5\tilde{\omega}^2) \le 0$$

Now, computing the second derivative $\ddot{V}_0$ (taking into account that v_d, ω_d are smooth) we find that it is finite, and so, by Barbalat's lemma (see Section 6.2.3), $\dot{V}_0$ is uniformly continuous and $\dot{V}_0 \to 0$ for $t \to \infty$. Then, it follows that $\varepsilon_1 \to 0$, $\varepsilon_3 \to 0$, $\tilde{v} \to 0$, and $\tilde{\omega} \to 0$. It can also easily be shown that $\varepsilon_2 \to 0$ by assuming that v_d and ω_d do not both go to zero simultaneously. It is again remarked that actually the above solution is based on the well known *backstepping control procedure.*

7.4 Sliding Mode Control of Mobile Robots

Here, the sliding mode control of Section 7.2.2 will be applied to the differential drive mobile robot ((7.66a) and (7.66b)), with kinematically compatible desired trajectory $[x_d, y_d, \phi_d]^T$ [12,17]. The tracking error ε, expressed in the local coordinate frame, is given by (see Eq. (7.69)):

$$\begin{bmatrix} \varepsilon_1 \\ \varepsilon_2 \\ \varepsilon_3 \end{bmatrix} = \begin{bmatrix} \cos\phi_d & \sin\phi_d & 0 \\ -\sin\phi_d & \cos\phi_d & 0 \\ 0 & 0 & 1 \end{bmatrix} \begin{bmatrix} x - x_d \\ y - y_d \\ \phi - \phi_d \end{bmatrix} \tag{7.78}$$

and satisfies the differential equations:

$$\begin{aligned} \dot{\varepsilon}_1 &= \dot{x}\cos\phi_d + \dot{y}\sin\phi_d + \omega_d\varepsilon_2 - v_d \\ \dot{\varepsilon}_2 &= -\dot{x}\sin\phi_d + \dot{y}\cos\phi_d - \omega_d\varepsilon_1 \\ \dot{\varepsilon}_3 &= \omega - \omega_d \end{aligned} \tag{7.79}$$

Without loss of generality, we will assume that $|\varepsilon_3| < \pi/2$. The system control inputs are $u_1 = \tau_r + \tau_l$ and $u_2 = \tau_r - \tau_l$.

Here, we have a two-dimensional sliding surface:

$$\mathbf{s} = \begin{bmatrix} s_1 \\ s_2 \end{bmatrix} \tag{7.80a}$$

and a two-component control of the type (7.22). The components of $\mathbf{s}$ are defined as:

$$s_1 = \dot{\varepsilon}_1 + \Lambda_1\varepsilon_1 \tag{7.80b}$$

$$s_2 = \dot{\varepsilon}_3 + \Lambda_2\varepsilon_3 + \Lambda_0|\varepsilon_2|\mathrm{sgn}\,(\varepsilon_3) \tag{7.80c}$$

Clearly, if $s_1 \to 0$, then $\varepsilon_1 \to 0$. If $s_2 \to 0$, then we have:

$$\dot{\varepsilon}_3 = -\Lambda_2\varepsilon_3 - \Lambda_0|\varepsilon_2|\mathrm{sgn}(\varepsilon_3) \tag{7.81}$$

Therefore, since $|\varepsilon_2|$ is bounded, we get the following conditions:

$$\begin{aligned} &\text{If } \varepsilon_3 < 0, \text{ then } \quad \dot{\varepsilon}_3 > 0 \\ &\text{If } \varepsilon_3 > 0, \text{ then } \quad \dot{\varepsilon}_3 < 0 \end{aligned} \tag{7.82}$$

Differentiating s_1 and s_2 in Eqs. (7.80b) and (7.80c), we get:

$$\begin{aligned} \dot{s}_1 &= \ddot{\varepsilon}_1 + \Lambda_1\dot{\varepsilon}_1 \\ \dot{s}_2 &= \ddot{\varepsilon}_3 + \Lambda_2\dot{\varepsilon}_3 + \Lambda_0|\varepsilon_2|'\,\mathrm{sgn}(\varepsilon_3) \end{aligned} \tag{7.83}$$

Equations (7.66a), (7.66b), (7.78), and (7.80a)–(7.80c) can be written as:

$$\dot{\mathbf{s}} = -\mathrm{H}\mathbf{s} - \boldsymbol{\Lambda}\ \mathrm{sgn}(\mathbf{s}) \tag{7.84}$$

where:

$$\mathrm{H} = \begin{bmatrix} \mathrm{H}_1 & 0 \\ 0 & \mathrm{H}_2 \end{bmatrix}, \quad \boldsymbol{\Lambda} = \begin{bmatrix} \Lambda_1 & 0 \\ 0 & \Lambda_2 \end{bmatrix}, \quad \mathrm{sgn}(\mathbf{s}) = \begin{bmatrix} \mathrm{sgn}(s_1) \\ \mathrm{sgn}(s_2) \end{bmatrix}$$

Now, as in the SISO case, we define the following candidate Lyapunov function:

$$V = \frac{1}{2}\mathbf{s}^{\mathrm{T}}\mathbf{s} = \frac{1}{2}s_1^2 + \frac{1}{2}s_2^2$$

Differentiating V, we get:

$$\begin{aligned} \dot{V} &= \mathbf{s}^{\mathrm{T}}\dot{\mathbf{s}} = \mathbf{s}^{\mathrm{T}}(-\mathrm{H}\mathbf{s} - \boldsymbol{\Lambda}\ \mathrm{sgn}(\mathbf{s})) \\ &= -\mathbf{s}^{\mathrm{T}}\mathrm{H}\mathbf{s} - s_1\Lambda_1\ \mathrm{sgn}(s_1) - s_2\Lambda_2\ \mathrm{sgn}(s_2) \\ &= -\mathbf{s}^{\mathrm{T}}\mathrm{H}\mathbf{s} - \Lambda_1|s_1| - \Lambda_2|s_2| \end{aligned}$$

Therefore, $\dot{V} \leq 0$ if H and $\boldsymbol{\Lambda}$ are selected as:

$$\mathrm{H}_1 > 0, \quad \mathrm{H}_2 > 0, \quad \Lambda_1 > 0, \quad \text{and} \quad \Lambda_2 > 0$$

The controller that satisfies the sliding condition (7.14) has the form of Eqs. (7.21) and (7.22). To avoid the undesired chattering effect, the function "sgn" of Eq. (7.15) is replaced by the "sat(·)" function given by Eq. (7.17). Therefore:

$$u_i = \hat{u}_i - k_i\ \mathrm{sat}(s_i/U) \quad (i = 1, 2) \tag{7.85}$$

where U is the thickness of the boundary layer.

The best continuous control law approximation $\hat{u}_i (i = 1, 2)$ that satisfies $\dot{s}_i = 0$ $(i = 1, 2)$ is found using the dynamic model of the robot (see Eqs. (7.67)–(7.68)):

$$\begin{aligned} \ddot{x} &= -(v \sin \phi)\dot{\phi} + (\beta_1 \cos \phi)u_1 \\ \ddot{y} &= (v \cos \phi)\dot{\phi} + (\beta_1 \sin \phi)u_1 \\ \ddot{\phi} &= \beta_2 u_2 \end{aligned}$$

together with Eqs. (7.79) and (7.80a)–(7.80c). The result is (see Eq. (7.21)):

$$\hat{u}_1 = (1/\hat{\beta}_1 \cos \varepsilon_3)[\dot{x}\dot{\varepsilon}_3 \sin \phi_{\mathrm{d}} - \dot{y}\dot{\varepsilon}_3 \cos \phi_{\mathrm{d}}] + \dot{\omega}_{\mathrm{d}}\varepsilon_2 + \omega_{\mathrm{d}}\dot{\varepsilon}_2 + \Lambda_1\dot{\varepsilon}_1 \tag{7.86a}$$

$$\hat{u}_2 = (1/\hat{\beta}_2)[-\dot{\omega}_{\mathrm{d}} + \Lambda_2\dot{\varepsilon}_3 + \Lambda_0(\mathrm{sgn}\ \varepsilon_3)\dot{\varepsilon}_2\ \mathrm{sgn}(\varepsilon_2)] \tag{7.86b}$$

The gains k_i ($i = 1, 2$) must be selected sufficiently large to assure that the sliding condition (7.24b) is satisfied, given the maximum bounds $|\hat{\beta}_i - \beta_i| = |\Delta\beta_i| \leq B_{i,\max}$ ($i = 1, 2$). After some calculation, we can verify that two gains which do the job are [17]:

$$k_1 = a_1[\mathrm{H}_1 s_1 + \Lambda_1 \ \mathrm{sat}(s_1/B) + p_1(B_{1,\max}/\hat{\beta}_1)]$$
$$k_2 = a_2[\mathrm{H}_2 s_2 + \Lambda_2 \ \mathrm{sat}(s_2/B) + p_2(B_{2,\max}/\hat{\beta}_2)]$$

where:

$$a_1 = 1/(\hat{\beta}_1 + B_{1,\max})\cos \varepsilon_3, \quad a_2 = 1/(\hat{\beta}_2 + B_{2,\max})$$

$$p_1 = \dot{x}\dot{\varepsilon}_3 \sin \phi_\mathrm{d} - \dot{y}\dot{\varepsilon}_3 \cos \phi_\mathrm{d} + \dot{\omega}_\mathrm{d}\varepsilon_2 + \omega_\mathrm{d}\dot{\varepsilon}_2 + \Lambda_1\dot{\varepsilon}_1$$

$$p_2 = -\dot{\omega}_\mathrm{d} + \Lambda_2\dot{\varepsilon}_3 + \Lambda_0(\mathrm{sgn}\ \varepsilon_3)\dot{\varepsilon}_2\mathrm{sgn}(\varepsilon_2)$$

7.5 Sliding Mode Control in Polar Coordinates

7.5.1 Modeling

The technique develops as in the previous section, with the basic difference that the kinematic model should be expressed in polar coordinates. Referring to the geometry of Figure 7.4, the robot's actual and desired kinematic equations:

$$\dot{x} = v \cos \phi, \quad \dot{y} = v \sin \phi, \quad \dot{\phi} = \omega; \quad \dot{x}_\mathrm{d} = v_\mathrm{d} \cos \phi_\mathrm{d}, \quad \dot{y}_\mathrm{d} = v_\mathrm{d} \sin \phi_\mathrm{d}, \quad \dot{\phi}_\mathrm{d} = \omega_\mathrm{d}$$

have the following polar coordinates form [12,13]:

$$\dot{x} = \begin{bmatrix} \dot{l} \\ \dot{\psi} \\ \dot{\phi} \end{bmatrix} = \begin{bmatrix} v\cos(\psi - \phi) \\ -(v/l)\sin(\psi - \phi) \\ \omega \end{bmatrix} \tag{7.87a}$$

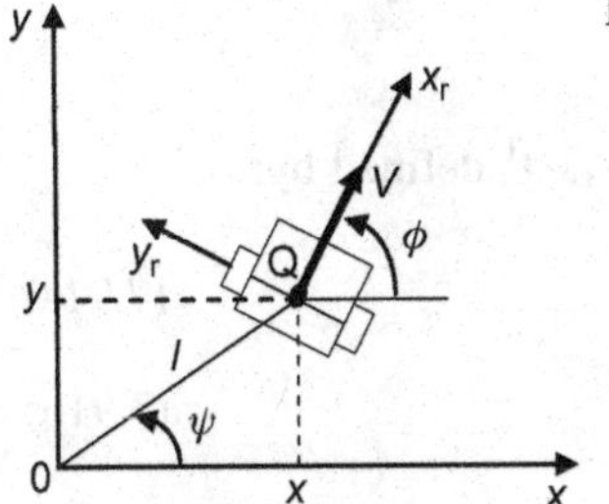

Figure 7.4 Geometry of polar coordinates.

$$\dot{x}_{\rm d} = \begin{bmatrix} \dot{l}_{\rm d} \\ \dot{\psi}_{\rm d} \\ \dot{\phi}_{\rm d} \end{bmatrix} = \begin{bmatrix} v_{\rm d}\cos(\psi_{\rm d}-\phi_{\rm d}) \\ -(v_{\rm d}/l_{\rm d})\sin(\psi_{\rm d}-\phi_{\rm d}) \\ \omega_{\rm d} \end{bmatrix} \tag{7.87b}$$

where:

$$l=\sqrt{x^2+y^2},\ \ \phi=tg^{-1}(y/x),\ \ l_{\rm d}=\sqrt{x_{\rm d}^2+y_{\rm d}^2},\ \ \phi_{\rm d}=tg^{-1}(y_{\rm d}/x_{\rm d}) \tag{7.87c}$$

We will work with the dynamic model (7.66b) enhanced with linear and rotational frictional terms and disturbance inputs, namely:

$$mr\dot{v}+c_1v+d_1=u_1 \tag{7.88a}$$

$$(Ir/2a)\dot{\omega}+c_2\omega+d_2=u_2 \tag{7.88b}$$

where c_1 and c_2 are linear friction coefficients and d_1, d_2 are unknown disturbances representing all uncertain inputs (e.g., due to slip). We assume that d_1 and d_2 are expressed in the form:

$$d_1=mr\overline{d}_1,\quad d_2=(Ir/2a)\overline{d}_2 \tag{7.89}$$

that is, they satisfy a matching condition similar to Eq. (7.32). To assure that $v_{\rm d}/l_{\rm d}$ takes finite values, we assume that $l_{\rm d}\geq l_{\rm d,min}>0$, where $l_{\rm d,min}$ is a selected minimum allowed value for $l_{\rm d}$. Also, for convenience, we assume $l\geq l_{\rm d,min}$. As usual, $l_{\rm d}, \phi_{\rm d}$, and $\psi_{\rm d}$ are assumed to have smooth first-order and second-order time derivatives. Finally, without loss of generality, we assume that:

$$l(t)>0,\quad l_{\rm d}(t)>0\ (\text{for } t>0),\quad l_{\rm d}(0)=0 \tag{7.90}$$

$$-\pi<\psi_{\rm d}<\pi,\quad -\pi<\psi<\pi,\quad -\pi<\phi_{\rm d}<\pi,\quad -\pi<\phi<\pi$$
$$||\psi_{\rm d}-\phi_{\rm d}|-(2k+1)\pi/2|\geq\eta\quad (k=0,1)$$

for all t, where η is a positive constant. (i.e., it is required the robot should not have a posture with heading angle tangential to any circle drawn around the world frame origin).

7.5.2 Sliding Mode Control

Here, we have a two-dimensional sliding surface $\mathbf{s}=[s_1,s_2]^{\rm T}$ defined by:

$$s_1=\dot{\tilde{l}}+\Lambda_1\tilde{l} \tag{7.91a}$$

$$s_2=\dot{\tilde{\phi}}+\Lambda_2\tilde{\phi}+|\tilde{\psi}|\ \mathrm{sgn}(\tilde{\phi}) \tag{7.91b}$$

where $\tilde{l}=l-l_{\rm d}$, $\tilde{\phi}=\phi-\phi_{\rm d}$ and $\tilde{\psi}=\psi-\psi_{\rm d}$.

If $s_1 \to 0$, then $\tilde{l} \to 0$ asymptotically. Now, since by Eq. (7.90), $|\tilde{\psi}|$ is bounded, $s_2 \to 0$ implies that:

If $\tilde{\phi} < 0$, then $\dot{\tilde{\phi}} > 0$
If $\tilde{\phi} > 0$, then $\dot{\tilde{\phi}} < 0$

Finally, if $\tilde{\phi} \to 0$ and $\dot{\tilde{\phi}} \to 0$, then $|\tilde{\psi}| \to 0$. To compute s_1 and s_2, we use the expressions for $\dot{\tilde{l}}$ and $\dot{\tilde{\phi}}$:

$$\dot{\tilde{l}} = \dot{l} - \dot{l}_{\rm d} = v\cos(\psi - \phi) - v_{\rm d}\cos(\psi_{\rm d} - \phi_{\rm d})$$
$$\dot{\tilde{\phi}} = \dot{\phi} - \dot{\phi}_{\rm d} = \omega - \omega_{\rm d}$$

Therefore, by Eqs. (7.91a) and (7.91b):

$$s_1 = v\cos(\psi - \phi) - v_{\rm d}\cos(\psi_{\rm d} - \phi_{\rm d}) + \Lambda_1(l - l_{\rm d}) \tag{7.92a}$$

$$s_2 = \omega - \omega_{\rm d} + \Lambda_2(\phi - \phi_{\rm d}) + (\psi - \psi_{\rm d})\text{sgn}(\phi - \phi_{\rm d}) \tag{7.92b}$$

To assure that $l \to l_{\rm d}$, $\phi \to \phi_{\rm d}$, and $\psi \to \psi_{\rm d}$ the sliding condition $\mathbf{s}^{\rm T}\dot{\mathbf{s}} < 0$, $\mathbf{s} = [s_1, s_2]^{\rm T}$ must hold. We therefore need the expressions for $\dot{s}_1$ and $\dot{s}_2$, which are found by differentiating Eqs. (7.92a) and (7.92b):

$$\begin{aligned}
\dot{s}_1 &= \dot{v}\cos(\psi - \phi) - \dot{v}_{\rm d}\cos(\psi_{\rm d} - \phi_{\rm d}) + v_{\rm d}\frac{\rm d}{{\rm d}t}\cos(\psi - \phi) \\
&\quad - v_{\rm d}\frac{\rm d}{{\rm d}t}\cos(\psi_{\rm d} - \phi_{\rm d}) + \Lambda_1\tilde{l} \\
&= \dot{v}\cos(\psi - \phi) - v\sin(\psi - \phi)(\dot{\psi} - \dot{\phi}) \\
&\quad - \dot{v}_{\rm d}\cos(\psi_{\rm d} - \phi_{\rm d}) + v_{\rm d}\sin(\psi_{\rm d} - \phi_{\rm d})(\dot{\psi}_{\rm d} - \dot{\phi}) + \Lambda_1\tilde{l} \\
&= \dot{v}\cos(\psi - \phi) - v\sin(\psi - \phi)\left[-\frac{v}{l}\sin(\psi - \phi)\right] + v\omega\cos(\psi - \phi) \\
&\quad - \dot{v}_{\rm d}\cos(\psi_{\rm d} + \phi_{\rm d}) + v_{\rm d}\sin(\psi_{\rm d} - \phi_{\rm d})\left[-\frac{v_{\rm d}}{l_{\rm d}}\sin(\psi_{\rm d} - \phi_{\rm d})\right] \\
&\quad - v_{\rm d}\omega_{\rm d}\cos(\psi_{\rm d} - \phi_{\rm d}) + \Lambda_1\tilde{l} \\
&= \dot{v}\cos(\psi - \phi) + F(x, v) - \dot{v}_{\rm d}\cos(\psi_{\rm d} - \phi_{\rm d}) - F_{\rm d}(x_{\rm d}, v_{\rm d}) + \Lambda_1\tilde{l}
\end{aligned} \tag{7.93a}$$

$$\dot{s}_2 = \dot{\omega} - \dot{\omega}_{\rm d} + \Lambda_2(\omega - \omega_{\rm d}) + \frac{\rm d}{{\rm d}t}|\tilde{\psi}|\,\text{sgn}(\tilde{\phi}) \tag{7.93b}$$

with:

$$F(x, v) = v\sin(\psi - \phi)[(v/l)\sin(\psi - \phi)] + v\omega\cos(\psi - \phi)$$

$$F(x_{\rm d}-v_{\rm d})=v_{\rm d}\sin(\psi_{\rm d}-\phi_{\rm d})[(v_{\rm d}/l_{\rm d})\sin(\psi_{\rm d}-\phi_{\rm d})]+v_{\rm d}\omega_{\rm d}\cos(\psi_{\rm d}-\phi_{\rm d})$$

Now, $\dot{v}$ and $\dot{\omega}$ are given by Eqs. (7.88a) and (7.88b), and so choosing u_1 and u_2 as:

$$u_1=mr\dot{v}_{\rm d}+c_1v+mr\upsilon_1 \tag{7.94a}$$

$$u_2=(Ir/2a)\dot{\omega}_{\rm d}+c_2\omega+(Ir/2a)\upsilon_2 \tag{7.94b}$$

we get the closed-loop equations:

$$\dot{v}=\dot{v}_{\rm d}+\upsilon_1-\overline{d}_1,\quad d_1=mr\overline{d}_1 \tag{7.95a}$$

$$\dot{\omega}=\dot{\omega}_{\rm d}+\upsilon_2-d_2,\quad d_2=(Ir/2a)\overline{d}_2 \tag{7.95b}$$

Introducing Eqs. (7.95a) and (7.95b) into Eqs. (7.93a) and (7.93b), and computing $\mathbf{s}^{\rm T}\dot{\mathbf{s}}$ we get:

$$\begin{aligned}\mathbf{s}^{\rm T}\dot{\mathbf{s}}&=s_1\dot{s}_1+s_2\dot{s}_2\\&=s_1[\dot{v}_{\rm d}\cos(\psi-\phi)+\upsilon_1\cos(\psi-\phi)\\&\quad-\overline{d}_1\cos(\psi-\phi)+F(x,v)-\dot{v}_{\rm d}\cos(\psi_{\rm d}-\phi_{\rm d})-F_{\rm d}(x_{\rm d},y_{\rm d})+\Lambda_1\tilde{l}]\\&\quad+s_2\left[\dot{\omega}_{\rm d}+\upsilon_2-\overline{d}_2-\dot{\omega}_{\rm d}+\Lambda_2\tilde{\phi}+\frac{\rm d}{{\rm d}t}|\tilde{\psi}|\ \mathrm{sgn}(\tilde{\phi})\right]\end{aligned}$$

Therefore, selecting υ_1 and υ_2 such that:

$$\begin{aligned}&\dot{v}_{\rm d}\cos(\psi-\phi)+\upsilon_1\cos(\psi-\phi)+F(x,v)-F_{\rm d}(x_{\rm d},v_{\rm d})\\&-\dot{v}_{\rm d}\cos(\psi_{\rm d}-\phi_{\rm d})+\Lambda_1\tilde{l}=-H_1s_1-K_1\ \mathrm{sgn}(s_1)\end{aligned}$$

and

$$\upsilon_2+\Lambda_2\tilde{\phi}+\frac{\rm d}{{\rm d}t}|\tilde{\psi}|\ \mathrm{sgn}(\tilde{\phi})=-H_2s_2-K_2\ \mathrm{sgn}(s_2)$$

that is:

$$\begin{aligned}\upsilon_1=&\frac{1}{\cos(\psi-\phi)}[-\dot{v}_{\rm d}\cos(\psi-\phi)-F(x,v)+F_{\rm d}(x_{\rm d},v_{\rm d})+\dot{v}_{\rm d}\cos(\psi_{\rm d}-\phi_{\rm d})]\\&-\Lambda_1\tilde{l}-H_1s_1-K_1\ \mathrm{sgn}(s_1)]\end{aligned} \tag{7.96a}$$

$$\upsilon_2=-\Lambda_2\tilde{\phi}-\mathrm{H}_2s_2-K_2\ \mathrm{sgn}(s_2)-\frac{\rm d}{{\rm d}t}|\tilde{\psi}|\ \mathrm{sgn}(\tilde{\phi}) \tag{7.96b}$$

yields:

$$\begin{aligned}\mathbf{s}^{\mathrm{T}}\dot{\mathbf{s}} &= -s_1[H_1 s_1 + K_1\ \mathrm{sgn}(s_1) - \overline{d}_1\cos(\psi-\phi)]\\ &= -s_2[H_2 s_2 + K_2\ \mathrm{sgn}(s_2) - \overline{d}_2]\\ &= -\mathbf{s}^T\mathrm{H}\mathbf{s} - [K_1|s_1| - \overline{d}_1 s_1\cos(\psi-\phi)\\ &\quad -[K_2|s_2| - \overline{d}_2 s_2]\end{aligned}$$

which is negative definite for:

$$\mathrm{H} = \begin{bmatrix}\mathrm{H}_1 & 0\\ 0 & \mathrm{H}_2\end{bmatrix} > 0,\quad K_1 > \overline{d}_1,\quad K_2 > \overline{d}_2 \tag{7.97}$$

7.6 Robust Control of Differential Drive Robot Using the Lyapunov Method

We consider the differential drive WMR with disturbances in both the torque/force input channels and the x, y velocities, that is:

$$\begin{aligned}&\dot{x} = (v + \overline{\mathrm{d}}_v)\cos\phi,\quad \dot{y} = (v + \overline{\mathrm{d}}_v)\sin\phi,\quad \dot{\phi} = \omega\\ &\dot{v} = -(c_1/mr)v + (1/mr)(u_1 - \mathrm{d}_1)\\ &\dot{\omega} = -(2ac_2/Ir)\omega + (2a/Ir)(u_2 - d_2)\end{aligned} \tag{7.98}$$

where the dynamic model ((7.88a) and (7.88b)) is considered. The model (7.98) can be written as:

$$\dot{\mathbf{x}} = \mathbf{f}(\mathbf{x}) + \mathbf{G}(\mathbf{x})\mathbf{u} + \mathbf{d},\quad \mathbf{u} = \begin{bmatrix}u_1 & u_2\end{bmatrix}^{\mathrm{T}} \tag{7.99a}$$

with $\mathbf{d} = [d_x, d_y, 0, d_1, d_2]^{\mathrm{T}}$ and:

$$\mathbf{x} = \begin{bmatrix}x\\ y\\ \phi\\ v\\ \omega\end{bmatrix},\quad \mathbf{f}(\mathbf{x}) = \begin{bmatrix}v\cos\phi\\ v\sin\phi\\ \omega\\ -(c_1/mr)v\\ -(2ac_2/Ir)\omega\end{bmatrix},\quad \mathbf{G}(\mathbf{x}) = \begin{bmatrix}0 & 0\\ 0 & 0\\ 0 & 0\\ 1/mr & 0\\ 0 & 2a/Ir\end{bmatrix} \tag{7.99b}$$

where $d_x = \overline{d}_v\cos\phi$ and $d_y = \overline{d}_v\sin\phi$. Defining state vectors $\mathbf{z}_1$ and $\mathbf{z}_2$ as:

$$\mathbf{z}_1 = \begin{bmatrix}x\\ y\end{bmatrix},\quad \mathbf{z}_2 = \begin{bmatrix}\dot{x}\\ \dot{y}\end{bmatrix} = \dot{\mathbf{z}}_1 \tag{7.100}$$

the system ((7.99a) and (7.99b)) can be written (after some standard algebraic manipulation) in the form [11]:

$$\begin{aligned}\dot{\mathbf{z}}_1 &= \mathbf{z}_2 \\ \dot{\mathbf{z}}_2 &= \mathbf{R}(\phi)[\mathbf{u} + \boldsymbol{\delta}(\mathbf{z}_2, \phi)] \\ \dot{\phi} &= \omega(\mathbf{z}_2, \phi)\end{aligned} \tag{7.101}$$

where $\boldsymbol{\delta}(\mathbf{z}_2, \phi)$ incorporates all the uncertainties of the system. The corresponding nominal model is obtained from Eq. (7.101) by setting to zero all the uncertain terms involved in $\boldsymbol{\delta}(\mathbf{z}_2, \phi)$, that is:

$$\begin{aligned}\dot{\mathbf{z}}_1 &= \mathbf{z}_2 \\ \dot{\mathbf{z}}_2 &= \mathbf{R}(\phi)[\mathbf{u} + \hat{\boldsymbol{\delta}}(\mathbf{z}_2, \phi)] \\ \dot{\phi} &= \omega(\mathbf{z}_2, \phi)\end{aligned} \tag{7.102}$$

where $\hat{\boldsymbol{\delta}}(\mathbf{z}_2, \phi)$ is the nominal part of $\delta(\mathbf{z}_2, \phi)$. The problem is to design a robust tracking controller for the system (7.101) using appropriately the Lyapunov method presented in Section 7.2.3. Defining the tracking errors:

$$\tilde{\mathbf{z}}_1(t) = \mathbf{z}_1(t) - \mathbf{z}_{1d}(t), \quad \tilde{\mathbf{z}}_2 = \dot{\tilde{\mathbf{z}}}_1(t) = \dot{\mathbf{z}}_1(t) - \dot{\mathbf{z}}_{1d}(t) \tag{7.103}$$

and using the computed torque control law:

$$\mathbf{u} = \mathbf{R}^{-1}(\phi)[\dot{\mathbf{z}}_{2d}(t) + \upsilon(t)] - \hat{\boldsymbol{\delta}}(\mathbf{z}_2, \phi) \tag{7.104}$$

the system (7.102) gives the following closed-loop error equation:

$$\dot{\varepsilon}(t) = \mathbf{A}\varepsilon(t) + \mathbf{B}\upsilon(t) \tag{7.105}$$

where $\varepsilon(t) = [\tilde{\mathbf{z}}_1^{\mathrm{T}}(t),\ \tilde{\mathbf{z}}_2^{\mathrm{T}}(t)]^{\mathrm{T}}$, $\upsilon(t)$ is the new control input, and:

$$\mathbf{A} = \begin{bmatrix}\mathbf{0} & \mathrm{I}_{2\times 2} \\ \mathbf{0} & \mathbf{0}\end{bmatrix} \in R^{4\times 4}, \quad \mathbf{B} = \begin{bmatrix}\mathbf{0} \\ \mathrm{I}_{2\times 2}\end{bmatrix} \in R^{4\times 2} \tag{7.106}$$

Thus, the control law (7.104), applied to the disturbed system (7.101), gives the linearized closed-loop error system:

$$\dot{\varepsilon}(t) = \mathbf{A}\varepsilon(t) + \mathbf{B}[\upsilon(t) + \zeta(\mathbf{z}_2, \phi)] \tag{7.107}$$

where $\zeta(\mathbf{z}_2, \phi)$ is the overall disturbance which is assumed to be bounded as:

$$||\zeta(\mathbf{z}_2, \phi)|| \leq \overline{\rho}(\varepsilon) < \infty \tag{7.108}$$

From now on, the robust tracking controller design is made in two steps [11]:

Step 1: Design a nominal stabilizing controller $\hat{\upsilon}(\varepsilon)$ for the error system (7.105).
Step 2: Design the robustifying control term $\upsilon_{\text{robust}}(\varepsilon)$ using Eq. (7.34).

7.6.1 Nominal Controller

Selecting a linear feedback controller:

$$\hat{\upsilon}(\varepsilon(t)) = -\mathbf{K}\varepsilon(t), \quad \mathbf{K} = \begin{bmatrix} K_1 & K_2 \end{bmatrix} \tag{7.109}$$

we obtain the closed-loop error system:

$$\dot{\varepsilon}(t) = \mathbf{A}_c\varepsilon(t), \quad \mathbf{A}_c = \mathbf{A} - \mathbf{BK} \tag{7.110}$$

To find the gain matrix $\mathbf{K}$ we use the Lyapunov function:

$$V(\varepsilon) = \frac{1}{2}\varepsilon^T\mathbf{P}\varepsilon$$

where $\mathbf{P}$ is a symmetric positive definite matrix given by (see Eq. (5.15)):

$$\mathbf{A}_c^T\mathbf{P} + \mathbf{PA}_c = -\mathbf{Q} \tag{7.111}$$

and $\mathbf{Q}$ is a selected positive definite matrix. This assures that $\mathbf{A}_c$ is a Hurwitz matrix (having eigenvalues with strictly negative real parts), and so $\varepsilon(t) \to 0$ exponentially as $t \to \infty$.

7.6.2 Robustifying Controller

We calculate $\upsilon_{\text{robust}}(\varepsilon)$ using (7.34). Here:

$$[\vartheta V(\varepsilon)/\vartheta\varepsilon]^T\mathbf{B} = \mathbf{B}^T\mathbf{P}\varepsilon(t)$$

Therefore (see (7.108)):

$$\upsilon_{\text{robust}}(\varepsilon) = \begin{cases} -\overline{\rho}(\varepsilon)\mathbf{B}^T\mathbf{P}\varepsilon(t)/||\mathbf{B}^T\mathbf{P}\varepsilon(t)||, & ||\mathbf{B}^T\mathbf{P}\varepsilon(t)|| \neq 0 \\ 0, & ||\mathbf{B}^T\mathbf{P}\varepsilon(t)|| = 0 \end{cases} \tag{7.112}$$

To summarize, the overall robust controller is given by the combination of the computed torque linearizing controller (7.104), the nominal stabilizing controller (7.109), and the robustifying controller (7.112).

Example 7.2

It is desired to design a robust kinematic tracking controller for the differential drive WMR using the Lyapunov-based method of Section 7.3.

Solution

The nonrobust kinematic controller was designed in Section 5.4.1 using the Lyapunov function (5.54):

$$V(\tilde{\mathbf{p}}_r) = \frac{1}{2}(\tilde{x}_r^2 + \tilde{y}_r^2) + (1 - \cos\tilde{\phi}_r)$$

for the kinematic error system (5.53)

$$\dot{\tilde{\mathbf{p}}}_r = \mathbf{f}(\tilde{\mathbf{p}}_r) + \mathbf{G}(\tilde{\mathbf{p}}_r)\mathbf{u}, \quad \mathbf{u} = [v, \omega]^T = [u_1, u_2]^T$$

where:

$$\tilde{\mathbf{p}}_r = \begin{bmatrix} \tilde{x}_r \\ \tilde{y}_r \\ \tilde{\phi}_r \end{bmatrix}, \quad \mathbf{f}(\tilde{\mathbf{p}}_r) = \begin{bmatrix} v_d \cos\tilde{\phi}_r \\ v_d \sin\tilde{\phi}_r \\ \omega_d \end{bmatrix}, \quad \mathbf{G}(\tilde{\mathbf{p}}_r) = \begin{bmatrix} \tilde{\mathbf{g}}_1 & \tilde{\mathbf{g}}_2 \end{bmatrix} = \begin{bmatrix} -1 & \tilde{y}_r \\ 0 & -\tilde{x}_r \\ 0 & -1 \end{bmatrix}$$

where $\tilde{x}_r, \tilde{y}_r$, and $\tilde{\phi}_r$ are the errors between the actual and desired trajectories of $x(t)$, $y(t)$, and $\phi(t)$, respectively.

The nominal controller that satisfies $\dot{V}(\tilde{\mathbf{p}}_r) \le 0$ is given by Eqs. (5.56a) and (5.56b):

$$\hat{\mathbf{u}} = \begin{bmatrix} \hat{v} \\ \hat{\omega} \end{bmatrix} = \begin{bmatrix} v_d \cos\tilde{\phi}_r + K_x \tilde{x}_r \\ \omega_d + v_d \tilde{y}_r + K_\phi \sin\tilde{\phi}_r \end{bmatrix}$$

Here, the partial derivative of $V(\tilde{\mathbf{p}}_r)$ with respect to $\tilde{\mathbf{p}}_r$ is:

$$\vartheta V/\vartheta \tilde{\mathbf{p}}_r = \begin{bmatrix} \tilde{x}_r & \tilde{y}_r & \sin\phi_r \end{bmatrix}^T$$

Therefore:

$$[\vartheta V/\vartheta \tilde{\mathbf{p}}_r]^T \mathbf{G}(\tilde{\mathbf{p}}_r) = \begin{bmatrix} -\tilde{x}_r & -\sin\tilde{\phi}_r \end{bmatrix}$$

Now, assume that the disturbances d_v and d_ω in v and ω satisfy the matching conditions:

$$\mathbf{d}_u = \begin{bmatrix} d_v \\ d_\omega \end{bmatrix} = \mathbf{G}(\tilde{\mathbf{p}}_r) \begin{bmatrix} \bar{d}_v \\ \bar{d}_\omega \end{bmatrix} = \mathbf{G}(\tilde{\mathbf{p}}_r)\bar{\mathbf{d}}_u$$

with:

$$\|\bar{\mathbf{d}}_u\| \le \bar{\rho}_u$$

Then, the robustifying control $u_{robust}(\tilde{\mathbf{p}}_r)$ is given by (see (7.34)):

$$\mathbf{u}_{robust}(\tilde{\mathbf{p}}_r) = \begin{cases} -\overline{\rho}_u \dfrac{\mathbf{G}^T(\tilde{\mathbf{p}}_r)[\vartheta V/\vartheta \tilde{\mathbf{p}}_r]}{||\mathbf{G}^T(\tilde{\mathbf{p}}_r)\vartheta V/\vartheta \tilde{\mathbf{p}}_r||}, & \left\|\mathbf{G}^T(\tilde{\mathbf{p}}_r)\dfrac{\vartheta V}{\vartheta \tilde{\mathbf{p}}_r}\right\| \neq 0 \\ 0, & ||\mathbf{G}^T(\tilde{\mathbf{p}}_r)\vartheta V/\vartheta \tilde{\mathbf{p}}_r|| = 0 \end{cases}$$

$$= \begin{cases} -\dfrac{\overline{\rho}_u}{\sqrt{\tilde{x}_r^2 + \sin^2 \tilde{\phi}_r}} \begin{bmatrix} -\tilde{x}_r \\ -\sin \tilde{\phi}_r \end{bmatrix}, & \sqrt{\tilde{x}_r^2 + \sin^2 \tilde{\phi}_r} \neq 0 \\ 0, & \sqrt{\tilde{x}_r^2 + \sin^2 \tilde{\phi}_r} = 0 \end{cases}$$

Therefore, the full robust controller is:

$$\mathbf{u}(\tilde{\mathbf{p}}_r) = \hat{\mathbf{u}}(\tilde{\mathbf{p}}_r) + \mathbf{u}_{robust}(\tilde{\mathbf{p}}_r)$$

The discontinuous controller can be smoothened as explained in Section 7.3. Further results and examples on WMR robust control can be found in Refs. [10,14,15].

References

[1] Marino R, Tomei P. Nonlinear control design: geometric, adaptive and robust. Upper Saddle River, NJ: Prentice Hall; 1995.

[2] Marino R. Adaptive control of nonlinear systems: basic results and applications. IFAC-Review Control 1997;21:55–66.

[3] Huang HC, Tsai CC. Adaptive trajectory tracking and stabilization for omnidirectional mobile robot with dynamic effect and uncertainties. Proceedings of 17th IFAC world congress. Seoul, Korea; July 6–11, 2008. p. 5383–88.

[4] Ashoorizad M, Barzarnimi R, Afshar A, Jouzdani J. Model reference adaptive path following for wheeled mobile robots. Proceedings of international conference on information and automation. Colombo, Sri Lanka (IEEE-ICIA2006); 2006. p. 289–94.

[5] Alicja M. A new universal adaptive tracking control law for nonholonomic wheeled mobile robots moving in R^3 space. Proceedings of IEEE international conference robotics and automation. Dedroit, MI; May 10–15,1999.

[6] Kuc TY, Baek SM, Park K. Adaptive learning controller for autonomous mobile robots. IEE Proc—Control Theory Appl 2001;148(1):49–54.

[7] Fetter Lages W, Hemerly EM. Adaptive linearizing control of mobile robots. In: Kopacek P, Pereira CE, editors. Intelligent manufacturing systems—a volume from IFAC workshop. Gramado-RS, Brazil: Pergamon Press; November 9–11, 1998. 2000. p. 23–9.

[8] Pourboghrat F, Karlsson MP. Adaptive control of dynamic mobile robots with nonholonomic constraints. Comput Electr Eng 2002;28:241–53.

[9] Gholipour A, Dehghan SM, Ahmadabadi MN. Lyapunov-based tracking control of nonholonomic mobile robot. Proceedings of 10th Iranian conference on electrical engineering. vol. 3. Tabriz, Iran; 2002. p. 262–69.

[10] Dong W, Kuhnert KD. Robust adaptive control of nonholonomic mobile robot with parameter and nonparameter uncertainties. IEEE Trans Robot 2005;21(2):261–6.
[11] Zhang Y, Hong D, Chung JH, Velinsky S. Dynamic model based robust tracking control of a differentially steered wheeled mobile robot. Proceedings of the American control conference, Philadelphia, PA; June 1988, p. 850–55.
[12] Yang JM, Kim JK. Sliding mode control for trajectory tracking of nonholonomic wheeled mobile robots. IEEE Trans Robot Autom 1999;15(3):578–87.
[13] Chwa D. Sliding-mode tracking control of nonholonomic wheeled mobile robots in polar coordinates. IEEE Trans Control Syst Technol 2004;12(4):637–44.
[14] Zhu X, Dong G, Hu D, Cai Z. Robust tracking control of wheeled mobile robots not satisfying nonholonomic constraints. Proceedings of the 6th international conference on intelligent systems design and applications (ISDA'06), 2006. p. 643–8.
[15] Dixon WE, Dawson DM, Zergeroglu E, Zhang F. Robust tracking and regulation control for mobile robots. Int J Robust Nonlinear Control 2000;10:199–216.
[16] Slotine JJ, Li W. Applied nonlinear control. Englewood Cliffs: Prentice Hall; 1991.
[17] Solea R, Filipescu A, Nunes U. Sliding mode control for trajectory tracking of a wheeled mobile robot in presence of uncertainties. Proceedings of 17th Asian control conference. Hong Kong, China; August 27–29, 2009. p. 1701–6.

8 Mobile Robot Control IV: Fuzzy and Neural Methods

8.1 Introduction

Fuzzy logic systems (or, simply, *fuzzy systems*, FSs) and *neural networks* (NNs) have found wide application in the identification, planning, and control of robotic and other complex nonlinear technological systems. This is because FSs and NNs are universal approximators, that is, they can approximate any nonlinear function (mapping) with any desired accuracy. They are two of the three field components of *computational intelligence*, with the third component being the field of genetic or evolutionary algorithms (GA). The NN concept was coined by Mc Culloch and Pitts in 1943 as a result of their study of the human brain cell, which they name "neuron." The big next step in NNs was done in 1949 by Hebb who coined the concept of *synaptic weights*. Fuzzy logic or fuzzy set theory in its present form was coined by the control scientist Lofti Zadeh in 1965, thus breaking down the classical two-valued (yes/no) Aristotelian logic. Genetic and evolutionary algorithms were developed in the 1950s and 1960s, and in 1973, Rechenberg introduced the so-called *evolution strategies* as a method of parameter optimization, using a population with two atoms: a parent and an off spring. Holland has extended this concept in 1975 to a multiple-member population introducing the operators of *crossover*, *inversion*, and *mutation*.

Fuzzy logic offers a unified approximate (linguistic) way of drawing conclusions from uncertain data using uncertain rules. NNs offer the possibility of learning and training either autonomously (*unsupervised learning*) or nonautonomously (*supervised learning*) or via evaluation of their performance (*reinforcement learning*) [1–7]. Pictorially, supervised learning (from a teacher), unsupervised learning (without a teacher), and reinforcement learning (with a critic) are shown in Figure 8.1.

In many practical cases (including mobile robots), we use combined *neurofuzzy systems* (NFSs) that provide better performance. NFSs are distinguished as follows:

- Cooperative NFS (the NN determines some sub-blocks of the FS, e.g., fuzzy rules, which are then used without the presence of the NN) (Figure 8.2A).
- Concurrent NFS (the NN and the FS work continuously together, where the NN preprocesses the inputs, or postprocesses the outputs, of the FS) (Figure 8.2B).
- Hybrid NFS (an FS that uses a heuristic learning algorithm, inspired by NN theory, to determine its parameters, i.e., fuzzy sets and fuzzy rules, via the input–output patterns).

Introduction to Mobile Robot Control. DOI: http://dx.doi.org/10.1016/B978-0-12-417049-0.00008-0

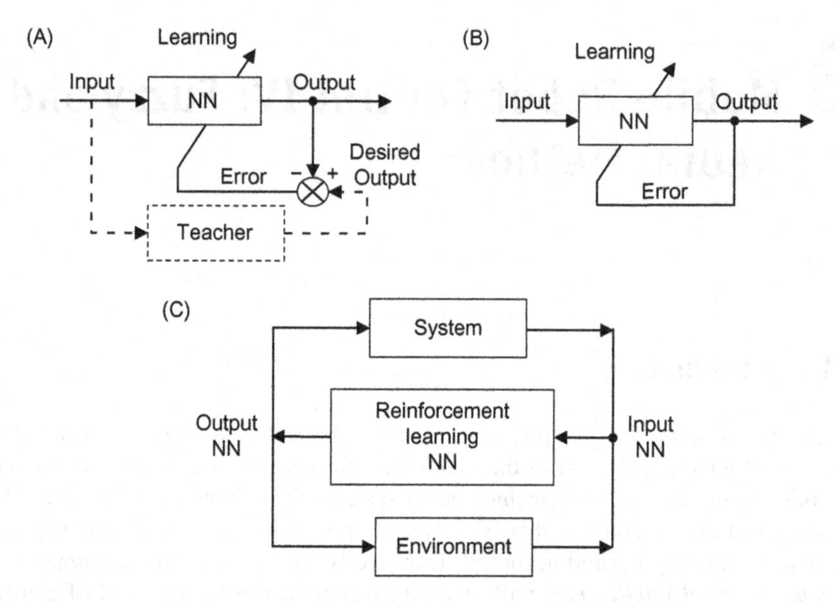

Figure 8.1 (A) Supervised learning NN, (B) unsupervised learning NN, and (C) reinforcement learning NN.

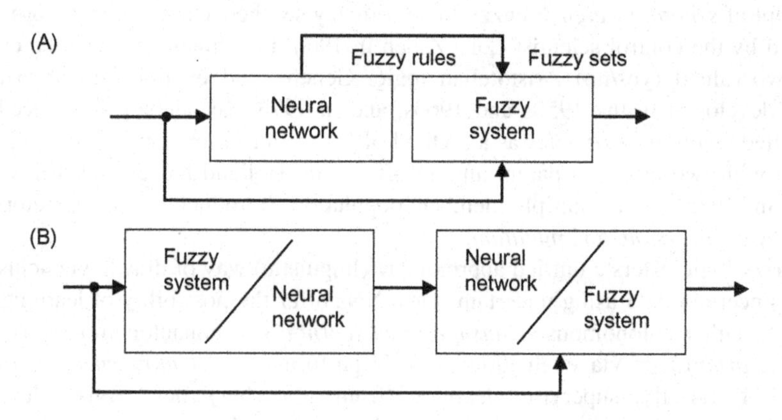

Figure 8.2 (A) Cooperative NFS, (B) concurrent neurofuzzy system.

The objectives of this chapter are the following:

- To provide a brief introduction to neural networks and FSs
- To derive and discuss the general structure of fuzzy and neural robot controllers
- To provide the details of mobile (nonholonomic) fuzzy tracker controller design
- To fuzzify the model-based sliding mode controller and apply it to mobile robots
- To solve the mobile adaptive tracking controller design problem using MLP and RBF neural networks.

8.2 Background Concepts

8.2.1 Fuzzy Systems

8.2.1.1 Fuzzy Sets

Fuzzy sets constitute an extension of the classical concept of set which is one of the foundations of the mathematics discipline. In a classical (or crisp) set X, only one of the following is true:

An element x belongs to X or does not belong to X, symbolically $x \in X$ or $x \notin X$.

This *dichotomy* was broken by the *fuzzy* sets coined by Zadeh [1]. Let $X = \{x_1, x_2, x_3, x_4, x_5\}$ be a classical set. The set X is called the *reference superset* (*universe of discourse*). Now, let $A = \{x_1, x_3, x_5\}$ be a classical subset of X. An equivalent representation of A is:

$$A = \{(x_1, 1), (x_2, 0), (x_3, 1), (x_4, 0), (x_5, 1)\}$$

which is an ordered set of pairs $(x, \mu_A(x))$, where x is the element of $x \in X$ of concern and $\mu_A(x)$ is the membership of x in the subset A, where:

$$\mu_A(x) = \left\{ \begin{array}{ll} 1 & \text{if} \quad x \in A \\ 0 & \text{if} \quad x \notin A \end{array} \right\}$$

That is, here we have $\mu_A{:}A \to \{0, 1\}$, where the set $\{0, 1\}$ has two elements, namely, 0 and 1. If we allow the membership function $\mu_A(x)$ to be:

$$\mu_A{:}A \to [0, 1]$$

where $[0, 1]$ is the full closed interval between 0 and 1 (i.e., $0 \le \mu_A(x) \le 1$), then we have the fuzzy subset A of X, defined as:

$$A = \{(x, \mu_A(x)) | x \in X, \mu_A(x){:}X \to [0, 1]\}$$

Another notation for the fuzzy set A is:

$$A = \mu_A(x_1)/x_1 + \mu_A(x_2)/x_2 + \cdots + \mu_A(x_n)/x_n$$

where the symbol "+" represents union of points, and the symbol "/" does not represent division.

Example 8.1

A fuzzy set with discrete points is:

$$\begin{aligned} A_1 &= \{(7, 0.1), (8, 0.5), (9, 0.8), (10, 1), (11, 0.8), (12, 0.5), (13, 0.1)\} \\ &= 0.1/7 + 0.5/8 + 0.8/9 + 1/10 + 0.8/11 + 0.5/12 + 0.1/13 \end{aligned}$$

A fuzzy set with continuous domain of elements x is:

$$A_2 = \{(x, \mu_A(x)) | x \in X, \mu_A(x) = 1/[1 + (x-10)^2]\}$$

Fuzzy Set Operations

The three fundamental operations of fuzzy sets are defined as extensions of the respective operations of classical sets, that is:

Intersection

$$C = A \cap B = \{(x, \mu_C(x)) | x \in X, \mu_C(x) = \min(\mu_A(x), \mu_B(x))\}$$

Union

$$D = A \cup B = \{(x, \mu_D(x)) | x \in X, \mu_D(x) = \max\{\mu_A(x), \mu_B(x)\}\}$$

Complement

$$A^c = \{(x, \mu_{A^c}(x)) | x \in X, \mu_{A^c}(x) = 1 - \mu_A(x)\}$$

It is easy to verify that the standard properties of sets hold also here (i.e., De Morgan, absorption, associativity, distributivity, idempotency).

Fuzzy Set Image

The image $f(A)$ of a fuzzy set A through the mapping (function) $f(\cdot)$ is the fuzzy set:

$$f(A) = \sum_Y \mu_A(x)/f(x)$$

For example, if $y = f(x_1) = f(x_2)$ for $x_1 \neq x_2$, we have:

$$\mu_A(x_1)/f(x_1) + \mu_A(x_2)/f(x_2) = \max\{\mu_A(x_1), \mu_A(x_2)\}/y$$

Fuzzy Inference

The *fuzzy inference* (or *fuzzy reasoning*) is an extension of the classical inference based on the *modus ponens* and *modus tollens* rules. Thus, we have:

Fuzzy modus ponens

Rule: IF $x = A$, THEN $y = B$
Fact: $x = A'$
Inference: $y = B'$
where A, A', B, and B' are fuzzy sets.

Fuzzy modus tollens

Rule: IF $x = A$, THEN $y = B$
Fact: $y = B'$
Inference: $x = A'$

Fuzzy Relations

Let X and Y be two reference supersets. Then, with the term fuzzy relation R, we mean a fuzzy set in the Cartesian product:

$$X \times Y = \{(x, y), x \in X, y \in Y\}$$

which has membership function $\mu_R(x, y)$:

$$\mu_R: X \times Y \rightarrow [0, 1]$$

For each pair (x, y) the membership function $\mu_R(x, y)$ represents the connection degree between x and y.

Zadeh's Max-Min Composition

On the basis of the above, we can formulate the rule of "max-min fuzzy composition" developed by Zadeh, which is as follows. Let the fuzzy sets A and B be:

$$A = \{(x, \mu_A(x)) | x \in X\}, \quad B = \{(y, \mu_B(y)) | y \in Y\}$$

and a fuzzy relation upon $X \times Y$, namely:

$$R = \{((x, y), \mu_R(x, y)) | (x, y) \in X \times Y\}$$

Then, if A is the input to R, the membership function of the output set B is given by the relation:

$$\mu_B(y) = \max_x \{\min[\mu_A(x), \mu_R(x, y)]\} \tag{8.1a}$$

or, symbolically:

$$B = A \circ R \tag{8.1b}$$

where "$\circ$" denotes the max-min operation.

If we are given a fuzzy rule:

IF x is A, THEN y is B we can find the corresponding fuzzy relation $R(x, y)$ using one of the following rules:

Mamdani's rule (minimum)

$$\mu_R(x_i, y_j) = \min\{\mu_A(x_i), \mu_B(y_j)\} \tag{8.2a}$$

Larsen's rule (product)

$$\mu_R(x_i, y_j) = \mu_A(x_i)\mu_B(y_j) \tag{8.2b}$$

Zadeh's arithmetic rule

$$\mu_R(x_i, y_j) = \min\{1, 1 - \mu_A(x_i) + \mu_B(y_j)\} \quad (8.2c)$$

Zadeh's maximum rule

$$\mu_R(x_i, y_j) = \max\{\min[\mu_A(x_i), \mu_B(y_j)], 1 - \mu_A(x_i)\} \quad (8.2d)$$

Example 8.2

Let $X = Y = \{1, 2, 3, 4\}$

$$A = \text{"}x\text{ small"} = \{(1, 1), (2, 0.6), (3, 0.2), (4, 0)\}$$

and

R = "x nearly equal to y" with fuzzy relation:

x/y	1	2	3	4
1	1	0.5	0	0
2	0.5	1	0.5	0
3	0	0.5	1	0.5
4	0	0	0.5	1

Then, the max-min rule $B = A \circ R$ gives:

$$\begin{aligned}\mu_B(y) &= \max_x\{\min\{\mu_A(x), \mu_R(x, y)\}\}\\ &= \{(1, 1), (2, 0.6), (3, 0.5), (4, 0.2)\}\end{aligned}$$

Obviously, the result can be interpreted as the fuzzy set "x=nearly small." Thus, in this case, the "fuzzy modus ponens" rule gives:

IF "x is small" AND "x is nearly equal to y," THEN "y is nearly small."

8.2.1.2 FSs Structure

The general structure of an FS (or fuzzy decision algorithm) involves the following four units (Figure 8.3) [2–7]:

1. A fuzzy rule base (FRB), that is, a base of IF-THEN rules
2. A fuzzy inference mechanism (FIM)
3. An input fuzzification unit (IFU)
4. An output defuzzification unit (ODU)

The FRB usually contains, besides the fuzzy or linguistic rules, a standard arithmetic database section. The fuzzy rules are provided by human experts or are derived through simulation. The IFU (*fuzzifier*) receives the nonfuzzy input values

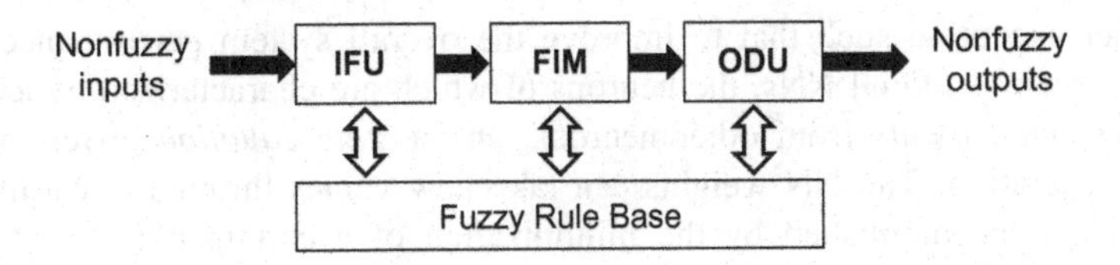

Figure 8.3 General structure of an FS.

and converts them into fuzzy or linguistic form. The FIM is the core of the system and involves the fuzzy inference logic (e.g., the max-min rule of Zadeh). Finally, the ODU (*defuzzifier*) converts the fuzzy results provided by FIM to nonfuzzy form using a defuzzification method.

The fuzzifier performs a mapping from the set of real input values $\mathbf{x} = [x_1, x_2, \ldots, x_n] \in X$ to the fuzzy subset A of the superset X.

Two possible choices of this mapping are:

$$\mu_A(x') = \begin{cases} 1 & \text{if} \quad x' = x \\ 0 & \text{if} \quad x' \neq x \end{cases} \quad \text{(Singleton fuzzifier)}$$

$$\mu_A(x') = \exp\left[\frac{(x'-x)^{\mathrm{T}}(x'-x)}{\sigma^2}\right] \quad \text{(Bell-type fuzzifier)}$$

The two most popular defuzzification methods are the following:

1. *Center of gravity (COG) method*: The defuzzified value w_0 is given by:

$$w_0 = \left[\sum_i w_i \mu_B(w_i)\right] \Big/ \left[\sum_i \mu_B(w_i)\right] \tag{8.3}$$

2. *Mean of maxima (MOM) method*: Here, the defuzzified output w_0 is equal to:

$$w_0 = \left[\sum_{j=1}^{m} w_j\right] \Big/ m \tag{8.4}$$

where w_j is the value that corresponds to the j maximum of the membership function $\mu_B(w)$.

8.2.2 Neural Networks

Neural networks are large-scale systems that involve a large number of special type nonlinear processors called *neurons*. Biological neurons are nerve cells which have a number of internal parameters called *synaptic weights*. The human brain consists of over 10 million neurons. The weights are adjusted adaptively according to the task under execution such that to improve the overall system performance. Here, we will discuss artificial NNs, the neurons of which are characterized by a *state,* a list of *weighted inputs* from other neurons, and a *state equation* governing their dynamic operation. The NN weights can take new values through a learning process which is accomplished by the minimization of a certain objective function through the gradient or the Newton–Raphson algorithm. The optimal values of the weights are stored as the strengths of the neurons' interconnections. The NN approach is suitable for systems or processes that cannot be modeled with concise and accurate mathematical models; typical examples being machine vision, pattern recognition, control systems, and human-based operations. The three primary features of NNs are (i) utilization of large amounts of sensory information, (ii) collective processing capability, and (iii) learning and adaptation capability [4]. Learning and control in neurocontrollers are achieved simultaneously, and learning continues as long as perturbations are present in the plant under control and/or its environment. The practical implementation of NNs was made possible by the recent developments in fast parallel methods (VLS, electro-optical, and others). The two NNs that are mostly suitable for decision and control purposes are the *multilayer perceptron* (MLP) and the *radial basis functions* (RBF) networks.

8.2.2.1 The Basic Artificial Neuron Model

This model is based on the McCulloch–Pitts model which has the form shown in Figure 8.4A. The neuron has a basic processing unit which consists of three elements:

1. A set of connection branches (synapses)
2. A linear summation node
3. An activation (nonlinear) function

Each connection branch has a weight (strength) which is positive if it has an *excitatory* role and negative if it has an *inhibitory* role. The summation node sums the input signals multiplied by the respective synaptic weights. Finally, the *activation function* (or as otherwise called) *squashing function* limits the allowable amplitude of the output signal to some finite value, typically in the normalized interval [0, 1] or, alternatively, in the interval [−1, 1]. The neuron model has also a threshold θ which is applied externally, and practically lowers the net input to the activation function.

From Figure 8.4A, it follows that the neuron is described by the following equations:

$$u = \sum_{i=1}^{n} w_i x_i$$

$$y = \sigma(z)$$

$$z = u - \theta, \quad \theta > 0$$

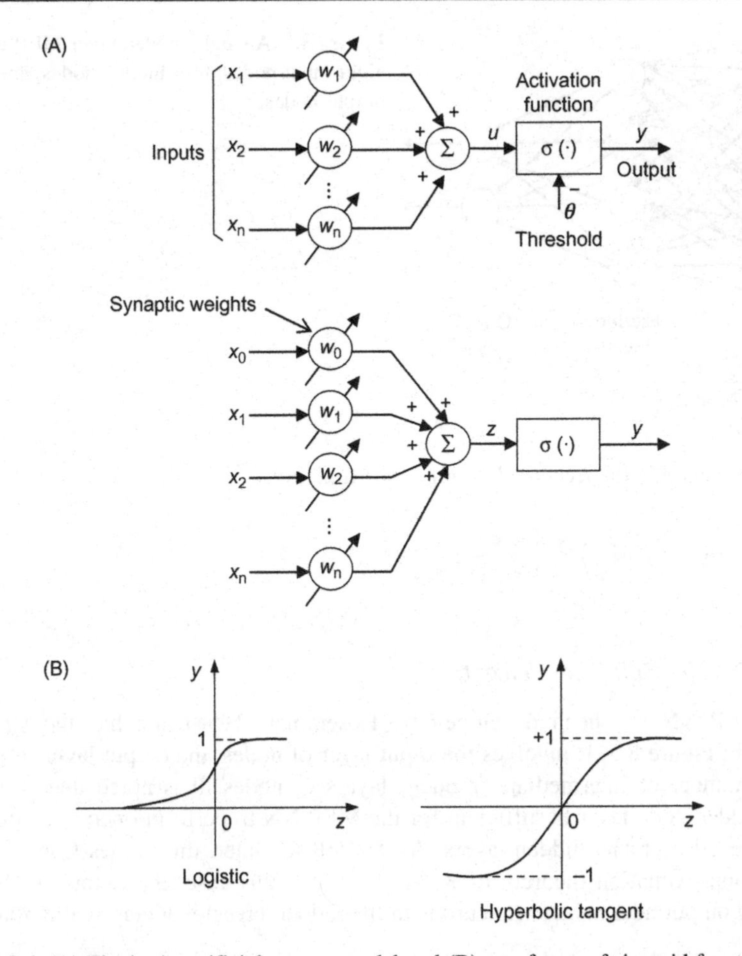

Figure 8.4 (A) The basic artificial neuron model and (B) two forms of sigmoid functions.

If the threshold $-\theta$ is regarded as a normal input $x_0 = -1$ with corresponding weight $w_0 = \theta$, then the neuron model takes the form:

$$y = \sigma(z), \quad z = \sum_{i=0}^{n} w_i x_i \tag{8.5}$$

The nonlinear activation function $\sigma(x)$ can be of the *on-off* or the *saturation* function type, or of the *sigmoid* function type with values either in the interval [0, 1] or in the interval [−1,1], as shown in Figure 8.4B.

The first sigmoid function is the *logistic function*:

$$y = \sigma(z) = \frac{1}{1 + e^{-z}}, \quad y \in [0, 1]$$

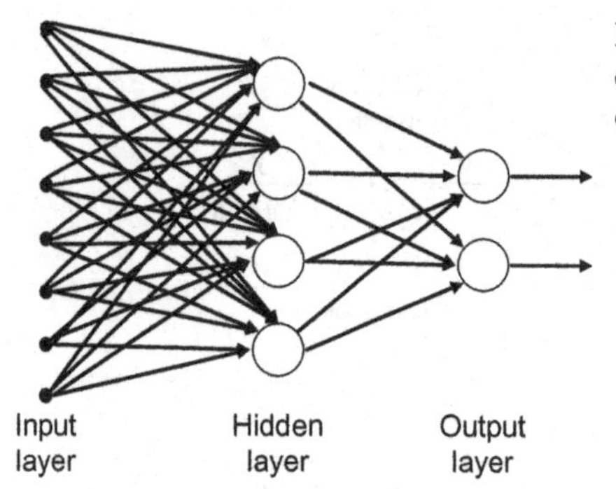

Figure 8.5 A single-hidden layer MLP with eight input nodes, four hidden nodes, and two output nodes.

and the second is the *hyperbolic tangential* function:

$$y = \sigma(z) = \tan\left(\frac{z}{2}\right) = \frac{1 - e^{-z}}{1 + e^{-z}}, \quad y \in [-1, 1]$$

8.2.2.2 The Multilayer Perceptron

The MLP NN has been developed by Rosemblat (1958) and has the structure shown in Figure 8.5. It involves the input layer of nodes, the output layer of nodes, and a number of intermediate (hidden) layers of nodes. It is noted that even one only hidden node layer is sufficient for the MLP NN to perform operations that can be achieved by many hidden layers. As we will see later, this comes from the universal approximation theorem of Kolmogorov. In this case, the output of the NN (with m output neurons and L neurons in the hidden layer) is given by the relation:

$$y_i = \sum_{j=1}^{L} \left[v_{ij}\sigma\left(\sum_{k=0}^{n} w_{jk}x_k \right) \right], \quad i = 1, 2, \ldots, m \tag{8.6}$$

where $x_k (k = 0, 1, 2, \ldots, n)$ are the NN inputs (including the thresholds), w_{jk} are the input-to-hidden-layer interconnection weights, and v_{ij} are the hidden-to-output-layer interconnection weights.

In compact form, Eq. (8.6) can be written as:

$$\mathbf{y} = \mathbf{V}^{\mathrm{T}}\boldsymbol{\sigma}(\mathbf{W}^{\mathrm{T}}\mathbf{x}) \tag{8.7}$$

where:

$$\mathbf{x} = [x_0, x_1, \ldots, x_n]^{\mathrm{T}}, \quad \mathbf{y} = [y_1, y_2, \ldots, y_m]^{\mathrm{T}}$$

$$\mathbf{V}^{\mathrm{T}} = \begin{bmatrix} \mathbf{v}_1^{\mathrm{T}} \\ \mathbf{v}_2^{\mathrm{T}} \\ \vdots \\ \mathbf{v}_m^{\mathrm{T}} \end{bmatrix}, \quad \boldsymbol{\sigma}(\mathbf{W}^{\mathrm{T}}\mathbf{x}) = \begin{bmatrix} \sigma(\mathbf{w}_1^{\mathrm{T}}\mathbf{x}) \\ \sigma(\mathbf{w}_2^{\mathrm{T}}\mathbf{x}) \\ \vdots \\ \sigma(\mathbf{w}_L^{\mathrm{T}}\mathbf{x}) \end{bmatrix}$$

$$\mathbf{v}_i^{\mathrm{T}} = [v_{i1}, v_{i2}, \ldots, v_{iL}], \quad \mathbf{w}_i^{\mathrm{T}} = [w_{i0}, w_{i1}, \ldots, w_{in}]$$

8.2.2.3 The Backpropagation Algorithm

The *backpropagation* (BP) algorithm is a supervisory learning algorithm which updates (adapts) the synaptic weights such that to minimize the mean square error between the desired and actual outputs after the presentation of each input vector (pattern) at the NN input layer. The errors are propagated via the hidden layers for the computation of the weight corrections in the same way as is done in the output layer.

Thus, the weights are updated so as to minimize the criterion:

$$E_p(t) = \frac{1}{2}\sum_k e_k^2(t), \quad e_k(t) = d_k(t) - y_k(t)$$

where:

$d_k(t)$ is the desired output (response) of the kth output neuron at time t.
$y_k(t)$ is the output neuron actually obtained after the presentation of each vector pattern $\mathbf{x}$ at the input layer at time t.

This means that the minimization of $E_p(t)$ must be done consecutively pattern by pattern. In the above criterion function, $E_p(t)$, $y_k(t)$, and $e_k(t)$ are given by:

$$y_k(t) = \sigma(z_k(t)), \quad e_k(t) = d_k(t) - \sigma_k(z_k(t)), \quad z_k = \sum_{i=0}^{n} w_{ki}(t)y_i(t)$$

The minimization of $E_p(t)$ can be done by the well-known gradient (steepest descent) rule:

$$\Delta w_{ki}(t) = -\gamma \frac{\partial E_p(t)}{\partial w_{ki}(t)}$$

for the p input pattern, where γ is the learning parameter.

Here:

$$\frac{\partial E_p(t)}{\partial w_{ki}(t)} = \frac{\partial E_p(t)}{\partial e_k(t)} \cdot \frac{\partial e_k(t)}{\partial y_k(t)} \cdot \frac{\partial y_k(t)}{\partial z_k(t)} \cdot \frac{\partial z_k(t)}{\partial w_{ki}(t)}$$

$$= e_k(t)(-1)\frac{\mathrm{d}\sigma(z_k(t))}{\mathrm{d}z_k(t)} y_i(t)$$

where, for the logistic function $\sigma(\cdot)$:

$$\frac{d\sigma(z_k(t))}{dz_k(t)} = \frac{d}{dz_k}\left[\frac{1}{1+e^{-z_k}}\right] = -\frac{(-e^{-z_k})}{(1+e^{-z_k})^2} = y_k(t)[1-y_k(t)]$$

Thus:

$$\Delta w_{ki}(t) = \gamma[d_k(t) - y_k(t)]y_k[1-y_k(t)]y_i(t) = \gamma\delta_k(t)y_i(t) \tag{8.8a}$$

where:

$$\delta_k(t) = [d_k(t) - y_k(t)][1-y_k(t)]y_k \tag{8.8b}$$

is the so-called *delta* of the BP algorithm. The index k extends over all the output neurons, and the index i over all the neurons of the last hidden layer.

The learning (weight updating) rule (8.8a) for the hidden layers takes the form:

$$\begin{aligned} \Delta w_{ji}(t) &= \gamma y_i(t)y_j(t)[1-y_j(t)]\sum_m \delta_m(t)w_{mj}(t) \\ \Delta w_{ji}(t) &= w_{ji}(t+1) - w_{ji}(t) \end{aligned} \tag{8.9}$$

where the index i refers to the neurons of the layer that lies behind the considered layer. To accelerate the convergence we add a "momentum" term $a[w_{ji}(t) - w_{ji}(t-1)]$, where a is a parameter in the interval $0 \leq a \leq 1$.

On the basis of the above, the BP learning algorithm involves the following steps:

Step 1: Select the initial weights and thresholds using small positive random values.
Step 2: Present the training input pattern vector $\mathbf{x}(t)$ and the desired output vector $\mathbf{d}(t)$:

$$\begin{aligned} \mathbf{x}(t) &= [x_0(t), x_1(t), \ldots, x_n(t)]^T \\ \mathbf{d}(t) &= [d_1(t), d_2(t), \ldots, d_m(t)]^T \end{aligned}$$

Step 3: Compute the actual outputs of all the neurons of the NN neuron-to-neuron in the forward direction using the current values of the synaptic weights, that is:

$$y_i(t) = \sigma_j(z_j(t)), \quad z_j(t) = \sum_i w_{ji}(t)y_i(t)$$

where $y_i(t)$ (the output of the ith neuron) is the ith input to the jth neuron and w_{ji} is the synaptic weight that connects the ith neuron with the jth neuron. For the neurons j of the first hidden layer, we have:

$$y_i(t) = x_i(t), \quad i = 1, 2, 3, \ldots, n$$

where $x_i(t)$ is the ith component of the input pattern vector $\mathbf{x}(t)$.

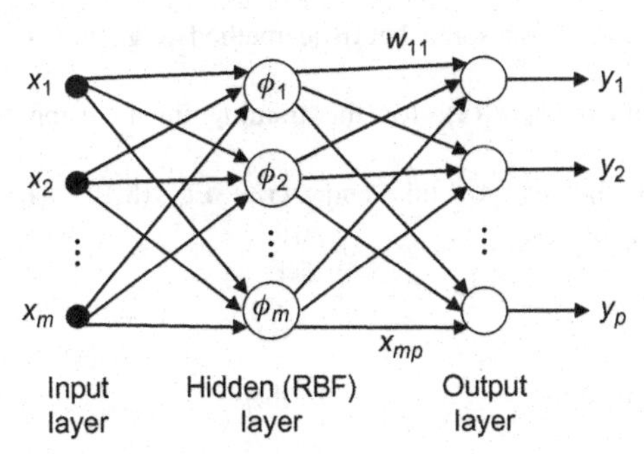

Figure 8.6 The RBF network.

Step 4: Update the synaptic weights starting from the output neurons and going backward toward the input layer, using the rule:

$$w_{ji}(t+1) = w_{ji}(t) + \gamma\delta_j(t)y_i(t) + a[w_{ji}(t) - w_{ji}(t-1)]$$

Step 5: Repeat the procedure from step 2 (until a desired accuracy is obtained or a maximum number of iterations is reached).

In the literature, there are available many variations or improvements of the above basic BP algorithm with faster convergence and acceptable computational effort.

8.2.2.4 The RBF Network

An RBF network approximates an input–output mapping by employing a linear combination of radially symmetric functions (Figure 8.6). The kth output y_k is given by:

$$y_k(\mathbf{x}) = \sum_{i=1}^{m} w_{ki}\phi_i(\mathbf{x}) \quad (k = 1, 2, \ldots, p) \tag{8.10}$$

where:

$$\phi_i(\mathbf{x}) = \phi(||\mathbf{x} - \mathbf{c}_i||) = \phi(r_i) = \exp\left(-\frac{r_i^2}{2\sigma_i^2}\right), \quad r_i \geq 0, \ \sigma_i \geq 0 \tag{8.11}$$

The RBF networks have always one hidden layer of computational nodes with nonmonotonic transfer functions $\phi(\cdot)$. Theoretical studies have shown that the choice of $\phi(\cdot)$ is not very crucial for the effectiveness of the network. In most cases, the Gaussian RBF given by Eq. (8.11) is used, where $\mathbf{c}_i$ and $\sigma_i(i = 1, 2, \ldots, m)$ are selected centers and widths, respectively. The training procedure of the RBF network involves the following steps:

Step 1: Group the training patterns in **M** subsets using some clustering algorithm (e.g., the k-means clustering algorithm) and select their centers $\mathbf{c}_i$.

Step 2: Select the widths, $\sigma_i (i = 1, 2, \ldots, m)$, using some heuristic method (e.g., the p nearest-neighbor algorithm).
Step 3: Compute the RBF activation functions, $\phi_i(\mathbf{x})$, for the training inputs using Eq. (8.11).
Step 4: Compute the weights by least squares. To this end, write Eq. (8.10) as $\mathbf{b}_k = \mathbf{A}\mathbf{w}_k \ (k = 1, 2, \ldots, p)$ and solve for $\mathbf{w}_k$, that is:

$$\mathbf{w}_k = \mathbf{A}^{\dagger}\mathbf{b}_k, \quad \mathbf{w}_k = [w_{k1}, \ldots, w_{km}]^{\mathrm{T}}$$

where $\mathbf{A}^{\dagger}$ is the generalized inverse of $\mathbf{A}$ given by:

$$\mathbf{A}^{\dagger} = (\mathbf{A}^{\mathrm{T}}\mathbf{A})^{-1}\mathbf{A}^{\mathrm{T}}$$

and $\mathbf{b}_k$ is the vector of the training values for the output k.

It is remarked that MLP NNs perform global matching to the input–output data, whereas in RBF NNs, this is done only locally, of course with better accuracy.

8.2.2.5 The Universal Approximation Property

Neural Network Approximator

The MLP neural network with (at least) one hidden layer has the universal approximation property.[1] This follows from the Kolmogorov theorem which states [4]:

Given any continuous function:

$$\mathbf{F}: [0, 1]^n \rightarrow \mathbf{F}(\mathbf{x}) = \mathbf{y}, \mathbf{y} \in R^m$$

where $\mathbf{I} = [0, 1]^n$ is the n-dimensional unit cube, $\mathbf{F}$ can be approximated (realized) exactly with a perceptron NN of three layers, n nodes in the input (source) layer $\{\mathbf{x} = [x_1, x_2, \ldots, x_n]^{\mathrm{T}}\}$, $L = 2n + 1$ nodes in the middle (hidden) layer, and m nodes in the output layer $\{\mathbf{y} = [y_1, y_2, \ldots, y_m]^{\mathrm{T}}\}$.

The nonlinear function $f(z)$ used in the nodes can be selected so as to satisfy the Lipschitz condition $|f(z_1) - f(z_2)| \leq c|z_1 - z_2|^{\mu}$ for $0 < \mu < 1$, where c is a constant. Any sigmoid type function $f(z) = \sigma(z)$ or RBF $f(z) = \phi(z)$ satisfies the Lipschitz condition.

In practice, the NN representation of $\mathbf{y} = \mathbf{F}(\mathbf{x})$, after a finite number of steps of weight updating (using the BP learning algorithm), is approximate, involving an error ε, that is:

$$\mathbf{V}^{\mathrm{T}}\sigma(\mathbf{W}^{\mathrm{T}}\mathbf{x}) = \mathbf{F}(\mathbf{x}) - \varepsilon \tag{8.12}$$

[1] By definition, an MLP has a nonconstant, bounded, and monotonically increasing continuous activation function.

for some number L of hidden neurons. Actually, for any positive number ε_0, the weights can be trained (e.g., by BP), and the number L of hidden neurons can be selected such that:

$$||\varepsilon|| < \varepsilon_0$$

for all $\mathbf{x} = \mathbf{I}$. The selection of $L \geq 2n + 1$ can be done in several ways, but it is still an open problem. Depending on the rate of change of $\mathbf{F}(\mathbf{x})$, one can determine the smallest L that assures a described accuracy ε_0.

Fuzzy Logic Universal Approximator

Consider a multi-input single-output (MISO) fuzzy logic system with singleton fuzzifier, COG defuzzifier, and Gaussian-type membership functions:

$$\mu_{A_i^j}(x_i) = \rho_i^j \exp\left[-\frac{1}{2}\left(\frac{x_i - \bar{x}_i^j}{\sigma^j}\right)^2\right] \tag{8.13}$$

where x_i is the ith component of the fuzzy input vector $\mathbf{x} = [x_1, x_2, \ldots, x_n]^{\text{T}}$, and ρ_i^j, $\bar{x}_i^j$, and σ_i^j are real-valued parameters with $0 \leq \rho_i^j \leq 1$. The system consists of m fuzzy rules $\mathbf{R}_j (j = 1, 2, \ldots, m)$ of the form $\mathbf{R}_j$: IF x_1 is A_1^j and x_2 is A_2^j and ... and x_n is A_n^j, THEN y is B^j, where $x_i (i = 1, 2, \ldots, n)$ and y are fuzzy (linguistic) variables represented by fuzzy membership functions $\mu_{A_i^j}(x_i)$ and $\mu_{B^j}(y)$, respectively, and y^j is the point in the output space y at which $\mu_{B^j}(y)$ takes its maximum value. If the membership functions of x_i are given by Eq. (8.13), then the output $F(\mathbf{x})$ of the system can be written as:

$$F(\mathbf{x}) = \frac{\sum_{j=1}^{m} \bar{y}_i^j \left(\prod_{i=1}^{m} \mu_{A_i^j}(x_i)\right)}{\sum_{j=1}^{m} \left(\prod_{i=1}^{m} \mu_{A_i^j}(x_i)\right)} \tag{8.14}$$

Indeed, consider *fuzzy basis functions* (FBFs) of the form:

$$\psi_j(\mathbf{x}) = \frac{\prod_{i=1}^{n} \mu_{A_i^j}(x_i)}{\sum_{j=1}^{m} \prod_{i=1}^{n} \mu_{A_i^j}(x_i)} \tag{8.15}$$

where $\mu_{A_i^j}(x_i)$ are the Gaussian functions (8.13). Then, the FS (8.14) can be written as an FBF expansion with $\bar{y}_i^j$ as a free coefficient of the form:

$$F(\mathbf{x}) = \sum_{j=1}^{m} \bar{y}_i^j \psi_j(\mathbf{x}) = \psi^{\text{T}}(\mathbf{x})\beta \tag{8.16}$$

where:

$$\psi(\mathbf{x}) = [\psi_1(\mathbf{x}), \ldots, \psi_m(\mathbf{x})]^{\mathrm{T}}$$

is the fuzzy basis vector, and

$$\beta = [\bar{y}^1, \bar{y}^2, \ldots, \bar{y}^m]^{\mathrm{T}}$$

is the parameter vector to be estimated. The output of a vector-valued MIMO function (FS) $\mathbf{F}(\mathbf{x})$ is expressed as:

$$\mathbf{F}(\mathbf{x}) = \Psi(\mathbf{x})\beta \tag{8.17}$$

where:

$$\mathbf{F}(\mathbf{x}) = \begin{bmatrix} F_1(\mathbf{x}) \\ \vdots \\ F_p(\mathbf{x}) \end{bmatrix}, \quad \Psi(\mathbf{x}) = \begin{bmatrix} \psi_1(\mathbf{x}) \\ \vdots \\ \psi_p(\mathbf{n}) \end{bmatrix} = \begin{bmatrix} \psi_{11}(\mathbf{x}) & \cdots & \psi_{1m}(\mathbf{x}) \\ \vdots & \ddots & \vdots \\ \psi_{p1}(\mathbf{x}) & \cdots & \psi_{pm}(\mathbf{x}) \end{bmatrix}$$

It is shown in the literature [8] that Eq. (8.14) (which is actually an *adaptive fuzzy system*) can uniformly approximate any real continuous function over a compact input set to any desired accuracy. Thus, Eq. (8.14) is a *fuzzy universal approximator.*

8.3 Fuzzy and Neural Robot Control: General Issues

8.3.1 Fuzzy Robot Control

The basic structure of a fuzzy robot control system is built using the FS of Figure 8.3, and has the form shown in Figure 8.7 [6].

Here, the FRB stores all the relevant knowledge (i.e., how to control the system) which eliminates the need to have available an analytical mathematical model of the robot. Two practical implementations of the basic control loop are shown in Figure 8.8A and B.

In the simple structure of Figure 8.8A, the input to the robot consists of delayed measured values of the control input and the output of the modeled robot. Thus, if $\mathbf{f}(\cdot)$ is the unknown nonlinear mapping that describes the unknown dynamics of the robot, and:

$$\mathbf{z}(k) = [\mathbf{y}(k-1), \mathbf{y}(k-2), \ldots, \mathbf{y}(k-n_y); \mathbf{u}(k-1), \ldots, \mathbf{u}(k-n_u)]^{\mathrm{T}}$$

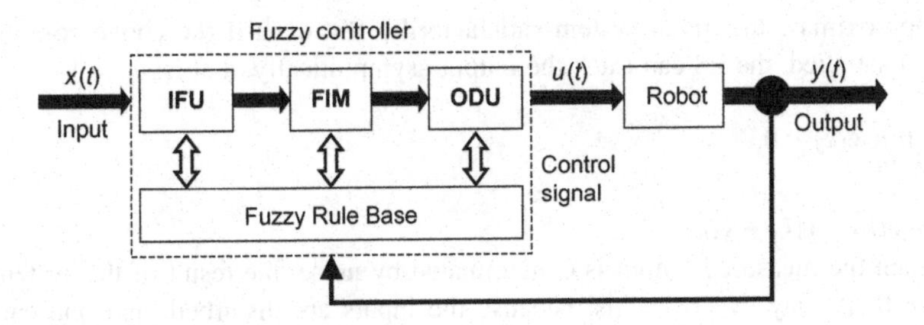

Figure 8.7 Basic fuzzy robot control loop.

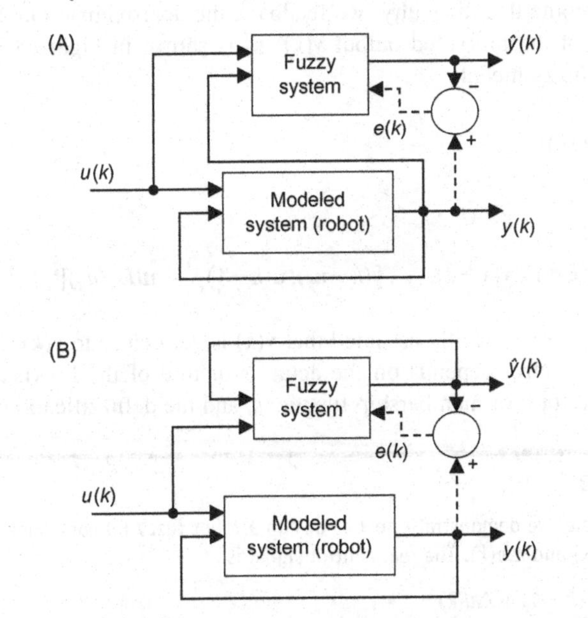

Figure 8.8 Two implementations of fuzzy robot control: (A) use of delayed input–output values and (B) use of output feedback.

is the information vector, then the training data are generated (modeled) by using the relation:

$$\mathbf{y}(k) = \mathbf{f}(\mathbf{z}(k))$$

Then, a fuzzy model of the robot has the form:

$$\hat{\mathbf{y}}(k) = \hat{\mathbf{f}}(\mathbf{z}(k))$$

where $\hat{\mathbf{f}}(\cdot)$ is the fuzzy approximation (description) of $\mathbf{f}(\cdot)$. Under the assumption that the measurements are correct (accurate), the delay degrees n_y and n_u have been estimated correctly, and that the control signal acts persistently, the fuzzy model can approximate the given system satisfactorily. But even if the above conditions are not satisfied, the FS can track the output asymptotically, that is:

$$\lim_{k \to \infty} \mathbf{e}(k) = \mathbf{0}$$

where $\mathbf{e}(k) = \mathbf{y}(k) - \hat{\mathbf{y}}(k)$.

When the measured output is contaminated by noise, the result of the system of Figure 8.8A may by erroneous, because the inputs are disturbed, and one cannot determine whether some output error is due to parametric uncertainty or input error. To overcome this difficulty, we feedback the approximate output $\hat{\mathbf{y}}(k)$ to the input (instead of the measured output $\mathbf{y}(k)$) as is shown in Figure 8.8B. Thus, we have now the fuzzy model:

$$\hat{\mathbf{y}}(k) = \hat{\mathbf{f}}(\mathbf{z}'(k))$$

where:

$$\mathbf{z}'(k) = [\hat{\mathbf{y}}(k-1), \hat{\mathbf{y}}(k-2), \ldots, \hat{\mathbf{y}}(k-n_y); \mathbf{u}(k-1), \ldots, \mathbf{u}(k-n_u)]^{\mathrm{T}}$$

Of course, here it is tacitly assumed that $\hat{\mathbf{y}}(k)$ is very close to $\mathbf{y}(k)$. In both cases, the success of control depends on the actual structure of the FS (i.e., the type of fuzzy rules, the form of membership functions, and the defuzzification method).

Example 8.3

In this example, we demonstrate how to design a robot fuzzy PD controller which receives the inputs $e(k)$ and $\Delta\mathbf{e}(k)$. The real control signal is:

$$u(k) = u(k-1) + \Delta u(k)$$

The fuzzy rules produce the incremental control signal $\Delta u(k)$ and are described in Table 8.1.

Table 8.1 Fuzzy Rule base of the Fuzzy Controller PD

	e(*k*)						
$\Delta e(k)$	**NB**	**NM**	**NS**	**AZ**	**PS**	**PM**	**PB**
NB	AZ	PS	PM	PB	PB	PB	PB
NM	NS	AZ	PS	PM	PB	PB	PB
NS	NM	NS	AZ	PS	PM	PB	PB
AZ	NB	NM	NS	AZ	PS	PM	PB
PS	NB	NB	NM	NS	AZ	PS	PM
PM	NB	NB	NB	NM	NS	AZ	PS
PB	NB	NB	NB	NB	NM	NS	AZ

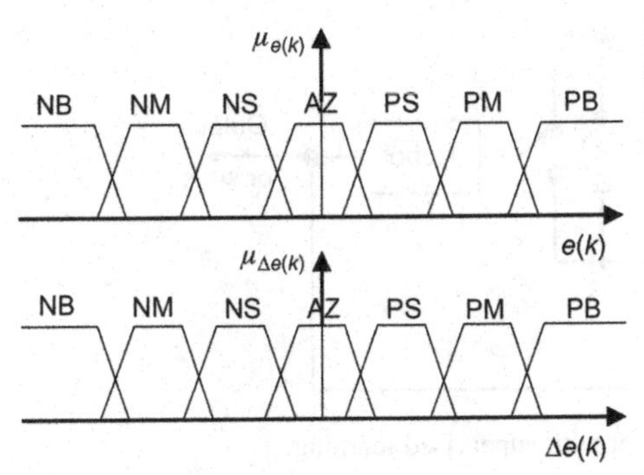

Figure 8.9 Fuzzy values (membership functions) of $e(k)$ and $\Delta e(k)$ of the fuzzy PI controller.

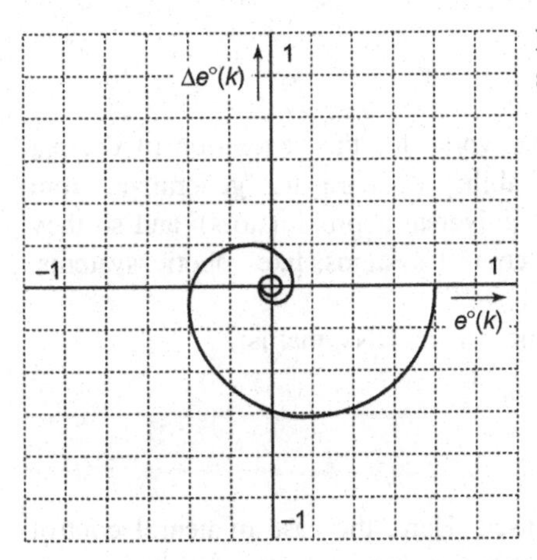

Figure 8.10 Trajectory of the error dynamics state $[e(k), \Delta e(k)]$ of the fuzzy PI controller.

Here, the symbols of the table for $e(k)$, $\Delta e(k)$, and $u(k)$ have the following meaning: NB, negative big; NS, negative small; NM, negative medium; AZ, almost zero; PB, positive big; PS, positive small; PM, positive medium.

These rules are pictorially illustrated in Figure 8.9.

Table 8.1 involves $7 \times 7 = 49$ rules, that is:

$R1$: IF $e(k) = \text{NB}$ AND $\Delta e(k) = \text{NB}$, THEN $\Delta u(k) = \text{AZ}$.
$R2$: IF $e(k) = \text{NB}$ AND $\Delta e(k) = \text{NM}$, THEN $\Delta u(k) = \text{NS}$
...
$R49$: IF $e(k) = \text{PB}$ AND $\Delta e(k) = \text{PB}$, THEN $\Delta u(k) = \text{AZ}$.

As we know, the tuning of conventional PD controllers is made by selecting the gains K_p and K_υ. But in fuzzy PD controllers, the tuning is done by selecting the position of the membership functions. The resulting form of the control action is an interpolation between several types of control for each subdivision of the input space. The performance of a fuzzy PD controller is illustrated pictorially in the phase plane, with axes $e(k)$ and $\Delta e(k)$. In Fig 8.10, the symbol "°" denotes normalized values in the interval $[-1,1]$.

The various regions of the phase plane correspond to positions in Table 8.1. Typically, the desired performance of a fuzzy controller is the one shown in Figure 8.10, that is, $(e(k), \Delta e(k))$ converges asymptotically to the equilibrium point (0,0).

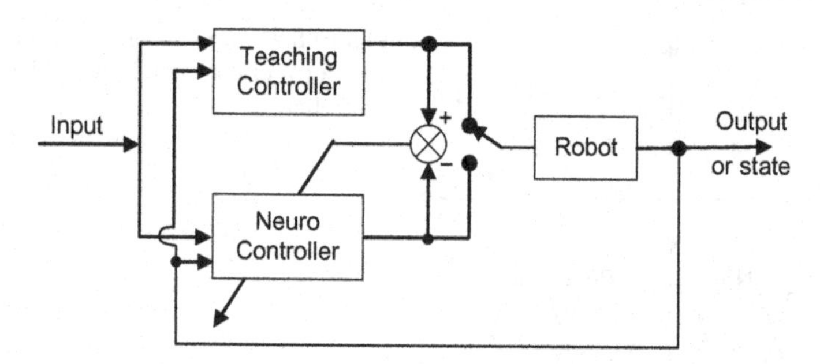

Figure 8.11 Structure of neurocontrolled robot with supervised learning.

8.3.2 Neural Robot Control

Neural control uses "well-defined" neural networks for the generation of desired control signals. Neural Networks have the ability to learn and generalize, from examples, nonlinear mappings (i.e., they are universal approximators), and so they are suitable for solving complex nonlinear control systems, like robotic systems, with high speed [5,9].

Neural control can be classified in the same way as NNs, that is:

- Neural control with supervised learning
- Neural control with unsupervised learning
- Neural control with reinforcement learning

In each case, the proper NN should be used. Here, the case of neural control with supervised learning, which is very popular for its simplicity, will be considered. The structure of supervised learning neurocontrol is as shown in Figure 8.11.

The teacher trains the neurocontroller via the presentation of control signal examples that can control the robot successfully. The teacher can be either a human controller or any classical, adaptive, or intelligent technological controller. The outputs or states are measured and sent to the teacher as well as to the neurocontroller. During the control period by the teacher, the control signal and outputs/states of the robotic system are sampled and stored for the training of the neural network. After the training period, the neurocontroller takes the control actions, and the teacher is deconnected from the system.

The most popular type of supervised neurocontrol is the direct inverse neurocontrol in which the NN learns successfully the robot inverse dynamics and is used directly as controller as shown in Figure 8.12A [9].

Another variation of the direct inverse neurocontroller is shown in Figure 8.12B which is called *specialized direct inverse neurocontrol*. Here, the NN is trained *on-line* and the error e of the closed-loop system is transmitted backward at each sampling instant. On the contrary, in the direct inverse neurocontrol scheme of Figure 8.12A, the training of the NN is performed "off-line."

Another neurocontrol scheme is the so-called indirect *neurocontrol* which uses two NNs as shown in Figure 8.13 [9].

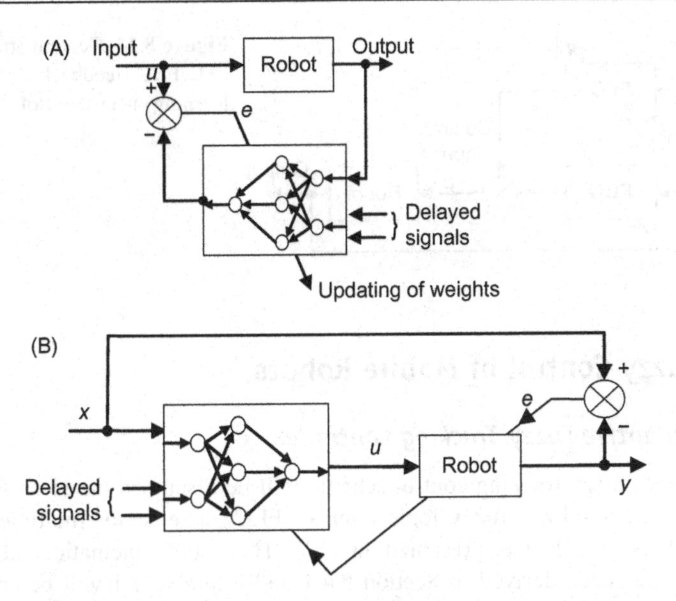

Figure 8.12 (A) Direct inverse neurocontroller and (B) specialized direct inverse neurocontroller.

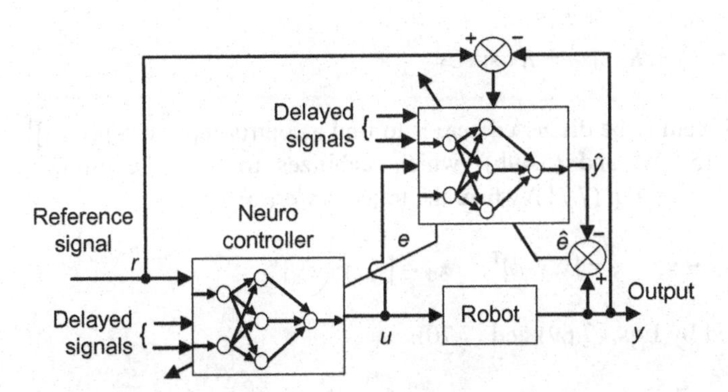

Figure 8.13 Indirect robot neurocontrol (use of an adaptive NN simulator and a NN controller).

The first NN is used as simulator of the robot, and the second NN as controller. The simulation NN can be trained either *off-line* (batch learning) or *on-line* using random inputs for the learning of the robot dynamics. All the above types of neurocontrol use NNs of the MLP type with BP learning.

The most general type of neurocontrol involves two NNs: the first is used as *feedforward controller* (FFC) and the second as *feedback controller* (FBC). The structure of this control scheme which is known as *feedback error learning neurocontroller* is shown in Figure 8.14.

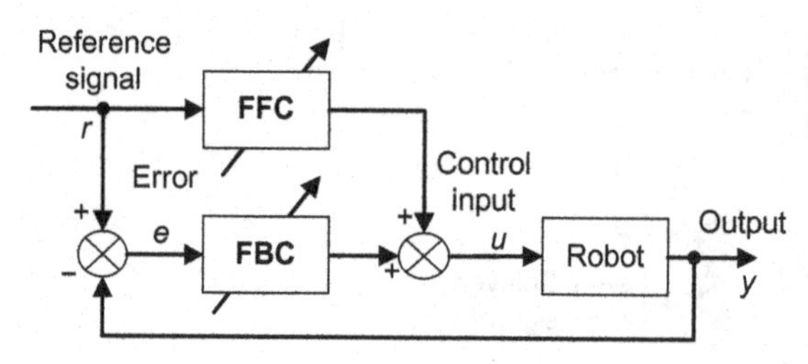

Figure 8.14 General structure of FFC-FBC (feedback error learning) neurocontrol.

8.4 Fuzzy Control of Mobile Robots

8.4.1 Adaptive Fuzzy Tracking Controller

Here, a direct fuzzy tracking control scheme will be discussed based on Figure 8.7 [10,11]. A decentralized fuzzy logic control (FLC) scheme for multiple wheeled mobile robots (WMRs) is presented in [12]. The robot kinematic and dynamic model that has been derived in Section 5.4.1 and Example 7.1 will be considered. For this system, we have derived the two-step (backstepping) controller. The kinematic controller is (see Eqs. (7.73a) and (7.73b)):

$$\begin{aligned} v_m &= v_d \cos \varepsilon_3 + K_1\varepsilon_1 \\ \omega_m &= \omega_d + K_2 v_d \varepsilon_2 + K_3 \sin \varepsilon_3 \end{aligned} \tag{8.18}$$

The problem to be discussed here is to find a control input $\mathbf{u} = [u_1, u_2]^T = [\tau_a, \tau_b]^T$ (see Eqs. (5.49a) and (5.49b)) which stabilizes to zero the errors: $\tilde{v} = v - v_m$, $\tilde{\omega} = \omega - \omega_m$ (see Eq. (7.71)) where the trajectory error:

$$\tilde{\mathbf{x}} = \mathbf{x}_d - \mathbf{x}, \quad \mathbf{x} = [x, y, \phi]^T, \quad \mathbf{x}_d = [x_d, y_d, \phi_d]^T$$

is described by Eqs. (7.69) and (7.70):

$$\boldsymbol{\varepsilon} = \begin{bmatrix} \varepsilon_1 \\ \varepsilon_2 \\ \varepsilon_3 \end{bmatrix} = \begin{bmatrix} \cos\phi & \sin\phi & 0 \\ -\sin\phi & \cos\phi & 0 \\ 0 & 0 & 1 \end{bmatrix} \begin{bmatrix} \tilde{x} \\ \tilde{y} \\ \tilde{\phi} \end{bmatrix} = \mathbf{E}(\tilde{\mathbf{x}})\tilde{\mathbf{x}} \tag{8.19a}$$

where:

$$\dot{\varepsilon}_1 = \omega\varepsilon_2 - v + v_d \cos \varepsilon_3, \quad \dot{\varepsilon}_2 = -\omega\varepsilon_1 + v_d \sin \varepsilon_3, \quad \dot{\varepsilon}_3 = \omega_d - \omega \tag{8.19b}$$

A simple design method is to use the Mamdani rule given by Eq. (8.2a). Here, we have two control inputs u_1, u_2 and two outputs v and ω. The fuzzy rules represent the mapping from $\tilde{v}$ and $\tilde{\omega}$ to the motor torques u_1 and u_2. For convenience, the reference superset (universe of discourse) is taken to be the interval [−1,1] for all fuzzy

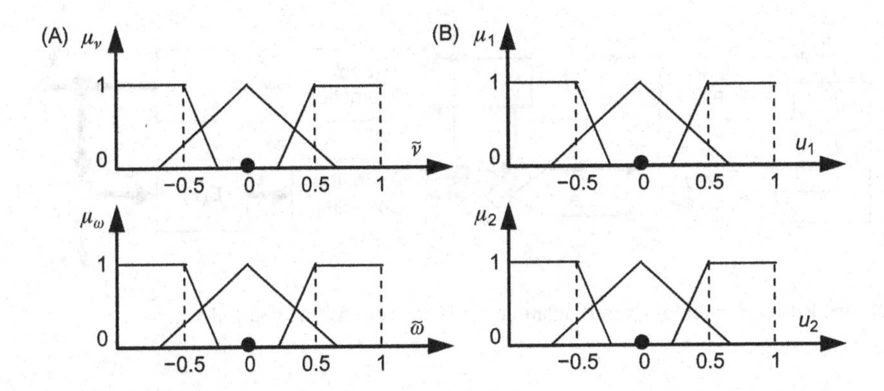

Figure 8.15 Membership functions $\mu(x)$. (A) Variables $\tilde{v}$ and $\tilde{\omega}$ and (B) variables u_1, u_2.

Table 8.2 Fuzzy Rule Base in Table form

	$\tilde{\omega}$		
$\tilde{v}$	**N**	**Z**	**P**
N	(N,N)	(N,Z)	(N,P)
Z	(Z,N)	(Z,Z)	(Z,P)
P	(P,N)	(P,Z)	(P,P)

variables. The membership functions of these variables are selected to have a triangular/trapezoidal form with three functions each, as shown in Figure 8.15 [11].

Clearly, we have the FRB given in Table 8.2, where the following notation is used for the linguistic values of $\tilde{v}, \tilde{\omega}, u_1$, and u_2: N, negative; Z, zero; P, positive.

The entries Table 8.2 represent the fuzzy values of the pairs (u_1, u_2) with corresponding fuzzy values (N,N), (N,Z), (N,P), and so on.

In linguistic form, the nine (3×3) fuzzy rules of the rule base of the FLC are the following:

Rule 1: IF $\tilde{v}$ is Z and $\tilde{\omega}$ is Z, THEN u_1 is Z and u_2 is Z.
Rule 2: IF $\tilde{v}$ is Z and $\tilde{\omega}$ is *P*, THEN u_1 is Z and u_2 is *P*.
Rule 3: IF $\tilde{v}$ is Z and $\tilde{\omega}$ is *N*, THEN u_1 is Z and u_2 is *N*.
Rule 4: IF $\tilde{v}$ is *P* and $\tilde{\omega}$ is *P*, THEN u_1 is *P* and u_2 is *P*.
Rule 5: IF $\tilde{v}$ is *P* and $\tilde{\omega}$ is *N*, THEN u_1 is *P* and u_2 is *N*.
Rule 6: IF $\tilde{v}$ is *P* and $\tilde{\omega}$ is Z, THEN u_1 is *P* and u_2 is Z.
Rule 7: IF $\tilde{v}$ is *N* and $\tilde{\omega}$ is *N*, THEN u_1 is *N* and u_2 is *N*.
Rule 8: IF $\tilde{v}$ is *N* and $\tilde{\omega}$ is *P*, THEN u_1 is *N* and u_2 is *P*.
Rule 9: IF $\tilde{v}$ is *N* and $\tilde{\omega}$ is Z, THEN u_1 is *N* and u_2 is Z.

The convenient deffuzification method is the COG/centroid of area (COA) given by Eq. (8.3). Clearly, the above controller is a fuzzy proportional controller.

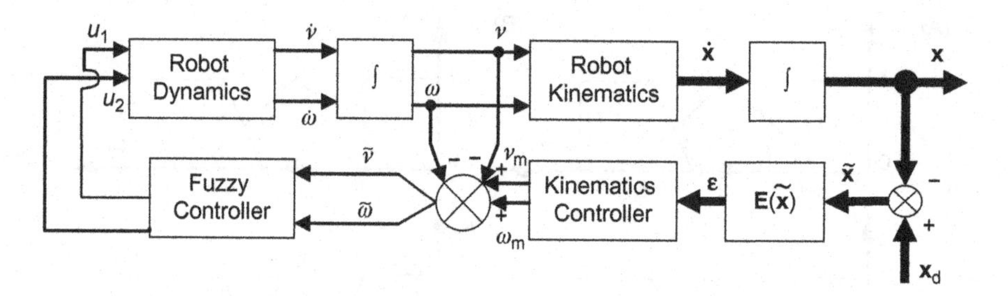

Figure 8.16 Structure of overall adaptive fuzzy tracking WMR controller.

The overall feedback tracking control system has the structure of Figure 5.11 with the dynamic controller ((5.59a) and (5.59b)) being replaced by the FLC designed on the basis of the rule base given in Table 8.2. The FLC accepts crisp values which are fuzzified in the IFU unit and provides crisp values for the robot inputs (torques) after the deffuzification process by the ODU unit (Figure 8.7). Therefore, the block diagram of the two-step controller, namely, the crisp kinematic controller Eq. (8.18) and the nine-rule FLC has the form of Figure 8.16 [11].

Some remarks about this control scheme are the following:

- The FLC performs, implicitly through the fuzzy rules, the robot dynamics identification, thus replacing both the deterministic dynamic controller of Section 5.4.2 (Eqs. (5.59a) and (5.59b)) and the adaptive controller of Section 7.3.1 which uses the parameter adaptation laws (7.40a), (7.40b), and (7.41).
- The kinematic controller remains a crisp controller. This does not reduce the applicability of the control scheme, since the kinematic controller (8.18) does not involve any unknown (or potentially unknown) parameter. All variables are known or measured.
- A better performance can be obtained if, instead of the present proportional fuzzy controller, we use a PD fuzzy controller such as the controller designed in Example 8.3.

Example 8.4

It is desired to describe fully the computation of the fuzzy output of a Mamdani fuzzy controller with inputs the error $e(t)$ and the change $\Delta e(t)$ of the error, with three linguistic values N, Z, P that are represented by triangular and trapezoidal membership functions.

Solution

A Mamdani-type fuzzy PD controller is described by the rule base (matrix) of Table 8.3, where the elements represent the changes Δu of the control signal u.

The linguistic values N (negative), Z (zero), and P (positive) for e, Δe, and Δu are as shown in Figure 8.17 with universes of discourse:

$$X_1 = \{e\} = [-3, 3], \quad X_2 = \{\Delta e\} = [-1, 1], \quad Y = \{\Delta u\} = [-6, 6]$$

We will compute the control action Δu when:

$$\{e, \Delta e\} = \{-2.5, 0.5\}$$

Table 8.3 Fuzzy PD Controller Rule Base

Δe	e		
	N	**Z**	**P**
N	N	N	Z
Z	N	Z	P
P	Z	P	P

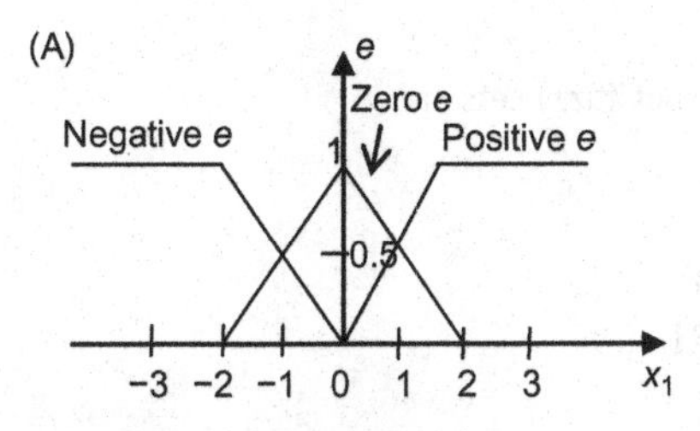

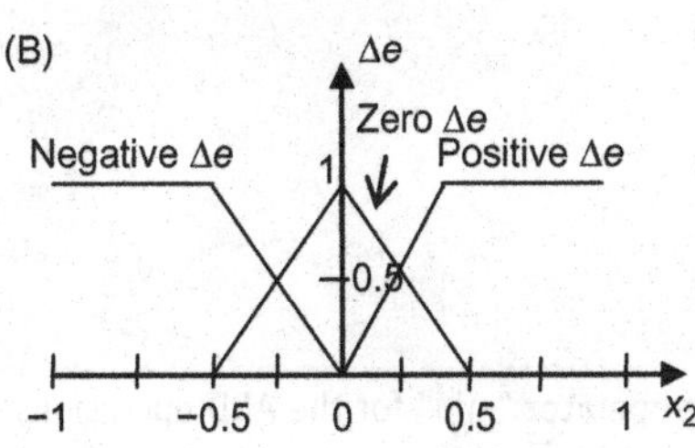

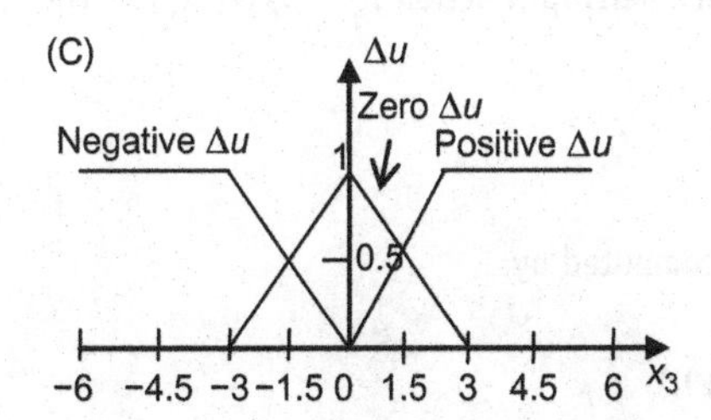

Figure 8.17 Fuzzy sets (membership functions) of e(A), Δe (B), and Δu(C).

Since we have a 3×3 FRB table, the controller consists of nine rules ($i = 1, 2, \ldots, 9$) of the form:

$\mathbf{R}_i$: IF $e = A_{i1}$ AND $\Delta e = A_{i2}$, THEN $\Delta u = B_i$.

where:

$\mathbf{R}_1$: A_{11} = negative e, A_{12} = negative Δe, B_1 = negative Δu;
$\mathbf{R}_2$: A_{21} = negative e, A_{22} = zero Δe, B_2 = negative Δu;
$\mathbf{R}_3$: A_{31} = negative e, A_{32} = positive Δe, B_3 = zero Δu;
$\mathbf{R}_4$: A_{41} = zero e, A_{42} = negative Δe, B_4 = negative Δu;

R$_5$: A_{51} = zero e, A_{52} = zero Δe, B_5 = zero Δu;
R$_6$: A_{61} = zero e, A_{62} = positive Δe, B_6 = positive Δu;
R$_7$: A_{71} = positive e, A_{72} = negative Δe, B_7 = zero Δu;
R$_8$: A_{81} = positive e, A_{82} = zero Δe, B_8 = positive Δu;
R$_9$: A_{91} = positive e, A_{92} = positive Δe, B_9 = positive Δu.

Using uniform partitioning of the universe of discourse $Y = [-6, 6]$ as shown in Figure 8.17, we have:

$$Y = [-6, \ -4.5, \ -3, \ -1.5, 0, 1.5, 3, 4.5, 6]$$

Therefore, the membership functions of the output fuzzy sets are:

$$\begin{aligned}
B_1(Y) &= [1 \quad 1 \quad 1 \quad 0.5 \quad 0 \quad 0 \quad 0 \quad 0 \quad 0] \\
B_2(Y) &= [1 \quad 1 \quad 1 \quad 0.5 \quad 0 \quad 0 \quad 0 \quad 0 \quad 0] \\
B_3(Y) &= [0 \quad 0 \quad 0 \quad 0.5 \quad 1 \quad 0.5 \quad 0 \quad 0 \quad 0] \\
B_4(Y) &= [1 \quad 1 \quad 1 \quad 0.5 \quad 0 \quad 0 \quad 0 \quad 0 \quad 0] \\
B_5(Y) &= [0 \quad 0 \quad 0 \quad 0.5 \quad 1 \quad 0.5 \quad 0 \quad 0 \quad 0] \\
B_6(Y) &= [0 \quad 0 \quad 0 \quad 0 \quad 0 \quad 0.5 \quad 1 \quad 1 \quad 1] \\
B_7(Y) &= [0 \quad 0 \quad 0 \quad 0.5 \quad 1 \quad 0.5 \quad 0 \quad 0 \quad 0] \\
B_8(Y) &= [0 \quad 0 \quad 0 \quad 0 \quad 0 \quad 0.5 \quad 1 \quad 1 \quad 1] \\
B_9(Y) &= [0 \quad 0 \quad 0 \quad 0 \quad 0 \quad 0.5 \quad 1 \quad 1 \quad 1]
\end{aligned}$$

Since the rules are of the Mamdani type, we use the operator "min" for the AND operations on the left-hand side of the rules, and so the membership function $\mu_i = \mu_{R_i}(x_1, x_2)$ of the left-hand side of rule R_i is equal to:

$$\mu_i = \min\{\mu_{Ai1}(x_1), \mu_{Ai2}(x_2)\}$$

Then, the output (conclusion) of this rule is computed by:

$$\begin{aligned}
\mu_{\Delta u,i}(y) &= \{\min(\mu_i, B_{i1}(y)), \ldots, \min(\mu_i, B_{19}(y))\} \\
&= \{\min[\mu_{A_{i1}}(x_1), \mu_{A_{i2}}(x_2), B_{i1}(y)], \ldots, \min[\mu_{A_{i1}}(x_1), \mu_{A_{i2}}(x_2), B_{i9}(y)]\}
\end{aligned}$$

From the data, we find:

$$\{x_1, x_2\} = \{e(k), \Delta e(k)\} = \{-2.5, 0.5\}$$

and so:

1. $\mu_{A_{11}}(-2.5) = 1, \quad \mu_{A_{12}}(0.5) = 0$

$$\mu_{\Delta u,1}(y) = [0 \quad 0 \quad 0 \quad 0 \quad 0 \quad 0 \quad 0 \quad 0 \quad 0]$$

2. $\mu_{A_{21}}(-2.5) = 1, \quad \mu_{A_{22}}(0.5) = 0$

$$\mu_{\Delta u,2}(y) = [0 \quad 0 \quad 0 \quad 0 \quad 0 \quad 0 \quad 0 \quad 0 \quad 0]$$

3. $\mu_{A_{31}}(-2.5) = 1, \quad \mu_{A_{32}}(0.5) = 1$

$$\mu_{\Delta u,3}(y) = [0 \quad 0 \quad 0 \quad 0.5 \quad 1 \quad 0.5 \quad 0 \quad 0 \quad 0]$$

4. $\mu_{A_{41}}(-2.5) = 1, \quad \mu_{A_{42}}(0.5) = 0$

$$\mu_{\Delta u,4}(y) = [0 \quad 0 \quad 0 \quad 0 \quad 0 \quad 0 \quad 0 \quad 0 \quad 0]$$

5. $\mu_{A_{51}}(-2.5) = 1, \quad \mu_{A_{52}}(0.5) = 0$

$$\mu_{\Delta u,5}(y) = [0 \quad 0 \quad 0 \quad 0 \quad 0 \quad 0 \quad 0 \quad 0 \quad 0]$$

6. $\mu_{A_{61}}(-2.5) = 1, \quad \mu_{A_{62}}(0.5) = 1$

$$\mu_{\Delta u,6}(y) = [0 \quad 0 \quad 0 \quad 0 \quad 0 \quad 0 \quad 0 \quad 0 \quad 0]$$

7. $\mu_{A_{71}}(-2.5) = 0, \quad \mu_{A_{72}}(0.5) = 0$

$$\mu_{\Delta u,7}(y) = [0 \quad 0 \quad 0 \quad 0 \quad 0 \quad 0 \quad 0 \quad 0 \quad 0]$$

8. $\mu_{A_{81}}(-2.5) = 1, \quad \mu_{A_{82}}(0.5) = 0$

$$\mu_{\Delta u,8}(y) = [0 \quad 0 \quad 0 \quad 0 \quad 0 \quad 0 \quad 0 \quad 0 \quad 0]$$

9. $\mu_{A_{91}}(-2.5) = 0, \quad \mu_{A_{92}}(0.5) = 0$

$$\mu_{\Delta u,9}(y) = [0 \quad 0 \quad 0 \quad 0 \quad 0 \quad 0 \quad 0 \quad 0 \quad 0]$$

Consequently, the total membership function which is equal with the "max" of $\mu_{\Delta u,i}(y)$ (union of fuzzy sets, $i = 1, 2, \ldots, 9$) is:

$$\mu_{\Delta u}(y) = [0 \quad 0 \quad 0 \quad 0.5 \quad 1 \quad 0.5 \quad 0 \quad 0 \quad 0]$$

Applying COG/COA defuzzification we find:

$$\Delta u = \frac{0.5 \times (-1.5) + 1 \times (0.0) + 0.5 \times (+1.5)}{0.5 + 1 + 0.5} = 0$$

Exercise: Repeat the above computation for $\{e, \Delta e\} = \{0.9, 0.2\}$ and $\{e, \Delta e\} = \{0.75, 0.75\}$.

8.4.2 Fuzzy Local Path Tracker for Dubins Car

8.4.2.1 The Problem

The kinematic model of Dubins car is found from the standard car-like model (2.52) by omitting the equation for the steering angle velocity $\dot{\psi}$, that is (see Figure 2.9):

$$\begin{bmatrix} \dot{x} \\ \dot{y} \\ \dot{\phi} \end{bmatrix} = \begin{bmatrix} \cos\phi \\ \sin\phi \\ (tg\,\psi)/D \end{bmatrix} v_1 \tag{8.20}$$

where, for notational simplicity, the index Q was dropped from $\dot{x}_Q$ and $\dot{y}_Q$ (since here it is not needed). Clearly, the Dubins car is a four-wheeled WMR under bounded curvature constraints, with drift, and forward motion. These constraints imply that the turning radius of the mobile robot is bounded (just like actual cars) and that without any input, the robot remains still [13]. Here, the following discretized form of the model (8.20), will be used:[2]

$$\Delta\phi = \kappa\Delta s \tag{8.21a}$$

$$\Delta x = (2/\kappa)\sin(\kappa(\Delta s)/2)\cos(\phi_0 + \kappa(\Delta s)/2) \tag{8.21b}$$

$$\Delta y = (2/\kappa)\sin(\kappa(\Delta s)/2)\sin(\phi_0 + \kappa(\Delta s)/2) \tag{8.21c}$$

where κ is the robot's curvature and Δs the covered distance in a control loop. The curvature is related to steering angle ψ by the following equation:

$$\kappa = (tg\,\psi)/D \tag{8.22}$$

which is obtained from the relation $y_1 = R\dot{\phi} = \dot{\phi}/\kappa$ and the third line of Eq. (8.20) for $\dot{\phi} \neq 0$.

The motion of the robot described by this model is as follows:

- At t_0, the robot has a curvature κ_0.
- The robot maintains the curvature for a path arc length Δs.
- At t_1, the robot has covered the distance Δs and instantly changes curvature to κ_1.
- The loop starts all over.

The control outputs are the curvature κ and the distance Δs. The control can be further simplified if the length Δs, covered at each control loop, is assumed to have a fixed constant value. Therefore, the only control output is the curvature. Geometrically, this model describes a motion consisting of connected arcs. Although this simplification eases the efforts to control the robot, it increases the

[2] Based on the approximations, $\Delta\phi/2 = \sin(\Delta\phi/2) = \sin(\kappa\Delta s/2)$ and $\phi = \phi_0 + \Delta\phi/2 = \phi_0 + \kappa\Delta s/2$ [14].

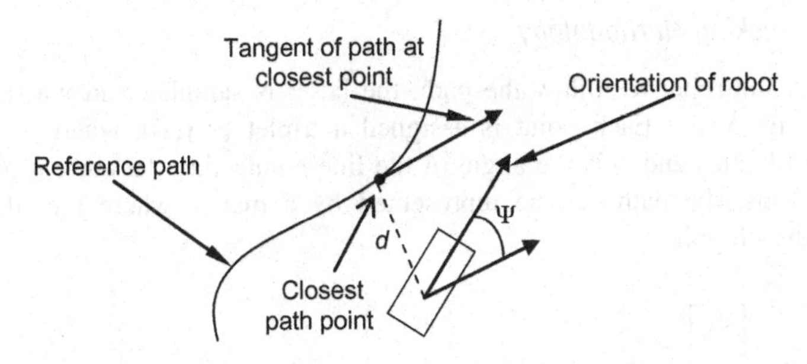

Figure 8.18 Standard control input variables for path tracking.

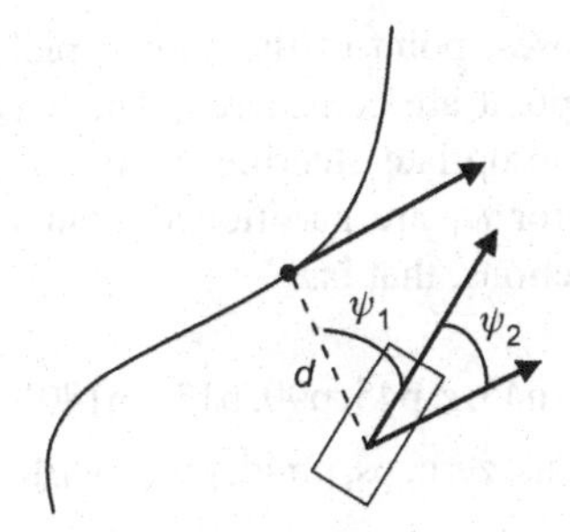

Figure 8.19 The control input variables for the fuzzy controller are the angular errors ψ_1 and ψ_2.

complexity of a physical implementation of the controller. Fixed Δs with a variable robot speed implies that the control period ΔT is also variable. The relation between Δs and ΔT is:

$$\Delta T = \Delta s / v_1$$

So, the controller must be a *multirate controller*. Now, suppose that we have a mobile robot that has to follow a reference path. In case the robot is misplaced, the path tracker must steer it back on course. Mathematically, this is equivalent to minimizing the orientation error ψ and the position error d as shown in Figure 8.18.

The majority of the path tracking controllers use one or both of these two variables as inputs and try to minimize them, thereby positioning the robot right on path. Here, use is made of a different set of variables as control inputs, which are more appropriate for the control philosophy of the present fuzzy control scheme [14,15]. This set consists of the orientation error ψ_2, the same as above, and the angular error ψ_1 (Figure 8.19).

In this case, in order for the robot to track the path, ψ_1 and ψ_2 must be zero. This set of variables allows us to define control actions for the whole universe of discourse, as ψ_1, ψ_2 are angles and $\psi_1, \psi_2 \in (-\pi, \pi]$.

8.4.2.2 Tracking Methodology

In order for the robot to follow the path, the latter is sampled under a fixed sampling spacing Δs_{path}. Each point is assigned a triplet (x, y, ϕ), where x, y are the point's coordinates and ϕ is the angle of the line connecting the current point with the next. Thus, the path can be represented by a matrix where the ith column describes the ith point:

$$\text{column } i: \begin{bmatrix} x_i \\ y_i \\ \phi_i \end{bmatrix}, \quad \text{column } i+1: \begin{bmatrix} x_{i+1} \\ y_{i+1} \\ \phi_{i+1} \end{bmatrix}, \quad \phi_i = \tan^{-1}\left(\frac{y_{i+1} - y_i}{x_{i+1} - x_i}\right) \tag{8.23}$$

As the robot moves, the closest point of the path is picked up, and the orientation and angular errors for this point are considered. These variables are presented to a fuzzy controller, and the appropriate steering command is issued. The orientation error ψ_2 and the angular error ψ_1 are partitioned in nine fuzzy sets with Gaussian (bell-type) membership functions, that is:

$$\psi_1 = \{\text{n180}, \text{n135}, \text{n90}, \text{n45}, z, \text{p45}, \text{p90}, \text{p135}, \text{p180}\}$$
$$\psi_2 = \{\text{nvb}, \text{nbig}, \text{nmid}, \text{ns}, \text{zero}, \text{ps}, \text{pmid}, \text{pbig}, \text{pvb}\}$$

where n180 denotes a bell-shaped function centered at 180° and so on. The symbols nvb, nb, and so on denote bell-type fuzzy numbers: nvb, negative very big; nb, negative big; nmid, negative medium; ns, negative small; z, zero; ps, positive small; pmed, positive medium; pb, positive big; pvb, positive very big (as shown in Figures 8.20 and 8.21).

The output variable, curvature κ, is partitioned in five sets with Gaussian membership functions as well (Figure 8.22).

The Mamdani inference scheme along with the "min" aggregation operator are suitable for the present problem. The rule base of the fuzzy logic controller consists of $9 \times 9 = 81$ IF-THEN rules. The philosophy of these rules is that of driving a car (robot) that sees road signs (closest path points) that show the right direction (angle ψ_2). So, according to where the sign is (angle ψ_1) and where it points to, the car is steered appropriately (curvature κ). Thus, a typical rule would have the form:

$R1$: IF ψ_1 is p45 AND ψ_2 is nb, THEN κ is pb,

which is shown pictorially in Figure 8.23.

If the closest point is at $\psi_1 = 45°$ and points to $\psi_2 = -135°$, the wheel is turned hard left. A special case that needs attention is when the closest point lies at $\psi_1 = 180°$, which is equivalent to $\psi_1 = -180°$.

Although $\psi_1 \in (-\pi, \pi)$ and thus ψ_1 never gets the value $-180°$, we speak in terms of fuzzy sets. The rules that apply to the fuzzy sets n180, p180 must be consistent with each other in terms of continuity. The sets n180, p180 essentially cover two cases that are "continuous" in the real world even if their mathematical values

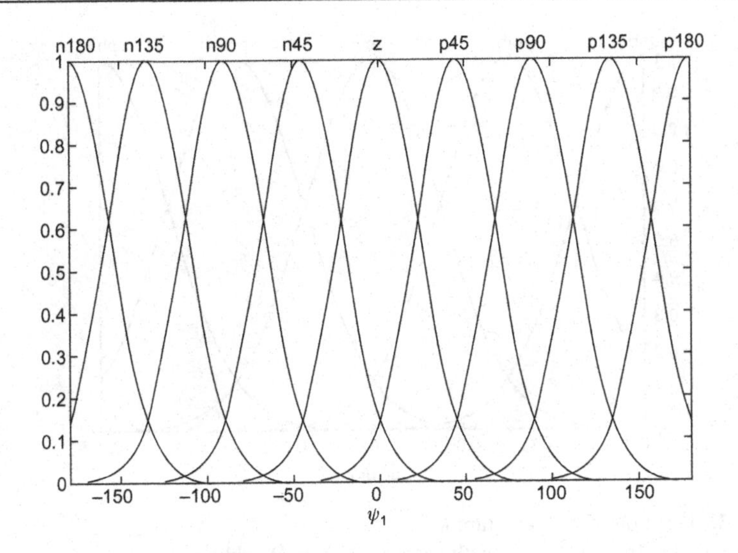

Figure 8.20 Partition of angular error ψ_1.
Source: Reprinted from Ref. [14], with permission from World Scientific.

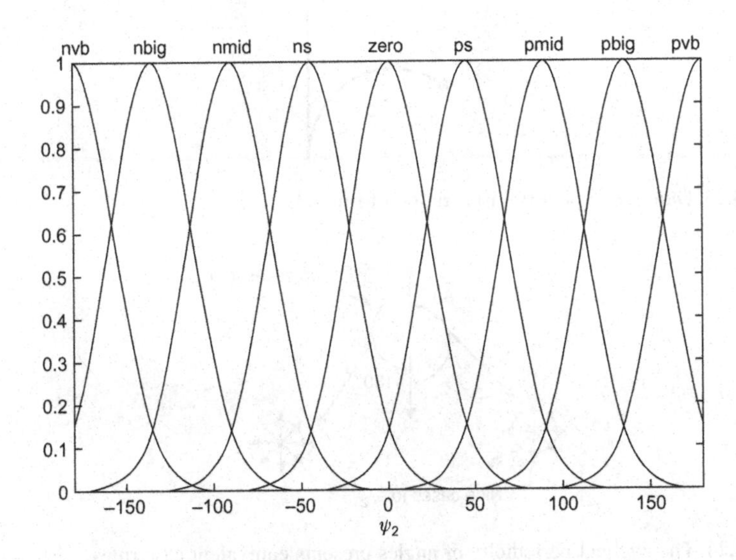

Figure 8.21 Partition of orientation error ψ_2.
Source: Reprinted from Ref. [14], with permission from World Scientific.

lie on the two opposing ends of the universe of discourse of ψ_1. This is a natural consequence of the cyclical periodicity of angles (see Figure 8.24). To better understand this, we must look at the FLC rule surface shown in Figure 8.25.

We observe that at $\psi_1, \psi_2 \approx -180°$, there is a spike. If the surface was flat, then when ψ_1 passes from $-180°$ to $180°$, the curvature would suddenly change

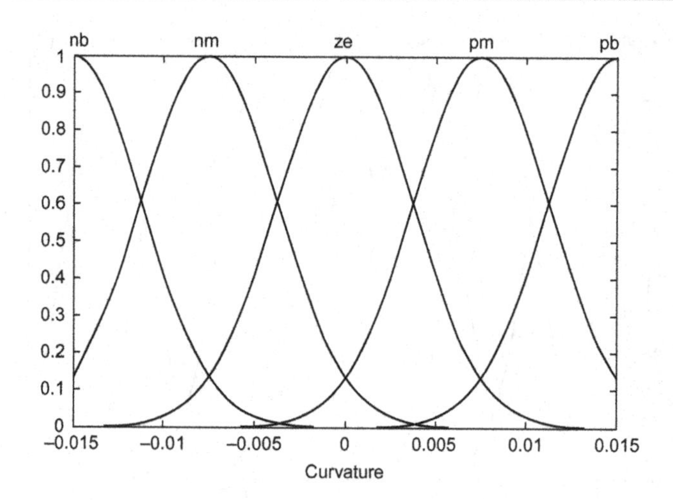

Figure 8.22 Partition of the curvature κ.
Source: Reprinted from Ref. [14], with permission from World Scientific.

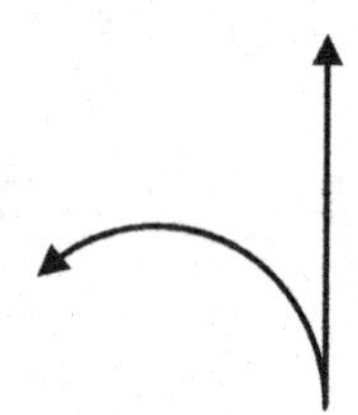

Figure 8.23 Qualitative pictorial illustration of rule $R1$.

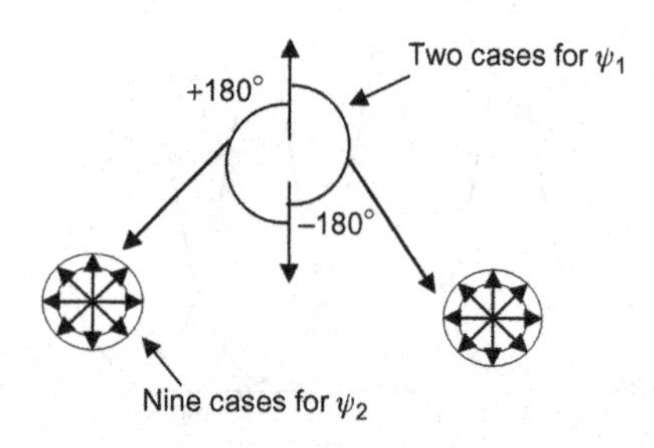

Figure 8.24 The cyclical periodicity of angles presents equivalent case rules.
Source: Reprinted from Ref. [14], with permission from World Scientific.

from a minimum negative to a positive value. This kind of behavior would cause an oscillatory motion of the robot, something which was experimentally confirmed. Therefore, in order to avoid this, we bias the rules there, hence introducing the spike on the FLC surface. The present path tracking controller was implemented on a SoC (System on Chip) consisting of a parameterized FLC IP core and a Xilinx Microblaze soft processor as the top level flow controller [16–18].

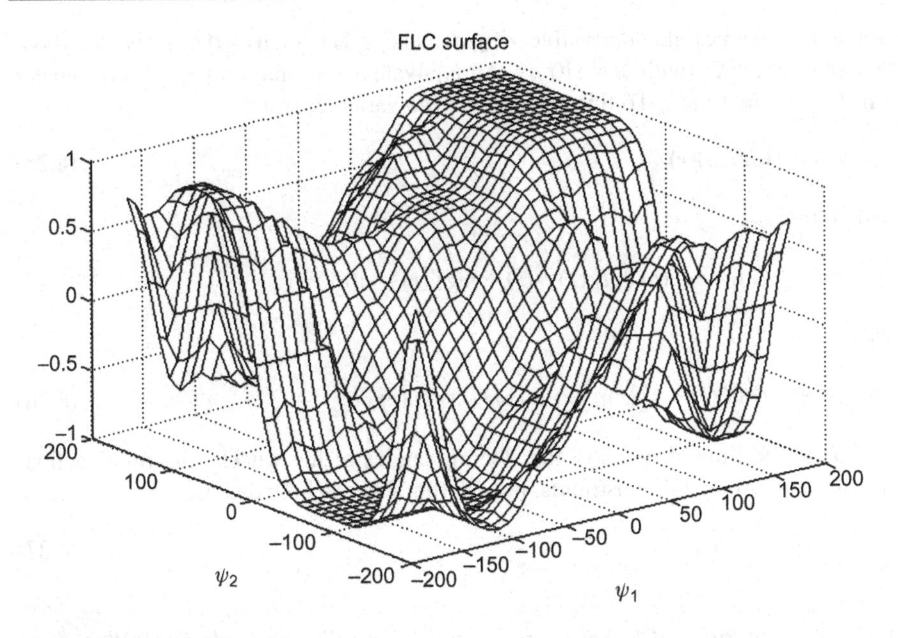

Figure 8.25 FLC rule surface.
Source: Reprinted from Ref. [14], with permission from World Scientific.

The Field Programmable Gate Array (FPGA) board (Spartan 3-1500) hosting the SoC was mounted to an ActivMedia Pioneer P3-DX8 robot which was used in the experiments. The experimental results were obtained using the zero-order Takagi-Sugeno inference scheme combined with triangular membership functions with first degree adjacent overlap (see also Section 13.10). A full account of mobile control using a scalable FPGA architecture is given in Ref. [19].

8.4.3 Fuzzy Sliding Mode Control

8.4.3.1 The Mobile Robot Model

This controller will be based on the controller presented in Section 7.2.2, and will be applied to a WMR that moves on a surface $g(x, y, z) = 0$ along a continuously differentiable path $\mathbf{p}(r) = [x(r), y(r), z(r)]$ [20,21]. The path $\mathbf{p}(r)$ describes the motion of the center of gravity of the robot with respect to a world coordinate frame. Under the assumption of conservative forces, the dynamic model of the mobile robot is:[3]

$$[m + I_{\mathrm{R}}(r)]\ddot{r}(t) + I'_{\mathrm{R}}(r)\dot{r}^2(t) + mgz'(r) = u(t) \tag{8.24}$$

[3] Followed from the Lagrange equation $d(\partial L/\partial \dot{r})/dt - \partial L/\partial r = u$, where $L(r, \dot{r}) = K - P = (1/2) I_{\mathrm{R}}(r)\dot{r}^2 + (1/2)m\dot{r}^2 - mgz(r)$ (see Eqs. (3.7)–(3.10)).

where $(\cdot)'$ denotes the derivative $d(\cdot)/dr$, and $I_R(r) = \langle \Omega(r), I\Omega(r) \rangle$ is the robot "reflected inertia" (with $\omega = \Omega(r)\dot{r}$). An equivalent formulation uses the parameter r instead of the time t. To this end, we use the variable:

$$v(r) = (dr/dt)(r) \tag{8.25}$$

and obtain:

$$[m + I_R(r)]v(r)v'(r) + I'_R(r)v^2(r) + mgz'(r) = u(r)$$

that is:

$$[m + I_R(r)]\dot{v}(r) + I'_R(r)v^2(r) + mgz'(r) = u(r) \tag{8.26}$$

Due to the limited range of the steering angles of the wheels, the floor inclination and the actuators' constraints, the steering force is bounded as:

$$-U_2 \leq u \leq U_1 \tag{8.27}$$

8.4.3.2 Similarity of Fuzzy Logic Controller and Sliding Mode Controller

The *sliding mode controller* (SMC) for a system of the type:

$$\ddot{x} = b(x) + a(x)u \tag{8.28}$$

where the gain function satisfies the inequality (7.24c):

$$1/\eta(x) \leq \hat{a}/a \leq \eta(x) \tag{8.29}$$

with $\eta = \eta(x)$ being the "gain margin" of the system, is given by Eq. (7.24d):

$$u = \hat{a}^{-1}\{\hat{u} - k\,\mathrm{sgn}(s)\} \tag{8.30a}$$

or

$$u = \hat{a}^{-1}\left\{\hat{u} - k\,\mathrm{sat}\left(\frac{s}{B}\right)\right\} \tag{8.30b}$$

where:

$$\mathrm{sat}(z) = \begin{cases} z & \text{if} \quad |z| \leq 1 \\ \mathrm{sgn}(z) & \text{if} \quad |z| > 1 \end{cases} \tag{8.31a}$$

$$\hat{u} = -b + \ddot{x}_d - \Lambda\dot{\tilde{x}} \tag{8.31b}$$

$$k \geq \eta(\rho_{max} + \gamma) + (\eta - 1)|\hat{u}| \tag{8.31c}$$

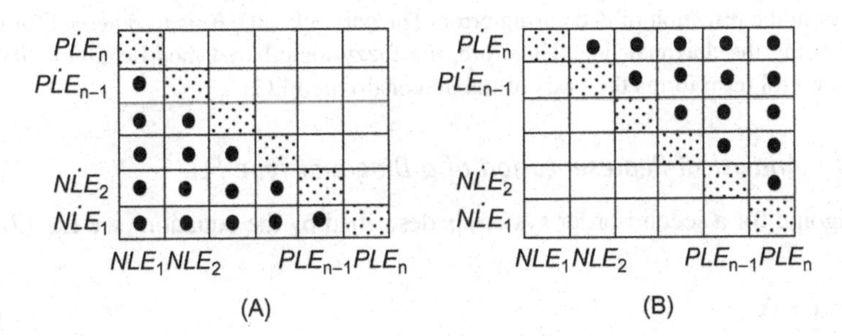

Figure 8.26 (A) Fuzzy regions below the diagonal and (B) fuzzy regions above the diagonal.

Now, consider a nonautonomous second-order single-input single-output nonlinear system. In the case of a diagonal-type FLC (similar to the one of Example 8.3, Table 8.1), regions where the controller output becomes zero lie on the diagonal which separates the fuzzy plane into two semi-planes. For all the fuzzy regions below (above) the diagonal, the controller output takes a positive (negative) fuzzy value with a magnitude that depends on the distance between this fuzzy region and a particular zero region on the diagonal, below (above) which the given fuzzy region is located. Figure 8.26A and B illustrate the set of all fuzzy regions below (above) the diagonal [20].

As distance between a "fuzzy region below (above) the diagonal and the diagonal" is defined the distance between the "center of this region and the center of the zero region below (above) which the given fuzzy region is located."

Looking at SMC (8.30)–(8.31), we see that the controller involves the following terms:

- A filtering term:

$$u_{\mathrm{filt}} = -\hat{a}^{-1}\Lambda\dot{\tilde{x}} \tag{8.32a}$$

which rejects unmodeled frequencies of the system.

- A feedforward term:

$$u_{\mathrm{ff}} = \hat{a}^{-1}\ddot{x}_{\mathrm{d}} \tag{8.32b}$$

- A compensation term:

$$u_{\mathrm{comp}} = -\hat{a}^{-1}b \tag{8.32c}$$

- A feedback control term:

$$u_{\mathrm{c}} = -\hat{a}^{-1}(k)\mathrm{sat}\left(\frac{s}{B}\right) \tag{8.32d}$$

which prevents the error state vector $\tilde{x} = x(t) - x_{\mathrm{d}}(t) = [\tilde{x}_1, \dot{\tilde{x}}_1]^{\mathrm{T}}$ from moving away from the sliding surface $s = 0$. The negative sign indicates that the control action takes place always in the direction of decreasing error. The part $-k\ \mathrm{sat}(s/B)$ is of diagonal form, with $s=0$ being the diagonal line. Therefore, the fuzzy-logic-based modification of the feedback control term forms the basis of the diagonal-type FLC.

8.4.3.3 Analytical Representation of a Diagonal-Type FLC

The diagonal for a second-order system is described by the equation (see Eq. (7.13)):

$$s=\dot{\tilde{x}}+\Lambda\tilde{x} \tag{8.33}$$

The rules of the diagonal-type FLC can be selected as follows:

1. The states $\tilde{x},\dot{\tilde{x}}$ are bounded as:

$$-\rho^0_{x,\max}\leq\tilde{x}\leq\rho^0_{x,\max}$$
$$-\rho^1_{x,\max}\leq\dot{\tilde{x}}\leq\rho^1_{x,\max}$$

2. The control signal u is bounded as:[4]

$$-\rho_{u,\max}\leq u\leq\rho_{u,\max}$$

3. The state $\tilde{x}$ and $\dot{\tilde{x}}$ that are located on the diagonal produce zero control signals.
4. The states $\tilde{x}$ and $\dot{\tilde{x}}$ that are located below the diagonal produce positive control signals.
5. The states $\tilde{x}$ and $\dot{\tilde{x}}$ that are located above the diagonal produce negative control signals.
6. The magnitude of the control signal $|u|$ increases when the distance from the diagonal increases, and vice versa. On the basis of the above, we can express the diagonal-type FLC by the formula:

$$u_{\mathrm{fuzzy}}=-K_{\mathrm{fuzzy}}(\tilde{x},\dot{\tilde{x}},\Lambda)\ \mathrm{sgn}(s) \tag{8.34}$$

with the condition:

$$K_{\mathrm{fuzzy}}(\tilde{x}_1,\dot{\tilde{x}}_1,\Lambda)\leq K_{\mathrm{fuzzy}}(\tilde{x}_2,\dot{\tilde{x}}_2,\Lambda)$$

for $|\Lambda\tilde{x}_1+\dot{\tilde{x}}_1|\leq|\Lambda\tilde{x}_2+\dot{\tilde{x}}_2|$, which means that the greater the distance of $(\tilde{x},\dot{\tilde{x}})$ from the sliding surface, the greater the control signal.

8.4.3.4 Reduced Complexity Sliding Mode Fuzzy Logic Controller

To design a *reduced-complexity sliding mode fuzzy logic controller* (RC-SMFLC), we start from the conditions that assure the error convergence to zero, namely:

IF $e(t)\dot{e}(t)<0$, THEN $x(t)\rightarrow x_{\mathrm{d}}(t)$.
that is, $e(t)=\tilde{x}E(t)=x(t)-x_{\mathrm{d}}(t)\rightarrow 0$.
IF $e(t)\dot{e}(t)>0$, THEN $x(t)$ deviates from $x_{\mathrm{d}}(t)$.

[4] Note that here $\tilde{x}$ is defined as $\tilde{x}(t)=e(t)=x(t)-x_{\mathrm{d}}(t)$. Therefore, if $\tilde{x}(t)\leq 0$, then the control action u should be positive, and if $\tilde{x}(t)>0$, the control action u should be negative.

As we can observe, these conditions could be derived by using the Lyapunov function $V=(1/2)e^2(t)$ for which $\dot{V}=e(t)\dot{e}(t)$. The sliding condition is then defined as:

$$\dot{V}(x,t)=e(t)\dot{e}(t)<0 \tag{8.35}$$

Here, we have two possible control actions: *increase* or *decrease*. Therefore, we have the following control rules:

1. IF $\mathrm{sgn}(e(t)\dot{e}(t))<0$, THEN the control action leads to convergence and should be maintained.
2. IF $\mathrm{sgn}(e(t)\dot{e}(t))>0$, THEN the control action leads to divergence and should be changed.

The increase (decrease) of the control signal is assured by the following rules:

1. *Increase*
   IF u_k is U_1, THEN u_{k+1} is $U_2,\ldots$; IF u_k is U_{n-1}, THEN u_{k+1} is U_n.
2. *Decrease*
   IF u_k is U_2, THEN u_{k+1} is $U_1,\ldots$; IF u_k is U_n, THEN u_{k+1} is U_{n-1}.

where $u_k=u(t)]_{t=k\Delta t}$ with Δt being a given time increment (sampling period) and $U_1,U_2,\ldots,U_n$ are the fuzzy subsets in which the fuzzy phase plane U of the control input is divided. In the RC-SMFLC case, $e=0$ plays the role of the diagonal. The properties of RC-SMFLC are analogous to the ones of a diagonal type, and can be expressed mathematically as:

$$u_{\mathrm{fuzzy}}=-K_{\mathrm{fuzzy}}(|s|\mathrm{sgn}(s)) \tag{8.36}$$

where s is the distance from the diagonal. If the membership functions have the same shape (e.g., triangular with the same width and slopes), then the control law produced is $u_{\mathrm{fuzzy}}=-K_{\mathrm{fuzzy}}\,\mathrm{sgn}(s)$. To overcome this problem, the width of the membership functions should be modified at every crossing of the diagonal $e=0$. The last two control signals u_{k-1},u_k are taken into account, namely, u_{k-1} is the last signal below (above) the diagonal; u_k is the last control signal above (below) the diagonal. As the diagonal is approached, the width of the membership functions is reduced, and so the gain K_{fuzzy} is reduced too. In this way, the control law takes the form:

$$u_{\mathrm{fuzzy}}=-K_{\mathrm{fuzzy}}(\mathrm{sgn}(e_k e_{k-1}))\mathrm{sgn}(s) \tag{8.37}$$

with $s=e\dot{e}$, where e_{k-1} is the error at the $(k-1)$th step of the algorithm and e_k the error at the kth step. On the basis of the above, the similarity between RC-SMFLC and the diagonal-type FLC or the conventional SMFLC becomes clear.

8.4.3.5 Application to the Mobile Robot

We write the model (8.26) as:

$$\dot{v}(r)+\frac{I'_{\mathrm{R}}(r)}{m+I_{\mathrm{R}}(r)}v^2(r)+\frac{mgz'(r)}{m+I_{\mathrm{R}}(r)}=\frac{1}{m+I_{\mathrm{R}}(r)}u(r)=u^*(r) \tag{8.38}$$

and assume that m, the moment of inertia $I(r)$, and the reflected inertia $I_R(r)$ are exactly known, the only parametric uncertainty being in the slope:

$$z'(r) = \partial z / \partial r$$

of the path.

The control signal that compensates the uncertainties and leads to closed-loop velocity stability has the form:

$$u^*(t) = \dot{v}_d(t) + \frac{I'_R(r)}{m + I_R(r)} v^2(t) + K_p^* e(t) + u_{\text{RC-SMFLC}}(t) \tag{8.39}$$

In the case of position control, the closed-loop stability can be established as follows: We use the robot model (8.24):

$$\ddot{r}(t) + \frac{I'_R(r)}{m + I_R(r)} \dot{r}^2(t) + \frac{mgz'(r)}{m + I_R(r)} = u^*(t) \tag{8.40a}$$

and the PD controller:

$$u^*(t) = \ddot{r}_d(t) + \frac{I'_R(r)}{m + I_R(r)} \dot{r}^2(t) + K_p e(t) + K_d \dot{e}(t) + u_{\text{RC-SMFLC}}(t) \tag{8.40b}$$

to get:

$$\ddot{r}_d(t) - \ddot{r}(t) + K_p e(t) + K_d \dot{e}(t) + u_{\text{RC-SMFLC}}(t) - \frac{mgz'(r)}{m + I_R(r)} = 0 \tag{8.41}$$

where $e(t) = r_d(t) - r(t)$. Thus:

$$\ddot{e}(t) + K_d \dot{e}(t) + K_p e(t) = \frac{mgz'(r)}{m + I_R(r)} - u_{\text{RC-SMFLC}}(t) \tag{8.42}$$

Therefore, if:

$$\lim_{t \to \infty} \left[\frac{mgz'(t)}{m + I_R(r)} - u_{\text{RC-SMLC}}(t) \right] = 0 \tag{8.43}$$

the gains K_p and K_d can be selected such as $e(t) \to 0$ as $t \to \infty$. The design of the RC-SMFLC controller guarantees that the condition (8.43) will be satisfied, and therefore, the closed-loop system will be asymptotically stable. The proof of stability in the case of velocity control is similar. The block diagram of the complete hybrid closed-loop system with the PD controller (gains K_p, K_d), and the RC-SMFLC controller is shown in Figure 8.27 [20].

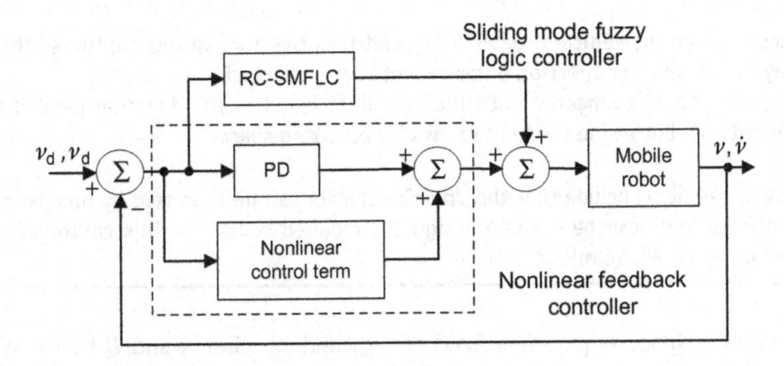

Figure 8.27 Structure of hybrid PD and RC-SMFLC control system.

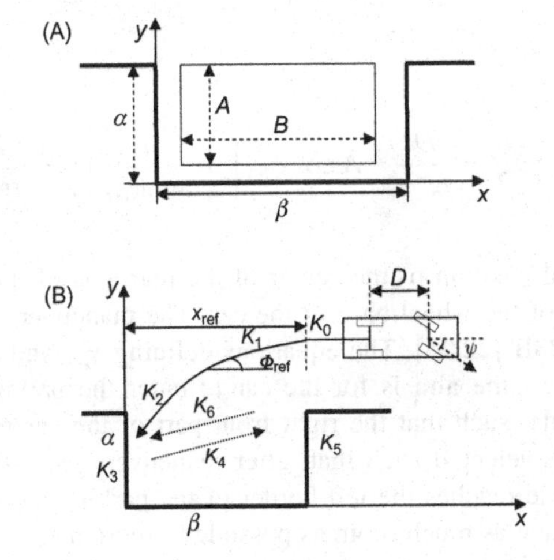

Figure 8.28 (A) Parking space and vehicle and (B) parameters of parking maneuver strategy.

Example 8.5

It is desired to provide a solution to the parallel car-parking problem using the RC-SMFLC controller.

Solution

Consider a car-like WMR which is to park in a restricted rectangular space of dimensions α and β (Figure 8.28A) [22].

The steps followed by a human driver for parallel parking are the following [22]:

Step 1: The vehicle is set parallel and ahead of the parking area.

Step 2: The vehicle backtracks to a certain point and then the front wheels are turned so that the vehicle moves toward the parking area until it reaches a certain angle of approach.

Step 3: As long as the vehicle stays inside the boundaries of the parking place, the vehicle continuous to backtrack.

Step 4: When the vehicle reaches the boundaries, the front wheels are turned the other way round, and the direction of movement is also changed.
Step 5: Step 3 is repeated until the vehicle is found in the direction parallel to the desired one, but inside the limitations of the parking space.

One can easily conclude that the driver's behavior can be expressed by linguistic rules, and so fuzzy logic can be used to design the required controller. This controller will be derived using RC-SMFLC rules.

The parking space is parameterized by ϕ_{ref} and x_{ref}. Let A and B be the dimensions of the vehicle, and assume that initially it is on the right of the parking place with $y_0 = h$ and $x_0 = 0$. The parameters ϕ_{ref} and x_{ref} are given by:

$$\sin \phi_{\text{ref}} = a/B$$

$$x_{\text{ref}} = \frac{1}{tg\,\phi_{\text{ref}}}\left[h + \frac{a}{2} - \frac{D}{tg\,\psi} - A\cos\phi_{\text{ref}}\right] + \frac{1}{\sin\phi_{\text{ref}}}\left[\frac{A}{2} + \frac{D}{tg\,\psi}\right]$$

where h is the initial position of the center of the rear axis of the car (i.e., $y_0 = h$) and D is the length of the wheel-base of the car. The maneuver strategy is demonstrated in Figure 8.28B [22,23]. The equations defining x_{ref} and ϕ_{ref} can be easily interpreted as follows: the aim is for the car to enter the parking place with the highest angle possible, such that the right front part of the car remains within the parking place. x_{ref} is selected such that, after maneuvers (K_0), (K_1), and (K_2) are completed, the vehicle reaches the left border of the parking area, with its left rear corner, in order to have as much room as possible to reorient.

The parameters x_{ref} and ϕ_{ref} are not always known in advance in real-world parking situations; however, empirical rules can be used to approximate them. In general, a genetic algorithm can be used to find the appropriate values of x_{ref} and ϕ_{ref}.

The parking maneuvers consist of the following steps: the vehicle backs up (K_0) until the back of the car reaches x_{ref}, it turns right (K_1) until the orientation overshoots ϕ_{ref}, and then it backs up (K_2) until the rear of the car touches the left or the lower border of the parking area. Finally, the vehicle reorients by repeating the following sequence: (i) if the rear part touches the border of the parking place (K_3), it drives forward and turns right (K_4); (ii) if the front part touches the border (K_5), or if the left rear corner of the car reaches the upper limit of the parking place, the vehicle backs up (K_6). As soon as the vehicle becomes parallel to the desirable axis ($\phi_d = 0°$), the parking maneuvers stop.

Every rule K_i can be implemented by a rule-based incremental controller. Standard parking algorithms need only to consider an increment $\Delta_v = 2v$ on the speed component of the input vector, and an increment $\Delta_\psi = 2\psi$ on the steering angle component of the input vector. Although this control approach does not explicitly minimize the number of maneuvers, it is rather effective. Actually, no increment on the speed component is considered, while the increment Δ_ψ on the steering angle is calculated via fuzzy inference. Additionally, we want to provide a controller that will not be strictly based on the rules given by a human expert. The controller will be able to determine smooth changes of the steering angle.

Referring to Figure 8.28B, the control strategy can be expressed as follows: the vehicle is set parallel and ahead of the parking place with its rear far ahead from point x_{ref}. The vehicle backtracks to the point x_{ref}, and then it turns the front wheels so as to move toward the parking place until it reaches the angle of approach ϕ_{ref}. As long as the vehicle stays inside the parking area, it continuous to backtrack using the modified RC-SMFLC algorithm. When it reaches the boundaries, some additional rules are used to accelerate the convergence to the desirable final position. The vehicle changes direction and moves forward turning its wheels driven by the controller output ψ. The goal is to set the vehicle in parallel to $\phi_{\text{d}} = 0°$ and inside the parking place. The error is $e = \phi - \phi_{\text{d}}$ and its derivative $\dot{e}$ is also calculated.

The present RC-SMFLC control algorithm involves the following rules:

R$_1$: IF $\text{sgn}(e(t)\dot{e}(t)) < 0$ AND the previous control action was to increase the control signal, THEN keep on increasing the control signal.
R$_2$: IF $\text{sgn}(e(t)\dot{e}(t)) < 0$ AND the previous control action was to decrease the control signal, THEN keep on decreasing the control signal.
R$_3$: IF $\text{sgn}(e(t)\dot{e}(t)) > 0$ AND the previous control action was to increase the control signal, THEN set $\psi = 0°$ and decrease ψ.
R$_4$: IF $\text{sgn}(e(t)\dot{e}(t)) > 0$ AND the previous control action was to decrease the control signal, THEN set $\psi = 0°$ and increase ψ.

To increase the flexibility in the vehicle's maneuvering, and to accelerate the convergence to the desirable angle $\phi_{\text{d}} = 0°$, the width of the fuzzy subsets describing the steering angle has to be modified each time the vehicle reaches the boundaries of the parking area. Thus, two more rules are employed:

R$_5$: IF the vehicle reaches the FRONT or the INSIDE boundary, THEN increase $\Delta\psi$ and change $\text{sgn}(u)$.
R$_6$: IF the vehicle reaches the REAR or the OUTSIDE boundary, THEN decrease $\Delta\psi$ and change $\text{sgn}(u)$.

An increase of $\Delta\psi$ means that the error vector $[e, \dot{e}]^{\text{T}}$ lies far from the desirable value $[0, 0]^{\text{T}}$, and so bigger changes of the control signal ψ are required to speed up convergence. A decrease of $\Delta\psi$ means that the error vector $[e, \dot{e}]^{\text{T}}$ lies near the desirable value $[0, 0]^{\text{T}}$, thus more subtle changes of the control signal ψ are required to speed up convergence (Figure 8.29) [22].

It is noted that unlike the conventional RC-SMFLC, described in Section 8.4.3.3, here the error e never changes sign. Also, between two consecutive control actions, the immediate change of sign of the control signal ψ is prohibited. Therefore, to switch from an *increase* to a *decrease* control action and vice versa, the steering angle ψ must first be set to $\psi = 0°$.

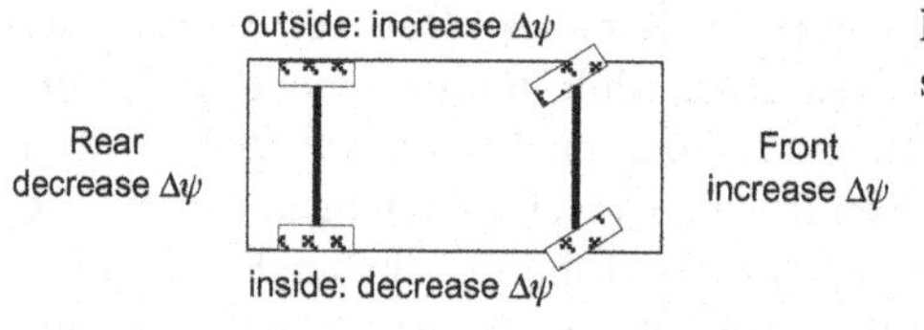

Figure 8.29 Adaptation of the fuzzy subsets describing steering angle.

8.5 Neural Control of Mobile Robots

In this section, the control of WMRs will be considered using the two types of neural networks presented in Section 8.2.2, namely:

- MLP network
- RBF network

In the literature, there are also used other types of NNs (wavelet networks, recurrent networks, self-organizing maps, local neural networks, etc.) or NFSs [24,25]. For all of them, the methodology is practically the same as that developed in this section.

8.5.1 Adaptive Tracking Controller Using MLP Network

The two-step (backstepping) procedure used in Section 8.4.1 for designing adaptive fuzzy controllers will be followed, namely:

- Kinematic controller design, which produces the auxiliary velocity control inputs $v_{\mathrm{m}}(t)$ and $\omega_{\mathrm{m}}(t)$ that assure asymptotic convergence of the robot trajectory $\mathbf{q}(t) = [x(t), y(t), \phi(t)]^{\mathrm{T}}$ to the desired one: $\mathbf{q}_{\mathrm{d}}(t) = [x_{\mathrm{d}}(t), y_{\mathrm{d}}(t), \phi_{\mathrm{d}}(t)]^{\mathrm{T}}$.
- Dynamic controller design, which produces the torque control inputs that assure asymptotic tracking of $v_{\mathrm{m}}(t)$ and $\omega_{\mathrm{m}}(t)$ which are used as reference inputs to the dynamic control subsystem.

The kinematic controller is (see Eq. (8.18)):

$$\begin{aligned} v_{\mathrm{m}} &= v_{\mathrm{d}} \cos \varepsilon_3 + K_1 \varepsilon_1 \\ \omega_{\mathrm{m}} &= \omega_{\mathrm{d}} + K_2 v_{\mathrm{d}} \varepsilon_2 + K_3 \sin \varepsilon_3 \end{aligned} \tag{8.44}$$

The dynamic controller was derived in Section 5.4.2 for the dynamic model ((5.49a) and (5.49b)) with exactly the known dynamic parameters (m, I, etc.), in Example 7.1 for the case of unknown constant parameters using the parameter adaptation law (7.77), and in Section 8.4.1 using the simple proportional fuzzy control law that has nine fuzzy control rules, as given in Table 8.2.

Here, the dynamic control job will be accomplished using a neural network controller in place of the fuzzy controller [26,27]. For more generality, we will consider the unconstrained dynamic model of the nonholonomic WMR described by Eqs. ((3.19a) and (3.19b)):

$$\overline{\mathbf{D}}(\mathbf{q})\dot{\mathbf{v}} + \overline{\mathbf{C}}(\mathbf{q}, \dot{\mathbf{q}})\mathbf{v} + \overline{\mathbf{g}}(\mathbf{q}) = \overline{\mathbf{E}}\boldsymbol{\tau}$$

omitting the gravitational term $\overline{\mathbf{g}}(\mathbf{q})$ which for the WMR is zero, and adding a linear friction term $\overline{\mathbf{B}}(\mathbf{v})$, that is:

$$\overline{\mathbf{D}}(\mathbf{q})\dot{\mathbf{v}} + \overline{\mathbf{C}}(\mathbf{q}, \dot{\mathbf{q}})\mathbf{v} + \overline{\mathbf{B}}(\mathbf{v}) = \overline{\boldsymbol{\tau}} \tag{8.45}$$

where $\overline{\boldsymbol{\tau}} = \mathbf{E}\boldsymbol{\tau}$ and $\mathbf{v} = [v, \omega]^{\mathrm{T}}$.

For convenience, the kinematic controller (8.44) is expressed as:

$$\mathbf{v}_{\mathrm{m}} = \mathbf{f}_{\mathrm{m}}(\boldsymbol{\varepsilon}_{\mathrm{m}}, \mathbf{v}_{\mathrm{d}}, \mathbf{K}), \quad \mathbf{K} = [K_1, K_2, K_3]^{\mathrm{T}} \tag{8.46}$$

Differentiating $\mathbf{v}_{\mathrm{m}}$ we get:

$$\begin{bmatrix} \dot{v}_{\mathrm{m}} \\ \dot{\omega}_{\mathrm{m}} \end{bmatrix} = \begin{bmatrix} \dot{v}_{\mathrm{d}} \cos \varepsilon_3 \\ \dot{\omega}_{\mathrm{d}} + K_2 \dot{v}_{\mathrm{d}} \varepsilon_2 \end{bmatrix} + \begin{bmatrix} K_1 & 0 & -v_{\mathrm{d}} \sin \varepsilon_3 \\ 0 & K_2 v_{\mathrm{d}} & K_3 \cos \varepsilon_3 \end{bmatrix} \begin{bmatrix} \dot{\varepsilon}_1 \\ \dot{\varepsilon}_2 \\ \dot{\varepsilon}_3 \end{bmatrix}$$

Now, if $v_{\mathrm{d}}(t) = v_{\mathrm{d}} = \text{const.}$ and $\omega_{\mathrm{d}}(t) = \omega_{\mathrm{d}} = \text{const.}$, then:

$$\begin{bmatrix} \dot{v}_{\mathrm{m}} \\ \dot{\omega}_{\mathrm{m}} \end{bmatrix} = \begin{bmatrix} K_1 & 0 & -v_{\mathrm{d}} \sin \varepsilon_3 \\ 0 & K_2 v_{\mathrm{d}} & K_3 \cos \varepsilon_3 \end{bmatrix} \dot{\boldsymbol{\varepsilon}} \tag{8.47}$$

and the feedback control input vector $\mathbf{u} = [u_1, u_2]^{\mathrm{T}}$ is selected as:

$$\mathbf{u} = \dot{\mathbf{v}}_{\mathrm{m}} + \mathbf{K}_4(\mathbf{v}_{\mathrm{m}} - \mathbf{v}) \tag{8.48}$$

where $\mathbf{K}_4$ is a diagonal positive definite matrix:

$$\mathbf{K}_4 = k_4 \mathbf{I}_{2 \times 2} \tag{8.49}$$

If $v_{\mathrm{d}}(t)$ and $\omega_{\mathrm{d}}(t)$ are varying, there is no change in the form of the control law (8.47). The two-step controller defined by Eqs. (8.46) and (8.48) assures asymptotic tracking of the desired state trajectory $\mathbf{q}_{\mathrm{d}}(t) = [x_{\mathrm{d}}(t), y_{\mathrm{d}}(t), \phi_{\mathrm{d}}(t)]^{\mathrm{T}}$. This can be shown as usual by using the extended Lyapunov function:

$$V = K_1(\varepsilon_1^2 + \varepsilon_2^2) + (2K_1/K_2)(1 - \cos \varepsilon_3) + (1/2K_4)[\varepsilon_4^2 + (K_1/K_2K_3)\varepsilon_5^2]$$

where ε_4 and ε_5 are the components of $\tilde{\mathbf{v}} = \mathbf{v}_{\mathrm{m}} - \mathbf{v}$.

We now proceed to show how to use the neural net for the design of the adaptive neurocontroller. Differentiating the velocity tracking error $\tilde{\mathbf{v}} = \mathbf{v}_{\mathrm{m}} - \mathbf{v}$ and using Eq. (8.45), we get the robot dynamics in terms of $\tilde{\mathbf{v}}$, namely:

$$\overline{\mathbf{D}}(\mathbf{q})\dot{\tilde{\mathbf{v}}} = -\overline{\mathbf{C}}(\mathbf{q}, \dot{\mathbf{q}})\tilde{\mathbf{v}} + \mathbf{F}(\mathbf{x}) - \overline{\boldsymbol{\tau}} \tag{8.50}$$

where:

$$\mathbf{F}(\mathbf{x}) = \overline{\mathbf{D}}(\mathbf{q})\dot{\mathbf{v}}_{\mathrm{m}} + \overline{\mathbf{C}}(\mathbf{q}, \dot{\mathbf{q}})\mathbf{v}_{\mathrm{m}} + \overline{\mathbf{B}}(\mathbf{v}) \tag{8.51}$$

The vector needed for the computation of $\mathbf{F}(\mathbf{x})$ is:

$$\mathbf{x} = \begin{bmatrix} v \\ v_{\mathrm{m}} \\ \dot{v}_{\mathrm{m}} \end{bmatrix}$$

which is available (measured). The function $\mathbf{F}(\mathbf{x})$ involves all the dynamic parameters of the robot (masses, moments of inertia, friction coefficient, etc.) that in practice are unknown or involve uncertainties.

From Eq. (8.50), we see that a proper feedback control law is:

$$\overline{\boldsymbol{\tau}} = \hat{\mathbf{F}}(\mathbf{x}) + \mathbf{K}_4\tilde{\mathbf{v}} - \boldsymbol{\mu} \tag{8.52}$$

where $\hat{\mathbf{F}}(\mathbf{x})$ is an estimate of $\mathbf{F}(\mathbf{x}), \mathbf{K}_4 = \mathrm{diag}[K_{41}, K_{42}]$ with $K_{41} > 0$, $K_{42} > 0$, and $\boldsymbol{\mu}$ is a robustifying term required to compensate any unmodeled structural disturbances.

Introducing Eq. (8.52) into Eq. (8.50), we get the closed-loop system:

$$\overline{\mathbf{D}}(\mathbf{q})\dot{\tilde{\mathbf{v}}} = -[\overline{\mathbf{C}}(\mathbf{q}, \dot{\mathbf{q}}) + \mathbf{K}_4]\tilde{\mathbf{v}} + \mathbf{F}(\mathbf{x}) - \hat{\mathbf{F}}(\mathbf{x}) + \boldsymbol{\mu} \tag{8.53}$$

Clearly, to assure that Eq. (8.53) is asymptotically stable at $\tilde{\mathbf{v}} = \mathbf{0}$, we have to select properly $\mathbf{K}_4$, $\hat{\mathbf{F}}(\mathbf{x})$, and $\boldsymbol{\mu}$. Here, the estimate $\hat{\mathbf{F}}(\mathbf{x})$ of $\mathbf{F}(\mathbf{x})$ is provided by a neural network approximator (8.12), namely:

$$\hat{\mathbf{F}}(\mathbf{x}) = \hat{\mathbf{V}}^{\mathrm{T}}\boldsymbol{\sigma}(\hat{\mathbf{W}}^{\mathrm{T}}\mathbf{x}) + \boldsymbol{\varepsilon} \tag{8.54}$$

in which case Eq. (8.52) becomes:

$$\overline{\boldsymbol{\tau}} = \hat{\mathbf{V}}^{\mathrm{T}}\boldsymbol{\sigma}(\hat{\mathbf{W}}^{\mathrm{T}}\mathbf{x}) + \mathbf{K}_4\tilde{\mathbf{v}} - \boldsymbol{\mu} + \boldsymbol{\varepsilon}$$

and the closed-loop system becomes:

$$\overline{\mathbf{D}}(\mathbf{q})\dot{\tilde{\mathbf{v}}} = -[\overline{\mathbf{C}}(\mathbf{q}, \dot{\mathbf{q}}) + \mathbf{K}_4]\tilde{\mathbf{v}} + \mathbf{V}^{\mathrm{T}}\boldsymbol{\sigma}(\mathbf{W}^{\mathrm{T}}\mathbf{x}) - \hat{\mathbf{V}}^{\mathrm{T}}\boldsymbol{\sigma}(\hat{\mathbf{W}}^{\mathrm{T}}\mathbf{x}) - \boldsymbol{\varepsilon} + \boldsymbol{\mu} \tag{8.55}$$

Now, we only have to select the neural network parameter updating (tuning) and the robustifying term $\boldsymbol{\mu}$. Choosing $\boldsymbol{\mu}(t)$ as:

$$\boldsymbol{\mu}(t) = -\overline{\mathbf{K}}_{\mu}\tilde{\mathbf{v}} \tag{8.56a}$$

we get:

$$\overline{\mathbf{D}}(\mathbf{q})\dot{\tilde{\mathbf{v}}} = -[\overline{\mathbf{C}}(\mathbf{q},\dot{\mathbf{q}}) + \mathbf{K}_4 + \mathbf{K}_\mu]\tilde{\mathbf{v}} + \mathbf{V}^{\mathrm{T}}\boldsymbol{\sigma}(\mathbf{W}^{\mathrm{T}}\mathbf{x}) - \hat{\mathbf{V}}^{\mathrm{T}}\boldsymbol{\sigma}(\hat{\mathbf{W}}^{\mathrm{T}}\mathbf{x}) - \varepsilon \tag{8.56b}$$

Then, we can show that updating the NN weights as (Section 8.5.3):

$$\dot{\hat{\mathbf{V}}} = \mathbf{A}\hat{\boldsymbol{\sigma}}\tilde{\mathbf{v}}^{\mathrm{T}} - \mathbf{A}\hat{\boldsymbol{\sigma}}'\hat{\mathbf{W}}^{\mathrm{T}}\mathbf{x}\tilde{\mathbf{v}}^{\mathrm{T}} - \lambda\mathbf{A}||\tilde{\mathbf{v}}||\hat{\mathbf{V}} \tag{8.57}$$

$$\dot{\hat{\mathbf{W}}} = \mathbf{B}\mathbf{x}(\hat{\boldsymbol{\sigma}}'^{\mathrm{T}}\hat{\mathbf{V}}\tilde{\mathbf{v}})^{\mathrm{T}} - \lambda\mathbf{B}||\tilde{\mathbf{v}}||\hat{\mathbf{W}} \tag{8.58}$$

where $\boldsymbol{\sigma}'$ is the hidden-layer gradient corresponding to $\boldsymbol{\sigma}$, $\mathbf{A}$ and $\mathbf{B}$ are positive definite design matrices, and $\lambda > 0$, stabilizes $\tilde{\mathbf{v}}$ to an invariant region around $\tilde{\mathbf{v}} = \mathbf{0}$. For the sigmoid function $\boldsymbol{\sigma}$, $\boldsymbol{\sigma}'$ is given by:

$$\boldsymbol{\sigma}'(\hat{\mathbf{z}}) = [\partial\boldsymbol{\sigma}(z)/\partial z]_{z=\hat{z}} = \mathrm{diag}\{\boldsymbol{\sigma}(\hat{\mathbf{W}}^{\mathrm{T}}\mathbf{x})\}[\mathbf{I} - \mathrm{diag}\{\boldsymbol{\sigma}(\hat{\mathbf{W}}^{\mathrm{T}}\mathbf{x})\}]$$

The block diagram of the overall adaptive neurocontroller is shown in Figure 8.30 [26,27].

Example 8.6

It is desired to design a fuzzy logic adaptive controller using the above neural network control scheme, along with the fuzzy universal approximator (8.14).

Solution

The analysis of the controller is exactly the same as that described in Section 8.5.1, up to the design of the controller (8.52), where $\hat{\mathbf{F}}(\mathbf{x})$ is an estimate of the dynamics function $\mathbf{F}(\mathbf{x})$. This estimate can be computed using Eq. (8.17) as:

$$\hat{\mathbf{F}}(\mathbf{x}) = \Psi(\mathbf{x})\hat{\beta} \tag{8.59}$$

where $\mathbf{x} = [v, v_m, \dot{v}_m]^{\mathrm{T}}, \Psi(\mathbf{x})$ is the FBF matrix with elements given by Eq. (8.13), and $\hat{\beta}$ is the approximator's parameter vector estimate that must be updated as the time passes. The adaptation law has the standard form:

$$\dot{\hat{\beta}} = \boldsymbol{\Gamma}\Psi^{\mathrm{T}}(\mathbf{x})\tilde{\mathbf{v}} - \lambda\boldsymbol{\Gamma}||\tilde{\mathbf{v}}||\hat{\beta}, \quad \tilde{\mathbf{v}} = \mathbf{v}_m - \mathbf{v} \tag{8.60}$$

Here, $\boldsymbol{\Gamma}$ is a selected positive constant matrix (e.g., $\boldsymbol{\Gamma} = \mathrm{diag}[\gamma_1, \gamma_2]$) and λ is a small positive constant. Using the standard Lyapunov stability theory, of adaptive control and adaptation laws, it follows that the estimator (8.60) combined with the controller (8.52) assures overall stability to $\tilde{\mathbf{v}} = 0$ (actually to a small vicinity of $\tilde{\mathbf{v}} = 0$). The structure of this adaptive controller is the same as that of NN controller with the neural estimator block being replaced by the fuzzy estimator (Figure 8.30).

8.5.2 Adaptive Tracking Controller Using RBF Network

The RBF NNs can be used in a way similar to MLP networks and the fuzzy universal approximator [28]. The approximation of the function $\mathbf{F}(\mathbf{x})$ in Eq. (8.51) is performed by the relation (8.10) for the jth component $y_j(\mathbf{x}) = F_j(\mathbf{x}),\ j = 1, 2, \ldots, p$:

$$F_j(\mathbf{x}) = \sum_{i=1}^{m} w_{ji}\phi_i(\mathbf{x}) \tag{8.61}$$

of $\mathbf{F}(\mathbf{x})$, where:

$$\phi_i(\mathbf{x}) = \phi(||\mathbf{x} - \mathbf{c}_i||) = \phi(r_i) = \exp(-r_i^2/2\sigma_i^2), \quad r_i \geq 0, \quad \sigma_i \geq 0 \tag{8.62}$$

The algorithm for updating the weights w_{ki}, the RBF centers c_i, and the widths σ_i involves the four steps described in Section 8.2.2.4. In general, the weight $w_i(k+1)$ and center $c_i(k+1)$ updating algorithm can be either the gradient descent law, or a second-order Newton-like updating law, which minimizes a squared error function:

$$J = \frac{1}{2}\sum_{k=1}^{n} \varepsilon^2(k) \tag{8.63a}$$

where:

$$\varepsilon(k) = y(k) - \sum_{i=1}^{m} w_i\phi(||x(k) - c_i||)^2 \tag{8.63b}$$

with input–output data $\{x(k), y(k)\}$.

Therefore, in the case of the gradient descent updating law, we have:

$$w_i(k+1) = w_i(k) - \eta_w \frac{\partial\varepsilon(k)}{\partial w_i(k)} \tag{8.64a}$$

$$c_i(k+1) = c_i(k) - \eta_c \frac{\partial\varepsilon(k)}{\partial c_i(k)} \tag{8.64b}$$

where η_w and η_c are design parameters that control the rate of convergence. Many alternative ways exist in the literature for updating (adapting) the RBF parameters and weights.

Overall, the RBF representation (8.61) is similar to the fuzzy universal approximator and can be used in the same manner for estimating $\mathbf{F}(\mathbf{x})$ required by the control law (8.52). This estimator is placed in the NN/FLC box of the overall control scheme shown in Figure 8.30.

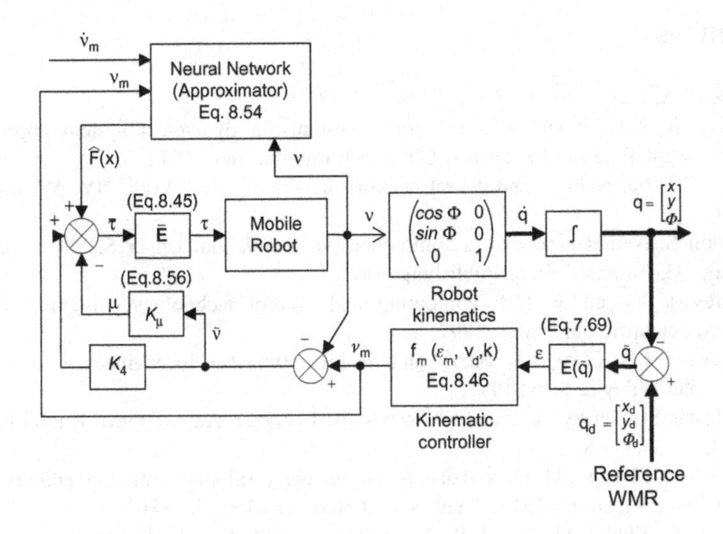

Figure 8.30 Structure of overall adaptive neurocontroller.

8.5.3 Appendix: Proof of Neurocontroller Stability

Here, a brief sketch of the proof of asymptotic convergence to an invariant set around the origin of the closed-loop error, obtained using the controller (Eqs. (8.52), (8.56a), (8.57), (8.58)), will be presented (omitting the lengthy calculations). We assume that the universal approximator Eq. (8.12) is valid with a given accuracy ε^* for all $\mathbf{x}$ in a compact set U_x, and consider the candidate Lyapunov function:

$$V_0 = \frac{K_2}{2}(\varepsilon_1^2 + \varepsilon_2^2) + \frac{1}{2}\varepsilon_3^2 + V_0', \quad V_0' = \frac{1}{2}\left[\tilde{\mathbf{v}}_{\mathrm{m}}^{\mathrm{T}}\overline{\mathbf{D}}\tilde{\mathbf{v}}_{\mathrm{m}} + tr\left\{\tilde{\mathbf{V}}^{\mathrm{T}}\mathbf{A}^{-1}\tilde{\mathbf{V}}\right\} + tr\left\{\tilde{\mathbf{W}}^{\mathrm{T}}\mathbf{B}^{-1}\tilde{\mathbf{W}}\right\}\right]$$

$$\tilde{\mathbf{V}} = \mathbf{V} - \hat{\mathbf{V}}, \tilde{\mathbf{W}} = \mathbf{W} - \hat{\mathbf{W}}$$

The derivative of V_0 is given by:

$$\dot{V}_0 = K_2(\varepsilon_1\dot{\varepsilon}_1 + \varepsilon_2\dot{\varepsilon}_2) + \varepsilon_3\dot{\varepsilon}_3 + \dot{V}_0'$$

Introducing $\dot{\varepsilon}_1, \dot{\varepsilon}_2$, and $\dot{\varepsilon}_3$ and computing $\dot{V}_0'$, we find (after the calculations) that the condition:

$$||\tilde{\mathbf{v}}_{\mathrm{m}}|| > \rho_{\mathrm{m}}$$

where ρ_{m}, a properly selected bound, assures that $\dot{V}_0 < 0$ outside the compact set U_x. According to the LaSalle invariant set stability condition (see Section 6.2.3), this means that $\tilde{\mathbf{v}}$ and $\tilde{\boldsymbol{\varepsilon}}$ (i.e., $\tilde{\mathbf{x}}$) converge to their corresponding invariant sets associated with U_x [26,27].

References

[1] Zadeh LA. Fuzzy sets. Inf Control 1965;8:338–53.

[2] Kosko B. Neural networks and fuzzy systems: a dynamical system approach to machine intelligence. Englewood Cliffs, NJ: Prentice Hall; 1992.

[3] Chen CH. Fuzzy logic and neural network handbook. New York, NY: McGraw-Hill; 1996.

[4] Haykin S. Neural networks: a comprehensive foundation. Upper Saddle River, New Jersey: Macmillan College Publishing; 1994.

[5] Tzafestas SG, editor. Soft computing and control technology. Singapore/London: World Scientific Publishers; 1997.

[6] Tsoukalas LH, Uhrig RE. Fuzzy and neural approaches in engineering. New York, NY: John Wiley & Sons; 1997.

[7] Tzafestas SG. Fuzzy systems and fuzzy expert control: An overview. Knowl Eng Rev 1994;9(3):229–68.

[8] Wang LX, Mendel JM. Fuzzy basis functions, universal approximation, and orthogonal least-squares learning. IEEE Trans Neural Networks 1992;3(5):807–14.

[9] Omatu S, Khalid M, Yusof R. Neuro-control and its applications. London/Berlin: Springer; 1996.

[10] Das T, Narayan Kar I. Design and implementation of a adaptive fuzzy logic-based controller for wheeled mobile robots. IEEE Trans Control Syst Technol 2006;14 (3):501–10.

[11] Castillo O, Aguilar LT, Cárdenas S. Fuzzy logic tracking control for unicycle mobile robots. Eng Lett 2006;13(2): [EL. 13-2-4:73-7].

[12] Driesen BJ, Feddema JT, Kwok KS. Decentralized fuzzy control of multiple nonholonomic vehicles. J Intell Robot Syst 1999;26:65–78.

[13] Balluchi A, Bicchi A, Balestrino A, Casalino G. Path tracking control for Dubins cars. Proceedings of 1996 IEEE international conference on robotics and automation, Minneapolis, MI; April 1996. p. 3123–28.

[14] Moustris G, Tzafestas SG. A robust fuzzy logic path tracker for non-holonomic mobile robots. J Artif Intell Tools 2005;14(6):935–65.

[15] Moustris G, Tzafestas SG. Switching fuzzy tracking control for the Dubins car. Control Eng Practice 2011;19(1):45–53.

[16] Deliparaschos KM, Moustris GP, Tzafestas SG. Autonomous SoC for fuzzy robot path tracking. Proceedings of the European control conference. Kos, Greece; July 2–5, 2007. p. 5471-78.

[17] Moustris GP, Deliparaschos KM, Tzafestas SG. Tracking control using the strip-wise affine transformation: an experimental SoC design. Proceedings of the European control conference. Budapest, Hungary; August 23–26, 2009. [paper MoC3.5.]

[18] Tzafestas SG, Deliparaschos KM, Moustris GP. Fuzzy logic path tracking control for autonomous non-holonomic robots: design of system on a chip. Rob Auton Syst 2010;58:1017–27.

[19] Moustris GP, Deliparaschos KM, Tzafestas SG. Feedback equivalence and control of mobile robots through a scalable FPGA architecture. In: Velenivov Topalov A, editor. Recent advances in mobile robotics. In Tech; 2011<www.interchopen.com/books>.

[20] Rigatos GG, Tzafestas CS, Tzafestas SG. Mobile robot motion control in partially unknown environments using a sliding-mode fuzzy-logic controller. Robot Autom Syst 2000;33:1–11.

[21] Kyriakopoulos KJ, Saridis GN. Optimal and Suboptimal motion planning for collision avoidance of mobile robots in non-stationary environments. J Intell Robot Syst 1995;11(3):223–67.

[22] Rigatos GG, Tzafestas SG, Evangelidis GJ. Reactive parking control of nonholonomic vehicles via a fuzzy learning automaton. IEE Proc Control Theory Appl 2001;148 (2):169–79.

[23] Luzeaux D. Parking maneuvers and trajectory tracking. Proceedings of the third international workshop on advanced motion control. Berkeley, CA: University of California; 1994.

[24] De Oliveira VM, De Pieri ER, Lages WF. Wheeled mobile robot using sliding modes and neural networks: learning and nonlinear models. Review Soc. Brasileria de Redes Neurais 2003;1(2):103–21.

[25] Oubbati M, Schanz M, Levi P. Kinematic and dynamic adaptive control of a nonholonomic mobile robot using a RNN. Proceedings of the IEEE symposium on computational intelligence in robotics and automation (CIRA'05). Espoo, Helsinki, Finland; June 27–30, 2005. p. 27–33.

[26] Lewis FL, Campos J, Selmic R. Neuro-fuzzy control of industrial systems with actuator nonlinearities. Philadelphia, PA: SIAM; 2002.

[27] Fierro R, Lewis FL. Control of a nonholonomic mobile robot: Backstepping kinematics into dynamics. J Robot Syst 1997;14(3):149–63.

[28] Bayar G, Konukseven EI, Koku AB. Control of a differentially driven mobile robot using radial basis function based neural networks. WSEAS Trans Syst Control 2008;3 (12):1002–13.

9 Mobile Robot Control V: Vision-Based Methods

9.1 Introduction

Vision is an important robotic sensor since it can be used for environment measurement without physical contact. *Visual robot control* or *visual servoing* is a feedback control methodology that uses one or more vision sensors (cameras) to control the motion of the robot. Specifically, the control inputs for the robot motors are produced by processing image data (typically, extraction of contours, features, corners, and other visual primitives). In robotic manipulators, the purpose of visual control is to control the *pose* of the robot's end-effector relative to a target object or a set of target features. In mobile robots, the vision controller's task is to control the *vehicle's pose* with respect to some landmarks. Tracking stability can be assured only if the vision sensing delays are sufficiently small and/or the dynamic model of the robot has a sufficient accuracy. Over the years many techniques were developed for compensating this delay of the visual system in robot control. A rich literature has been oriented to the control of nonholonomic systems in order to handle various challenging problems associated with vision-based control.

The objectives of this chapter are as follows:

- To present the basic aspects of visual servoing
- To discuss the position-based and image-based visual-control problems
- To apply visual servoing to a number of selected mobile robot control problems
- To study the mobile robot visual servoing using omnidirectional vision

9.2 Background Concepts

9.2.1 Classification of Visual Robot Control

Visual robot controllers (VRCs) depend on whether the vision system provides set-points as input to the robot joint controllers, or computes directly the joint level inputs, and whether the error signal is determined in task space coordinates or directly in terms of image features.

Therefore, VRCs are classified in the following four categories [1–8]:

- *Dynamic look-and-move system*—Here the control structure is hierarchical in which the vision system provides set-point inputs to the joint controllers and so the robot is internally controlled using joint feedback.

Introduction to Mobile Robot Control. DOI: http://dx.doi.org/10.1016/B978-0-12-417049-0.00009-2

- *Direct visual servo control*—Here the robot joint controller is eliminated and replaced by a visual servo controller which directly computes the inputs of the joints, and stabilizes the robot using only vision signals. Actually, most implemented VRCs are of the look-and-move type because internal feedback with a high sampling rate provides the visual controller with an accurate axis dynamic model. Also, look-and-move control separates the kinematics singularities of the system from the visual controller, and bypasses the low sampling rates at which the direct visual control can work.
- *Position-based visual robot control (PBVRC)*—Here, use is made of features extracted from the image and used together with a geometric model of the target and the available camera model to determine the pose of the target with respect to the camera. Thus, the feedback loop is closed using the error in the estimated pose space.
- *Image-based visual robot control (IBVRC)*—Here, direct computation of the control signals is performed using the image features. IBVRC reduces the computational time, does not need image interpretation, and eliminates the errors of sensors' modeling and camera calibration. But its implementation is more difficult due to the complex nonlinear dynamics of robots.

The PBVRC and IBVRC control schemes have the structure shown in Figure 9.1.

9.2.2 Kinematic Transformations

Kinematic (homogeneous) transformations were discussed in Section 2.2.2. The task space of a robot (fixed or mobile) is denoted by C_s and describes the set of positions and orientations attainable by a mobile robot or by the end-effector of a fixed robot. Robotic tasks are typically specified with respect to specific coordinate frames (e.g., the camera frame, the target/object frame). As seen in Section 2.2.2,

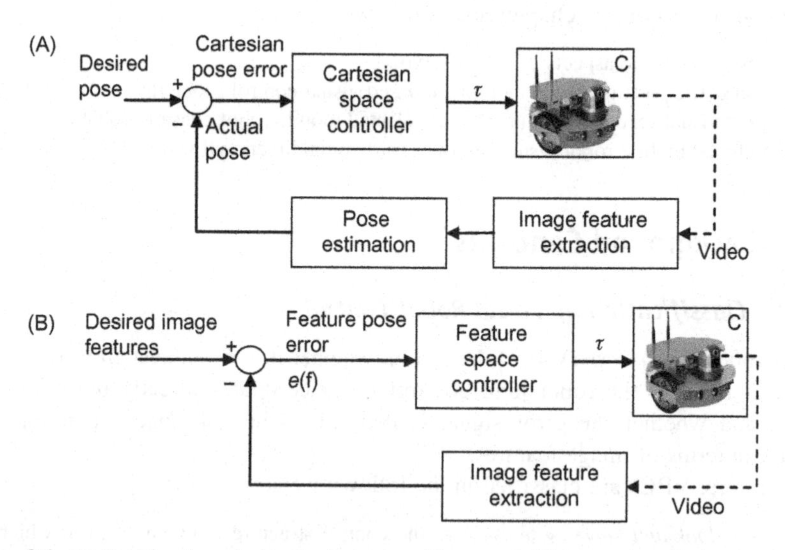

Figure 9.1 (A) Position-based visual robot control loop and (B) image-based visual robot control loop.

the notation $\mathbf{A}_i^j$ represents both a coordinate transformation or a pose, involving a rotation matrix $\mathbf{R}_i^j$ and a translation $\mathbf{p}_i^j$ (or $\mathbf{d}_i^j$) (see Eqs. (2.9) and (2.10)). A multiple coordinate transformation is used to obtain a desired multistep change of coordinates as shown by Eq. (2.18).

In vision-based robot control, the following coordinate frames are typically needed:

- the target coordinate frame $\mathbf{A}_t$ attached to the object (target),
- the world coordinate frame $\mathbf{A}_0$ (attached to a given fixed point of the workspace),
- the camera coordinate frame $\mathbf{A}_{ci}$ attached to the ith camera,
- the robot frame $\mathbf{A}_r$ (attached to a fixed point of the robot body).

Consider a fixed-robot manipulator working in a 3D space. The motion of its end-effector is described in world coordinates by a translational velocity $\mathbf{v}(t)$ and an angular velocity $\boldsymbol{\omega}(t)$, where:

$$\begin{aligned}\mathbf{v}(t) &= [v_x, v_y, v_z]^{\mathrm{T}} \\ \boldsymbol{\omega}(t) &= [\omega_x, \omega_y, \omega_z]^{\mathrm{T}}\end{aligned} \tag{9.1}$$

Let $\mathbf{p} = [x,y,z]^{\mathrm{T}}$ be a point rigidly attached to the end-effector, where x,y,z are the world coordinates of $\mathbf{p}$. Then, $\dot{\mathbf{p}}$ is given by:

$$\dot{\mathbf{p}} = \boldsymbol{\omega} \times \mathbf{p} + \mathbf{v} \tag{9.2a}$$

where $\boldsymbol{\omega} \times \mathbf{p}$ is the cross product of $\boldsymbol{\omega}$ and $\mathbf{p}$, that is:

$$\boldsymbol{\omega} \times \mathbf{p} = \begin{bmatrix} \omega_y z - y\omega_z \\ \omega_z x - z\omega_x \\ \omega_x y - x\omega_y \end{bmatrix} \tag{9.2b}$$

The combined velocity vector $\dot{\mathbf{r}}$, where:

$$\dot{\mathbf{r}}(t) = \begin{bmatrix} \mathbf{v} \\ \boldsymbol{\omega} \end{bmatrix} \tag{9.3}$$

is called the *velocity screw* (or *velocity twist*) of the robotic end-effector. In compact form, Eqs. (9.2a) and (9.2b) can be written as:

$$\dot{\mathbf{p}}(t) = \mathbf{J}_0(\mathbf{p})\dot{\mathbf{r}}, \quad \mathbf{J}_0(\mathbf{p}) = [\mathbf{I}_{3\times 3} \vdots \mathbf{S}(\mathbf{p})] \tag{9.4a}$$

where $\mathbf{S}(\mathbf{p})$ is the *skew symmetric* matrix:

$$\mathbf{S}(\mathbf{p}) = \begin{bmatrix} 0 & z & -y \\ -z & 0 & x \\ y & -x & 0 \end{bmatrix} \tag{9.4b}$$

The platform of mobile robots is moving with a linear velocity $\mathbf{v}(t) = [v_x, v_y, 0]^T$ and angular velocity $\boldsymbol{\omega}(t) = [0,0,\omega]^T$ $(\omega = \omega_z)$. Therefore Eq. (9.2b) becomes:

$$\boldsymbol{\omega} \times \mathbf{p} = \begin{bmatrix} -y\omega \\ x\omega \\ 0 \end{bmatrix}$$

and (9.2a) gives:

$$\begin{bmatrix} \dot{x} \\ \dot{y} \\ \dot{z} \end{bmatrix} = \begin{bmatrix} -y\omega \\ x\omega \\ 0 \end{bmatrix} + \begin{bmatrix} v_x \\ v_y \\ 0 \end{bmatrix} \quad \text{or} \quad \begin{bmatrix} \dot{x} \\ \dot{y} \end{bmatrix} = \begin{bmatrix} v_x - y\omega \\ v_y + x\omega \end{bmatrix}$$

with $z = \text{constant} = 0$. Embedding the relation $\dot{\phi} = \omega$ in the above equation, we get:

$$\begin{bmatrix} \dot{x} \\ \dot{y} \\ \dot{\phi} \end{bmatrix} = \begin{bmatrix} 1 & 0 & -y \\ 0 & 1 & x \\ 0 & 0 & 1 \end{bmatrix} \dot{\mathbf{r}}, \quad \dot{\mathbf{r}} = \begin{bmatrix} v_x \\ v_y \\ \omega \end{bmatrix} \tag{9.4c}$$

If the wheeled mobile robot (WMR) involves a steering angle ψ, then the corresponding equation $\dot{\psi} = \omega_\psi$ must be added to Eq. (9.4c) as a fourth line.

9.2.3 Camera Visual Transformations

As discussed in Section 4.5.1, each camera involves a lens that forms a 2D projection of the scene on the image plane where the sensor is located. Due to this projection, *depth information is lost*, and so each point on the image plane corresponds to a ray in 3D space. Depth information may be obtained from multiple cameras, multiple views with a single camera, or knowledge of the geometric relation between several feature points on the target.

Here, the perspective projection model (presented in Section 4.5.2.4) will be used. Two other projection models are the scaled *orthographic projection* and the *affine projection*. The geometry of the perspective projection model is shown in Figure 9.2 (see Figure 4.10).

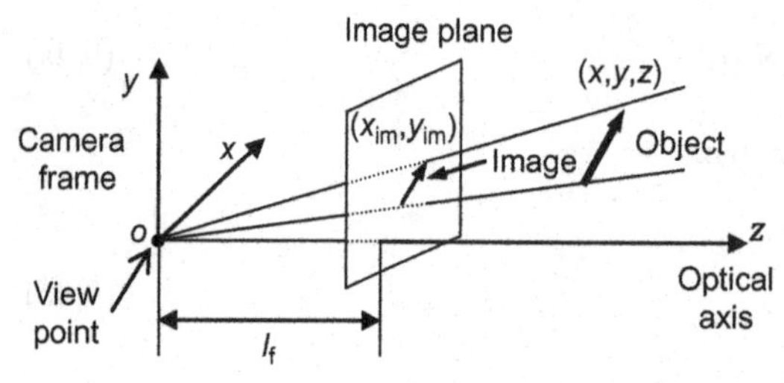

Figure 9.2 Geometry of camera lens system (l_f = focal length).

In perspective projection, a point $\mathbf{p} = [x,y,z]^{\mathrm{T}}$ whose coordinates are expressed with respect to the camera coordinate frame $\mathbf{A}_{\mathrm{c}}$, projects onto the image plane point $\mathbf{f} = [x_{\mathrm{im}}, y_{\mathrm{im}}]^{\mathrm{T}}$ given by:

$$\mathbf{f}(x,y,z) = \begin{bmatrix} x_{\mathrm{im}} \\ y_{\mathrm{im}} \end{bmatrix} = \frac{l_{\mathrm{f}}}{z}\begin{bmatrix} x \\ y \end{bmatrix} \tag{9.5}$$

If the point $\mathbf{p}$ is expressed with coordinates in an arbitrary frame $\mathbf{A}_{\mathrm{a}}$, then these coordinates must first be transformed to the camera coordinate frame.

Image feature is any structural feature that can be extracted from an image (e.g., a contour edge, corner). Usually, an image feature corresponds to the projection of a physical feature of an object (e.g., the end-effector, a goal object), onto the camera image plane. *Image feature parameter* is defined to be any real-valued numerical quantity which can be determined from one or more image features. An *image parameter vector* $\mathbf{f}$ is a vector with components $f_i (i = 1,2,\ldots,k)$ the image feature parameters, that is:

$$\mathbf{f} = [f_1, f_2, \ldots, f_k]^{\mathrm{T}} \in F \tag{9.6}$$

where F is the image parameter vector space. The mapping M from the position and orientation of the WMR or the fixed-robot end-effector to the perspective image feature parameters, that is:

$$M = C_{\mathrm{s}} \rightarrow F \tag{9.7}$$

can be found using the projective geometry of the camera. Here, the *perspective projection geometry* is used:

$$\mathbf{f} = [x_{\mathrm{im}}, y_{\mathrm{im}}]^{\mathrm{T}}$$

where x_{im} and y_{im} are given by Eq. (9.5). The actual form of Eq. (9.7) partly depends on the relative configuration of the camera and the mobile robot or end-effector. The two typical camera configurations are shown in Figure 9.3:

- Onboard camera configuration
- Fixed camera in the workspace (e.g., on the ceiling)

Clearly, in Figure 9.3 we have:

$$\mathbf{A}_0^{\mathrm{t}} = \mathbf{A}_0^{\mathrm{r}} \mathbf{A}_{\mathrm{r}}^{\mathrm{c}} \mathbf{A}_{\mathrm{c}}^{\mathrm{t}} \tag{9.8a}$$

$$\mathbf{A}_0^{\mathrm{t}} = \mathbf{A}_0^{\mathrm{c}} \mathbf{A}_{\mathrm{c}}^{\mathrm{t}} \tag{9.8b}$$

Another camera configuration is that in which the camera is not fixed in the workspace but mounted on another robot or pan-tilt head so as to observe the visually controlled robot from the best position. In all cases, camera calibration is needed before the execution of the visual-control task.

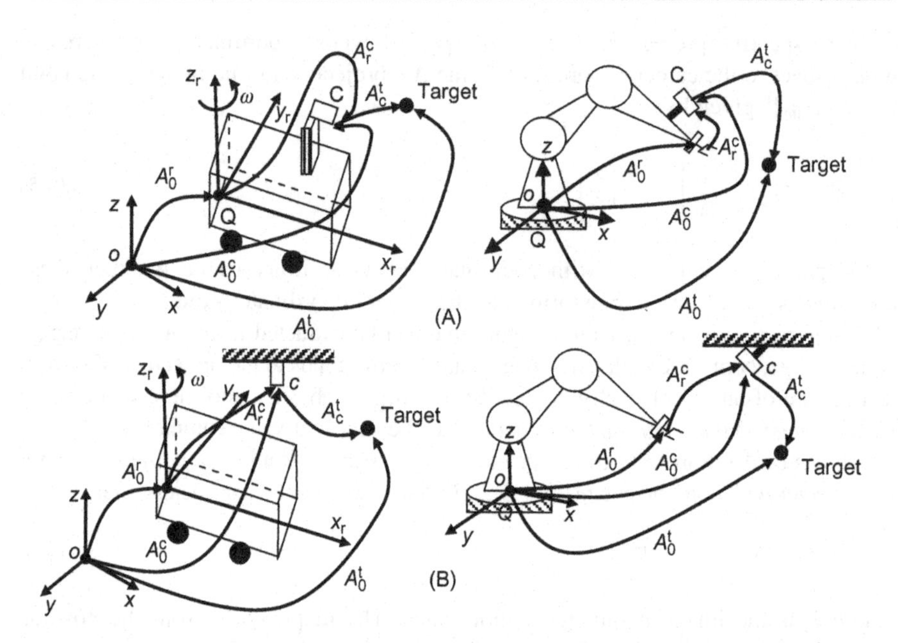

Figure 9.3 Coordinate frames used in vision-based robot control: (A) onboard/eye-in-hand configuration and (B) fixed camera configuration.

9.2.4 Image Jacobian Matrix

The discussion that follows is referred to the end-effector of a fixed robot, but it also covers the mobile robots' local coordinate frame. Given a feature parameter vector: $\mathbf{f} = [f_1, f_2, \ldots, f_k]^{\mathrm{T}} \in F$, $\dot{\mathbf{f}}(i = 1,2,\ldots,k)$ is the corresponding vector of rates of change $\dot{f}_i(i = 1,2,\ldots,k)$ of the feature parameters. Let $\mathbf{r}$ be the end-effector coordinate vector in some parameterization of the task (configuration) space C_s and $\dot{\mathbf{r}}$ the corresponding velocity, that is, the screw of the end-effector:

$$\dot{\mathbf{r}} = \begin{bmatrix} \mathbf{v} \\ \boldsymbol{\omega} \end{bmatrix} \tag{9.9}$$

The *image Jacobian* $\mathbf{J}_{\mathrm{im}}$ is defined to be the transformation of $\dot{\mathbf{r}}$ to $\dot{\mathbf{f}}$, that is:

$$\dot{\mathbf{f}} = \mathbf{J}_{\mathrm{im}}(\mathbf{r})\dot{\mathbf{r}} \tag{9.10}$$

where:

$$\mathbf{J}_{\mathrm{im}}(\mathbf{r}) = \left[\frac{\partial \mathbf{f}}{\partial \mathbf{r}}\right] = \begin{bmatrix} \partial f_1(r)/\partial r_1 \ldots \partial f_1(r)/\partial r_m \\ \cdots \\ \partial f_k(r)/\partial r_1 \ldots \partial f_k(r)/\partial r_m \end{bmatrix} \tag{9.11}$$

where m is the dimension of the space C_s. The image Jacobian matrix is also known as *interaction matrix* or *feature sensitivity matrix*. Equation (9.10) provides the feature parameters changes corresponding to a change of the robot pose. In VRC, we need to determine the end-effector velocity $\dot{\mathbf{r}}$ required to obtain a certain desired value $\dot{\mathbf{f}}_d$ of $\dot{\mathbf{f}}$.

In the following we consider point features, but other cases (such as lines, contours, areas) can be handled by properly extending the present analysis.

To compute $\mathbf{J}_{im}(\mathbf{r})$ we work as follows. Let $\mathbf{p} = [x,y,z]^T$ a point expressed with respect to the camera frame. This point is assumed to be fixed in the world coordinate frame. Then, from Eq. (9.5) we know that the image parameter vector $\mathbf{f}$ is:

$$\mathbf{f} = \begin{bmatrix} x_{im} \\ y_{im} \end{bmatrix} = \frac{l_f}{z}\begin{bmatrix} x \\ y \end{bmatrix} \tag{9.12}$$

Differentiating Eq. (9.12) we get:

$$\begin{aligned} \dot{\mathbf{f}} &= \begin{bmatrix} \dot{x}_{im} \\ \dot{y}_{im} \end{bmatrix} = \begin{bmatrix} l_f/z & 0 & -x_{im}/z \\ 0 & l_f/z & -y_{im}/z \end{bmatrix}\dot{\mathbf{p}} \\ &= \mathbf{J}_c(x_{im}, y_{im}, l_f)\dot{\mathbf{p}} \end{aligned} \tag{9.13a}$$

where:

$$\mathbf{J}_c = \begin{bmatrix} l_f/z & 0 & -x_{im}/z \\ 0 & l_f/z & -y_{im}/z \end{bmatrix} \tag{9.13b}$$

Now, $\dot{\mathbf{p}}(t)$ is given by Eqs. (9.4a) and (9.4b). Therefore, combining Eqs. (9.4a) and (9.4b) and Eqs. (9.13a) and (9.13b) we get:

$$\begin{aligned} \dot{\mathbf{f}} &= \mathbf{J}_c(x_{im}, y_{im}, z, l_f)\mathbf{J}_O(\mathbf{p})\dot{\mathbf{r}} \\ &= \mathbf{J}_{im}(x_{im}, y_{im}, z, l_f)\dot{\mathbf{r}} \end{aligned} \tag{9.14}$$

where:

$$\begin{aligned} \mathbf{J}_{im}(x_{im}, y_{im}, z, l_f) &= \mathbf{J}_c(x_{im}, y_{im}, z, l_f)\mathbf{J}_O(\mathbf{p}) \\ &= \begin{bmatrix} \dfrac{l_f}{z} & 0 & -\dfrac{x_{im}}{z} & -\dfrac{x_{im}y_{im}}{l_f} & \dfrac{l_f^2 + x_{im}^2}{l_f} & -y_{im} \\ 0 & \dfrac{l_f}{z} & -\dfrac{y_{im}}{z} & -\dfrac{l_f^2 + y_{im}^2}{l_f} & \dfrac{x_{im}y_{im}}{l_f} & x_{im} \end{bmatrix} \end{aligned} \tag{9.15}$$

This expresses the image plane velocity of a point in terms of the velocity of the end-effector with respect to the camera. The image Jacobian Eq. (9.15) depends on the distance z of the end-effector (or the target point being imaged, in general). If the target is the end-effector, this distance can be calculated using information from the camera calibration, and the forward kinematics of the robot. It is remarked that if the image feature parameters are point coordinates, then the rates $\dot{x}_{\text{im}}$ and $\dot{y}_{\text{im}}$ are image plane velocities.

Example 9.1

It is desired to derive the image Jacobian matrix of a unicycle-type WMR with a pinhole onboard camera and a target with three feature points in the camera field of view.

Solution

We will apply the procedure described in Section 9.2.4. Consider a camera-equipped WMR as shown in Figure 9.4, where the target S is identifiable via the three point features D,E,F on the (x,y) plane [9,10].

Here, we have the following coordinate frames:

$O_w(x_w,y_w)$ is world coordinate frame with origin at O_w,
$Q(x_r,y_r)$ is local coordinate frame with origin at Q,
$C(x_c,y_c)$ is camera coordinate frame with origin at C,
$S(x_s,y_s)$ is target coordinate frame with origin at S.

The relative position and orientation of these coordinate frames is as shown in Figure 9.4.

The coordinates x_m^c, y_m^c of the three features points m = {D,E,F} in the camera frame C are:

$$x_m^c = x_s^c + x_m^s \cos\phi_s^c - y_m^s \sin\phi_s^c \tag{9.16}$$

$$y_m^c = y_s^c + x_m^s \sin\phi_s^c + y_m^s \cos\phi_s^c$$

where x_s^c, y_s^c are the position coordinates, and ϕ_s^c the orientation angle, of S with respect to the camera frame C. The feature points D,E, and F, represented by $(x_D^c, y_D^c), (x_E^c, x_E^c)$ and

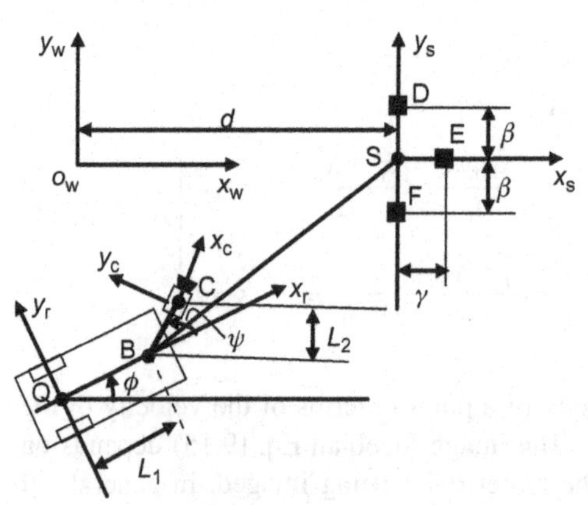

Figure 9.4 WMR with rotating onboard camera (The target frame $S(x_s, y_s)$ is translated along the axis $O_w x_w$ by a distance d).

(x_F^c, y_F^c) in the camera frame, are mapped on the image plane (which here is 1D, i.e., a line) through the forward perspective transformation:

$$f_m = l_f \frac{y_m^c}{x_m^c} \quad (m = D, E, F) \tag{9.17}$$

where f_m is the image of (x_m^c, y_m^c), x is the "depth" variable (replacing the depth variable z of Eq. (9.12)). Therefore, in our case the sensory data vector is:[1]

$$\mathbf{f} = [f_D, f_E, f_F, \psi]^T \tag{9.18a}$$

The full position and orientation vector of the target frame S with respect to the camera frame C is:

$$\mathbf{x}_s^c = [x_s^c, y_s^c, \phi_s^c, \psi_s^c]^T \tag{9.18b}$$

Differentiating Eq. (9.17) we get (for m = D,E,F):

$$\begin{aligned} \dot{f}_m &= l_f \frac{\dot{y}_m^c}{x_m^c} - l_f \frac{y_m^c}{(x_m^c)^2} \dot{x}_m^c \\ &= \left[-\frac{f_m}{x_m^c}, \frac{l_f}{x_m^c}, 0 \quad 0 \right] \dot{\mathbf{x}}_m^c \end{aligned} \tag{9.19a}$$

where:

$$\dot{\mathbf{x}}_m^c = [\dot{x}_m^c, \dot{y}_m^c, \dot{\phi}_m^c, \dot{\psi}_m^c]^T$$

Now, by Eq. (9.4c) we have:

$$\dot{\mathbf{x}}_m^c = \begin{bmatrix} 1 & 0 & -y_m^c & 0 \\ 0 & 1 & x_m^c & 0 \\ 0 & 0 & 1 & 0 \\ 0 & 0 & 0 & 1 \end{bmatrix} \dot{\mathbf{x}}_s^c \tag{9.19b}$$

where the relations $\dot{\phi}_m^c = \dot{\phi}_s^c$ and $\dot{\psi}_m^c = \dot{\psi}_s^c = \dot{\psi}$ were used. Therefore by Eqs. (9.18a) and (9.18b) and Eqs. (9.19a) and (9.19b) we get:

$$\dot{\mathbf{f}} = \mathbf{J}_{im}^c \dot{\mathbf{x}}_s^c \tag{9.20}$$

where:

$$\mathbf{J}_{im}^c = \begin{bmatrix} -f_D/x_D^c & l_f/x_D^c & (l_f^2 + f_D^2)/l_f & 0 \\ -f_E/x_E^c & l_f/x_E^c & (l_f^2 + f_E^2)/l_f & 0 \\ -f_F/x_F^c & l_f/x_F^c & (l_f^2 + f_F^2)/l_f & 0 \\ 0 & 0 & 0 & 1 \end{bmatrix} \tag{9.21}$$

[1] The camera angular deviation ψ is assumed to be measured by a proper sensor.

We will now find the relation between the state vector velocity:

$$\dot{\mathbf{x}} = [\dot{x}_Q, \dot{y}_Q, \dot{\phi}, \dot{\psi}]^T \tag{9.22a}$$

and the velocity vector:

$$\dot{\mathbf{x}}_s^c = [\dot{x}_s^c, \dot{y}_s^c, \dot{\phi}_s^c, \dot{\psi}]^T \tag{9.22b}$$

of the target frame S in the camera frame C. To this end, we first write the equations for x_Q, y_Q and ϕ from Figure 9.4:

$$\begin{aligned} x_Q &= -x_s^c \cos\phi_s^c - y_s^c \sin\phi_s^c \\ &\quad - L_1\cos(\phi_s^c + \psi) - L_2\cos\phi_s^c + d \\ y_Q &= x_s^c \sin\phi_s^c - y_s^c\cos\phi_s^c + L_1\sin(\phi_s^c+\psi) + L_2\sin\phi_s^c \\ \phi &= -(\phi_s^c + \psi) \end{aligned} \tag{9.23}$$

Differentiating Eq. (9.23) and using the notation $\dot{\psi} = \omega_\psi$, we get the relation:

$$\dot{\mathbf{x}}_s^c = \mathbf{J}_0(\phi, \psi, L_1, L_2)\dot{\mathbf{x}} \tag{9.24a}$$

where:

$$\mathbf{J}_0 = \begin{bmatrix} -\cos(\phi+\psi) & -\sin(\phi+\psi) & L_1\sin\psi & 0 \\ \sin(\phi+\psi) & -\cos(\phi+\psi) & -(L_1\cos\psi + L_2) & -L_2 \\ 0 & 0 & -1 & -1 \\ 0 & 0 & 0 & 1 \end{bmatrix} \tag{9.24b}$$

Combining (9.20) and (9.24a) we obtain the overall image Jacobian relation from the state vector velocity $\dot{\mathbf{x}}$ to the feature rate of change vector $\dot{\mathbf{f}}$:

$$\dot{\mathbf{f}} = \mathbf{J}_{im}\dot{\mathbf{x}} \tag{9.25a}$$

where:

$$\mathbf{J}_{im} = \mathbf{J}_{im}^c \mathbf{J}_0 \tag{9.25b}$$

9.3 Position-Based Visual Control: General Issues

As it is evident from Figure 9.3, the origin Q of the local coordinate frame Q $x_r y_r$ of a WMR (see, e.g., Figure 2.7A) plays the role of the origin of the end-effector (flange) coordinate frame of a fixed robot. Therefore, for more generality, our analysis will be presented for a fixed or mobile manipulator's end-effector.

Point-to-point positioning—Some point $\mathbf{p}_e$ on the robot with end-effector coordinates has to reach a fixed stationary point **s** visible in the scene.

Pose-based motion control—Here, the end-effector positioning tasks are directly determined in terms of a known object pose, which can be defined in terms of stationing points with respect to the object pose.

A positioning task is represented by an error $\mathbf{e}(t)$ from the configuration (task) space C_s to the screw space R^6. A positioning task is completed when the end-effector pose $\mathbf{x}_e$ satisfies $\mathbf{e}(\mathbf{x}_e) = \mathbf{0}$. Actually, the error function restricts some number $d < 6$ degrees of freedom of the robot, where d is called the *degree of constraint*. The error function can be regarded as a virtual kinematic constraint between the end-effector and the target.

9.3.1 Point-to-Point Positioning

Consider a fixed camera. Then, the error $\mathbf{e}_p$ in world coordinates is given by:

$$\mathbf{e}_p = \mathbf{x}_e(\mathbf{p}_e) - \mathbf{s} \tag{9.26}$$

where $\mathbf{x}_e(\mathbf{p}_e)$ expresses the position of the point $\mathbf{p}_e$ in world coordinates, and is actually the variable to be controlled. We assume that an estimate $\hat{\mathbf{s}}_c$ of $\mathbf{s}$ is provided by a calibrated camera with respect to the camera coordinate frame. Then, using the camera pose $\hat{\mathbf{x}}_c$ in world coordinates, we have:

$$\hat{\mathbf{s}} = \hat{\mathbf{x}}_c(\hat{\mathbf{s}}_c) \tag{9.27}$$

From Eqs. (9.26) and (9.27) it follows that, if no disturbance exists, the proportional negative feedback law:

$$\begin{aligned}\mathbf{u} &= -\mathbf{K}\mathbf{e}_p \\ &= -\mathbf{K}[\mathbf{x}_e(\mathbf{p}_e) - \hat{\mathbf{x}}_c(\hat{\mathbf{s}}_c)]\end{aligned} \tag{9.28}$$

with $\mathbf{K}$ a positive definite gain matrix, drives the equilibrium state to a value where $\mathbf{e}_p = \mathbf{0}$. In practice, $\mathbf{x}_e$ may also be subject to errors. In these cases, $\mathbf{x}_e$ in Eq. (9.28) must be replaced by an estimate $\hat{\mathbf{x}}_e$ of $\mathbf{x}_e$, which may lead to positioning errors. The case where the cameras are mounted on board and calibrated to the end-effector can be treated in the same way.

9.3.2 Pose-Based Motion Control

Here, we work as above defining the positioning tasks in terms of the object pose. We use a desired stationing pose $\mathbf{x}_{e,d}$ of the end-effector with respect to the target coordinate frame. If $\mathbf{e}_{pose}$ is the positioning error of the actual end-effector pose $\mathbf{x}_e$ from the desired one, then again a proportional feedback controller:

$$\begin{aligned}\mathbf{u} &= -\mathbf{K}\mathbf{e}_{pose} \\ &= -\mathbf{K}(\mathbf{x}_e - \mathbf{x}_{e,d}), \quad \mathbf{K} > 0\end{aligned} \tag{9.29}$$

stabilizes to zero the error $\mathbf{e}_{\text{pose}}$. Clearly, the problem in the above control laws is the estimation of the variables that are used to parameterize the feedback (i.e., $\mathbf{p}_e$,$\mathbf{s}$). This issue will be discussed later.

9.4 Image-Based Visual Control: General Issues

9.4.1 Use of the Inverse Jacobian

The Jacobian Eq. (9.14) provides the rate of change $\dot{\mathbf{f}}$ of the image feature parameters, perceived in the image plane, using the screw vector $\dot{\mathbf{r}}$ of translational and angular velocities of the end-effector. But visual robot control applications require the inverse, that is, to determine $\dot{\mathbf{r}}$ from $\dot{\mathbf{f}}$. This can be done by solving Eq. (9.14) for $\dot{\mathbf{r}}$, but the solution is not always unique. If $\mathbf{J}_{\text{im}}(\mathbf{r})$ is invertible ($k = m$) the solution is exact and unique given by:

$$\dot{\mathbf{r}} = \mathbf{J}_{\text{im}}^{-1}(\mathbf{r})\dot{\mathbf{f}} \tag{9.30}$$

If $k \neq m$, the inverse Jacobian does not exist. Therefore, we use the (least squares) generalized Jacobian matrix given by Eqs. (2.8a) and (2.8b), that is:

$$\mathbf{J}_{\text{im}}^{\dagger} = (\mathbf{J}_{\text{im}}^{T}\mathbf{J}_{\text{im}})^{-1}\mathbf{J}_{\text{im}}^{T} \text{ when } k > m \tag{9.31a}$$

$$\mathbf{J}_{\text{im}}^{\dagger} = \mathbf{J}_{\text{im}}^{T}(\mathbf{J}_{\text{im}}\mathbf{J}_{\text{im}}^{T})^{-1} \text{ when } k < m \tag{9.31b}$$

If there are more feature parameters than the task degrees of freedom, that is, $k > m$, the algebraic system (9.14) is *overdetermined*. If $k < m$ the algebraic system (9.14) is *underdetermined* (i.e., there are some components of the object that cannot be observed, because there are not enough features to determine uniquely the object velocity $\dot{\mathbf{r}}$). Thus, the proper generalized inverse Jacobian is given by Eq. (9.31b).

If as control vector $\mathbf{u}$ we use the vector $\dot{\mathbf{r}}$, then in the nonsingular case (9.30) we have:

$$\mathbf{u} = \mathbf{J}_{\text{im}}^{-1}(\mathbf{r})\dot{\mathbf{f}} \tag{9.32}$$

Therefore, defining the error function $\mathbf{e}(\mathbf{f})$ as:

$$\mathbf{e}(\mathbf{f}) = \mathbf{f}_{\text{d}} - \mathbf{f} \tag{9.33}$$

where $\mathbf{f}_{\text{d}}$ is a desired feature parameter vector to be reached, a convenient proportional (resolved-rate) control law is:

$$\mathbf{u} = \mathbf{J}_{\text{im}}^{-1}(\mathbf{r})\mathbf{K}\mathbf{e}(\mathbf{f}) \tag{9.34}$$

where $\mathbf{K}$ is a constant positive definite gain matrix of appropriate dimensionality (usually diagonal). This control law guarantees asymptotic convergence of $\mathbf{e}(\mathbf{f})$ to zero. To verify this, we select the Lyapunov function:

$$V(\mathbf{e}(\mathbf{f})) = \frac{1}{2}\mathbf{e}^{\mathrm{T}}(\mathbf{f})\mathbf{e}(\mathbf{f}) > 0 \tag{9.35}$$

The derivative of V is given by:

$$\dot{V}(\mathbf{e}(\mathbf{f})) = \mathbf{e}^{\mathrm{T}}(\mathbf{f})\dot{\mathbf{e}}(\mathbf{f}) \tag{9.36}$$

where, using Eqs. (9.32)–(9.34):

$$\begin{aligned}\dot{\mathbf{e}} &= -\dot{\mathbf{f}} = -\mathbf{J}_{\mathrm{im}}(\mathbf{r})\mathbf{u} \\ &= -\mathbf{J}_{\mathrm{im}}(\mathbf{r})[\mathbf{J}_{\mathrm{im}}^{-1}(\mathbf{r})\mathbf{K}\mathbf{e}(\mathbf{f})] \\ &= -\mathbf{K}\mathbf{e}(\mathbf{f})\end{aligned} \tag{9.37}$$

Thus:

$$\dot{V}(\mathbf{e}(\mathbf{f})) = -\mathbf{e}^{\mathrm{T}}(\mathbf{f})\mathbf{K}\mathbf{e}(\mathbf{f}) < 0 \tag{9.38}$$

for $\mathbf{K}$ positive definite. This proves asymptotic stability of $\mathbf{e}(\mathbf{f})$. If $k \neq m$ we use the control law (9.34) with $\mathbf{J}_{\mathrm{im}}^{-1}$ being replaced by Eq. (9.31a) or Eq. (9.31b) as the case may be. The asymptotic stability of the error dynamics of $\mathbf{e}_{\mathrm{p}}$ (see Eq. (9.28)) or $\mathbf{e}_{\mathrm{pose}}$ (see Eq. (9.29)) can be proved in the same way using the proper Lyapunov functions.

9.4.2 Use of the Transpose-Extended Jacobian

An alternative method to image visual-control design is to use the *transpose-extended Jacobian* $\mathbf{J}_0^{\mathrm{T}}$ instead of the *inverse image Jacobian* $\mathbf{J}_{\mathrm{im}}^{-1}$ (or $\mathbf{J}_{\mathrm{im}}^{\dagger}$). This method bypasses the problem of inverting the Jacobian, but actually, if a physical singularity occurs the control may be erroneous, although it will not fail computationally. The accuracy of this method is better if larger gains can be used.

Consider the Lagrange dynamic model of a robotic manipulator:

$$\mathbf{D}(\mathbf{q})\ddot{\mathbf{q}} + \mathbf{C}(\mathbf{q}, \dot{\mathbf{q}})\dot{\mathbf{q}} + \mathbf{g}(\mathbf{q}) = \boldsymbol{\tau} \tag{9.39}$$

and the *extended Jacobian* $\mathbf{J}_0(\mathbf{q})$ relating the rate of change $\dot{\mathbf{f}}$ of the image feature parameter vector $\mathbf{f}$ with the joint variable velocity vector $\dot{\mathbf{q}}$, that is:

$$\dot{\mathbf{f}} = \mathbf{J}_0(\mathbf{q})\dot{\mathbf{q}} \tag{9.40}$$

The control law is assumed to have the form:

$$\boldsymbol{\tau} = \mathbf{J}_0^{\mathrm{T}}(\mathbf{q})\mathbf{K}_{\mathrm{p}}\mathbf{e}(\mathbf{f}) - \mathbf{K}_{v}\dot{\mathbf{q}} + \mathbf{g}(\mathbf{q}) \tag{9.41}$$

where $\mathbf{e(f)} = \mathbf{f}_\mathrm{d} - \mathbf{f}$ and $\mathbf{K}_\mathrm{p}, \mathbf{K}_\upsilon$ are symmetric positive definite gain matrices.

Using the control law (9.41) in the above robot model, we get the autonomous closed-loop system (for which $\dot{\mathbf{q}} = \mathbf{0}$ is an equilibrium point):

$$\mathbf{D(q)}\ddot{\mathbf{q}} + \mathbf{C}(\mathbf{q}, \dot{\mathbf{q}})\dot{\mathbf{q}} = \mathbf{J}_0^\mathrm{T}(\mathbf{q})\mathbf{K}_\mathrm{p}\mathbf{e(f)} - \mathbf{K}_\upsilon\dot{\mathbf{q}}$$

which assures that the image-based visual robot control objective:

$$\lim_{t \to \infty} \mathbf{e(f)} = \mathbf{0}$$

is achieved.

This can be established by using the Lyapunov function:

$$V(\mathbf{q}, \dot{\mathbf{q}}) = \frac{1}{2}\dot{\mathbf{q}}^\mathrm{T}\mathbf{D(q)}\dot{\mathbf{q}} + \frac{1}{2}\mathbf{e}^\mathrm{T}(\mathbf{f})\mathbf{K}_\mathrm{p}\mathbf{e(f)}$$

The time derivative of V along the trajectories of the closed-loop system is found to be:

$$\dot{\mathbf{V}}(\mathbf{q}, \dot{\mathbf{q}}) = -\dot{\mathbf{q}}^\mathrm{T}\mathbf{K}_\upsilon\dot{\mathbf{q}}, \quad \mathbf{K}_\upsilon > 0 \tag{9.42}$$

where use was made of the property $\dot{\mathbf{f}} = \mathbf{J}_\mathrm{O}(\mathbf{q})\dot{\mathbf{q}}$, and the skew symmetricity property:

$$\dot{\mathbf{q}}^\mathrm{T}\left[\frac{1}{2}\dot{\mathbf{D}}(\mathbf{q}) - \mathbf{C}(\mathbf{q}, \dot{\mathbf{q}})\right]\dot{\mathbf{q}} = 0$$

We see in Eq. (9.42) that $\dot{\mathbf{V}}(\mathbf{q}, \dot{\mathbf{q}})$ is negative which, as usual, proves that the error $\mathbf{e(f)}$ tends to zero asymptotically.

9.4.3 Estimation of the Image Jacobian Matrix

The main advantage of position-based control is that the desired tasks can be described in terms of the Cartesian pose, as it is the common practice in robotics. The principal disadvantage is that the feedback is closed using estimated values of quantities that are functions of system calibration parameters. Therefore, in many cases the sensitivity of the controller to calibration errors is very large [11,12]. Pose-based methods provide a generic approach to visual-based control, but often they are computationally very expensive due to the computation time needed for solving the orientation problem involved. However, Kalman-filter-based estimation methods provide a good and fast solution via the use of microprocessor and field-programmable logic gate array technologies [13–16].

In image-based visual control, the key issue is to get accurate estimates of the image Jacobian matrix, despite the possibly large uncertainties in focal length l_f

(*intrinsic camera parameter*), hand–eye/vehicle–eye calibration (*extrinsic camera parameters*), depth z of point features, etc.

The estimation of $\mathbf{J}_{\text{im}}$ can be split in two parts: estimation of the robot Jacobian part $\mathbf{J}_0(\mathbf{p})$ (see Eq. (9.4a)) and estimation of the visual interaction matrix part $\mathbf{J}_c(x_{\text{im}}, y_{\text{im}}, l_f, z)$ (see Eq. (9.13b)). In the closed loop, the error dynamics of features becomes (see Eq. (9.37)):

$$\dot{\mathbf{e}}(\mathbf{f}) = -\mathbf{J}_{\text{im}}(\mathbf{r})\hat{\mathbf{J}}_{\text{im}}^{\dagger}(r)\mathbf{K}\mathbf{e}(\mathbf{f})$$

where $\hat{\mathbf{J}}_{\text{im}}(\mathbf{r})$ is the estimate of $\mathbf{J}_{\text{im}}(\mathbf{r})$. In the ideal case we have $\mathbf{J}_{\text{im}}\mathbf{J}_{\text{im}}^{\dagger} = \mathbf{I}$, and in reality (due to the estimation error) we have $\mathbf{J}_{\text{im}}\hat{\mathbf{J}}_{\text{im}}^{\dagger} \neq \mathbf{I}$. Clearly, a sufficient condition for local convergence is:

$$\mathbf{J}_{\text{im}}\hat{\mathbf{J}}_{\text{im}}^{\dagger} > 0$$

Now, let us consider the problem of estimating the depth z for each considered point feature. This can be solved using a dynamic state estimator of the type:

$$\dot{\hat{\mathbf{x}}} = \mathbf{F}(\hat{\mathbf{x}}, \mathbf{f})\mathbf{u} + \mathbf{g}(\hat{\mathbf{x}}, \mathbf{f}, \mathbf{u})$$

where:

$$\mathbf{x} = [x_{\text{im}}, y_{\text{im}}, 1/z]^{\text{T}}$$

is the actual state, and

$$\hat{\mathbf{x}} = [\hat{x}_{\text{im}}, \hat{y}_{\text{im}}, 1/\hat{z}]^{\text{T}}$$

is the estimated state, with:

$$\mathbf{f} = [x_{\text{im}}, y_{\text{im}}]^{\text{T}}$$

being the measured output. This estimator assures that $||\mathbf{x}(t) - \hat{\mathbf{x}}|| \to 0$ as $t \to \infty$, if:

- the camera has linear velocity different than zero,
- the linear velocity is not aligned with the projection ray of the point feature under consideration,
- there are persistent excitation conditions (i.e., the system is state observable).

Figure 9.5A shows the structure of the nonlinear dynamic state estimator (e.g., extended Kalman filter),[2] and Figure 9.5B shows how this estimator can be integrated into an image-based visual robot control system [15,16].

[2] The Kalman filter is discussed in Section 12.2.3 and the extended Kalman filter in Section 12.8.2

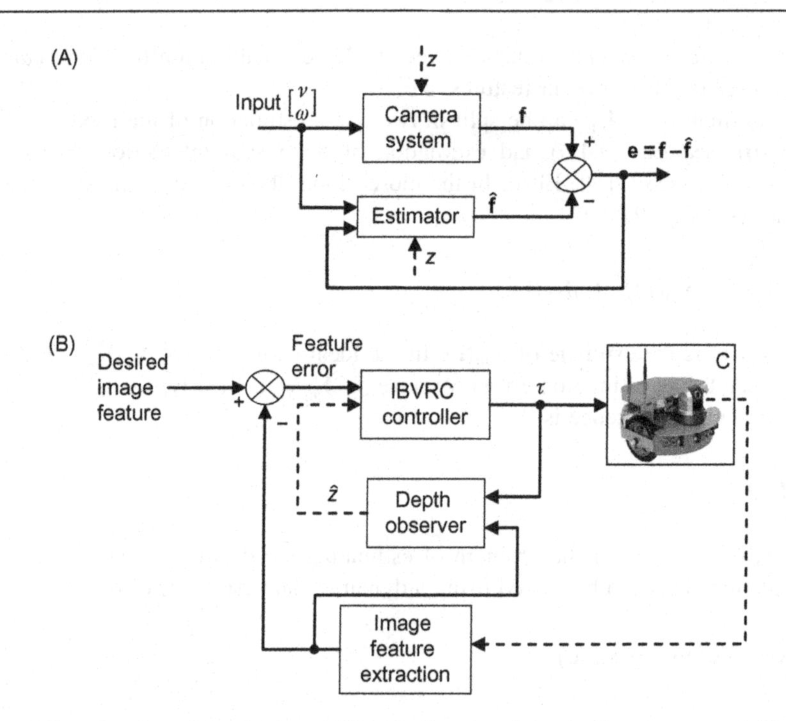

Figure 9.5 Estimation of the depth z: (A) block diagram of the depth estimator and (B) integration of the estimator into an image-based VRC.

Here, it is useful to discuss an important property of the Jacobian matrix $\mathbf{J}_{im}$ and how it can be exploited. Looking at Eq. (9.14) we see that it can be written as:

$$\dot{\mathbf{f}} = \mathbf{J}_{im,v}(x_{im},y_{im},z)\mathbf{v} + \mathbf{J}_{im,\omega}(x_{im},y_{im},l_f)\boldsymbol{\omega}$$

where $\mathbf{J}_{im,v}$ contains the first three columns of $\mathbf{J}_{im}$ and depends on both the image coordinates (x_{im},y_{im}) and the depth z, while $\mathbf{J}_{im,\omega}$ contains the last three columns which are only functions of (x_{im},y_{im}) and do not depend on the depth z. This means that errors in z merely cause a scaling of the matrix $\mathbf{J}_{im,v}$ which can be easily compensated for through fairly simple control procedures. The above property constitutes the core of the so-called *partitioned estimation and control methods*.

The camera velocity $[\mathbf{v}^T,\boldsymbol{\omega}^T]^T$ has six degrees of freedom, but only two values (x_{im} and y_{im}) are observed in the image. This means that the matrix $\mathbf{J}_{im} \in R^{2\times 6}$ has a null space of dimension 4, that is, the equation:

$$\mathbf{J}_{im}(x_{im},y_{im},l_f,z)\boldsymbol{\alpha} = \mathbf{0}$$

has a solution vector $'\alpha'$ that lies on a 4D subspace R^4. Actually, it can be verified that the null space of $\mathbf{J}_{\text{im}}$ in Eq. (9.15) is spanned by the following four vectors:

$$\xi_1 = [x_{\text{im}}, y_{\text{im}}, l_{\text{f}}, 0, 0, 0]^{\text{T}}$$
$$\xi_2 = [0, 0, 0, x_{\text{im}}, y_{\text{im}}, l_{\text{f}}]^{\text{T}}$$
$$\xi_3 = [x_{\text{im}} y_{\text{im}} z, -(x_{\text{im}}^2 + l_{\text{f}}^2) z, l_{\text{f}} y_{\text{im}} z, -l_{\text{f}}^2, 0, l_{\text{f}} x_{\text{im}}]^{\text{T}}$$
$$\xi_4 = [l_{\text{f}}(x_{\text{im}}^2 + y_{\text{im}}^2 + l_{\text{f}}^2) z, 0, -x_{\text{im}}(x_{\text{im}}^2 + y_{\text{im}}^2 + l_{\text{f}}^2) z, l_{\text{f}} x_{\text{im}} y_{\text{im}}, -(x_{\text{im}}^2 + l_{\text{f}}^2) z, l_{\text{f}}^2 x_{\text{im}}]^{\text{T}}$$

The first vector corresponds to motion of the camera frame along the projection ray that contains the point $\mathbf{p} = [x, y, z]^{\text{T}}$, and the second vector corresponds to rotation of the camera frame about a projection ray which contains $\mathbf{p}$.

9.5 Mobile Robot Visual Control

Vision-based methods have been applied to solve several robot control problems, pose-based and image-based. The basic tool in all these methods is the robot and image Jacobian matrices. As we already know the treatment of nonholonomic WMRs is more challenging due to the fact that continuous controllers cannot exponentially stabilize them to a desired pose [17–21]. In Chapters 5 through 8 we have presented several ways for treating the pose control problem.

In the present chapter, we will treat several particular problems including pose stabilization, path following, wall following, target vehicle following, and keeping a landmark in the camera's field of view.

9.5.1 Pose Stabilizing Control

The general pose-based visual-control loop of a mobile robot is shown in Figure 9.1. The controller involves two parts:

1. a standard (nonvision-based) controller,
2. a pose estimator based on image feature measurements provided by one or more cameras.

Here, we will illustrate this scheme as applied to the unicycle-type WMR of Figure 9.4, where it is desired to asymptotically stabilize to zero the pose:

$$\mathbf{x} = [x_{\text{Q}}, y_{\text{Q}}, \phi, \psi]^{\text{T}} \tag{9.43}$$

using measurements from the onboard camera and a sensor (e.g., encoder) that measures ψ.

The kinematic performance of the unicycle-type WMR:

$$\dot{x}_{\text{Q}} = v \cos \phi, \quad \dot{y}_{\text{Q}} = v \sin \phi, \quad \dot{\phi} = \omega \tag{9.44}$$

is described by several equivalent affine models of the form of Eq. (6.1), for example, chain model, Brockett integrator model.

As an example, we will use the chain model:

$$\begin{aligned}
\dot{z}_1 &= u_1 \\
\dot{z}_2 &= u_2 \\
\dot{z}_3 &= z_2 u_1 \\
u_1 &= \omega \\
u_2 &= v - z_3 u_1
\end{aligned} \tag{9.45}$$

which was derived via the variable transformation (2.62):

$$\begin{aligned}
&\mathbf{z}' = \mathbf{F}(\mathbf{x}'), \quad \mathbf{z}' = [z_1, z_2, z_3], \quad \mathbf{x}' = [x_Q, y_Q, \phi]^T \\
&z_1 = \phi, \quad z_2 = x_Q \cos\phi + y_Q \sin\phi, \quad z_3 = x_Q \sin\phi - y_Q \cos\phi
\end{aligned} \tag{9.46a}$$

with inverse:

$$\mathbf{x}' = \mathbf{F}^{-1}(\mathbf{z}'), \quad v = u_2 + z_3 u_1, \quad \omega = u_1 \tag{9.46b}$$

$$x_Q = z_2 \cos z_1 + z_3 \sin z_1, \quad y_Q = z_2 \sin z_1 - z_3 \cos z_1, \quad \phi = z_1$$

For this model, we have derived the stabilizing controller (6.81a) using the invariant manifolds method:

$$\mathbf{u} = \begin{bmatrix} u_1 \\ u_2 \end{bmatrix} = \mathbf{u}(\mathbf{z}', t) \tag{9.47}$$

To ensure that the three point landmarks (see Figure 9.4) remain in the camera's field of view, during the mobile robot's movement, the error z_4:

$$\begin{aligned}
&z_4(\mathbf{x}) = \psi - \theta(x_Q, y_Q, \phi) \\
&\theta = tg^{-1}\left(\frac{y_Q + L_1 \sin\phi}{x_Q + L_1 \cos\phi - d}\right) - \phi
\end{aligned} \tag{9.48a}$$

must be stabilized to zero [9]. The time evolution of z_4 is described by the relation:

$$\begin{aligned}
\dot{z}_4 &= \omega_\psi - \left(\frac{\partial\theta}{\partial\mathbf{x}'}\right)\left(\frac{\partial\mathbf{F}^{-1}}{\partial\mathbf{z}'}\right)\dot{\mathbf{z}}' \\
&= \omega_\psi - \left(\frac{\partial\theta}{\partial\mathbf{x}'}\right)\left(\frac{\partial\mathbf{F}^{-1}}{\partial\mathbf{z}'}\right)\begin{bmatrix} 1 & 0 \\ 0 & 1 \\ z_2 & 0 \end{bmatrix}\begin{bmatrix} u_1 \\ u_2 \end{bmatrix}
\end{aligned} \tag{9.48b}$$

where the function $\mathbf{x}' = \mathbf{F}^{-1}(\mathbf{z}')$ is defined by Eq. (9.46b).

Having available a feedback stabilizing controller $\mathbf{u}(\mathbf{z}')=[u_1,u_2]^{\mathrm{T}}$ for the chained model (9.45), the unicycle controller $[v,\omega]^{\mathrm{T}}$ is obtained as:

$$v(\mathbf{x}')=u_2+z_3u_1=v(\mathbf{F}(\mathbf{x}'),t) \tag{9.49a}$$

$$\omega(\mathbf{x}')=\omega(\mathbf{F}(\mathbf{x}'),t) \tag{9.49b}$$

Now, to ensure that $z_4(t)$ tends to zero, asymptotically, we must select the control signal $\omega_\psi=\dot{\psi}$ in Eq. (9.48b) such that:

$$\dot{z}_4=-k_4z_4 \text{ with } k_4>0$$

Therefore, we select ω_ψ as:

$$\omega_\psi(\mathbf{x})=-k_4z_4(\mathbf{x})+\mathbf{G}(\mathbf{x}')\mathbf{u}(\mathbf{F}(\mathbf{x}'),t) \tag{9.50a}$$

where:

$$\mathbf{G}(\mathbf{x}')=\left(\frac{\partial\theta}{\partial\mathbf{x}'}\right)\left(\frac{\partial\mathbf{F}^{-1}}{\partial\mathbf{z}'}\right)\begin{bmatrix}1 & 0\\ 0 & 1\\ z_2 & 0\end{bmatrix},\quad \mathbf{x}=\begin{bmatrix}\mathbf{x}'\\ \psi\end{bmatrix} \tag{9.50b}$$

The overall state-feedback stabilizing controller $\mathbf{U}(\mathbf{x})$ of the WMR is:

$$\mathbf{U}(\mathbf{x})=\begin{bmatrix}v(\mathbf{x}')\\ \omega(\mathbf{x}')\\ \omega_\psi(\mathbf{x})\end{bmatrix} \tag{9.51}$$

where $v(\mathbf{x}'),\omega(\mathbf{x}')$ and $\omega_\psi(\mathbf{x})$ are given by Eqs. (9.49a), (9.49b), (9.50a), and (9.50b). This completes the design of the nonvision-based part of the controller which assumes that the state vector (pose) $\mathbf{x}=[x_Q,y_Q,\phi,\psi]^{\mathrm{T}}$ is available.

An estimate of $\mathbf{x}$ can be constructed using the camera measurements of the feature points D, E, and F (see Figure 9.4) through the image Jacobian relation (9.25a) and (9.25b). Assuming that the initial pose is sufficiently close to the desired pose $\mathbf{x}_\mathrm{d}=\mathbf{0}$, the image Jacobian relation (9.25a) and (9.25b) gives:

$$\Delta\mathbf{f}=\mathbf{J}_{\mathrm{im}}(\mathbf{x}_\mathrm{d})\Delta\mathbf{x},\quad \mathbf{J}_{\mathrm{im}}=\mathbf{J}_{\mathrm{im}}^{\mathrm{c}}\mathbf{J}_\mathrm{O} \tag{9.52a}$$

where:

$$\Delta\mathbf{f}=\mathbf{f}-\mathbf{f}_\mathrm{d},\quad \Delta\mathbf{x}=\mathbf{x}-\mathbf{x}_\mathrm{d}=\mathbf{x} \tag{9.52b}$$

and $\mathbf{f}_\mathrm{d}$ are the camera data at the desired pose. From Eqs. (9.52a) and (9.52b) we get the vision-based estimate $\hat{\mathbf{x}}$ of $\mathbf{x}$ as:

$$\hat{\mathbf{x}}=\mathbf{J}_{\mathrm{im}}^{-1}(\mathbf{x}_\mathrm{d})(\mathbf{f}-\mathbf{f}_\mathrm{d}) \tag{9.53}$$

which, if introduced into Eq. (9.51), gives the overall pose-based visual controller:

$$\mathbf{U}(\hat{\mathbf{x}}) = [v(\hat{\mathbf{x}}), \omega(\hat{\mathbf{x}}), \omega_{\psi}(\hat{\mathbf{x}})]^{\mathrm{T}} \tag{9.54}$$

The block diagram of the closed-loop vision-based pose control system is shown in Figure 9.6.

Exactly the same procedure can be applied using other types of stabilizing controllers, and also to the car-like WMR described by Eqs. (2.66)–(2.69) with the controller (6.101).

9.5.2 Wall Following Control

Wall following is actually a particular case of path following. Consider a car-like WMR which uses an omnidirectional camera as a sensor without use of odometry. As described in Section 4.5.8, an omnidirectional camera provides 360° field of view, and there exists a proper geometric mapping from the image plane or any other desired plane [17,22–25].

The kinematic model of the robot in terms of path variables is (Figure 9.7):

$$\begin{aligned}\dot{s} &= v_1 \cos \phi_{\mathrm{p}} \\ \dot{\mathrm{d}} &= v_1 \sin \phi_{\mathrm{p}} \\ \dot{\phi}_{\mathrm{p}} &= v_1 (tg\psi)/D \\ \dot{\psi} &= v_2\end{aligned} \tag{9.55}$$

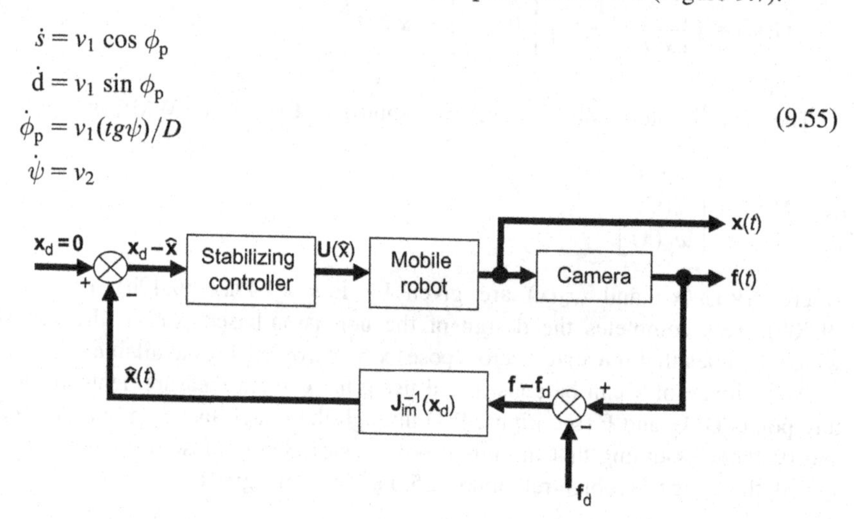

Figure 9.6 General structure of the pose-based visual-control system.

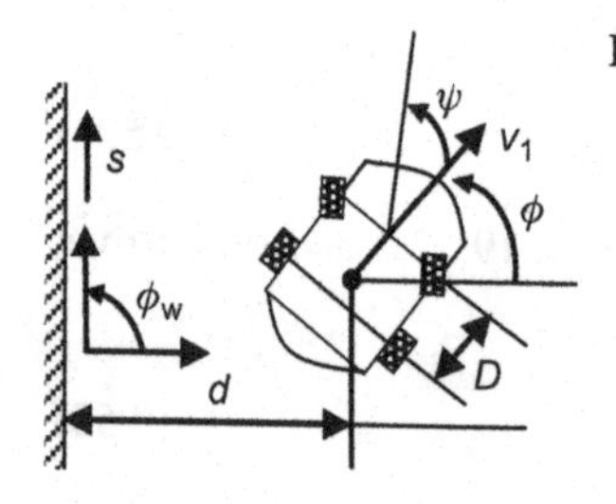

Figure 9.7 WMR wall following ($\phi_{\mathrm{w}} = \pi/2, \phi_{\mathrm{p}} = \pi/2 - \phi$).

where $\phi_{\mathrm{p}}=\phi_{\mathrm{w}}-\phi, \quad \phi_{\mathrm{w}}=\pi/2$. The robot is to follow the wall with a piecewise constant velocity $v_1(t)$.

The output $z(t)$ of the system and its derivative are $z(t)=d(t)$ and $\dot{z}(t)=\dot{\mathrm{d}}(t)$. Then:

$$\ddot{\mathrm{d}}=v_1(\cos\phi_{\mathrm{p}})\dot{\phi}_{\mathrm{p}}=(v_1^2/D)(\cos\phi_{\mathrm{p}})tg\psi=u \tag{9.56}$$

If the desired distance to the wall is d_0 (a constant), then the feedback control law for u is:

$$u=\ddot{\mathrm{d}}_0+K_v(\dot{\mathrm{d}}_0-\dot{\mathrm{d}})+K_{\mathrm{p}}(\mathrm{d}_0-\mathrm{d}),\quad K_{\mathrm{p}}>0,\quad K_v>0 \tag{9.57}$$

from which we obtain the feedback steering angle control law:

$$\psi=tg^{-1}\left[\frac{D}{v_1^2\cos\phi_{\mathrm{p}}}\{K_{\mathrm{p}}(\mathrm{d}_0-\mathrm{d})-K_v v_1\sin\phi_{\mathrm{p}}\}\right] \tag{9.58}$$

This control law requires the measurements of the angle $\phi_{\mathrm{p}}=\pi/2-\phi$, the distance d, and the linear velocity $v_1=\dot{\mathrm{d}}/\sin\phi_{\mathrm{p}}$. The first two variables can be measured by "wall detection" and "obstacle detection" sensing, which is performed by processing the image data. For example, one can apply a Sobel gradient[3] to the original image of the omnidirectional image, considering the resulting edges in the image to be the features of interest. Then, assuming that the ground is planar, the distance to the nearest feature in the sector of interest can be determined from its relative elevation angle of the mirror. This gives a range map of all the obstacles at frame rate [14,17].

The velocity v_1 can be estimated by numerically differentiating the respective distance provided by the range measurement (see next section). A better velocity estimator can be constructed using an extended Kalman filter, discussed in Section 12.8.2.

9.5.3 Leader–Follower Control

Consider a WMR that follows a *leader* mobile robot moving in an arbitrary trajectory with unknown velocity (Figure 9.8A). The problem is the WMR (called the *follower*) to keep a desired distance l_{d} to the leader vehicle while pointing to it (i.e., $\phi_{\mathrm{d}}=0$) [18]. Actually, the *leader–follower* problem is a special case of the *formation control* where a convoy of robots exists [26].

The robot has a fixed mounted camera on the robot center, which looks ahead and captures the image of a pattern mounted on the leader WMR, with four marks lying on the corners of a square with known side L (Figure 9.8B). The positions of the pattern marks on the image are $(x_i,y_i), i=\mathrm{A,B,C,D}$ (in pixels), and are

[3] The Sobel gradient method detects the edges by looking for the maximum and minimum in the first derivative of the image. An edge has the one-dimensional shape of a ramp and calculating the derivative of the image can highlight its location (see References [7–9] of Ch.4).

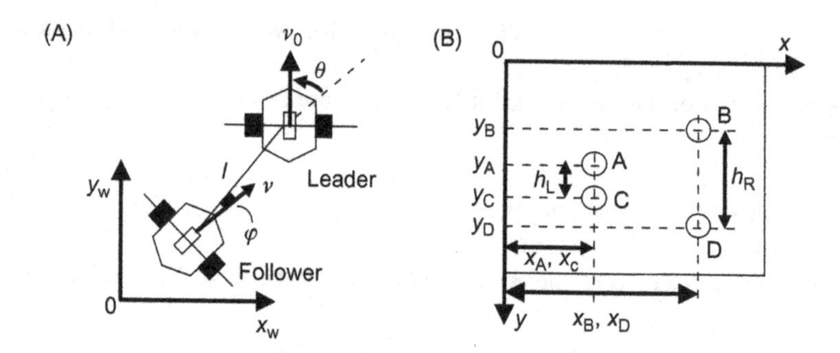

Figure 9.8 (A) Leader and follower positions and (B) image of the pattern's marks.

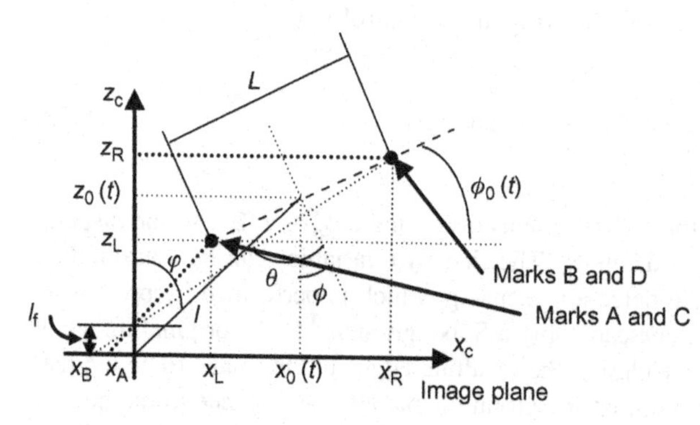

Figure 9.9 Horizontal projection (image) plane of the vision system (location of the leader vehicle).

considered as the image features. From these features, which are measured by the camera, we can compute the pose:

$$\mathbf{x}_0 = [x_0, y_0, \phi_0]$$

of the leading (target) vehicle in the camera coordinate frame $O_c(C, x_c, z_c)$. Referring to Figure 9.9 [18], which shows the horizontal projection of the vision system, the components of $\mathbf{x}_0$ are found to be:

$$x_0 = \frac{x_R + x_L}{2}, \quad z_0 = \frac{z_R + z_L}{2}, \quad \cos\varphi_0 = \frac{x_R - x_L}{L} \tag{9.59}$$

We use reverse perspective imaging projection and obtain the following (see Figure 4.10A):

$$\begin{aligned} x_L &= \left(\frac{z_L - l_f}{l_f}\right) x_A, \quad x_R = \left(\frac{z_R - l_f}{l_f}\right) x_B \\ z_L &= l_f\left(1 + \frac{L}{h_L}\right), \quad z_R = l_f\left(1 + \frac{L}{h_R}\right) \end{aligned} \tag{9.60}$$

where $x_A = x_C$, $x_B = x_D$, h_L and h_R are as shown in Figure 9.8B. Using Eqs. (9.59) and (9.60) we find the relative pose of the follower WMR with respect to the leader WMR:

$$tg\phi = \frac{z_R - z_L}{x_R - x_L}, \quad tg\varphi = \frac{x_0}{z_0}$$
$$\theta = \phi + \varphi, \quad l = \sqrt{x_0^2 + z_0^2} \tag{9.61}$$

The control objective is to make: $l \to l_d$ and $\varphi \to \varphi_d$ as $t \to \infty$ using feedback control laws for v and ω.

To this end, we derive the dynamic equations for the errors:

$$e_l = l_d - l \text{ and } e_\varphi = \varphi_d - \varphi \tag{9.62}$$

Projecting the velocities of the leader and follower (v_0 and v respectively) on the line that connects them, we get the dynamic equation for $\dot{e}_l$:

$$\dot{e}_l = -v_0 \cos\theta + v \cos e_\varphi \tag{9.63a}$$

Similarly, taking into account that the angle error velocity $\dot{e}_\varphi$ has three components, that is, the angular velocity ω of the follower WMR, and the rotational effects (interactions) of both WMRs, we get:

$$\dot{e}_\varphi = \omega + (v_0/l)\sqrt{1-\cos^2\theta} + (v/l)\sqrt{1-\cos^2 e_\varphi} \tag{9.63b}$$

To stabilize e_l and e_φ asymptotically to zero, we must select v and ω in Eqs. (9.63a) and (9.63b) such that the closed-loop equations are:

$$\dot{e}_l = -K_l e_l, \quad \dot{e}_\varphi = -K_\varphi e_\varphi, \quad K_l > 0, \quad K_\varphi > 0 \tag{9.64}$$

From Eqs. (9.63a), (9.63b), and (9.64) we get the control laws:

$$v(t) = (1/\cos e_\varphi)(-K_l e_l + v_0 \cos\theta)$$
$$\omega(t) = -K_\varphi e_\varphi - (1/l)(v_0 \operatorname{sen}\theta + v \operatorname{sen} e_\varphi) \tag{9.65}$$

where by definition $\operatorname{sen} a = \sqrt{1-\cos^2 a}$.

As usual, selecting the Lyapunov function $V = (1/2)(e_l^2 + e_\varphi^2)$, we obtain $\dot{V} = e_l\dot{e}_l + e_\varphi\dot{e}_\varphi = -(K_l e_l^2 + K_\varphi e_\varphi^2) < 0$ which assures that e_l and e_φ are asymptotically stabilized to zero.

From Eq. (9.65) we see that the controllers require the knowledge of v_0, the linear velocity of the leader robot. This can be estimated from the camera visual data. A simple estimation way is to approximate $\dot{e}$ as $\dot{e}_d \simeq (l_k - l_{k-1})/T$ where

$l_k = l(kT)$, $k = 1,2,3,\ldots$, and T is a suitable sampling period. Then, Eq. (9.63a), if solved for v_0, gives:

$$\hat{v}_0 = (1/\cos\theta)[v \cos e_\varphi - (l_k - l_{k-1})/T] \tag{9.66}$$

which uses visual range data for the distance l.

It is noted that for successful control, the follower robot should have weaker curvature constraint, that is:

$$\kappa_{\text{follower}} \geq \kappa_{\text{leader}}$$

where $\kappa = 1/R = \omega/v$ (R = the instantaneous curvature radius of the robots).

9.6 Keeping a Landmark in the Field of View

The problem is to develop a control law that enables a unicycle-type WMR to move on a piecewise smooth trajectory, while keeping a landmark in the camera's field of view [27]. The point landmark is located at the origin, the camera optical center is above the origin of the robot's local frame origin, and the camera optical axis is parallel to the robot axis x_r.

The robot's kinematic model is:

$$\dot{x} = v\cos\phi, \quad \dot{y} = v\sin\phi, \quad \dot{\phi} = \omega \tag{9.67}$$

The robot must be pointed toward the landmark, in order to keep the landmark lying at the origin within the field of view of the camera. This is so if (Figure 9.10) [27]:

$$\phi + \psi = \pi + tg^{-1}(y/x) = \pi + \theta \tag{9.68}$$

The field of view of the camera ranges from $-\tilde{\psi}$ to $+\tilde{\psi}$, as measured from the robot platform x_r-axis. If $-\tilde{\psi} < \psi < \tilde{\psi}$, then the landmark is visible in the image.

The two steps of the controller are [27]:

- rotation of the robot until the image of the point landmark at the origin is on the edge of the image (i.e., $\psi = \pm\tilde{\psi}$),
- forward or backward motion of the robot while keeping the image point on the edge of the image.

This results in a curved path, called *T-curve*, which is described by:

$$\rho = \rho_0 \exp\{(\theta_0 - \theta)/tg\tilde{\psi}\} \tag{9.69}$$

where (ρ_0, θ_0) is a point via which the T-curve passes. We have two T-curves one at $\psi = -\tilde{\psi}$ and the other at $\psi = \tilde{\psi}$. Respecting the constraint $\rho_{\min} < \rho < \rho_{\max}$, by

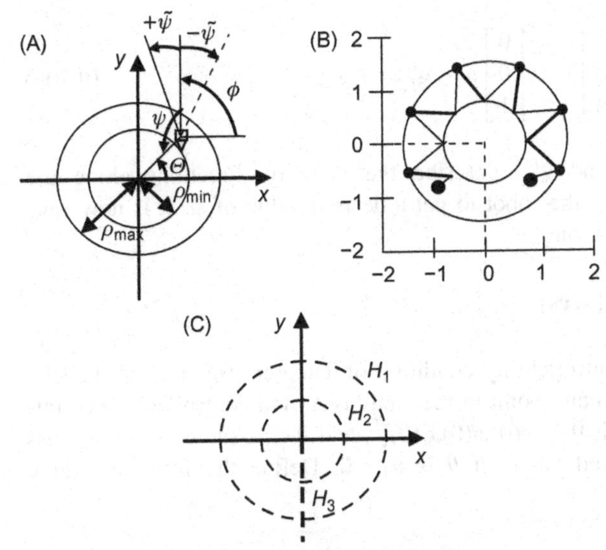

Figure 9.10 (A) Geometry of robot with the camera, (B) an S-curve for $\psi = \pi/6$, and (C) switching surfaces H_1, H_2, and H_3.

alternating between these two T-curves, the robot can move within an *annulus A* centered on the landmark. This set of T-curves is called an *S-curve*. The S-curve allows the robot to move between any two points in the annulus while keeping the landmark in the field of view. This can be shown using the reachability and controllability concept of affine systems. As shown in Example 6.5, using Lie brackets, the unicycle system is locally reachable from everywhere and controllable. This was done using the two standard unicycle fields:

$$\mathbf{g}_1^0 = [\cos\phi, \sin\phi, 0]^{\mathrm{T}}, \quad \mathbf{g}_2 = [0,0,1]^{\mathrm{T}}$$

The question is whether this is also true when the robot motion is constrained to the two available T-curves, and find a proper relevant stabilizing controller. Referring to Figure 9.10A, the polar kinematic model of the robot is:

$$\begin{bmatrix} \dot{\rho} \\ \dot{\theta} \\ \dot{\phi} \end{bmatrix} = \begin{bmatrix} \cos(\theta-\phi) \\ (1/\rho)\sin(\theta-\phi) \\ 0 \end{bmatrix} v + \begin{bmatrix} 0 \\ 0 \\ 1 \end{bmatrix} \omega = \mathbf{g}_1^1 v + \mathbf{g}_2 \omega$$

Now, using the constraint $\phi + \psi = \pi + \theta$ (which assures that the robot points toward the landmark), and $\psi = \pm\tilde{\psi}$ (at the edge of the image), we get: $\cos(\theta-\phi) = -\cos(\psi)$, $\sin(\theta-\phi) = \sin(\psi)$, $\dot{\phi} = \dot{\theta}$ and so the above polar model splits in two models, one valid at $\psi = -\tilde{\psi}$ and the other at $\psi = \tilde{\psi}$:

$$\begin{bmatrix} \dot{\rho} \\ \dot{\theta} \\ \dot{\theta} \end{bmatrix} = \begin{bmatrix} -\cos(\tilde{\psi}) \\ (1/\rho)\sin(\tilde{\psi}) \\ (1/\rho)\sin(\tilde{\psi}) \end{bmatrix} v + \begin{bmatrix} 0 \\ 0 \\ 1 \end{bmatrix} \omega = \mathbf{g}_{1,1}^1 v + \mathbf{g}_2 \omega \tag{9.70a}$$

$$\begin{bmatrix} \dot{\rho} \\ \dot{\theta} \\ \dot{\theta} \end{bmatrix} = \begin{bmatrix} -\cos(-\tilde{\psi}) \\ (1/\rho)\sin(-\tilde{\psi}) \\ (1/\rho)\sin(-\tilde{\psi}) \end{bmatrix} v + \begin{bmatrix} 0 \\ 0 \\ 1 \end{bmatrix} \omega = \mathbf{g}_{1,2}^{1} v + \mathbf{g}_2 \omega \tag{9.70b}$$

The vector fields $\mathbf{g}_{1,1}^{1}$ and $\mathbf{g}_{1,2}^{1}$ describe the velocity directions along the T-curves. The field $\mathbf{g}_2$ rotates the robot to get a desired value of ϕ. It is now easy to show that the field distribution:

$$\Delta = \{\mathbf{f}_1, \mathbf{f}_2, \mathbf{f}_3\} = \{\mathbf{g}_{1,1}^{1}, \mathbf{g}_{1,2}^{1}, \mathbf{g}_2\}$$

satisfies the reachability–controllability condition of Theorem 6.8, and so the system is locally reachable from any point in the annulus A, and controllable. The initial posture of the robot is $\mathbf{X}(0) = [\rho(0), \theta(0), \phi(0)]^{\mathrm{T}}$, and it is assumed (without loss of generality) that the desired value of θ is $\theta_{\mathrm{d}} = 0$. Define the following three switching surfaces:

$$\begin{aligned} H_1 &= \{\mathbf{X} = [\rho, \theta, \phi]^{\mathrm{T}} \quad \text{with} \quad \rho = \rho_{\max}\} \\ H_2 &= \{\mathbf{X} = [\rho, \theta, \phi]^{\mathrm{T}} \quad \text{with} \quad \rho = \rho_{\min}\} \\ H_3 &= \{\mathbf{X} = [\rho, \theta, \phi]^{\mathrm{T}} \quad \text{with} \quad \theta = \theta_{\mathrm{d}}\} \end{aligned} \tag{9.70c}$$

If $\theta(0) > \theta_{\mathrm{d}}$ (i.e., if the robot must move clockwise), then the robot starts following the flow of the vector field:

$$\mathbf{f}_0 = [\tilde{v}\cos\tilde{\psi}, -(1/\rho)\tilde{v}\sin\tilde{\psi}, -(1/\rho)\tilde{v}\sin\tilde{\psi}]^{\mathrm{T}}, \quad \tilde{v} > 0 \quad \text{(a constant velocity)}$$

The state variable ρ increases and the state variable $\theta(t)$ decreases. When the robot contacts the surface H_1, then the system switches to the vector field:

$$\mathbf{f}_{H1} = [-\tilde{v}\cos(-\tilde{\psi}),\ (1/\rho)\tilde{v}\sin(-\tilde{\psi}),\ (1/\rho)\tilde{v}\sin(-\tilde{\psi})]^{\mathrm{T}}$$

which decreases both $\rho(t)$ and $\theta(t)$. If the robot state contacts the surface H_2 the controller switches to follow the vector field:

$$\mathbf{f}_{H2} = \mathbf{f}_0$$

If the state contacts H_3, then the controller is switched to follow the vector field:

$$\begin{aligned} \mathbf{f}_{H3} &= [-k\cos(0)[\rho(t) - \rho_{\mathrm{d}}], -(1/\rho)\tilde{v}\sin(0), -(1/\rho)\tilde{v}\sin(0)]^{\mathrm{T}} \\ &= [-k(\rho(t) - \rho_{\mathrm{d}}), 0, 0]^{\mathrm{T}} \end{aligned}$$

where the gain k is positive. A similar switching control procedure is applied when $\theta(0) < \theta_{\mathrm{d}}$.

Defining a Lyapunov function $V_\theta = (1/2)e_\theta^2$, where $e_\theta(t) = \theta(t) - \theta_d$, it is easy to show that along the field $\mathbf{f}_{H1}, \dot{V}_\theta = e_\theta(\tilde{v}/\rho)\sin(-\tilde{\psi}) < 0$ with $\rho_{\min} < \rho < \rho_{\max}$. The same is true when following the field $\mathbf{f}_{H2}$. Thus, during the motion along the fields $\mathbf{f}_{H1}$ and $\mathbf{f}_{H2}$, θ converges to θ_d, in finite time. When, for $\theta(t) = \theta_d$, the state is on H_3, the system switches to follow the field $\mathbf{f}_{H3}$. Defining a new Lyapunov function $V = V_\theta + (1/2)e_\rho^2$, where $e_\rho = \rho(t) - \rho_d$, we find that along the flow $\mathbf{f}_{H3}$, that is, along $\dot{\rho} = -ke_\rho$ and $\dot{\theta} = 0$, we have $\dot{V} = -ke_\rho^2 < 0$. Therefore, when $\theta(t) = \theta_d$, we have $\rho(t) \rightarrow \rho_d$ along the field $\mathbf{f}_{H3}$. Finally, at this point the system can be controlled to go asymptotically to the desired orientation ϕ_d, following the vector field $\mathbf{g}_2 = [0,0,\omega]^T$.

Overall, the above switching control scheme assures that any initial pose $\mathbf{X}(0) = [\rho(0),\theta(0),\phi(0)]^T$ is driven asymptotically to any desired goal pose $\mathbf{X}_d = [\rho_d,\theta_d,\phi_d]^T$, as $t \rightarrow \infty$.

The controller needs a *single point landmark*, the availability of the *distance to the origin* (measured by the camera), and the *ability to switch instantly* between the T-curves. It assures that while the robot is moving within the task space (i.e., within the annulus defined by $\rho_{\min}$ and $\rho_{\max}$), the landmark is kept in the camera's field of view.

The vision-based control algorithm is as follows [27]:

Step 1: The orientation angle ϕ is increased until the landmark lies on the left edge of the image.
Step 2: When $\psi = \tilde{\psi}$, the robot moves backward and steered so as the image of the landmark is kept on the left edge.
Step 3: When $\rho = \rho_{\max}$, ϕ is decreased to move the landmark on the right edge of the image.
Step 4: When $\psi = -\tilde{\psi}$, the robot moves forward and steered so as the landmark remains on the right edge of the image.
Step 5: When the robot is on the radial line with $\theta = \theta_d$ the robot is turned so as $\psi = 0$, and driven toward ρ_d.

An analogous sequence is applied when $-\pi < \theta(0) \leq \theta_d$, with the robot moving counterclockwise around the workspace. The switching surfaces are reached when $\rho = \rho_{\min}$, $\rho = \rho_{\max}$ and $\theta = \theta_d$ (see Figure 9.10C).

The vision-based implementation of this method together with experimental results for the case where the landmark is a square of coplanar points is provided in Ref. [27]. In the experiments, an image of the landmark taken at the goal position $\mathbf{x}_d$ and knowledge of the feature point at the goal were assumed to be available. Therefore, the homography between images of planar points could be used to estimate the robot state $\mathbf{x}(t)$. The robot started from the position $\mathbf{x}(0) = [\rho,\ \theta,\ \phi]^T = [2.75,\ \pi/3,\ 5\pi/6]^T$ and the goal position was $\mathbf{x}_d = [-1.2,\ 0,\ 0]^T$. The application of the above switching controller has led to a bounded and periodic distance error $e_\rho = \rho(t) - \rho_d$ and an angle error $e_\theta = \theta(t) - \theta_d$ decreasing asymptotically to zero with time. Then, the remaining error in $\rho(t)$ was regulated to zero. These results showed the ability of the switching controller to keep the landmark in the camera field of view. The details of the pose reconstruction using the homography between images are given in Ref. [27]. A more general homography-based control methodology for WMRs with nonholonomic and field-of-view constraints is presented in Ref. [28]. In this methodology, the control laws are directly expressed in terms of the individual entries in the homography matrix (i.e., the pose parameters are not estimated using the homography between images as in other approaches). The methodology is applied for the development of specific control laws for the three standard types of paths, namely, circular, straight line segments, and logarithmic spirals. The control law appropriate to each case is selected via tomography decomposition prior to initiating the navigation.

Example 9.2

Throughout the book we have seen that selecting the controller such that the closed-loop dynamics of the error e, between the desired and actual variable under control, has the form:

$$\dot{e}(t) = -Ke(t), \quad K>0$$

assures that $e(t)$ goes to zero asymptotically as $t \to \infty$. To avoid the saturation of the controller, in many practical cases, we multiply the gain K by the hyperbolic tangential function $tgh(\mu e(t))$ instead of $e(t)$ [18].

(a) Show that using the modified control law, convergence of the error to zero is still assured.

(b) Investigate the robustness of the modified controller against measurement errors, in the mobile robot leader–follower controller of Section 9.5.3.

Solution

(a) In the present case, the closed-loop error dynamic equation becomes:

$$\dot{e}(t) = -K\, tgh(\mu e(t)) \tag{9.71}$$

To show that $e(t) \to 0$, we select as usual the Lyapunov function:

$$V = (1/2)e^2$$

Then, we find:

$$\begin{aligned} \dot{V} &= e\dot{e} = -K\, etgh(\mu e) \\ &= -K' x tgh(x) \end{aligned} \tag{9.72}$$

where $x = \mu e$ and $K' = K/\mu$. The function $tgh(x)$ is defined as:

$$y = tgh(x) = (e^x - e^{-x})/(e^x + e^{-x})$$

and has the plot of Figure 9.11.

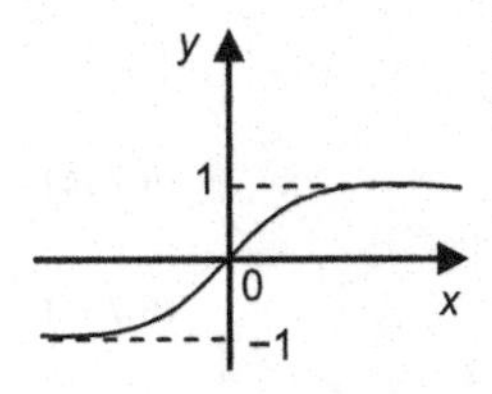

Figure 9.11 Graphical representation of the hyperbolic tangential function.

We observe that $y = tgh(x)$ has the following properties:

$$tgh(0) = 0$$
$$tgh(x) > 0 \quad \text{for} \quad x > 0$$
$$tgh(x) < 0 \quad \text{for} \quad x < 0$$

Therefore, $xtgh(x) > 0$ for $x \neq 0$, which implies that $\dot{V}$ in Eq. (9.72) has the properties as:

$$\dot{V}(0) = 0$$
$$\dot{V}(e) < 0 \quad \text{for} \quad e \neq 0$$

This assures that the feedback controller (9.71) stabilizes asymptotically to zero the error $e(t)$.

(b) As seen in Section 9.5.3 [18], the controllers in Eq. (9.65) assume the availability of the velocity v_0 of the leader robot which can be estimated by the vision system. Let $\hat{v}_0$ be the estimate, and $\tilde{v}_0 = v_0 - \hat{v}_0$ the estimation error. Then:

$$v_0 = \hat{v}_0 + \tilde{v}_0 \tag{9.73}$$

Using Eq. (9.73) in Eqs. (9.63a) and (9.63b) we get:

$$\begin{aligned} \dot{e}_l &= -(\hat{v}_0 + \tilde{v}_0)\cos\theta + v\cos e_\varphi \\ &= (-\hat{v}_0\cos\theta + v\cos e_\phi) - \tilde{v}_0\cos\theta \end{aligned} \tag{9.74a}$$

$$\begin{aligned} \dot{e}_\varphi &= \omega + [(\hat{v}_0 + \tilde{v}_0)/l]\sqrt{1-\cos^2\theta} + (v/l)\sqrt{1-\cos^2 e_\varphi} \\ &= \omega + (\hat{v}_0/l)\sqrt{1-\cos^2\theta} + (v/l)\sqrt{1-\cos^2 e_\varphi} + (\tilde{v}_0/l)\sqrt{1-\cos^2\theta} \end{aligned} \tag{9.74b}$$

Therefore, selecting v and ω such as:

$$-\hat{v}_0\cos\theta + v\cos e_\varphi = -F_l(e_l) \tag{9.75a}$$

$$\omega + (\hat{v}_0/l)\sqrt{1-\cos^2\theta} + (v/l)\sqrt{1-\cos^2 e_\varphi} = -F_\varphi(e_\varphi) \tag{9.75b}$$

where:

$$F_l(e_l) = K_l tgh(\mu_l e_l), \quad F_\varphi(e_\varphi) = K_\varphi tgh(\mu_\varphi e_\varphi) \tag{9.75c}$$

we get, from Eqs. (9.74a) and (9.74b):

$$\dot{e}_l = -F(e_l) - \tilde{v}_0 \cos\theta \tag{9.75d}$$

$$\dot{e}_\varphi = -F(e_\varphi) + (\tilde{v}_0/l)\sqrt{1-\cos^2\theta} \tag{9.75e}$$

Now, if we select the Lyapunov function:

$$V = V_l + V_\varphi, \quad V_l = (1/2)e_l^2, \quad V_\varphi = (1/2)e_\varphi^2$$

we get:

$$\dot{V}_l = e_l\dot{e}_l = -[e_l F(e_l) + e_l\tilde{v}_0 \cos\theta] \tag{9.76a}$$

$$\dot{V}_\varphi = e_\varphi\dot{e}_\varphi = -[e_\varphi F(e_\varphi) - e_\varphi(\tilde{v}_0/l)\sqrt{1-\cos^2\theta}] \tag{9.76b}$$

From Eq. (9.76a) we see that $\dot{V}_l < 0$ if:

$$e_l F_l(e_l) > |e_l||\tilde{v}_0 \cos\theta| \tag{9.77}$$

When $t \to \infty$, we have $|F_l(e_l)| \to K_l$ and so Eq. (9.77) gives $K_l > |\tilde{v}_0|$. For small e_l (in the linear region) we obtain:

$$\begin{aligned} F_l(e_l) &= F_l(0) + F_l'(0)e_l \\ &= K_l\mu_l \sec^2(0)e_l = K_l\mu_l e_l \end{aligned}$$

Therefore, Eq. (9.77) gives:

$$K_l\mu_l > |\tilde{v}_0|$$

which implies that $e(t)$ converges to a neighborhood of the origin with radius:

$$\sigma_l = |\tilde{v}_0|/K_l\mu_l \tag{9.78a}$$

In the same way, from Eq. (9.76b) it follows that $\dot{V}_\varphi < 0$ if $e_\varphi F(e_\varphi) > |e_\varphi||\tilde{v}_0/l|$, which in the saturation state ($t \to \infty$) becomes:

$$K_\varphi > |\tilde{v}_0|/l$$

For e_φ in the linear region, we find (after some calculation) that e_φ tends to a neighborhood of zero with radius [18]:

$$\sigma_\varphi = \frac{|\tilde{v}|K_l\mu_l}{K_\varphi(K_l\mu_l l - |\tilde{v}_0|)} \tag{9.78b}$$

for $K_l\mu_l l > |\tilde{v}_0|$.

9.7 Adaptive Linear Path Following Visual Control

9.7.1 Image Jacobian Matrix

Here, an adaptive vision-based control scheme for straight path following will be presented using a camera mounted on the center of mass of the WMR [21,29]. The image plane $[x_{\text{im}}, y_{\text{im}}]$ is parallel to the ground and the camera points downwards. The coordinate frames needed are $O_w x_w y_w z_w$ (world coordinate frame), $O_r x_r y_r z_r$ (local WMR frame), $O_c x_c y_c z_c$ (camera frame), $O_p x_p y_p z_p$ (path frame), and $O_{\text{im}} x_{\text{im}} y_{\text{im}} z_{\text{im}}$ (image frame). Figure 9.12 shows the coordinate frames O_c and O_{im}, and the image feature parameterization [21,29].

Consider a straight line path L_G on the ground parameterized with the parameters k_0 and λ_0 as:

$$y_w^L = k_0 x_w^L + \lambda_0 \tag{9.79}$$

The camera image of L_G is also a straight line described by a similar equation:

$$y_{\text{im}}^L = k_{\text{im}} x_{\text{im}}^L + \lambda_{\text{im}} \tag{9.80a}$$

using the parameters k_{im} and λ_{im}, or by the polar form equations:

$$x_{\text{im}}^L = \rho \cos \theta, \quad y_{\text{im}}^L = \rho \sin \theta \tag{9.80b}$$

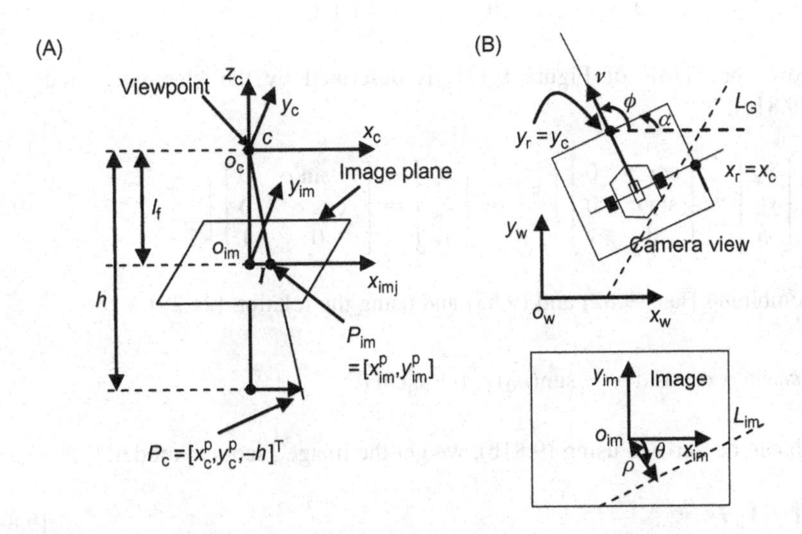

Figure 9.12 (A) Coordinate frames and (B) image feature parameterization.

with parameters ρ and θ. The polar relations in Eq. (9.80b) can also be written in the form of Eq. (9.80a) as:

$$x_{\text{im}}^{\text{L}} \cos\theta + y_{\text{im}}^{\text{L}} \sin\theta = \rho$$

Working with the polar coordinates ρ and θ, which are found from $[x_{\text{w}}, y_{\text{w}}, z_{\text{w}}]^{\text{T}}$ and Eq. (9.80a), using the related perspective transformation $x_{\text{im}}^{\text{L}} = \mu x_{\text{w}}^{\text{L}}, y_{\text{im}}^{\text{L}} = \mu y_{\text{w}}^{\text{L}}$ from the camera to the ground, we find that ρ and θ are given by (see Figure 9.12B):

$$\rho = \mu \frac{k_0 x_{\text{w}}^{\text{L}} + \lambda_0 - y_{\text{w}}^{\text{L}}}{\sqrt{1+k_0^2}} \text{sgn}(k_0) \tag{9.81a}$$

$$\theta = -a - tg^{-1}(1/k_0) \tag{9.81b}$$

$$a = \phi - 90° \tag{9.81c}$$

where the parameter μ is defined by:

$$\mu = l_{\text{f}}/h \quad \text{(aspect ratio)}$$

with l_{f} being the camera's focal length, and h the distance of the camera from the ground. Differentiating Eqs. (9.81a) and (9.81b) with respect to time we get:

$$\begin{bmatrix} \dot{\rho} \\ \dot{\theta} \end{bmatrix} = \begin{bmatrix} \dfrac{\mu k_0 \text{sgn}(k_0)}{\sqrt{1+k_0^2}} & -\dfrac{\mu\,\text{sgn}(k_0)}{\sqrt{1+k_0^2}} & 0 \\ 0 & 0 & -1 \end{bmatrix} \begin{bmatrix} \dot{x}_w \\ \dot{y}_w \\ \dot{\alpha} \end{bmatrix} \tag{9.82}$$

Now, the WMR of Figure 9.12B is described by the kinematic model (see Eq. (9.81c)):

$$\begin{bmatrix} \dot{x}_{\text{w}} \\ \dot{y}_{\text{w}} \\ \dot{\phi} \end{bmatrix} = \begin{bmatrix} \cos\phi & 0 \\ \sin\phi & 0 \\ 0 & 1 \end{bmatrix} \begin{bmatrix} v \\ \omega \end{bmatrix} \text{ or } \begin{bmatrix} \dot{x}_{\text{w}} \\ \dot{y}_{\text{w}} \\ \dot{\alpha} \end{bmatrix} = \begin{bmatrix} -\sin\alpha & 0 \\ \cos\alpha & 0 \\ 0 & 1 \end{bmatrix} \begin{bmatrix} v \\ \omega \end{bmatrix} \tag{9.83}$$

Combining Eqs. (9.82) and (9.83) and using the relation [21,29]:

$$k_0 \sin\alpha + \cos\alpha = -\text{sgn}(\kappa_0)\sqrt{1+k_0^2}\sin\theta$$

which can be verified using (9.81b), we get the image Jacobian model:

$$\dot{\mathbf{f}} = \mathbf{J}_{\text{im}}\dot{\mathbf{r}} \tag{9.84a}$$

$$\mathbf{f}=\begin{bmatrix}\rho\\ \theta\end{bmatrix},\quad \dot{\mathbf{r}}=\begin{bmatrix}v\\ \omega\end{bmatrix},\quad \mathbf{J}_{\text{im}}=\begin{bmatrix}\mu\sin\theta & 0\\ 0 & -1\end{bmatrix} \tag{9.84b}$$

If the camera is placed elsewhere, $\mathbf{J}_{\text{im}}$ takes the form:

$$\mathbf{J}_{\text{im}}=\begin{bmatrix}\mu\sin\theta & h(\theta)\\ 0 & -1\end{bmatrix} \tag{9.85a}$$

$$h(\theta)=\mu l\sin(\zeta+\theta-\gamma) \tag{9.85b}$$

where l is the distance between the vehicle's center of mass and the center of camera, ζ is the angle between x_{r} and x_{im}, and γ is the angle between x_{r} and $\vec{QO}_{\text{c}}$(see Figure 9.12).

9.7.2 The Visual Controller

The image Jacobian model in Eqs. (9.84a) and (9.84b) describes the kinematics of the entire system including the robot and the camera, and will be used in the design of the controller. The controller design develops in three steps (backstepping procedure), as described in Sections 5.4 and 7.3, namely:

- Design of the kinematic controller
- Design of the dynamic controller
- Design of the adaptive controller

embedding a parameter adaptation scheme to the dynamic controller.

9.7.2.1 Kinematic Controller

It is assumed that the feature vector $\mathbf{f}=[\rho,\theta]^{\text{T}}$ is measured by the camera. For properly approaching the path, we use an approach angle $\psi(\rho)$ defined as [29,30]:

$$\psi(\rho)=-\operatorname{sign}(v)\frac{e^{2k_{\psi}\rho}-1}{e^{2k_{\psi}\rho}+1}\theta_{\text{a}},\quad \psi(0)=0 \tag{9.86}$$

where $v\neq 0$ is a given linear velocity, $k_{\psi}>0$ is a constant gain, and $\theta_{\text{a}}\leq\pi/2$. Clearly, when ρ is away from zero, $\psi(\rho)$ is approximately equal to θ_{α}. Then using the Lyapunov function:

$$V_1=(1/2)(\theta-\psi)^2 \tag{9.87a}$$

we find that the feedback kinematic law:

$$\omega_{\theta}=-K_{\theta}(\theta-\psi)+\dot{\psi},\quad K_{\theta}>0 \tag{9.87b}$$

gives:

$$\begin{aligned}\dot{V}_1 &= (\theta - \psi)(\dot{\theta} - \dot{\psi}) = (\theta - \psi)(\omega_\theta - \dot{\psi}) \quad (\omega_\theta = \dot{\theta}) \\ &= -K_\theta(\theta - \psi)^2 < 0\end{aligned} \tag{9.87c}$$

for $K_\theta > 0$, which implies that in the closed-loop ρ and θ converge asymptotically to zero. Now, since by Eq. (9.81b) $\dot{\theta} = \omega_\theta = -\dot{a} = -\omega$, in terms of ω the controller (9.87b) becomes:

$$\omega = \mathrm{K}_\theta(\theta - \psi) - \dot{\psi} \tag{9.88}$$

9.7.2.2 Dynamic Controller

The dynamic model of the WMR is (see, e.g., Eq. (7.66b)):

$$\dot{v} = (1/mr)u_1, \quad \dot{\omega} = (2a/Ir)u_2 \tag{9.89}$$

where $u_1 = \tau_r + \tau_l$, $u_2 = \tau_r - \tau_l$ and τ_r, τ_l are the right wheel and left wheel motor torques. The problem is to design feedback control laws for u_1 and u_2 so as to drive $v(t)$ to zero, and $\omega(t)$ to $\omega^*(t)$, where:

$$\omega^*(t) = K_\theta(\theta - \psi) - \dot{\psi} \tag{9.90}$$

as specified by Eq. (9.88). Working in the usual way, we use the candidate Lyapunov functions:

$$V_v = \frac{1}{2}(v - v_d)^2, \quad V_\omega = \frac{1}{2}(\omega - \omega^*)^2$$

Differentiating V_v with respect to time gives:

$$\dot{V}_v = (v - v_d)(\dot{v} - \dot{v}_d) = (v - v_d)\left[\frac{1}{mr}u_1 - \dot{v}_d\right]$$

Therefore, selecting u_1 as:

$$u_1 = mr[\dot{v}_d - K_v(v - v_d)] \tag{9.91}$$

gives $\dot{V}_v = -K_v(v - v_d)^2$ which is negative for $K_v > 0$. Consequently, $v(t)$ converges to v_d asymptotically.

Similarly:

$$\begin{aligned}\dot{V}_\omega &= (\omega - \omega^*)(\dot{\omega} - \dot{\omega}^*) \\ &= (\omega - \omega^*)[(2a/Ir)u_2 - K_\theta(\dot{\theta} - \dot{\psi}) + \ddot{\psi}]\end{aligned} \tag{9.92}$$

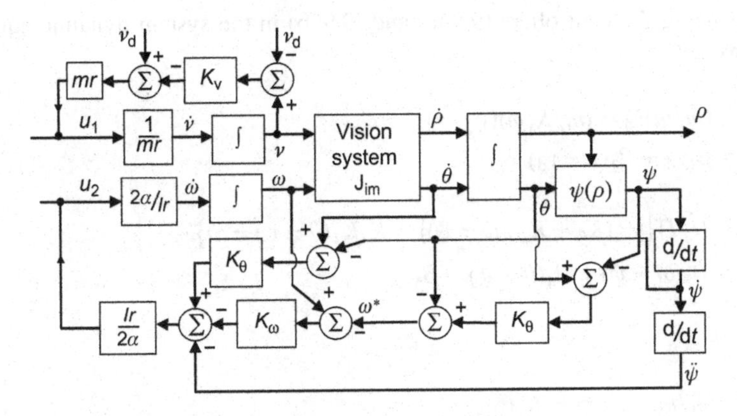

Figure 9.13 Block diagram of vision-based path following system.

which, selecting u_2 as:

$$u_2 = (Ir/2a)[K_\theta(\dot{\theta} - \dot{\psi}) - K_\omega(\omega - \omega^*) - \ddot{\psi}] \tag{9.93}$$

becomes:

$$\dot{V}_\omega = -K_\omega(\omega - \omega^*)^2 < 0 \text{ for } K_\omega > 0 \tag{9.94}$$

Thus, the feedback control law in Eq. (9.93) assures that $\omega \to \omega^*$, asymptotically as $t \to \infty$. The controllers in Eqs. (9.91) and (9.93) need the availability of θ, ψ (i.e., ρ), and v which are measured by the vision system and the robot position and velocity sensors. The block diagram of the closed-loop vision-based system is shown in Figure 9.13.

9.7.2.3 Adaptive Controller

The design of the adaptive vision-based controller is performed as described in Section 7.3.1. Assuming that only m and I are unknown, the controllers in Eqs. (9.91) and (9.93) are written as:

$$u_1 = \hat{m}r[\dot{v}_d - K_v(v - v_d)] \tag{9.95a}$$

$$u_2 = (\hat{I}r/2a)\{K_\theta(\dot{\theta} - \dot{\psi}) - K_\omega[\omega - K_\theta(\theta - \psi) + \dot{\psi}] - \ddot{\psi}\} \tag{9.95b}$$

where (see Eqs. (9.84a), (9.84b), and (9.86)):

$$\dot{\psi} = s_1(t)\mu, \quad s_1(t) = (\vartheta\psi/\vartheta\rho)(\sin\theta)v \tag{9.96a}$$

$$\ddot{\psi} = s_1(t)\dot{\mu} + s_2(t)\mu^2, \quad s_2(t) = (\vartheta s_1/\vartheta\rho)(\sin\theta)v \tag{9.96b}$$

Introducing the controllers (9.95a) and (9.95b) in the system dynamic equations (9.89) we get:

$$\begin{aligned}\dot{v} &= (\hat{m}/m)\dot{v}_\mathrm{d} - (\hat{m}/K_v m)(v - v_\mathrm{d}) \\ &= \beta_1 \dot{v}_\mathrm{d} + \beta_2(v - v_\mathrm{d})\end{aligned} \tag{9.97a}$$

$$\begin{aligned}\dot{\omega} &= (\hat{I}/I)\{-(K_\theta + K_\omega)(\omega + \dot{\psi})\} + K_\theta K_\omega(\theta - \psi) - \ddot{\psi}\} \\ &= \beta_3(\omega + \dot{\psi}) + \beta_4(\theta - \psi) + \beta_5 \ddot{\psi}\end{aligned} \tag{9.97b}$$

where:

$$\beta_1 = \hat{m}/m, \quad \beta_2 = -K_v \beta_1 \tag{9.98a}$$

$$\beta_3 = -(K_\theta + K_\omega)(\hat{I}/I), \quad \beta_4 = K_\theta K_\omega(\hat{I}/I), \quad \beta_5 = -(\hat{I}/I) \tag{9.98b}$$

The dynamic relations in Eqs. (9.97a) and (9.97b) with parameters $\beta_i (i = 1,2,3,4,5)$ are similar to Eqs. (7.36a) and (7.36b) with parameters $\beta_i (i = 1,2,3,4)$. Therefore, the adaptation laws can be derived in the same way and have the general form of Eqs. (7.40a), (7.40b), and (7.41). If the aspect ratio μ is also unknown it can be included in the parameters to be estimated, in which case the functions $\dot{\psi}$ and $\ddot{\psi}$ in Eqs. (9.96a) and (9.96b) are replaced by their estimates:

$$\hat{\dot{\psi}} = s_1(t)\hat{\mu} \tag{9.99}$$

$$\hat{\ddot{\psi}} = s_1(t)\dot{\hat{\mu}} + s_2(t)\hat{\mu}^2 \tag{9.100}$$

The general procedure remains the same. The reader may also derive, as an exercise, an adaptive vision-based neurocontroller following the results of Section 8.5.

Example 9.3

(a) Derive the image Jacobian of a differential-drive WMR assuming that the rigidly mounted camera optical axis z is identical with the direction of the linear velocity v.

(b) Extend this matrix to include as an extra feature the actual measured distance ρ between the camera and the target (landmark) where ρ replaces z.

Solution

(a) According to the requirement about the camera optical axis z the configuration of the WMR and camera is as shown in Figure 9.14A [31].

We have the following coordinate frames:

$O_w x_w y_w z_w$ (world coordinate frame),

$Q\, x_r y_r z_r$ (local coordinate frame).

$C\, x_c y_c z_r$ (camera coordinate frame with z_c-axis identical to z_r).

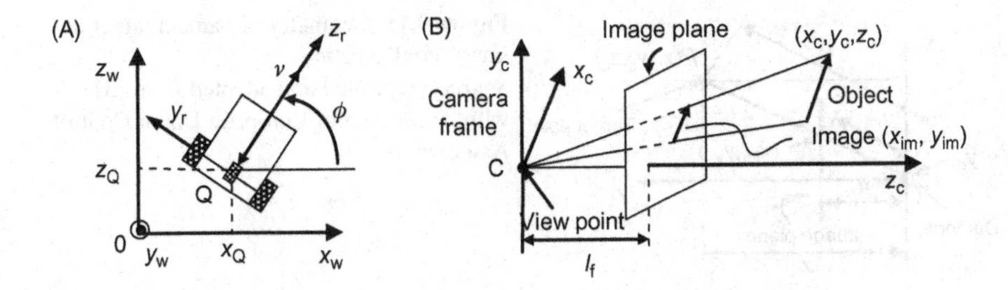

Figure 9.14 (A) Coordinate frames and (B) perspective projection geometry.
Source: Reprinted and adapted from [31], with permission from European Union Control Association.

In the present case, the velocity vectors $\mathbf{v}(t)$ and $\boldsymbol{\omega}(t)$ in Eq. (9.1) are:

$$\mathbf{v}(t) = [v_x, v_z]^{\mathrm{T}}, \quad \boldsymbol{\omega}(t) = \omega_y$$

Therefore, only the columns of $\mathbf{J}_{im}$ in Eq. (9.15), which correspond to v_x, v_z, and ω_y should be retained, that is:

$$J_{im}(x_{im}, y_{im}, z, l_f) = \begin{bmatrix} \dfrac{l_f}{z} & -\dfrac{x_{im}}{z} & \dfrac{l_f^2 + x_{im}^2}{l_f} \\ 0 & -\dfrac{y_{im}}{z} & \dfrac{x_{im} y_{im}}{l_f} \end{bmatrix} \tag{9.101a}$$

that is:

$$\dot{\mathbf{f}} = \begin{bmatrix} \dot{x}_{im} \\ \dot{y}_{im} \end{bmatrix} = \mathbf{J}_{im}(x_{im}, y_{im}, z, l_f) \begin{bmatrix} \dot{x}_Q \\ \dot{z}_Q \\ \dot{\phi} \end{bmatrix} \tag{9.101b}$$

where:

$$\dot{x}_Q = v \cos \phi, \quad \dot{z}_Q = v \sin \phi, \quad \dot{\phi} = \omega_y \tag{9.101c}$$

(b) We assume that the actual distance ρ of the landmark from the camera is measured by a range-finder camera [31,32]. Referring to Figure 9.15 we find:

$$\rho_{im} / \rho = l_f / z$$

that is:

$$z = \rho l_f / \rho_{im}, \quad \rho_{im} = \sqrt{x_{im}^2 + y_{im}^2 + l_f^2} \tag{9.102}$$

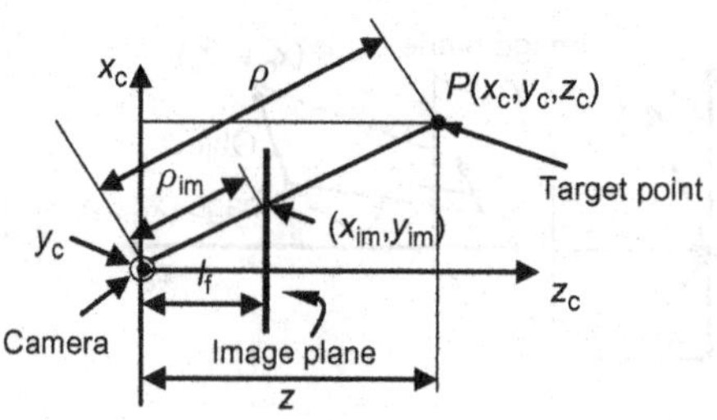

Figure 9.15 Geometry of camera target and image configuration.
Source: Reprinted and adapted from [31], with permission of European Union Control Association.

Therefore, $\mathbf{J}_{im}$ in Eq. (9.101a) becomes:

$$\mathbf{J}_{im} = \begin{bmatrix} \dfrac{\rho_{im}}{\rho} & -\dfrac{x_{im}\rho_{im}}{\rho l_f} & \dfrac{l_f^2 + x_{im}^2}{l_f} \\ 0 & -\dfrac{y_{im}\rho_{im}}{\rho l_f} & \dfrac{x_{im}y_{im}}{l_f} \end{bmatrix} \tag{9.103}$$

Now, if we use ρ_{im} as an extra feature, the rate of change $\dot{\rho}_{im}$ should be expressed in terms of $[\dot{x}_Q,\dot{y}_Q,\dot{\phi}]^T$, and added as a third component to the feature vector velocity $\dot{\mathbf{f}}$.

From Eq. (9.102) we obtain:

$$\begin{aligned} \dot{\rho}_{im} &= \frac{1}{\rho_{im}}(x_{im}\dot{x}_{im} + y_{im}\dot{y}_{im}) \\ &= \frac{1}{\rho_{im}}[x_{im},y_{im}]\begin{bmatrix} \dot{x}_{im} \\ \dot{y}_{im} \end{bmatrix} = \frac{1}{\rho_{im}}[x_{im},y_{im}]\mathbf{J}_{im}\begin{bmatrix} \dot{x}_Q \\ \dot{y}_Q \\ \dot{\phi} \end{bmatrix} \\ &= [x_{im}/\rho, -(x_{im}^2 + y_{im}^2)/\rho l_f, x_{im}\rho_{im}/l_f]\dot{\mathbf{x}} \end{aligned} \tag{9.104}$$

Therefore, the increased image Jacobian relation is:

$$\dot{\mathbf{f}} = \bar{\mathbf{J}}_{im}\dot{\mathbf{x}} \tag{9.105}$$

where $\dot{\mathbf{f}} = [\dot{x}_{im},\dot{y}_{im},\dot{\rho}_{im}]^T$, $\dot{\mathbf{x}} = [\dot{x}_Q,\dot{y}_Q,\dot{\phi}]^T$

$$\bar{\mathbf{J}}_{im} = \begin{bmatrix} \dfrac{\rho_{im}}{\rho} & -\dfrac{x_{im}\rho_{im}}{\rho l_f} & \dfrac{l_f^2 + x_{im}^2}{l_f} \\ 0 & -\dfrac{y_{im}\rho_{im}}{\rho l_f} & \dfrac{x_{im}y_{im}}{l_f} \\ \dfrac{x_{im}}{\rho} & -\dfrac{x_{im}^2 + y_{im}^2}{\rho l_f} & \dfrac{x_{im}\rho_{im}}{l_f} \end{bmatrix} \tag{9.106}$$

This is the desired *extended image Jacobian matrix* of the WMR.

9.8 Image-Based Mobile Robot Visual Servoing

In Sections 9.5–9.7, the problems of pose stabilization, wall following, vehicle leader–follower, etc. were treated using position-based visual control (Figure 9.1A), where the vision system provides estimates of the parameters that are needed to implement conventional controllers. Here, we will examine the problem of pose control (parking, docking) using image-based visual control where the controller is designed and implemented employing directly the visual data as shown in Figure 9.1B [33,34].

The main problem for the visual control of differential-drive WMR (and any other nonholonomic vehicle) is to keep the target always visible while the robot is moving. Two methods for keeping the target (landmarks) in the camera field of view were presented in Sections 9.5.1 and 9.6. The method of Section 9.5.1 belongs to the general method which uses a range-finder camera mounted on a pan-tilt head, and controls the pan-angle so as to keep the features in the camera field of view. Since the camera rotates independently of the platform in order to track the target features, there occurs a difference angle ψ between the orientations of the camera and the platform as shown in Figure 9.16 (see also Figure 9.4). Therefore, a special care is required to control $\psi(t)$. This needs the computation of the coordinates of the landmarks on the image plane.

To implement the closed-loop system, the images are monitored continuously while the robot is moving, using the camera vision system. Assuming that the image Jacobian matrix for each feature point $\mathbf{f}$ is given by Eq. (9.106), and inverting Eq. (9.105), we get:

$$\dot{\mathbf{x}} = \bar{\mathbf{J}}_{\text{im}}^{\dagger}\dot{\mathbf{f}}$$

Typically, for reducing the camera measurement error (or noise) effects, we use more feature points than the number n of degrees of motion of the WMR (i.e., $k > n$). Here, a minimum of four feature points will be used (since $n = 3$). Since $\bar{\mathbf{J}}_{\text{im}}$ is full rank (rank $\bar{\mathbf{J}}_{\text{im}} = n$), $\bar{\mathbf{J}}_{\text{im}}^{\dagger}$ is given by Eq. (9.31a):

$$\bar{\mathbf{J}}_{\text{im}}^{\dagger} = (\bar{\mathbf{J}}_{\text{im}}^{T}\bar{\mathbf{J}}_{\text{im}})^{-1}\bar{\mathbf{J}}_{\text{im}}^{T}$$

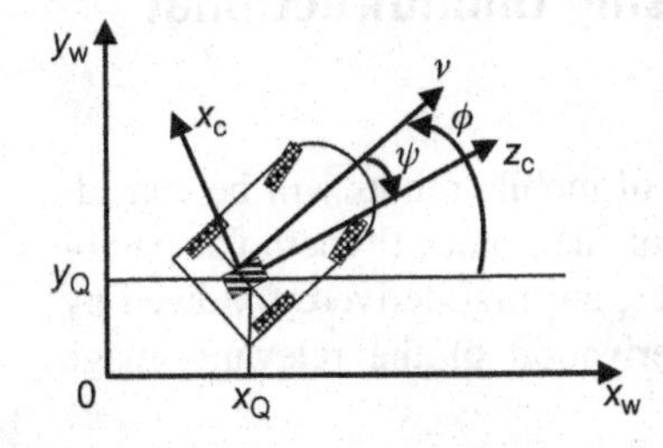

Figure 9.16 Differential-drive WMR with a camera mounted on a pan-tilt head.

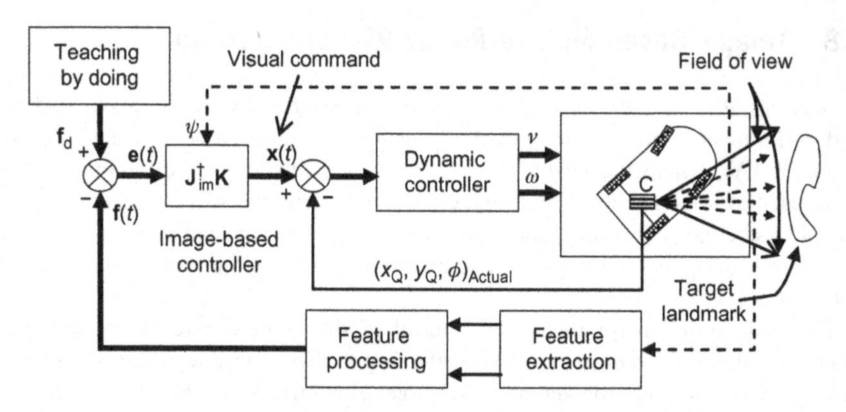

Figure 9.17 Complete kinematic and dynamic visual servoing of the mobile robot (including the feedback loop for the camera orientation deviation $\psi(t)$).

The feedback control law is:

$$\dot{\mathbf{x}} = \mathbf{J}_{\mathrm{im}}^{\dagger}\mathbf{K}\mathbf{e}(\mathbf{f}), \quad \mathbf{e}(\mathbf{f}) = \mathbf{f}_{\mathrm{d}} - \mathbf{f}$$

where $\mathbf{f}_{\mathrm{d}}$ is the desired feature vector (corresponding to the desired pose of the robot). The desired vector $\mathbf{f}_{\mathrm{d}}$ is determined before the application of the control by generating and storing a trajectory that leads from the starting pose to the desired pose employing the typical robotics technique of "*teaching-by-doing*." The block diagram of the complete image-based WMR visual servoing system is shown in Figure 9.17, where a dynamic controller is also in an inner loop.

The orientation angle difference ψ between the platform and the rotating camera is included in the image-based controller by using the concept of an infinite "*virtual image plane*," which is obtained by rotating the physical image plane by the varying angle ψ (which is continuously measured by the camera) [33,34]. In this way, the platform and the camera point in the same direction, and the controller gains $\mathbf{J}_{\mathrm{im}}^{\dagger}\mathbf{K}$ can be used in the standard way by introducing to $\mathbf{J}_{\mathrm{im}}^{\dagger}$ the transformed feature values corresponding to the virtual image plane. This is conceptually shown in Figure 9.17 with the dotted line ψ from the camera to the image-based controller.

9.9 Mobile Robot Visual Servoing Using Omnidirectional Vision

In this section, the visual servoing design problem of mobile robots will be considered using omnidirectional vision. The equations of the conics (hyperbola, parabola, ellipse), that are needed in catadioptric cameras, are first derived, followed by the catadioptric projection geometry, and the derivation of the relevant image Jacobian used for the visual servoing.

9.9.1 General Issues: Hyperbola, Parabola, and Ellipse equations

Catadioptric vision systems are composed by combinations of hyperbolic, parabolic, and elliptic mirrors and cameras (lenses). These systems are distinguished according to whether they have a unique *effective viewpoint* or not. Most fish-eye lenses do not possess this unique viewpoint property, but a hyperbolic mirror in front of a perspective camera, a parabolic mirror in front of an orthographic camera, or an elliptic mirror in front of a perspective camera have a single effective viewpoint.

Actually, a catadioptric camera is equivalent (up to distortion) to a perspective camera, and the lines in space, along which the image is constant, intersect at a single effective viewpoint (Figure 9.18A).

It was shown that "a necessary and sufficient condition for a catadioptric camera to have a single effective viewpoint is that the mirror's cross section is a conic section" (i.e., a hyperbola, parabola, ellipse, or circle shown in Figure 9.19A–D) [25].

Hyperbola—The hyperbola (Figure 9.20A) is defined to be a conic section that represents the locus of all points Σ in the plane for which the distances $d_1 = (F_1\Sigma)$ and $d_2 = (F_2\Sigma)$ from two fixed points (the foci) F_1 and F_2, separated by a distance $2c$, have a given constant difference $d_2 - d_1 = \lambda$. When the point Σ is on the left

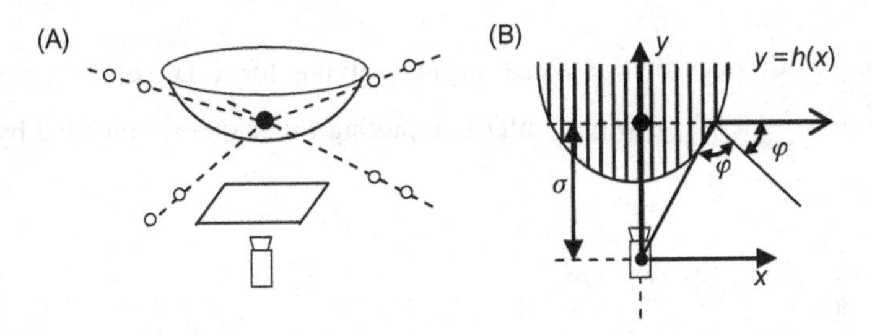

Figure 9.18 The single effective viewpoint property of catadioptric systems.

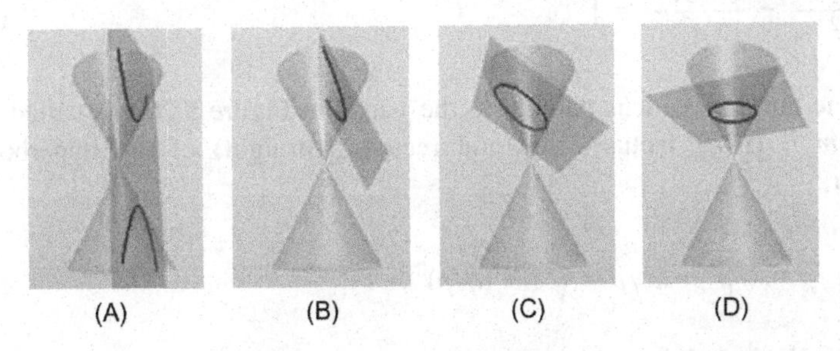

Figure 9.19 Conic sections (conics). Conics are generated by the intersections of a plane with one or two nappes of a double-napped cone: (A) hyperbola, (B) parabola, (C) ellipse, and (D) circle.
Source: http://math2.org/math/algebra/conics.htm.

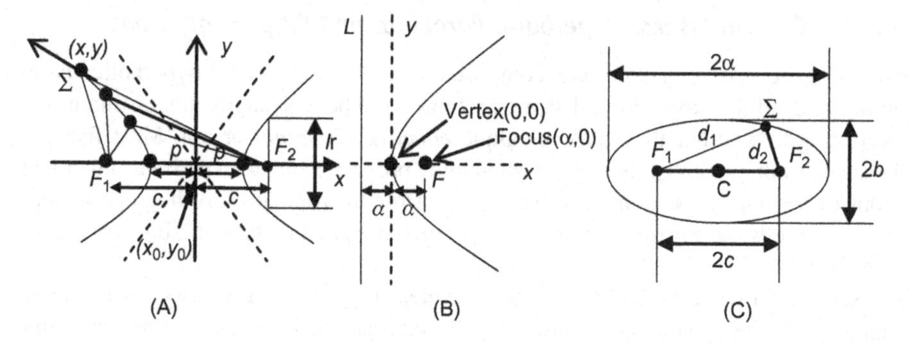

Figure 9.20 Geometry of conics on the plane: (A) hyperbola, (B) parabola, and (C) ellipse.

vertex we find that $\lambda = (c+p) - (c-p) = 2p$. Let x,y be the coordinates of Σ in a Cartesian frame centered at $(x_0,y_0) = (0,0)$. Then, by the hyperbola definition we get (see Figure 9.20A):

$$\sqrt{(x+c)^2 + y^2} - \sqrt{(x-c)^2 + y^2} = 2p \text{ or } \sqrt{(x+c)^2 + y^2} = 2p + \sqrt{(x-c)^2 + y^2}$$

Squaring both sides of this equation and solving for $\sqrt{(x-c)^2 + y^2}$ we obtain $\sqrt{(x-c)^2 + y^2} = cx/p - p$ which, after completing the squares, gives the hyperbola equation:

$$\frac{x^2}{p^2} - \frac{y^2}{q^2} = 1, \quad q^2 = c^2 - p^2 \tag{9.107a}$$

which, when $(x_0,y_0) \neq (0,0)$, becomes:

$$\frac{(x-x_0)^2}{p^2} - \frac{(y-y_0)^2}{q^2} = 1 \tag{9.107b}$$

The line through a focus parallel to the y axis in Figure 9.20A is called the *lactus rectum* l_r (from lactus = side and rectum = straight) of the hyperbola. The parameter:

$$e = c/p = \sqrt{p^2 + q^2}/p = \sqrt{1 + (q/p)^2} > 1 \tag{9.107c}$$

is called the hyperbola's *eccentricity*.

Parabola—A parabola is the locus of all points in the plane equidistant from a given line L (called *directrix* since it shows the direction of the conic section) and a given point F(the focus) not on the line (Figure 9.20B). The distance between this line and the focus is equal to $p = 2a$, where a is the distance of the vertex from the line L. For a parabola opening to the right with vertex at (0,0) as in Figure 9.20B, the parabola's Cartesian equation is:

$$\sqrt{(x-a)^2 + y^2} = x + a$$

which reduces to:

$$y^2 = 4ax \tag{9.108a}$$

For $x = a$ we get $y^2 = 4a^2$, that is, $y = \pm 2a$ and so the *lactus rectum* l_r is equal to:

$$l_r = 2|y| = 4a \tag{9.108b}$$

If the *vertex* is at $(x_0,y_0) \neq (0,0)$, then the parabola equation is:

$$(y-y_0)^2 = 4a(x - x_0) \tag{9.108c}$$

For a parabola opening upwards the equation is:

$$x^2 = 4ay \tag{9.108d}$$

Ellipse—An ellipse is the locus of all points Σ in the plane for which the sum of their distances d_1 and d_2 from two fixed points F_1 and F_2 (the foci) separated by a distance $2c$ is a given positive constant $2a$, that is, $d_1 + d_2 = 2a$, where a is called the *semimajor axis* (Figure 9.20C).

Assuming that C is at $(x_0,y_0) = (0,0)$, the foci points are $F_1(-c,0)$ and $F_2(c,0)$. Therefore, from the ellipse definition we get:

$$\sqrt{(x+c)^2 + y^2} + \sqrt{(x-c)^2 + y^2} = 2a$$

which leads to $\sqrt{(x+c)^2 + y^2} = 2a - \sqrt{(x-c)^2 + y^2}$ or $(x+c)^2 + y^2 = 4a^2 + (x-c)^2 + y^2 - 4a\sqrt{(x-c)^2 + y^2}$. Solving the last equation for $\sqrt{(x-c)^2 + y^2}$, and squaring gives $(x-c)^2 + y^2 = [a-(c/a)x]^2$, from which we get the ellipse equation:

$$\frac{x^2}{a^2} + \frac{y^2}{b^2} = 1, \quad b^2 = a^2 - c^2 \quad (b<a) \tag{9.109a}$$

When $(x_0,y_0) \neq (0,0)$, the ellipse equation is:

$$(x-x_0)^2/a^2 + (y-y_0)^2/b^2 = 1 \tag{9.109b}$$

The circle is a special case of ellipse, when F_1 and F_2 are at C, and so $d_1 = d_2$ and $a = b = r$ (circle radius).

9.9.2 Catadioptric Projection Geometry

A basic result in central catadioptric projection is the following theorem [35]:

> *Catadioptric projection with a single effective viewpoint (called central catadioptric projection) is equivalent to projection to a sphere followed by projection to a plane from a point.* ■

The three central catadioptric vision cases are illustrated in Figure 9.21.

To establish the above theorem, we will derive the projection equations for each of the above projection models, starting from the spherical mirror [35].

Consider the projection of a point $\Sigma(x,y,z)$ of the world space to a unit sphere centered at O(0,0,0), and then to an image plane $z = -\xi$ (Figure 9.22).

From Figure 9.22 we see that the world point $\Sigma(x,y,z)$ is projected via the sphere to two antipodal points $\Sigma_1(x/\rho, y/\rho, z/\rho)$ and $\Sigma_2(-x/\rho, -y/\rho, -z/\rho)$ on the sphere where $\rho = \sqrt{x^2 + y^2 + z^2}$. These points are then projected to the points Σ_1' and Σ_2' on the image plane $z = -\xi$ from the point N(0,0,d). This is expressed by:

$$\sigma_{1,d,\xi}(x,y,z) = \left[\frac{x(d+\xi)}{d\rho - z}, \frac{y(d+\xi)}{d\rho - z}, -\xi\right]^{\mathrm{T}} \tag{9.110a}$$

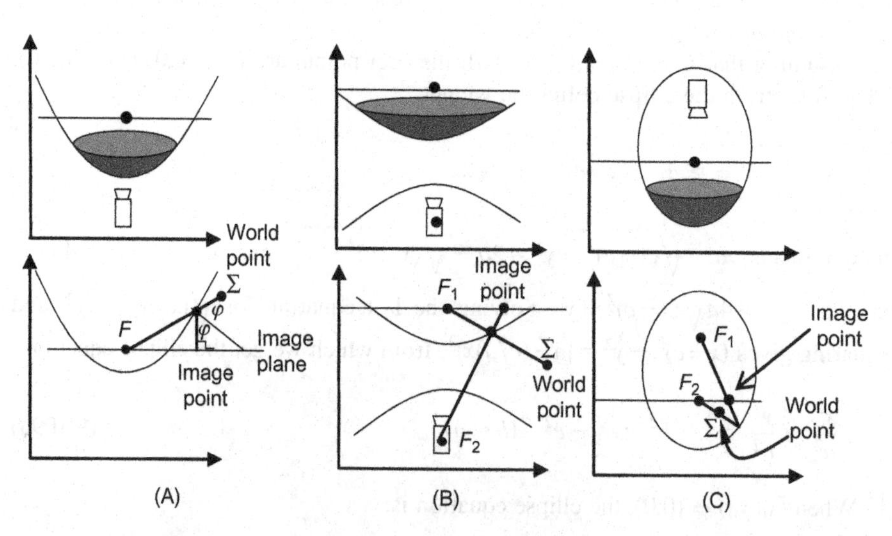

Figure 9.21 Central catadioptric solutions and projection models: (A) parabolic mirror and orthographic camera, (B) hyperbolic mirror and perspective camera, and (C) elliptic mirror and perspective camera.

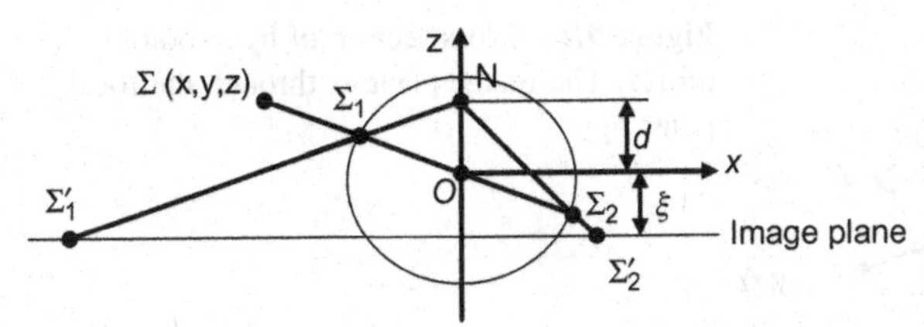

Figure 9.22 A unit spherical mirror centered at (0,0,0). A point $\Sigma(x,y,z)$ is projected to the points Σ_1 and Σ_2 which are then projected to the image plane points Σ_1' and Σ_2'.

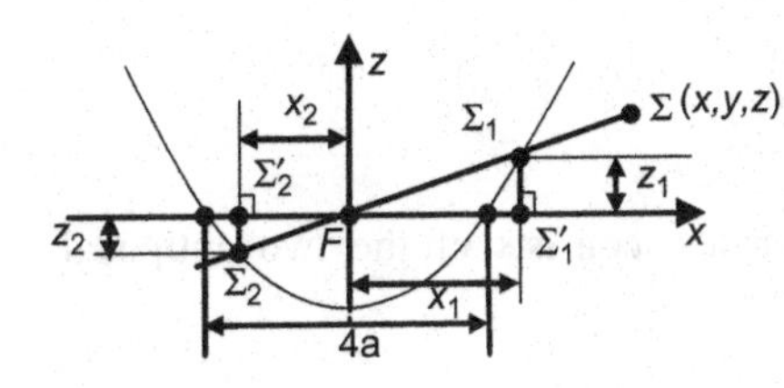

Figure 9.23 Parabolic mirror. The image plane is via the focal point F.

for point Σ_1, and:

$$\sigma_{2,d,\xi}(x,y,z) = \left[-\frac{x(d+\xi)}{d\rho+z}, -\frac{y(d+\xi)}{d\rho+z}, -\xi \right]^{\mathrm{T}} \tag{9.110b}$$

for point Σ_2. If the projection is on the plane $z = -\beta$, then the relation between the two projections is:

$$\sigma_{i,d,\xi}(x,y,z) = \left(\frac{d+\xi}{d+\beta} \right) \sigma_{i,d,\beta}(x,y,z) \tag{9.110c}$$

that is, they differ only by a scaling factor $(d+\xi)/(d+\beta)$. Therefore, if ξ is not given it can be taken as $\xi = 1$. When $d = 1$ and $\xi = 0$ the point of projection N is the North pole, and if $d = 0$ and $\xi = 1$ we get perspective projection:

$$\sigma_{i,0,1}(x,y,z) = \left[\frac{x}{z}, \frac{y}{z} \right]^{\mathrm{T}} \tag{9.110d}$$

Parabolic mirror—Now, consider a parabolic mirror (Figure 9.23). The points Σ_1 and Σ_2 are orthographically projected to the points Σ_1' and Σ_2' on the image plane, that is, any line (e.g., a ray of light) incident with the focus F is reflected such that it is perpendicular to the image plane. Therefore, the projection of Σ is equivalent to its central projection followed by standard orthographic projection.

A paraboloid surface opening upwards (in the z direction) is described by:

$$z = \frac{1}{4a}(x^2 + y^2) - a \tag{9.111}$$

where a is its focal length, its axis is assumed to be the z-axis, and its focus is located at the origin.

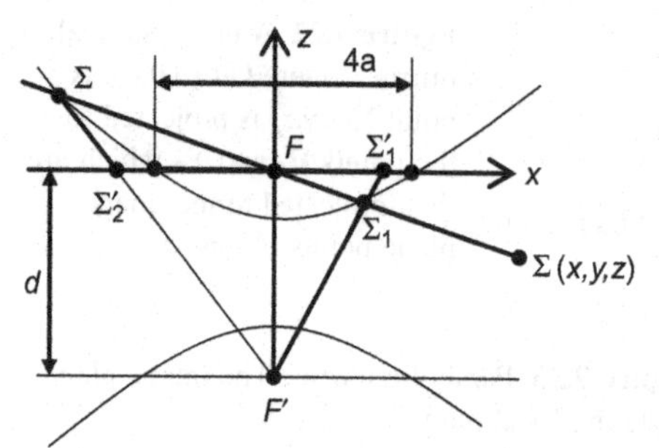

Figure 9.24 Cross section of hyperboloid mirror. The image plane is through the focal point F.

The projection of Σ to the paraboloid surface consists of the two antipodal points Σ_1 and Σ_2.

$$\Sigma_1\left(\frac{2ax}{\rho-z},\frac{2ay}{\rho-z},\frac{2az}{\rho-z}\right),\quad \Sigma_2\left(-\frac{2ax}{\rho+z},-\frac{2ay}{\rho+z},-\frac{2az}{\rho+z}\right) \tag{9.112}$$

where $\rho=\sqrt{x^2+y^2+z^2}$. These points are then orthographically projected to the plane $z=0$ giving the points Σ_1' and Σ_2':

$$\Sigma_1'\left(\frac{2ax}{\rho-z},\frac{2ay}{\rho-z}\right),\quad \Sigma_2'\left(-\frac{2ax}{\rho+z},-\frac{2ay}{\rho+z}\right) \tag{9.113}$$

Hyperbolic Mirror—Consider the hyperbolic mirror of Figure 9.24, where the image plane is again through the focal point F. The 3D point Σ of the world is projected to the antipodal points Σ_1 and Σ_2 which are then perspectively projected to Σ_1' and Σ_2' from the second focal point F', that is, here, rays incident with one of the focal points of the mirror are reflected into rays incident with the second focal point.

The focus F is at the origin $(x,y,z)=(0,0,0)$ and the focus F' is at $(x,y,z)=(0,0,-d)$. The surface of the hyperboloid is described by:

$$\frac{(z+d/2)^2}{p^2}-\frac{x^2+y^2}{q^2}=1 \tag{9.114a}$$

where:

$$p=\frac{1}{2}\left(\sqrt{d^2+4a^2}-2a\right),\quad q=\sqrt{a\sqrt{d^2+4a^2}-2a^2} \tag{9.114b}$$

The perspective projection (image) points Σ_1' and Σ_2' of Σ_1 and Σ_2 via the point (focus) $F'(0,0,-d)$ are:

$$\Sigma_1':\left(\frac{2xad/\sqrt{d^2+4a^2}}{\frac{d}{\sqrt{d^2+4a^2}}\rho-z},\frac{2yad/\sqrt{d^2+4a^2}}{\frac{d}{\sqrt{d^2+4a^2}}\rho-z}\right) \tag{9.115a}$$

$$\Sigma_2': \left(-\frac{2xad/\sqrt{d^2+4a^2}}{\frac{d}{\sqrt{d^2+4a^2}}\rho + z}, -\frac{2yad/\sqrt{d^2+4a^2}}{\frac{d}{\sqrt{d^2+4a^2}}\rho + z} \right) \tag{9.115b}$$

Ellipsoid Mirror—The surface equation of an ellipsoid with foci at (0,0,0) and $(0,0,-d)$ (see Figure 9.21C) and *lactus rectum*, $4a$, is:

$$\frac{(z+d/2)^2}{p^2} + \frac{(x^2+y^2)}{q^2} = 1 \tag{9.116a}$$

where:

$$p = \frac{1}{2}\left(\sqrt{d^2+4a^2} + 2a\right), \quad q = \sqrt{a\sqrt{d^2+4a^2} + 2a^2} \tag{9.116b}$$

Here, the projection (image) points Σ_1' and Σ_2' are:

$$\Sigma_1': \left(\frac{2xad}{\rho d + z\sqrt{d^2+4a^2}}, \frac{2yad}{\rho d + z\sqrt{d^2+4a^2}} \right) \tag{9.117}$$

$$\Sigma_2': \left(-\frac{2xad}{\rho d - z\sqrt{d^2+4a^2}}, -\frac{2yad}{\rho d - z\sqrt{d^2+4a^2}} \right) \tag{9.118}$$

Equations (9.115a), (9.115b), (9.117), and (9.118) show that the ellipsoid mirror gives the same reflection about $z=0$ as the hyperboloid mirror. Comparing Eqs. (9.110a), (9.110b), (9.113), (9.115a), (9.115b), (9.117), and (9.118) the validity of the catadioptric projection theorem, stated at the beginning of this section, is directly established. This theorem provides a unified geometric model of catadioptric projection, using the projective (image) plane induced by the projection point $\sigma_{1,d,\xi}(x,y,z)$ or $\sigma_{2,d,\xi}(x,y,z)$ (see Eqs. (9.110a) and (9.110b)).

Actually, the central catadioptric imaging can be modeled by projecting the scene on the surface of the sphere and then reprojecting these points on the image plane from a new projection point N (see Figure 9.22). This is illustrated in Figure 9.25 which shows that the projections (images) Σ_1' and Σ_2' of the point Σ to the catadioptric image plane Π via the parabolic mirror and orthographic projection, and via the spherical mirror through the point N and perspective projection are coincident. As an exercise, the reader can verify geometrically that this is true using the parabola and circle properties.

To describe formally this general catadioptric projection we draw the model shown in Figure 9.26, where F_c corresponds to the projection point N of Figure 9.22 [36,37].

Specifically, the mapping of 3D world points to points in the catadioptric image plane involves three stages as shown in Figure 9.27.

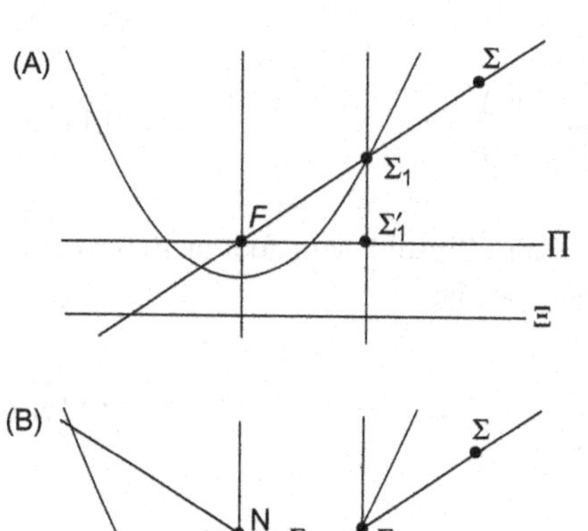

Figure 9.25 Parabolic orthographic image point Σ_1' (A) is coincident with spherical perspective image point Σ_2'(B).

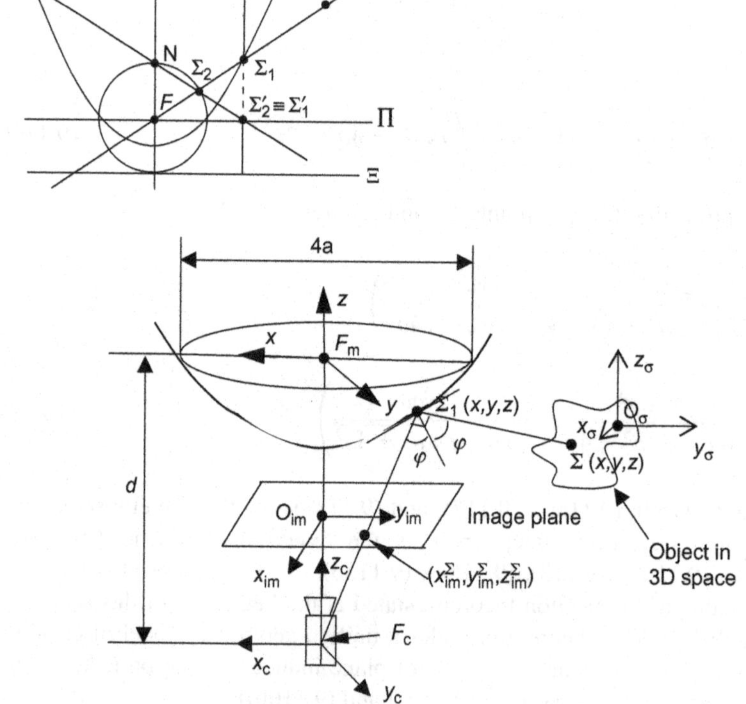

Figure 9.26 Central catadioptric projection of a point Σ of the 3D world.

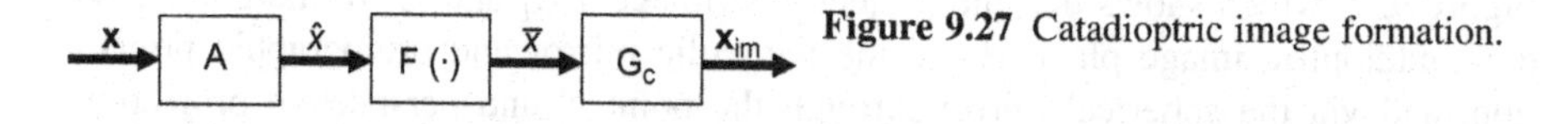

Figure 9.27 Catadioptric image formation.

The world point $\mathbf{x} = [x,y,z,1]^T$ is projected to the catadioptric image point $\mathbf{x}_{im} = [x_{im},y_{im},z_{im}]^T$. Each visible point Σ can be associated with a projective ray $\hat{\mathbf{x}}$ joining the point with the effective viewpoint of the system. The vectors $\hat{\mathbf{x}}$ and $\mathbf{x}$ are related by:

$$\hat{\mathbf{x}} = \mathbf{A}\mathbf{x} \tag{9.119a}$$

where $\mathbf{A}$ is a typical 3×4 projection matrix:

$$\mathbf{A} = \mathbf{R}[\mathbf{I}_{3\times3} \vdots - \mathbf{s}_0] \tag{9.119b}$$

with $\mathbf{R}$ being the rotation matrix between world and mirror coordinate systems, and $\mathbf{s}_0$ being the world's frame origin. Assuming, without loss of generality, that the world and sensor coordinate systems are the same, then:

$$\mathbf{A} = \begin{bmatrix} 1 & 0 & 1 \vdots & 0 \\ 0 & 1 & 0 \vdots & 0 \\ 0 & 0 & 1 \vdots & 0 \end{bmatrix} \tag{9.119c}$$

We regard the projective ray $\hat{\mathbf{x}}$ as a point in an oriented projective plane P^2 which is transformed to a point $\overline{\mathbf{x}} \in P^2$ by a nonlinear transformation $\mathbf{F}(\hat{\mathbf{x}})$. Then, the image point $\mathbf{x}_{\text{im}}$ is obtained via the transformation:

$$\mathbf{x}_{\text{im}} = \mathbf{G}_c\overline{\mathbf{x}}, \quad \mathbf{G}_c = \mathbf{Q}_c\mathbf{R}_c\mathbf{T}_c \tag{9.120}$$

where $\mathbf{Q}_c$ involves the camera intrinsic parameters, $\mathbf{R}_c$ is the rotation matrix between the camera and the mirror, and $\mathbf{T}_c$ depends on the shape of the mirror. The parameters d' and ξ' of this general model for parabolic, hyperbolic, elliptic, and perspective (spherical) systems are shown in Table 9.1, where $4a$ is the lactus rectum of the mirror and d the distance between the foci of the camera and mirror.

These parameters are found by comparing the spherical projection equation (9.110a) and (9.110b) and the parabolic, hyperbolic, and elliptic mirror projection equations (9.113), (9.115a), and (9.117). The function $\mathbf{F}(\mathbf{x})$ and the matrix $\mathbf{T}_c$ are given by:

$$\mathbf{F}(\mathbf{x}) = \begin{bmatrix} \dfrac{x}{z + d'\sqrt{x^2 + y^2 + z^2}} \\ \dfrac{y}{z + d'\sqrt{x^2 + y^2 + z^2}} \\ 1 \end{bmatrix} \tag{9.121a}$$

Table 9.1 Unified Catadioptric Model Parameters d' and ξ'

Mirror	d′	ξ′	Point
Parabolic	1	$2a - 1$	$(x, y, z, 1)$
Hyperbolic	$d/\sqrt{d^2 + 2a^2}$	$(2a - 1)d/\sqrt{d^2 + 2a^2}$	$(x, y, z, 1)$
Elliptic	$d/\sqrt{d^2 + 2a^2}$	$(2a - 1)d/\sqrt{d^2 + 2a^2}$	$(x, y, -z, 1)$
Perspective	0	1	$(x/z, y/z, 1)$

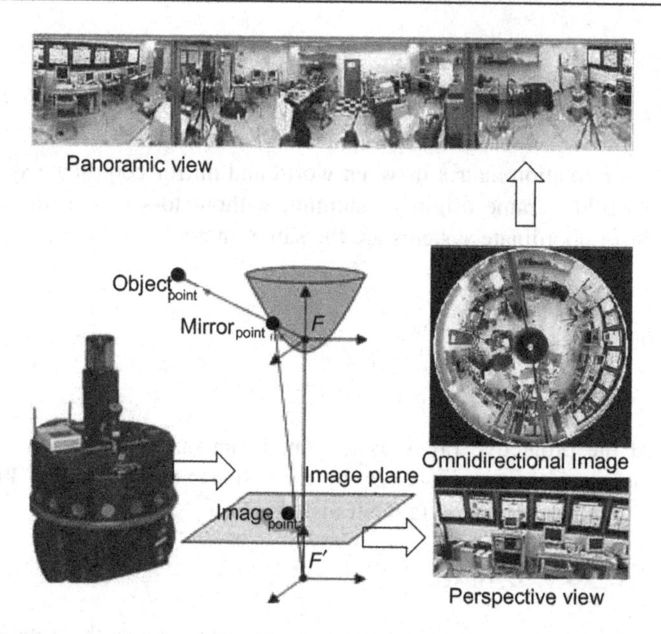

Figure 9.28 A hyperboloidal catadioptric vision system and examples of generated transformed images.

or, equivalently:

$$\mathbf{F}(\mathbf{x}) = \begin{bmatrix} x \\ y \\ z + d'\sqrt{x^2 + y^2 + z^2} \end{bmatrix} \tag{9.121b}$$

and

$$\mathbf{T}_c = \begin{bmatrix} \xi' + d' & 0 & 0 \\ 0 & \xi' + d' & 0 \\ 0 & 0 & 1 \end{bmatrix} \tag{9.121c}$$

Detailed derivations and extensions of the above results to more general models can be found in Refs. [38–43].

An example of hyperboloidal omnidirectional image and the corresponding panoramic (cylindrical) and perspective images, falling in the general catadioptric camera model of Figures 9.26 and 9.27, is shown in Figure 9.28.

Example 9.4

It is desired to derive the catadioptric image Jacobian matrix for a set of n points (features) Σ_i:$\mathbf{x}_{\sigma,1}, \mathbf{x}_{\sigma,2}, \ldots, \mathbf{x}_{\sigma,n}$ of an object shown in Figure 9.26.

Solution

Denote by $\mathbf{x}_\sigma$ an arbitrary point of the object in the local coordinate frame $O_\sigma(x_\sigma, y_\sigma, z_\sigma)$. If $\mathbf{R}$ is the rotation matrix between the frame O_σ and the mirror's reference frame $F(x,y,z)$, and $\mathbf{s}$ is the position vector of O_σ in mirror's coordinates then the homogeneous transformation from $F(x,y,z)$ to $O_\sigma(x_\sigma, y_\sigma, z_\sigma)$ can be expressed as:

$$\mathbf{x} = \mathbf{R}\mathbf{x}_\sigma + \mathbf{s} \tag{9.122}$$

The point $\mathbf{x}$ is projected to the point $\mathbf{x}_{\mathrm{im}_i} = \mathbf{F}_i(\mathbf{x})$ of the catadioptric image plane as shown by Eq. (9.121a). The motion of the object is described by the screw vector:

$$\dot{\mathbf{r}} = \begin{bmatrix} \mathbf{v} \\ \boldsymbol{\omega} \end{bmatrix}$$

and so the 3D velocity of the point Σ_i due to the object motion is:

$$\dot{\mathbf{x}} = \mathbf{J}_0\dot{\mathbf{r}} \tag{9.123}$$

Now, if $\mathbf{J}_i$ is the Jacobian matrix of the first two lines of the function $\mathbf{f}_i(\mathbf{x})$ in Eq. (9.121a), then the image Jacobian $\mathbf{J}_{\mathrm{im}}$ from $\dot{\mathbf{x}}_{\sigma,i}$ to $\dot{\mathbf{r}}$ is given by:

$$\begin{bmatrix} \dot{\mathbf{x}}_{\sigma,1} \\ \dot{\mathbf{x}}_{\sigma,2} \\ \vdots \\ \dot{\mathbf{x}}_{\sigma,n} \end{bmatrix} = \mathbf{J}_{\mathrm{im}}\dot{\mathbf{r}} = \begin{bmatrix} \mathbf{J}_{\mathrm{im}}^1 \\ \mathbf{J}_{\mathrm{im}}^2 \\ \vdots \\ \mathbf{J}_{\mathrm{im}}^n \end{bmatrix} \dot{\mathbf{r}} \tag{9.124a}$$

where:

$$\mathbf{J}_{\mathrm{im}}^i = \mathbf{J}_i\mathbf{J}_0 \tag{9.124b}$$

From Eq. (9.121a) we find that [37]:

$$\mathbf{J}_i = \frac{1}{\rho(z+\zeta\rho)^2} \begin{bmatrix} \rho z + \zeta(y^2+z^2) & -\zeta xy & -x(\rho+\zeta z) \\ \zeta xy & -[\rho z + \zeta(x^2+z^2)] & y(\rho+\zeta z) \end{bmatrix} \tag{9.125a}$$

where $\zeta = d'$ and $\rho = \sqrt{x^2+y^2+z^2}$. The matrix $\mathbf{J}_0$ is equal to $\mathbf{J}_0 = [\mathbf{I}_{3\times 3} \vdots \mathbf{S}(\mathbf{x})]$, where $\mathbf{S}(\mathbf{x})$ is the skew symmetric matrix associated with $\mathbf{x}$, that is (see (9.4a) and (9.4b)):

$$\mathbf{S}(\mathbf{x}) = \begin{bmatrix} 0 & z & -y \\ -z & 0 & x \\ y & -x & 0 \end{bmatrix} \tag{9.125b}$$

Using the above expressions for $\mathbf{J}_i$ and $\mathbf{J}_0$, Eq. (9.124b) gives:

$$\mathbf{J}_{\mathrm{im}}^i = [(\mathbf{J}_{\mathrm{im}}^i)_1 \vdots (\mathbf{J}_{\mathrm{im}}^i)_2]$$

$$
(\mathbf{J}^i_{\text{im}})_1 = \begin{bmatrix} x_{\sigma,i}\, y_{\sigma i} & \dfrac{(1+x^2_{\sigma,i})\mu - y^2_{\sigma,i}\zeta}{\mu+\zeta} & y_{\sigma,i} \\ \dfrac{(1+y^2_{\sigma,i})\mu - x^2_{\sigma,i}\zeta}{\mu+\zeta} & x_{\sigma,i} y_{\sigma,i} & -x_{\sigma,i} \end{bmatrix}
$$

$$
(\mathbf{J}^i_{\text{im}})_2 = \begin{bmatrix} \dfrac{1+x^2_{\sigma,i}[1-\zeta(\mu+\zeta)]y^2_{\sigma,i}}{\rho(\mu+\zeta)} & \dfrac{x_{\sigma,i} y_{\sigma,i}\zeta}{\rho} & \dfrac{-x_{\sigma i}\mu}{\rho} \\ \dfrac{-x_{\sigma,i} y_{\sigma,i}\zeta}{\rho} & \dfrac{1+x^2_{\sigma,i}+y^2_{\sigma,i}[1-\zeta(\mu+\zeta)]}{\rho(\mu+\zeta)} & \dfrac{-y_{\sigma,i}\mu}{\rho} \end{bmatrix}
\tag{9.126}
$$

where $\mu = \sqrt{1+(x^2_{\sigma,i}+y^2_{\sigma,i})(1-\zeta^2)}$. We can verify that when $\zeta = 0$ (perspective projection), we get the standard Jacobian matrix of perspective cameras. In fact, in this case Eq. (9.121a) becomes the standard perspective transformation $\mathbf{f}_i = [x/z, -y/z]$, and $\mathbf{J}_i$ takes the form:

$$
\mathbf{J}_i = \mathbf{J}_{a,i}\mathbf{J}_{c,i} \tag{9.127a}
$$

$$
\mathbf{J}_{a,i} = \begin{bmatrix} \dfrac{z(\rho z + \zeta(y^2+z^2))}{\rho(z+\zeta\rho)^2} & \dfrac{\zeta xyz}{\rho(z+\zeta\rho)^2} \\ \dfrac{\zeta xyz}{\rho(z+\zeta\rho)^2} & \dfrac{z(\rho z + \zeta(x^2+z^2))}{\rho(z+\zeta\rho)^2} \end{bmatrix} \tag{9.127b}
$$

$$
\mathbf{J}_{c,i} = \begin{bmatrix} 1/z & 0 & -x/z^2 \\ 0 & -1/z & y/z^2 \end{bmatrix} \tag{9.127c}
$$

For $\zeta = 0$, the matrix $\mathbf{J}_{a,i}$ reduces to the identity matrix, and so:

$$
\mathbf{J}_i = \begin{bmatrix} 1/z & 0 & -x/z^2 \\ 0 & -1/z & y/z^2 \end{bmatrix} \tag{9.127d}
$$

which is the standard Jacobian matrix (see Eqs. (9.13a) and (9.31b)). The above Jacobian matrix can be used directly for visual servoing based on catadioptric omnidirectional vision.

9.9.3 Omnidirectional Vision-Based Mobile Robot Visual Servoing

Having available the catadioptric image Jacobian matrix, the visual servoing of a fixed or mobile robot can be performed in the usual way (see, e.g., Section 9.8), that is, using the feedback control law (see Eq. (9.34)):

$$
\mathbf{u} = \dot{\mathbf{r}} = \mathbf{J}^{\dagger}_{\text{im}}\mathbf{K}\mathbf{e}_\sigma, \quad \mathbf{e}_\sigma = \mathbf{x}_{\sigma,\text{d}} - \mathbf{x}_\sigma
$$

where $\mathbf{e}_\sigma = [\mathbf{e}_{\sigma,1}^{\mathrm{T}}, \mathbf{e}_{\sigma,2}^{\mathrm{T}}, \ldots, \mathbf{e}_{\sigma,n}^{\mathrm{T}}]^{\mathrm{T}}$, and:

$$\mathbf{J}_{\mathrm{im}} = \begin{bmatrix} \mathbf{J}_{\mathrm{a},1} & 0 & \cdots & 0 \\ 0 & \mathbf{J}_{\mathrm{a},2} & \cdots & 0 \\ \cdots & \cdots & \cdots & \cdots \\ \cdots & \cdots & \cdots & \cdots \\ 0 & 0 & \cdots & \mathbf{J}_{\mathrm{a},n} \end{bmatrix} \begin{bmatrix} \mathbf{J}_{\mathrm{c},1} \\ \mathbf{J}_{\mathrm{c},2} \\ \vdots \\ \mathbf{J}_{\mathrm{c},n} \end{bmatrix}$$

with $\mathbf{J}_{\mathrm{a},i}$ given by Eq. (9.127b) and $\mathbf{J}_{\mathrm{c},i}$ by Eq. (9.127c), $i = 1,2,\ldots,n$. The generalized inverse $\mathbf{J}_{\mathrm{im}}^{\dagger}$ is given by Eqs. (9.31a) and (9.31b). To get a realizable solution, the number n of object points (features) must be larger than the number of the task degrees of freedom in which case we use Eq. (9.31a).

Now, consider the case of mobile robot image-based visual servoing using a perspective camera-hyperbolic mirror catadioptric sensor. Since the extraction of target feature parameters, needed to implement catadioptric visual feedback, is computationally very demanding, a feasible approach is to consider a small pixel region that corresponds to the target. However, for a moving target we need to continuously update the coordinates of the target, in order to make possible a real-time computation of the target parameters. Using a convenient landmark on the target (e.g., a ■ drawn on it) the target visual tracking can be done in two steps:

Step 1: Localization of the target in the entire image (this is done off line).
Step 2: Real-time visual tracking with initial position coordinates of the target the coordinates found in step 1.

To apply the real-time tracking we can use a discrete time affine model describing the target movement between two successive times t and $t + \Delta t$ which involves a translation and rotation displacement specified by a vector $\mathbf{d}$ and a rotation matrix $\mathbf{R}$. Since the affine transformation cannot be applied directly to the omnidirectional image, this image is first transformed to a perspective image $(x_{\mathrm{im}}, y_{\mathrm{im}})$ described by the parameters l_{fp}, ϕ_{a}, and ϕ_{e} (see Example 4.1, Eqs. (4.5a), (4.5b), (4.6a), and (4.6b)). The updated pose $(\phi_{\mathrm{new}}, \psi_{\mathrm{new}})$ is computed using the relations:

$$\phi_{\mathrm{new}} = \phi + tg^{-1}\left(\frac{\mathrm{d}y_{\mathrm{im}}}{l_{\mathrm{f}}}\right), \quad \psi_{\mathrm{new}} = \psi + tg^{-1}\left(\frac{\mathrm{d}x_{\mathrm{im}}}{l_{\mathrm{f}}}\right)$$

where $\mathrm{d}x_{\mathrm{im}}$ and $\mathrm{d}y_{\mathrm{im}}$ are the horizontal and vertical translation, respectively, of the target center. In this way, we assure that the target remains always at the center of the perspective image. Clearly, this process would be the same if a virtual perspective camera mounted on a pan-tilt mechanism is used, with its focus always placed at the hyperbola focus (see Section 9.8). Now, the standard image-based visual servoing procedure (described in Section 9.8) can be applied. Choosing the target region, being tracked, to be a square, and the target features to be the coordinates in the perspective image of the four corners ($i = 1,2,3,4$) of this square, then the feature vector $\mathbf{f}$ is (see Figure 4.16):

$$\mathbf{f} = [x_{\mathrm{im},1}, x_{\mathrm{im},2}, x_{\mathrm{im},3}, x_{\mathrm{im},4}, y_{\mathrm{im},1}, y_{\mathrm{im},2}, y_{\mathrm{im},3}, y_{\mathrm{im},4}]^{\mathrm{T}}$$

The motion of the perspective camera frame is described by the vector (see Eq. (9.9)):

$$\dot{\mathbf{r}}_{\mathrm{c}} = \begin{bmatrix} \mathbf{v}_{\mathrm{c}} \\ \boldsymbol{\omega}_{\mathrm{c}} \end{bmatrix} = [v_x^{\mathrm{c}}, v_y^{\mathrm{c}}, v_z^{\mathrm{c}}, \omega_x^{\mathrm{c}}, \omega_y^{\mathrm{c}}, \omega_z^{\mathrm{c}}]^{\mathrm{T}}$$

The relation of $\dot{\mathbf{f}}$ and $\dot{\mathbf{r}}_{\mathrm{c}}$ is given by Eq. (9.9):

$$\dot{\mathbf{f}} = \mathbf{J}_{\mathrm{im}}(\mathbf{r})\dot{\mathbf{r}}_{\mathrm{c}}$$

where $\dot{\mathbf{f}}$ has eight rows and six columns. Assuming that the square target region is centered at the center of the image and has dimension 2η, the feature vector $\mathbf{f}_T$ becomes:

$$\mathbf{f}_{\mathrm{T}} = [-\eta, \eta, \eta, -\eta, -\eta, \eta, \eta, -\eta]^{\mathrm{T}}$$

and the Jacobian matrix $\mathbf{J}_{\mathrm{im}}]_{\mathbf{f}\equiv\mathbf{f}_T}$ has four 2×6 blocks of the form of Eq. (9.15), where we can put $z = 1$ without loosing convergence. The image-based feedback control law is as usual:

$$\dot{\mathbf{r}} = \mathbf{J}_{\mathrm{im}}^{\dagger}\mathbf{K}\mathbf{e}_{\mathrm{T}}(\mathbf{f}_{\mathrm{T}}), \quad \mathbf{e}_{\mathrm{T}}(\mathbf{f}_{\mathrm{T}}) = \mathbf{f}_{\mathrm{T,d}} - \mathbf{f}_{\mathrm{T}}$$

where $\mathbf{f}_{\mathrm{T,d}}$ is the desired feature vector. The dimensionality of $\mathbf{J}$ can be reduced by taking into account the fact that here we have $v_y^{\mathrm{c}} = 0$, since the robot and the camera move only on a horizontal plane (see Figure 4.16). Considering a synchrodrive WMR, the control vector is $\mathbf{u} = [v_z^{\mathrm{p}}, \omega_\psi^{\mathrm{p}}]$, where v_z^{p} is the translational speed, and ω_ψ^{p} the steering speed of the robot platform. Here, we can also assume that $\omega_x^{\mathrm{c}} = \omega_z^{\mathrm{c}} = 0$. Therefore, $\dot{\mathbf{r}}_{\mathrm{c}}$ becomes:

$$\dot{r}_{\mathrm{c}} = [v_x^{\mathrm{c}}, v_z^{\mathrm{c}}, \omega_y^{\mathrm{c}}]$$

Example 9.5

Derive the kinematic transformation of a 3D point, that moves on a panoramic (cylindrical) image plane, to the perspective camera image plane.

Solution

Consider a hyperboloidal mirror-perspective camera vision system of the form shown in Figure 9.26. The corresponding virtual cylindrical image plane has the form of Figure 9.29.

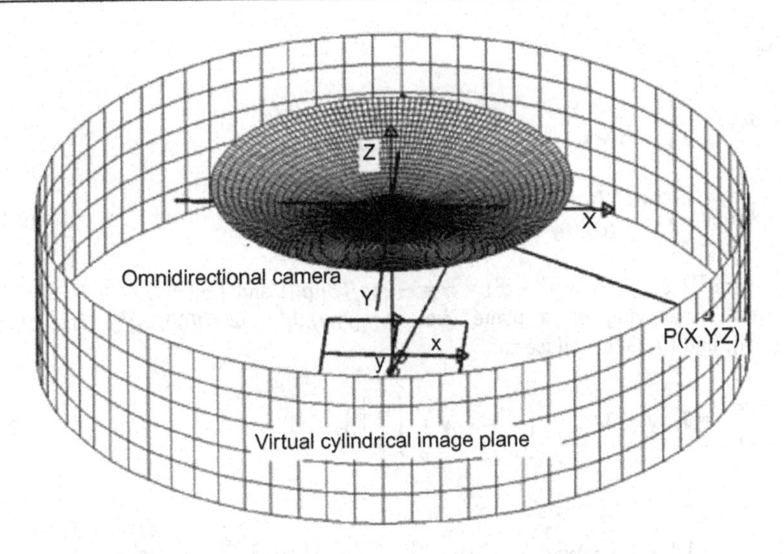

Figure 9.29 Virtual cylindrical image plane.
Source: Reprinted from [23], with permission from International Journal of Control, Automation and Systems.

The world point $P(X,Y,Z)$ is mapped to the point (x,y) of the image plane after reflection of the light ray originated from P at the mirror. The focal points of the mirror and camera are denoted by F_m and F_c, respectively. The hyperboloidal mirror surface equation is:

$$(Z+c)^2/p^2 - (X^2+Y^2)/q^2 = 1, \quad c = \sqrt{p^2+q^2} \tag{9.128}$$

In terms of the parameters p,q,c and l_f (the camera focal length), the coordinates of the image point of $P(X,Y,Z)$ are given by (see Eqs. (9.114a), (9.114b), and (9.121a)):

$$\begin{bmatrix} x \\ y \end{bmatrix} = \mu \begin{bmatrix} X \\ Y \end{bmatrix}, \quad \mu = \frac{l_f(p^2-c^2)}{(p^2+c^2)Z - 2pc\sqrt{X^2+Y^2+Z^2}} \tag{9.129}$$

The projection of a point P of the cylindrical surface to the image plane is given by:

$$\lambda \mathbf{x} = \mathbf{A}\mathbf{p} \tag{9.130}$$

where $\mathbf{x} = [x,y,0]^{\mathrm{T}}, \mathbf{A} = \mathrm{diag}[1,1,0]$, $\mathbf{p} = [X,Y,Z]^{\mathrm{T}}$, and $\lambda = 1/\mu$. As we know, the motion of a 3D space point $\mathbf{p} = [X,Y,Z]^{\mathrm{T}}$ which is related to the camera motion is described by the linear velocity $\mathbf{v}(t)$ and angular velocity $\boldsymbol{\omega}(t)$ as shown in Eq. (9.2a) and (9.2b), that is, $\dot{\mathbf{p}} = -(\boldsymbol{\omega} \times \mathbf{p} + \mathbf{v})$ where $\mathbf{v} = [v_x,v_y,v_z]^{\mathrm{T}}$ and $\boldsymbol{\omega} = [\omega_x,\omega_y,\omega_z]^{\mathrm{T}}$. Therefore, differentiating Eq. (9.130) with respect to time we get $\lambda\dot{\mathbf{x}} + \dot{\lambda}\mathbf{x} = \mathbf{A}\dot{\mathbf{p}}$ which gives:

$$\begin{bmatrix} \dot{x} \\ \dot{y} \end{bmatrix} = \mathbf{J}_{0,v}\mathbf{v} + \mathbf{J}_{0,\omega}\boldsymbol{\omega} \tag{9.131}$$

where:

$$\mathbf{J}_{0,v} = \frac{1}{\lambda}\begin{bmatrix} -1+\gamma\lambda x^2 & \gamma\lambda xy & (\alpha/\beta)x+\gamma xZ \\ \gamma\lambda xy & -1+\gamma\lambda y^2 & (\alpha/\beta)y+\gamma yZ \end{bmatrix} \tag{9.132a}$$

$$\mathbf{J}_{0,\omega} = \begin{bmatrix} (\alpha/\beta)xy & -Z/\lambda-(\alpha/\beta)x^2 & y \\ Z/\lambda+(\alpha/\beta)y^2 & -(\alpha/\beta)xy & -x \end{bmatrix} \tag{9.132b}$$

with $\alpha = p^2 + c^2$, $\beta = l_f(p^2 - c^2)$, $\gamma = -2pc/(\beta\|\mathbf{p}\|)$ and $\|\mathbf{p}\| = \sqrt{X^2+Y^2+Z^2}$. Since the WMR is moving on a plane (i.e., $\mathbf{v} = [v_x, v_y, 0]^T$, $\omega = [0,0,\omega_z]^T$), Eqs. (9.131), (9.132a), and (9.132b) reduce to:

$$\begin{bmatrix} \dot{x} \\ \dot{y} \end{bmatrix} = \mathbf{J}_{0,v}\mathbf{v}' + \mathbf{J}_{0,\omega}\omega \quad \left(\omega = \omega_z, \mathbf{v}' = \begin{bmatrix} v_x \\ v_y \end{bmatrix}\right) \tag{9.133}$$

where:

$$\mathbf{J}_{0,v} = \frac{1}{\lambda}\begin{bmatrix} -1+\gamma\lambda x^2 & \gamma\lambda xy \\ \gamma\lambda xy & -1+\gamma\lambda y^2 \end{bmatrix}, \quad \mathbf{J}_{0,\omega} = \begin{bmatrix} y \\ -x \end{bmatrix} \tag{9.134}$$

Example 9.6

It is desired to find the image Jacobian matrix of a camera WMR system, where the camera is fixed to the ceiling above the WMR such as the $x_c y_c$ camera plane is parallel to the image plane $x_{im} y_{im}$.

Solution

We consider the system configuration shown in Figure 9.30 [44].

We have the following coordinate frames:

$O_w x_w y_w z_w$ is world coordinate frame.
$O_c x_c y_c z_c$ is camera coordinate frame.
$O_{im} x_{im} y_{im} z_{im}$ is image plane coordinate frame.

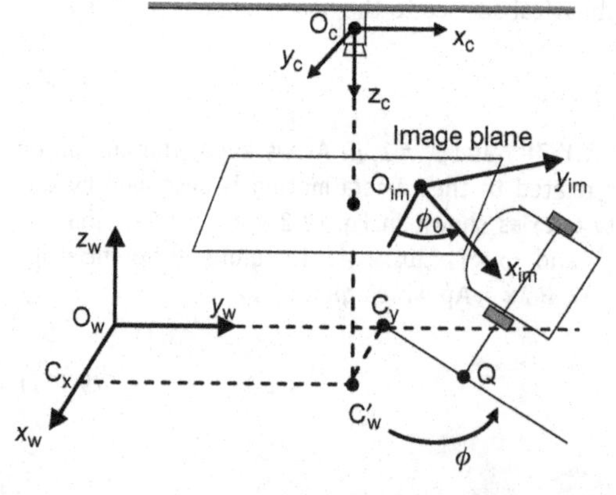

Figure 9.30 Geometry of the vision and robot system.

Let $C'_w(c_x, c_y)$ be the point where the camera optical axis crosses the $O_w x_w y_w$ (robot) plane (assumed parallel to the image plane), $(c_{x,\mathrm{im}}, c_{y,\mathrm{im}})$ the coordinates of the image of O_c in the image plane, $Q(x_Q, y_Q)$ the position of the WMR in the world coordinate frame, and $(x_{Q,\mathrm{im}}, y_{Q,\mathrm{im}})$ the image of (x_Q, y_Q).

Then, it is easy to find that the camera perspective model is:

$$\mathbf{x}_{Q,\mathrm{im}} = \mathbf{\Lambda}\mathbf{R}(\phi_0)[\mathbf{x}_Q - \mathbf{c}_w] + \mathbf{c}_{\mathrm{im}} \tag{9.135a}$$

where:

$$\mathbf{x}_{Q,\mathrm{im}} = \begin{bmatrix} x_{Q,\mathrm{im}} \\ y_{Q,\mathrm{im}} \end{bmatrix}, \mathbf{x}_Q = \begin{bmatrix} x_Q \\ y_Q \end{bmatrix}, \mathbf{c}_w = \begin{bmatrix} c_x \\ c_y \end{bmatrix}, \mathbf{c}_{\mathrm{im}} = \begin{bmatrix} c_{x,\mathrm{im}} \\ c_{y,\mathrm{im}} \end{bmatrix} \tag{9.135b}$$

and $\mathbf{\Lambda} = \mathrm{diag}[\lambda_1, \lambda_2]$ with λ_1, λ_2 being constants that are specified by the vision system parameters (depth, focal length, scaling factors in the x_{im} and y_{im} axis respectively). The rotation matrix $\mathbf{R}(\phi_0)$ is given by:

$$\mathbf{R}(\phi_0) = \begin{bmatrix} \cos\phi_0 & \sin\phi_0 \\ -\sin\phi_0 & \cos\phi_0 \end{bmatrix} \tag{9.136}$$

where ϕ_0 is the angle between the x_{im} axis and the x_w axis, positive in the anticlockwise direction.

Carrying out the algebraic calculations in Eq. (9.135a), using Eq. (9.136), we get [44]:

$$\mathbf{f} = \mathbf{G}\mathbf{x} + \mathbf{F} \tag{9.137a}$$

where:

$$\mathbf{f} = [\mathbf{x}_{Q,\mathrm{im}}^{\mathrm{T}}, \phi]^{\mathrm{T}}, \mathbf{x} = [\mathbf{x}_Q^{\mathrm{T}}, \phi]^{\mathrm{T}} \tag{9.137b}$$

$$\mathbf{G} = \begin{bmatrix} \lambda_1\cos\phi_0 & \lambda_1\sin\phi_0 & 0 \\ -\lambda_2\sin\phi_0 & \lambda_2\cos\phi_0 & 0 \\ 0 & 0 & 1 \end{bmatrix}, \quad \mathbf{F} = \begin{bmatrix} -c_x\lambda_1\cos\phi_0 & -c_y\lambda_1\sin\phi_0 + c_{x,\mathrm{im}} \\ c_x\lambda_2\sin\phi_0 & -c_y\lambda_2\cos\phi_0 + c_{y,\mathrm{im}} \end{bmatrix} \tag{9.137c}$$

Differentiating Eq. (9.137a) with respect to time we obtain:

$$\dot{\mathbf{f}} = \mathbf{J}_{\mathrm{im}}\dot{\mathbf{r}}, \quad \dot{\mathbf{r}} = [v, \omega]^{\mathrm{T}} \tag{9.138a}$$

where $\mathbf{J}_{\mathrm{im}}$ is the image Jacobian matrix:

$$\mathbf{J}_{\mathrm{im}} = \begin{bmatrix} \lambda_1\cos(\phi - \phi_0) & 0 \\ \lambda_2\sin(\phi - \phi_0) & 0 \\ 0 & 1 \end{bmatrix} \tag{9.138b}$$

which represents the kinematic model of the camera robot system, from the WMR screw vector $\dot{\mathbf{r}}$ to the vision feature rate vector $\dot{\mathbf{f}}$.

Example 9.7

The problem is to find a vision-based control law $\boldsymbol{\tau}(t)$ which drives the system trajectory $\mathbf{x}(t) = [x_Q, y_Q, \phi]^T$ of a differential-drive robot to a desired trajectory $\mathbf{x}_d(t) = [x_{Q,d}, y_{Q,d}, \phi_d]^T$ and is robust against a bounded unknown disturbance $\mathbf{d}(t)$ with:

$$||\mathbf{d}(t)|| \leq d_{max} \tag{9.139}$$

Solution

We will apply a two-step backstepping procedure, assuming as usual that the desired trajectory satisfies the same kinematic equations as the WMR [44,45]. Working with the camera robot configuration of Figure 9.30, the desired trajectory satisfies the kinematic model (9.138a) of the robot trajectory on the image plane:

$$\dot{\mathbf{f}}_d = \mathbf{J}_{im}\dot{\mathbf{r}}, \quad \dot{\mathbf{r}} = [v, \omega]^T \tag{9.140a}$$

where $\mathbf{f}_d = [x_{Q,d}, y_{Q,d}, \phi_d]^T$ and:

$$\mathbf{J}_{im} = \begin{bmatrix} \lambda_1 \cos(\phi_d - \phi_0) & 0 \\ \lambda_2 \sin(\phi_d - \phi_0) & 0 \\ 0 & 1 \end{bmatrix} \tag{9.140b}$$

Kinematic Controller

Without loss of generality we assume that $\lambda_1 = \lambda_2 = \lambda (0 < \lambda_{min} \leq \lambda \leq \lambda_{max})$, which is the unknown vision system parameter. The feature vector error:

$$\tilde{\mathbf{f}} = \mathbf{f}_d - \mathbf{f} \tag{9.141}$$

expressed in the WMR (local) coordinate frame is (see Eqs. (5.51) and (7.69)):

$$\boldsymbol{\varepsilon} = \mathbf{E}\tilde{\mathbf{f}}, \quad \boldsymbol{\varepsilon} = [\varepsilon_1, \varepsilon_2, \varepsilon_3]^T \tag{9.142a}$$

where:

$$\mathbf{E} = \begin{bmatrix} \cos(\phi - \phi_0) & \sin(\phi - \phi_0) & 0 \\ -\sin(\phi - \phi_0) & \cos(\phi - \phi_0) & 0 \\ 0 & 0 & 1 \end{bmatrix} \tag{9.142b}$$

Therefore, since $\mathbf{E}$ is invertible, $\boldsymbol{\varepsilon} \to \mathbf{0}$ if and only if $\tilde{\mathbf{f}} \to \mathbf{0}$. Now, in analogy to Eq. (7.70), using Eqs. (9.138a), (9.138b), (9.140a), (9.140b), (9.142a), and (9.142b) we get the following error dynamics:

$$\begin{aligned} \dot{\varepsilon}_1 &= \omega\varepsilon_2 - \lambda v + \lambda v_d \cos \varepsilon_3 \\ \dot{\varepsilon}_2 &= -\omega\varepsilon_1 + \lambda v_d \sin \varepsilon_3 \\ \dot{\varepsilon}_3 &= \omega_d - \omega \end{aligned} \tag{9.143}$$

Therefore, working as in Example 7.1 (Eqs. (7.71) and (7.72)), we choose the following candidate Lyapunov function [44,45]:

$$V = \frac{1}{2}(\varepsilon_1^2 + \varepsilon_2^2 + \lambda\varepsilon_3^2) \tag{9.144}$$

Differentiating V with respect to time and introducing Eq. (9.143) into the result we find:

$$\begin{aligned}\dot{V} &= \varepsilon_1\dot{\varepsilon}_1 + \varepsilon_2\dot{\varepsilon}_2 + \lambda\varepsilon_3\dot{\varepsilon}_3\\ &= \varepsilon_1(\omega\varepsilon_2 - \lambda v_{\mathrm{m}} - \lambda\tilde{v} + \lambda v_{\mathrm{d}}\cos\varepsilon_3)\\ &\quad + \varepsilon_2(-\omega\varepsilon_1 + \lambda v_{\mathrm{d}}\sin\varepsilon_3) + \lambda\varepsilon_3(\omega_{\mathrm{d}} - \omega_{\mathrm{m}} - \tilde{\omega})\\ &= \varepsilon_1\lambda(-v_{\mathrm{m}} + v_{\mathrm{d}}\cos\varepsilon_3) + \varepsilon_3\lambda\left[\varepsilon_2 v_{\mathrm{d}}\left(\frac{\sin\varepsilon_3}{\varepsilon_3}\right) + \omega_{\mathrm{d}} - \omega_{\mathrm{m}}\right] - \lambda\varepsilon_1\tilde{v} - \lambda\varepsilon_3\tilde{\omega}\end{aligned} \tag{9.145}$$

Therefore, selecting:

$$\begin{aligned}&-v_{\mathrm{m}} + v_{\mathrm{d}}\cos\varepsilon_3 = -K_1\varepsilon_1\\ &\varepsilon_2 v_{\mathrm{d}}\left(\frac{\sin\varepsilon_3}{\varepsilon_3}\right) + \omega_{\mathrm{d}} - \omega_{\mathrm{m}} = -K_3\varepsilon_3\end{aligned}$$

that is:

$$v_{\mathrm{m}} = K_1\varepsilon_1 + v_{\mathrm{d}}\cos\varepsilon_3 \tag{9.146a}$$

$$\omega_{\mathrm{m}} = \omega_{\mathrm{d}} + K_3\varepsilon_3 + \varepsilon_2 v_{\mathrm{d}}\left(\frac{\sin\varepsilon_3}{\varepsilon_3}\right) \tag{9.146b}$$

we get:[4]

$$\dot{V} = -\lambda K_1\varepsilon_1^2 - \lambda K_3\varepsilon_3^2 - \lambda\varepsilon_1\tilde{v} - \lambda\varepsilon_3\tilde{\omega} \tag{9.146c}$$

The first two terms of $\dot{V}$ are negative, and so, for asymptotic trajectory tracking and parameter convergence, we have to ensure that $\tilde{v}$ and $\tilde{\omega}$ tend asymptotically to zero. This will be done by the dynamic adaptive controller that follows.

Adaptive Controller

We will use the dynamic model of Eq. (3.29) enhanced with the external disturbance $\mathbf{d}(t)$:

$$\overline{\mathbf{D}}\mathbf{v} = \overline{\mathbf{E}}\boldsymbol{\tau} + \mathbf{d}(t) \tag{9.147a}$$

[4] This is a valid controller since for $\varepsilon_3 \to 0, (\sin\varepsilon_3)/\varepsilon_3 \to 1$.

where:

$$\mathbf{v} = \begin{bmatrix} v \\ \omega \end{bmatrix}, \mathbf{d} = \begin{bmatrix} d_1 \\ d_2 \end{bmatrix}, \bar{\mathbf{D}} = \begin{bmatrix} m & 0 \\ 0 & \mathrm{I} \end{bmatrix}, \bar{\mathbf{E}} = \frac{1}{r}\begin{bmatrix} 1 & 1 \\ 2a & -2a \end{bmatrix} \tag{9.147b}$$

Using the velocity error vector:

$$\tilde{\mathbf{v}} = \begin{bmatrix} \tilde{v} \\ \tilde{\omega} \end{bmatrix} = \begin{bmatrix} v - v_{\mathrm{m}} \\ \omega - \omega_{\mathrm{m}} \end{bmatrix} = \mathbf{v} - \mathbf{v}_{\mathrm{m}} \tag{9.148}$$

the model of Eq. (9.147a) can be written as:

$$\bar{\mathbf{D}}\tilde{\mathbf{v}} = -\mathbf{M}\boldsymbol{\beta} + \bar{\mathbf{E}}\boldsymbol{\tau} + \mathbf{d}(t) \tag{9.149a}$$

where the relation:

$$\begin{gathered} \bar{\mathbf{D}}\mathbf{v}_{\mathrm{m}} = \begin{bmatrix} m & 0 \\ 0 & I \end{bmatrix}\begin{bmatrix} v_{\mathrm{m}} \\ \omega_{\mathrm{m}} \end{bmatrix} = \begin{bmatrix} v_{\mathrm{m}} & 0 \\ 0 & \omega_{\mathrm{m}} \end{bmatrix}\begin{bmatrix} \beta_1 \\ \beta_2 \end{bmatrix} = \mathbf{M}\boldsymbol{\beta} \\ \mathbf{M} = \begin{bmatrix} v_{\mathrm{m}} & 0 \\ 0 & \omega_{\mathrm{m}} \end{bmatrix}, \boldsymbol{\beta} = \begin{bmatrix} \beta_1 \\ \beta_2 \end{bmatrix} = \begin{bmatrix} m \\ I \end{bmatrix} \end{gathered} \tag{9.149b}$$

was used. The matrix $\mathbf{M}$ is a known as *linear regressor matrix* that multiplies the unknown parameter vector $\boldsymbol{\beta}$ (see Eq. (3.102)).

To design the dynamic adaptive controller, we add to V in Eq. (9.144) an extra Lyapunov function term V' defined as:

$$V' = \frac{1}{2}\tilde{\mathbf{v}}^{\mathrm{T}}\bar{\mathbf{D}}\tilde{\mathbf{v}} + \frac{1}{2}\tilde{\boldsymbol{\beta}}^{\mathrm{T}}\boldsymbol{\Gamma}^{-1}\tilde{\boldsymbol{\beta}} + \frac{1}{2\gamma_3}\tilde{\lambda}^2 \tag{9.150}$$

where:

$$\begin{gathered} \tilde{\boldsymbol{\beta}} = \boldsymbol{\beta} - \hat{\boldsymbol{\beta}}, \boldsymbol{\beta} = [\beta_1, \beta_2]^T = \begin{bmatrix} m & I \end{bmatrix}^T, \tilde{\lambda} = \lambda - \hat{\lambda} \\ \boldsymbol{\Gamma} = \mathrm{diag}[\gamma_1, \gamma_2], \gamma_1 > 0, \gamma_2 > 0, \gamma_3 > 0 \end{gathered} \tag{9.151}$$

We have seen that the time derivative of V along the trajectory of the system described by Eq. (9.143), with the feedback controls of Eqs. (9.146a) and (9.146b) leads to the expression (9.146c). Therefore, the time derivative of the total Lyapunov function $V = V + V'$ along the trajectories of the system in Eqs. (9.143), (9.149a), and (9.149b) is given by:

$$\begin{aligned}\dot{V}_0 =& -\lambda K_1\varepsilon_1^2-\lambda K_3\varepsilon_3^2-\lambda\varepsilon_1\tilde{v}-\lambda\varepsilon_3\tilde{\omega}\\&+\tilde{\mathbf{v}}^{\mathrm{T}}(-\mathbf{M}\boldsymbol{\beta}+\overline{\mathbf{E}}\boldsymbol{\tau}+\mathbf{d})+\tilde{\boldsymbol{\beta}}\boldsymbol{\Gamma}^{-1}\dot{\tilde{\boldsymbol{\beta}}}+\frac{1}{\gamma_3}\tilde{\lambda}\dot{\tilde{\lambda}}\\=& -\lambda K_1\varepsilon_1^2-\lambda K_3\varepsilon_3^2-(\hat{\lambda}+\tilde{\lambda})\varepsilon_1\tilde{v}-(\hat{\lambda}+\tilde{\lambda})\varepsilon_3\tilde{\omega}\\&+\tilde{\mathbf{v}}^T(-\mathbf{M}\hat{\boldsymbol{\beta}}-\mathbf{M}\tilde{\boldsymbol{\beta}}+\mathbf{E}\boldsymbol{\tau}+\mathbf{d})-\tilde{\boldsymbol{\beta}}\boldsymbol{\Gamma}^{-1}\dot{\hat{\boldsymbol{\beta}}}-\frac{1}{\gamma_3}\tilde{\lambda}\dot{\hat{\lambda}}\\=& -\lambda K_1\varepsilon_1^2-\lambda K_3\varepsilon_3^2+\tilde{\mathbf{v}}^T\left(-\mathbf{M}\hat{\boldsymbol{\beta}}-\hat{\lambda}\begin{bmatrix}\varepsilon_1\\\varepsilon_2\end{bmatrix}+\overline{\mathbf{E}}\boldsymbol{\tau}+\mathbf{d}\right)\\&-\tilde{\boldsymbol{\beta}}^{\mathrm{T}}(\boldsymbol{\Gamma}^{-1}\dot{\hat{\boldsymbol{\beta}}}+\mathbf{M}^{\mathrm{T}}\tilde{\mathbf{v}})-\tilde{\lambda}\left(\frac{1}{\gamma_3}\dot{\hat{\lambda}}+\tilde{\mathbf{v}}^{\mathrm{T}}\begin{bmatrix}\varepsilon_1\\\varepsilon_2\end{bmatrix}\right)\end{aligned} \tag{9.152}$$

where the relations $\boldsymbol{\beta}=\hat{\boldsymbol{\beta}}+\tilde{\boldsymbol{\beta}}$, $\lambda=\hat{\lambda}+\tilde{\lambda}$, and $\mathbf{v}=\hat{\mathbf{v}}+\tilde{\mathbf{v}}$ were used. Now, selecting the control vector $\boldsymbol{\tau}$ and the updating laws of the parameter estimates as:

$$\overline{\mathbf{E}}\boldsymbol{\tau}=\mathbf{M}\hat{\boldsymbol{\beta}}+\hat{\lambda}\begin{bmatrix}\varepsilon_1\\\varepsilon_2\end{bmatrix}-\mathbf{K}_{\mathrm{a}}\tilde{\mathbf{v}}-\mathbf{u}_{\mathrm{robust}} \tag{9.153}$$

$$\dot{\hat{\boldsymbol{\beta}}}=-\boldsymbol{\Gamma}\mathbf{M}^{\mathrm{T}}\tilde{\mathbf{v}} \tag{9.154}$$

$$\dot{\hat{\lambda}}=-\gamma_3[\varepsilon_1,\varepsilon_3]\tilde{\mathbf{v}} \tag{9.155}$$

$$\mathbf{K}_{\mathrm{a}}=\begin{bmatrix}K_{\mathrm{a}1}&0\\0&K_{\mathrm{a}2}\end{bmatrix}, K_{\mathrm{a}1}>0, K_{\mathrm{a}2}>0$$

we get:

$$\dot{V}_0=-\lambda_1K_1\varepsilon_1^2-\lambda K_3\varepsilon_3^2-\tilde{\mathbf{v}}^{\mathrm{T}}\mathbf{K}_{\mathrm{a}}\tilde{\mathbf{v}}+\tilde{\mathbf{v}}^{\mathrm{T}}(\mathbf{d}-\mathbf{u}_{\mathrm{robust}}) \tag{9.156}$$

The first three terms of $\dot{V}_0$ are nonpositive. Therefore, to guarantee that $\dot{V}_0\le 0$ the robustifying control term $\mathbf{u}_{\mathrm{robust}}$ must be selected such as:

$$\tilde{\mathbf{v}}^T(\mathbf{d}-\mathbf{u}_{\mathrm{robust}})\le 0 \tag{9.157}$$

Given the bound $d_{\max}$ of the disturbance $\mathbf{d}(t)$ (see Eq. (9.139)), the condition (9.157) is satisfied by selecting:

$$\mathbf{u}_{\mathrm{robust}}=d_{\max}\,\mathrm{sgn}\,\tilde{\mathbf{v}},\quad \mathrm{sgn}\,\tilde{\mathbf{v}}=\begin{bmatrix}\mathrm{sgn}\,\tilde{v}\\\mathrm{sgn}\,\tilde{\omega}\end{bmatrix}$$

which, by Eq. (9.139) implies:

$$\tilde{\mathbf{v}}^{\mathrm{T}}(\mathbf{d} - d_{\max} \operatorname{sgn} \tilde{\mathbf{v}}) \leq ||\tilde{\mathbf{v}}||(||\mathbf{d}|| - d_{\max}) \leq 0$$

To summarize, the adaptive robust trajectory controller for the differential-drive WMR is given by:

$$v_{\mathrm{m}} = K_1\varepsilon_1 + v_{\mathrm{d}} \cos \varepsilon_3 \tag{9.158a}$$

$$\omega_{\mathrm{m}} = \omega_{\mathrm{d}} + K_3\varepsilon_3 + \varepsilon_2 v_{\mathrm{d}}\left(\frac{\sin \varepsilon_3}{\varepsilon_3}\right) \tag{9.158b}$$

$$\tau = \overline{\mathbf{E}}^{-1}\left(\mathbf{M}\hat{\beta} + \hat{\lambda}\begin{bmatrix}\varepsilon_1 \\ \varepsilon_2\end{bmatrix} - \mathbf{K}_{\mathrm{a}}\tilde{\mathbf{v}} - d_{\max} \operatorname{sgn} \tilde{\mathbf{v}}\right) \tag{9.158c}$$

$$\dot{\hat{\beta}} = -\Gamma\mathbf{M}^{\mathrm{T}}\tilde{\mathbf{v}} \tag{9.158d}$$

$$\dot{\hat{\lambda}} = -\gamma_3[\varepsilon_1, \varepsilon_2]\tilde{\mathbf{v}} \tag{9.158e}$$

This controller assures that $V_0(t)$ is a nonincreasing function that converges to a limiting value $\overline{V}_0 \geq 0$. Thus, $\varepsilon(t), \tilde{\mathbf{v}}(t)$, $\tilde{\beta}$, and $\tilde{\lambda}$ are all bounded. Now, assuming that $v_{\mathrm{d}}, \dot{v}_{\mathrm{d}}, \omega_{\mathrm{d}}, \dot{\omega}_{\mathrm{d}}$ are bounded, it follows that $\dot{\varepsilon}_1, \dot{\varepsilon}_2, \dot{\varepsilon}_3, \tilde{v}, \tilde{\omega}$ are all bounded. Then, $\varepsilon_1, \varepsilon_3, \tilde{v}$, and $\tilde{\omega}$ are uniformly continuous, and so by Barbalat lemma $\varepsilon_1 \to 0$, $\varepsilon_3 \to 0$, $\tilde{v} \to 0$, and $\tilde{\omega} \to 0$. It only remains to show that $\varepsilon_2 \to 0$. To this end, we consider the closed-loop equation for ε_3, namely:

$$\dot{\varepsilon}_3 = -K_3\varepsilon_3 - (\varepsilon_2/\varepsilon_3)v_{\mathrm{d}} \sin \varepsilon_3 - \tilde{\omega}$$

Then, from $\varepsilon_3 \to 0$, $\tilde{\omega} \to 0$, and $(\sin \varepsilon_3)/\varepsilon_3 \to 1$, and assuming that $\min|v_{\mathrm{d}}| \to \eta > 0$, $t \to \infty$, it follows that $\varepsilon_2 \to 0$. Therefore, overall, we have established that $\varepsilon_1 \to 0$, $\varepsilon_2 \to 0$, $\varepsilon_3 \to 0$, $\tilde{v} \to 0$, $\tilde{\omega} \to 0$, and $\hat{\beta}, \hat{\lambda}$ bounded (not necessarily equal to the true values β and λ).

To overcome the chattering that may occur due to the signum robustifying term $\mathbf{u}_{\mathrm{robust}} = d_{\max} \operatorname{sgn} \tilde{\mathbf{v}}$, we replace the signum function by the saturation function:

$$\operatorname{sat} \tilde{\mathbf{v}} = \begin{cases} \tilde{\mathbf{v}}/U, & \text{if } ||\tilde{\mathbf{v}}|| \leq U(t) \\ \operatorname{sgn} \tilde{\mathbf{v}}, & \text{if } ||\tilde{\mathbf{v}}|| > U(t) \end{cases}$$

where $U(t)$ is the width of the boundary layer used (see Eq. (7.17)). By selecting $U(t)$ such as $\int_0^\infty U(\tau)d\tau \leq \delta$, where δ is a nonnegative constant we can show that the saturation robustifying term leads again to $\varepsilon_1 \to 0$, $\varepsilon_2 \to 0$, $\varepsilon_3 \to 0$, $\tilde{v} \to 0$, $\tilde{\omega} \to 0$, and bounded $\hat{\beta}$ and $\hat{\lambda}$ (the proof is left as an exercise).

Remark A similar controller can be derived by selecting, in analogy to Eq. (7.72), the Lyapunov function:

$$V = \frac{1}{2}(\varepsilon_1^2 + \varepsilon_2^2) + (\lambda/K_2)(1 - \cos\varepsilon_3), \quad \lambda > 0,\ K_2 > 0$$

Then, differentiating V along the trajectory of the system of Eq. (9.143), we find that selecting:

$$v_{\mathrm{m}} = K_1\varepsilon_1 + v_{\mathrm{d}}\cos\varepsilon_3 \tag{9.159a}$$

$$\omega_{\mathrm{m}} = \omega_{\mathrm{d}} + K_2 v_{\mathrm{d}}\varepsilon_2 + K_3\sin\varepsilon_3 \tag{9.159b}$$

gives

$$\dot{V} = -K_1\varepsilon_1^2 - \lambda(K_3/K_2)(\sin\varepsilon_3)^2 - \varepsilon_1\lambda\tilde{v} - (\lambda/K_2)(\sin\varepsilon_3)\tilde{\omega}$$

which is analogous to Eq. (9.146c). Therefore, adding a second Lyapunov function term V', given by Eq. (9.150), and working as above we find the kinematic controller (9.159a) and (9.159b), and the dynamic adaptive-robust trajectory tracking controller:

$$\begin{aligned} \boldsymbol{\tau} &= \overline{\mathbf{E}}^{-1}\left(\mathbf{M}\hat{\boldsymbol{\beta}} + \hat{\lambda}\begin{bmatrix}\varepsilon_1 \\ \sin\varepsilon_3\end{bmatrix} - \mathbf{K}_{\mathrm{a}}\tilde{\mathbf{v}} - d_{\max}\,\mathrm{sgn}\,\tilde{\mathbf{v}}\right) \\ \dot{\hat{\boldsymbol{\beta}}} &= -\boldsymbol{\Gamma}\mathbf{M}^T\tilde{\mathbf{v}} \\ \dot{\hat{\lambda}} &= -\gamma_3[\varepsilon_1, \sin\varepsilon_3]\tilde{\mathbf{v}} \end{aligned} \tag{9.160}$$

Like the controller of Eqs. (9.158a)–(9.158e), we again achieve asymptotic convergence of $\varepsilon_1, \varepsilon_2, \varepsilon_3$ to zero, and of $\hat{\boldsymbol{\beta}}, \hat{\lambda}$ to bounded values (possibly different than the true values of $\boldsymbol{\beta}$ and λ).

References

[1] Corke P. Visual control of robot manipulators: a review. In: Hashimoto K, editor. Visual servoing. Singapore: Word Scientific; 1993. p. 1–31.

[2] Espiau B, Chaumette F, Rives P. A new approach to visual servoing in robotics. IEEE Trans Robot Autom 1992;8:313–26.

[3] Hutchinson S, Hager G, Corke P. A tutorial on visual servo control. IEEE Trans Robot Autom 1996;12:651–70.

[4] Haralik RM, Shapiro LG. Computer and robot vision. Reading, MA: Addison Wesley; 1993.

[5] Cherubini A, Chaumette F, Oriolo G. An image-based visual servoing scheme for following paths with nonholonomic mobile robots. In: Proceedings of international conference on control automation. Robotics and Vision. Hanoi, Vietnam; December 17–20, 2008. p. 108–113.

[6] Cherubini A, Chaumette F, Oriolo G. A position-based visual servoing scheme for following paths with nonholonomic robots. In: Proceedings of IEEE/RSJ international conference on intelligent robots and systems (IROS 2008). Nice, France; September 2008. p. 1648–54.

[7] Chaumette F, Hutchinson S. Visual servo control tutorial, parts I and II. IEEE Robot Autom Mag 2007;13(14):82–90, 14(1):109–118

[8] Burshka D, Hager G. Vision-based control of mobile robots. In: Proceedings of 2001 IEEE international conference on robotics and automation. Seoul, Korea; May 21–26, 2001. p. 1707–13.

[9] Tsakiris D, Samson C, Rives P. Vision-based time-varying stabilization of a mobile manipulator. In: Proceedings of fourth international conference on control, automation, robotics and vision (ICARV'96). Westin Stamford, Singapore; December 3–6, 1996.

[10] Tsakiris D, Samson C, Rives P. Vision-based time-varying robot control. In: Proceedings of ERNET workshop. Darmstadt, Germany; September 9–10, 1996. p. 163–72.

[11] Huang TS, Netravali NA. Motion and structure from feature correspondences: a review. Proc IEEE 1994;82(2):252–68.

[12] Kumar R. Robust methods for estimating pose and a sensitivity analysis. CVGIP: Image Underst 1994;3:313–42.

[13] Wilson W. Visual servo control of robots using Kalman filter estimates of robot pose relative to workpieces. In: Hashimoto K, editor. Visual servoing. Singapore: World Scientific; 1994. p. 71–104.

[14] Horswill I. Polly: a vision-based artificial agent. In: Proceedings of eleventh national conference on artificial intelligence (AAAI'93). Washington, DC: MIT Press; July 11–15, 1993. p. 824–29.

[15] Rives P, Espiau B. Closed-loop recursive estimation of 3D features for a mobile vision system. In: Proceedings of IEEE international conference on robotics and automation (ICRA 1987). Rayleigh, NC; April 1987. p. 1436–44.

[16] Qian J, Su J. Online estimation of image Jacobian matrix by Kalman–Bucy filter for uncalibrated stereo vision feedback. In: Proceedings of international conference on robotics and automation (ICRA 2002). 2002. p. 562–7.

[17] Das AK, Fierro R, Kumar V, Southall B, Spletzer J, Taylor CJ. Real-time vision-based control of a nonholonomic mobile robot. In: Proceedings of 2001 IEEE international conference on robotics and automation. Seoul, Korea; May 21–26, 2001. p. 1714–19.

[18] Carelli R., Soria CM, Morales B. Vision-based tracking control for mobile robots. In: Proceedings of twelveth international conference on advanced robotics (ICAR'05). Seatle, WA; July, 18–20, 2005. p. 148–52.

[19] Maya-Mendez M, Morin P, Samson C. Control of a nonholonomic mobile robot via sensor-based target tracking and pose estimation. In: Proceedings of 2006 IEEE/RSJ international conference on intelligent robots and systems. Beijing, China; October 9–15, 2009. p. 5612–18.

[20] Dixon WE, Dawson DM, Zergeroglu E. Adaptive tracking control of a wheeled mobile robot via an uncalibrated camera system. IEEE Trans Syst Man Cybern Cybern 2001;31(3):341–52.

[21] Soetanto D, Lapierre L, Pascoal A. Adaptive, non-singular path-following control of dynamic wheeled robots. In: Proceedings of fourtysecond IEEE conference on decision and control. Maui, HI; December 2003. p. 1765–70.

[22] Abdelkader HH, Mezouar Y, Andreff N, Martinet P. Image-based control of mobile robot with central catadioptric cameras. In: Proceedings of 2005 IEEE international conference on robotics and automation. Barcelona, Spain; April 2005. p. 3533–8.
[23] Kim J, Suga Y. An omnidirectional vision-based moving obstacle detection in mobile robot. Int J Control Autom Syst 2007;5(6):663–73.
[24] Shakernia O, Vidal R, Sastry S. Infinitesimal motion estimation from multiple central panoramic views. In: Proceedings of IEEE international workshop on motion and video computing. December, 2002. p. 229–34.
[25] Baker S, Nayar S. A theory of single-viewpoint catadioptric image formation. Int J Comput Vis 1999;35(2):1–22.
[26] Das AK, Fierro R, Kumar V, Ostrowski JP, Spletzer J, Taylor CJA. Vision-based formation control framework. IEEE Trans Robot Autom 2002;18(5):813–25.
[27] Gans NR, Hutchinson SA. A stable vision-based control scheme for nonholonomic vehicles to keep a landmark in the field of view. In: Proceedings of 2007 international conference on robotics and automation. Rome, Italy; April 10–14, 2007. p. 2196–200.
[28] Lopez-Nicolas G, Gans NR, Bhattacharya S, Sagues C, Guerrero JJ, Hutchinson S. Homography-based control scheme for mobile robots with nonholonomic and field-of-view constraints. IEEE Trans Syst Man Cybern Cybern 2010;40(4):1115–27.
[29] Lapierre LP, Soetanto DJ, Pascoal A. Adaptive vision-based path following control of a wheeled robot. In: Proceedings of IEEE Mediterranean control conference (MED 2002). Lisbon, July 9–12, 2002. p. 1–6.
[30] Micaelli A, Samson C. Trajectory tracking for unicycle-type and two-steering—wheels mobile robots. INRIA Technical Report No. 2097, France; 1993.
[31] Lietmann T, Bornstedt B, Lohmann B. Visual servoing for a non-holonomic mobile robot using a range-image camera. In: Proceedings of European control conference (ECC'01). Porto, Portugal; September 4–7, 2001.
[32] Augustin B, Lietmann T, Lohmann B. Image-based visual servoing of a non-holonomic mobile platform using a pan-tilt head. In: Proceedings of eight IEEE international conference on methods and models in automation and robotics (MMAR 2002). Szczecin, Poland; 2002, p. 941–6.
[33] Grigorescu S, Macesanu G, Cocias T, Puiu D, Moldoveanu F. Robust camera pose and scene structure analysis for service robotics. Robot Auton Syst 2011;58:1–13.
[34] Kragic D, Christensen HI. Advances in robot vision. Robot Auton Syst 2005;52:1–3.
[35] Geyer C, Daniilidis K. A unifying theory for central panoramic systems and practical implementations. In: Vernon D, editor. Proceedings of European Conference on Computer Vision (ECCV'2000). Dublin: Springer; 2000. p. 445–61.
[36] Okamoto Jr J, Grassi Jr V. Visual servo control of a mobile robot using omnidirectional vision. In: Proceedings of mechatronics 2002, University of Twente; June 24–26, 2002. p. 413–22.
[37] Barreto J, Martin F, Horaud R. Visual servoing/tracking using central catadioptric images. Expl Rob Springer Tracks Adv Robot 2003;5:245–54.
[38] Baker S, Nayar S. A theory of catadioptric image formation. In: Proceedings of 1997 international conference on computer vision (ICCV'97). 1998. p. 35–42.
[39] Barreto JP, Araujo H. Geometric properties of central catadioptric line images and their application to calibration. IEEE Trans Pattern Anal Mach Intell 2005;27(8):1327–33.
[40] Barreto J.P. General central projection systems: modeling, calibration and visual servoing [Ph.D. thesis]. University of Coimbra; 2003.
[41] Geyer C.M. Catadioptric projective geometry: theory and applications [Ph.D. dissertation]. University of Pennsylvania; 2003.

[42] Gong X. Omnidirectional vision for an autonomous surface vehicle [Ph.D. dissertation]. Virginia Polytechnic Institute and State University; 2008.
[43] Barreto JP, Araujo H. Issues on the geometry of central catadioptric image formation. Proc Comput Vis Pattern Recog 2001;:422–7.
[44] Yang F, Wang C. Adaptive tracking control for uncertain dynamic nonholonomic mobile robots based on visual servoing. J Control Theory Appl 2012;10(1):56–63.
[45] Liang Z, Wang C. Robust exponential stabilization of nonholonomic wheeled mobile robots with unknown visual parameters. J Control Theory Appl 2011;9(2):295–301.

10 Mobile Manipulator Modeling and Control

10.1 Introduction

Mobile manipulators (MMs) are robotic systems consisting of articulated arms (manipulators) mounted on holonomic or nonholonomic mobile platforms. They provide the dexterity of the former and the workspace extension of the latter. They are able to reach and work over objects that are initially outside the working space of the articulated arm. Therefore, MMs are appealing for many applications, and today they constitute the main body of service robots [1–27]. One of the principal and most challenging problems in MMs research is to design accurate controllers for the entire system. Due to the strong interaction and coupling of the mobile platform subsystem and the manipulator arm(s) mounted on the platform, a proper coordination between the respective controllers is needed [18,23]. However, in the literature there are also available unified control design methods that treat the whole system using the full-state MM model [3,10,22,24]. In either case the control methods studied in this book (computed-torque control, feedback linearizing control, robust sliding-mode or Lyapunov-based control, adaptive control, and visual-based control) can be employed and combined.

The objectives of this chapter are as follows:

- To present the Denavit–Hartenberg method of articulated robots' kinematic modeling, and provide a practical method for inverse kinematics
- To study the general kinematic and dynamic models of MMs
- To derive the kinematic and dynamic model of a differential drive, and a three-wheel omnidirectional MM, with two-link and three-link mounted articulated arms, respectively
- To derive a computed-torque controller for the above differential-drive MM, and a sliding-mode controller for the three-wheel omnidirectional MM
- To discuss some general issues of MM visual-based control, including a particular indicative example of hybrid coordinated (feedback/open-loop) visual control and a panoramic visual servoing example.

10.2 Background Concepts

Here, the following aspects which will be used in the chapter are considered:

- The Denavit–Hartenberg direct manipulator kinematics method
- A general method for inverse manipulator kinematics

Introduction to Mobile Robot Control. DOI: http://dx.doi.org/10.1016/B978-0-12-417049-0.00010-9

- The manipulability measure concept
- The two-link manipulator's direct and inverse kinematic model and manipulability measure.

10.2.1 The Denavit–Hartenberg Method

The *Denavit–Hartenberg* method provides a systematic procedure for determining the position and orientation of the end-effector of an n-joint robotic manipulator, that is, for computing $\mathbf{X}^0$ in (2.18).

Consider a link i that lies between the joints i and $i+1$ of a robot as shown in Figure 10.1.

Each link is described by the distance a_i between the (possibly nonparallel) axes i and $i+1$ of the two joints, and the rotation angle α_i from the axis i to the axis $i+1$ with respect to the common normal of the two axes. Each joint (prismatic or rotary) is driven by a motor (translational or rotational) that produces the motion of link i. In overall, the robotic arm has n joints and $n+1$ links. The functional relation of the end-effector and the displacements of the joints can be found using the convention of parameters shown in Figure 10.2, called the *Denavit–Hartenberg parameters*.

These parameters refer to the relative position of the coordinate frames $O_{i-1}x_{i-1}y_{i-1}z_{i-1}$ and $O_ix_iy_iz_i$, and are the following:

- The length a_i of the common normal Σ_iO_i
- The distance d_i between the origin of O_{i-1} and the point Σ_i
- The angle α_i between the joint i (i.e., the axis z_{i-1}) and the axis z_i in the positive (clockwise) direction
- The angle θ_i between the axis x_{i-1} and the common normal (i.e., rotation about the axis z_{i-1}) in the positive direction.

On the basis of the above, the transfer from the frame $O_{i-1}x_{i-1}y_{i-1}z_{i-1}$ to the frame $O_ix_iy_iz_i$ can be done in four steps:

Step 1: Rotation of frame $i-1$ about the axis z_{i-1} by an angle θ_i.
Step 2: Translation of frame $i-1$ along the axis z_{i-1} by d_i.
Step 3: Translation of the rotated axis x_{i-1} (which now coincides with x_i) along the common normal by a_i.
Step 4: Rotation about x_i by an angle α_i.

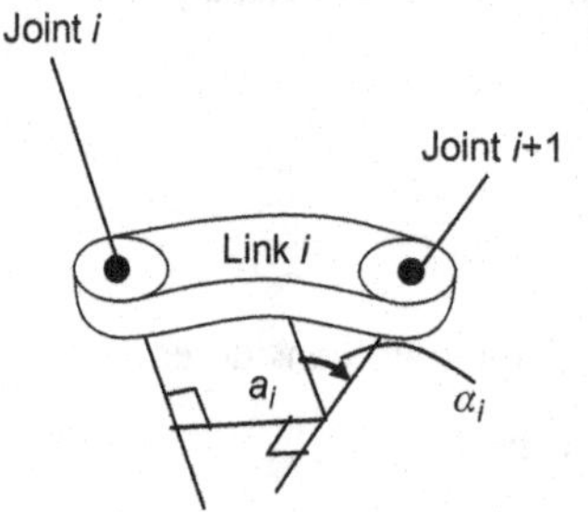

Figure 10.1 A robotic link between joints i and i+1.

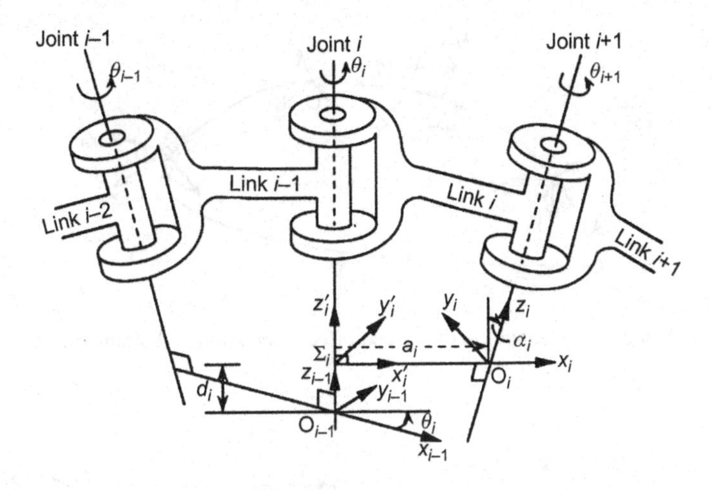

Figure 10.2 Denavit–Hartenberg robotic parameters.

Denoting by $\mathbf{A}_i^*$ the result of steps 3 and 4 and by $\mathbf{A}_*^{i-1}$ the result of steps 1 and 2, the overall result of steps 1 through 4 is given by:

$$\mathbf{A}_i^{i-1} = \mathbf{A}_i^*\mathbf{A}_*^{i-1} = \begin{bmatrix} \cos\theta_i & -\sin\theta_i\cos\alpha_i & \sin\theta_i\sin\alpha_i & a_i\cos\theta_i \\ \sin\theta_i & \cos\theta_i\cos\alpha_i & -\cos\theta_i\sin\alpha_i & a_i\sin\theta_i \\ 0 & \sin\alpha_i & \cos\alpha_i & d_i \\ 0 & 0 & 0 & 1 \end{bmatrix} \tag{10.1}$$

where $\mathbf{A}_i^{i-1}$ gives the position and orientation of the frame i with respect to the frame $i-1$. The first three columns of $\mathbf{A}_i^{i-1}$ contain the direction cosines of the axes of frame i, whereas the fourth column represents the position of frame O_i.

In general, the displacement of joint i is denoted as q_i, where:

$q_i = \theta_i$ for revolute joint,
$q_i = d_i$ for prismatic joint.

The position and orientation of link i with respect to link $i-1$ is a function of q_i, that is, $\mathbf{A}_i^{i-1}(q_i)$.

The kinematic equation of a robotic arm gives the position and orientation of the last link with respect to the coordinate frame of the base, and obviously contains all generalized variables $q_1, q_2, \ldots, q_n$ of the joints. Figure 10.3 shows pictorially the consecutive coordinate frames from the base up to the end-effector of the serial robotic kinematic chain.

According to Eq. (2.18), the matrix $\mathbf{T}$ given by:

$$\mathbf{T} = \mathbf{A}_1^0(q_1)\mathbf{A}_2^1(q_2)\ldots\mathbf{A}_n^{n-1}(q_n) \tag{10.2}$$

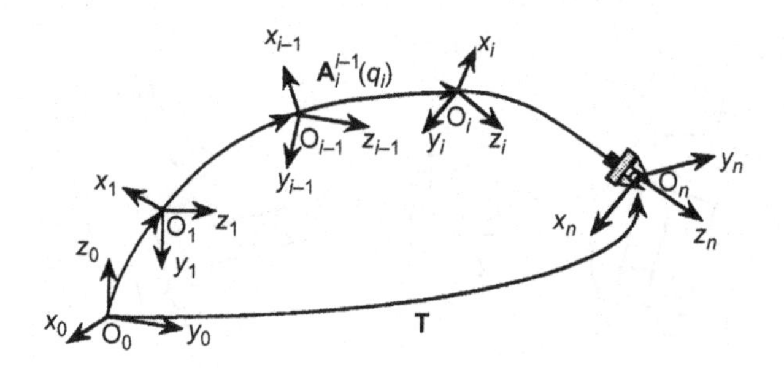

Figure 10.3 Pictorial representation of the end-effector position and orientation by the 4×4 matrix **T**.

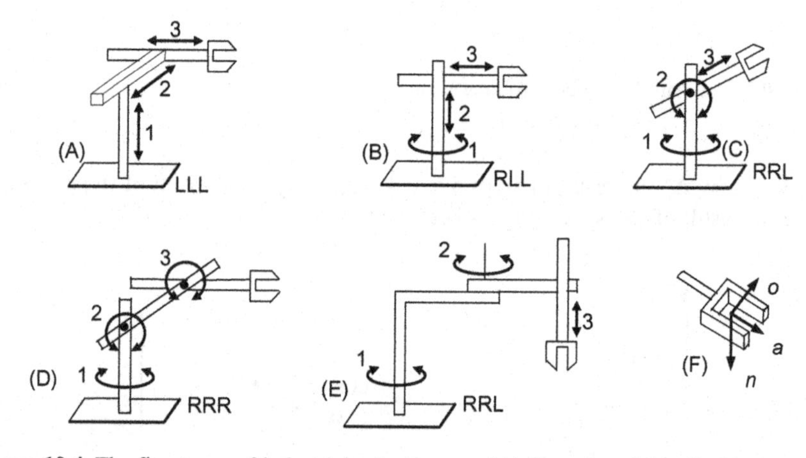

Figure 10.4 The five types of industrial robotic arms: (A) Cartesian, (B) cylindrical, (C) spherical (polar), (D) articulated (anthropomorphic), (E) SCARA robot, and (F) end-effector coordinate frame. L denotes linear (translational) motion, and R denotes rotational motion.

represents the position and orientation of the end-effector (which is the final link with respect to the base). It is now easy to determine **T** for all types of robots (Figure 10.4) using Eq. (10.2).

10.2.2 Robot Inverse Kinematics

In the direct kinematics problem we are finding **T** knowing the values of $q_1, q_2, \ldots, q_n$. In the inverse kinematics problem we do the converse, that is, given **T** we determine $q_1, q_2, \ldots, q_n$ by solving (10.2) with respect to $q_i (i = 1, 2, \ldots, n)$.

The direct kinematics Eq. (10.2) can be written in the vectorial form:

$$\mathbf{X} = \mathbf{f}(\mathbf{q}) \tag{10.3}$$

where $\mathbf{X}$ is the six-dimensional vector:

$$\mathbf{X}=\begin{bmatrix}\mathbf{p}\\ \cdots\\ \boldsymbol{\psi}\end{bmatrix},\quad \mathbf{p}=\begin{bmatrix}x\\ y\\ z\end{bmatrix},\quad \boldsymbol{\psi}=\begin{bmatrix}\psi_x\\ \psi_y\\ \psi_z\end{bmatrix} \tag{10.4}$$

of the end-effector's position $\mathbf{p}$ and orientation $\boldsymbol{\psi}$, $\mathbf{f}$ is a six-dimensional nonlinear column vectorial function, and

$$\mathbf{q}=[q_1,q_2,\ldots,q_n]^T \tag{10.5}$$

Therefore, the inverse kinematics equation is:

$$\mathbf{q}=\mathbf{f}^{-1}(\mathbf{X}) \tag{10.6}$$

where $\mathbf{f}^{-1}(\cdot)$ denotes the usual inverse function of $\mathbf{f}(\cdot)$.

A straightforward practical method for inverting the kinematic Eq. (10.2) is the following. We start from:

$$\mathbf{T}=\mathbf{A}_1^0(\mathbf{q}_1)\mathbf{A}_2^1(\mathbf{q}_2)\ldots\mathbf{A}_6^5(\mathbf{q}_6)$$

and obtain the following sequence of equations:

$$\begin{aligned}
(\mathbf{A}_1^0)^{-1}\mathbf{T} &= \mathbf{T}_6^1\\
(\mathbf{A}_2^1)^{-1}(\mathbf{A}_1^0)^{-1}\mathbf{T} &= \mathbf{T}_6^2\\
(\mathbf{A}_3^2)^{-1}(\mathbf{A}_2^1)^{-1}(\mathbf{A}_1^0)^{-1}\mathbf{T} &= \mathbf{T}_6^3\\
(\mathbf{A}_4^3)^{-1}(\mathbf{A}_3^2)^{-1}(\mathbf{A}_2^1)^{-1}(\mathbf{A}_1^0)^{-1}\mathbf{T} &= \mathbf{T}_6^4\\
(\mathbf{A}_5^4)^{-1}(\mathbf{A}_4^3)^{-1}(\mathbf{A}_3^2)^{-1}(\mathbf{A}_2^1)^{-1}(\mathbf{A}_1^0)^{-1}\mathbf{T} &= \mathbf{T}_6^5
\end{aligned} \tag{10.7}$$

The elements of the left-hand sides of these equations are functions of the elements of $\mathbf{T}$ and the first $i-1$ variables of the robot. The elements of the right-hand sides are constants or functions of the variables $q_i, q_{i+1}, \ldots, q_6$. From each matrix equation we get 12 equations, that is, one equation for each of the elements of the four vectors $\mathbf{n}, \mathbf{o}, \mathbf{a}$, and $\mathbf{p}=\mathbf{x}_0$. From these equations we can determine the values of $q_i(i=1,2,\ldots,6)$ of the robot.

Although the solution of the direct kinematics problem is unique, it is not so for the inverse kinematics problem, because of the presence of trigonometric functions. In some cases, the solution can be found analytically, but, in general, the solution can only be found approximately using some approximate numerical method and the computer. Also, if the robot has more than six degrees of freedom (i.e., if we have a redundant robot), there are infinitely many solutions to $q_i(i=1,2,\ldots,n;\ n>6)$ that lead to the same position and orientation of the end-effector.

10.2.3 Manipulability Measure

An important factor that determines the ease of arbitrarily changing the position and orientation of the end-effector by a manipulator is the so-called *manipulability measure*. Other factors are the size and geometric shape of the workspace envelope, the accuracy and repeatability, the reliability and safety, etc. Here, we will examine the manipulability measure, which was fully studied by Yoshikawa.

Given a manipulator with n degrees of freedom, the direct kinematics equation is given by Eq. (10.3):

$$\mathbf{X} = \mathbf{f}(\mathbf{q})$$

where $\mathbf{X}$ is the m-dimensional vector ($m = 6$) of the end-effector's position $\mathbf{p} = [x, y, z]^T$ and orientation vector $\boldsymbol{\psi} = [\psi_x, \psi_y, \psi_z]^T$. Differentiating this relation we get the well-known differential kinematics model:

$$\dot{\mathbf{X}} = \mathbf{J}(\mathbf{q})\dot{\mathbf{q}}$$

where $\mathbf{J}(\mathbf{q})$ is the manipulator's Jacobian. This relation relates the velocity $\dot{\mathbf{q}} = [\dot{q}_1, \dot{q}_2, \ldots, \dot{q}_n]^T$ of the manipulator joints with the velocity of the end-effector, that is, the *screw* of the robot.

The set of all end-effector velocities that are realizable by joint velocities such as:

$$||\dot{\mathbf{q}}|| = \sqrt{\dot{q}_1^2 + \dot{q}_2^2 + \cdots + \dot{q}_n^2} \leq 1$$

is an *ellipsoid* in the m-dimensional Euclidean space where m is the dimensionality of $\dot{\mathbf{X}}$. The maximum speed of motion of the end-effector is along the major axis, and the minimum speed along the minor axis of this ellipsoid is called *manipulability ellipsoid*. One can show that the manipulability ellipsoid is the set of all $\mathbf{v} = \dot{\mathbf{X}}$ that satisfy:

$$\mathbf{v}^T(\mathbf{J}^\dagger)^T\mathbf{J}^\dagger\mathbf{v} \leq 1$$

for all $\mathbf{v}$ in the range of $\mathbf{J}$. Indeed, by using the relation $\dot{\mathbf{q}} = \mathbf{J}^\dagger\mathbf{v} + (\mathbf{I} - \mathbf{J}^\dagger\mathbf{J})\mathbf{k}$ ($\mathbf{k}$ = arbitrary constant vector) and the equality $(\mathbf{I} - \mathbf{J}^\dagger\mathbf{J})^T\mathbf{J}^\dagger = \mathbf{0}$, we get

$$\begin{aligned} ||\dot{\mathbf{q}}||^2 = \dot{\mathbf{q}}^T\dot{\mathbf{q}} &= \mathbf{v}^T(\mathbf{J}^\dagger)^T\mathbf{J}^\dagger\mathbf{v} + 2k^T(\mathbf{I} - \mathbf{J}^\dagger\mathbf{J})^T\mathbf{J}^\dagger\mathbf{v} \\ &\quad + \mathbf{k}^T(\mathbf{I} - \mathbf{J}^\dagger\mathbf{J})^T(\mathbf{I} - \mathbf{J}^\dagger\mathbf{J})\mathbf{k} \\ &\geq \mathbf{v}^T(\mathbf{J}^\dagger)^T\mathbf{J}^\dagger\mathbf{v} \end{aligned}$$

Thus, if $||\dot{\mathbf{q}}|| \leq 1$, then $\mathbf{v}^T(\mathbf{J}^\dagger)^T\mathbf{J}^\dagger\mathbf{v} \leq 1$. Conversely, if we choose an arbitrary $\hat{\mathbf{v}}$ such that $\hat{\mathbf{v}}^T(\mathbf{J}^\dagger)^T\mathbf{J}^\dagger\hat{\mathbf{v}} \leq 1$, then there exists a vector $\hat{\mathbf{z}}$ such that $\hat{\mathbf{v}} = \mathbf{J}\hat{\mathbf{z}}$, that is, $\hat{\dot{\mathbf{q}}} = \mathbf{J}^\dagger\hat{\mathbf{v}}$. Then, we find that:

$$\mathbf{J}\hat{\dot{\mathbf{q}}} = \mathbf{J}\mathbf{J}^\dagger\hat{\mathbf{v}} = \mathbf{J}\mathbf{J}^\dagger\mathbf{J}\hat{\mathbf{z}} = \mathbf{J}\hat{\mathbf{z}} = \hat{\mathbf{v}}$$

and

$$||\dot{\mathbf{q}}|| = \hat{\mathbf{v}}(\mathbf{J}^{\dagger})^{\mathrm{T}}\mathbf{J}^{\dagger}\hat{\mathbf{v}} \leq 1$$

In nonsingular configurations, the manipulability ellipsoid is given by:

$$\mathbf{v}^{\mathrm{T}}(\mathbf{J}^{-1})^{\mathrm{T}}\mathbf{J}^{-1}\mathbf{v} \leq 1$$

The manipulability measure w of the manipulator is defined as:

$$w = \sqrt{\det \mathbf{J}(\mathbf{q})\mathbf{J}^{\mathrm{T}}(\mathbf{q})}$$

which for $m = n$ reduces to:

$$w = |\det \mathbf{J}(\mathbf{q})|$$

In this case, the set of all velocities that are implemented by a velocity $\dot{\mathbf{q}}$ of the joints such that:

$$|\dot{q}_i| \leq 1, \quad i = 1, 2, \ldots, m$$

is a *parallelepiped* in m-dimensional space with a volume of $2^m w$. This means that the measure w is proportional to the volume of the parallelepiped, a fact that provides a physical representation of the manipulability measure.

10.2.4 The Two-Link Planar Robot

10.2.4.1 Kinematics

Consider the planar robot of Figure 10.5.

The kinematic model of this robot can be found by simple trigonometric calculation. Referring to Figure 10.5 we readily obtain the direct kinematic model:

$$\begin{aligned} x(\theta_1, \theta_2) &= l_1 \cos \theta_1 + l_2 \cos(\theta_1 + \theta_2) \\ y(\theta_1, \theta_2) &= l_1 \sin \theta_1 + l_2 \sin(\theta_1 + \theta_2) \end{aligned} \tag{10.8a}$$

that is:

$$\mathbf{p} = \mathbf{f}(\boldsymbol{\theta}), \quad \mathbf{p} = \begin{bmatrix} x \\ y \end{bmatrix}, \quad \boldsymbol{\theta} = \begin{bmatrix} \theta_1 \\ \theta_2 \end{bmatrix} \tag{10.8b}$$

To derive the inverse kinematic model $\boldsymbol{\theta} = \mathbf{f}^{-1}(\mathbf{p})$ we work on Figure 10.5B. Thus, using the cosine rule we find:

$$x^2 + y^2 = l_1^2 + l_2^2 - 2l_1 l_2 \cos(180^\circ - \theta_2)$$

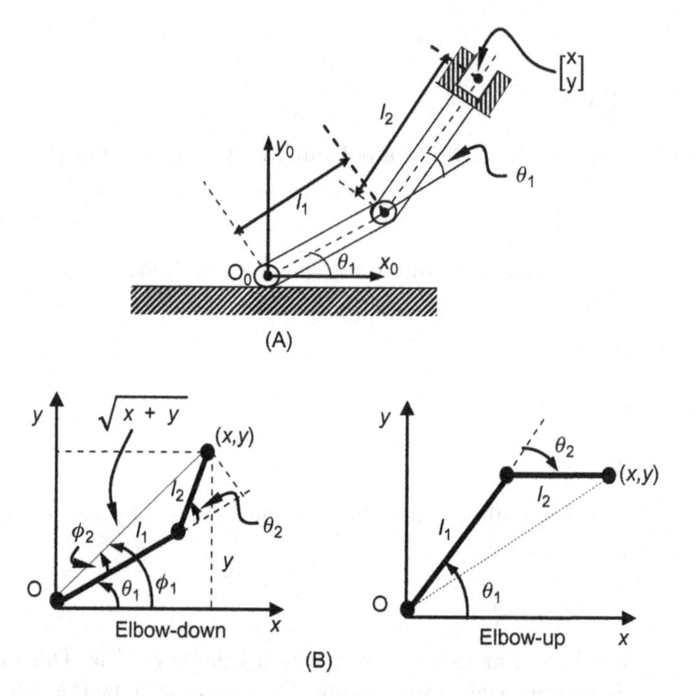

Figure 10.5 (A) Planar two degrees of freedom robot and (B) geometry for finding the inverse kinematic model (elbow-down, elbow-up).

from which the angle θ_2 is found to be:

$$\theta_2 = \arccos[(x^2 + y^2 - l_1^2 - l_2^2)/2l_1l_2] \tag{10.9a}$$

The angle θ_1 is equal to:

$$\theta_1 = \phi_1 - \phi_2$$

where:

$$\tan\phi_1 = \frac{y}{x}, \quad \tan\phi_2 = \frac{l_2 \sin\theta_2}{l_1 + l_2\cos\theta_2}$$

Therefore:

$$\theta_1 = \arctan\left(\frac{y}{x}\right) - \arctan\left(\frac{l_2\sin\theta_2}{l_1 + l_2\cos\theta_2}\right) \tag{10.9b}$$

Actually, we have two configurations that lead to the same position **p** of the end-effector, viz., *elbow-down* and *elbow-up* as shown in Figure 10.5. When $(x,y) = (0,0)$, that can be obtained if $l_1 = l_2$, the ratio y/x is not defined.

If $\theta_2 = 180°$, the base $(x, y) = (0, 0)$ can be reached for all θ_1. Finally, when a point is out of the workspace, the inverse kinematic problem has no solution.

The differential kinematics equation is:

$$\mathbf{dp} = \mathbf{J}\,\mathrm{d}\boldsymbol{\theta}, \quad \mathbf{J} = \begin{bmatrix} \partial x/\partial\theta_1 & \partial x/\partial\theta_2 \\ \partial y/\partial\theta_1 & \partial y/\partial\theta_2 \end{bmatrix}$$

Here, the Jacobian matrix $\mathbf{J}$ is obtained by differentiating Eq. (10.8a), that is:

$$\mathbf{J} = \begin{bmatrix} J_{11} & J_{12} \\ J_{21} & J_{22} \end{bmatrix} = \begin{bmatrix} -l_1 \sin\theta_1 - l_2 \sin(\theta_1 + \theta_2) & -l_2 \sin(\theta_1 + \theta_2) \\ l_1 \cos\theta_1 + l_2 \cos(\theta_1 + \theta_2) & l_2 \cos(\theta_1 + \theta_2) \end{bmatrix} \tag{10.10}$$

The inverse of $\mathbf{J}$ is found to be:

$$\begin{aligned} \mathbf{J}^{-1} &= \frac{1}{J_{11}J_{22} - J_{21}J_{12}} \begin{bmatrix} J_{22} & -J_{12} \\ -J_{21} & J_{11} \end{bmatrix} \\ &= \frac{1}{l_1 l_2 \sin\theta_2} \begin{bmatrix} l_2 \cos(\theta_1 + \theta_2) & l_2 \sin(\theta_1 + \theta_2) \\ -l_1 \cos\theta_1 - l_2 \cos(\theta_1 + \theta_2) & -l_1 \sin\theta_1 - l_2 \sin(\theta_1 + \theta_2) \end{bmatrix} \end{aligned} \tag{10.11}$$

Thus, the inverse differential kinematics equation of the robot is:

$$\frac{\mathrm{d}\boldsymbol{\theta}}{\mathrm{d}t} = \mathbf{J}^{-1} \frac{\mathbf{dp}}{\mathrm{d}t} \tag{10.12}$$

The singular (degenerate) configurations occur when $\det \mathbf{J} = J_{11}J_{22} - J_{21}J_{22} = l_1 l_2 \sin\theta_2 = 0$, that is, when $\theta_2 = 180°$ or $\theta_2 = 0°$. These two configurations correspond, respectively to the origin $(0, 0)$ and to the full extension (i.e., when the robot end-effector is at the boundary of the workspace) (see Figure 10.5B).

10.2.4.2 Dynamics

To derive the dynamic model of the robot we apply directly the Lagrange method. Consider the notation of Figure 10.6.

The symbols θ_1 and θ_2 have the usual meaning, m_1 and m_2 are the masses of the two links (concentrated at their centers of gravity), and l_1 and l_2 are the lengths of the links. The symbol $l_{c\,i}$ denotes the distance of the ith-link's center of gravity (COG) from the axis of the joint i, and $\tilde{I}_i$ denotes the moment of inertia of link i with respect to the axis passing via the COG of this link and is perpendicular to the plane xy (parallel to the axis z). Here, $q_1 = \theta_1$ and $q_2 = \theta_2$, and the kinematic and potential energies of the links 1 and 2 are given by:

$$K_1 = \frac{1}{2} m_1 l_{c1}^2 \dot{\theta}_1^2 + \frac{1}{2} \tilde{I}_1 \dot{\theta}_1^2, \quad P_1 = mgl_{c1}S_1 \tag{10.13a}$$

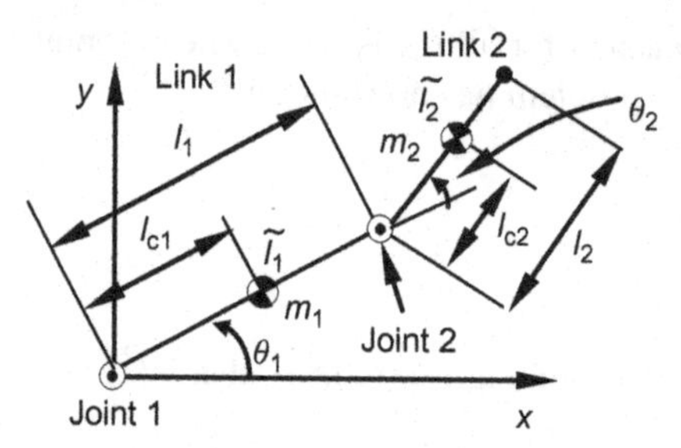

Figure 10.6 Two-link planar robot.

$$K_2 = \frac{1}{2}m_2\dot{s}_2^{\mathrm{T}}\dot{s}_2 + \frac{1}{2}\tilde{I}_2(\dot{\theta}_1^2 + \dot{\theta}_1\dot{\theta}_2), \quad P_2 = m_2 g(l_1 S_1 + l_{c2}S_{12}) \tag{10.13b}$$

where $S_1 = \sin\theta_i (i = 1,2)$, $C_i = \cos\theta_i$ $(i = 1,2)$, $s_2 = [s_{2x}, s_{2y}]^{\mathrm{T}}$ is the position vector of the COG of link 2 with:

$$s_{2x} = l_1 C_1 + l_{c2}C_{12}, \quad s_{2y} = l_1 S_1 + l_{c2}S_{12}$$

and $C_{ij} = \cos(\theta_i + \theta_j)$, $S_{ij} = \sin(\theta_i + \theta_j)$.

Using the Lagrangian function of the robot, $L = \mathrm{K}_1 + \mathrm{K}_2 - P_1 - P_2$, we find the equations:[1]

$$D_{11}\ddot{\theta}_1 + D_{12}\ddot{\theta}_2 + h_{122}\dot{\theta}_2^2 + 2h_{112}\dot{\theta}_1\dot{\theta}_2 + g_1 = \tau_1, \; D_{21}\ddot{\theta}_1 + D_{22}\ddot{\theta}_2 + h_{211}\dot{\theta}_1^2 + g_2 = \tau_2 \tag{10.14}$$

with:

$$\begin{aligned}
&D_{11} = m_1 l_{c1}^2 + \tilde{I}_1 + m_2(l_1^2 + l_{c2}^2 + 2l_1 l_{c2}C_2) + \tilde{I}_2 \\
&D_{12} = D_{21} = m_2(l_{c2}^2 + l_1 l_{c2}C_2) + \tilde{I}_2, \quad D_{22} = m_2 l_{c2}^2 + \tilde{I}_2 \\
&h_{122} = h_{112} = -h_{211} = -m_2 l_1 l_{c2} S_2 \\
&g_1 = m_1 g l_{c1} C_1 + m_2 g(l_1 C_1 + l_{c2}C_{12}), \quad g_2 = m_2 g l_{c2} C_{12}
\end{aligned} \tag{10.15}$$

where τ_1 and τ_2 are the external torques applied to joints 1 and 2. The coefficient D_{ii} is the *effective inertia* of the joint i, D_{ij} is the *coupling inertia* of joints i and j, h_{ijj} is the *coefficient of centrifugal force*, h_{ijk} $(j \neq k)$ is the *coefficient of Coriolis acceleration* of the joint i due to the velocities of the joints j and k, and g_i $(i = 1,2)$ represent the torques due to gravity. The dynamic relations in Eqs. (10.14) and (10.15) can be written in the standard compact form:

$$\mathbf{D}(\boldsymbol{\theta})\ddot{\boldsymbol{\theta}} + \mathbf{h}(\boldsymbol{\theta},\dot{\boldsymbol{\theta}}) + \mathbf{g}(\boldsymbol{\theta}) = \boldsymbol{\tau} \tag{10.16a}$$

[1] If there is friction τ_f in the motors of the joints (e.g., $\tau_f = -\beta\dot{\theta}_i$, β = friction coefficient), then τ_i should be replaced by $\tau_i' = \tau_i - \beta\dot{\theta}_i$.

$$\mathbf{D}(\boldsymbol{\theta}) = \begin{bmatrix} D_{11} & D_{12} \\ D_{21} & D_{22} \end{bmatrix}, \quad \mathbf{g}(\boldsymbol{\theta}) = \begin{bmatrix} g_1 \\ g_2 \end{bmatrix}, \quad \boldsymbol{\tau} = \begin{bmatrix} \tau_1 \\ \tau_2 \end{bmatrix} \tag{10.16b}$$

$$\mathbf{h}(\boldsymbol{\theta},\dot{\boldsymbol{\theta}}) = \text{col}\left[\sum_{j=1}^{2}\sum_{k=1}^{2}\left(\frac{\partial D_{ij}}{\partial \theta_k} - \frac{1}{2}\frac{\partial D_{jk}}{\partial \theta_i}\right)\dot{\theta}_j\dot{\theta}_k\right] \tag{10.16c}$$

where col$[h_i]$ denotes a column vector with elements $h_i (i = 1,2)$. In the special case, where the link masses m_1 and m_2 are assumed to be concentrated at the end of each link, we have $l_{c1} = l_1$ and $l_{c2} = l_2$. It is easy to verify the properties described by Eqs. (3.12) and (3.13) and the antisymmetricity of $\dot{\mathbf{D}} - 2\mathbf{C}$, where K is the total kinetic energy $K = K_1 + K_2$ of the robot.

10.2.4.3 Manipulability Measure

If we use the position $[x, y]^T$ of the end-effector as the vector $v = \dot{X}$, then (see (10.10)):

$$\mathbf{J} = \begin{bmatrix} -l_1 \sin\theta_1 - l_2 \sin(\theta_1 + \theta_2) & -l_2 \sin(\theta_1 + \theta_2) \\ l_1 \cos\theta_1 + l_2 \cos(\theta_1 + \theta_2) & l_2 \cos(\theta_1 + \theta_2) \end{bmatrix}$$

and so the manipulability measure w is:

$$w = |\det \mathbf{J}| = l_1 l_2 |\sin\theta_2|$$

Thus, the optimal configurations of the manipulator, at which w is maximum, are: $\theta_2 = \pm 90°$ for any l_1, l_2 and θ_1. If the lengths l_1 and l_2 are specified under the condition of constant total length (i.e., $l_1 + l_2 = \text{const.}$), then the manipulability measure takes its maximum when $l_1 = l_2$ for any θ_1 and θ_2.

10.3 MM Modeling

The total kinematic and dynamic models of an MM are complex and strongly coupled, combining the models of the mobile platform and the fixed robotic manipulator [1,3,7]. Today, MMs are in use with differential-drive, tricycle, car-like, and omnidirectional platforms that offer maximum maneuverability.

10.3.1 General Kinematic Model

Consider the MM of Figure 10.7 which has a differential-drive platform and a multilink robotic manipulator.

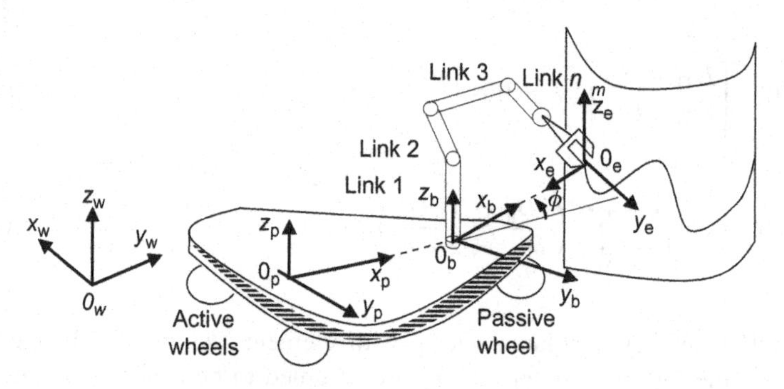

Figure 10.7 Geometric features of an MM with differential-drive platform.

Here, we have the following four coordinate frames:

$O_w x_w y_w z_w$ is the world coordinate frame,
$O_p x_p y_p z_p$ is the platform coordinate frame,
$O_b x_b y_b z_b$ is the manipulator base coordinate frame,
$O_e x_e y_e z_e$ is the end-effector coordinate frame.

Then, the manipulator's end-effector position/orientation with respect to $O_w x_w y_w z_w$ is given by (see Eq. (10.2)):

$$\mathbf{T} = \mathbf{A}_p^w \mathbf{A}_b^p \mathbf{A}_e^b \tag{10.17}$$

where:

$\mathbf{A}_p^w$ is the transformation matrix from O_w to $\mathbf{O}_p$,
$\mathbf{A}_b^p$ is the transformation matrix from O_p to O_b,
$\mathbf{A}_e^b$ is the transformation matrix from O_b to O_e.

In vectorial form, the end-effector position/orientation vector $\mathbf{x}_e^w$ in world coordinates has the form:

$$\mathbf{x}_e^w = \mathbf{F}(\mathbf{q}) \tag{10.18}$$

where:

$$\mathbf{q} = [\mathbf{p}^T, \boldsymbol{\theta}^T]^T$$

$\mathbf{p} = [x, y, \phi]^T$ (Platform configuration)

$\boldsymbol{\theta} = [\theta_1, \theta_2, \ldots, \theta_{n_m}]^T$ (Manipulator configuration)

Therefore, differentiating Eq. (10.18) with respect to time we get:

$$\dot{\mathbf{x}}_e^w = \left(\frac{\partial \mathbf{F}}{\partial \mathbf{p}}\right)\dot{\mathbf{p}} + \left(\frac{\partial \mathbf{F}}{\partial \boldsymbol{\theta}}\right)\dot{\boldsymbol{\theta}} \tag{10.19}$$

where $\dot{\mathbf{p}}$ is given by the kinematic model of the platform (see (6.1)):

$$\dot{\mathbf{p}} = \mathbf{G}(\mathbf{p})\mathbf{u}_{\mathrm{p}}, \quad \mathbf{u}_{\mathrm{p}} \in R^2 \tag{10.20a}$$

and $\dot{\boldsymbol{\theta}}$ by the manipulator's kinematic model. Assuming that the manipulator is constraint-free, we can write:

$$\dot{\boldsymbol{\theta}} = \mathbf{u}_{\mathrm{m}} \tag{10.20b}$$

where $\mathbf{u}_{\mathrm{m}}$ is the vector of manipulator's joint commands. Combining Eqs. (10.19), (10.20a), and (10.20b) we get the overall kinematic model of the MM:

$$\begin{aligned}\dot{\mathbf{x}}_{\mathrm{e}}^{\mathrm{w}}(t) &= \left(\frac{\partial \mathbf{F}}{\partial \mathbf{p}}\right)\mathbf{G}(\mathbf{p})\mathbf{u}_{\mathrm{p}} + \left(\frac{\partial \mathbf{F}}{\partial \boldsymbol{\theta}}\right)\mathbf{u}_{\mathrm{m}} \\ &= \mathbf{J}(\mathbf{q})\mathbf{u}(t)\end{aligned} \tag{10.21a}$$

where:

$$\begin{aligned}\mathbf{u}(t) &= [\mathbf{u}_{\mathrm{p}}^{\mathrm{T}}(t), \mathbf{u}_{\mathrm{m}}^{\mathrm{T}}(t)]^{\mathrm{T}} \in R^{2+n_{\mathrm{m}}} \\ \mathbf{J}(\mathbf{q}) &= [\mathbf{J}_{\mathrm{p}}(\mathbf{q})\mathbf{G}(\mathbf{q}) \vdots \mathbf{J}_{\mathrm{m}}(\boldsymbol{\theta})] \\ \mathbf{J}_{\mathrm{p}}(\mathbf{q}) &= \partial \mathbf{F}/\partial \mathbf{p} \\ \mathbf{J}_{\mathrm{m}}(\mathbf{q}) &= \partial \mathbf{F}/\partial \boldsymbol{\theta}\end{aligned} \tag{10.21b}$$

Equation (10.21a) represents the total kinematic model of the MM from the inputs to the end-effector (task) variables.

Actually, in the present case the system is subject to the nonholonomic constraint:

$$\mathbf{M}(\mathbf{p})\dot{\mathbf{p}} = 0, \quad \mathbf{M}(\mathbf{p}) = [-\sin\phi, \cos\phi, 0, \ldots, 0] \tag{10.22}$$

where $\dot{\mathbf{p}}$ cannot be eliminated by integration. Therefore, $\mathbf{J}(\mathbf{q})$ should involve a row representing this constraint. Since the dimensionality $2 + n_{\mathrm{m}}$ of the control input vector $\mathbf{u}(t)$ is less than the total number $3 + n_{\mathrm{m}}$ of variables (degrees of freedom) to be controlled, the system is always *underactuted*.

10.3.2 General Dynamic Model

The Lagrange dynamic model of the MM is given by Eq. (3.16):

$$\mathbf{D}(\mathbf{q})\ddot{\mathbf{q}} + \mathbf{C}(\mathbf{q},\dot{\mathbf{q}})\dot{\mathbf{q}} + \mathbf{g}(\mathbf{q}) + \mathbf{M}^{\mathrm{T}}(\mathbf{q})\boldsymbol{\lambda} = \mathbf{E}\boldsymbol{\tau} \tag{10.23}$$

where $\mathbf{M(q)}$ is given by Eq. (10.22). This model contains two parts, namely:

Platform Part

$$\mathbf{D}_p(\mathbf{q}_p, \mathbf{q}_m)\ddot{\mathbf{q}}_p + \mathbf{C}_p(\mathbf{q}_p, \mathbf{q}_m, \dot{\mathbf{q}}_p, \dot{\mathbf{q}}_m) = \mathrm{E}_p\boldsymbol{\tau}_p - \mathrm{M}^T(\mathbf{q}_p)\boldsymbol{\lambda} - \mathbf{D}_p(\mathbf{q}_p, \mathbf{q}_m)\ddot{\mathbf{q}}_m$$

Manipulator Part

$$\mathbf{D}_m(\mathbf{q}_m)\ddot{\mathbf{q}}_m + \mathbf{C}_m(\mathbf{q}_p, \mathbf{q}_m; \dot{\mathbf{q}}_p) = \boldsymbol{\tau}_m - \mathbf{D}_m(\mathbf{q}_p, \mathbf{q}_m)\ddot{\mathbf{q}}_p$$

where the index p refers to the platform and m to the manipulator, and the symbols have the standard meaning. Appling the technique of Section 3.2.4 we eliminate the nonholonomic constraint $\mathbf{M}(\mathbf{q}_p)\dot{\mathbf{q}}_p = \mathbf{0}$, and get the reduced (unconstrained) model of the form of Eqs. (3.19a) and (3.19b):

$$\overline{\mathbf{D}}(\mathbf{q})\dot{\mathbf{v}} + \overline{\mathbf{C}}(\mathbf{q},\dot{\mathbf{q}})\mathbf{v} + \overline{\mathbf{g}}(\mathbf{q}) = \overline{\mathbf{E}}\boldsymbol{\tau} \tag{10.24}$$

where:

$$\mathbf{q} = [\mathbf{q}_p^T, \mathbf{q}_m^T]^T$$

and $\overline{\mathbf{D}}(\mathbf{q})$, $\overline{\mathbf{C}}(\mathbf{q},\dot{\mathbf{q}})$, and $\overline{\mathbf{g}}(\mathbf{q})$ are given by the combination of the respective terms in the platform and manipulator parts (the details of derivation are straightforward and are left as exercise).

10.3.3 Modeling a Five Degrees of Freedom Nonholonomic MM

Here, the above general methodology will be applied to the MM of Figure 10.8 which consists of a differential-drive mobile platform and a two-link planar manipulator [3,22].

10.3.3.1 Kinematics

Without loss of generality, the COG G is assumed to coincide with the rotation point Q (midpoint of the two wheels), that is, $b = 0$. The nonholonomic constraint of the platform is:

$$-\dot{x}_Q \sin\phi + \dot{y}_Q \cos\phi = 0$$

This constraint, when expressed at the base $O_b(x_b, y_b)$ of the manipulator, becomes:

$$-\dot{x}_b \sin\phi + \dot{y}_b \cos\phi + l_b\dot{\phi} = 0$$

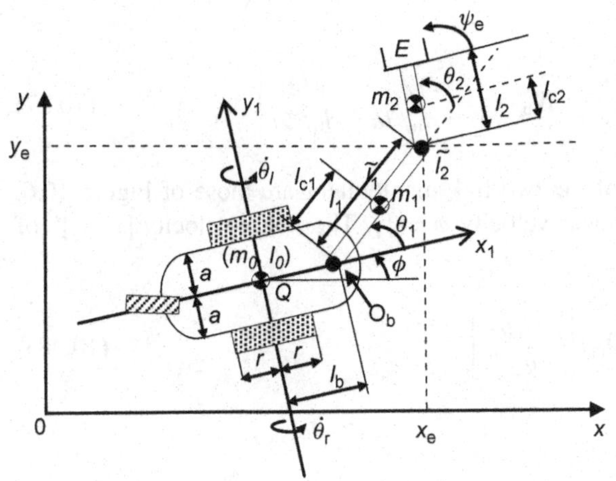

Figure 10.8 The five degrees of freedom MM. Platform's mass and moment of inertia m_0 and I_0.

where l_b is the distance between the points $G(Q)$ and O_b. The kinematic equations of the wheeled mobile robot (WMR) platform (at the point O_b) are:

$$\begin{aligned}
\dot{x}_b &= \left(\frac{r}{2}\cos\phi + \frac{rl_b}{2a}\sin\phi\right)\dot{\theta}_1 + \left(\frac{r}{2}\cos\phi - \frac{rl_b}{2a}\sin\phi\right)\dot{\theta}_r \\
\dot{y}_b &= \left(\frac{r}{2}\sin\phi - \frac{rl_b}{2a}\cos\phi\right)\dot{\theta}_1 + \left(\frac{r}{2}\sin\phi + \frac{rl_b}{2a}\cos\phi\right)\dot{\theta}_r \\
\dot{\phi} &= \frac{r}{2a}(\dot{\theta}_r - \dot{\theta}_1)
\end{aligned} \tag{10.25}$$

which are written in the Jacobian form:

$$\dot{\mathbf{p}} = \mathbf{J}\dot{\boldsymbol{\theta}}, \quad \dot{\mathbf{p}} = [\dot{x}_b, \dot{y}_b, \dot{\phi}]^T, \quad \dot{\boldsymbol{\theta}}^T = [\dot{\theta}_1, \dot{\theta}_r]^T$$

where:

$$\mathbf{J} = \begin{bmatrix} \mathbf{J}_b \\ \vdots \\ -r/2a \quad r/2a \end{bmatrix}$$

$$\begin{aligned}
\mathbf{J}_b &= \begin{bmatrix} \frac{r}{2}\cos\phi + \frac{rl_b}{2a}\sin\phi & \frac{r}{2}\cos\phi - \frac{rl_b}{2a}\sin\phi \\ \frac{r}{2}\sin\phi - \frac{rl_b}{2a}\cos\phi & \frac{r}{2}\sin\phi + \frac{rl_b}{2a}\cos\phi \end{bmatrix} \\
&= \mathbf{R}(\phi)\mathbf{W}_b
\end{aligned} \tag{10.26}$$

with:

$$\mathbf{R}(\phi) = \begin{bmatrix} \cos\phi & -\sin\phi \\ \sin\phi & \cos\phi \end{bmatrix}, \quad \mathbf{W}_\mathrm{b} = \begin{bmatrix} r/2 & r/2 \\ -rl_\mathrm{b}/2a & rl_\mathrm{b}/2a \end{bmatrix} \tag{10.27}$$

The kinematic parameters of the two-link manipulator are those of Figure 10.6, where now its first joint has linear velocity $\dot{\phi} + \dot{\theta}_1$. The linear velocity $[\dot{x}_\mathrm{e}, \dot{y}_\mathrm{e}]^T$ of the end-effector is given by:

$$\begin{bmatrix} \dot{x}_\mathrm{e} \\ \dot{y}_\mathrm{e} \end{bmatrix} = \begin{bmatrix} \dot{x}_\mathrm{b} \\ \dot{y}_\mathrm{b} \end{bmatrix} + \mathbf{R}(\phi)\mathbf{J}_\mathrm{m}(\boldsymbol{\theta}_\mathrm{m}) \begin{bmatrix} \dot{\phi} + \dot{\theta}_1 \\ \dot{\theta}_2 \end{bmatrix} \tag{10.28a}$$

where ($\boldsymbol{\theta}_\mathrm{m} = [\theta_1, \theta_2]^\mathrm{T}$):

$$\begin{aligned} \mathbf{J}_\mathrm{m}(\boldsymbol{\theta}_\mathrm{m}) &= \begin{bmatrix} -l_1 \sin\theta_1 - l_2 \sin(\theta_1 + \theta_2) & -l_2 \sin(\theta_1 + \theta_2) \\ l_1 \cos\theta_1 + l_2 \cos(\theta_1 + \theta_2) & l_2 \cos(\theta_1 + \theta_2 \end{bmatrix} \\ &= \begin{bmatrix} J_{\mathrm{m},11} & J_{\mathrm{m},12} \\ J_{\mathrm{m},21} & J_{\mathrm{m},22} \end{bmatrix} \end{aligned} \tag{10.28b}$$

is the Jacobian matrix of the manipulator with respect to its base coordinate frame (see Eq. (10.10)). Here, $\dot{\phi}$ is given by the third part of Eq. (10.25), that is, $\dot{\phi} = (r/2a)(-\dot{\theta}_1 + \dot{\theta}_\mathrm{r})$. Thus:

$$\begin{bmatrix} \dot{\phi} + \dot{\theta}_1 \\ \dot{\theta}_2 \end{bmatrix} = \mathbf{S} \begin{bmatrix} \dot{\theta}_1 \\ \dot{\theta}_\mathrm{r} \end{bmatrix}, \quad \mathbf{S} = \begin{bmatrix} 1 - r/2a & r/2a \\ 0 & 1 \end{bmatrix} \tag{10.29a}$$

and so:

$$\begin{bmatrix} \dot{x}_\mathrm{e} \\ \dot{y}_\mathrm{e} \end{bmatrix} = [\mathbf{R}(\phi)\mathbf{W}_\mathrm{b} + \mathbf{R}(\phi)\mathbf{J}_\mathrm{m}(\boldsymbol{\theta}_\mathrm{m})\mathbf{S}] \begin{bmatrix} \dot{\theta}_1 \\ \dot{\theta}_\mathrm{r} \end{bmatrix} \tag{10.29b}$$

The overall kinematic equation of the MM is:

$$\dot{\mathbf{p}}_0 = \mathbf{J}_0 \dot{\boldsymbol{\theta}}_0 \tag{10.30a}$$

where:

$$\dot{\mathbf{p}}_0 = \begin{bmatrix} \dot{x}_\mathrm{e} \\ \dot{y}_\mathrm{e} \\ \vdots \\ \dot{x}_\mathrm{b} \\ \dot{y}_\mathrm{b} \end{bmatrix}, \quad \dot{\boldsymbol{\theta}}_0 = \begin{bmatrix} \dot{\theta}_1 \\ \dot{\theta}_\mathrm{r} \\ \vdots \\ \dot{\theta}_1 \\ \dot{\theta}_2 \end{bmatrix}, \quad \mathbf{J}_0 = \begin{bmatrix} \mathbf{R}(\phi)[\mathbf{W}_\mathrm{b} + \mathbf{J}_\mathrm{m}(\boldsymbol{\theta}_\mathrm{m})\mathbf{S}] \\ \cdots\cdots\cdots\cdots \\ \mathbf{R}(\phi)\mathbf{W}_\mathrm{b} \end{bmatrix} \tag{10.30b}$$

If the mobile platform of the MM is of the tricycle or car-like type shown in Figure 2.8 or Figure 2.9, its position/orientation is described by four variables $\dot{x}_Q, \dot{y}_Q, \dot{\phi}$, and $\dot{\psi}$, and the kinematic Eq. (2.45) or Eq. (2.53) is enhanced as in Eq. (10.25) with the l_b terms. The overall kinematic model of the MM is found by the same procedure.

10.3.3.2 Dynamics

Assuming that the moment of inertia of the passive wheel is negligible, the Lagrangian of the MM is found to be:

$$\begin{aligned} L = & \frac{1}{2}m_0(\dot{x}_Q^2 + \dot{y}_Q^2) + \frac{1}{2}I_0\dot{\phi}^2 + \frac{1}{2}m_1(\dot{x}_A^2 + \dot{y}_A^2) \\ & + \frac{1}{2}I_1(\dot{\phi}+\dot{\theta}_1)^2 + \frac{1}{2}m_2(\dot{x}_B^2 + \dot{y}_B^2) + \frac{1}{2}I_2(\dot{\phi} + \dot{\theta}_1 + \dot{\theta}_2) \end{aligned} \tag{10.31}$$

where:

m_0, I_0 are mass and moment of inertia of the platform,
m_1, m_2, I_1, I_2 are masses and moments of inertia of links 1 and 2, respectively,
$\dot{x}_Q, \dot{y}_Q$ are x, y components of the velocity of point Q.
$\dot{x}_A, \dot{y}_A, \dot{x}_B, \dot{y}_B$ are x, y velocities of links 1 and 2, respectively.

Using the Lagrangian (10.31) we derive the dynamic model of Eq. (10.23), with the nonholonomic constraint:

$$\mathbf{M}(\mathbf{q}) = [-\sin\phi, \cos\phi, l_b, 0, 0] \tag{10.32}$$

Now, using the transformation $\dot{\mathbf{q}}(t) = \mathbf{B}(\mathbf{q})\mathbf{v}(t)$ with $\mathbf{B}^T(\mathbf{q})\mathbf{M}^T(\mathbf{q}) = \mathbf{0}$, where:

$$\mathbf{q} = [x_b, y_b, \phi, \theta_1, \theta_2]^T$$
$$\mathbf{v} = [v_1, v_2, v_3, v_4]^T = [\dot{\theta}_1, \dot{\theta}_r, \dot{\theta}_1, \dot{\theta}_2]^T$$

and $\mathbf{B}$ is selected as:

$$\mathbf{B} = \begin{bmatrix} \mathbf{B}' & \mathrm{O} \\ \mathrm{O} & \mathrm{I} \end{bmatrix}, \quad \mathbf{B}' = \begin{bmatrix} b'_{11} & b'_{12} \\ b'_{21} & b'_{22} \\ b'_{31} & b'_{32} \end{bmatrix} \tag{10.33}$$

where I is the 2×2 unit matrix, and

$$\begin{aligned} b'_{11} &= r(\cos\phi)/2 + l_b r(\sin\phi)/2a \\ b'_{12} &= r(\cos\phi)/2 - l_b r(\sin\phi)/2a \\ b'_{21} &= r(\sin\phi)/2 - l_b r(\cos\phi)/2a \\ b'_{22} &= r(\sin\phi)/2 + l_b r(\cos\phi/2a \\ b'_{31} &= -r/2a, \quad b'_{32} = r/2a \end{aligned} \tag{10.34}$$

the constrained Lagrangian model is reduced, as usual, to the unconstrained form (10.24).

Example 10.1

It is desired to investigate the manipulability measure of an MM and compute it for a five degrees of freedom MM.

Solution

The manipulability measure is given by:

$$w = \sqrt{\det \mathbf{J}(\mathbf{q})\mathbf{J}^{\mathrm{T}}(\mathbf{q})}$$

or $w = |\det \mathbf{J}|$, if $\mathbf{J}$ is square. Let us define by $\sigma_1, \sigma_2, \ldots, \sigma_m$, the m larger numbers $\sqrt{\lambda_i}, i = 1, 2, \ldots, n$, where λ_i is the ith eigenvalue of the matrix $\mathbf{J}^{\mathrm{T}}\mathbf{J}$. Then, the above definition of w reduces to:

$$w_1 = \sigma_1\sigma_2\ldots\sigma_m, \quad \sigma_1 \geq \sigma_2 \geq \cdots > \sigma_m \geq 0$$

The numbers $\sigma_i (i = 1, 2, \ldots, m)$ are known as *singular values* of $\mathbf{J}$, and form the matrix $\boldsymbol{\Sigma}$:

$$\boldsymbol{\Sigma} = \begin{bmatrix} \sigma_1 & 0 & & \vdots & \\ & \ddots & & & \mathbf{0} \\ 0 & & \sigma_m & \vdots & \end{bmatrix}$$

which defines the *singular value decomposition* $\mathbf{J} = \mathbf{U}\boldsymbol{\Sigma}\mathbf{V}^{\mathrm{T}}$ of $\mathbf{J}$ with $\mathbf{U}$ and $\mathbf{V}$ being orthogonal matrices of dimensionality m and n respectively. The principal axes of the manipulability ellipsoid are $\sigma_1\mathbf{u}_1, \sigma_2\mathbf{u}_2, \ldots, \sigma_m\mathbf{u}_m$ where $\mathbf{u}_i$ are the columns of $\mathbf{U}$. Some other definitions of the manipulability measure are:

(i) $w_2 = \sigma_m/\sigma_1$: The ratio of the minimum and maximum singular values (radii) of the manipulator ellipsoid. This provides only qualitative information about the manipulability of the robot. If $\sigma_m = \sigma_1$, then the ellipsoid is a sphere and the robot has the same manipulability in all directions.

(ii) $w_3 = \sigma_m$: The minimum radius which gives the upper bound of the velocity of the end-effector motion in any direction.

(iii) $w_4 = (\sigma_1\sigma_2\ldots\sigma_m)^{1/m}$: The radius of the sphere that has the same volume of the manipulability ellipsoid.

(iv) $w_5 = \sqrt{1 - \sigma_m^2/\sigma_1^2}$: The generalized concept of eccentricity of an ellipse.

Consider the manipulability definition via $\mathbf{J}(\mathbf{q})$. In the present case, the total Jacobian matrix of the MM is given by Eqs. (10.21a) and (10.21b), and represents the transformation from:

$$\mathbf{u}(t) = \begin{bmatrix} \mathbf{u}_{\mathrm{p}}(t) \\ \mathbf{u}_{\mathrm{m}}(t) \end{bmatrix}, \quad \mathbf{u}_{\mathrm{p}}(t) = \begin{bmatrix} v(t) \\ \omega(t) \end{bmatrix}, \quad \mathbf{u}_{\mathrm{m}}(t) = \begin{bmatrix} \theta_1 \\ \theta_2 \\ \vdots \\ \theta_m \end{bmatrix}$$

to

$$\mathbf{x}_{\mathrm{e}} = [\dot{x}_{\mathrm{e}}, \dot{y}_{\mathrm{e}}, \dot{\psi}]^T$$

where x_e, y_e are the position coordinates, and ψ_e the orientation angle of the end-effector. Then:

$$w = \sqrt{\det \mathbf{J}(\mathbf{q})\mathbf{J}^T(\mathbf{q})}$$

and the MM manipulability ellipsoid is defined by:

$$||\mathbf{u}|| = \left\| \begin{bmatrix} \mathbf{u}_p(t) \\ \mathbf{u}_m(t) \end{bmatrix} \right\| \leq 1$$

Unfortunately, in most cases it is not possible to decouple $\mathbf{u}_p(t)$ and $\mathbf{u}_m(t)$ in computing the ellipsoid $||\mathbf{u}|| \leq 1$. Now, consider the five-link MM of Figure 10.8. It is straightforward to find that:

$$\mathbf{J} = \begin{bmatrix} J_{11} & J_{12} & J_{13} & J_{14} \\ J_{21} & J_{22} & J_{23} & J_{24} \\ 0 & 0 & 0 & 1 \end{bmatrix}$$

$$\begin{aligned}
J_{11} &= \cos\phi \\
J_{12} &= -l_b \sin\phi - l_1 \sin(\phi + \theta_1) - l_2 \sin(\phi + \theta_1 + \theta_2) \\
J_{13} &= -l_1 \sin(\phi + \theta_1) - l_2 \sin(\phi + \theta_1 + \theta_2) \\
J_{14} &= -l_2 \sin(\phi + \theta_1 + \theta_2) \\
J_{21} &= \sin\phi \\
J_{22} &= l_b \cos\phi + l_1 \cos(\phi + \theta_1) + l_2 \cos(\phi + \theta_1 + \theta_2) \\
J_{23} &= l_1 \cos(\phi + \theta_1) + l_2 \cos(\phi + \theta_1 + \theta_2) \\
J_{24} &= l_2 \cos(\phi + \theta_1 + \theta_2)
\end{aligned}$$

Consider the manipulability measure w_5 which gives information on the shape of the ellipses. As w_5 tends to zero, that is, $\sigma_m/\sigma_1 \to 1$ the ellipsoid tends to a sphere and the attainable end-effector speeds tend to be the same in all directions. Actually, they can be equal (isotropic) when $w_5 = 0$, for a given bounded velocity control signal.

10.3.4 Modeling an Omnidirectional MM

Here, the kinematic and dynamic model of the MM shown in Figure 10.9 will be derived. This MM involves a three-wheel omnidirectional platform (with orthogonal wheels) and a 3D manipulator [8,11].

10.3.4.1 Kinematics

As usual, the kinematic model of the MM is found by combining the kinematic models of the platform and the manipulator.

The workspace of the robot is described by the world coordinate frame $O_w x_w y_w z_w$. The platform's coordinate frame $O_p x_p y_p z_p$ has its origin at the COG of

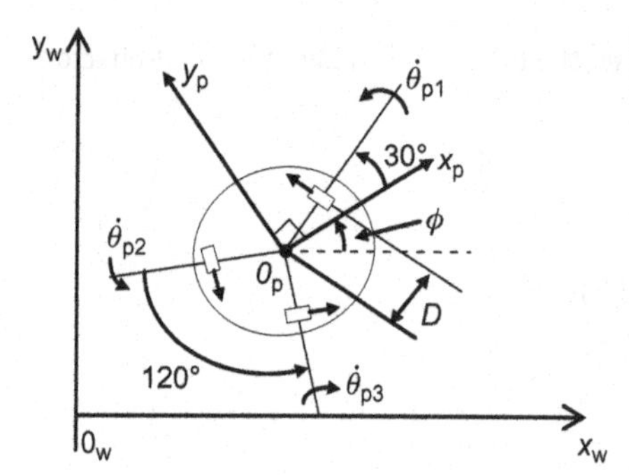

Figure 10.9 Geometric features of the omnidirectional platform.

the platform, and the wheels have a radius r, an angle 120° between them, and a distance D from the platform's COG. The kinematic model of the robot was derived in Section 2.4. For the platform shown in Figure 10.9, the model given by Eqs. (2.74a) and (2.74b) becomes:

$$\begin{bmatrix} \dot{\theta}_{p1} \\ \dot{\theta}_{p2} \\ \dot{\theta}_{p3} \end{bmatrix} = \frac{1}{r} \begin{bmatrix} -1/2 & \sqrt{3}/2 & D \\ -1/2 & -\sqrt{3}/2 & D \\ 1 & 0 & D \end{bmatrix} \begin{bmatrix} \dot{x}_p \\ \dot{y}_p \\ \dot{\phi} \end{bmatrix}$$

that is,

$$\dot{\boldsymbol{\theta}}_r = \mathbf{J}_0^{-1} \dot{\mathbf{p}}_p \tag{10.35}$$

where $\dot{\theta}_{pi} (i = 1, 2, 3)$ are the angular velocities of the wheels, ϕ the rotational angle between x_w and x_p axes, and $\dot{x}_p, \dot{y}_p$ are the components of the platform's velocity expressed in the platform's (moving) coordinate system. The coordinate transformation matrix between the world and the platform coordinate frames is:

$$\mathbf{R}_p^w = \begin{bmatrix} \cos\phi & -\sin\phi & 0 \\ \sin\phi & \cos\phi & 0 \\ 0 & 0 & 1 \end{bmatrix} \tag{10.36}$$

Thus, the relation between $\mathbf{v}_p^w = [\dot{x}_p^w, \dot{y}_p^w, \dot{\phi}]^T$ and $\dot{\boldsymbol{\theta}}_r = [\dot{\theta}_{p1}, \dot{\theta}_{p2}, \dot{\theta}_{p3}]^T$ is

$$\dot{\boldsymbol{\theta}}_r = \mathbf{J}_0^{-1} \mathbf{R}_p^w \mathbf{v}_p^w \tag{10.37}$$

Inverting Eq. (10.37) we get the direct differential kinematic model of the robot from $\dot{\boldsymbol{\theta}}_r$ to $\mathbf{v}_p^w$ as:

$$\mathbf{v}_p^w = \mathbf{J}_p(\phi) \dot{\boldsymbol{\theta}}_r \tag{10.38a}$$

where:

$$\mathbf{J}_{\mathrm{p}}(\phi) = (\mathbf{J}_0^{-1}\mathbf{R}_{\mathrm{p}}^{\mathrm{w}})^{-1} = \frac{r}{3}\begin{bmatrix} -\cos\phi - \sqrt{3}\sin\phi & -\cos\phi + \sqrt{3}\sin\phi & 2\cos\phi \\ -\sin\phi + \sqrt{3}\cos\phi & -\sin\phi - \sqrt{3}\cos\phi & 2\sin\phi \\ 1/D & 1/D & 1/D \end{bmatrix} \tag{10.38b}$$

Now, the platform's kinematic model of Eq. (10.38a) will be integrated with the manipulator's kinematic model (see Figure 10.10).

From Figure 10.9 we get:

$$\begin{aligned} x_{\mathrm{e}}^{\mathrm{w}} &= x_{\mathrm{p}}^{\mathrm{w}} + [l_2\cos\theta_2 + l_3\cos(\theta_2+\theta_3)]\cos(\phi+\theta_1) \\ y_{\mathrm{e}}^{\mathrm{w}} &= y_{\mathrm{p}}^{\mathrm{w}} + [l_2\cos\theta_2 + l_3\cos(\theta_2+\theta_3)]\sin(\phi+\theta_1) \\ z_{\mathrm{e}}^{\mathrm{w}} &= r + H + l_1 + l_2\sin\theta_2 + l_3\sin(\theta_2+\theta_3) \end{aligned} \tag{10.39}$$

where H is the height of the platform. Now, differentiating Eq. (10.39) and using Eqs. (10.38a) and (10.38b) we get:

$$\dot{\mathbf{p}}_{\mathrm{e}} = \mathbf{J}_{\mathrm{e}}(\phi,\theta_1,\theta_2,\theta_3)\dot{\mathbf{q}} \tag{10.40}$$

where: $\dot{\mathbf{q}} = [\dot{\theta}_{\mathrm{p}1}, \dot{\theta}_{\mathrm{p}2}, \dot{\theta}_{\mathrm{p}3}, \dot{\theta}_1, \dot{\theta}_2, \dot{\theta}_3]^{\mathrm{T}}$ and $\mathbf{J}_{\mathrm{e}}$ is a 3×6 matrix with columns:

$$\mathbf{J}_{\mathrm{e}1} = \begin{bmatrix} -\frac{r}{3}(C_\Phi + \sqrt{3}S_\Phi) - \frac{r}{3D}(l_2C_2S_{\Phi1} + l_3C_{23}S_{\Phi1}) \\ \frac{r}{3}(-S_\Phi + \sqrt{3}C_\Phi) + \frac{r}{3D}(l_2C_2C_{\Phi1} + l_3C_{23}C_{\Phi1}) \\ 0 \end{bmatrix}, \mathbf{J}_{\mathrm{e}4} = \begin{bmatrix} -l_2C_2S_{\Phi1} - l_3C_{23}S_{\Phi1} \\ l_2C_2C_{\Phi1} + l_3C_{23}C_{\Phi1} \\ 0 \end{bmatrix}$$

$$\mathbf{J}_{\mathrm{e}2} = \begin{bmatrix} \frac{r}{3}(-C_\Phi + \sqrt{3}S_\Phi) - \frac{r}{3D}(l_2C_2S_{\Phi1} + l_3C_{23}S_{\Phi1}) \\ -\frac{r}{3}(S_\Phi + \sqrt{3}C_\Phi) + \frac{r}{3D}(l_2C_2C_{\Phi1} + l_3C_{23}C_{\Phi1}) \\ 0 \end{bmatrix}, \mathbf{J}_{\mathrm{e}5} = \begin{bmatrix} -l_2S_2C_{\Phi1} - l_3S_{23}C_{\Phi1} \\ -l_2S_2S_{\Phi1} - l_3S_{23}S_{\Phi1} \\ l_2C_2 + l_3C_{23} \end{bmatrix}$$

$$\mathbf{J}_{\mathrm{e}3} = \begin{bmatrix} \frac{2}{3}rC_\Phi - \frac{r}{3D}(l_2C_2S_{\Phi1} + l_3C_{23}S_{\Phi1}) \\ \frac{2}{3}rS_\Phi + \frac{r}{3D}(l_2C_2C_{\Phi1} + l_3C_{23}C_{\Phi1}) \\ 0 \end{bmatrix}, \mathbf{J}_{\mathrm{e}6} = \begin{bmatrix} -l_3S_{23}C_{\Phi1} \\ -l_3S_{23}S_{\Phi1} \\ l_3C_{23} \end{bmatrix} \tag{10.41}$$

The notations used in the above expressions are:

$$
\begin{aligned}
S_\Phi &= \sin\phi & C_\Phi &= \cos\phi \\
S_i &= \sin\theta & C_i &= \cos\theta \quad (i=1,\ldots,3) \\
S_{\Phi 1} &= \sin(\phi+\theta_1) & C_{\Phi 1} &= \cos(\phi+\theta_1) \\
S_{23} &= \sin(\theta_2+\theta_3) & C_{23} &= \cos(\theta_2+\theta_3)
\end{aligned}
$$

10.3.4.2 Dynamics

To find the dynamic model of the MM, the local coordinate frames are employed according to the Denavit–Hartenberg convention. Figure 10.10 shows all the coordinate frames involved. Using the above notations, we obtain the following rotational matrices:

$$
R_1^{\mathrm{p}} = \begin{bmatrix} C_1 & 0 & S_1 \\ S_1 & 0 & -C_1 \\ 0 & 1 & 0 \end{bmatrix}, \quad R_2^1 = \begin{bmatrix} C_2 & -S_2 & 0 \\ S_2 & C_2 & 0 \\ 0 & 0 & 1 \end{bmatrix}, \quad R_3^2 = \begin{bmatrix} C_3 & -S_3 & 0 \\ S_3 & C_3 & 0 \\ 0 & 0 & 1 \end{bmatrix}
$$

$$
\mathbf{R}_1^{\mathrm{w}} = \mathbf{R}_{\mathrm{p}}^{\mathrm{w}}\mathbf{R}_1^{\mathrm{p}} = \begin{bmatrix} C_{\Phi 1} & 0 & S_{\Phi 1} \\ S_{\Phi 1} & 0 & -C_{\Phi 1} \\ 0 & 1 & 0 \end{bmatrix}
$$

$$
\mathbf{R}_2^{\mathrm{w}} = \mathbf{R}_1^{\mathrm{w}}\mathbf{R}_2^1 = \begin{bmatrix} C_2 C_{\Phi 1} & -S_2 C_{\Phi 1} & S_{\Phi 1} \\ C_2 S_{\Phi 1} & -S_2 S_{\Phi 1} & -C_{\Phi 1} \\ S_2 & C_2 & 0 \end{bmatrix}
$$

$$
\mathbf{R}_3^{\mathrm{w}} = \mathbf{R}_2^{\mathrm{w}}\mathbf{R}_3^2 = \begin{bmatrix} C_{23} C_{\Phi 1} & -S_{23} C_{\Phi 1} & S_{\Phi 1} \\ C_{23} S_{\Phi 1} & -S_{23} S_{\Phi 1} & -C_{\Phi 1} \\ S_{23} & C_{23} & 0 \end{bmatrix}
$$

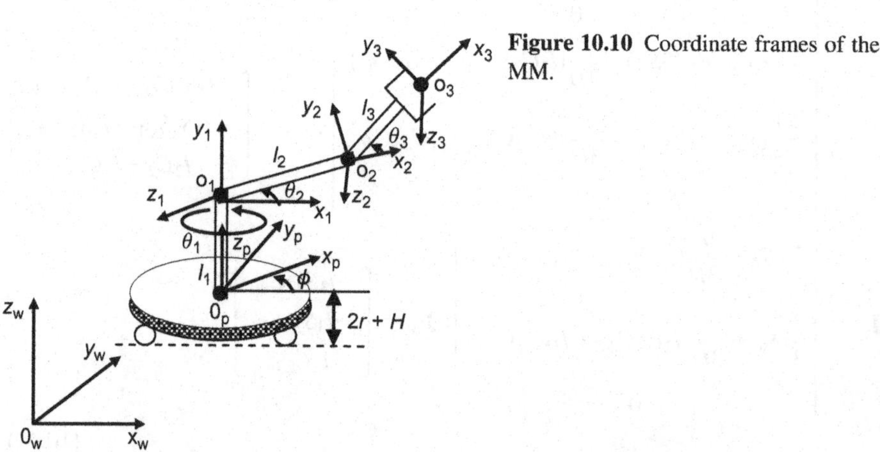

Figure 10.10 Coordinate frames of the MM.

Now, for each rigid body composing the MM, the kinetic and potential energies are found. Then the Lagrangian function (L) is built. Finally, the equations of the generalized forces are calculated using the Lagrangian function:

$$\frac{d}{dt}\left(\frac{\partial L}{\partial \dot{\theta}_{pi}}\right) - \frac{\partial L}{\partial \theta_{pi}} = \tau_{pi} \quad (i = 1, \ldots 3) \quad \frac{d}{dt}\left(\frac{\partial L}{\partial \dot{\theta}_{i}}\right) - \frac{\partial L}{\partial \theta_{i}} = \tau_{i} \quad (i = 1, \ldots 3)$$

where τ_{pi} $(i = 1, \ldots, 3)$ are the generalized torques for the wheel actuators that drive the platform, and τ_i $(i = 1, \ldots, 3)$ are the generalized torques for the joint actuators that drive the links of the manipulator. The parameters of the MM are defined in Table 10.1:

Then, the differential equations of the dynamic model are expressed in compact form as:

$$\mathbf{D(q)\ddot{q} + h(q,\dot{q}) + g(q) = \tau} \tag{10.42}$$

where:

$$\mathbf{h(q,\dot{q}) = B(q)\dot{q} \cdot \dot{q} + C(q)\dot{q}^2}$$

$$\mathbf{D(q)} = \begin{bmatrix} {}^1a_1 & \cdots & {}^1a_6 \\ \vdots & \ddots & \vdots \\ {}^6a_1 & \cdots & {}^6a_6 \end{bmatrix}; \quad \mathbf{B(q)} = \begin{bmatrix} {}^1b_1 & \cdots & {}^1b_{15} \\ \vdots & \ddots & \vdots \\ {}^6b_1 & \cdots & {}^6b_{15} \end{bmatrix}; \quad \mathbf{C(q)} = \begin{bmatrix} {}^1c_1 & \cdots & {}^1c_6 \\ \vdots & \ddots & \vdots \\ {}^6c_1 & \cdots & {}^6c_6 \end{bmatrix}$$

Table 10.1 Parameters of the Omnidirectional MM

m_r: Mass of each lateral orthogonal wheel	l_2: Length of link 2
r: Radius of the wheels	l_{c2}: Distance of the COG of link 2 from the second joint
m_p: Mass of the platform	I_{xx}^2: Moment of inertia of link 2 with respect to x-axis
D: Distance of the platform's COG from each assembly	I_{zz}^2: Moment of inertia of link 2 with respect to z-axis
g: Gravitational constant (9.8062 m/s^2)	m_{l_3}: Mass of link 3
m_{l_1}: Mass of link 1	l_3: Length of link 3
I_{xx}^1: Moment of inertia of link 1 with respect to x-axis	l_{c3}: Distance of the COG of link 3 from the third joint
I_{zz}^p: Moment of inertia of the platform with respect to z-axis	I_{xx}^3: Moment of inertia of link 3 with respect to x-axis
m_{l_2}: Mass of link 2	I_{zz}^3: Moment of inertia of link 3 with respect to z-axis

$$
\begin{aligned}
\mathbf{g}(\mathbf{q}) &= [{}^1g, {}^2g, {}^3g, {}^4g, {}^5g, {}^6g]^{\mathrm{T}} \\
\ddot{\mathbf{q}} &= [\ddot{\theta}_{\mathrm{p}1}, \ddot{\theta}_{\mathrm{p}2}, \ddot{\theta}_{\mathrm{p}3}, \ddot{\theta}_1, \ddot{\theta}_2, \ddot{\theta}_3]^{\mathrm{T}} \\
\dot{\mathbf{q}}\,\dot{\mathbf{q}} &= [\dot{\theta}_{\mathrm{p}1}\dot{\theta}_{\mathrm{p}2}, \ldots, \dot{\theta}_{\mathrm{p}1}\dot{\theta}_3;\ \dot{\theta}_{\mathrm{p}2}\dot{\theta}_{\mathrm{p}3}, \ldots, \dot{\theta}_{\mathrm{p}2}\dot{\theta}_3;\ \dot{\theta}_{\mathrm{p}3}\dot{\theta}_1, \ldots, \dot{\theta}_{\mathrm{p}3}\theta_3, \dot{\theta}_1\dot{\theta}_2, \dot{\theta}_1\dot{\theta}_3, \dot{\theta}_2\dot{\theta}_3]^{\mathrm{T}} \\
\dot{\mathbf{q}}^2 &= [\dot{\theta}_{\mathrm{p}1}^2, \dot{\theta}_{\mathrm{p}2}^2, \dot{\theta}_{\mathrm{p}3}^2, \dot{\theta}_1^2, \dot{\theta}_2^2, \dot{\theta}_3^2]^{\mathrm{T}} \\
\boldsymbol{\tau} &= [\tau_{\mathrm{p}1}, \tau_{\mathrm{p}2}, \tau_{\mathrm{p}3}, \tau_1, \tau_2, \tau_3]^{\mathrm{T}}
\end{aligned}
\tag{10.43}
$$

with ${}^ja_i, {}^jb_k, c_i^j$ and ${}^ig$ $(i, j = 1, 2, \ldots, 6; k = 1, 2, \ldots, 15)$ being defined via intermediate coefficients [8].

10.4 Control of MMs

10.4.1 Computed-Torque Control of Differential-Drive MM

Here, the control of the five degrees of freedom MM studied in Section 10.3.3 will be considered [22]. We use the reduced dynamic model of Eq. (10.24) with parameters given by Eqs. (10.32)–(10.34). The vector $\mathbf{q}$ of generalized variables is:

$$\mathbf{q} = [x_{\mathrm{b}}, y_{\mathrm{b}}, \phi, \theta_1, \theta_2]^{\mathrm{T}} \tag{10.44a}$$

and the vector $\mathbf{v}$ is:

$$\mathbf{v} = [v_1, v_2, v_3, v_4]^{\mathrm{T}} = [\dot{\theta}_1, \dot{\theta}_{\mathrm{r}}, \dot{\theta}_1, \dot{\theta}_2]^{\mathrm{T}} = \boldsymbol{\theta}_0^{\mathrm{T}} \tag{10.44b}$$

To convert $\mathbf{v} = \boldsymbol{\theta}_0^{\mathrm{T}}$ to the Cartesian coordinates velocity vector:

$$\dot{\mathbf{p}}_0 = [\dot{x}_{\mathrm{e}}, \dot{y}_{\mathrm{e}}, \dot{x}_{\mathrm{b}}, \dot{y}_{\mathrm{b}}]^{\mathrm{T}} \tag{10.44c}$$

we use the Jacobian relation (10.30a) and (10.30b).

$$\dot{\mathbf{p}}_0 = \mathbf{J}_0\dot{\boldsymbol{\theta}}_0 = \mathbf{J}_0\mathbf{v} \tag{10.44d}$$

where the 4×4 Jacobian matrix $\mathbf{J}_0$ is invertible. Differentiating Eq. (10.44d) we get:

$$\ddot{\mathbf{p}}_0 = \dot{\mathbf{J}}_0\mathbf{v} + \mathbf{J}_0\dot{\mathbf{v}}$$

which, if solved for $\dot{\mathbf{v}}$, gives:

$$\dot{\mathbf{v}} = \mathbf{J}_0^{-1}(\ddot{\mathbf{p}}_0 - \dot{\mathbf{J}}_0\mathbf{v}) \tag{10.45}$$

Now, introducing Eq. (10.45) into Eq. (10.24), and premultiplying by $(\mathbf{J}_0^{-1})^{\mathrm{T}}$ gives the model:

$$\mathbf{D}^*\ddot{\mathbf{p}}_0 + \mathbf{F}^*\dot{\mathbf{p}}_0 + \mathbf{G}^* = \mathbf{E}^*\boldsymbol{\tau} \tag{10.46}$$

where:

$$\begin{aligned} \mathbf{D}^* &= \mathbf{J}_0^{-1^{\mathrm{T}}}\overline{\mathbf{D}}\mathbf{J}_0^{-}1 \\ \mathbf{F}^* &= \mathbf{J}_0^{-1^{\mathrm{T}}}(\overline{\mathbf{C}} - \overline{\mathbf{D}}\mathbf{J}_0^{-}1\dot{\mathbf{J}}_0)\mathbf{J}_0^{-1} \\ \mathbf{G}^* &= \mathbf{J}_0^{-1^{\mathrm{T}}}\overline{\mathbf{g}} \\ \mathbf{E}^* &= \mathbf{J}_0^{-1^{\mathrm{T}}}\overline{\mathbf{E}} \end{aligned} \tag{10.47}$$

If the desired path for $\mathbf{p}_0(t)$ is $\mathbf{p}_{0,\mathrm{d}}(t)$, and the error is defined as $\tilde{\mathbf{p}}_0 = \mathbf{p}_{0,\mathrm{d}} - \mathbf{p}_0$, we select as usual the computed-torque law:

$$\mathbf{E}^*\boldsymbol{\tau} = \mathbf{D}^*\mathbf{u} + \mathbf{F}^*\dot{\mathbf{p}}_0 + \mathbf{G}^* \tag{10.48}$$

Introducing Eq. (10.48) into Eq. (10.46), under the assumption that the robot parameters are precisely known, gives:

$$\ddot{\mathbf{p}}_0 = \mathbf{u} \tag{10.49a}$$

Now, we can use the linear feedback control law:

$$\mathbf{u} = \ddot{\mathbf{p}}_{0,\mathrm{d}} + \mathbf{K}_{\upsilon}\dot{\tilde{\mathbf{p}}}_0 + \mathbf{K}_{\mathrm{p}}\tilde{\mathbf{p}}_0, \quad \tilde{\mathbf{p}}_0 = \mathbf{p}_{0,\mathrm{d}} - \mathbf{p}_0 \tag{10.49b}$$

In this case, the dynamics of the closed-loop error system is described by:

$$\ddot{\tilde{\mathbf{p}}}_0 + \mathbf{K}_{\upsilon}\dot{\tilde{\mathbf{p}}}_0 + \mathbf{K}_{\mathrm{p}}\tilde{\mathbf{p}}_0 = \mathbf{0}$$

Therefore, selecting properly $\mathbf{K}_{\mathrm{p}}$ and $\mathbf{K}_{\upsilon}$ we can get the desired performance specifications.

10.4.2 Sliding-Mode Control of Omnidirectional MM

We work with the dynamic model of the MM, given by Eq. (10.42) which due to the omnidirectionality of the platform does not involve any nonholonomic constraint [8,11]. Applying computed-torque control:

$$\boldsymbol{\tau} = \mathbf{D}(\mathbf{q})\mathbf{u} + \mathbf{h}(\mathbf{q},\dot{\mathbf{q}}) + \mathbf{g}(\mathbf{q}) \tag{10.50a}$$

we arrive, as usual, to the linear dynamic system:

$$\ddot{\mathbf{q}} = \mathbf{u} \tag{10.50b}$$

and the feedback controller:

$$\mathbf{u} = \ddot{\mathbf{q}}_d + \mathbf{K}_v\dot{\tilde{\mathbf{q}}} + \mathbf{K}_p\tilde{\mathbf{q}} \tag{10.50c}$$

that leads to an asymptotically stable error dynamics.

If $\mathbf{D}, \mathbf{h}$, and $\mathbf{g}$ are subject to uncertainties, and we know only their approximations $\hat{\mathbf{D}},\hat{\mathbf{h}},\hat{\mathbf{g}}$, then the computed-torque controller is:

$$\hat{\boldsymbol{\tau}} = \hat{\mathbf{D}}(\mathbf{q})\mathbf{u} + \hat{\mathbf{h}}(\mathbf{q},\dot{\mathbf{q}}) + \hat{\mathbf{g}}(\mathbf{q}) \tag{10.51a}$$

and the system dynamics is given by:

$$\ddot{\mathbf{q}} = (\mathbf{D}^{-1}\hat{\mathbf{D}})\mathbf{u} + \mathbf{D}^{-1}(\hat{\mathbf{h}} - \mathbf{h}) + \mathbf{D}^{-1}(\hat{\mathbf{g}} - \mathbf{g}) \tag{10.51b}$$

In this case, some appropriate robust control technique must be used. A convenient controller is the sliding-mode controller described in Section 7.2.2. This controller involves a nominal term (based on the computed-torque control law), and an additional term that faces the imprecision of the dynamic model.

The sliding condition for a multiple-input multiple-output system is a generalization of Eq. (7.14):

$$\frac{1}{2}\frac{d}{dt}\mathbf{s}^{\mathrm{T}}(x,t)\mathbf{s}(x,t) \le -\eta(\mathbf{s}^{\mathrm{T}}\mathbf{s})^{1/2}, \eta > 0 \tag{10.52a}$$

with:

$$\mathbf{s} = \dot{\tilde{\mathbf{q}}} + \boldsymbol{\Lambda}\tilde{\mathbf{q}} = \dot{\mathbf{q}} - \dot{\mathbf{q}}_r \tag{10.52b}$$

where $\dot{\mathbf{q}}_r = \dot{\mathbf{q}}_d - \boldsymbol{\Lambda}\tilde{\mathbf{q}}$ and $\boldsymbol{\Lambda}$ is a *Hurwitz* (stable) matrix. The condition (10.52a) guarantees that the trajectories point towards the surface $\mathbf{s} = 0$ for all $t > 0$. This can be shown by choosing a Lyapunov function of the form:

$$V(t) = \frac{1}{2}\mathbf{s}^{\mathrm{T}}\mathbf{D}\mathbf{s}$$

where $\mathbf{D}$ is the inertia matrix of the MM. Differentiating $V(t)$ we get:

$$\dot{V}(t) = \mathbf{s}^{\mathrm{T}}(\boldsymbol{\tau} - \mathbf{D}\ddot{\mathbf{q}}_r - \mathbf{h} - \mathbf{g}) \tag{10.53a}$$

The control law is now defined as:

$$\boldsymbol{\tau} = \hat{\boldsymbol{\tau}} - \mathbf{k}\,\mathrm{sgn}(\mathbf{s}) \tag{10.53b}$$

where $\mathbf{k}\,\mathrm{sgn}(\mathbf{s})$ is a vector with components $k_i\,\mathrm{sgn}(s_i)$. Furthermore, the $\hat{\tau}$ term is the control law part which could make $\dot{V}(t)=0$, if there is no dynamic imprecision inside the estimated dynamic model, that is, according to Eq. (10.53a):

$$\hat{\tau}=\hat{\mathbf{D}}\ddot{\mathbf{q}}+\hat{\mathbf{h}}+\hat{\mathbf{g}}$$

where $\hat{\mathbf{D}}, \hat{\mathbf{h}}$, and $\hat{\mathbf{g}}$ are the available estimates of $\mathbf{D}, \mathbf{h}$, and $\mathbf{g}$. Now, calling $\mathbf{D}, \mathbf{h}$, and $\mathbf{g}$ the real matrices of the robot, we define the matrices:

$$\tilde{\mathbf{D}}=\hat{\mathbf{D}}-\mathbf{D}, \quad \tilde{\mathbf{h}}=\hat{\mathbf{h}}-\mathbf{h}, \quad \tilde{\mathbf{g}}=\hat{\mathbf{g}}-\mathbf{g}$$

as the bounds on the modeling errors. Then, it is possible to choose the components k_i of the vector $\mathbf{k}$ such that:

$$k_i \geq ||[\tilde{\mathbf{D}}(q)\ddot{\mathbf{q}}_{\mathrm{r}}+\tilde{\mathbf{h}}(\mathbf{q},\dot{\mathbf{q}})+\tilde{\mathbf{g}}(\mathbf{q})]_i||+\eta_i, \eta_i>0 \tag{10.54a}$$

If this control mode is used, the condition:

$$\dot{V}(t) \leq -\sum_{i=1}^{n} \eta_i |s_i| \leq 0 \tag{10.54b}$$

is verified to hold. This means that the sliding surface $\mathbf{s}=0$ is reached in a finite time, and that once on the surface, the trajectories remain on the surface and, therefore, tend to $\mathbf{q}_{\mathrm{d}}(t)$ exponentially.

Example 10.2

The end-effector of a five degrees of freedom planar MM is dragged by a human operator to follow a desired trajectory. The problem is:

(a) To derive a nonlinear controller that drives the mobile platform such that the end-effector follows this trajectory with a desired (preferred) manipulator configuration.
(b) To extend the method of (a) and find a state controller that achieves the trajectory tracking by completely compensating the platform–manipulator dynamic interaction.

Solution
(a) Clearly, the MM must track the desired trajectory with the manipulator fixed at the desired configuration which here is selected as the one that maximizes the manipulator's manipulability measure (see Section 10.2.4.3):

$$\begin{aligned} w &= \sqrt{\det \mathbf{J}(\mathbf{q})\mathbf{J}^{\mathrm{T}}(\mathbf{q})} \\ &= |\det \mathbf{J}| = l_1 l_2 |\sin\theta_2| \end{aligned}$$

The optimal configurations of the manipulator for which w is maximum are $\theta_2=\pm 90^\circ$ for any l_1, l_2, and θ_1. Here we choose the desired angles $\theta_{2\mathrm{d}}=+90^\circ$ and $\theta_{1\mathrm{d}}=-45^\circ$ as

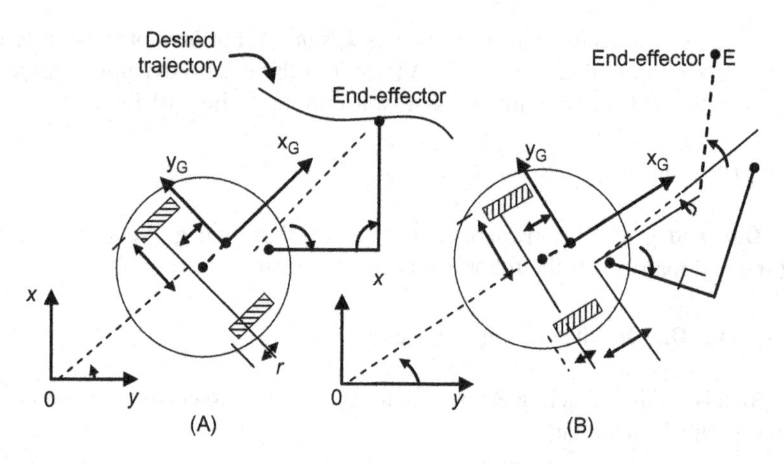

Figure 10.11 (A) The MM at the desired manipulator configuration $\theta_{1d} = -45°$ and $\theta_{2d} = +90°$. The point G is the COG of the platform with world coordinates x_G and y_G. The world coordinates of the manipulator base point O_b are x_b and y_b. (B) While the MM is moving to achieve trajectory tracking, the platform moves to bring the manipulator into the preferred configuration.

shown in Figure 10.11A. If $l_1 = l_2$ the choice of $\theta_{1d} = -45°$ implies that the end-effector is located on a line parallel to the WMR axis which passes through the point O_b.

The coordinates x_{rG}, y_{rG} (in the platform coordinate frame) of the manipulator end-effector at the desired (optimal) configuration are constant, being equal to:

$$\begin{aligned}
x_{rG} &= x_{bG} + l_1 \cos\theta_{1d} + l_2 \cos(\theta_{1d} + \theta_{2d}) \\
&= x_{bG} + (\sqrt{2}/2)l_1 + (\sqrt{2}/2)l_2 \\
y_{rG} &= y_{bG} + l_1 \sin\theta_{1d} + l_2 \sin(\theta_{1d} + \theta_{2d}) \\
&= y_{bG} - (\sqrt{2}/2)l_1 + (\sqrt{2}/2)l_2
\end{aligned}$$

where x_{bG} and y_{bG} are the coordinates of the manipulator base O_b in the platform coordinate frame Gx_Gy_G. It is remarked that any trajectory tracking error will result in getting the manipulator out of the desired configuration, thus reducing the value of the manipulability measure. We will apply the affine-systems methodology of Section 6.3.2 [23,24].

Since the manipulator is kept at the desired configuration, its dynamics is ignored. Therefore, we need only the dynamic model of the platform which in reduced form (with the nonholonomic constraint eliminated) has the affine state-space form of Eqs. (6.43a) and (6.43b):

$$\dot{\mathbf{x}}(t) = \mathbf{f}(\mathbf{x}) + \mathbf{g}(\mathbf{x})\mathbf{u}(t), \quad \mathbf{u} = \boldsymbol{\tau} = [\tau_r, \tau_l]^T$$

$$\mathbf{f}(\mathbf{x}) = \begin{bmatrix} \mathbf{Bv} \\ \vdots \\ -\overline{\mathbf{D}}^{-1}\overline{\mathbf{C}}\mathbf{v} \end{bmatrix}, \quad \mathbf{g}(\mathbf{x}) = \begin{bmatrix} \mathbf{0} \\ \vdots \\ \overline{\mathbf{D}}^{-1} \end{bmatrix}$$

where $\mathbf{B}, \overline{\mathbf{C}}$, and $\overline{\mathbf{D}}$ are given by Eq. (6.41b) and:

$$\mathbf{x} = \begin{bmatrix} \mathbf{q} \\ \vdots \\ \mathbf{v} \end{bmatrix}, \quad \mathbf{q} = [x, y, \theta_r, \theta_l]^T, \quad \mathbf{v} = [v_1, v_2]^T$$

Applying the computed-torque-like controller given by Eq. (6.44a):

$$\mathbf{u}(t) = \mathbf{F}(\mathbf{x}) + \mathbf{G}(\mathbf{x})\upsilon(t), \quad \mathbf{F}(\mathbf{x}) = \overline{\mathbf{C}}\mathbf{v}, \quad \mathbf{G}(\mathbf{x}) = \overline{\mathbf{D}}$$

where $\upsilon(t)$ is a new control vector, the above model takes the form (6.45a)–(6.45c):

$$\dot{\mathbf{x}}(t) = \mathbf{f}_c(\mathbf{x}) + \mathbf{g}_c(\mathbf{x})\upsilon(t)$$

where:

$$\mathbf{f}_c(\mathbf{x}) = \begin{bmatrix} \mathbf{B}v \\ \vdots \\ \mathbf{0} \end{bmatrix}, \quad \mathbf{g}_c(\mathbf{x}) = \begin{bmatrix} \mathbf{0} \\ \vdots \\ \mathrm{I}_{2\times 2} \end{bmatrix}$$

We have seen in Section 6.3.2 that if the output components y_1 and y_2 are taken to be the coordinates of the wheels' axis midpoint Q, the above model can be input–output linearizable (and decoupled) by using a dynamic nonlinear state-feedback controller, but it cannot be linearized by any static state-feedback controller. However, as explained in Example 6.8 it is possible to decouple the inputs and outputs using a static feedback controller if y_1 and y_2 are the coordinates x_r, y_r of a reference point C in front of the vehicle. To find the general conditions under which a differential-drive WMR can be input–output decoupled by static state-feedback control, we assume that the output vector components y_1 and y_2 are the world coordinates x_r and y_r of an arbitrary reference point, that is:

$$\mathbf{y} = \mathbf{h}(\mathbf{x}) = \begin{bmatrix} x_r \\ y_r \end{bmatrix}$$

We now express x_r and y_r in the platform's coordinate frame with origin at the platform COG G. The result is:

$$x_r = x_G + x_{rG}\cos\phi - y_{rG}\sin\phi$$
$$y_r = y_G + x_{rG}\sin\phi + y_{rG}\cos\phi$$

where x_{rG} and y_{rG} are the coordinates of the reference point in the platform coordinate frame.

Working as in Example 6.8 we find, via successive differentiation of $\mathbf{y}$, the dynamic model of $\mathbf{y}$, namely:

$$\ddot{\mathbf{y}} = \dot{\mathbf{H}}(\mathbf{x})\mathbf{v} + \mathbf{H}(\mathbf{x})\upsilon$$

where $\mathbf{H(x)}$ is the matrix:

$$\mathbf{H(x)} = \begin{bmatrix} H_{11} & H_{12} \\ H_{21} & H_{22} \end{bmatrix}$$

with ($\rho = r/2a$):

$$\begin{aligned} H_{11} &= \rho[(a - y_{rG})\cos\phi - (b + x_{rG})\sin\phi] \\ H_{12} &= \rho[(a + y_{rG})\cos\phi + (b + x_{rG})\sin\phi] \\ H_{21} &= \rho[(a - y_{rG})\sin\phi + (b + x_{rG})\cos\phi] \\ H_{22} &= \rho[(a + y_{rG})\sin\phi - (b + x_{rG})\cos\phi] \end{aligned}$$

The determinant of $\mathbf{H(x)}$ is found to be:

$$\det \mathbf{H(x)} = -r^2(b + x_{rG})/2a$$

which is different than zero if $x_{rG} \neq -b$, that is, if the reference point is not a point of the WMR wheel axis. Therefore, if $x_{rG} \neq -b$, the matrix $\mathbf{H(x)}$ is nonsingular (i.e., it is a decoupling matrix) and can be inverted.[2] As a result, choosing the state-feedback control law:

$$\upsilon(t) = \mathbf{H}^{-1}(\mathbf{x})[\mathbf{w} - \dot{\mathbf{H}}(\mathbf{x})\mathbf{v}]$$

where $\mathbf{v} = [v_1, v_2]^T = [\dot{\theta}_r, \dot{\theta}_l]^T$, and $\mathbf{w}(t)$ is the new control input vector:

$$\mathbf{w}(t) = \begin{bmatrix} w_1 \\ w_2 \end{bmatrix}$$

we get the decoupled system $\ddot{y}_1 = w_1, \ddot{y}_2 = w_2$.

The fact that for decouplability the reference point must not lie on the WMR wheel axis is due to the nonholonomicity property which implies that a point on the wheel axis has instantaneously one degree of freedom, whereas all other points have instantaneously two degrees of freedom.

On the basis of the above, it follows that the overall static feedback controller that leads to a linear input–output decoupled system consists of the computed control-like part:

$$\mathbf{u}(t) = \bar{\mathbf{C}}\mathbf{v} + \bar{\mathbf{D}}\upsilon(t) \tag{10.55a}$$

and the decoupling part:

$$\upsilon(t) = \mathbf{H}^{-1}(\mathbf{w} - \dot{\mathbf{H}}\mathbf{v}) \tag{10.55b}$$

[2] Note that the computed-torque control technique applied in Section 5.5 for the reference point C that does not belong to the common wheel axis (i.e., $b \neq 0$) is a special case of this general input–output linearizing and decoupling static feedback controller.

with the result being $\ddot{\mathbf{y}} = \mathbf{w}$, that is:

$$\ddot{y}_1 = w_1, \ \ddot{y}_2 = w_2 \tag{10.55c}$$

Now, the asymptotic tracking of the desired trajectory $\mathbf{y}_d(t) = [y_{id}(t), y_{2d}(t)]^T$ can be achieved by a linear PD controller as usual.

(b) In this case, we have to drive the MM at the desired trajectory and at the same time achieve the desired manipulator configuration. Therefore, we need the dynamic models of both the platform and the manipulator, which in combined reduced (unconstrained) form are given by Eq. (10.24):

$$\overline{\mathbf{D}}(\mathbf{q})\dot{\mathbf{v}} + \overline{\mathbf{C}}(\mathbf{q},\dot{\mathbf{q}})\mathbf{v} = \overline{\mathbf{E}}\boldsymbol{\tau}$$

where $\overline{\mathbf{D}}(\mathbf{q}), \overline{\mathbf{C}}(\mathbf{q},\dot{\mathbf{q}})$ are given by the combination of the respective terms of the platform and manipulator, and:

$$\mathbf{q} = [\mathbf{q}_p^T, \mathbf{q}_m^T]^T, \boldsymbol{\tau} = [\tau_r, \tau_l, \tau_1, \tau_2]^T$$

This model can be written in the state-space affine form:

$$\dot{\mathbf{x}}(t) = \mathbf{f}_c(\mathbf{x}) + \mathbf{g}_c(\mathbf{x})\upsilon(t)$$

where:

$$\mathbf{x} = [\mathbf{q}_p^T, \mathbf{q}_m^T, \mathbf{v}^T, \dot{\mathbf{q}}_m^T]^T$$

$$\mathbf{f}_c(\mathbf{x}) = \begin{bmatrix} \mathbf{Bv} \\ \dot{\mathbf{q}}_m \end{bmatrix}, \mathbf{g}_c(\mathbf{x}) = \begin{bmatrix} \mathbf{0} \\ \mathbf{0} \\ \mathbf{I} \end{bmatrix}$$

Here, we have four inputs, $\boldsymbol{\tau} \in R^4$ and so we can have input–output decoupling of up to four outputs. Therefore, we will select four outputs: $\mathbf{y} \in R^4$. The manipulator end-effector cannot track the desired trajectory alone without the help of the mobile platform, because it may be overstretched reaching the boundary of its workspace. Therefore, while the manipulator is moving to track as much as possible the desired trajectory, the mobile platform should be controlled such that to bring the manipulator into the preferred configuration.

The first two components of the output vector $\mathbf{y}$ are selected to be the coordinates x_E, y_E of the point E in the platform coordinate frame (see Figure 10.11B), that is:

$$y_1 = x_{bG} + l_1 \cos\theta_1 + l_2\cos(\theta_1 + \theta_2)$$
$$y_2 = y_{bG} + l_1 \sin\theta_1 + l_2\sin(\theta_1 + \theta_2)$$

To assure that the platform controller moves the platform so as to bring always the manipulator to the desired configuration, we choose the other two output components as:

$$y_3 = x_b + [l_1\cos\theta_{1d} + l_2\cos(\theta_{1d} + \theta_{2d})]\cos\phi$$
$$y_4 = y_b + [l_1\sin\theta_{1d} + l_2\sin(\theta_{1d} + \theta_{2d})]\sin\phi$$

which are the coordinates of the reference point R of Figure 10.11 in the world coordinate frame. Clearly, the desired values y_{3d} and y_{4d} of y_3 and y_4 must be set to the actual location of the end-effector position E such that to bring R to E, that is, the manipulator configuration into the preferred one.

Thus, the overall output vector **y** is:

$$\mathbf{y} = \mathbf{h}(\mathbf{x}) = \begin{bmatrix} y_1 \\ y_2 \\ y_3 \\ y_4 \end{bmatrix} = \begin{bmatrix} x_E \\ y_E \\ x_R \\ x_R \end{bmatrix}$$

From this point, the controller design proceeds as before. Differentiating twice the output **y** we get:

$$\ddot{\mathbf{y}} = \dot{\mathbf{H}}(\mathbf{x})\mathbf{v}_M + \mathbf{H}(\mathbf{x})\upsilon$$

where $\mathbf{v}_M = [\mathbf{v}^T, \dot{\mathbf{q}}_p^T]^T$, and:

$$\mathbf{H}(\mathbf{x}) = \begin{bmatrix} \mathbf{H}_1 & \vdots & \mathbf{H}_2 \end{bmatrix}$$

$$\mathbf{H}_1 = \begin{bmatrix} \rho(a\cos\phi - 2l\sin\phi) & \rho(a\cos\phi + 2l\sin\phi) \\ \rho(a\sin\phi + 2l\cos\phi) & \rho(a\sin\phi) - 2l\cos\phi \\ 0 & 0 \\ 0 & 0 \end{bmatrix}$$

$$\mathbf{H}_2 = \begin{bmatrix} 0 & 0 \\ 0 & 0 \\ -l_1\sin\theta_1 - l_2\sin(\theta_1+\theta_2) & -l_2\sin(\theta_1+\theta_2) \\ l_1\cos\theta_1 + l_2\cos(\theta_1+\theta_2) & l_2\cos(\theta_1+\theta_2) \end{bmatrix}$$

Therefore, selecting $\upsilon(t)$ as:

$$\upsilon(t) = \mathbf{H}^{-1}(\mathbf{w} - \dot{\mathbf{H}}\mathbf{v}_M)$$

we get the linear decoupled system:

$$\ddot{y}_1 = w_1, \ddot{y}_2 = w_2, \ddot{y}_3 = w_3, \ddot{y}_4 = w_4$$

which can be controlled by a diagonal PD state-feedback controller as usual with given desired natural frequency ω_n and damping ratio ζ. As an exercise, the reader is advised to solve the problem by taking into consideration the WMR maneuverability. The above controllers for the problems (a) and (b) were tested by simulation in Refs. [23,24] showing very satisfactory performance (see Section 13.9.2). It is noted that the tracking control problem studied in Section 10.4.1 is a special case of the present problem (b) in which the maximum (or other desired) manipulability measure condition is relaxed.

10.5 Vision-Based Control of MMs

10.5.1 General Issues

As described in Section 9.4, the image-based visual control uses the image Jacobian (or interaction) matrix. Actually, the inverse or generalized inverse of the Jacobian matrix or its generalized transpose is used, in order to determine the *screw (or velocity twist)* **r** of the end-effector which assures the achievement of a desired task. In MMs where the platform is omnidirectional (and does not involve any nonholonomic constraint), the method discussed in Section 9.4 is directly applicable. However, if the MM platform is nonholonomic, special care is required for both determining and using the corresponding image Jacobian matrix. This can be done by combining the results of Sections 9.2, 9.4, and 10.3.

The Jacobian matrix is defined by Eq. (9.11) and relates the end-effector screw:

$$\dot{\mathbf{r}} = \begin{bmatrix} \mathbf{v} \\ \boldsymbol{\omega} \end{bmatrix} \tag{10.56a}$$

with the feature vector rate of change:

$$\dot{\mathbf{f}} = [f_1, f_2, \ldots, f_k]^{\mathrm{T}} \tag{10.56b}$$

that is:

$$\dot{\mathbf{f}} = \mathbf{J}_{\mathrm{im}}(\mathbf{r})\dot{\mathbf{r}} \tag{10.56c}$$

The relation of $\dot{\mathbf{r}}$ to $\dot{\mathbf{p}}$, where $\mathbf{p}$ is the position vector $[x, y, z]^{\mathrm{T}}$ rigidly attached to the end-effector, is given by Eq. (9.4a):

$$\dot{\mathbf{p}}(t) = \mathbf{J}_{\mathrm{O}}(\mathbf{p})\dot{\mathbf{r}}(t), \quad \mathbf{J}_{\mathrm{O}}(\mathbf{p}) = [\mathbf{I}_{3\times 3} \vdots \mathbf{S}(\mathbf{p})] \tag{10.57}$$

where $\mathbf{S}(\mathbf{p})$ is the skew symmetric matrix of Eq. (9.4b). The matrix $\mathbf{J}_{\mathrm{O}}(\mathbf{p})$ is the *robot Jacobian part*. The relation of $\dot{\mathbf{p}}$ to $\dot{\mathbf{f}}$ is given by Eq. (9.13a):

$$\dot{\mathbf{f}} = \mathbf{J}_{\mathrm{c}}(x_{\mathrm{im}}, y_{\mathrm{im}}, l_{\mathrm{f}})\dot{\mathbf{p}} \tag{10.58a}$$

where:

$$\mathbf{J}_{\mathrm{c}} = \begin{bmatrix} l_{\mathrm{f}}/z & 0 & -x_{\mathrm{im}}/z \\ 0 & l_{\mathrm{f}}/z & -y_{\mathrm{im}}/z \end{bmatrix} \tag{10.58b}$$

The matrix $\mathbf{J}_{\mathrm{c}}(x_{\mathrm{im}}, y_{\mathrm{im}}, l_{\mathrm{f}})$ is the *camera interaction part*. Combining Eqs. (10.57) and (10.58a) we get the expression of $\mathbf{J}_{\mathrm{im}}(r)$, namely,

$$\mathbf{J}_{\mathrm{im}}(\mathbf{r}) = \mathbf{J}_{\mathrm{c}}(x_{\mathrm{im}}, y_{\mathrm{im}}, l_{\mathrm{f}})\mathbf{J}_{\mathrm{O}}(\mathbf{p}) \tag{10.59}$$

as given in Eq. (9.15) for the case of two image plane features x_{im} and y_{im}.

Now, we will examine the case where the task features are the components of the vector:

$$\mathbf{f} = [x_{\mathrm{im},1}, y_{\mathrm{im},1}, x_{\mathrm{im},2}, y_{\mathrm{im},2}, \ldots, x_{\mathrm{im},k/2}, y_{\mathrm{im},k/2}]^{\mathrm{T}} \in R^k$$

In this case, the *camera interaction part* $\mathbf{J}_{\mathrm{c}}$, in Eq. (10.58b), is given by:

$$\bar{\mathbf{J}}_{\mathrm{c}} = \begin{bmatrix} \mathbf{J}_{\mathrm{c},1} \\ \mathbf{J}_{\mathrm{c},2} \\ \vdots \\ \mathbf{J}_{\mathrm{c},k/2} \end{bmatrix}, \quad \mathbf{J}_{\mathrm{c},i} = \begin{bmatrix} l_{\mathrm{f}}/z_i & 0 & -x_{\mathrm{im},i}/z_i \\ 0 & l_{\mathrm{f}}/z_i & -y_{\mathrm{im},i}/z_i \end{bmatrix} \tag{10.60a}$$

for $i = 1, 2, \ldots, k/2$. Therefore, the overall Jacobian matrix $\bar{\mathbf{J}}_{\mathrm{im}}$ is given by:

$$\bar{\mathbf{J}}_{\mathrm{im}} = \bar{\mathbf{J}}_{\mathrm{c}}\mathbf{J}_{\mathrm{O}} = \begin{bmatrix} \mathbf{J}_{\mathrm{c},1}\mathbf{J}_{\mathrm{O}} \\ \mathbf{J}_{\mathrm{c},2}\mathbf{J}_{\mathrm{O}} \\ \vdots \\ \mathbf{J}_{\mathrm{c},k/2}\mathbf{J}_{\mathrm{O}} \end{bmatrix}$$

$$= \begin{bmatrix} \frac{l_{\mathrm{f}}}{z_1} & 0 & \frac{-x_{\mathrm{im},1}}{z_1} & -\frac{x_{\mathrm{im},1}y_{\mathrm{im},1}}{l_{\mathrm{f}}} & \frac{l_{\mathrm{f}}^2 + x_{\mathrm{im},1}^2}{l_{\mathrm{f}}} & -y_{\mathrm{im},1} \\ 0 & \frac{l_{\mathrm{f}}}{z_1} & -\frac{y_{\mathrm{im},1}}{z_1} & -\frac{l_{\mathrm{f}}^2 + y_{\mathrm{im},1}^2}{l_{\mathrm{f}}} & \frac{x_{\mathrm{im},1}y_{\mathrm{im},1}}{l_{\mathrm{f}}} & x_{\mathrm{im},1} \\ \cdots & \cdots & \cdots & \cdots & \cdots & \\ \cdots & \cdots & \cdots & \cdots & \cdots & \\ \frac{l_{\mathrm{f}}}{z_{k/2}} & 0 & -\frac{x_{\mathrm{im},k/2}}{z_{k/2}} & -\frac{x_{\mathrm{im},k/2}y_{\mathrm{im},k/2}}{l_{\mathrm{f}}} & \frac{l_{\mathrm{f}}^2 + x_{\mathrm{im},k/2}^2}{l_{\mathrm{f}}} & -y_{\mathrm{im},k/2} \\ 0 & \frac{l_{\mathrm{f}}}{z_{k/2}} & -\frac{y_{\mathrm{im},k/2}}{z_{k/2}} & -\frac{l_{\mathrm{f}}^2 + y_{\mathrm{im},k/2}^2}{l_f} & \frac{x_{\mathrm{im},k/2}y_{\mathrm{im},k/2}}{l_{\mathrm{f}}} & x_{\mathrm{im},k/2} \end{bmatrix} \tag{10.60b}$$

The control of the MM needs the control of the platform and the manipulator mounted on it. Two ways to do this are the following:

1. Control the platform and the arm separately, and then treat the existing coupling between them.
2. Control the platform and the arm jointly with the coupling kinematics and dynamics included in the overall model used (full-state model).

Example 10.3

Outline a method for MM visual-based point stabilization, taking into consideration the coupling that exists between the platform and the manipulator.

Solution

We will consider the problem of stabilizing an MM in a desired configuration using measurements of the manipulator (arm) joint positions and the measures (feature measurements) provided by an onboard camera mounted on the end-effector. This problem involves the following requirements:

1. To reduce asymptotically to zero the feature errors provided by the camera measurements.
2. To move the platform such that, during the arm control for reducing the feature errors, to maintain the manipulator in the nonsingular configuration space.
3. To steer the platform such that to have the manipulator to the desired configuration, while the arm is controlled to keep constant the camera measures.

These requirements can be satisfied using a *hybrid control scheme* that merges a feedback and an open-loop control strategy as follows [18].

Feedback Control

The feedback control U_f drives the camera measurement errors asymptotically to zero, and, at the same time, maintains the manipulator far from singularities. If $\mathbf{e}_f = \mathbf{f} - \mathbf{f}_d$ is the error between the actual and desired image features, then $\dot{\mathbf{e}}_f = \dot{\mathbf{f}}$, and the manipulator position $\mathbf{q}_m(t)$ must be such that $\mathbf{e}_f(t) \to \mathbf{0}$ as $t \to \infty$, asymptotically, that is, such that:

$$\dot{\mathbf{e}}_f(t) = -\mathbf{K}_f \mathbf{e}_f(t) \tag{10.61}$$

where $\mathbf{K}_f = \mathrm{diag}[k_f, k_f, \ldots, k_f]$, $k_f > 0$ is a matrix gain determining the rate of convergence.

The kinematic model of the MM from the input velocities to the features' rate of variation $\dot{\mathbf{f}}$ has the form of Eq. (10.21a), that is:

$$\dot{\mathbf{f}}(t) = \mathbf{J}_{im,p}(\mathbf{f},\mathbf{q})\mathbf{u}_p + \mathbf{J}_{im,m}(\mathbf{f},\mathbf{q}_m)\dot{\mathbf{q}}_m(t) \tag{10.62}$$

where $\mathbf{q}_m = [q_{m,1}, q_{m,2}, \ldots, q_{m,n_m}]^T$ is the vector of the manipulator joint positions, $\mathbf{q} = [\mathbf{q}_p^T, \mathbf{q}_m^T]^T$ ($\mathbf{q}_p = [x, y, \phi]^T$ is the platform position/orientation vector, $\mathbf{J}_{im,p}$ and $\mathbf{J}_{im,m}$ are the platform and manipulator image Jacobian matrices, respectively, and $\mathbf{u}_p = [v, \omega]^T$ is the platform control vector (linear velocity, v, and angular velocity, ω). The image Jacobian matrices $\mathbf{J}_{im,p}$ and $\mathbf{J}_{im,m}$ consist of the robot part and camera part as shown in Eq. (10.59). Note that if we use three landmark points the camera part is a 6×6 matrix, and if these landmarks are not aligned the matrix is nonsingular (see Eqs. (10.60a), (10.60b), and (9.15)).

To get Eq. (10.61), $\dot{\mathbf{q}}_m(t)$ in Eq. (10.62) must be selected as:

$$\dot{\mathbf{q}}_m(t) = -\mathbf{J}_{im,m}^{\dagger}(\mathbf{f},\mathbf{q}_m)[\mathbf{J}_{im,p}(\mathbf{f},\mathbf{q})\mathbf{u}_p + \mathbf{K}_f \mathbf{e}_f] \tag{10.63}$$

Now if the desired position of the manipulator is $\mathbf{q}_m^d$, we get $\mathbf{e}_m = \mathbf{q}_m - \mathbf{q}_m^d$ and $\dot{\mathbf{e}}_m = \dot{\mathbf{q}}_m$. Therefore, selecting the platform control vector $\mathbf{u}_p$ as:

$$\mathbf{u}_p(t) = (\mathbf{J}_{im,m}^{\dagger}\mathbf{J}_{im,p})^{\dagger}[\mathbf{K}_p\mathbf{e}_m - \mathbf{J}_{im,m}^{\dagger}\mathbf{K}_f\mathbf{e}_f] \tag{10.64}$$

we obtain:

$$\dot{\mathbf{e}}_m = -\mathbf{K}_p\mathbf{e}_m \tag{10.65}$$

where the gain $\mathbf{K}_p = \text{diag}[k_p, k_p, \ldots, k_p]$, $k_p > 0$ must be suitably selected such that to have the desired rate of convergence to zero.

To keep the arm far from singular configurations, the above two controllers must be suitably coordinated. To this end, $\mathbf{K}_f = \text{diag}[k_f, k_f, \ldots, k_f]$ in Eq. (10.64) is modified as:

$$k_f(\mathbf{q}_m) = \begin{cases} \sigma||\mathbf{e}_m|| & \text{if} \quad |\mathbf{e}_m| < \eta \\ 0 & \text{if} \quad |\mathbf{e}_m| \geq \eta \end{cases} \tag{10.66}$$

where η is a threshold ensuring the avoidance of singular configurations, and $\sigma(\cdot)$ is a suitable strictly decreasing function such as:

$$\sigma(||\mathbf{e}_m||) = \sigma_0 \, e^{-\mu||\mathbf{e}_m||^2} \tag{10.67}$$

with proper coefficients σ_0 and μ. Indeed, for manipulator configurations far from the desired configuration (where also singular configurations may occur), we have $k_f \rightarrow 0$ and $\dot{\mathbf{e}}_f \rightarrow \mathbf{0}$. Therefore, the manipulator is forced to compensate only the motion of the platform which is controlled to reduce $||\mathbf{e}_m||$ according to Eq. (10.64).

Open-Loop Control

The open-loop control U_0 is to steer the platform so as to have the manipulator to the desired configuration $\mathbf{q}_m^d$, while the arm is controlled to maintain constant the camera measures.

The hybrid control scheme is as follows:

$$U(t) = \begin{cases} U_f(f, f_d, q_m, q_m^d, t) & \text{for} \quad kT \leq t \leq (k+1)T \\ U_0(q_m, q_d, t) & \text{for} \quad (k+1)T < t \leq (k+2)T \end{cases} \tag{10.68}$$

where T is a selected time period, and $k = 0, 2, 4, \ldots$. Given an initial configuration $\mathbf{q}_m(t_0) = \mathbf{q}_{m,0}$, the open-loop control must assure that $\mathbf{q}_m(t_1) = \mathbf{q}_m^d$, and $\mathbf{q}_m(t)$ belongs to the nonsingular configuration space for all $t \in [t_0, t_1]$, where $t_0 = (k+1)\text{T}$ and $t_1 = (k+2)\text{T}$. Setting $\mathbf{K}_f = \mathbf{0}$ in Eq. (10.63) (i.e., no feature feedback control) we get:

$$\dot{\mathbf{q}}_m(t) = -\mathbf{J}_{im,m}^{\dagger}(\mathbf{q}_m)\mathbf{J}_{im,p}(\mathbf{f},\mathbf{q})\mathbf{u}_p \tag{10.69}$$

Now, for a given $\mathbf{u}_p(\tau), t_0 \leq \tau \leq t$, we have:

$$\mathbf{q}_m(t) = \mathbf{q}_0 - \int_{t_0}^{t} \mathbf{J}_{im,m}^{-1}(\mathbf{q}(\tau))\mathbf{J}_{im,p}(\mathbf{f},\mathbf{q})\mathbf{u}_p(\tau)d\tau \tag{10.70}$$

and so the joint trajectory is fully determined by $\mathbf{q}_0$ and $\mathbf{u}_p(\tau), t_0 \leq \tau \leq t$. Therefore, as long as the arm is controlled using Eq. (10.69), the open-loop control problem is to compute an open-loop platform control $\mathbf{u}_p^0(\mathbf{q}_{m,0}, t)$ such that $\mathbf{q}_m(t_1) = \mathbf{q}_m^d$, and $\mathbf{q}_m(t)$ is a nonsingular configuration for all $t_0 \le t \le t_1$. Actually, many such control sequences can be found. For the planar case, it can be verified that the sequence:

$$u_p(t) = \begin{cases} \left[0, \frac{1}{T_1}\operatorname{arctg}\left(\frac{y_{pf}}{x_{pf}}\right)\right]^T, & 0 < t^* \le T_1 \\ \left[\frac{1}{T_2 - T_1}\operatorname{arctg}\left(\sqrt{x_{pf}^2 + y_{pf}^2}\right), 0\right], & T_1 < t^* \le T_2 \\ \left[0, \frac{1}{T - T_2}\arcsin(\xi)\right]^T, & T_2 < t^* \le T \end{cases} \tag{10.71}$$

does the job. Here, $[x_{pf}, y_{pf}, \phi_{pf}]^T$ is the platform final configuration, $t^* = t - (k+1)T$, $0 < T_1 < T_2 < T$, and:

$$\xi = \begin{bmatrix} \cos(\phi_{bf}) \\ \sin(\phi_{bf}) \end{bmatrix} \times \begin{bmatrix} \cos[\operatorname{arctg}(y_{pf}/x_{pf})] \\ \sin[\operatorname{arctg}(y_{pf}/x_{pf})] \end{bmatrix}$$

with the assumption that the platform coordinate frame coincides with the world coordinate frame [18].

10.5.2 Full-State MM Visual Control

The vision-based control of mobile platforms was studied in Section 9.5, where several representative problems (pose stabilizing control, wall following control, etc.) were considered. Here we will consider the pose stabilizing control of MMs combining the platform pose and manipulator/camera pose control [3,14]. Actually, the results of Section 9.5.1 concern the case of stabilizing the platform's pose (x_Q, y_Q, ϕ) and the camera's pose ψ. The camera pose control can be regarded as the stabilizing pose control of a one-link manipulator (using the notation $\theta_1 = \psi$). Therefore, the vision-based control of a general MM can be derived by direct extension of the controller derived in Section 9.5.1. This controller is given by Eqs. (9.49a), (9.49b), (9.50a), and (9.50b), and stabilizes to zero the total pose vector:

$$\mathbf{x} = [x_Q, y_Q, \phi, \psi]^T$$

on the basis of feature measurements:

$$\mathbf{f} = [f_D, f_E, f_F]^T$$

of three nonaligned target feature points D, E, and F, provided by an onboard camera, and a sensor measuring ψ (see Figure 9.4). This controller employs the image Jacobian relation (9.25a) and (9.25b):

$$\dot{\mathbf{f}} = \mathbf{J}_{im}\dot{\mathbf{x}}, \quad \mathbf{J}_{im} = \mathbf{J}_{im}^c \mathbf{J}_0$$

where $\mathbf{J}_{im}^c$ and $\mathbf{J}_0$ are given by Eqs. (9.21) and (9.24b), respectively.

Using three feature points as in Figure 9.4, the camera part $\mathbf{J}_{\mathrm{im}}^{\mathrm{c}}$ of the Jacobian $\mathbf{J}_{\mathrm{im}}$ has the same form as in Eq. (9.21). But the robot part $\mathbf{J}_0$ of the image Jacobian is different depending on the number and type of the manipulator links.

For example, if we consider the five-link MM of Figure 10.8, the Jacobian matrix of the overall system (including the coupling between the platform and the manipulator) is given by Eqs. (10.30a) and (10.30b), where the wheel motor velocities $\dot{\theta}_{\mathrm{l}}$ and $\dot{\theta}_{\mathrm{r}}$ are equivalently used as controls instead of $v = (r/2)(\dot{\theta}_{\mathrm{r}} + \dot{\theta}_{\mathrm{l}})$ and $\omega = (r/2a)(\dot{\theta}_{\mathrm{r}} - \dot{\theta}_{\mathrm{l}})$. Now, the controller design proceeds in the usual way as described in Sections 9.3 and 9.4 for both cases of position-based and image-based control.

Example 10.4

We consider a five degrees of freedom MM consisting of a four degrees of freedom articulated robotic manipulator, a one degree of freedom linear slide, a fixed camera, and a stationary spherical mirror as shown in Figure 10.12. Assuming that two fictitious landmarks are mounted on the end-effector, the problem is to derive an appropriate image Jacobian that can be used for 3D visual servoing of the MM.

Solution

The system has the structure of Figure 9.3B with the following coordinate frames:

- $O_r(x_r, y_r, z_r)$ is the robot coordinate frame,
- $O_c(x_c, y_c, z_c)$ is the camera coordinate frame,
- $O_m(x_m, y_m, z_m)$ is the mirror coordinate frame.

The mirror frame is mapped to the camera frame by the homogenous transformation:

$$\mathbf{A}_m^c = \left[\begin{array}{c|c} \mathbf{I}_{3\times3} & \mathbf{d} \\ \hline 0 & 1 \end{array}\right], \quad \mathbf{d} = \begin{bmatrix} 0 \\ 0 \\ -d \end{bmatrix} \tag{10.72}$$

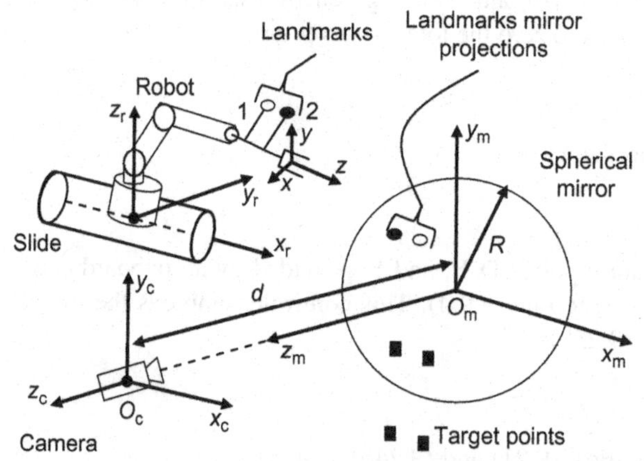

Figure 10.12 Structure of MM/spherical catadioptric vision system. The optical axis of the camera passes through the center O_m.

where d is the distance of the mirror and camera. Similarly, the camera frame can be transformed to the robot frame by T_c^r. The relationship between the landmarks $\mathbf{L}_{r1}^{m}, \mathbf{L}_{r2}^{m}$ and their mirror projections (reflections) $\mathbf{L}_{m1}^{m}$ and $\mathbf{L}_{m2}^{m}$ represented in the spherical mirror frame is found using the spherical convex mirror reflection rule as [25,26]:

$$\mathbf{L}_{mi}^{m} = \mu_i \mathbf{L}_{ri}, \quad \mu_i = R/(2||\mathbf{L}_{ri}^{m}|| - R) \quad (i = 1, 2) \tag{10.73}$$

where R is the spherical mirror radius. Using Eqs. (10.72) and (10.73) one finds that the transformation from the landmarks and their mirror reflections in $0_m(x_m, y_m, z_m)$ to those in the camera frame $0_c(x_c, y_c, z_c)$ are:

$$\mathbf{L}_{ri}^{c} = \mathbf{A}_m^c \mathbf{L}_{ri}^{m}, \quad \mathbf{L}_{mi}^{c} = \mathbf{A}_m^c \mathbf{L}_{mi}^{m} \quad (i = 1, 2) \tag{10.74}$$

from which we obtain that the landmarks' coordinates to its mirror reflection, represented in the camera frame, are given by (compare with the geometry of Figure 9.22):

$$x_{mi}^{c} = \left(\frac{R}{\mu_i}\right) x_{ri}^{c}, \quad y_{mi}^{c} = \left(\frac{R}{\mu_i}\right) y_{ri}^{c}, \quad z_{mi}^{c} = \left(\frac{R}{\mu_i}\right)(z_{ri}^{c} + d) - d \tag{10.75}$$

Differentiating Eq. (10.75), we find the transformation $\mathbf{T}_i = [T_{jk}]$ of the velocity screw of the landmarks to their mirror reflections, represented in the camera frame, namely:

$$\dot{\mathbf{x}}_{mi}^{c} = \mathbf{T}_i \dot{\mathbf{x}}_{ri}^{c} \quad (i = 1, 2) \tag{10.76}$$

where:

$$\mathbf{x}_{mi}^{c} = [x_{mi}^{c}, y_{mi}^{c}, z_{mi}^{c}]^{\mathrm{T}}, \quad \mathbf{x}_{ri}^{c} = [x_{ri}^{c}, y_{ri}^{c}, z_{ri}^{c}]^{\mathrm{T}}$$

$$T_{11} = \frac{R}{2}\left(\frac{1}{\lambda_i} - \frac{x_{ri}^2}{\lambda_i^3}\right), \quad T_{22} = \frac{R}{2}\left(\frac{1}{\lambda_i} - \frac{y_{ri}^2}{\lambda_i^3}\right), \quad T_{33} = \frac{R}{2}\left(\frac{1}{\lambda_i} - \frac{(z_{ri}^{c} + d)^2}{\lambda_i^3}\right)$$

$$T_{12} = T_{21} = -\frac{R}{2}\left(\frac{x_{ri}^{c} y_{ri}^{c}}{\lambda_i^3}\right), \quad T_{13} = T_{31} = -\frac{R x_{ri}^{c}(z_{ri}^{c} + d)}{2\lambda_i^3}, \quad T_{23} = T_{32} = -\frac{R y_{ri}^{c}(z_{ri}^{c} + d)}{2\lambda_i^3}$$

with $\lambda_i = 1/||L_{ri}^{m}||^3 \quad (i = 1, 2)$.

As we know (see Figures 9.21 and 9.22) in a spherical catadioptric vision system, the camera is a typical perspective camera which, in general, is described by:

$$\begin{bmatrix} u_i \\ v_i \\ 1 \end{bmatrix} = \begin{bmatrix} \lambda_u & 0 & u_{\mathrm{offset}} \\ 0 & \lambda_v & v_{\mathrm{offset}} \\ 0 & 0 & 1 \end{bmatrix} \lambda_i \begin{bmatrix} x_i \\ y_i \\ z_i \end{bmatrix} \tag{10.77}$$

with u_i, v_i being the image coordinates in the 2D image plane, $\lambda_i = 1/z_i$, $\lambda_u = l_f k_u$, $\lambda_v = l_f k_v$ where l_f is the camera focal length, k_u and k_v are scaling factors that correspond to the effective pixels' size in the horizontal and vertical directions, and $u_{\mathrm{offset}}, v_{\mathrm{offset}}$ are offset parameters that represent the principal point of the image in the pixel frame (typically at or near the image center).

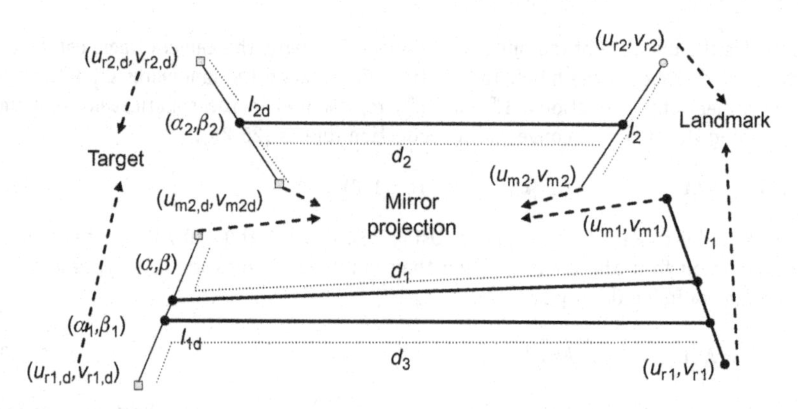

Figure 10.13 Landmarks, target points, and their images on the image plane.

Now, let (u_{r1}, v_{r1}) and (u_{r2}, v_{r2}) be the 2D image plane coordinates of the landmarks 1 and #2 mounted on the end-effector. Similarly, let (u_{m1}, v_{m1}) and (u_{m2}, v_{m2}) be the mirror reflections of the landmarks on the image plane (Figure 10.13) [25]. Thus, actually we have eight features:

$$\mathbf{f} = [u_{r1}, v_{r1}; u_{r2}, v_{r2}; u_{m1}, v_{m1}; u_{m2}, v_{m2}]$$

By bringing these features to their desired values in the image plane where:

$$\mathbf{f}_{\mathrm{d}} = [u_{r1,d}, v_{r1,d}; u_{r2,d}, v_{r2,d}; u_{m1,d}, v_{m1,d}; u_{m2,d}, v_{m2,d}]$$

(simultaneously) we can have 3D visual control of the end-effector.

Since the landmarks on the end-effector are considered as geometric points, they do not possess any roll motion. Therefore, in the present case, the above eight features can be reduced to five features, still allowing 3D visual servoing. These five features can be found using any possible morphology. Referring to Figure 10.13 one can choose the following five features: l_1, l_2, d_1, d_2, and d_3 where l_1 and l_2 are the distances between the landmarks and their images, d_1 and d_2 are the distances between the middle points of the line segments.

$((u_{r1}, v_{r1}) - (u_{m1}, v_{m1}))$ and $((u_{r1,d}, v_{r1,d}) - (u_{m1,d}, v_{m1,d}))$, and d_3 is the distance between the one-thirds of the above line segments [25,26]. From the geometry of Figure 10.13 we find that:

$$l_1(u_{m1}, u_{r1}; v_{m1}, v_{r1}) = \sqrt{(u_{m1} - u_{r1})^2 + (v_{m1} - v_{r1})^2}$$

$$l_2(u_{m2}, u_{r2}; v_{m2}, v_{r2}) = \sqrt{(u_{m2} - u_{r2})^2 + (v_{m2} - v_{m2})^2}$$

$$d_1(u_{m1}, u_{r1}; v_{m1}, v_{m2}) = \sqrt{\left[\left(\frac{u_{m1}+u_{r1}}{2} - \alpha_1\right)^2 + \left(\frac{v_{m1}+v_{r1}}{2} - \beta_1\right)^2\right]}$$

$$d_2(u_{m2}, u_{r2}; v_{m2}, v_{r2}) = \sqrt{\left(\frac{u_{m2}+u_{r2}}{2} - \alpha_2\right)^2 + \left(\frac{v_{m2}+v_{r2}}{2} - \beta_2\right)^2}$$

$$d_3(u_{m1}, u_{r1}; v_{m1}, v_{m2}) = \sqrt{\left(\frac{u_{m1}+2u_{r1}}{3} - \alpha\right)^2 + \left(\left(\frac{v_{m1}+2v_{r1}}{3}\right)^2 - \beta\right)^2}$$

where (α_1, β_1) and (α_2, β_2) are the 2D image coordinates of the midpoint of the line segment connecting the desired positions of the landmarks and their images (reflections), respectively, and (α, β) represent the 2D image coordinates of the one-third of the line segment that connects the desired position of landmark 1 and its reflection, projected onto the 2D image plane. Clearly:

$$\begin{aligned}
l_1(u_{m1}, u_{r1}; v_{m1}, v_{r1}) &= l_1(u_{r1}, u_{m1}; v_{r1}, v_{m1}) \\
l_2(u_{m2}, u_{r2}; v_{m2}, v_{r2}) &= l_2(u_{r2}, u_{m2}; v_{r2}, v_{m2}) \\
d_1(u_{m1}, u_{r1}; v_{m1}, v_{r1}) &= d_1(u_{r1}, u_{m1}; v_{r1}, v_{m1}) \\
d_2(u_{m2}, u_{r2}; v_{m2}, v_{r2}) &= d_2(u_{r2}, u_{m2}; v_{r2}, v_{m2}) \\
d_3(u_{m1}, u_{r1}; v_{m1}, v_{r1}) &\neq d_3(u_{r1}, u_{m1}; v_{r1}, v_{m1})
\end{aligned} \tag{10.78}$$

This means that the values of the features l_1, l_2, d_1, and d_2 are invariant in terms of the image information obtained on the real landmarks or their reflections. This is not so for the feature d_3, and so one can use d_3 to distinguish the landmarks from their projections on the image plane. Indeed it was shown in Ref. [27] that with the above 5D feature vector:

$$\mathbf{f} = [l_1, d_1, d_3, l_2, d_2]^{\mathrm{T}}$$

the mirror reflection of a landmark projected onto the image plane will always remain closer to the center of the image plane than from the landmark. Thus, a real landmark can be distinguished from its mirror reflection without the need of any image tracking. It is noted that special care has to be taken if a real landmark and its mirror reflection are seen to be the same on the image plane. In this case, the singularity can be avoided by moving the robot around the singular configuration.

We now have to study the motion of the landmarks in the 3D space. To this end, we consider as end-effector (tip) point the midpoint between the landmarks in the robot's base coordinate frame. The relevant geometry is shown in Figure 10.14 [25].

In Figure 10.14, let us assume that $0_1(x_{\mathrm{r}}^{\mathrm{r}}, y_{\mathrm{r}}^{\mathrm{r}}, z_{\mathrm{r}}^{\mathrm{r}})$ is the midpoint between the landmarks $(x_{\mathrm{r}1}^{\mathrm{r}}, y_{\mathrm{r}1}^{\mathrm{r}}, z_{\mathrm{r}1}^{\mathrm{r}})$ and $(x_{\mathrm{r}2}^{\mathrm{r}}, y_{\mathrm{r}2}^{\mathrm{r}}, z_{\mathrm{r}2}^{\mathrm{r}})$ where the upper index r denotes position coordinate in the robot coordinate frame. Here, ϕ is the angle between the x-axis of the robot coordinate frame and the projection on the xy-plane of the line segment between the points $(x_{\mathrm{r}1}^{\mathrm{r}}, y_{\mathrm{r}1}^{\mathrm{r}}, z_{\mathrm{r}1}^{\mathrm{r}})$ and $(x_{\mathrm{r}2}^{\mathrm{r}}, y_{\mathrm{r}2}^{\mathrm{r}}, z_{\mathrm{r}2}^{\mathrm{r}})$.

The angle θ is the angle between the line segment defined by the above two points and the xy-plane of the robot coordinate frame.

Denoting by D the distance between the landmarks in the world coordinate frame we find that:

$$\begin{aligned}
x_{\mathrm{r}1}^{\mathrm{r}} &= x_{\mathrm{r}}^{\mathrm{r}} - (D/2)\cos\theta\cos\phi \\
x_{\mathrm{r}2}^{\mathrm{r}} &= x_{\mathrm{r}}^{\mathrm{r}} + (D/2)\cos\theta\cos\phi \\
y_{\mathrm{r}1}^{\mathrm{r}} &= y_{\mathrm{r}}^{\mathrm{r}} - (D/2)\cos\theta\sin\phi \\
y_{\mathrm{r}2}^{\mathrm{r}} &= y_{\mathrm{r}}^{\mathrm{r}} + (D/2)\cos\theta\sin\phi \\
z_{\mathrm{r}1}^{\mathrm{r}} &= z_{\mathrm{r}}^{\mathrm{r}} - (D/2)\sin\theta \\
z_{\mathrm{r}2}^{\mathrm{r}} &= z_{\mathrm{r}}^{\mathrm{r}} + (D/2)\sin\theta
\end{aligned} \tag{10.79}$$

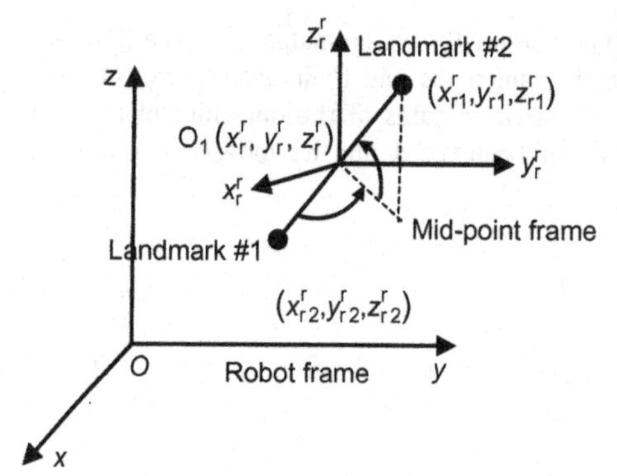

Figure 10.14 Geometry for defining the positions of the two landmarks in 3D space.

The image Jacobian matrix which relates the velocity screw:

$$\dot{\mathbf{r}} = [\mathbf{v}^{\mathrm{T}}, \boldsymbol{\omega}^{\mathrm{T}}]^{\mathrm{T}}, \quad \mathbf{v} = [\dot{x}_r, \dot{y}_r, \dot{z}_r]^{\mathrm{T}}, \quad \boldsymbol{\omega} = [\dot{\varphi}, \dot{\theta}] \tag{10.80}$$

of the landmark midpoint coordinate frame (with origin at $O_1(x_r^r, y_r^r, z_r^r)$) and the image feature rate vector:

$$\dot{\mathbf{f}} = [\dot{l}_1, \dot{d}_1, \dot{d}_3, \dot{l}_2, \dot{d}_2] \tag{10.81}$$

can be found, as usual, by differentiating Eqs. (10.78) and (10.79), namely:

$$\dot{\mathbf{f}} = \mathbf{J}_{\mathrm{im}}\dot{\mathbf{r}} \tag{10.82}$$

Now, defining the feature error $\mathbf{e}(\mathbf{f}) = \mathbf{f}_d - \mathbf{f}$, the control law that assures asymptotic convergence of $\mathbf{e}(\mathbf{f})$ to zero is the resolved-rate control law given by Eq. (9.34):

$$\mathbf{u} = \mathbf{J}_{\mathrm{im}}^{-1}\mathbf{K}\mathbf{e}(\mathbf{f}) \quad (\mathbf{u} = \dot{\mathbf{r}} = \text{control vector}) \tag{10.83}$$

where $\mathbf{K}$ is a constant positive definite gain matrix (usually diagonal). Of course, if $\mathbf{J}_{\mathrm{im}}$ is not exactly known we can use in Eq. (10.83) an estimate $\hat{\mathbf{J}}_{\mathrm{im}}$ which can be found as explained in Section 9.4.3 using the estimates of the camera-intrinsic and -extrinsic parameters. Here, the intrinsic parameters include the two scaling factors λ_u, λ_v and the two offset terms u_{offset} and v_{offset}. The extrinsic parameter is the distance d of the camera frame from the mirror's center. Therefore, the vector $\boldsymbol{\xi}$ that has to be estimated is:

$$\boldsymbol{\xi} = [u_{\mathrm{offset}}, v_{\mathrm{offset}}, \lambda_u, \lambda_v, d]^{\mathrm{T}}$$

while the measurement vector $\mathbf{z}$ is:

$$\mathbf{z} = [u_{m1}, v_{m1}, u_{r1}, v_{r1}; u_{m2}, v_{m2}, u_{r2}, v_{r2}]^{\mathrm{T}}$$

References

[1] Padois V, Fourquet JY, Chiron P. Kinematic and dynamic model-based control of wheeled mobile manipulators: a unified framework for reactive approaches. Robotica 2007;25(2):157−73.

[2] Seelinger M, Yoder JD, Baumgartner ET, Skaar BR. High-precision visual control of mobile manipulators. IEEE Trans Robot Autom 2002;18(6):957−65.

[3] Tzafestas CS, Tzafestas SG. Full-state modeling, motion planning and control of mobile manipulators. Stud Inform Control 2001;10(2):109−27.

[4] Kumara P, Abeygunawardhana W, Murakami T. Control of two-wheel mobile manipulator on a rough terrain using reaction torque observer feedback. J Autom Mobile Robot Intell Syst 2010;4(1):56−67.

[5] Chung JH, Velinsky SA. Robust interaction control of a mobile manipulator: dynamic model based coordination. J Intell Robot Syst 1999;26(1):47−63.

[6] Li Z, Chen W, Liu H. Robust control of wheeled mobile manipulators using hybrid joints. Int J Adv Rob Syst 2008;5(1):83−90.

[7] Yamamoto Y, Yun X. Effect of the dynamic interaction on coordinated control of mobile manipulators. IEEE Trans Rob Autom 1996;12(5):816−24.

[8] Tzafestas SG, Melfi A, Krikochoritis T. Omnidirectional mobile manipulator modeling and control: analysis and simulation. Syst Anal Model Control 2001;40:329−64.

[9] Mazur A, Szakiel D. On path following control of non-holonomic mobile manipulators. Int J Appl Math Comput Sci 2009;19(4):561−74.

[10] Meghadari A, Durali M, Naderi D. Investigating dynamic interaction between one d.o.f. manipulator and vehicle of a mobile manipulator. J Intell Robot Syst 2000;28:277−90.

[11] Watanabe K, Sato K, Izumi K, Kunitake Y. Analysis and control for an omnidirectional mobile manipulator. J Intell Robot Syst 2000;27(1−2):3−20.

[12] De Luca A, Oriolo G, Giordano PR. Image-based visual servoing schemes for nonholonomic mobile manipulators. Robotica 2007;25:131−45.

[13] Burshka D, Hager G. Vision-based control of mobile robots. In: Proceedings of 2001 IEEE international conference robotics and automation. Seoul, Korea; May 21−26, 2001.

[14] Tsakiris D, Samson C, Rives P. Vision-based time-varying stabilization of a mobile manipulator. In: Proceedings of fourth international conference on control, automation, robotics and vision (ICARCV'96). Westin Stanford, Singapore; December 3−6, 1996. p. 1−5.

[15] Yu Q, Chen Ming I. A general approach to the dynamics of nonholonomic mobile manipulator systems. Trans ASME 2002;124:512−21.

[16] DeLuca A, Oriolo G, Giordano PR. Kinematic control of nonholonomic mobile manipulators in the presence of steering wheels. In: Proceedings of 2010 IEEE international conference on robotics and automation. Anchorage, AK; May 3−8, 2010. p. 1792−98.

[17] Phuoc LM, Martinet P, Kim H, Lee S. Motion planning for nonholonomic mobile manipulator based visual servo under large platform movement errors at low velocity. In: Proceedings of 17th IFAC world congress. Seoul, Korea; July 6−11, 2008. p. 4312−17.

[18] Gilioli M, Melchiori C. Coordinated mobile manipulator point-stabilization using visual-servoing techniques. In: Proceedings of IEEE/RSJ international conference on intelligent robots and systems. Lausanne, CH; 2002. p. 305−10.

[19] Ma Y, Kosecha J, Sastry S. Vision guided navigation for nonholonomic mobile robot. IEEE Trans Robot Autom 1999;15(3):521−36.

[20] Zhang Y. Visual servoing of a 5-DOF mobile manipulator using an omnidirectional vision system. Technical Report RML-3-2, University of Regina, 2006.

[21] Bayle B, Fourquet JY, Renaud M. Manipulability analysis for mobile manipulators. In: Proceedings of 2001 IEEE international conference on robotics and automation. Seoul, Korea; May 21−26, 2001. p. 1251−56.

[22] Papadopoulos E, Poulakakis J. Trajectory planning and control for mobile manipulator systems. In: Proceedings of eighth IEEE Mediterranean conference on control and automation. Patras, Greece; July 17−19, 2000.

[23] Yamamoto Y, Yun X. Coordinating locomotion and manipulation of a mobile manipulator. In: Proceedings of 31st IEEE conference on decision and control. Tucson, AZ; December, 1992. p. 2643−48.

[24] Yamamoto Y, Yun X. Modeling and compensation of the dynamic interaction of a mobile manipulator. In: Proceedings of IEEE conference on robotics and automation. San Diego, CA; May 8−13, 1994. p. 2187−92.

[25] Zhang Y, Mehrandezh M. Visual servoing of a 5-DOF mobile manipulator using a panoramic vision system. In: Proceedings of 2007 Canadian conference on electrical and computer engineering. Vancouver, BC, Canada; April 22−26, 2007. p. 453−56.

[26] Zhang Y, Mehrandezh M. Visual servoing of a 5-DOF mobile manipulator using a catadioptric vision system. In: Proceedings of SPIE, the international society for optical engineering; 2007. p. 6−17.

[27] Zhang Y. Visual servoing of a 5-DOF mobile manipulator using panoramic vision system [M.Sc. thesis]. Regina, Canada, Faculty of Engineering, University of Regina, 2007.

11 Mobile Robot Path, Motion, and Task Planning

11.1 Introduction

Robot planning is concerned with the general problem of figuring out how to move to get from one place to another and how to perform a desired task. As a whole, it is a wide research field in itself. Actually, the term planning means different things to different scientific communities. The three categories of planning in robotics are [1–32]:

1. Path planning
2. Motion planning
3. Task planning

Planning represents a class of *problem solving* which is an interdisciplinary area of system theory and artificial intelligence (AI) [12,19]. A *general problem solver* is basically a search program in which the problem to be solved is defined in terms of a given initial state and a desired goal state. This program guides the search by evaluating the current state and operator (rule) ordering, that is, by applying only the operators that promise the best movement in the search space. The principal problem-solving method is the *means-ends analysis*, which consists in repeated reduction of the difference between the goal state and the current state. This needs a special feedback control strategy. An example of the use of means-ends analysis in AI problem solving is the well-known *Towers of Hanoi* puzzle problem. The general problem-solving methodology has limited value for specific problems that can be solved only by skilled experts such as fault diagnosis, decision support, and robot planning.

The objectives of this chapter are as follows:

- To present the general conceptual definition of robot path planning, motion planning, and task planning
- To investigate the path planning problem of mobile robots, including the basic operations and classification of methods
- To study in some detail the model-based mobile robot path planning, including configuration space and road map planning methods
- To discuss mobile robot motion planning presenting the vector fields method and the analytical parameterized method

Introduction to Mobile Robot Control. DOI: http://dx.doi.org/10.1016/B978-0-12-417049-0.00011-0

- To show how global and local path planning can be integrated such that to achieve the desired features of path smoothness and trapping avoidance
- To outline the basic issues of fixed and mobile robot task planning, including plan representation and generation, and the three phases of task planning, namely, world modeling, task specification, and robot program synthesis.

11.2 General Concepts

Path planning is a general capability embedded in all kinds of robots (robotic manipulators, mobile robots/manipulators, humanoid robots, etc.). In a broad sense, robot path planning is concerned with the determination of how a robot will move and maneuver in a workspace or environment in order to achieve its goals. The path planning problem involves computing a collision-free path between a start position and a goal position. Very often, besides the obstacle avoidance, the robot must also satisfy some other requirements or optimize certain performance criteria. Path planning is distinguished according to the knowledge available about the environment (i.e., fully known/structured environment, partially known environment, and fully unknown/unstructured environment). In most practical cases, the environment is only *partially known*, where the robot, prior to path planning and navigation, has already knowledge of some areas within the workspace (i.e., areas likely to pose local minima problems). The nature of an obstacle is described via its configuration which may be *convex* shaped or *concave* shaped or both. The status of an obstacle may be *static* (when its position and orientation relative to a known fixed coordinate frame is invariant in time), or *dynamic* (when either its position), or orientation or both change relative to the fixed coordinate frame change.

Path planning may be either *local* or *global*. Local path planning is performed while the robot is moving, taking data from local sensors. In this case, the robot has the ability to generate a new path in response to the changes of the environment. Global path planning can be performed only if the environment (obstacles, etc.) is static and perfectly known to the robot. In this case, the path planning algorithm produces a complete path from the start point to the goal point before the robot starts its motion.

Motion planning is the process of selecting a motion and the corresponding inputs such that to assure that all constraints (obstacle avoidance, risk avoidance, etc.) are satisfied. Motion planning can be considered as a set of computations which provide subgoals or set points for the control of the robot. These computations and the resulting plans are based on a suitable model of the robot and the environment in which it is moved. The process by which the robot executes (follows) the planned motion is the control process studied in Chapters 5–10.

We recall that the motion of a robot can be described in three different spaces:

1. Task or Cartesian space
2. Joints' (motors') space
3. Actuators' space

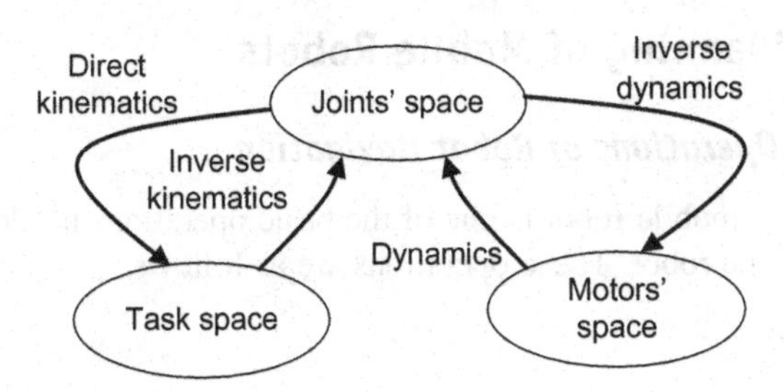

Figure 11.1 Robot motion spaces and their relationships.

The relation of these spaces is pictorially shown in Figure 11.1.

The description of a motion in the task space, that is, the specification of a reference point (e.g., the tip of the end effector) is typically made in the Cartesian world coordinate frame.

The specification of the end effector or mobile robot position is not always sufficient to determine the positions of all the links. For this reason, we use the *joints' space* which is the Cartesian product of the allowable ranges of all the degrees of freedom (usually a subset of n). Finally, for each robot motion that is compatible with the kinematic and dynamic constraints, there must exist at least one set of forces/torques that produce that motion. The forces/torques of the actuators (motors) that generate all the allowable motions define the *motors' space*. A basic issue in robot motion planning is the presence of obstacles and the resulting requirement to find an *obstacle-free* path. This is the *path planning* or *path finding* problem. The need for motion planning comes from the fact that there is a very large number (theoretically, an infinite number) of motions via which the robot can go to a goal pose and execute a desired task. Also, for a given motion, there may be more than one inputs (forces/torques) of the motors that produce the desired motion.

Task planning involves three phases [8]:

1. World modeling
2. Task specification
3. Robot program synthesis

World modeling: A world model must involve a geometric description of robots and objects in the environment (which is usually embodied in a CAD system), a physical description of objects (e.g., mass and inertia of parts), a kinematic description of robot body and linkages, and a description of robot features (e.g., joint limits, maximum possible or allowed acceleration, and sensor features).

Task specification: The task planner usually receives the tasks as a sequence of models of the world state at several steps of the task's execution. Actually, a task specification is a model of the world accompanied by a sequence of changes in the positions of the model components.

Robot program synthesis: This is the most important phase for the task planner to be successful. The program synthesized must include grasp commands, motion properties, sensor commands, and error tests. This implies that the program must be in the robot-level programming language of the robot.

11.3 Path Planning of Mobile Robots

11.3.1 Basic Operations of Robot Navigation

Path planning of a mobile robot is one of the basic operations needed to implement the navigation of the robot. These operations are as follows:

- Self-localization
- Path planning
- Map building and map interpretation

Robot localization provides the answer to the question "where am I?" The *path planning* operation provides the answer to the question "how should I get to where I' am going?" Finally, the *map building/interpretation* operation provides the geometric representation of the robot's environment in notations suitable for describing locations in the robot's reference frame. Vision-based navigation employs optical sensors including laser-based range finders and CCD cameras by which the visual features needed for the localization in the robot's environment are extracted.

Actually, to date, there is no generic method for mobile robot positioning (localization). The specific techniques that exist are divided into two categories:

1. Relative localization methods
2. Absolute localization methods

Because no single, globally good localization method is available, designers of autonomous guided vehicles (AGVs) and autonomous mobile robots (AMRs) usually employ some combination of methods, one from each category.

Relative localization is performed by odometry or inertial navigation. The first uses encoders to measure wheel rotation and/or steering angle. Inertial navigation employs gyroscopes (or accelerometers in some cases) to measure the rate of rotation and the angular acceleration.

Absolute localization uses the following:

- *Active beacons*, where the absolute position of the mobile robot is computed by measuring the direction of incidence of three or more transmitted beacons. The transmitters use light or radio frequencies and are placed at known positions in the environment.
- *Recognition of artificial landmarks*, which are placed at known locations in the environment and are designed so as to provide maximal detectability even under bad environmental conditions.
- *Recognition of natural landmarks*, that is, distinctive features of the environment, which must be known in advance. This method has lower reliability than the artificial landmarks method.
- *Model matching*, that is, comparison of the information received from on-board sensors and a map of the environment. The absolute location of the robot can be estimated if the sensor-based features match the world model map. The robot navigation maps are

distinguished in *geometric maps* and *topological maps*. The first category represents the world in a global coordinate frame, whereas the second category represents the world as a network of arcs and nodes.

11.3.2 Classification of Path Planning Methods

Path planning is a robotics field on its own. Its solution gives a feasible collision-free path for going from one place to another. Humans do path planning without thinking how it is done. If there is an obstacle ahead that has not been there before, humans just pass it. Very often, the human needs to change his/her pose in order to go through a narrow passage. This is a simple type of the so-called *piano-mover's problem*. If the obstacle blocks the way completely, humans just use another way. To perform all the above operations, a robot must be equipped with suitable high-level intelligence capabilities.

A very broad classification of free (obstacle-avoiding) path planning involves three categories, which include six distinct strategies. These are the following:

1. *Reactive control* ("Wander" routine, circumnavigation, potential fields, motor schemas)
2. *Representational world modeling* (certainty grids)
3. *Combinations of both* (vector field histogram)

In many cases, the above techniques do not assure that a path is found that passed obstacles although it exits, and so they need a higher level algorithm to assure that the mobile robot does not end up in the same position over and over again. In practice, it may be sufficient that the robot detects that it is "stuck" despite the fact that a feasible path way exists, and calls for help. In indoor applications, a maneuver for avoiding an obstacle is a good action. Outdoor situations are more complex, and more advanced perception techniques are needed (e.g., for distinguishing a small tree from an iron pole).

A research topic receiving much attention over the years is the *piano-mover's problem*, which is well known to most people that tried a couch or big table through a narrow door. The object has to be tilted and moved around through the narrow door. One of the first research works on this problem is described in Latombe [1].

On the basis of the way the information about the robot's environment is obtained, most of the path planning methods can be classified into two categories:

1. Model-based approach
2. Model-free approach

In the first category, all the information about the robot's workspace are prelearned, and the user specifies the geometric models of objects and a description of them in terms of these models. In the model-free approach, some of the information about the robot's environment is obtained via sensors (e.g., vision, range, touch sensors). The user has to specify all the robotic motions needed to accomplish a task.

11.4 Model-Based Robot Path Planning

The obstacles that may exist in a robotic work environment are distinguished into *static obstacles* and *moving obstacles*. Therefore, two types of path finding problems have to be solved, namely:

- Path planning among stationary obstacles
- Path planning among moving obstacles

The path planning methodology for stationary obstacles is based on the configuration space concept, and is implemented by the so-called *road map planning methods* discussed below.

The path planning problem for the case of moving obstacles is decomposed into two subproblems:

1. Plan a path to avoid collision with static obstacles.
2. Plan the velocity along the path to avoid collision with moving obstacles.

This combination constitutes the *robot motion planning* [2–5].

11.4.1 Configuration Space

Configuration space is a representation in which path planning for both manipulator robots and (most of) mobile robots is performed. For example, if the mobile system is a free-flying rigid body (i.e., a body that can move freely in space in any direction without any kinematics constraint), then six configuration parameters are required to determine fully its position, namely, x, y, z and three directional (Euler) angles. Path planning finds a path in this six-dimensional space. But, actually, a robot is not a free-flying robot. Its possible motion depends on its kinematic structure, for example, a unicycle-like robot has three configuration parameters: x, y, and ϕ. As we have seen, very often, these parameters are not independent, for example, the robot may or may not be able to turn on the spot (change ϕ while keeping x and y fixed), or be able to move sideway. A robotic arm which has n rotational joints needs n configuration parameters to specify its configuration in space, in addition to constraints such as the minimum or maximum values of each angular join. For example, a typical car-like robotic manipulator has 10 configuration parameters (4 for the mobile platform with the trailer, and 6 for the arm), whereas a certain humanoid robot such ASIMO or HRP may have 52 configuration parameters (2 for the head, 7 for each arm, 6 for each leg, and 12 for each hand that involves 4 fingers with 3 articulations each).

Now, given a robot with n configuration parameters moving in a certain environment, we define the following:

- The configuration $\mathbf{q}$ of the robot, that is, an n-tuple of real numbers that specifies the n parameters needed to determine the position of the robot in physical space.
- The configuration space CS of the robot, that is, the set of values that its configuration $\mathbf{q}$ may take.

- The free configuration space CS_{free}, that is, the subset of CS of configurations that are not in collision with the obstacles existing in the robot's environment.

It is noted that the degrees of freedom of a mobile robot are its control variables (a robot arm or a humanoid robot has as many degrees of freedom as configuration parameters, but a differential drive robot has three configuration parameters and only two degrees of freedom).

From the above definitions, it follows that path planning is the problem of finding a path in the free configuration space CS_{free}, between an initial and a final configuration. Thus, if CS_{free} could be determined explicitly, then path planning is reduced to a search for a path in this n-dimensional continuous space. Actually, the explicit definition of CS_{free} is a computationally difficult problem (its computational complexity increases exponentially with the dimension of CS), but there are available efficient probabilistic methods that solve this path planning problem in reasonable time [11]. As we have seen before, these techniques must involve two operations:

1. *Collision checking* (i.e., check whether a configuration $\mathbf{q}$ or a path between two configurations lies entirely in CS_{free})
2. *Kinematic steering* (i.e., find a path between two configurations $\mathbf{q}_0$ and $\mathbf{q}_f$ in CS that satisfies the kinematic constraints, without taking into account obstacles)

Example 11.1 Two-link planar manipulator

Consider the two-link planar manipulator of Figure 11.2A which has two configuration parameters θ_1 and θ_2 (i.e., its CS is two-dimensional).

An obstacle in an n-dimensional configuration space CS is represented by a *slice projection* which is defined by a range of values for one of the defining parameters of CS and an $(n-1)$-dimensional volume. The approximation of the full obstacle is built as the union of a number of $(n-1)$-dimensional slice projections, each for a different range of values of the same joint parameter [8]. In the present case (Figure 11.2A), the obstacles are approximated by a set of θ_2 ranges (which are shown by dark lines) for a set of values of θ_1. A sample path of a two-link robot in another environment with obstacles is shown in Figure 11.2B.

11.4.2 Road Map Path Planning Methods

The robot navigation maps which are used to represent the environment can be a continuous geometric description or a decomposition-based geometric map or a topological map. These maps must be converted to discrete maps appropriate for the path algorithm under implementation. This conversion (or decomposition) can be done by four general methodologies, namely [1,21]:

1. Road maps
2. Cell decomposition
3. Potential fields
4. Vector field histograms

In the following sections, a short description of these methodologies is given.

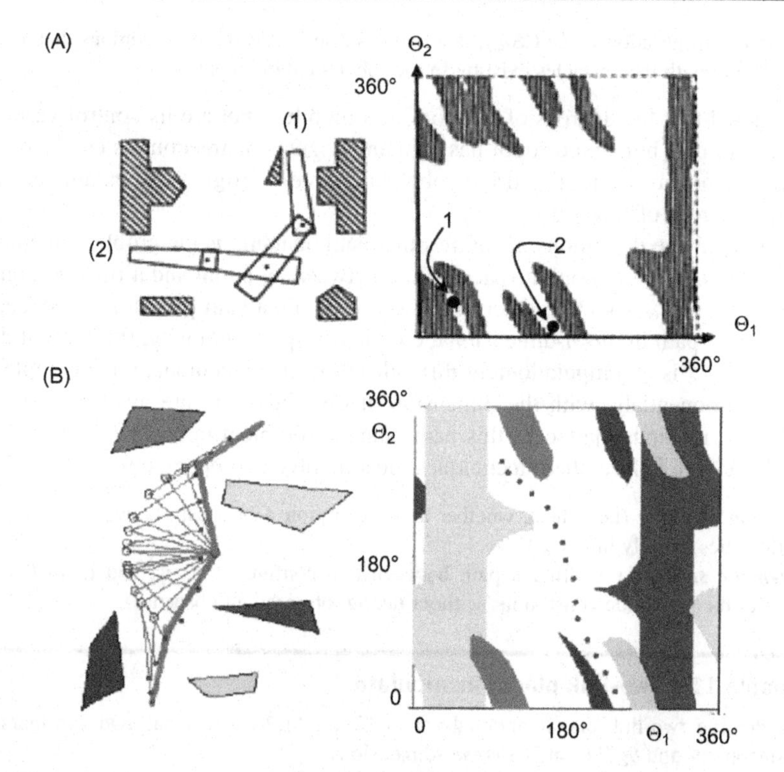

Figure 11.2 Configuration space of a two-link planar robot. (A) The manipulator with the obstacles, and the CS with obstacles approximated by a set of one-dimensional slice projections (shown in dark). (B) Another two-link manipulator in an environment with obstacles, and a possible path in CS_{free}.
Source: http://robotics.stanford.edu/~latombe/cs326/2009/class3/class3.htm; http://www.cs.cmu.edu/~motionplanning/lecture/Chap3-Config-Space_howie.pdf.

11.4.2.1 Road Maps

The basic idea is to capture the connectivity of CS_{free} with a road map (graph or network) of one-dimensional curves. After its construction, a road map is used as a network of path (road) segment for the planning of the robot motion. The main goal here is to construct a set of roads that as a whole enable the robot to reach any position in its CS_{free}. This is actually a hard problem.

The path planning problem is stated as follows:

Given input configurations $\mathbf{q}_{start}$ and $\mathbf{q}_{goal}$ and the set B of obstacles.
Find a path in CS_{free} connecting $\mathbf{q}_{start}$ and $\mathbf{q}_{goal}$.

The basic steps of the path planning algorithm are as follows:

Step 1: Build a road map in CS_{free} (the road map nodes are free or semifree configurations; two nodes are connected by an edge if the robot can easily move between them).

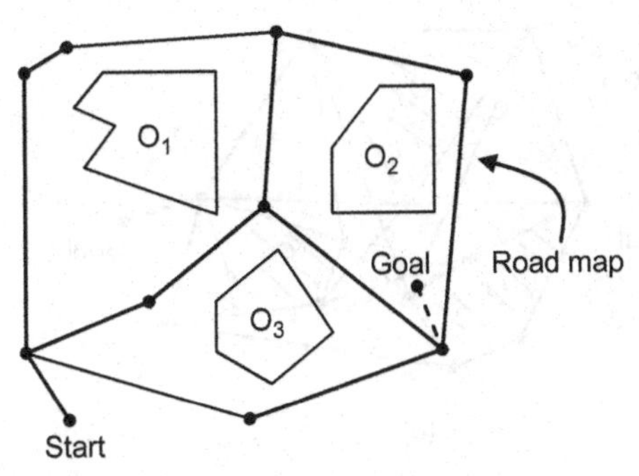

Figure 11.3 A road map example in an environment with obstacles O_1, O_2, O_3.

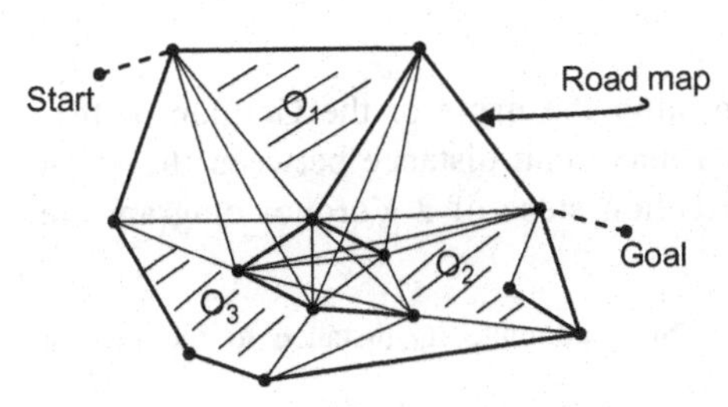

Figure 11.4 A visibility graph example.

Step 2: Connect $\mathbf{q}_{\text{start}}$ and $\mathbf{q}_{\text{goal}}$ to road map nodes υ_{start} and υ_{goal}.
Step 3: Find a path in the road map between υ_{start} and υ_{goal}, which gives directly a path in CS_{free}.

A road map example is shown in Figure 11.3.
The methods for building road maps are distinguished into:

- Traditional deterministic methods (they are suitable only for low-dimensional CSs, they build CS_{free}, and they are complete).
- Modern probabilistic methods (they do not build CS_{free}, they apply to both low- and high-dimensional CSs, but they are not complete).

Visibility graph of CS: A visibility graph for a polygonal CS is an undirected graph G where the nodes in G correspond to vertices of the polygonal obstacles, under the condition that the nodes can be connected by a straight line that lies fully in CS_{free} or by the edges of the obstacles (i.e., under the condition that the nodes can see each other including the initial and goal positions as vertices as well). An example of visibility graph is shown in Figure 11.4.

Mobile path planning based on visibility graphs is popular because of its simplicity. There are efficient algorithms with complexity $O(n^2)$, where n is the number of vertices of the objects or $O(E + n \log n)$, where E is the number of edges in G [17,22]. Visibility graphs are really suitable only for two-dimensional CS. It is noted that they are also methods for constructing "reduced" visibility graphs where not all edges are needed. A visibility graph and a reduced visibility graph (corresponding to it) are shown in Figure 11.5.

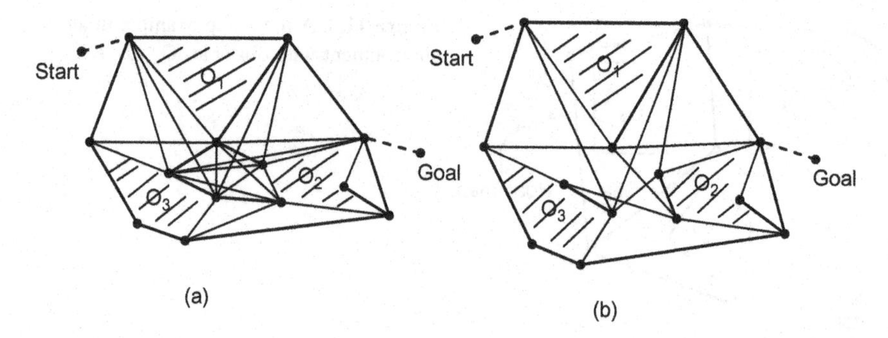

Figure 11.5 (A) A visibility graph. (B) An associated reduced visibility graph.

Voronoi diagram: A Voronoi diagram, after the name of the German mathematician who coined it in 1908, ensures a maximum distance between the robot and the obstacles in the map. The construction steps of a Voronoi diagram are as follows:

Step 1: For each point in the free space CS_{free}, compute its distance to the nearest obstacle.
Step 2: Plot that distance as a vertical height. The height increases as the point is moving away from the obstacle.
Step 3: At equidistant points from two or more obstacles, this distance plot has sharp ridges.
Step 4: Construct the Voronoi diagram by the union of the edges formed by these sharp ridges.

Mathematically, the Voronoi diagram V is a polygonal region defined as:

$$V = \{\mathbf{q} \in CS_{free}: |\text{near}(\mathbf{q})| > 1, \ \ CS = R^2\}$$

where:

$\text{near}(\mathbf{q}) = \{\mathbf{p} \in \delta: ||\mathbf{q} - \mathbf{p}|| = \text{clearance}(\mathbf{q})\}$
$\text{clearance}(\mathbf{q}) = \min\{||\mathbf{q} - \mathbf{p}||: \mathbf{p} \in \delta\} \text{ for } \mathbf{q} \in CS_{free}\}$
$\delta = \partial(CS_{free})$, the boundary of CS_{free}

We see that, actually, near($\mathbf{q}$) is the set of boundary points of CS_{free} that minimize the distance to $\mathbf{q}$. The Voronoi diagram V consists of all points in CS_{free} with at least two nearest neighbors in the CS_{free} boundary δ. The Voronoi diagram is a finite collection of straight line segments and parabolic segments (called arcs), where:

- *Straight arcs* are defined by two vertices or two edges of the set of obstacles (i.e., the set of points equally close to two points or two line segments is a line).
- *Parabolic arcs* are defined by one vertex and one edge of the obstacle set (i.e., the set of points equally close to a point (focus) and a line (directrics) is a parabolic segment).

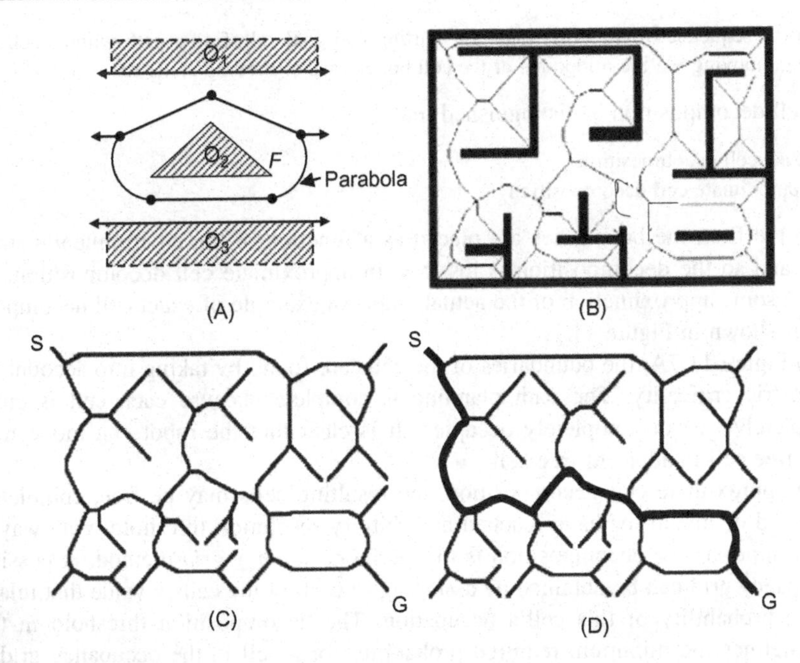

Figure 11.6 (A) An example of Voronoi diagram. (B) Environment (map) of an office floor. (C) Voronoi diagram of the floor. (D) A safe path found by the fast marching method.
Source: Reprinted from Ref. [33] with the courtesy of S. Garrido and permission from CSC Press.

A construction procedure of V based on the above definition involves the following basic steps:

Step 1: Compute all arcs (for all vertex-vertex, edge-edge, vertex-edge pairs).
Step 2: Compute all intersection points (dividing arcs into segments).
Step 3: Keep the segments which are closest only to the vertices/edges that defined them.

A simple Voronoi diagram example for a terrain with two parallel walls and a triangular object between them is shown in Figure 11.6A. The Voronoi diagram of an office floor that has the map of Figure 11.6B is shown in Figure 11.6C, and a safe path from S to G found by the *fast marching* path planning method is depicted in Figure 11.6D [33–36].

11.4.2.2 Cell Decomposition

Cell decomposition is concerned with the discrimination between cells (i.e., connected geometric areas) in CS that are free (i.e., belong to CS_{free}) or are occupied by objects. The cells have a predefined resolution. After the step of determining the cells, the path planning procedure proceeds as follows:

- Determine the open cells that are adjacent and construct a connectivity graph G.
- Determine the cells that contain the starting and goal configurations and search for a path in G that joins the start and goal cell.

- In the sequence of cells that joins the starting and goal cell, find a path within each cell (e.g., moving via the midpoints of the cell boundaries or moving along the walls).

Cell decomposition is distinguished into:

- Exact cell decomposition
- Approximate cell decomposition

In the first, the boundaries are placed as a function of the environment's structure, and so the decomposition is lossless. In approximate cell decomposition, we obtain some approximation of the actual map. An example of exact cell decomposition is shown in Figure 11.7.

In Figure 11.7A, the boundaries of the cells are found by taking into account the geometric criticality. The path planning is complete because each cell is either completely free or completely occupied. It is clear that the robot can move from each free cell to adjacent free cells.

In approximate cell decomposition, the resulting cells may be free, completely occupied or mixed, or have reached an arbitrary resolution threshold. One way to apply approximate decomposition is to use the *occupancy grid* method. A possible occupancy grid can be obtained by assigning to each of the cells a value that relates to the probability of this cell's occupation. The decomposition threshold in this case defines the minimum required probability for a cell in the occupancy grid to be deemed occupied. Actually, the idea of "approximate" is to fuse neighboring

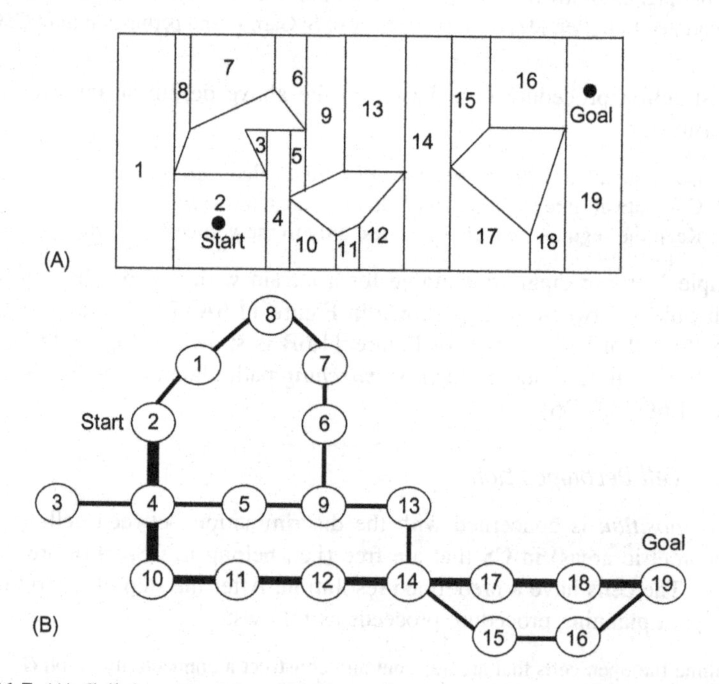

Figure 11.7 (A) Cell decomposition. (B) Network representation and a possible path from the start node to the goal node.

free cells into larger cells to allow a fast path determination. In two-dimensional workspaces, the approximate cell decomposition operation consists in recursively subdividing each cell which contains a mix of free and obstructed space into cells. Because of this property, this method is known as *quadtree method.* The recursion is ended when each cell is found to contain entirely free or obstructed space, or when the maximum desired resolution is reached. The height of the decomposition is the maximum allowable level of recursion, and specifies the resolution of the decomposition.

Figure 11.8 shows an application example of the approximate cell decomposition method. The workspace contains three obstacles and the robot has to move from S to G.

The result of approximate decomposition is a drastic reduction of the number of cells to be considered. In an example, a high resolution map that contains 250,000 cells, with a crude decomposition of height 4 was reduced to just 109 cells [37]. A problem that has to be faced here is to determine *cell adjacency* (i.e., to find which cells share a common border or edge with another one). Actually, many techniques are available to solve the adjacency problem.

One of them is to use *tesseral* (or *quad tesseral*) *addressing* which has the ability to map every part of a 2D (or *n*D) spatial domain into 1D sequence, and when stored with attributes in a database, each address can perform as a single key to data [37,38]. To generate tesseral addresses, the positive quadrant of 2D-Cartesian space is quartered to give parent tiles with labeling as shown in Figure 11.9A. This process can be continued with new tesseral addresses generated by always appending to the right of the parent addresses, until a desired depth is reached.

Another solution to global path planning, via decomposition, is to use local node refinement, path nodes refinement, and curve parametric interpolation [39].

A quad tesseral address can be stored in a quadtree structure. For example, the address of Figure 11.9A (right) can be stored by the quadtree structure of

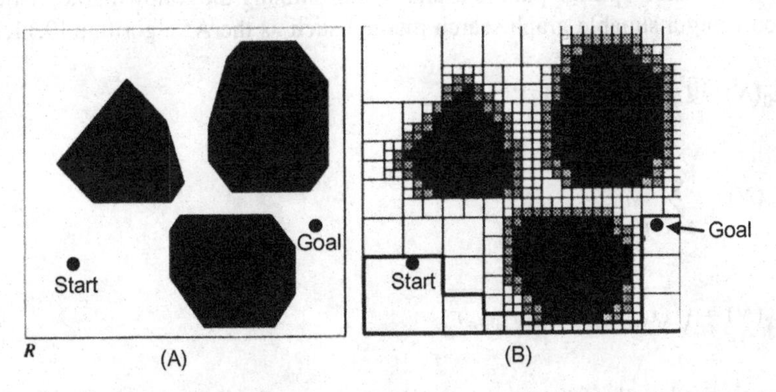

Figure 11.8 (A) A simple three-obstacle path planning problem. (B) An obstacle-free path based on approximate cell decomposition.
Source: http://www-cs-faculty.stanford.edu/~eroberts/courses/soco/1998-99/robotics/basicmotion.html.

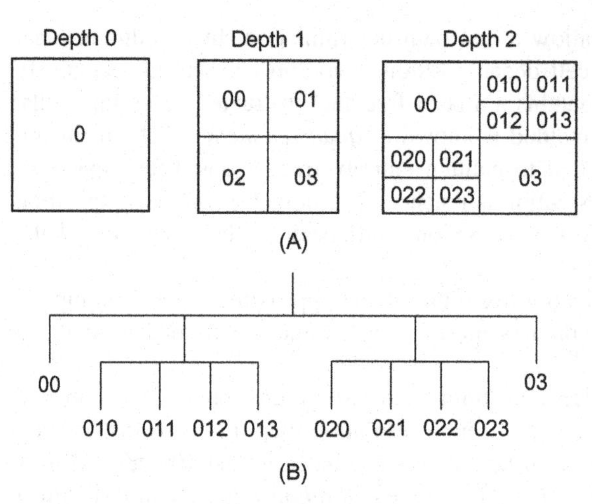

Figure 11.9 (A) 2D tesseral decomposition (addressing). (B) Quadtree structure of the above depth-2 address.

Figure 11.9B, which if traversed from left to right produces a linearization. This important property of tesseral addressing facilitates the storage, comparison, and translation of groups of addresses.

In global path planning, the environment is assumed to be *a priori* known, in which case, a *distance optimal path* from the robot's current position to the goal can be found. This path consists of a sequence of waypoints whose proximity to one another is specified by the decomposition resolution and the configuration of the obstacles in the environment. In a changing environment, one must combine the above global path planning technique with a waypoint driven local path planning method. A distance-optimal global path planning method which uses tesseral addressing is presented in Ref. [37]. In this method, use is made of the connectivity graph to produce *nodes* that represent physical locations in the environment and *arcs* that represent the ability to avoid the obstacles during the motion along the path. The distance-optimal path is found by minimizing the following heuristic cost function using a suitable graph search method such as the A* algorithm [9,11,19]:

$$L_0(N) = L_s(N) + L_g(N)$$

$$L_s(N) = \sum_{i=s+1}^{N} a_i$$

$$L_g(N) = \sqrt{(x_g - x_n)^2 + (y_g - y_n)^2}$$

where a_i is the length of the arc (n_{i-1}, n_i) between two adjacent nodes $i-1$ and i in a sequence $(n_s, n_{s+1}, \ldots, n_{N-1}, n_N)$ of nodes that connect the node s to node N, and $L_g(N)$ is the distance from node N to the goal node G. Clearly, $L_g(N)$ is an underestimate of the optimal cost from N to G.

To carry out the minimization, the so-called *tesseral*[1] *arithmetic* is used. For example, in this arithmetic, addition or subtraction is *tile translation* to the *right* (addition of tesseral 1, binary 01) or to the *left* (subtraction of tesseral 1). Correspondingly, translation *down* and *up* is equivalent to the addition or subtraction of tesseral 2 (binary 10), respectively. It is remarked that actually the resulting optimal path depends on the connectivity graph employed between the initial robot's position and the goal, that is, it is not globally optimal. The computation time of the tesseral arithmetic algorithms is of order $O(4n)$. A critical review of spatial reasoning using quad tesseral representation is provided in Ref. [38]. This representation allows to carry out spatial reasoning (i.e., manipulation of 2D and 3D objects) as if in only one dimension.

11.4.2.3 Potential Fields

The potential field method for path planning is very attractive because of its simplicity and elegance. The idea of potential field is borrowed from nature, for example, a charge particle navigating a magnetic field, or a small ball rolling on a hill. We know that depending on the strength of the field, or the slope of the hill, the particle, or the ball, will go to the source of the field (the magnet), or the valley of this example. In robotics, we can emulate the same phenomenon by developing an artificial potential field that will attract the robot to the goal. In conventional path planning, we calculate the relative position of the robot to the goal, and then apply the necessary forces that will drive the robot to the goal.

If the robot's environment is free of obstacles, we simply create an *attractive field* that goes to the goal. The potential field is defined over the entire free space, and at each time increment, we compute the potential field at the robot position, and then calculate the induced force by this field. The result is that the robot will move according to this force. If the robot's space has an obstacle, we make it generate a *repulsive field* in the surrounding space, which when the robot approaches the obstacle, pushes the robot away from it. If we want to make the robot going to the goal and avoiding the obstacle, then we superimpose an attractive field around the goal and a repulsive field around each obstacle in the robot's space. The above concepts are illustrated in Figure 11.10A–D, where the path of the robot from a certain starting point to the goal is also shown (Figure 11.10E).

In the simplest situation, a mobile robot can be considered to be a *point robot*, in which case the robot's orientation ϕ is not included in the calculations, the potential field is two-dimensional, and the robot's position is simply $\mathbf{q} = [x, y]^T$. In general, a mobile robot has $\mathbf{q} = [x, y, \phi]^T$, and a robotic manipulator has $\mathbf{q} = [q_1, q_2, \ldots, q_n]^T$. Mathematically, the potential field method develops as follows [1,5] (see also [24,25] for more formulations). Define:

- $U(\mathbf{q})$ the total potential field, which is equal to the sum of the attractive potential field $U_{att}(\mathbf{q})$ and the repulsive potential field $U_{rep}(\mathbf{q})$.

[1] The term *tesseral* comes from the Greek word τέσσερα (tessera=four).

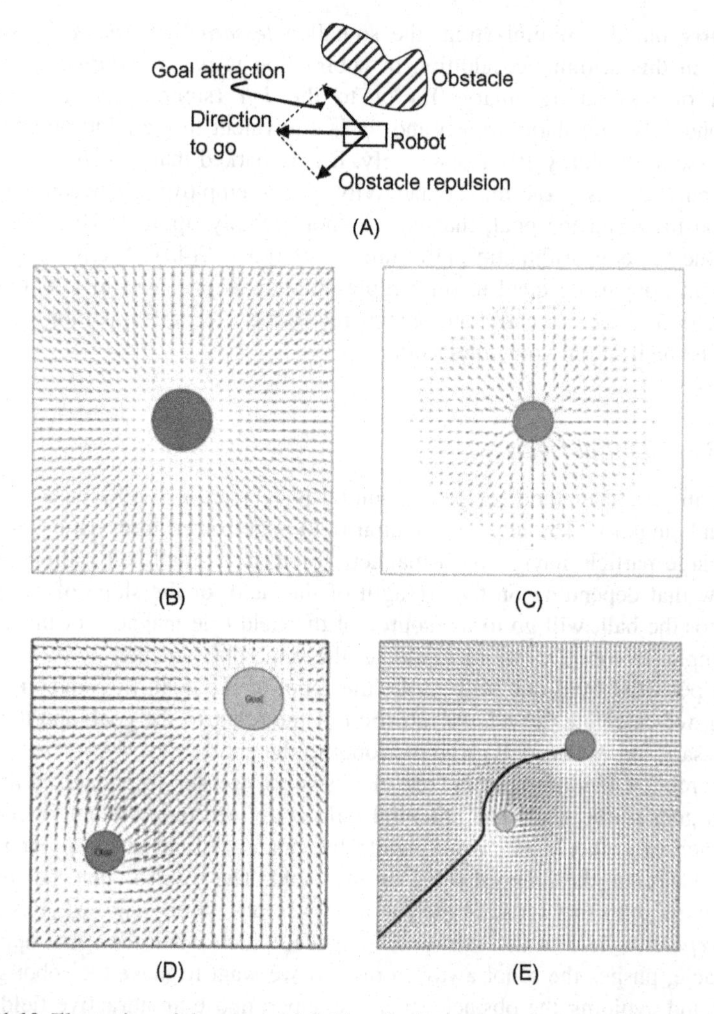

Figure 11.10 Illustration of the potential field path planning method. (A) Goal attraction, obstacle repulsion, and direction to go. (B) Attractive field to the goal. (C) Repulsive field around an obstacle. (D) Combination of the above two fields. (E) A possible robot path. *Source*: http://www.cs.mcgill.ca/~hsafad/robotics/index.html.

- $\mathbf{F}_{att}(\mathbf{q})$ the attractive force, $\mathbf{F}_{rep}(\mathbf{q})$ the repulsive force, and $\mathbf{F}(\mathbf{q})$ the total force applied to the robot.

Then, we have:

$$U(\mathbf{q}) = U_{att}(\mathbf{q}) + U_{rep}(\mathbf{q})$$

$$\mathbf{F}(\mathbf{q}) = \mathbf{F}_{att}(\mathbf{q}) + \mathbf{F}_{rep}(\mathbf{q})$$

where:

$$\mathbf{F}(\mathbf{q}) = -\nabla U(\mathbf{q}), \mathbf{F}_{\text{att}}(\mathbf{q}) = -\nabla U_{\text{att}}(\mathbf{q}), \mathbf{F}_{rep}(\mathbf{q}) = -\nabla U_{\text{rep}}(\mathbf{q})$$

with $\nabla[\cdot] = [\partial/\partial q_1, \partial/\partial q_2, \ldots, \partial/\partial q_n]^{\mathrm{T}}$ being the gradient operator.

Now, suppose that $U_{\text{att}}(\mathbf{q})$ is quadratic:

$$U_{\text{att}}(\mathbf{q}) = \frac{1}{2}||\mathbf{q} - \mathbf{q}_{\text{goal}}||^2 = \frac{1}{2}\sum_{i=1}^{n}(q_i - q_{i,\text{goal}})^2$$

In this case, we have:

$$\begin{aligned}
\nabla U_{\text{att}}(\mathbf{q}) &= \frac{1}{2}(2||\mathbf{q} - \mathbf{q}_{\text{goal}}||\nabla||\mathbf{q} - \mathbf{q}_{\text{goal}}||) \\
&= ||\mathbf{q} - \mathbf{q}_{\text{goal}}||\frac{(\mathbf{q} - \mathbf{q}_{\text{goal}})}{||\mathbf{q} - \mathbf{q}_{\text{goal}}||} \\
&= \mathbf{q} - \mathbf{q}_{\text{goal}}
\end{aligned}$$

Therefore:

$$\mathbf{F}_{\text{att}}(\mathbf{q}) = -(\mathbf{q} - \mathbf{q}_{\text{goal}})$$

which shows that $F_{\text{att}}(\mathbf{q})$ tends linearly to 0 with distance to goal. This is good for stability but tends to infinity when the distance goes far from goal.

If we define $U_{\text{att}}(\mathbf{q})$ as the norm of $\mathbf{q} - \mathbf{q}_{\text{goal}}$, that is:

$$U_{\text{att}}(\mathbf{q}) = ||\mathbf{q} - \mathbf{q}_{\text{goal}}||$$

then:

$$\begin{aligned}
\nabla_{\text{att}} U(\mathbf{q}) &= \nabla\left\{\sum_{i=1}^{n}(q_i - q_{i,\text{goal}})^2\right\}^{1/2} \\
&= \frac{1}{2}\left\{\sum_{i=1}^{n}(q_i - q_{i,\text{goal}})^2\right\}^{-1/2}\nabla\left[\sum_{i=1}^{n}(q_i - q_{i,\text{goal}})^2\right] \\
&= (\mathbf{q} - \mathbf{q}_{\text{goal}})/\left[\sum_{i=1}^{n}(q_i - q_{i,\text{goal}})^2\right]^{1/2} \\
&= \frac{(\mathbf{q} - \mathbf{q}_{\text{goal}})}{||\mathbf{q} - \mathbf{q}_{\text{goal}}||}
\end{aligned}$$

In this case:

$$\mathbf{F}_{\text{att}}(\mathbf{q}) = -(\mathbf{q} - \mathbf{q}_{\text{goal}})/||\mathbf{q} - \mathbf{q}_{\text{goal}}||$$

which is singular (unstable) at the goal, and tends to 1 far from the goal.

Very often, in practice, we use a composite attractive potential field of the form:

$$U_{\text{att}}(\mathbf{q}) = \begin{cases} \frac{1}{2}\lambda||\mathbf{q} - \mathbf{q}_{\text{goal}}||^2 & \text{if} \quad ||\mathbf{q} - \mathbf{q}_{\text{goal}}|| < \varepsilon \\ \mu||\mathbf{q} - \mathbf{q}_{\text{goal}}|| & \text{if} \quad ||\mathbf{q} - \mathbf{q}_{\text{goal}}|| \geq \varepsilon \end{cases}$$

where λ and μ are scaling factors and ε is a selected distance from the goal. The repulsive potential field is typically selected as:

$$U_{\text{rep}}(\mathbf{q}) = 1/||\mathbf{q} - \mathbf{q}^*||$$

where $\mathbf{q}^*$ is the closest point to the obstacle.

Therefore:

$$\mathbf{F}_{\text{rep}}(\mathbf{q}) = -\nabla U_{\text{rep}}(\mathbf{q}) = (\mathbf{q} - \mathbf{q}^*)/||\mathbf{q} - \mathbf{q}^*||^2$$

For each additional obstacle, we add a corresponding repulsive potential field. If an obstacle is non-convex, we triangulate it into multiple convex obstacle and weigh the separate fields of the object in case the summed repulsive fields are greater than that of the original obstacle. The main problem with potential-field-based path planning is that the robot may be trapped to a local minimum of the field. Actually, several methods exist for avoiding this trapping. Specifically, when the robot goes into a local minimum position, we can correct this situation in one of the following ways [27]:

- Backtrack from the local minimum and use an alternative strategy to avoid the local minimum.
- Perform certain random movements, hoping that they will help escaping the local minimum.
- Use more complex potential fields that are local minimum free (e.g., harmonic potential fields).
- Apply an extra force $\mathbf{F}_{\text{vfs}}$ which is called *virtual free space force*.

Of course, all the above methods assume that the robot can detect that it is trapped, which is also a difficult problem on its own. The virtual free space force is proportional to the amount of free space around the robot [28], and helps to pull the robot away from the local minimum region. In other words, the virtual force drags the robot outside the local minimum, and so the robot can begin again using the potential field planner. Fortunately, it is unlikely the robot to be trapped again to the same local minimum. The above virtual force concept and its effect is illustrated pictorially in Figure 11.11A–C, where the robot is moved by the total force:

$$\mathbf{F} = \mathbf{F}_{\text{att}} + \mathbf{F}_{\text{rep}} + \mathbf{F}_{\text{vfs}}$$

A hybrid virtual force field method that integrates the virtual force field concept with the virtual obstacle and virtual goal concept is presented in Ref. [40].

One way to detect that the robot is trapped is to employ an open-loop position estimator which estimates the current position of the robot. If the current position does not change for a predefined period of time (time threshold), then the virtual force is generated and applied to the robot for pulling it out from the local minimum. Obviously, due to the repulsive force, the robot velocity is decreased as it approaches an obstacle. A Java source code of a path planning algorithm that uses $\mathbf{F}_{\text{vfs}}$ is provided in Ref. [27].

Three other problems that may be encountered in potential-field-based path planning are the following [29]:

1. The robot cannot pass between closely spaced obstacles.
2. Oscillations occur in the presence of obstacle disturbances.
3. Oscillations occur in narrow passages.

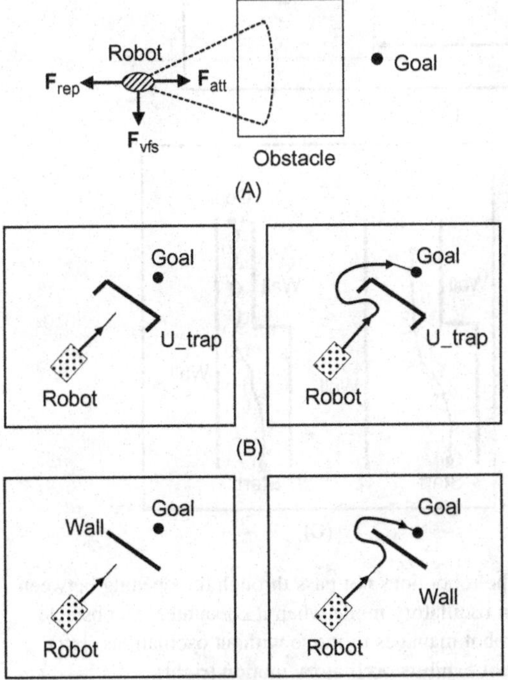

Figure 11.11 (A) The virtual free space force $\mathbf{F}_{\text{vfs}}$ drags the robot outside the local minimum area in CS_{free}. (B) $\mathbf{F}_{\text{vfs}}$ untraps the robot from a U-trap. (C) Untrapping from a wall trap.

The first situation is illustrated in Figure 11.12A. The two obstacles exert the repulsive forces $\mathbf{F}_{rep,1}$ and $\mathbf{F}_{rep,2}$ giving a total (resultant) repulsive force $\mathbf{F}_{rep}$. Thus, the robot moves under the influence of the goal attractive force $\mathbf{F}_{att}$ and the total obstacles' repulsive force $\mathbf{F}_{rep}$. Their sum:

$$\mathbf{F}_{total} = \mathbf{F}_{att} + \mathbf{F}_{rep}$$

in this case gets the robot away from the passage that leads toward the robot. Of course, depending on the relative magnitude of $\mathbf{F}_{att}$ and $\mathbf{F}_{rep}$, the robot may be moved toward the goal passing through the opening between the obstacles.

In Figure 11.12B, a robot is forced to move alongside a wall which obstructed its path. At a certain point, the wall has a discontinuity which causes the robot to move in an oscillatory mode. Figure 11.12C (right) shows the oscillatory behavior caused when the robot moves to a narrow corridor. This is due to that the robot is subject to two repulsive forces from opposite sides, simultaneously. If the corridor

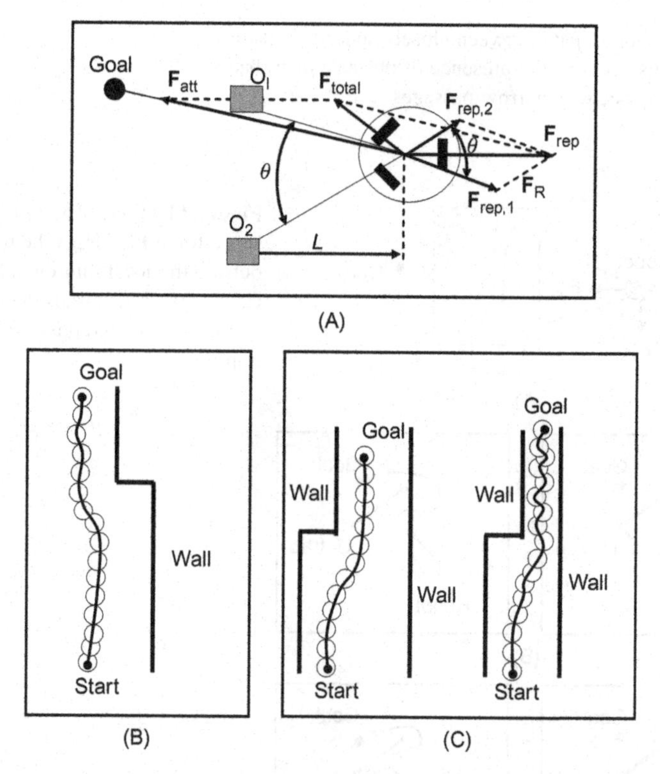

Figure 11.12 (A) A situation in which the robot does not pass through the opening between obstacles. (B) The robot motion enters an oscillatory mode when it encounters an obstacle disturbance. (C) In a wide corridor, the robot manages to move without oscillations (left), but if the corridor is very narrow, the robot exhibits oscillatory motion (right).

is sufficiently wide (Figure 11.12C, left), the robot may manage to get a stable (nonoscillatory) motion. All the above phenomena have been studied analytically by stability theory, and have been observed in practical or simulated experiments [29]. To overcome these limitations of the potential field approach, Koren and Borenstein developed the so-called *vector field histogram* (VFH) method [30], which was further improved in Ref. [31].

11.4.2.4 Vector Field Histograms

In this method, the obstacle-free path planning is performed with the aid of an intermediate data structure about the local obstacle distribution, called *polar histogram* which is an array of, say, 72 (5° wide) *angular sectors* (Figure 11.13). To take into account the robot changing position and the new sensor readings, the polar histogram is totally updated and rebuilt every, say, 30 ms (sampling period). The method involves two steps [30]:

1. The histogram grid is reduced to one-dimensional polar histogram which is built around the robot's instantaneous location (Figure 11.15A). Each sector in the polar histogram involves a value that represents the *polar obstacle density* (POD) in this direction.
2. The most suitable sector from among all polar histogram sectors with a low POD is selected, and the robot moves in that direction.

To implement these steps, a window (called *active window)* moves with the robot, overlying a square region of cells (e.g., 33 × 33 cells) in the histograms. All

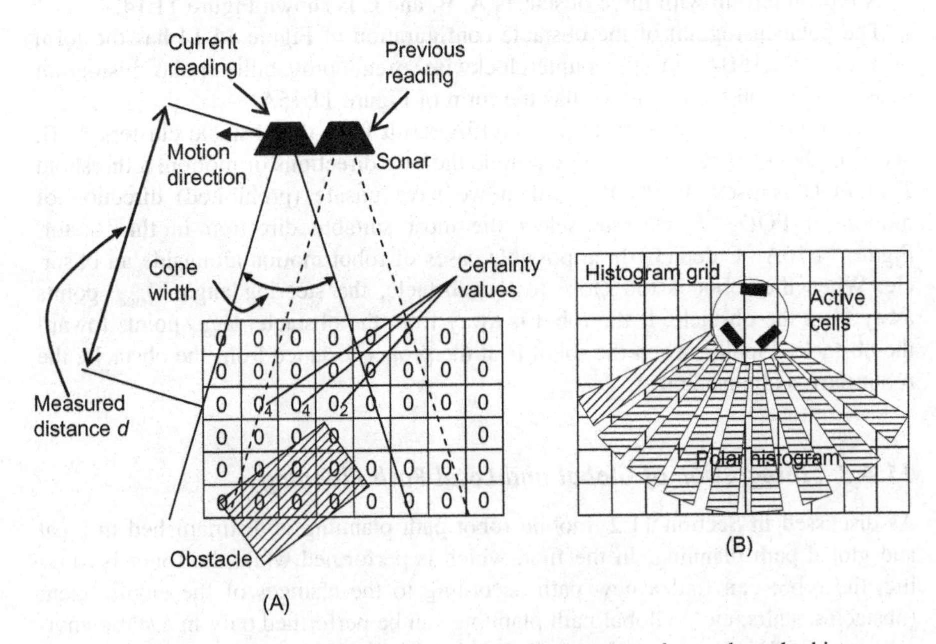

Figure 11.13 (A) Histogram grid. (B) The active cells are mapped onto the polar histogram.

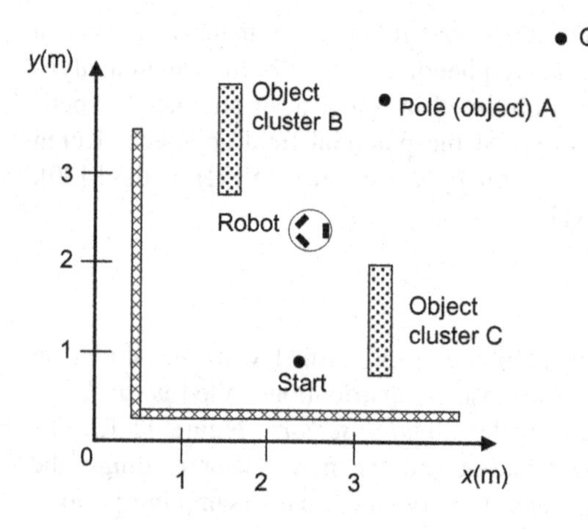

Figure 11.14 A mobile robot moving in a terrain with three obstacles (or clusters of obstacles) A, B, and C.

cells that lie on the moving window, each time, are called *active cells*. Actually, the cell that lies on the sonar axis and corresponds to the measured distance d found by each range reading is incremented and increases the *certainty value* (CV) of the cell (Figure 11.13A). The contents of each active cell in the histogram are mapped onto the corresponding polar histogram sector (say the kth sector), which gives a value H_k to this sector (Figure 11.13A). The value is higher if there are many cells with high CV in one sector. Obviously, this value can be regarded as the POD in the direction of sector k.

A typical terrain with three obstacles A, B, and C is shown Figure 11.14.

The polar histogram of the obstacle configuration of Figure 11.14 has the form of Figure 11.15B, and its counterclockwise pseudoprobability polar histogram (from A to C) on the H-k plane has the form of Figure 11.15A.

The peaks A, B, and C in Figure 11.15A result from the obstacle clusters A, B, and C in the histogram grid. To determine the safe directions of motion, a threshold T in POD is used. If POD $> T$, then we have unsafe (prohibited) direction of motion. If POD $< T$, we can select the most suitable direction in this sector. Figure 11.16A–C depicts three possible cases of robot motion alongside an obstacle. When the robot is too close to the obstacle, the steering angle ψ_{steer} points away from the obstacle. If the robot is away from the obstacle, ψ_{steer} points toward the obstacle. Finally, when the robot is at the proper distance from the obstacle, the robot moves alongside it.

11.4.3 Integration of Global and Local Path Planning

As discussed in Section 11.2, mobile robot path planning is distinguished in local and global path planning. In the first, which is performed while the robot is moving, the robot can find a new path according to the changes of the environment (obstacles, stairs, etc.). Global path planning can be performed only in a static environment which is known to the robot. In this case, the path planning algorithm produces a complete path from the start point to the goal before the robot starts its

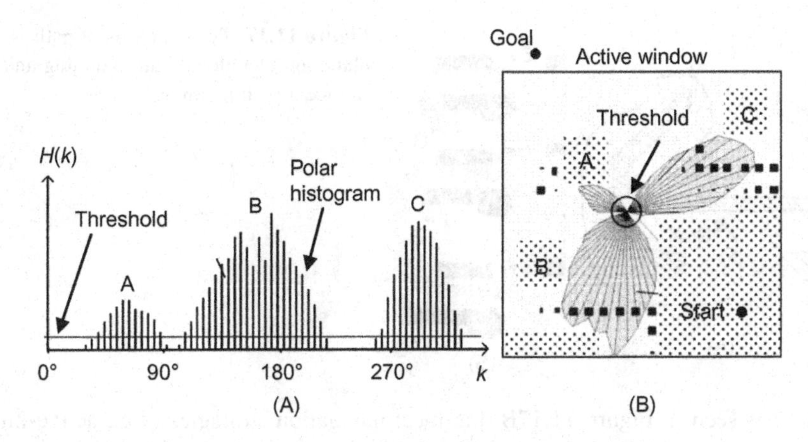

Figure 11.15 (A) The one-dimensional polar histogram (pseudoprobability distribution) corresponding to obstacles A, B, and C of Figure 11.14 in the counterclockwise direction from the x-axis. (B) The polar form of (A).
Source: Reprinted from Ref. [30], with permission from Institute of Electrical and Electronic Engineers.

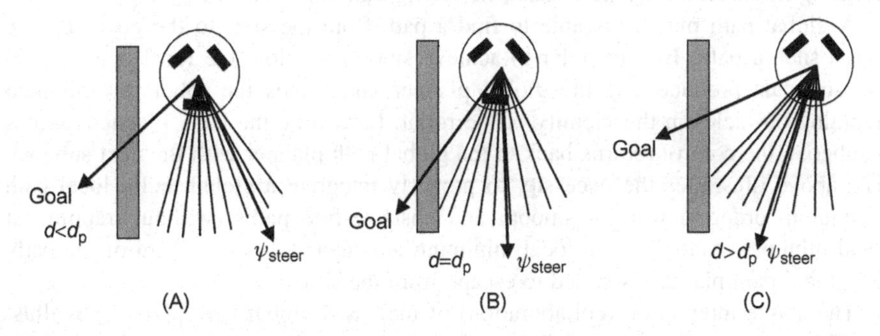

Figure 11.16 (A) Direction of motion when the robot obstacle distance d is smaller than the proper (desired) distance d_p. (B) Motion alongside the obstacle ($d = d_p$). (C) Direction of motion when $d > d_p$.

motion. A property that characterizes a path is its *smoothness* which is defined as the maximum curvature of the path measured over both its segments and the entire path. A motion planning method that takes into account the path curvature is presented in Section 11.5.3. The relative advantages and disadvantages of a typical global and local path planning can be seen in Figure 11.17.

The global path might be produced, for example, by the approximate cell decomposition method, but this could be done by including segments with high curvature values (e.g., the sharp turn shown in Figure 11.17A at the original motion steps of the navigation). This is a result of the nature of global path planning and is caused by the digitization of the space cells. But a global path method (e.g., cell decomposition) has an excellent capability in driving the robot out of an H-shaped obstacle which is known to be one of the most difficult cases (Figure 11.17A).

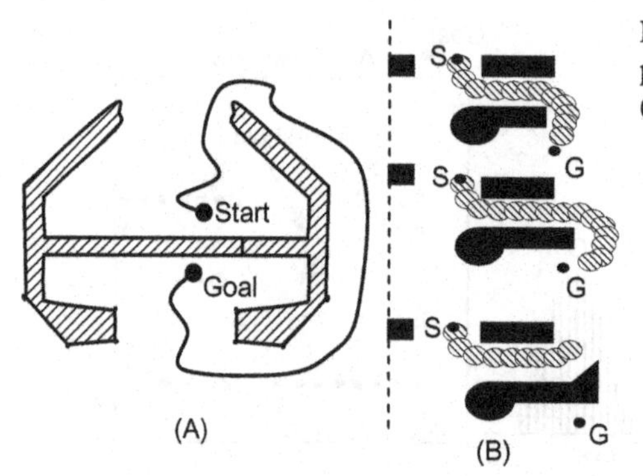

Figure 11.17 Typical cases of path planning: (A) global path planning and (B) local path planning.

As it is seen in Figure 11.17B, the local navigation strategies (e.g., active kinematic histograms [41]) present a series of simpler problems. The local planner manages to reach the goal in the first two cases, but in the last case, when the obstacle fully obstructs the way to the goal, the method fails to find a path although one exists, and the robot stays trapped in local minima in front of the obstacle. In this case, all the paths produced are sufficiently smooth, and the robot is traveling even adapting its speed according to obstacle configuration.

A global path planner is able to find a path from the start to the goal, if there exists such a path. But, in order to achieve smooth motion, the local planner gets the subgoals produced by the global planner and drives the robot toward them avoiding obstacles in the vicinity of the robot. Each time the local planner reaches a subgoal, the control returns back to the global path planner with the next subgoal. The above illustrates the necessity to properly integrate a global and a local path planner in order to warrant smooth and obstacle-free paths (without trapping at local minima). Actually, if a local minimum is detected at some point of the path, the global path planner is called to escape from the situation.

The above integration (collaboration) of local and global path planning is illustrated in pseudocode form as follows [42]:

Go to goal *(start position, final goal)*
While not at final goal

next subgoal = Global Path Planner (final goal)

Local path planner *(next subgoal)*
While not at *next subgoal* and *not blocked in local minima*

Drive the robot toward *next subgoal*

If blocked at local minima

Call recursively the go to goal algorithm with *start operant* the current position and *goal operant* the final goal.

A fuzzy algorithm which implements global and local navigation integration and uses potential fields is presented in Section 13.10. Another integrated (hybrid) path planner for indoor environments is presented in Ref. [32]. This hybrid planner

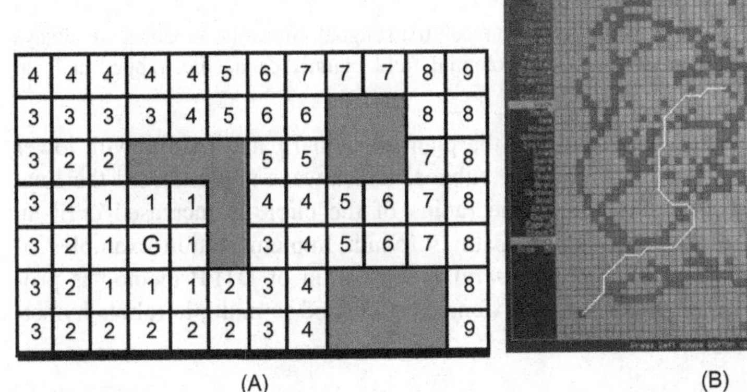

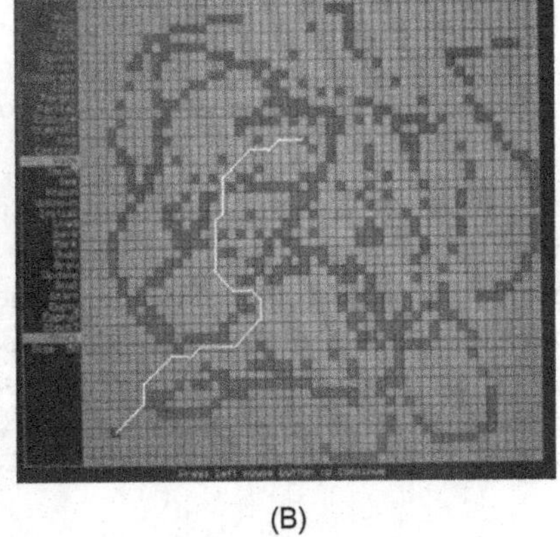

Figure 11.18 (A) The DT path planning scheme. (B) A typical minimal length path found by the steepest descend and DT.
Source: Reprinted from Ref. [43], with permission from International Journal of Computer Science and Applications.

combines the so-called *distance transform path planner* (DTPP) and the potential field planner. Unlike most planners, the distance transform (DT) planner treats the task of path planning by finding paths from the goal location back to the start location. This path planner propagates a distance wave front through all free space grid cells in the environment from the goal cell as shown in Figure 11.18A. Figure 11.18B shows a minimal length DT path [43].

Actually, the robot has to build the local map on the move, and at the same time constantly recompute the *distance map* and the path using locally available sensory data. DTPP employs an occupancy grid-based map of the workspace to compute its distance map. At the initialization stage, the goal cell and obstacle cells are assigned *values* that represent their distance from the goal. These distance values are propagated flowing around the obstacles. An algorithm similar to raster scanning iterates until the values of all cells are stabilized, in which case distance values are assigned to the rest of the cells. Obviously, the obstacle cells must be given very high values, and are passed over in the raster scan, and the raster scans are repeated until no further changes occur. The free configuration space cells should be treated in the same way. It was shown that the resulting DT is independent of any start point and represents a potential field without local minima [32]. Thus, a globally minimum distance path from any start point in free space can be determined via the standard gradient/steepest descend technique (see section 7.2.1). For a nonpoint robot, the obstacles are grown as usual by the *maximum effective radius* of the robot to convert the path planning to that of a point (dimensionless) robot.

Omitting the details, the steps of DTPP are as follows [32]:

Step 1: Using DT, create an optimal global path from the start point to the final goal.

Step 2: If not at the final goal, create a circle centered at the robot current position with reasonable radius depending on the configuration of the environment.

Step 3: Select the next subgoal as the intersection of the circumference of the circle and the planned global path.

Step 4: Use the potential field planner to reach the subgoal. Since the subgoals are always on the obstacle-free global path, the potential field planner cannot be trapped at local minima.

If there are two intersections along the planned global path, the one with lower distance value is selected such that the subgoal will always move toward the final goal. If no intersections are found, the radius of the circle is increased until an intersection with the planned global path is found. Implementation examples of DTPP are provided in Ref. [32]. A useful modification of DTPP method is provided in Ref. [44]. This modification extends the method to multiple robots by setting the initial exploration directions.

11.4.4 Complete Coverage Path Planning

Complete coverage path planning produces a path in which the robot sweeps all areas of free space in an environment. It is needed very often in practice (e.g., in autonomous room vacuum, security robots, lawn mowers). In the complete coverage DTPP, the robot moves away from the goal keeping track of the cell it has visited. The robot moves to a cell that has a smaller distance from the goal, only if it has visited all the neighboring cells that lie further away from the goal. The typical DTPP algorithm for complete coverage is as follows) [45,46]:

Set the start cell to current cell
Set all cells to not visited
Loop
Find unvisited neighboring cell with highest DT
If no neighbor cell is found then
Mark as visited and stop at Goal
If Neighbor cell DT <= Current cell DT then
Mark as visited and stop at Goal
Set current cell to neighboring cell
End Loop

A one-obstacle environment with corresponding DT values is shown in Figure 11.19A, and a complete coverage path for this environment is depicted in Figure 11.19B.

A modified complete coverage DT path planning, called *path transform path planning* (PTPP) [45] propagates a weighted sum of the distance from the goal and a measure of the discomfort of moving too close to obstacles. Thus actually, a PTPP is a form of DTPP without the trapping to local minima possibility. The steps of the PTPP are the following:

Step 1: Invert the DT transform into an *obstacle transform* (OT), where the obstacle cells become the goals. This implies that for each free cell, the minimal distance from the center of the free space to the boundary of an obstacle cell is obtained.

13	12	11	10	9	8	7	7	7	7	7	7	7	7
13	12	11	10	9	8	7	6	6	6	6	6	6	6
S	12	11	10	9	8	7	6	5	5	5	5	5	5
									4	4	4	4	4
9	8	7	6	5	4	3	3	3	3	3	3	3	4
9	8	7	6	5	4	3	2	2	2	2	2	3	4
9	8	7	6	5	4	3	2	1	1	1	2	3	4
9	8	7	6	5	4	3	2	1	G	1	2	3	4
9	8	7	6	5	4	3	2	1	1	1	2	3	4
9	8	7	6	5	4	3	2	2	2	2	2	3	4

(A)

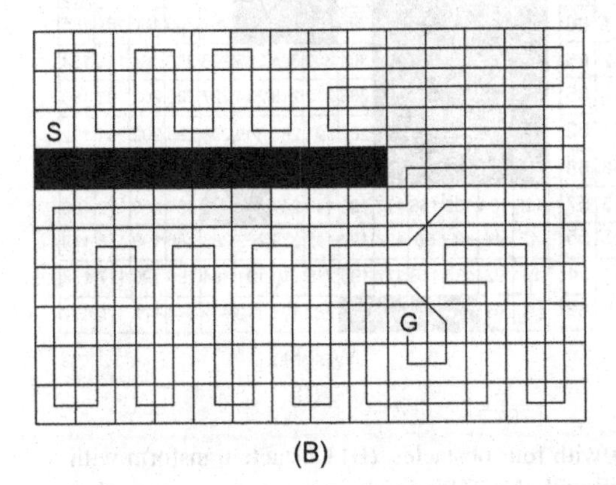

(B)

Figure 11.19 (A) An environment with a single obstacle. (B) A complete coverage DT path from S to G. *Source*: Reprinted from Ref. [45], with permission from Japan Robot Association.

Step 2: Define a cost function transform, called the *path transform* (V_{PT}), of the form:

$$V_{\mathrm{PT}}(c) = \min_{p \in P} \left\{ L(p) + \sum_{c_i \in p} \lambda\, O(c_i) \right\}$$

where P is the set of all possible paths to the goal, c_i is the ith cell in P, and p is a single path in P. The function $L(p)$ is the length of the path p to the goal, and the function $O(c_i)$ is a cost function produced by using the values of OT. The constant $\lambda \geq 0$ is a weight that specifies how strongly the PT will avoid obstacles. The minimization of $L(p) + \sum \lambda\, O(c_i)$ is done by the standard steepest descent method, without the possibility to be locked at some minimum. This is because all costs of paths to the goal from each cell are computed.

Figure 11.20 depicts an environment with four obstacles and distances shown (A), a path found with the obstacle transform (B), and two paths found by path transform with weight constants $\lambda_1 < \lambda_2$ (C and D) [45,46].

Figure 11.21A shows the actual path generated by PTPP implemented on the AMROS simulator [47] in a 7 m × 6 m room with a 1.5 m × 1.0 m obstacle at the room's center.

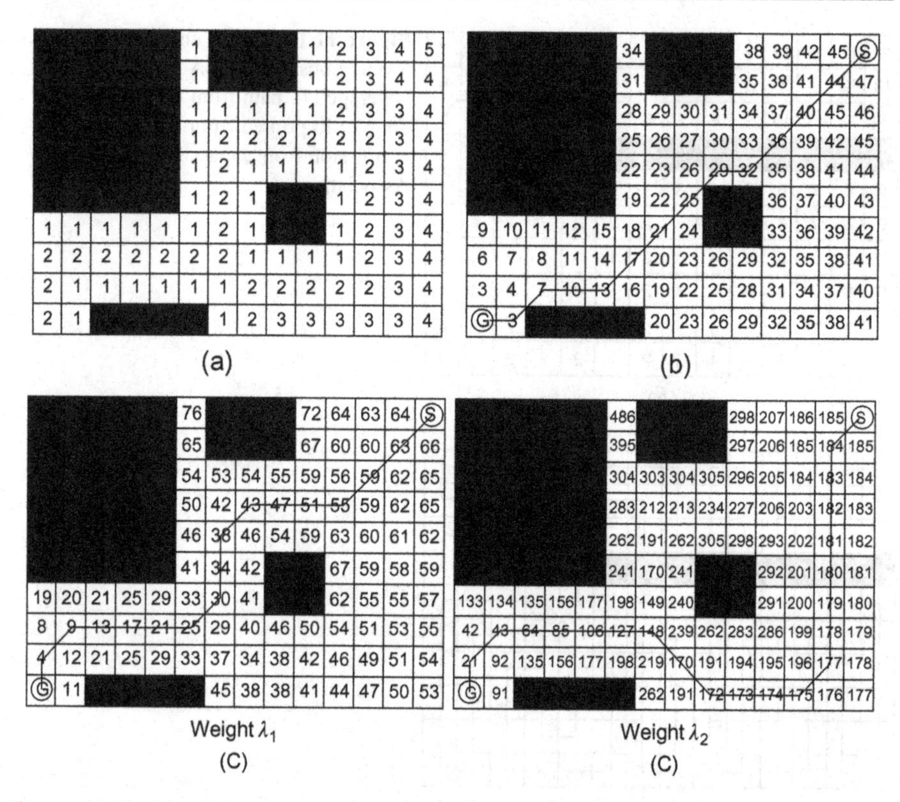

Figure 11.20 (A) DT for an environment with four obstacles. (B) Obstacle transform with the corresponding path. (C, D) Paths found with the PTPP for the same environment, where each path corresponds to the minimum propagated path cost to the goal. The path (C) corresponds to a lighter cost function than (D), that is, $\lambda_1 < \lambda_2$.
Source: Reprinted from Ref. [45], with permission from Japan Robot Association.

Another class of complete coverage path planning methods is based on the spiral algorithm as shown in Figure 11.21B [48].

If no obstacle on the right turn right
ELSE
If no obstacle in the front
Move forward
ELSE
If no obstacle on the left
Turn left
OTHERWISE
Terminate the algorithm

In this algorithm, previously covered cells and presently occupied cells are regarded as obstacles. The above basic spiral complete coverage path planning algorithm has been improved by many authors. For example, in Ref. [48], the complete coverage algorithm represents the environment as a union of robot-sized cells

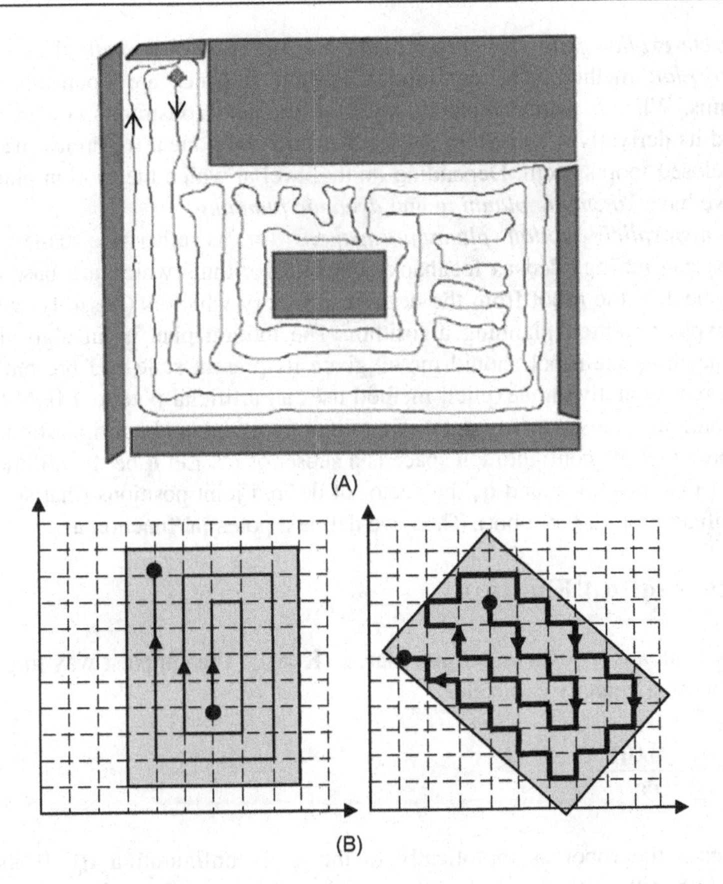

Figure 11.21 (A) A simulated complete coverage path for the Yamabico robot using PTPP with estimated position correction. (B) Spiral complete coverage path planning for initial robot orientation parallel (left) and nonparallel (right) to the wall.
Source: (A) Reprinted from Ref. [45], with permission from Japan Robot Association.

and then uses a spiral scheme. The overall path planner links the basic spiral paths using the constrained inverse DT.

11.5 Mobile Robot Motion Planning

11.5.1 General Online Method

The motion planning algorithms are distinguished into:

- Explicit
- Nonexplicit

according to the underlying path planning scheme.

Discrete explicit path planning schemes are the *road map* method and the *cell decomposition* method. The continuous explicit schemes are open-loop control algorithms. When a path compatible with the problem constraints is selected, this path and its derivatives constitute the feedforward component (reference trajectory) for the closed-loop system. Depending on the level at which the motion planning is made, we have *kinematic planning* and *dynamic planning*.

The *nonexplicit motion planning methods* (or as otherwise called, *online methods*) can be regarded as feedback control algorithms which are based on the displacement of the robot from the desired trajectory which is (possibly) generated by an explicit motion planning algorithm. The motion plan is an algorithm that determines how the robot should move, given its present state and present knowledge. A representative nonexplicit method uses an artificial potential field function in the configuration space which takes a minimum at the goal configuration.

Suppose that the configuration space is a subset of R^n. Let $\mathbf{q}$ be the n-dimensional vector of joint positions, and $\mathbf{q}_\mathrm{d}$ the vector of desired joint positions (that specify the goal configuration of the robot). Then, we define a potential function as:

$$V_0(\mathbf{q}) = (\mathbf{q}-\mathbf{q}_\mathrm{d})^\mathrm{T}\mathbf{K}(\mathbf{q}-\mathbf{q}_\mathrm{d}) \tag{11.1a}$$

where $\mathbf{K}$ is an $n \times n$ positive definite matrix ($\mathbf{K}>0$). The simplest way to go to the goal is to use a velocity controller:

$$\dot{\mathbf{q}} = -\frac{\partial V_0(\mathbf{q})}{\partial \mathbf{q}} \tag{11.1b}$$

which leads the robot asymptotically to the goal configuration $\mathbf{q}_\mathrm{d}$. If there are obstacles, then the velocity $\dot{\mathbf{q}}$ should move the robot away from them. Let $d(\mathbf{q}, E_i)$ be a function that gives the distance of the robot's point that lies in a smaller distance from the obstacle E_i. We construct a new potential function $V_{E_i}(\mathbf{q})$ as:

$$V_{E_i}(\mathbf{q}) = -k_i/d(\mathbf{q}, E_i)^r \tag{11.2}$$

where r is a suitable integer and k_i a positive constant, and instead of Eq. (11.1a) we use a total potential function:

$$V(\mathbf{q}) = V_0(\mathbf{q}) + V_{E_i}(\mathbf{q})$$

Then, the controller:

$$\dot{\mathbf{q}} = -\frac{\partial V(\mathbf{q})}{\partial \mathbf{q}} = -\frac{\partial V_0(\mathbf{q})}{\partial \mathbf{q}} - \frac{\partial V_{E_i}(\mathbf{q})}{\partial \mathbf{q}} \tag{11.3}$$

leads the robot to the goal $\mathbf{q}_\mathrm{d}$ while simultaneously moves it away from the obstacle E_i. Clearly, for each obstacle $E_i (i = 1, 2, \ldots)$, we must add a new $V_{E_i}(\mathbf{q})$.

Example 11.2

The problem is to derive the local motion planning equations for a WMR using potential fields and polar representation.

Solution

For the purposes of local motion planning, we only need the local (relative) polar coordinates depicted in Figure 11.22 [26].

These coordinates are as follows:

- Distance d_{GR} and angle ϕ_{GR} of the goal relative to the robot
- Distance d_{OR} and angle ϕ_{OR} of the obstacle relative to the robot
- Distance d_{OG} and angle ϕ_{OG} of the obstacle relative to the goal
- Angle ϕ_γ between the robot axis direction and the direction in which the robot should go to the goal
- Angle ϕ_δ between the robot-goal and the obstacle-goal direction lines

From the geometry of Figure 11.22, we find the relations:

$$d_{OR} = (d_{GR}^2 + d_{OG}^2 - 2d_{GR}d_{OG}\cos\phi_\delta)^{1/2}$$
$$\phi_\delta = (\phi_{GR} - \phi_\gamma) - (\pi - \phi_{OG})$$
$$\phi_{OR} = \phi^* \,\mathrm{sgn}(\phi_\gamma) + \phi_{GR}$$

where:

$$\cos\phi^* = \left(\frac{d_{GR}^2 + d_{OR}^2 - d_{OG}^2}{2d_{GR}d_{OR}}\right)$$

The above relations give the distance d_{OR} and angle ϕ_{OR} of the obstacle relative to the robot in terms of d_{OG} and ϕ_{OG} which are independent of the robot position and direction of motion.

Now, applying the general potential-field-based motion planning method of Section 11.5.1, we define a potential function:

$$V(\mathbf{q}) = V_{att}(\mathbf{q}) + V_{rep}(\mathbf{q})$$

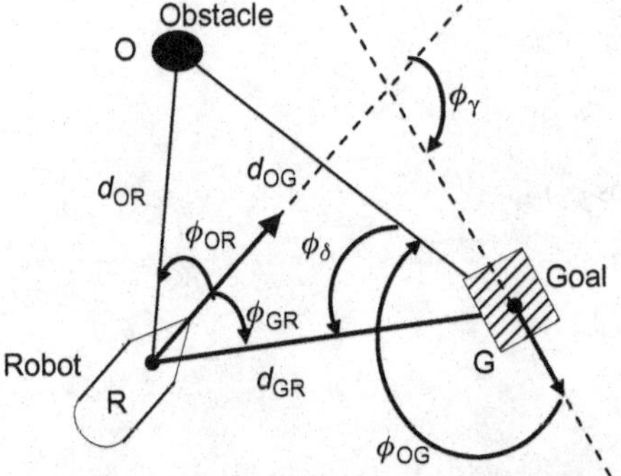

Figure 11.22 Geometry of local WMR motion planning in the presence of an obstacle.

which involves the goal-attraction term:

$$V_{att}(\mathbf{q}) = \begin{cases} \frac{1}{2}\lambda||\mathbf{q} - \mathbf{q}_{goal}||^2 = \frac{1}{2}\lambda d_{GR}^2 & \text{when} \quad d_{GR} < \varepsilon \\ \mu||\mathbf{q} - \mathbf{q}_{goal}|| = \mu d_{GR} & \text{when} \quad d_{GR} > \varepsilon \end{cases}$$

and an obstacle-repulsion term:

$$V_{rep}(\mathbf{q}) = 1/||\mathbf{q} - \mathbf{q}^*_{obstacle}|| = 1/d_{OR}$$

with λ, μ being scaling factors, $\varepsilon > 0$ being a distance from the goal, and $\mathbf{q}^*_{obstacle}$ the closest point to the obstacle. For each additional obstacle O_i, a corresponding repulsive potential field $V_{rep,i}(\mathbf{q})(i = 1, 2\ldots, M)$ should be added, where M is the number of obstacles. In this case, the total repulsive potential is equal to:

$$V_{rep}(\mathbf{q}) = \sum_{i=1}^{M} V_{rep,i}(\mathbf{q})$$

The attractive and repulsive forces applied to the robot are:

$$F_{att}(\mathbf{q}) = -\partial V_{att}(\mathbf{q})/\partial \mathbf{q}, \quad F_{rep}(\mathbf{q}) = -\partial V_{rep}(\mathbf{q})/\partial \mathbf{q}$$

And the total force is:

$$F(\mathbf{q}) = F_{att}(\mathbf{q}) + F_{rep}(\mathbf{q})$$

Clearly, since here we consider only local motion planning, we can assume that $d_{GR} < \varepsilon$ for some appropriate value of the distance ε that specifies the local area around the robot. Therefore:

$$\begin{aligned} V_{att}(d_{GR}) &= (1/2)\lambda d_{GR}^2 \\ V_{rep}(d_{OR}) &= (d_{GR}^2 + d_{OG}^2 - 2d_{GR}d_{OG}\cos\phi_\delta)^{-1/2} \\ \phi_\delta &= \phi_{GR} - (\phi_\gamma + \pi - \phi_{OG}) \end{aligned}$$

and

$$\begin{aligned} \partial V_{att}(d_{GR})/\partial d_{GR} &= \lambda d_{GR} \\ \partial V_{rep}/\partial d_{GR} &= -(V_{rep})^3(d_{GR} - d_{OG}\cos\phi_\delta) \\ \partial V_{rep}/\partial \phi_{GR} &= -(V_{rep})^3 d_{GR}d_{OG}\sin\phi_\delta \end{aligned}$$

since $\partial\phi_\delta/\partial\phi_{GR} = 1$.

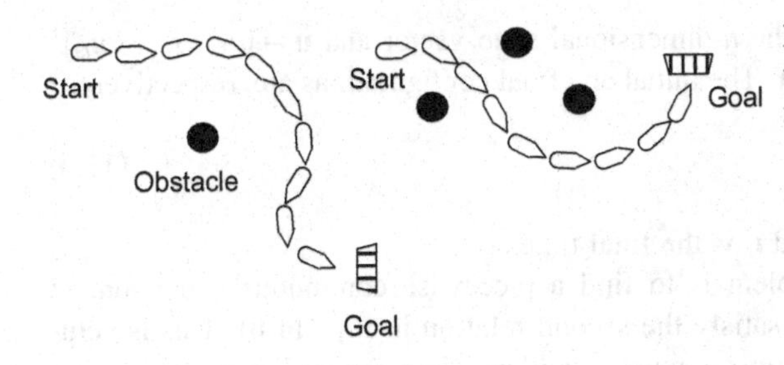

Figure 11.23 Robot paths with the above potential-based local motion planning.

These gradients are now used in the robot motion (velocity) planning controller (11.3). The result is:

$$v_{\mathrm{GR}} = \dot{d}_{\mathrm{GR}} = -\frac{\partial V_{\mathrm{att}}}{\partial d_{\mathrm{GR}}} - \frac{\partial V_{\mathrm{rep}}}{\partial d_{\mathrm{GR}}}$$

$$= -\lambda d_{\mathrm{GR}} + (V_{\mathrm{rep}})^3 (d_{\mathrm{GR}} - d_{\mathrm{OG}} \cos \phi_\delta)$$

$$\omega_{\mathrm{GR}} = \dot{\phi}_{\mathrm{GR}} = (V_{\mathrm{rep}})^3 d_{\mathrm{GR}} d_{\mathrm{OG}} \sin \phi_\delta$$

Two examples of the potential field path followed by a robot in a terrain with one or three obstacles are shown in Figure 11.23.

The potential field motion planning method enhanced with neural network learning was studied in Ref. [26].

11.5.2 Motion Planning Using Vector Fields

This method is particularly useful for the motion planning of nonholonomic WMRs. Consider the differential drive WMR of Figure 2.7 which is described by the affine kinematic model (2.37a):

$$\begin{bmatrix} \dot{x}_Q \\ \dot{y}_Q \\ \dot{\phi} \end{bmatrix} = \begin{bmatrix} (r/2)\cos\phi \\ (r/2)\sin\phi \\ r/2a \end{bmatrix} u_1 + \begin{bmatrix} (r/2)\cos\phi \\ (r/2)\sin\phi \\ -r/2a \end{bmatrix} u_2 \tag{11.4}$$

where $u_1 = \dot{\theta}_r$ and $u_2 = \dot{\theta}_l$. As we already know, although using the two inputs u_1 and u_2 in Eq. (11.4), the robot can go to any desired point of the terrain, and due to the nonholonomicity, the vehicle cannot follow all the possible trajectories on the plane.

Any motion plan (program) should include the constraint Eq. (11.4). A solution to this is as follows [23].

Let a robot be described by:

$$\dot{\mathbf{x}} = \mathbf{f}_1(\mathbf{x})u_1 + \mathbf{f}_2(\mathbf{x})u_2 + \cdots + \mathbf{f}_m(\mathbf{x})u_m \tag{11.5}$$

where $\mathbf{x}=[x_1,x_2,\ldots,x_n]^{\mathrm{T}}$ is the n-dimensional state vector and $\mathbf{u}=[u_1,u_2,\ldots,u_m]^{\mathrm{T}}$ the m-dimensional input vector. The initial and final configurations are, respectively:

$$\mathbf{x}(t_0)=\mathbf{x}_0, \quad \mathbf{x}(t_\mathrm{f})=\mathbf{x}_\mathrm{f} \tag{11.6}$$

where t_0 is the initial time and t_f is the final time.

The *motion planning* problem is to find a piecewise continuous and bounded input vector $\mathbf{u}(t)$ such that to satisfy the second relation in Eq. (11.6). This is actually an *open-loop final state (pose) control problem.* Here, we will examine it using the vector-field and Lie bracket theory. The Lie bracket of the vector fields $\mathbf{f}_i$ and $\mathbf{f}_j(i,j=1,2,\ldots,m)$ in the configuration space CS is defined as in Eq. (6.12), that is:

$$[\mathbf{f}_i,\mathbf{f}_j](\mathbf{x})=\frac{\partial \mathbf{f}_j}{\partial \mathbf{x}}\mathbf{f}_i-\frac{\partial \mathbf{f}_i}{\partial \mathbf{x}}\mathbf{f}_j \tag{11.7}$$

It is an antisymmetric operator that returns a vector field and provides a measure of how the flows $\partial \mathbf{f}/\partial \mathbf{x}$ that correspond to the vector fields $\mathbf{f}_i$ and $\mathbf{f}_j$ are interchanged. Actually, the Lie bracket represents the infinitesimal motions that result when we have a forward flow by $\mathbf{f}_i$ and then by $\mathbf{f}_j$ and after that a backward flow by $-\mathbf{f}_i$ and then by $-\mathbf{f}_j$. Clearly, if this input sequence is applied to a linear system, it will lead to a zero net motion. But in nonlinear systems, we can generate new motion directions by simply transposing vector fields through the Lie bracket (see Figure 6.3).

This can be done by applying the Lie bracket (operator) to each of the new motion directions and each of the directions that has been used in their generation.

Let $\boldsymbol{\Delta}_0$ be the *span* distribution of the initial input vector fields $\mathbf{f}_1,\mathbf{f}_2,\ldots,\mathbf{f}_m$. We denote the sequential application of the Lie bracket to the input vector fields as:

$$\boldsymbol{\Delta}_i=\boldsymbol{\Delta}_{i-1}+\mathrm{span}\{[\alpha,\beta],\ \alpha\in\boldsymbol{\Delta}_0,\ \beta\in\boldsymbol{\Delta}_{i-1}\}$$

The criterion for checking controllability is the following (see also Theorem 6.9):

"The system is controllable if and only if we have $\boldsymbol{\Delta}_k=R^n$ for some k." Applying this criterion to our differential drive WMR Eq. (11.4), we have:

$$\mathbf{f}_1=\begin{bmatrix}(r/2)\cos\phi\\(r/2)\sin\phi\\r/2a\end{bmatrix},\quad \mathbf{f}_2=\begin{bmatrix}(r/2)\cos\phi\\(r/2)\sin\phi\\-r/2a\end{bmatrix} \tag{11.8}$$

The Lie bracket of $\mathbf{f}_1$ and $\mathbf{f}_2$ is:

$$\mathbf{f}_3=[\mathbf{f}_1,\mathbf{f}_2]=\begin{bmatrix}-(r^2/2a)\sin\phi\\(r^2/2a)\cos\phi\\0\end{bmatrix} \tag{11.9}$$

Because these vector fields cover all the allowable motion directions on the plane, the system is *controllable*. This criterion can be used for the generation of desired motions (i.e., of successfully controlling robots with nonholonomic constraints along a motion plan). Indeed, suppose that we want to find a motion for a Lie bracket (system) $[\mathbf{f}_i, \mathbf{f}_j]$ of first order (Eq. (11.9)). This can be done by using high-frequency sinusoid inputs of the form [23,49].

$$u_i = \xi_i(t) + \sqrt{\omega}\xi_{ij}(t)\sin \omega t \tag{11.10a}$$

$$u_j = \xi_j(t) + \sqrt{\omega}\xi_{ji}(t)\cos \omega t \tag{11.10b}$$

In the limit $\omega \to \infty$, we get the motion:

$$\dot{\mathbf{x}} = \mathbf{f}_i(\mathbf{x})\xi_i(t) + \mathbf{f}_j(\mathbf{x})\xi_j(t) + \frac{1}{2}\xi_{ij}(t)\xi_{ji}(t)\left[\mathbf{f}_i, \mathbf{f}_j\right](\mathbf{x}) \tag{11.11}$$

This means that using the above u_i and u_j allows the robot to follow the direction of the Lie bracket $[\mathbf{f}_i, \mathbf{f}_j]$ as if it was one of the initial controlled directions.

11.5.3 Analytic Motion Planning

The analytic obstacle avoidance motion planning approach consists in describing the WMR trajectory parametrically and determining the optimal values of the parameters used [13]. Here, we will use sinusoid parameterization. Let $\mathbf{x}_0$ be an initial configuration of the mobile robot M that moves on the configuration space $\mathrm{CS} = R^2$, which involves a number of stationary obstacles $B_i (i = 1, 2, \ldots, n)$. The problem is to find a trajectory that determines a sequence of configurations of M from $\mathbf{x}_0$ to a desired configuration (pose) $\mathbf{x}_\mathrm{d}$ such that the robot avoids collisions with the obstacles B_i.

To solve this problem, we express the plane trajectory by the Cartesian parametric equations $x = x(t)$ and $y = y(t)$ which represent the position of the robot M as a function of the parameters from $t = 0$ to $t = t_\mathrm{f} = 1$ at the end of the trajectory. Such a curve must be sufficiently smooth to assure the continuity of the curvature $\kappa(t)$, which is defined as:

$$\kappa(t) = \frac{\mathrm{d}\phi}{\mathrm{d}s} = \frac{\mathrm{d}\phi/dt}{\mathrm{d}s/dt} = \frac{\dot{\phi}(t)}{\sqrt{\dot{x}^2 + \dot{y}^2}} \tag{11.12a}$$

where ϕ is the tangential angle and s is the curve length. The derivative $\dot{\phi}(t)$ can be found using the relation:

$$tg\ \phi(t) = \frac{\mathrm{d}y}{\mathrm{d}x} = \frac{\mathrm{d}y/dt}{\mathrm{d}x/dt} = \frac{\dot{y}(t)}{\dot{x}(t)} \tag{11.12b}$$

From Eq. (11.12b), we find:

$$\begin{aligned}\frac{\mathrm{d}}{\mathrm{d}t} tg\, \phi(t) &= \frac{1}{\cos^2 \phi(t)} \dot{\phi}(t) \\ &= (1 + tg^2\, \phi(t))\dot{\phi}(t) \\ &= (1 + \dot{y}^2/\dot{x}^2)\dot{\phi}(t)\end{aligned}$$

that is:

$$\begin{aligned}\dot{\phi}(t) &= \left(\frac{\dot{x}^2}{\dot{x}^2 + \dot{y}^2}\right) \frac{\mathrm{d}}{\mathrm{d}t} tg\, \phi(t) \\ &= \left(\frac{\dot{x}^2}{\dot{x}^2 + \dot{y}^2}\right) \frac{\mathrm{d}}{\mathrm{d}t} \left(\frac{\dot{y}(t)}{\dot{x}(t)}\right) \\ &= \frac{\dot{x}\ddot{y} - \dot{y}\ddot{x}}{\dot{x}^2 + \dot{y}^2}\end{aligned} \tag{11.13}$$

Therefore, Eq. (11.12a) gives:

$$\kappa(t) = \frac{\dot{x}(t)\ddot{y}(t) - \dot{y}(t)\ddot{x}(t)}{[\dot{x}^2(t) + \dot{y}^2(t)]^{3/2}} \tag{11.14}$$

It follows that for $\kappa(t)$ to be continuous, the functions $x(t)$ and $y(t)$ must be differentiable at least up to order 3.

Some typical forms of parametric curves $\{x(t), y(t)\}$ used for path/motion planning are as follows:

Cartesian polynomials

$$x(t) = \sum_{i=0}^{n} a_i t^i, \quad y(t) = \sum_{i=0}^{n} b_i t^i \quad (a_i, b_i \in R)$$

Cubic spirals

$$\dot{\phi}(t) = \frac{1}{2}At^2 + Bt + C = \kappa(t)$$

$$\phi(t) = \frac{1}{6}At^3 + \frac{1}{2}Bt^2 + Ct + D$$

This class minimizes the integral of the square of the curvature's derivative.

Finite sums of sinusoids

$$x(t) = \sum_{i=1}^{n}[A_i \cos(i\omega t) + B_i \sin(i\omega t)] \tag{11.15a}$$

$$y(t) = \sum_{i=1}^{n}[C_i \cos(i\omega t) + D_i \sin(i\omega t)] \tag{11.15b}$$

For $n = \infty$, these functions can approximate any continuous function.

The optimization criterion is typically the length of the curve $c(t) = c(x(t), y(t))$ which is given by:

$$J = \int_0^1 [\dot{x}^2(t) + \dot{y}^2(t)]dt \tag{11.16}$$

The *optimal trajectory* minimizes J. Any trajectory that satisfies the inequality $J < J_{max}$, where J_{max} is a predetermined (allowed) upper bound, is defined as *sub-optimal trajectory*.

In the following, we will assume a given number n of sinusoids and a given frequency ω. Thus, the parameter vector $\boldsymbol{\theta}$ to be selected has dimensionality $4n$:

$$\boldsymbol{\theta} = [A_1, B_1, C_1, D_1, A_2, B_2, C_2, D_2, A_3, B_3, C_3, D_3, \ldots, A_n, B_n, C_n, D_n]^{\mathrm{T}} \tag{11.17}$$

Differentiating Eqs. (11.15a) and (11.15b), we get:

$$\dot{x}(t) = \sum_{i=1}^{n}[-(i\omega)A_i \sin(i\omega t) + (i\omega) B_i \cos(i\omega t)] \tag{11.18a}$$

$$\dot{y}(t) = \sum_{i=1}^{n}[-(i\omega)C_i \sin(i\omega t) + (i\omega)D_i \cos(i\omega t)] \tag{11.18b}$$

Therefore, the above quadratic function J is a square function of the components of $\boldsymbol{\theta}$, independent of the parameter t.

Now, suppose that the robot is to go from the initial position $x(0) = 0$, $y(0) = 0$ with orientation ϕ_0, to the goal (final) position $x(1) = x_{\mathrm{f}}$, $y(1) = y_{\mathrm{f}}$ with orientation ϕ_{f}. The corresponding conditions for A_i, B_i, C_i, and D_i are found by replacing the above initial and final conditions in Eqs. (11.15a), (11.15b), (11.18a), and (11.18b), namely:

$$x(0) = 0 \text{ gives } \sum_{i=1}^{n} A_i = 0$$

$$y(0) = 0 \text{ gives } \sum_{i=1}^{n} C_i = 0$$

$$x(1) = x_f \text{ gives } \sum_{i=1}^{n} [A_i \cos(i\omega) + B_i \sin(i\omega)] = x_f$$

$$y(1) = y_f \text{ gives } \sum_{i=1}^{n} [C_i \cos(i\omega) + D_i \sin(i\omega)] = y_f$$

$$\frac{\dot{y}(0)}{\dot{x}(0)} = tg\ \phi_0 \text{ gives } \frac{\sum_{i=1}^{n} [i\omega D_i]}{\sum_{i=1}^{n} [i\omega B_i]} = tg\ \phi_0,$$

$$\frac{\dot{y}(1)}{\dot{x}(1)} = tg\ \phi_f \text{ gives } \frac{\sum_{i=1}^{n} [-i\omega C_i \sin(i\omega) + i\omega D_i \cos(i\omega)]}{\sum_{i=1}^{n} [-i\omega A_i \sin(i\omega) + i\omega B_i \cos(i\omega)]} = tg\ \phi_f$$

The obstacles that exist in the task space can also be represented by parametric curves. For example, an object with the shape of an ellipse is described by:

$$(x(t) - x_*)^2/a + (y(t) - y_*)^2/b = 1 \tag{11.19}$$

An edge of a polygonic object is described by a straight line segment:

$$y(t) = ax(t) + b$$

Therefore, one way to guarantee obstacle avoidance is to assure that the path and objects parametric curves do not intersect. This can be done if we equate the expression of $c(t)$ with any boundary of the obstacle (which is another parametric function with parameter s) and solve the resulting algebraic system with respect to t and s. The two curves intersect if:

$$0 \le t \le 1 \quad \text{and} \quad 0 \le s \le 1$$

Another method is to consider the constrained minimization of the function J with the constraint:

$$g(\theta) \le 0 \tag{11.20}$$

that expresses the obstacle-free task space. For example, for the elliptic obstacle, this obstacle avoidance constraint is:

$$[x(t) - x_*]^2/a + [y(t) - y_*]^2/b - 1 \le 0 \tag{11.21}$$

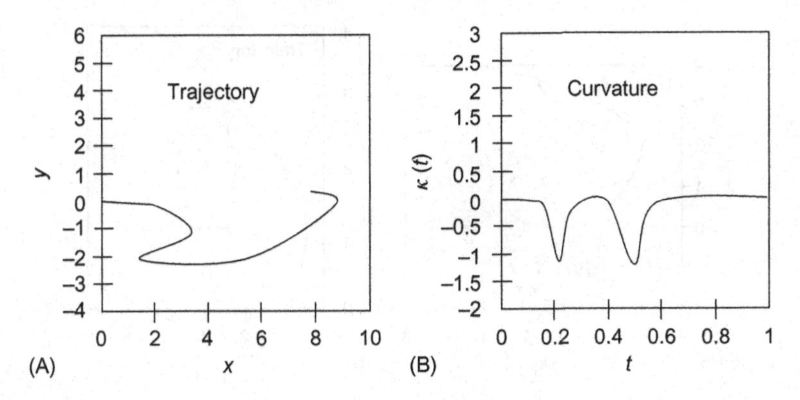

Figure 11.24 Obstacle-free case ($n = 2$, $\omega = \pi$). (A) Trajectory and (B) curvature.

The difficulty of this technique is that the constraints (11.20) and (11.21) depend on the parameter t, whereas the optimization problem requires their expression in terms of $\boldsymbol{\theta}$. One way to face this difficulty is to apply Eq. (11.21) for a certain number of values (e.g., equidistant values) of t. This is done in the simulation results that follow, where the optimization was performed with the aid of Matlab's optimization toolbox (function "constr"). The input data used are:

$$(x_0, y_0) = (0, 0), \quad \phi_0 = 0, \quad (x_f, y_f), \quad \phi_f, \quad \omega = \pi$$

Case 1 *Obstacle-free space with* $n = 2$, $\omega = \pi$
Initial parameter value $\boldsymbol{\theta}_0 = [0\,0\,0\,0\,0\,0\,0\,0]^T$
Final position/orientation $x_f = 9$, $y_f = 1$, $\phi_f = 3\pi/4$
The optimal value of J obtained is $J^0 = 14.28$. The resulting trajectory $\{x(t), y(t)\}$ and curvature $\kappa(t)$ are shown in Figure 11.24 [13].
We see that the curvature does not take large values. Thus, this trajectory may be appropriate for a WMR with nonholonomic or hard constraints (e.g., constraints on the steering angle).

Case 2 *Obstacle nonfree space with* $n = 2$, $\omega = \pi$
Here:

$$(x_0, y_0) = (0, 0), \quad (x_f, y_f) = (8, 4), \quad \phi_0 = \phi_f = 0$$

The initial parameter value is again $\boldsymbol{\theta}_0 = [0\,0\,0\,0\,0\,0\,0\,0]^T$.
The resulting trajectories for an elliptic and hexagonal obstacle are shown in Figure 11.25.

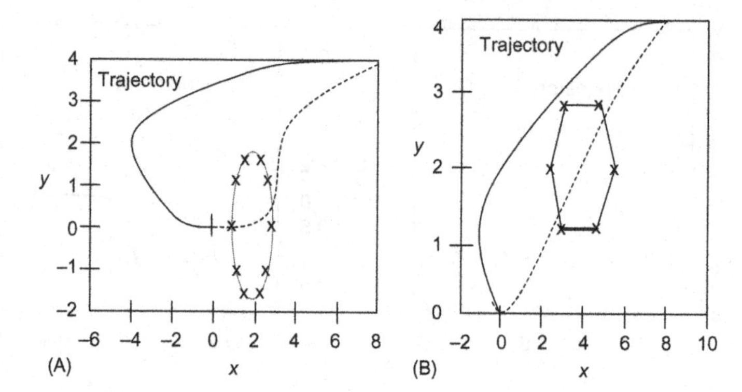

Figure 11.25 Path generation for obstacle avoidance. (A) Elliptic obstacle and (B) hexagonal obstacle.

The trajectory length for the elliptic obstacle is $L = 10.37$ and for the hexagonal $L = 8.75$. In both cases, the obstacle-avoiding path is shown by the continuous line, while the obstacle-free path is shown by the dotted line.

Example 11.3

Examine the motion planning problem in the joint space for a single joint of a fixed or mobile robot to go from an initial angular position θ_0 at time $t = 0$, to a final position θ_f at $t = t_f$. No obstacle is present in the robot's configuration space.

Solution

Each joint of the robot is driven (actuated) by a motor (see Example 5.1). The simplest way to plan the motion is to consider that the acceleration and deceleration periods of the motor are equal. We assume that at $t = 0$, the motor is stationing (zero velocity and zero acceleration). At time $t = 0^+$, we apply a constant acceleration $+a_0$, and we assume that the maximum velocity allowed is v_{max}. When reaching the maximum velocity at time t_1, the motor is moving with constant speed up to time $t_f - t_1$, and then decelerates with deceleration $-a_0$. During the acceleration period $[0, t_1 = 2T]$, we have:

$$\theta(t) = \frac{1}{2}a_0 t^2, \quad \theta_1 = \theta(t_1) = \frac{1}{2}a_0 t_1^2, \quad v_{max} = a t_1$$

During the constant velocity period $[t_1, t_2 = t_f - t_1]$, we have:

$$\theta(t) = \theta_1 + v_{max}(t - t_1), \quad \theta_2 = \theta(t_2) = \theta_1 + v_{max}(t_2 - t_1)$$

Finally, during the deceleration period $[t_2, t_f]$ we have:

$$\theta(t) = \theta_2 + v_{max}(t - t_2) - \frac{1}{2}a_0(t - t_2)^2$$

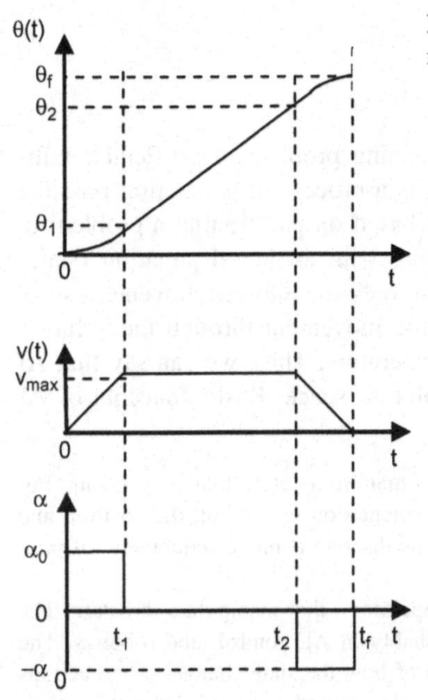

Figure 11.26 Typical plots for the point-to-point motion planning case.

The final position $\theta_f = \theta(t_f)$ is equal to:

$$\theta_f = \theta_2 + v_{max}(t_f - t_2) - \frac{1}{2}a_0(t_f - t_2)^2$$

$$= \theta_2 + v_{max}t_1 - \frac{1}{2}a_0 t_1^2 = \theta_2 - \theta_1 + v_{max}t_1$$

$$= v_{max}(t_2 - t_1) + v_{max}t_1 = v_{max}t_2$$

The time t_1 (switching from constant acceleration $+a_0$ to zero acceleration) and the time t_2 (switching from zero acceleration to deceleration $-a_0$) are given by:

$$t_1 = v_{max}/a_0 \quad \text{and} \quad t_2 = \theta_f/v_{max} \tag{11.22}$$

The total time of motion is equal to:

$$t_f = t_1 + t_2 = \frac{v_{max}}{a_0} + \frac{\theta_f}{v_{max}} \tag{11.23}$$

Given a_0, v_{max}, and θ_f, the switching times t_1 and t_2 can be automatically computed using the above formulas (11.22). Figure 11.26 shows the acceleration, velocity, and position plots for the constant acceleration/deceleration case.

11.6 Mobile Robot Task Planning

11.6.1 General Issues

Robot task planning belongs to the general planning problem of artificial intelligence, called *AI planning* [6–12]. AI planning is a process of generating possible multistage solutions to a specific problem. It is based on partitioning a problem in smaller, simpler, quasi-independent steps starting with an initial situation (state) and reaching a specified goal state by performing only the allowed movement steps along the solution path. This is equivalent to the movement through the solution space using the permitted state transformation operators. Thus, we can say that AI planning is a specific search process in the solution space. Basic concepts of AI planning are the following:

- *State space*: A space that involves all possible situations (states) that may occur. For example, a state could represent the position and orientation of a robot, the position and velocity of a car, and so on. The state space may be discrete (finite or countably infinite) or continuous (uncountably infinite).
- *Actions or operators*: A plan generates actions (operators) that manipulate the state. The terms actions and operators are used indistinguishably in AI, control, and robotics. The planning formulation must include a specification of how the state changes when actions (or controls) are applied. In discrete time, this may be defined as a state-valued function, and in continuous time as an ordinary differential equation. In some cases, actions could be selected by nature and are not under the control of the decision maker.
- *Initial and goal states*: A planning problem must involve some initial state, and specify the goal state in a set of goal states. The actions must be selected so as to drive the system from the initial state to the desired goal state.
- *Criterion*: This encodes the desired outcome of a plan in terms of the state and actions that are performed. The two basic criteria used in planning are *feasibility* (i.e., find a plan that causes the arrival at a goal state, regardless of its efficiency) and *optimality* (i.e., find a feasible plan that optimizes the system performance in some desired way, in addition to arriving at a goal state). Achieving optimality is much more challenging than simply assuring feasibility.
- *Plan*: A plan imposes a specific strategy or behavior on a decision maker. It may simply specify a sequence of actions to be taken, independently of how complicated this sequence is.
- *Algorithm*: This is difficult to be precisely and uniquely defined. In theoretical computer science, algorithm is a *Turing machine* (i.e., a finite-state machine with a special head that can read and write along an infinite piece of tape). Using the Turing machine as a model for algorithms implies that the physical world must be first properly modeled and written on tape before the algorithm can make decisions. But if changes take place in the world during the execution of the algorithm, it is not clear what may happen (e.g., in case a mobile robot is moving in a cluttered environment where humans are walking around). Thus, an online algorithm model is more appropriate in these cases (which relies on special sensory information).
- *Planner*: A planner simply produces a plan and may be a machine or a human. If the planner is a machine, it is generally considered to be a planning algorithm (a Turing machine or an online program). In many cases, the planners are humans who develop

plan(s) that can work in all situations. The three ways of employing a generated plan are: *execution* by a machine (e.g., a robot) or a simulator, (ii) *refinement* (i.e., improvement to a better plan), and (iii) *hierarchical inclusion* (i.e., packaging it as an action in a higher level plan).

Representative examples of robot task planning problems are the following:

- Automotive and other assembly planning
- Sealing cracks in automotive assembly
- Parking cars and trailers
- Mobile robot navigation
- Mobile manipulator service planning

Questions that have to be addressed in task planning are: What is a plan? How is a plan represented? How is it computed? What is expected to achieve? How is its quality evaluated? and Who or what is going to use it? Answers to these questions can be found in AI textbooks.

11.6.2 *Plan Representation and Generation*

11.6.2.1 *Plan Representation*

Plans can be represented and generated as:

- State-space problems
- Action-ordering problems

State-Space Representation

It is very familiar in systems theory and automata theory. When applied to *plan representation,* the state vectors represent the states of the application domain of the plan and the actions or operators. The state-space representation is the best way to recognize a solution of the problem at hand. In addition, state-space representations can be easily converted to equivalent computer programs, and are directly usable when searching algorithms are applied to problem solving. In this case, the solution to a problem in state space provides a path that connects the initial and goal state in a feasible way.

Clearly, the planning process in the state space is actually a search process, driven by the requirement of achieving a goal (i.e., it is essentially a goal-driven search process). Nodes of search trees can be viewed as possible states of the world presented as some possible partial plans, and the goal pursuing process, as a partial-plan search process.

Action-Ordering Representation

This representation involves a number of actions in the order in which they should be executed, along with the associated restrictions to be compiled. For example, it might be a simple list of actions under execution in the order they appear on the list. The action-ordering representation focuses on actions to be performed and not on conditions that could be affected by the individual actions. It is therefore a plan representation very different than the state-space representation. Actually, no states outside the list are used for problem representation. Essentially, action-ordering representation is a behavioral representation, because it contains no explicit notion of state, although one can associate with each action-ordering plan a certain number of possible states to be generated by the actions. Two major advantages of action-ordering plan representation are as follows:

1. Capability to represent parallel activities, which is very important when the action-ordering list is only partially ordered leaving room for reordering actions.
2. Easy implementation in a computer (e.g., as a directed graph structure), where the nodes represent the actions and the arcs the corresponding ordering relations.

Partially ordered plans allow parallel actions, and are called *parallel plans*. Totally, ordered plans require sequential actions, and are called *serial plans*. Serial plans are equivalent to state-space transition diagrams with single-operator labeled arcs. Alternative methods for formulating robotic planning problems are *predicate calculus*, *tick lists*, and *triangle tables*.

11.6.2.2 Plan Generation

As we saw before, planning is a problem of performing a search through a state space. At each planning stage, several computational steps are required before its execution. The computations provide a selection guide for the next move in the search space, that is, for the path construction that leads from the initial state to the goal state. Each plan node in the state space involves partial-plan information, and so the full set of nodes between the initial and the goal state represents the total plan. By searching through possible plan states, the planner *generates the plan*. A basic issue in planning is to select a suitable modeling method or knowledge representation describing the task domain. Some methods commonly used are as follows:

- Production rules (forward/backward)
- Predicate logic
- Procedural nets
- Object or schema-based methods
- AND/OR graphs
- Petri nets

Usually, if the problem is too complicated and the state space to be searched is very large, the total planning problem is decomposed into a number of simple subprocesses that can be performed separately. When solving a planning problem, the planning system must take into account the existing constraints which must be properly formulated, propagated, and satisfied.

Constraint formulation helps in specifying the interactions among individual solutions to different goals. *Constraint propagation* creates new constraints out of old ones and helps to refine the solution path in screening its nonpromising links. Finally, *constraint satisfaction* must be assured by the generated plan.

The least-commitment method helps to refine only those parts of the plan that will not be abandoned later.

The plan generation methods so far discussed produce plans that are linear sequences of complete subplans and are known as *linear planning*. On the contrary, *nonlinear planners* can generate plans whose individual subplans, needed to achieve the given conjunctive goals, are generated in parallel.

11.6.3 World Modeling, Task Specification, and Robot Program Synthesis

We now discuss a little more the three phases of robot task planning, namely, *world modeling*, *task specification*, and *program synthesis*.

11.6.3.1 World Modeling

World modeling is concerned with the geometric and physical description of the objects in the workspace (including the robot itself) and the representation of the assembly state of the objects. A geometric model provides the spatial information (dimensions, volume, shape, etc.) of the object in the workspace. The typical method used for modeling 3D objects is the *constructive solid geometry* (CSG) where the objects are defined as combinations or constructions via regularized set operations (union, intersection, etc.) of primitive objects (cube, parallelpiped, cylinder). The primitives can be represented in many ways, such as:

- A set of points and edges
- A set of surfaces
- Generalized cylinders
- Cell decomposition
- A procedure name that calls to other procedures representing other objects.

In the *AUTOPASS* assembly planner, the world state is represented by a graph, the nodes of which represent objects and edges represent relationships, such as [6]:

- *Attachment* (rigid, nonrigid, or conditional attachment of objects)
- *Constraints* (relationships that represent physical constraints, translational or rotational, between objects)
- *Assembly component* (which indicates that the subgraph linked by this edge is an assembly part that can be referenced as an object).

11.6.3.2 Task Specification

This can be done using a high-level specification language. For example, an assembly task can be described as a sequence of states of the world model. The states must be provided by the configurations of all the objects in the workspace (as it will be done later in Example 11.4). Another way of task description is to use a sequence of symbolic operations on the objects, including proper spatial constraints to eliminate possible ambiguities. Most robot-oriented languages use this type of task specification. AUTOPASS uses this kind of specification, but it uses a more detailed syntax, which divides its assembly related statements into three groups:

1. *State change statement* (i.e., an assembly operation description)
2. *Tools statement* (i.e., description of the kind of tools that must be used)
3. *Fastener statement* (i.e., description of fastening operation)

11.6.3.3 Robot Program Synthesis

This is the most difficult and important phase of task planning. The major steps in this phase are as follows:

- Grasp planning
- Motion planning
- Plan checking

Grasp planning specifies how the way the object is grasped affects all subsequent operations. The way the robot can grasp an object depends on the geometry of the object being grasped and the other objects present in the workspace. To be useful, a grasping configuration must be feasible and stable. The robot must be able to reach the object without any collision with other objects in the workspace. Also, the object once grasped must be stable during subsequent operations of the robot.

Motion planning specifies how the robot must move the object to its destination and complete the operation. The steps for successful motion planning are as follows:

- *Guarded departure* from the current configuration
- *Collision-free motion* to the desired configuration

Plan checking is made to ensure that the plan accomplishes the desired sequence of tasks, and if not to attempt an alternative plan.

Example 11.4 *(Blocks world problem)*

We are given three blocks A, B, and C in the initial configuration shown in Figure 11.27A. The problem is for the robotic manipulator to move the blocks to the final configuration shown in Figure 11.27B through the application of the following forward production rules (F-rules):

1. pick up (X)
2. put down (X)

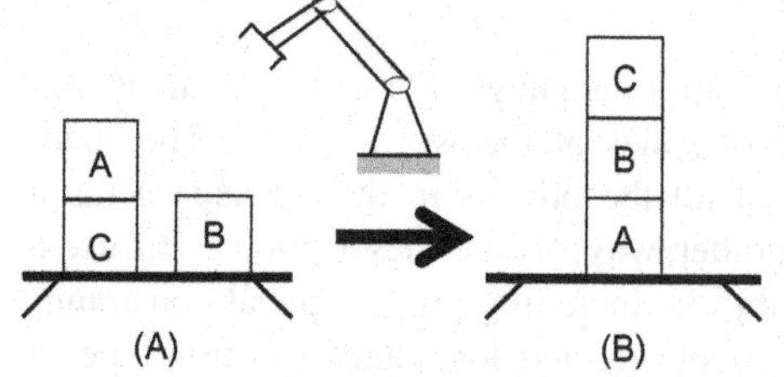

Figure 11.27 A "blocks world" problem. (A) Initial state and (B) final state.

3. stack (X,Y)
4. unstack (X,Y)

that represent possible robot actions.

Solution
The initial block configuration (arrangement) can be represented by the conjunction of the following statements:

clear (B)	Block B has a clear top
clear (A)	Block A has a clear top
on (A,C)	Block A is on block C
on table (B)	Block B is on the table
on table (C)	Block C is on the table
handempty	The robot hand is empty

Robot actions change one state, or arrangement, of the world to another. Given that the final state is to be achieved using the above four F-rules, the triangle table representation of the planning process is as shown in Figure 11.28.

Each column of the table describes the new situation after an operator (action) (e.g., *unstuck, put down*) has been carried out, while *handempty* and *handholding* mean that the

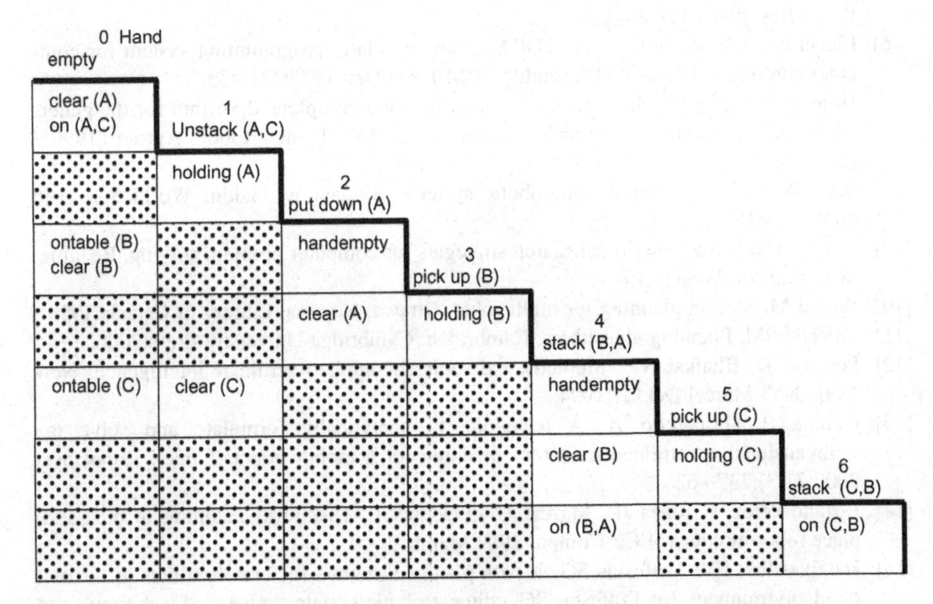

Figure 11.28 The triangle table representation of the planning process that solves the problem of Figure 11.15 (the numbers above the columns indicate the sequence of robot actions).

robot hand is empty or it is holding a block, respectively. Each column represents the preconditions of the next F-rule to be applied.

In our case, the *plan sequence* of F-rules is the following:

- *handempty*
- *unstack* (A,C)
- *put down* (A)
- *pick up* (B)
- *stack* (B,A)
- *pick up* (C)
- *stack* (C,B)

with the preconditions shown in the previous (left) column of each F-rule being applied.

References

[1] Latombe JC. Robot motion planning. Boston, MA: Kluwer; 1991.

[2] Lozano-Perez T. Spatial planning: a configuration space approach. IEEE Trans Comput 1983;32(2):108–20.

[3] Erdmann M, Lozano-Perez T. On multiple moving obstacles. Algorithmica 1987;2 (4):477–521.

[4] Fugimura K. Motion planning in dynamic environments. Berlin/Tokyo: Springer; 1991.

[5] Khatib O. Real-time obstacle avoidance for manipulators and mobile robots. Int J Robot Res 1986;5(1):90–8.

[6] Lieberman LL, Wesley M. AUTOPASS: an automatic programming system for computer controlled mechanical assembly. IBM J Res Dev 1977;321–33.

[7] Homemde Mello LS, Sanderson AC. A correct and complete algorithm for the generation of mechanical assembly sequences. IEEE Trans Robot Autom 1990;7 (2):228–40.

[8] Sheu PCY, Xue Q. Intelligent robotic systems. Singapore/London: World Scientific Publishers;1993.

[9] Pearl J. Heuristics: intelligent search strategies for computer problem solving. Reading, MA: Addison-Wesley;1984.

[10] Bonert M. Motion planning for multi-robot. Ottawa: National Library of Canada;1999.

[11] LaValle SM. Planning algorithms. Cambridge: Cambridge University Press;2006.

[12] Popovic D, Bhatkar VP. Methods and tools for applied artificial intelligence. New York, NY: Marcel Dekker;1994.

[13] Gallina P, Gasparetto A. A technique to analytically formulate and solve the 2-dimensional constrained trajectory planning for a mobile robot. J Intell Robot Syst 2000;27(3):237–62.

[14] Lozano-Pérez T, Jones JL, Mazers E, O'Donnel P. Task-level planning of pick-and-place robot motions. IEEE Comput 1989;March:21–9.

[15] Hatzivasiliou FV, Tzafestas SG. A path planning method for mobile robots in a structured environment. In: Tzafestas SG, editor. Robotic systems: advanced techniques and applications. Dordrecht/Boston: Kluwer;1992.

[16] Kant K, Zuckler S. Toward efficient trajectory planning: the path velocity decomposition. Int J Robot Res 1986;5:72–89.

[17] Canny J. The complexity of robot motion planning. Cambridge, MA: MIT Press;1988.
[18] Garcia E, De Santos PG. Mobile-robot navigation with complete coverage of unstructured environment. Robot Auton Syst 2004;46:195−204.
[19] Russel S, Norwig P. Artificial intelligence: a modern approach. Upper Saddle River, NJ: Prentice Hall;2003.
[20] Stentz A. Optimal and efficient path planning for partially known environments. Proceedings of IEEE conference on robotics and automation. San Diego, CA; May 1994. p. 3310−17.
[21] Amato N. Randomized motion planning, Part 1, roadmap methods, course notes. University of Padova;2004.
[22] Welzl E. Constructing the visibility graph for n line segments in $O(n^2)$ time. Inf Process Lett 1985;20:161−71.
[23] Kumar V, Zefran M, Ostrowski J. Motion planning and control of robots.In:Nof S. editor.Handbook of Industrial Robotics. New York: Wiley and Sons:1999, p.295−315.
[24] Wang Y, Chirikjian GS. A new potential field method for robot path planning. Proceedings of 2000 IEEE conference robotics and automation. San Francisco, CA; April 2000. p. 977−82.
[25] Pimenta LCA. Robot navigation based on electrostatic field computation. IEEE Trans Magn 2006;42(4):1459−62.
[26] Engedy I, Horvath G. Artificial neural network based local motion planning of a wheeled mobile robot. Proceedings of 11th international symposium on computational intelligence and informatics. Budapest, Hungary; November 2010. p. 213−18.
[27] Safadi H. Local path planning using virtual potential field. Report COMP 765: spatial representation and mobile robotics—project. School of Computer Science, McGill University, Canada; April 2007.
[28] Ding FG, Jiang P, Bian XQ, Wang HJ. AUV local path planning based on virtual potential field. Proceedings of IEEE international conference on mechatronics and automation. Niagara Falls, Canada; 2005. 4, p. 1711−16.
[29] Koren Y, Borenstein J. Potential field methods and their inherent limitations for mobile robot navigation. Proceedings of IEEE conference on robotics and automation. Sacramento, CA; April 2005. p. 1398−1404.
[30] Borenstein J, Koren Y. The vector field histogram: fast obstacle avoidance for mobile robots. IEEE J Robot Autom 1991;7(3):278−88.
[31] Ulrich I. Borenstein journal of VFH*: local obstacle avoidance with look-ahead verification. Proceedings of IEEE international conference on robotics and automation. San Francisco, CA; May 2000.
[32] Wang LC, Yong LS, Ang Jr MR. Hybrid of global path planning and local navigation implemented on a mobile robot in indoor environment. Proceedings of IEEE international symposium on intelligent control. October 2002. p. 821−26.
[33] Garrido S, Moreno L, Blanco D, Jurewicz P. Path planning for mobile robot navigation using Voronoi diagram and fast marching. Int J Robot Autom 2011;2(1):42−64.
[34] Sethian JA. Theory, algorithms, and applications of level set methods for propagating interfaces. Acta Numerica, Cambridge: Cambridge University Press; 1996. p. 309−95.
[35] Sethian JA. Level set methods. Cambridge: Cambridge University Press;1996.
[36] Garrido S, Moreno L, Blanco D. Exploration of a cluttered environment using Voronoi transform and fast marching. Robot Auton Syst 2008;56(12):1069−81.
[37] Arney T. An efficient solution to autonomous path planning by approximate cell decomposition. Proceedings of international conference on information and automation for sustainability (ICIAFS 07). Colombo, Sri Lanka; December 4−6, 2007. p. 88−93.

[38] Coenen FP, Beattle B, Shave MJR, Bench-Capon TGM, Diaz GM. Spatial reasoning using the quad tesseral representation. J Artif Intell Rev 1998;12(4):321–43.
[39] Katevas NI, Tzafestas SG, Pnevmatikatos CG. The approximate cell decomposition with local node refinement global path planning method: path nodes refinement and curve parametric interpolation. J. Intell Robot Syst 1998;22:289–314.
[40] Olunloyo VOS, Ayomoh MKO. Autonomous mobile robot navigation using hybrid virtual force field concept. Eur J Sci Res 2009;31(2):204–28.
[41] Katevas NI, Tzafestas SG. The active kinematic histogram method for path planning of non-point non-holonomically constrained mobile robots. Adv Robot 1998;12 (4):375–95.
[42] Katevas NI, Tzafestas SG, Matia F. Global and local strategies for mobile robot navigation. In: Katevas N, editor. Mobile robotics in healthcare. Amsterdam, the Netherlands: IOS Press;2001.
[43] Jarvis R. Intelligent robotics: past, present and future. Int J Comput Sci Appl 2008;5 (3):23–35.
[44] Taylor T, Geva S, Boles WW. Directed exploration using modified distance transform. Proceedings of international conference on digital imaging computing: techniques and applications. Cairus, Australia; December 2012. p. 208–16.
[45] Zelinsky A, Jarvis RA, Byrne JC, Yuta S. Planning paths of complete coverage of an unstructured environment by a mobile robot. Proceedings of international symposium on advanced robotics. Tokyo, Japan; November 1993.
[46] Zelinsky A, Yuta S. A unified approach to planning, sensing and navigation for mobile robots. Proceedings of international symposium on experimental robotics. Kyoto, Japan; October 1993.
[47] Kimoto K, Yuta S.A Simulator for programming the behavior of an autonomous sensor-based mobile robot. Proceedings of international conference on intelligent robots and systems (IROS '92). Raleigh, NC; July 1992.
[48] Choi Y-H, Lee T-K, Baek S-H, Oh S-Y. Online complete coverage path planning for mobile robots based on linked spiral paths using constrained inverse distance transform. Proceedings of IEEE/RSJ international conference on intelligent robots and systems. St. Louis, MO; 2009. p. 5688–5712.
[49] Murray RM, Sastry SS. Nonholonomic motion planning: steering using sinusoids. IEEE Transactions on Automatic Control 1993;38(5):700–715.

12 Mobile Robot Localization and Mapping

12.1 Introduction

As described in Section 11.3.1, localization and mapping are two of the three basic operations needed for the navigation of a robot. Very broadly, other fundamental capabilities and functions of an integrated robotic system from the task/mission specification to the motion control/task execution (besides path planning) are the following:

- Cognition of the task specification
- Perception of the environment
- Control of the robot motion

A pictorial illustration of the interrelations and organization of the above functions in a working wheel-driven mobile robot (WMR) is shown in Figure 12.1.

The robot must have the ability to perceive the environment via its sensors in order to create the proper data for finding its location (localization) and determining how it should go to its destination in the produced map (path planning). The desired destination is found by the robot through processing of the desired task/mission command with the help of the cognition process. The path is then provided as input to the robot's motion controller which drives the actuators such that the robot follows the commanded path. The path planning operation was discussed in Chapter 11. Here we will deal with the localization and map building operations. Specifically, the objectives of this chapter are as follows:

- To provide the background concepts on stochastic processes, Kalman estimation, and Bayesian estimation employed in WMR localization
- To discuss the sensor imperfections that have to be taken care in their use
- To introduce the relative localization (dead reckoning) concept, and present the kinematic analysis of dead reckoning in WMRs
- To study the absolute localization methods (trilateration, triangulation, map matching)
- To present the application of Kalman filters for mobile robot localization, sensor calibration, and sensor fusion
- To treat the simultaneous localization and mapping (SLAM) problem via the extended Kalman filter (EKF), Bayesian estimation, and particle filter (PF)

Introduction to Mobile Robot Control. DOI: http://dx.doi.org/10.1016/B978-0-12-417049-0.00012-2

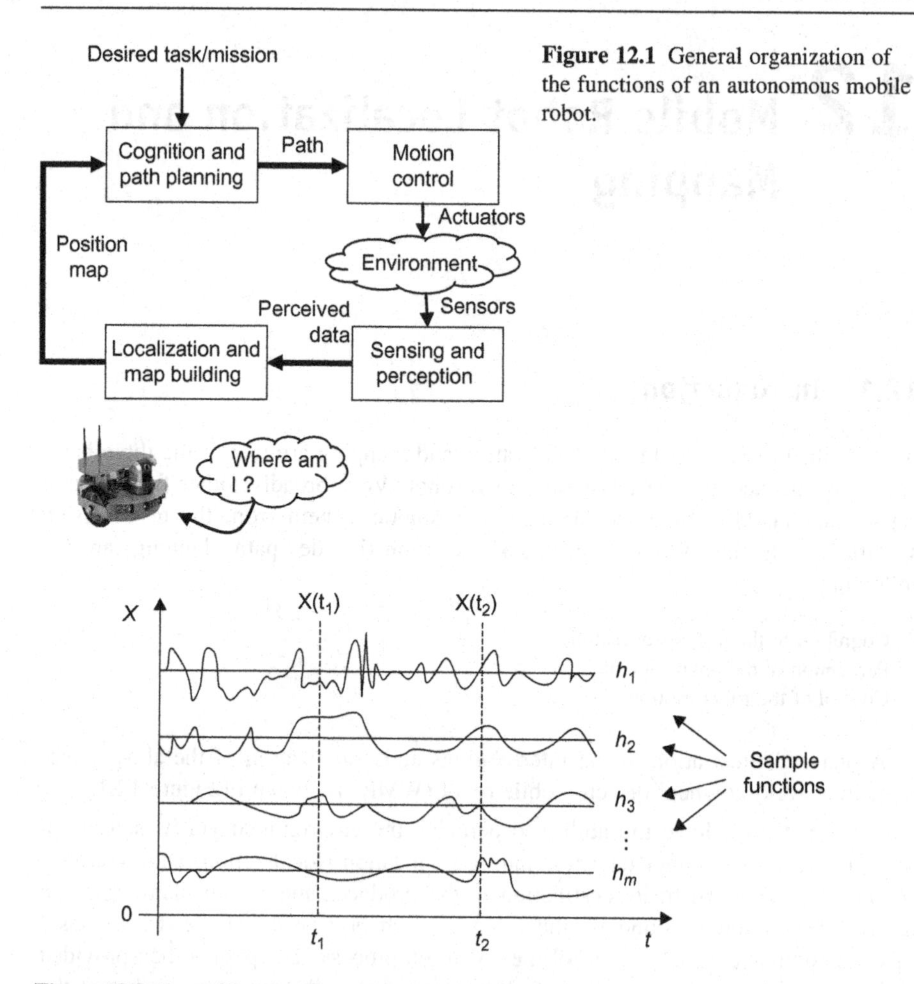

Figure 12.1 General organization of the functions of an autonomous mobile robot.

Figure 12.2 Ensemble representation of a stochastic process.

12.2 Background Concepts

Stochastic modeling, Kalman filtering, and Bayesian estimation techniques are important tools for robot localization. Here, an introduction to the above concepts and techniques, which will be used in the chapter, is provided [1–3].

12.2.1 Stochastic Processes

Stochastic process is defined to be a collection (or ensemble) of time functions $X(h, t)$ of random variables corresponding to an infinitely countable set of experiments h, with an associated probability description, for example, $p(x, t)$ (see Figure 12.2). At time t_1, $X(t_1)$ is a random variable with probability $p(x, t_1)$.

Similarly, $X(t_2)$ is a random variable with probability $p(x, t_2)$. Here, t_1 is a random variable over the ensemble every time, and so we can define first and higher-order statistics for the variables $t_1, t_2, \ldots$.

First-order statistics is concerned only with a single random variable $X(t)$ and is expressed by a probability distribution $P(x, t)$ and its density $p(x, t) = \mathrm{d}P(x, t)/dt$ for the continuous-time case, or its distribution function $P(x_i, t)$ for the discrete-time case.

Second-order statistics is concerned with two random variables $X(t_1)$ and $X(t_2)$ at two distinct time instances t_1 and t_2. In the continuous-time case, we have the following probability functions for $x(t_1) = x_1$ and $x(t_2) = x_2$.

$$P(x(t_1), x(t_2){:}t_1, t_2) \quad \text{(Joint distribution)}$$

$$p(x(t_1), x(t_2){:}t_1, t_2) = \mathrm{d}P/\mathrm{d}x_1\mathrm{d}x_2 \quad \text{(Joint density)}$$

$$p(x_2, t_2) = \int_{-\infty}^{\infty} p(x_1, x_2{:}t_1, t_2)\mathrm{d}x_1 \quad \text{(Marginal density)}$$

$$p(x_1, t_1 | x_2, t_2) = p(x_1, x_2{:}t_1, t_2)/p(x_2, t_2) \quad \text{(Conditional density)}$$

Using the above probability functions we get the first- and second-order averages (moments) as follows:

$$\bar{x}(t) = E[X(t)] = \int_{-\infty}^{\infty} xp(x, t)\mathrm{d}x \quad \text{(Mean value)}$$

$$R_{xx}(t_1, t_2) = \int_{-\infty}^{\infty}\int_{-\infty}^{\infty} x_1x_2p(x_1, x_2{:}t_1, t_2)\mathrm{d}x_1\mathrm{d}x_2 \quad \text{(Autocorrelation)}$$

$$\begin{aligned} C_{xx}(t_1, t_2) &= E([x(t_1) - \bar{x}(t_1)][x(t_2) - \bar{x}(t_2)]) \\ &= R_{xx}(t_1, t_2) - \bar{x}(t_1)\bar{x}(t_2) \end{aligned} \quad \text{(Autocovariance)}$$

$$C_{xx}(t, t) = \sigma_x^2 \quad \text{(Autocovariance)}$$

The sample statistics time averages of order n over the sample functions of the continuous-time process $X(t)$ are defined as:

$$\langle X^n \rangle = \lim_{T\to\infty} \frac{1}{2T}\int_{-T}^{+T} X(t)^n \mathrm{d}t$$

and the time averages of the discrete-time process X_i as:

$$E[X_i^n] = \lim_{N\to\infty} \sum_{i=1}^{N} X_i^n P_x(x_i)$$

Stationarity: A stochastic process is said to be *stationary* if all its marginal and joint density functions do not depend on the choice of the time origin. If this is not so, then the process is called *nonstationary*.

Ergodicity: A stochastic process is said to be *ergodic* if its ensemble moments (averages) are equal to its corresponding sample moments, that is, if:

$$E[X^n] = \int_{-\infty}^{\infty} x^n p(x)\mathrm{d}x = \lim_{T\to\infty}\frac{1}{2T}\int_{-T}^{T} X(t)^n \mathrm{d}t = \langle X^n \rangle$$

Ergodic stochastic processes are always stationary. The converse does not always hold.

Stationarity and Ergodicity in the Wide Sense: Motivated by the normal (Gaussian) density, which is completely described by its *mean* and *variance*, in practice we usually employ only first- and second-order moments. If a process is stationary or ergodic up to second-order moments, then it is called a stationary or ergodic process in the *wide sense*.

Markov Process: A Markov process is the process in which for $t_1 > t_2 \cdots > t_n$ the following property *(called Markovian property)* is true:

$$\begin{aligned}\mathrm{Prob}\{X(t_1) \le x_1 | X(t_2) = x_2, \ldots, X(t_n) = x_n\} \\ = \mathrm{Prob}\{X(t_1) \le x_1 | X(t_2) = x_2\}\end{aligned} \tag{12.1}$$

This means that the p.d.f. of the state variable $X(t_1)$ at some time instant t_1 depends only on the state $X(t_2)$ at the immediately previous time t_2 (not on more previous times).

12.2.2 Stochastic Dynamic Models

We consider an n-dimensional linear discrete-time dynamic process described by:

$$\mathbf{x}_{k+1} = \mathbf{A}_k\mathbf{x}_k + \mathbf{B}_k\mathbf{w}_k, \quad \mathbf{x}_k \in R^n, \quad \mathbf{w}_k \in R^r \tag{12.2a}$$

$$\mathbf{z}_k = \mathbf{C}_k\mathbf{x}_k + \mathbf{v}_k, \quad \mathbf{v}_k \in R^m \tag{12.2b}$$

where $\mathbf{A}_k$, $\mathbf{B}_k$, $\mathbf{C}_k$ are matrices of proper dimensionality depending on the discrete-time index k, and $\mathbf{w}_k$, $\mathbf{v}_k$ are stochastic processes (the input disturbance and measurement noise, respectively) with properties:

$$\begin{aligned} &E[\mathbf{w}_k] = \mathbf{0}, \quad E[\mathbf{v}_k] = \mathbf{0} \\ &E[\mathbf{w}_k\mathbf{w}_j^{\mathrm{T}}] = \mathbf{Q}_k\delta_{kj}, \quad E[\mathbf{v}_k\mathbf{v}_j^{\mathrm{T}}] = \mathbf{R}_k\delta_{kj} \\ &E[\mathbf{w}_k\mathbf{v}_j^{\mathrm{T}}] = \mathbf{0}\end{aligned} \tag{12.3}$$

where δ_{kj} is the Kronecker delta defined as $\delta_{kk} = 1$, $\delta_{kj} = 0$ $(k \neq j)$.

The initial state $\mathbf{x}_0$ is a random (stochastic) variable, such that:

$$E[\mathbf{v}_k\mathbf{x}_0^{\mathrm{T}}] = E[\mathbf{w}_k\mathbf{x}_0^{\mathrm{T}}] = E[\mathbf{w}_k\mathbf{v}_j^{\mathrm{T}}] = \mathbf{0} \tag{12.4}$$

The above properties imply that the processes $\mathbf{w}_k$ and $\mathbf{v}_k$ and the random variable $\mathbf{x}_0$ are statistically independent. If they are also Gaussian distributed, then the model is said to be a discrete-time *Gauss–Markov model* (or *Gauss–Markov chain*), since as can be easily seen, the process $\{x_k\}$ is Markovian. The continuous-time counterpart of the Gauss–Markov model has an analogous differential equation representation.

12.2.3 Discrete-Time Kalman Filter and Predictor

We consider a discrete-time Gauss–Markov system of the type described by Eqs. (12.2)–(12.4), with available measurements $\{z(1), z(2), \ldots, z(k)\}$. Let $\hat{\mathbf{x}}(k+1|k)$ and $\hat{\mathbf{x}}(k+1|k+1)$ be the estimate of $\mathbf{x}(k+1)$ with measurements up to $z(k)$ and $z(k+1)$, respectively. Then, the discrete-time Kalman filter for the stochastic system (12.2a) with measurement process (12.2b) is described by the following recursive equations:

$$\hat{\mathbf{x}}(k+1|k+1) = \mathbf{A}(k)\hat{\mathbf{x}}(k|k) + \mathbf{K}(k+1)[\mathbf{z}(k+1) - \mathbf{C}(k+1)\mathbf{A}(k)\hat{\mathbf{x}}(k|k)] \tag{12.5}$$

$$\mathbf{K}(k+1) = \boldsymbol{\Sigma}(k+1|k)\mathbf{C}^{\mathrm{T}}(k+1)[\mathbf{C}(k+1)\boldsymbol{\Sigma}(k+1|k)\mathbf{C}^{\mathrm{T}}(k+1) + \mathbf{R}(k+1)]^{-1} \tag{12.6}$$

$$\boldsymbol{\Sigma}(k+1|k+1) = \boldsymbol{\Sigma}(k+1|k) - \mathbf{K}(k+1)\mathbf{C}(k+1)\boldsymbol{\Sigma}(k+1|k) \tag{12.7}$$

$$\boldsymbol{\Sigma}(k+1|k) = \mathbf{A}(k)\boldsymbol{\Sigma}(k|k)\mathbf{A}^{\mathrm{T}}(k) + \mathbf{B}(k)\mathbf{Q}(k)\mathbf{B}^{\mathrm{T}}(k), \quad \boldsymbol{\Sigma}(0|0) = \boldsymbol{\Sigma}_0 \tag{12.8}$$

where $\boldsymbol{\Sigma}(j|i)$ is the covariance matrix of the estimate $\hat{x}(j)$ on the basis of data up to time i:

$$\boldsymbol{\Sigma}(k|k) = E[\tilde{\mathbf{x}}(k|k)\tilde{\mathbf{x}}^{\mathrm{T}}(k|k)] \tag{12.9a}$$

$$\begin{aligned} \boldsymbol{\Sigma}(k+1|k) &= E[\tilde{\mathbf{x}}(k+1|k)\tilde{\mathbf{x}}^{\mathrm{T}}(k+1|k)] \\ \tilde{\mathbf{x}}(k+1|k) &= \mathbf{x}(k+1) - \hat{\mathbf{x}}(k+1|k) \end{aligned} \tag{12.9b}$$

and the notation $\mathbf{z}(k) = \mathbf{z}_k$, $\mathbf{Q}(k) = \mathbf{Q}_k$, etc. is used. These equations can be represented in the block diagram (Figure 12.3).

An alternative expression of $\mathbf{K}(k+1)$ is obtained by applying the matrix inversion Lemma, and is the following:

$$\mathbf{K}(k+1) = \boldsymbol{\Sigma}(k+1|k+1)\mathbf{C}^{\mathrm{T}}(k+1)\mathbf{R}^{-1}(k+1) \tag{12.10}$$

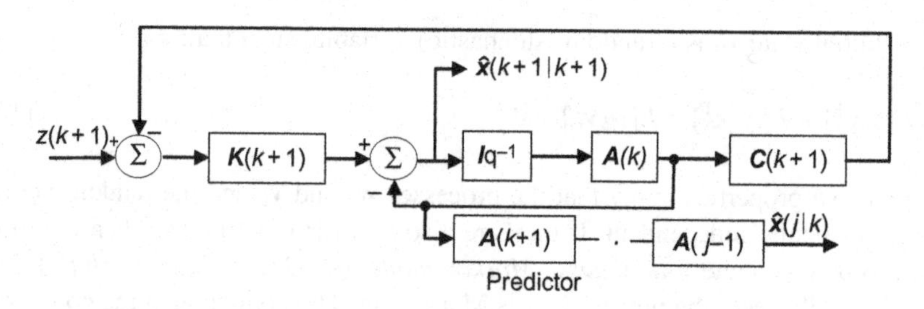

Figure 12.3 Block diagram representation of the Kalman filter and predictor ($\mathbf{I}$ = unit matrix, q^{-1} is the delay operator $q^{-1}\mathbf{x}(k+1)=\mathbf{x}(k|k)$).

The initial conditions for the state estimate Eq. (12.5) and the covariance equation (12.7) are:

$$\hat{\mathbf{x}}(0|0)=\mathbf{0},\ \mathbf{\Sigma}(0|0)=\mathbf{\Sigma}_0 \quad \text{(positive definite)} \tag{12.11}$$

State Prediction: Using the filtered estimate $\hat{\mathbf{x}}(k|k)$ of $\mathbf{x}(k)$ at the discrete-time instant k, we can compute a predicted estimate $\hat{\mathbf{x}}(j|k)$, $j>k$, $k=0,1,2,\ldots$ of the state $\mathbf{x}(j)$ on the basis of measurements up to the instant k. This is done by using the equations:

$$\hat{\mathbf{x}}(k+1|k)=\mathbf{A}(k)\hat{\mathbf{x}}(k|k) \tag{12.12a}$$

$$\hat{\mathbf{x}}(k+2|k)=\mathbf{A}(k+1)\hat{\mathbf{x}}(k+1|k)=\mathbf{A}(k+1)\mathbf{A}(k)\hat{\mathbf{x}}(k|k) \tag{12.12b}$$

$$\hat{\mathbf{x}}(k+3|k)=\mathbf{A}(k+2)\hat{\mathbf{x}}(k+2|k)=\mathbf{A}(k+2)\mathbf{A}(k+1)\mathbf{A}(k)\hat{\mathbf{x}}(k|k) \tag{12.12c}$$

and so on. Therefore:

$$\hat{\mathbf{x}}(j|k)=\mathbf{A}(j-1)\mathbf{A}(j-2)\ldots\mathbf{A}(k+1)\mathbf{A}(k)\hat{\mathbf{x}}(k|k), \quad j>k \tag{12.13}$$

12.2.4 Bayesian Learning

Formally, the probability of event A occurring, given that the event B occurs, is given by:

$$\text{Prob(A|B)}=\frac{\text{Prob(A and B)}}{\text{Prob(B)}}$$

and similarly:

$$\text{Prob(B|A)}=\frac{\text{Prob(A and B)}}{\text{Prob(A)}}$$

For notational simplicity, the above formulas are written as:

$$P(\mathrm{A}|\mathrm{B}) = \frac{P(\mathrm{A},\mathrm{B})}{P(\mathrm{B})}, \quad P(\mathrm{B}|\mathrm{A}) = \frac{P(\mathrm{A},\mathrm{B})}{P(\mathrm{A})} \tag{12.14}$$

where $P(\mathrm{A},\mathrm{B})$ is the probability of A and B occurring jointly.

Combining the two formulas (12.14), we get the so-called *Bayes formula*:

$$P(\mathrm{A}|\mathrm{B}) = \frac{P(\mathrm{B}|\mathrm{A})P(\mathrm{A})}{P(\mathrm{B})} \tag{12.15a}$$

Now, if $\mathrm{A} = \mathrm{H}$ where H is a hypothesis (about the truth or cause), and $\mathrm{B} = \mathrm{E}$ where E represents observed symptoms or measured data or evidence, Eq. (12.15a) gives the so-called *Bayes updating (or learning) rule*:

$$P(\mathrm{H}|\mathrm{E}) = \frac{P(\mathrm{E}|\mathrm{H})P(\mathrm{H})}{P(\mathrm{E})} \tag{12.15b}$$

where $P(\mathrm{E}) = P(\mathrm{E}|\mathrm{H})P(\mathrm{H}) + P(\mathrm{E}|\mathrm{notH})P(\mathrm{notH})$ and $P(\mathrm{notH}) = 1 - P(\mathrm{H})$. Here, $P(\mathrm{H})$ is the "*prior*" probability of the hypothesis H (before any evidence is obtained), $P(\mathrm{E})$ is the probability of the evidence, and $P(\mathrm{E}|\mathrm{H})$ is the probability that the evidence is true when H holds. The term $P(\mathrm{H}|\mathrm{E})$ is the "posterior" probability (i.e., the "*updated*" probability) of H after the evidence is obtained. Consider now the case that two evidences E_1 and E_2 about H are observed one after the other. Then, Eq. (12.15b) gives:

$$P(\mathrm{H}|\mathrm{E}_1) = \frac{P(\mathrm{E}_1|\mathrm{H})P(\mathrm{H})}{P(\mathrm{E}_1)}$$

$$P(\mathrm{H}|\mathrm{E}_1,\mathrm{E}_2) = \frac{P(\mathrm{E}_2|\mathrm{H})P(\mathrm{H}|\mathrm{E}_1)}{P(\mathrm{E}_2)} = \frac{P(\mathrm{E}_2|\mathrm{H})P(\mathrm{E}_1|\mathrm{H})P(\mathrm{H})}{P(\mathrm{E}_1)P(\mathrm{E}_2)}$$

Clearly, the above equation says that $P(\mathrm{H}|\mathrm{E}_1,\mathrm{E}_2) = P(\mathrm{H}|\mathrm{E}_2,\mathrm{E}_1)$, that is, the order in which the evidence (data) are observed does not influence the resulting posterior probability of H given E_1 and E_2.

Example 12.1

It is desired to derive the Kalman filter using the least-squares estimation approach.

Solution

The Kalman filter described by Eqs.(12.5)–(12.11) can be derived using several approaches, viz., (i) orthogonality principle, (ii) Gaussian Bayesian technique, and (iii) the least-squares approach (various formulations). Here, the least-squares estimator (3.103) will be used which was derived for the measurement system (3.102). This estimator uses all available data $\mathbf{y} = [\mathbf{y}_1, \mathbf{y}_2, \ldots, \mathbf{y}_m]^{\mathrm{T}}$ at once, and provides an off-line estimate of $\boldsymbol{\xi}$. The derivation of the Kalman filter (12.5)–(12.11) will be done by converting Eq. (3.103) to recursive form that updates (improves) the estimate $\boldsymbol{\xi}_k$ (found on the basis of k measurements $\mathbf{y}_1, \ldots, \mathbf{y}_k$) using the next, $(k+1)$th, measurement $\mathbf{y}_{k+1}$. To this end, we define:

$$\mathbf{M}_{k+1} = \left[\frac{\mathbf{M}_k}{\boldsymbol{\mu}_{k+1}^{\mathrm{T}}}\right], \quad \mathbf{y}_{k+1} = \left[\frac{\mathbf{y}_k}{y_{k+1}}\right]$$
$$\boldsymbol{\Sigma}_k = \left[\mathbf{M}_k^{\mathrm{T}}\mathbf{M}_k\right]^{-1}, \quad \boldsymbol{\Sigma}_{k+1} = \left[\mathbf{M}_{k+1}^{\mathrm{T}}\mathbf{M}_{k+1}\right]^{-1} \tag{12.16a}$$

The matrices $\boldsymbol{\Sigma}_{k+1}$ and $\boldsymbol{\Sigma}_k$ are related as follows:

$$\begin{aligned}\boldsymbol{\Sigma}_{k+1} &= \left(\left[\frac{\mathbf{M}_k}{\boldsymbol{\mu}_{k+1}^{\mathrm{T}}}\right]^T\left[\frac{\mathbf{M}_k}{\boldsymbol{\mu}_{k+1}^{\mathrm{T}}}\right]\right)^{-1} = \left([\mathbf{M}_k^{\mathrm{T}}|\boldsymbol{\mu}_{k+1}]\left[\frac{\mathbf{M}_k}{\boldsymbol{\mu}_{k+1}^{\mathrm{T}}}\right]\right)^{-1} \\ &= (\mathbf{M}_k^{\mathrm{T}}\mathbf{M}_k + \boldsymbol{\mu}_{k+1}\boldsymbol{\mu}_{k+1}^{\mathrm{T}})^{-1} = (\boldsymbol{\Sigma}_k^{-1} + \boldsymbol{\mu}_{k+1}\boldsymbol{\mu}_{k+1}^{\mathrm{T}})^{-1}\end{aligned}$$

Therefore:

$$\boldsymbol{\Sigma}_k^{-1} = \boldsymbol{\Sigma}_{k+1}^{-1} - \boldsymbol{\mu}_{k+1}\boldsymbol{\mu}_{k+1}^{\mathrm{T}} \tag{12.16b}$$

Now, using Eqs. (3.103) and (12.16a) we get:

$$\hat{\boldsymbol{\xi}}_k = \boldsymbol{\Sigma}_k\mathbf{M}_k^{\mathrm{T}}\mathbf{y}_k$$

$$\hat{\boldsymbol{\xi}}_{k+1} = \boldsymbol{\Sigma}_{k+1}(\mathbf{M}_k^{\mathrm{T}}\mathbf{y}_k + \boldsymbol{\mu}_{k+1}y_{k+1})$$

To express $\hat{\boldsymbol{\xi}}_{k+1}$ in terms of $\hat{\boldsymbol{\xi}}_k$ we must eliminate $\mathbf{M}_k^{\mathrm{T}}\mathbf{y}_k$ from the above equations. The first of these equations gives $\mathbf{M}_k^{\mathrm{T}}\mathbf{y}_k = \boldsymbol{\Sigma}_k^{-1}\hat{\boldsymbol{\xi}}_k$. Thus, the second equation can be written as:

$$\begin{aligned}\hat{\boldsymbol{\xi}}_{k+1} &= \boldsymbol{\Sigma}_{k+1}(\boldsymbol{\Sigma}_k^{-1}\hat{\boldsymbol{\xi}}_k + \boldsymbol{\mu}_{k+1}y_{k+1}) \\ &= \boldsymbol{\Sigma}_{k+1}[(\boldsymbol{\Sigma}_{k+1}^{-1} - \boldsymbol{\mu}_{k+1}\boldsymbol{\mu}_{k+1}^{\mathrm{T}})\hat{\boldsymbol{\xi}}_k + \boldsymbol{\mu}_{k+1}y_{k+1}] \\ &= \hat{\boldsymbol{\xi}}_k + \boldsymbol{\Sigma}_{k+1}\boldsymbol{\mu}_{k+1}(y_{k+1} - \boldsymbol{\mu}_{k+1}^{\mathrm{T}}\hat{\boldsymbol{\xi}}_k)\end{aligned} \tag{12.17a}$$

Equation (12.17a) provides the desired *recursive least-squares estimator*. The new estimate $\hat{\boldsymbol{\xi}}_{k+1}$ is a function of the old estimate $\hat{\boldsymbol{\xi}}_k$, the new data pair $[\boldsymbol{\mu}_{k+1}^{\mathrm{T}}, y_{k+1}]$ and the matrix $\boldsymbol{\Sigma}_{k+1}$. Actually, the new estimate $\hat{\boldsymbol{\xi}}_{k+1}$ is equal to the old estimate $\hat{\boldsymbol{\xi}}_k$ plus a correction term equal to an *adaptation* (or *learning*) *gain vector* $\boldsymbol{\Sigma}_{k+1}\boldsymbol{\mu}_{k+1}$ multiplied by the prediction (learning) error or *innovation process* $\tilde{y}_{k+1} = y_{k+1} - \boldsymbol{\mu}_{k+1}^{\mathrm{T}}\hat{\boldsymbol{\xi}}_k$. For this reason Eq. (12.17a) is also called *"least-squares learning equation or rule."* We now have to find how $\boldsymbol{\Sigma}_{k+1}$ can be found from $\boldsymbol{\Sigma}_k$.

From Eq. (12.16b) we obtain:[1]

$$\begin{aligned}\boldsymbol{\Sigma}_{k+1} &= (\boldsymbol{\Sigma}_k^{-1} + \boldsymbol{\mu}_{k+1}\boldsymbol{\mu}_{k+1}^{\mathrm{T}})^{-1} \\ &= \boldsymbol{\Sigma}_k - \boldsymbol{\Sigma}_k\boldsymbol{\mu}_{k+1}(\mathbf{I} + \boldsymbol{\mu}_{k+1}^{\mathrm{T}}\boldsymbol{\Sigma}_k\boldsymbol{\mu}_{k+1})^{-1}\boldsymbol{\mu}_{k+1}^{\mathrm{T}}\boldsymbol{\Sigma}_k \\ &= \boldsymbol{\Sigma}_k - \frac{\boldsymbol{\Sigma}_k\boldsymbol{\mu}_{k+1}\boldsymbol{\mu}_{k+1}^{\mathrm{T}}\boldsymbol{\Sigma}_k}{1 + \boldsymbol{\mu}_{k+1}^{\mathrm{T}}\boldsymbol{\Sigma}_k\boldsymbol{\mu}_{k+1}}\end{aligned} \tag{12.17b}$$

for $k = 0, 1, 2, \ldots, m-1$ (since $\boldsymbol{\mu}_{k+1}^{\mathrm{T}}\boldsymbol{\Sigma}_k\boldsymbol{\mu}_{k+1}$ is a scalar). The full recursive least-squares estimation algorithm given by Eqs. (12.17a) and (12.17b) is initiated using selected initial values $\boldsymbol{\Sigma}_0$ and $\hat{\boldsymbol{\xi}}_0$. One may also start from $k = n$, by using the first n measurements to get $\hat{\boldsymbol{\xi}}_n$ and $\boldsymbol{\Sigma}_n$ directly:

$$\boldsymbol{\Sigma}_n = [\mathbf{M}_n^{\mathrm{T}}\mathbf{M}_n]^{-1}, \quad \hat{\boldsymbol{\xi}}_n = \boldsymbol{\Sigma}_n\mathbf{M}_n^{\mathrm{T}}\mathbf{y}_n$$

Now, using in Eqs. (12.17a) and (12.17b) the following variable substitutions:

$$\begin{aligned}&\hat{\boldsymbol{\xi}}_{k+1} = \hat{\mathbf{x}}(k+1|k+1), \quad \hat{\boldsymbol{\xi}}_k = \hat{\mathbf{x}}(k+1|k) = \mathbf{A}(k)\hat{\mathbf{x}}(k|k), \quad \boldsymbol{\Sigma}_k = \boldsymbol{\Sigma}(k+1|k), \\ &\boldsymbol{\Sigma}_{k+1} = \boldsymbol{\Sigma}(k+1|k+1), \quad \boldsymbol{\mu}_{k+1}^{\mathrm{T}} = \mathbf{C}(k+1) \text{ and } \mathbf{y}_k = \mathbf{z}(k)\end{aligned}$$

we get the Kalman filter Eqs. (12.5)–(12.11).

The initial conditions for the state estimate Eq. (12.5) and the covariance equation (12.7) are:

$$\hat{\mathbf{x}}(0|0) = \mathbf{0}, \boldsymbol{\Sigma}(0|0) = \boldsymbol{\Sigma}_0 \quad \text{(positive definite)}$$

To see how the Kalman filter works, we consider a scalar time-invariant discrete-time Gauss–Markov system:

$$\begin{aligned}x(k+1) &= Ax(k) + w(k) \\ z(k) &= x(k) + v(k)\end{aligned} \quad (k = 0, 1, 2, \ldots)$$

with $A = 1$, $Q = 25$, $R = 15$, and $\Sigma_0 = 100$. The Kalman filter Eqs. (12.5)–(12.11) give:

$$\hat{x}(k+1|k+1) = A\hat{x}(k|k) + K(k+1)[z(k+1) - A\hat{x}(k|k)], \hat{x}(0|0) = 0$$

$$\Sigma(k+1|k) = A^2\Sigma(k|k) + Q$$

$$K(k+1) = [A^2\Sigma(k|k) + Q]/[A^2\Sigma(k|k) + Q + R]$$

$$\Sigma(k+1|k+1) = R[A^2\Sigma(k|k) + Q]/[A^2\Sigma(k|k) + Q + R], \quad \Sigma(0|0) = \Sigma_0$$

[1] Using the following matrix inversion lemma: $(\mathbf{A}+\mathbf{BC})^{-1} = \mathbf{A}^{-1} - \mathbf{A}^{-1}\mathbf{B}(\mathbf{I}+\mathbf{CA}^{-1}\mathbf{B})^{-1}\mathbf{CA}^{-1}$ with $\mathbf{A} = \boldsymbol{\Sigma}_k^{-1}$, $\mathbf{B} = \boldsymbol{\mu}_{k+1}$, and $\mathbf{C} = \boldsymbol{\mu}_{k+1}^{\mathrm{T}}$.

Since $\Sigma(k|k) \geq 0$, the second equation tells us that $\Sigma(k+1|k) \geq Q$, that is, the one-step prediction accuracy is at minimum equal to the variance of the input disturbance $w(k)$. From the third equation, we see that $0 \leq K(k+1) \leq 1 (k = 0, 1, 2, \ldots)$. The fourth equation implies that $\Sigma(k+1|k+1) = RK(k+1)$. Thus, $0 \leq \Sigma(k+1|k+1) \leq R$. This means that if $\Sigma(0|0) \gg R$, the use of the first measurement $z(1)$ gives $\Sigma(1|1) \leq R \ll \Sigma_0$, that is, it provides a drastic reduction in the estimation error. Using the given values $A = 1$, $Q = 25$, $R = 15$, and $\Sigma_0 = \Sigma(0|0) = 100$, the variance equations give the results as shown in Table 12.1.

The steady-state value of $\Sigma(k|k)$ is found by setting $\Sigma(k+1|k+1) = \Sigma(k|k) = \overline{\Sigma}$, in which case we obtain $\overline{\Sigma}^2 + 25\overline{\Sigma} - 375 = 0$. Since $\overline{\Sigma} \geq 0$, the acceptable solution is $\overline{\Sigma} = 10.55$. Thus, $\overline{K} = K(k+1) = 0.703$, and

$$\begin{aligned}\hat{x}(k+1|k+1) &= \hat{x}(k|k) + 0.703[z(k+1) - \hat{x}(k|k)] \\ &= 0.297\hat{x}(k|k) + 0.703z(k+1) \quad (k = 4, 5, 6, \ldots)\end{aligned}$$

12.3 Sensor Imperfections

The primary prerequisite for an accurate localization is the availability of reliable high-resolution sensors. Unfortunately, the practically available real-life sensors have several imperfections that are due to different causes. As we saw in Chapter 4, the usual sensors employed in robot navigation are range finders via sonar, laser and infrared technology, radar, tactile sensors, compasses, and GPS. Sonars have very low spatial bandwidth capabilities and are subject to noise due to wave scattering, and so the use of laser range sensing is preferred. But despite laser sensors possess a much wider bandwidth, they are still affected by noise. Furthermore, lasers have a restricted field of view, unless special measures are taken (e.g., incorporation of rotating mirrors in the design).

Since both sonar-based and laser-based navigation have the above drawbacks, robotic scientists have strengthened their attention to vision sensors and camera-based systems. Vision sensors can offer a wide field of view, millisecond sampling rates, and are easily applicable for control purposes. Some of the drawbacks of vision are the lack of depth information, image occlusion, low resolution, and the need for recognition and interpretation. However, despite the relative advantages

Table 12.1 Evolution of the Optimal Filter

k	$\Sigma(k\|k-1)$	$K(k)$	$\Sigma(k\|k)$
0	–	–	100
1	125	0.893	13.40
2	38.4	0.720	10.80
3	35.8	0.704	10.57
4	35.6	0.703	10.55

and disadvantages, the use of monocular cameras as a sensor for navigation leads to a competitive selection.

In general, the sensor imperfections can be grouped in: *sensor noise* and *sensor aliasing* categories [4–8]. The sensor noise is primarily caused by the environmental variations that cannot be captured by the robot. Examples of this in vision systems are the illumination conditions, the blooming, and the blurring. In sonar systems, if the surface accepting the emitted sound is relatively smooth and angled, much of the signal will be reflected away, failing to produce a return echo. Another source of noise in sonar systems is the use of multiple sonar emitters (16–40 emitters) that are subject to echo interference effects. The second imperfection of robotic sensors is the *aliasing*, that is, the fact that sensor readings are not unique. In other words, the mapping from the environmental states to the robot's perceptual inputs is many-to-one (not one-to-one). The sensor aliasing implies that (even if no noise exists) the available amount of information is in most cases not sufficient to identify the robot's position from a single sensor reading. The above issues suggest that in practice special sensor signal processing/fusion techniques should be employed to minimize the effect of noise and aliasing, and thus get an accurate estimate of the robot position over time. These techniques include *probabilistic and information theory methods* (Bayesian estimators, Kalman filters, EKFs, PFs, Monte Carlo estimators, information filters, fuzzy/neural approximators, etc).

12.4 Relative Localization

Relative localization is performed by *dead reckoning*, that is, by the measurement of the movement of a wheeled mobile robot (WMR) between two locations. This is done repeatedly as the robot moves and the movement measurements are added together to form an estimate of the distance traveled from the starting position. Since the individual estimates of the local positions are not exact, the errors are accumulated and the absolute error in the total movement estimate increases with traveled distance. The term *dead reckoning* comes from the sailing days term "*deduced reckoning*" [4].

For a WMR, the dead reckoning method is called "*odometry*," and is based on data obtained from incremental wheel encoders [9].

The basic assumption of odometry is that wheel revolutions can be transformed into linear displacements relative to the floor. This assumption is rarely ideally valid because of wheel slippage and other causes. The errors in odometric measurement are distinguished in:

- *Systematic errors* (e.g., due to unequal wheel diameters, misalignment of wheels, actual wheel base is different than nominal wheel base, finite encoder resolution, encoder sampling rate).
- *Non-systematic errors* (e.g., uneven floors, slippery floors, overacceleration, nonpoint contact with the floor, skidding/fast turning, internal and external forces).

Systematic errors are cumulative and occur principally in indoor environment. Nonsystematic errors are dominating in outdoor environments.

Two simple models for representing the systematic odometric errors are:
Error due to unequal wheel diameter:

$$e_{\mathrm{d}} = d_{\mathrm{R}}/d_{\mathrm{L}}$$

Error due uncertainty about the effective wheel base:

$$e_{\mathrm{b}} = b_{\mathrm{actual}}/b_{\mathrm{nominal}}$$

where d_{R}, d_{L} are the actual diameters of the right and left wheels, respectively, and b is the wheel base. It has been verified that either e_{d} or e_{b} can cause the same final error in both position and orientation measurement. To estimate the nonsystematic errors, we use the worst possible case, that is, we estimate the biggest possible disturbance. A good measure of nonsystematic errors is provided by the *average absolute orientation errors* e_θ clockwise and counterclockwise directions, namely $e_{\theta,\mathrm{cw}}^{\mathrm{avrg}}$ and $e_{\theta,\mathrm{ccw}}^{\mathrm{avrg}}$, respectively. Some ways for reducing systematic odometry errors are:

- *Addition of auxiliary wheels* (a pair of "knife-edge," nonload-bearing encoder wheels; Figure 12.4).
- Use of a trailer with encoder wheels.
- Careful systematic calibration of the WMR.

Methods for reducing nonsystematic odometry errors include:

- *Mutual referencing* (use of two robots that measure their positions mutually).
- *Correction internal position error* (two WMRs mutually correct their odometry errors).
- *Use of mechanical or solid-state gyroscopes.*

Clearly, more accurate odometry reduces the requirements on absolute position updates, but does not eliminate completely the necessity for periodic updating of the absolute position.

12.5 Kinematic Analysis of Dead Reckoning

The robot state (position and orientation) at instant k is $\mathbf{x}(k) = [x(k), y(k), \phi(k)]^{\mathrm{T}}$. The dead-reckoning (odometry)-based localization of WMRs is to estimate $\mathbf{x}(k+1)$ at instant $k+1$, and linear and angular position increments $\Delta l(k)$, and

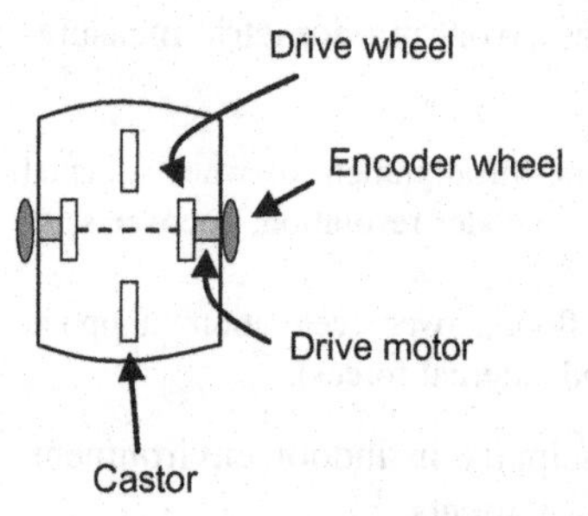

Figure 12.4 Pictorial illustration of the use of encoder wheels for a differential drive WMR.

$\Delta\phi(k) = \phi(k+1) - \phi(k)$, respectively, between the time instants k and $k+1$, where $\Delta l(k) = (\Delta x^2(k) + \Delta y^2(k))^{1/2}$, $\Delta x(k) = x(k+1) - x(k)$, and $\Delta y(k) = y(k+1) - y(k)$. Assuming that $\Delta\phi(k)$ is very small, we have the following approximations:

$$x(k+1) = x(k) + \Delta l(k)\cos(\phi(k) + \Delta\phi(k)/2)$$
$$y(k+1) = y(k) + \Delta l(k)\sin(\phi(k) + \Delta\phi(k)/2)$$
$$\phi(k+1) = \phi(k) + \Delta\phi(k)$$

To determine the position and angle increments from the wheel movements (i.e., the encoder values), we must use the kinematic equations of the WMR under consideration. We therefore investigate the localization problem for differential drive, tricycle drive, synchrodrive, Ackerman steering, and omnidirectional drive.

12.5.1 Differential Drive WMR

In this case, two optical encoders are sufficient. They supply the increments $\Delta l_1(k)$ and $\Delta l_2(k)$ of the left and right wheels. If $2a$ is the distance of the two wheels, and $R(k)$ the instantaneous curvature radius, we have:

$$\Delta l_1(k) = [R(k) - a]\Delta\phi(k)$$
$$\Delta l_2(k) = [R(k) + a]\Delta\phi(k)$$

and therefore:

$$\Delta l_2(k) - \Delta l_1(k) = 2a\Delta\phi(k)$$

Now

$$\begin{aligned}\Delta l(k) &= R(k)\Delta\phi(k) = \Delta l_1(k) + a\Delta\phi(k)\\ &= \Delta l_1(k) + \frac{1}{2}[\Delta l_2(k) - \Delta l_1(k)]\\ &= \frac{1}{2}[\Delta l_1(k) + \Delta l_2(k)]\end{aligned} \qquad (12.18a)$$

and

$$\Delta\phi(k) = \frac{1}{2a}[\Delta l_2(k) - \Delta l_1(k)] \qquad (12.18b)$$

These formulas give $\Delta l(k)$ and $\Delta\phi(k)$ in terms of the encoders' values $\Delta l_1(k)$ and $\Delta l_2(k)$.

12.5.2 Ackerman Steering

We work using the robot's diagram of Figure 1.28 denoted by D the length of the vehicle (from the rotation axis of the back wheels to the axis of the front axis). We have:

$$ctg\phi_{\rm i} = L/D, \quad ctg\phi_{\rm o} = (L+2a)/D$$
$$ctg\phi_{\rm o} - ctg\phi_{\rm i} = 2a/D, \quad ctg\phi_{\rm o} - ctg\phi_{\rm s} = a/D$$

where $\phi_{\rm s}$ is the actual steering angle of the vehicle (i.e., the angle of the virtual wheel located in the middle of the two front wheels), and $\phi_{\rm o}$, $\phi_{\rm i}$ are the steering angles of the outer and inner wheel, respectively.

Working as in the differential drive case, we find:

$$\Delta l(k) = \frac{1}{2}[\Delta l_1(k) + \Delta l_2(k)] \tag{12.19a}$$

$$\Delta\phi(k) = \frac{1}{D}[\Delta l_2(k) - \Delta l_1(k)] \tag{12.19b}$$

Here, we also find:

$$\Delta\phi(k) = \frac{1}{2D} tg\phi_{\rm s}(k)[\Delta l_1(k) + \Delta l_2(k)] \tag{12.20}$$

Two encoders are sufficient to provide $\Delta l(k)$ and $\Delta\phi(k)$ using the two relations (12.19a) and (12.19b). To use Eq. (12.20) we need a third encoder. But the result can be used to decrease the errors of the other two encoders.

12.5.3 Tricycle Drive

Here, the solution is the same as that for the Ackerman steering:

$$\Delta l(k) = \frac{1}{2}[\Delta l_1(k) + \Delta l_2(k)], \quad \Delta\phi(k) = \frac{1}{D}[\Delta l_2(k) - \Delta l_1(k)] \tag{12.21}$$

$$\Delta\phi(k) = \frac{1}{2D} tg\phi_{\rm s}(k)[\Delta l_1(k) + \Delta l_2(k)] \tag{12.22}$$

12.5.4 Omnidirection Drive

For the three-wheel case, we have the following relations:

$$v_1 = v_x + \omega R \tag{12.23a}$$

$$v_2 = -v_x \sin 30^\circ - v_y \cos 30^\circ + \omega R \tag{12.23b}$$

$$v_3 = -v_x \sin 30^\circ + v_y \cos 30^\circ + \omega R \tag{12.23c}$$

12.6 Absolute Localization

12.6.1 General Issues

Landmarks and *beacons* form the basis for absolute localization of a WMR [10]. Landmarks may or may not have identities. If the landmarks have identities, no *a priori* knowledge about the navigating robot's position is needed. Otherwise, if landmarks have no identities some rough knowledge of the position of the robot must be available. If natural landmarks are employed, then the presence or not of identities depends on the environment. For example, in an office building, one of the rooms may have a large three-glass window, which can be used as a landmark with identity.

The landmarks (beacons) are distinguished in:

- active artificial landmarks,
- passive artificial landmarks,
- natural landmarks.

An active landmark emits some kind of signal to enable easy detection. In general, the detection of a passive landmark needs more of the sensor. To simplify the detection of passive landmarks, we may use an active sensor together with a passive landmark that responds to this activeness. For example, use of bar codes on walls and visual light to detect them for building topological maps. It is preferable that natural landmarks can be employed with sustainable robustness of the navigational system. They can provide increased flexibility to the adaptation to new environments with minimum requirement of engineering design. For example, natural landmarks in indoor environments can be chairs, desks, and tables.

According to Borenstein [4], localization using active landmarks is distinguished in:

- trilateration,
- triangulation.

12.6.2 Localization by Trilateration

In *trilateration*, the location of a WMR is determined using distance measurements to known active beacon, B_i. Typically, three or more transmitters are used and one receiver on the robot. The locations of transmitters (beacons) must be known. Conversely, the robot may have one transmitter onboard and the receivers can be placed at known positions in the environment (e.g., on the walls of a room). Two examples of localization by trilateration are the beacon systems that use ultrasonic sensors (Figure 12.5) and the GPS system (Figure 4.24b).

Ultrasonic-based trilateration localization systems are appropriate for use in relatively small area environments (because of the short range of ultrasound), where no obstructions that interfere with the wave transmission exist. If the environment is large, the installation of multiple networked beacons throughout the operating area

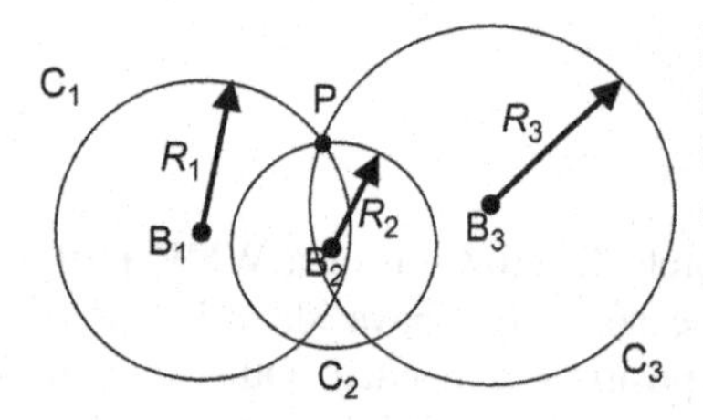

Figure 12.5 Graphical illustration of localization by trilateration. The robot P lies at the intersection of the three circles C_i centered at the beacons B_i with radii $R_i (i = 1, 2, 3)$.

is of increased complexity, a fact that reduces the applicability a sonar-based trilateration. The two alternative designs are:

- Use of a single transducer transmitting from the robot with many receivers at fixed positions.
- Use of a single listening receiver on the robot, with multiple fixed transmitters serving as beacons (this is analogous to the GPS concept).

The position $P(x, y)$ of the robot in the world coordinate system can be found by least-squares estimation as follows. In Figure 12.5, let $(x_i, y_i), i = 1, 2, 3, \ldots, N$ $(N \geq 3)$ be the known world coordinates of the beacons $B_i, i = 1, 2, 3, \ldots, N$ and $R_i, i = 1, 2, 3, \ldots, N$ the corresponding robot–beacon distances (radii of the intersecting circles). Then, we have:

$$(x_i - x)^2 + (y_i - y)^2 = R_i^2 \quad (i = 1, 2, 3, \ldots, N)$$

Expanding and rearranging gives:

$$x^2 + y^2 + (-2x_i)x + (-2y_i)y + x_i^2 + y_i^2 = R_i^2$$

To eliminate the squares of the unknown variables x and y, we pairwise subtract the above equations, for example, we subtract the kth equation from the ith equation to get:

$$2(x_k - x_i)x + 2(y_k - y_i)y = b_i$$

where:

$$b_i = R_i^2 - R_k^2 + (x_k^2 + y_k^2) - (x_i^2 - y_i^2)$$

Overall, we obtain the following overdetermined system of linear equations with two unknowns:

$$\mathbf{Ax} = \mathbf{b}, \quad \mathbf{x} = [x, y]^{\mathrm{T}}, \quad \mathbf{b} = [b_1, b_2, \ldots, b_N]^{\mathrm{T}}$$

where $\mathbf{A}$ is the matrix:

$$\mathbf{A} = 2\begin{bmatrix} (x_2 - x_1) & (y_2 - y_1) \\ (x_3 - x_1) & (y_3 - y_1) \\ \vdots & \vdots \\ (x_N - x_{N-1}) & (y_N - y_{N-1}) \end{bmatrix}$$

The solution of this linear algebraic system is given by (see Eqs. (2.7) and (2.8a)):

$$\mathbf{x} = \mathbf{A}^{\dagger}\mathbf{B}, \quad \mathbf{A}^{\dagger} = (\mathbf{A}^{\mathrm{T}}\mathbf{A})^{-1}\mathbf{A}^{\mathrm{T}}$$

under the assumption that the matrix $\mathbf{A}^{\mathrm{T}}\mathbf{A}$ is invertible.

An alternative way of computing $\mathbf{x} = [x, y]^{\mathrm{T}}$ is to apply the general *iterative least-squares* technique for nonlinear estimation. To this end, we expand the nonlinear function:

$$F_i = (x_i - x)^2 + (y_i - y)^2 - R_i^2 = 0 \quad (i = 1, 2, \ldots, N)$$

in Taylor series about an *a priori* position estimate $\hat{\mathbf{x}}_q = [\hat{x}_q, \hat{y}_q]^{\mathrm{T}} (q = 0)$, and keep only first-order terms, namely:

$$\mathbf{F}(\mathbf{x}) = \mathbf{F}(\hat{\mathbf{x}}_q) + \mathbf{A}(\hat{\mathbf{x}}_q)\Delta\mathbf{x}_q = 0, \quad q = 0, 1, 2, \ldots.$$

where:

$$\Delta\mathbf{x}_q = \mathbf{x} - \hat{\mathbf{x}}_q, \mathbf{F}(\mathbf{x}) = [F_1(\mathbf{x}), F_2(\mathbf{x}), \ldots, F_N(\mathbf{x})]^{\mathrm{T}}$$

and $\mathbf{A}(\hat{\mathbf{x}}_q)$ is the Jacobian matrix of $\mathbf{F}(\mathbf{x})$ evaluated at $\mathbf{x} = \hat{\mathbf{x}}_q$:

$$\mathbf{A}(\hat{\mathbf{x}}_q) = \left[\frac{\partial F(\mathbf{x})}{\partial \mathbf{x}}\right]_{\mathbf{x}=\hat{\mathbf{x}}_q}$$

$$= 2\begin{bmatrix} (\hat{x}_q - x_1) & (\hat{y}_q - y_1) \\ (\hat{x}_q - x_2) & (\hat{y}_q - y_2) \\ \vdots & \vdots \\ (\hat{x}_q - x_N) & (\hat{y}_q - y_N) \end{bmatrix}$$

Therefore, we get the following iterative equation for updating the estimate $\hat{\mathbf{x}}_q$:

$$\hat{\mathbf{x}}_{q+1} = \hat{\mathbf{x}}_q - \mathbf{A}^{\dagger}(\hat{\mathbf{x}}_q)\mathbf{F}(\hat{\mathbf{x}}_q), q = 0, 1, 2, \ldots.$$

where $\hat{\mathbf{x}}_0$ is known.

The iteration is repeated until $||\hat{\mathbf{x}}_{q+1} - \hat{\mathbf{x}}_q||$ becomes smaller than a predetermined small number ε or the number of iterations arrives at a maximum number $q_{\max} = Q$.

In practice, due to measurement errors in the positions of the beacons and the radii of the circles, the circles do not intersect at a single point but overlap in a small region. The position of the robot is somewhere in this region. Each pair of circles gives two intersection points, and so with three circles (beacons) we have six intersection points. Three of these points are clustered together more closely than the other three points. A simple procedure for determining the smallest intersection (clustering) region is as follows:

- Compute the distance between each pair of circle intersection points.
- Select the two closest intersection points as the initial cluster.
- Compute the centroid of this cluster (The x,y coordinates of the centroid are obtained by the averages of the x,y coordinates of the points in the cluster).
- Find the circle intersection point which is closest to the cluster centroid.
- Add this intersection point to the cluster and compute again the cluster centroid.
- Continue in the same manner until N intersection points have been added to the cluster, where N is the number of beacons (circles).
- The location of the robot is given by the centroid of the final cluster.

As an exercise, the reader may consider the case of two beacons (circles) and compute geometrically the positions of their intersection points. Also, the conditions under which the solution is nonsingular must be determined.

12.6.3 Localization by Triangulation

In this method, there are three or more *active beacons* (transmitters) mounted at known positions in the workspace (Figure 12.6). A rotating sensor mounted on the robot R registers the angles θ_1, θ_2, and θ_3 at which the robot "sees" the active beacons relative to the WMR's longitudinal axis x_r. Using these three readings, we can compute the position (x, y, ϕ) of the robot, where x stands for X_0 and y stands for Y_0.

The active beacons must be sufficiently powerful in order to be able to transmit the appropriate signal power in all directions. The two design alternatives of triangulation are:

1. Rotating transmitter and stationary receivers
2. Rotating transmitter–receiver and stationary reflectors

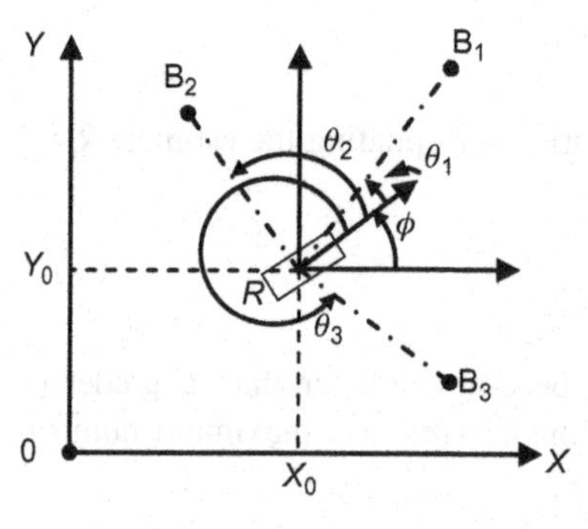

Figure 12.6 Triangulation method. A rotating sensor head measures the angles θ_1, θ_2, and θ_3 between the three active beacons B_1, B_2, and B_3 and the robot's longitudinal axis.

It must be noted that when the observed angles are small, or the observation point is very near to a circle containing the beacons, the triangulation results are very sensitive to small angular errors. The three-point triangulation method works efficiently only when the robot lies inside the boundary of the triangle formed by the three beacons. The geometric circle intersection shows large errors when the three beacons and the robot all lie on or are very close to the same circle. In case the Newton–Raphson method is used to determine the robot's location, no solution can be provided if the initial guess of the robot's position/orientation is in the exterior of a certain bound. A variation of the triangular method creates a virtual beacon as follows: a range or angle obtained from a beacon at time t can be used at time $t+\Delta t$, provided that the cumulative movement vector, recorded since the reading was obtained, is added to the position vector of the beacon. This generates a new virtual beacon.

Referring to Figure 12.6, the basic trigonometric equation, which relates x,y, and ϕ with the measured angles $\theta_i(i=1,2,3)$ of the beacons, is:

$$tg(\phi+\theta_i)=(y_i-y)/(x_i-x)$$

where x_i, y_i are the coordinates of the beacon i. Three beacons are sufficient for computing (x,y,ϕ) under the condition that the measurements are noise free. Then, we will have a nonlinear algebraic system of three equations with three unknowns which can be solved by approximate numerical techniques such as the Newton–Raphson technique and its variations. But if the measurements are noisy (as it typically happens in practice), we need more beacons. In this case, we get an overdetermined noisy nonlinear algebraic system which can be solved by the iterative least-squares technique, in analogy to the above-mentioned trilateration method [11].

The results obtained are more accurate if the artificial beacons are optimally placed (e.g., if they are about 120° apart in the three-beacon case). Otherwise, the robot position and orientation may have large variations with respect to an optimal value. Moreover, if all beacons are identical, it is very difficult to identify which beacon has been detected. Finally, mismatch is likely to occur if the presence of obstacles in the environment obscures one or more beacons. A natural way to overcome the problem of distinguishing one beacon from another is to manually initialize and recalibrate the robot position when the robot is lost. However, such a manual initialization/recalibration is not convenient in actual applications. Therefore, many automated initialization/recalibration techniques have been developed for use in robot localization by triangulation. One of them uses a self-organizing neural network (Kohonen network) which is able to recognize and distinguish the beacons [12] (see also Section 12.7).

12.6.4 Localization by Map Matching

Map-matching-based localization is a method in which a robot uses its sensors to produce a map of its local environment [13]. When we talk about a map in real life, we usually imagine a geometrical map like the map of a country or a town with names of places and streets. To find our way in a map of a city using street names, we walk up to a street corner, read the name (i.e., the landmark) and compare the name with what is written in the map. Now, if we know where we wish to go, we plan a path to this goal. In our way, we use natural landmarks (e.g., street corners, buildings, streets, blocks) to keep track of where we are along the planned path. If we lose track we have to revert the reading signs and matching the street names to the map. If a road is closed, dead-end, or one-way street, then we plan an alternative path. This methodology is in principle used by WMRs in map generation and matching, of course with landmarks suitable for the robot's environment in each case.

The robot compares its local map to a global map already stored in the computer's memory. If a match is found, then the robot can determine its actual position and orientation in the environment. The prestored map may be a CAD model of the environment, or it can be created from prior sensor data.

The basic structure of a map-based localization system is shown in Figure 12.7.

The benefits of the map-based positioning method are the following:

- No modification of the environment is required (only structured natural landmarks are needed).
- Generation of an updated map of the environment can be made.
- The robot can learn a new environment and enhance positioning accuracy via exploration.

Some drawbacks are as follows:

- A sufficient number of stationary, easily distinguishable, features are needed.
- The accuracy of the map must be sufficient for the tasks at hand.
- A substantial amount of sensing and processing power is required.

Maps are distinguished as follows:

Topological maps: They are suitable for sensors that provide information about the identity of landmarks. A real-life topological map example is the subway map which provides information only on how you get from one station (place) to another (no information

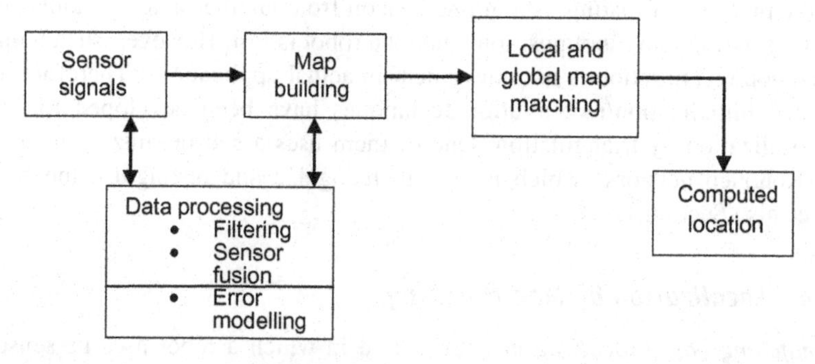

Figure 12.7 Structure of map-based localization (search can be reduced if the initial position guess is close to the true position).

about distances is given). To enable easy orientation when changing trains, color codings are provided. Topological maps model the world by a network of arcs and nodes.
Geometrical maps: These maps can be constructed from topological maps if the proper equipment for measuring distance and angle is available. Geometrical maps suit sensors that provide geometrical information (e.g., range), like sonar and laser range finders. In general, it is easier to use geometrical maps in conjunction with sensors of this type, than using topological maps, since less advanced perception techniques are needed. In extreme situations, the sensor measurements can be directly matched to the stored map.

Currently, used maps are often CAD drawings or they have been measured by hand. The human is deciding on which parts of the environment should be included in the map. It is also up to the human to decide which sensor is the most appropriate to use (taking of course into account its price), and how it will respond to the environment. Ideally, it is desired that the WMR is able to construct maps autonomously by itself. In this direction, a solid methodology called *simultaneous localization and mapping* has been developed which will be examined in the following text.

12.7 Kalman Filter-Based Localization and Sensor Calibration and Fusion

The Kalman filter described by Eqs. (12.5)–(12.11) can be used for three purposes, namely:

- Robot localization
- Sensor calibration
- Sensor fusion

12.7.1 Robot Localization

The steps of the localization algorithm that is based on Kalman filter are the following:

Step 1—One-Step Prediction
The predicted robot position $\hat{\mathbf{x}}(k+1|k)$ at time $k+1$, given the known filtered position $\hat{\mathbf{x}}(k|k)$ at time k, is given by Eq. (12.12a), that is:

$$\hat{\mathbf{x}}(k+1|k) = \mathbf{A}(k)\hat{\mathbf{x}}(k|k) \tag{12.24}$$

where $\mathbf{A}(k)$ represents the kinematic equation of the robot. The corresponding position uncertainty is given by the covariance matrix $\mathbf{\Sigma}(k+1|k)$ described by Eq. (12.8), that is:

$$\mathbf{\Sigma}(k+1|k) = \mathbf{A}(\mathbf{k})\mathbf{\Sigma}(k|k)\mathbf{A}^{\mathrm{T}}(k) + \mathbf{B}(k)\mathbf{Q}(k)\mathbf{B}^{\mathrm{T}}(k) \tag{12.25}$$

with $\mathbf{\Sigma}(0|0) = \mathbf{\Sigma}_0$ (a known value). The respective prediction $\hat{\mathbf{z}}(k+1|k)$ of the sensors' measurements is given by:

$$\hat{\mathbf{z}}(k+1|k) = \mathbf{C}(k+1)\hat{\mathbf{x}}(k+1|k)$$

Step 2—Sensor Observation

At this step, a new set of sensor measurements $\mathbf{z}(k+1)$ at time $k+1$ is taken.

Step 3—Matching

The measurement innovation process:

$$\begin{aligned}\tilde{\mathbf{z}}(k+1|k) &= \mathbf{z}(k+1) - \hat{\mathbf{z}}(k+1|k) \\ &= \mathbf{z}(k+1) - \mathbf{C}(k+1)\hat{\mathbf{x}}(k+1|k) \\ &= \mathbf{z}(k+1) - \mathbf{C}(k+1)\mathbf{A}(k)\hat{\mathbf{x}}(k|k)\end{aligned} \tag{12.26}$$

$$\begin{aligned}\mathbf{S}(k+1|k) &= E\{\tilde{\mathbf{z}}(k+1|k)\tilde{\mathbf{z}}^T(k+1|k)\} \\ &= \mathbf{C}(k+1)\boldsymbol{\Sigma}(k+1|k)\mathbf{C}^{\mathrm{T}}(k+1) + \mathbf{R}(k)\end{aligned} \tag{12.27}$$

is constructed.

Step 4—Position Estimation

Calculate the Kalman filter gain matrix $\mathbf{K}(k+1)$ using Eq. (12.10), that is:

$$\mathbf{K}(k+1) = \boldsymbol{\Sigma}(k+1|k+1)\mathbf{C}^{\mathrm{T}}(k+1)\mathbf{R}^{-1}(k+1) \tag{12.28}$$

and the updated position estimate $\hat{\mathbf{x}}(k+1|k+1)$ is given by Eqs. (12.5)–(12.8), that is:

$$\hat{\mathbf{x}}(k+1|k+1) = \hat{\mathbf{x}}(k+1|k) + \mathbf{K}(k+1)\tilde{\mathbf{z}}(k+1|k) \tag{12.29}$$

$$\boldsymbol{\Sigma}(k+1|k+1) = \boldsymbol{\Sigma}(k+1|k) - \mathbf{K}(k+1)\mathbf{C}(k+1)\boldsymbol{\Sigma}(k+1|k) \tag{12.30}$$

where $\tilde{\mathbf{z}}(k+1|k)$ and $\boldsymbol{\Sigma}(k+1|k)$ are given by Eqs. (12.26) and (12.25), respectively. The initial value $\boldsymbol{\Sigma}(0|0) = \boldsymbol{\Sigma}_0$ of the estimate's covariance, as well as the values of $\mathbf{Q}(k)$ and $\mathbf{R}(k)$ needs to be determined experimentally (before the application of the method) using the available information about the sensors and their models. Typically, $\mathbf{Q}(k) = \mathbf{Q} = \text{const.}$ and $\mathbf{R}(k) = \mathbf{R} = \text{const.}$ Taking into account that the measurement (sensor observation) vector $\mathbf{z}(k)$ may involve data from several sensors (usually of different type), we can say that the Kalman filter provides also a sensor fusion and integration method (see Example 12.5).

In all cases, it is advised to check if the new measurements are statistically independent from the previous measurements and disregard them if they are statistically dependent, since they do not actually provide new information. This can be done at the matching step on the basis of the innovation process (or residual) $\tilde{\mathbf{z}}(k+1|k)$ and its covariance $\mathbf{S}(k+1|k)$ given by Eqs. (12.26) and (12.27). The process $\tilde{\mathbf{z}}(k+1|k)$ expresses the difference between the new measurement $\mathbf{z}(k+1)$ and its expected value $\mathbf{C}(k+1)\mathbf{A}(k)\hat{\mathbf{x}}(k|k)$ according to the system model.

To this end, we use the so-called *normalized innovation squared* (NIS) defined as:

$$\tilde{\mathbf{z}}_{\mathrm{NIS}}(k+1|k) = \tilde{\mathbf{z}}^{\mathrm{T}}(k+1|k)\mathbf{S}^{-1}(k+1|k)\tilde{\mathbf{z}}(k+1|k)$$

The process $\tilde{\mathbf{z}}_{\mathrm{NIS}}(k+1|k)$ follows a *chi-square* (χ^2) distribution with m degrees of freedom (more precisely when the innovation process $\tilde{\mathbf{z}}(k+1|k)$ is Gaussian). Therefore, we can read from the chi-square, the bounding values (confidence interval) for a χ^2 random variable with m degrees of freedom, and discard a measurement, if its $\tilde{\mathbf{z}}_{\mathrm{NIS}}(k+1|k)$ is outside the desired confidence interval. More specifically, if the criterion:

$$\tilde{\mathbf{z}}_{\mathrm{NIS}}(k+1|k) \leq \chi^2$$

is satisfied, then the measurement $\mathbf{z}(k+1)$ is considered to be independent from $\mathbf{z}(k)$ and so it is included in the filtering process. Measurements that do not satisfy the above criterion are ignored. The random variable χ^2 is defined as follows. Consider two random variables (sensor signals) x and y which are independent if the probability distribution of x is not affected by the presence of y. Then, the chi-square variable is given by:

$$\chi^2 = \sum_{ij} \frac{(f_{ij} - e_{ij})^2}{e_{ij}}$$

where f_{ij} is the observed frequency of events belonging to both the ith category of x and jth category of y, and e_{ij} is the corresponding expected frequency of events if x and y are independent. The use of chi-square (or contingency table) is described in books on statistics (see, e.g., in http://onlinestatbook.com/stat_sim/chisq_theory/index.html).

12.7.2 Sensor Calibration

The Kalman filter provides also a method for sensor calibration, which is based on the minimization of the error between the actual measurements and the simulated measurement of the sensor with the parameters at hand. The parameters that have to be calibrated may be *intrinsic* or *extrinsic*. Intrinsic parameters cannot be directly observed, but extrinsic parameters can. For instance, the focal distance, the location of the optical center, and the distortion of a camera are intrinsic parameters, but the position of the camera is an extrinsic parameter.

To perform sensor calibration with the aid of the Kalman filter, we use the parameters under calibration, as the state vector of the model, in place of the robot's position/orientation. The prediction step gives simulated values of the parameters at the present state (i.e., with their present values). The state observation step provides a set of sensor captions taken at different robot positions, and not during the motion of the robot (as we did for robot localization, i.e., for position estimation and mapping). The use of Kalman filters for sensor calibration is general and independent of the nature of the sensor (range, vision, etc.). The only things that change are the state vector $\mathbf{x}(k)$ and the measurement vector with the associated matrices $\mathbf{A}(k)$, $\mathbf{B}(k)$, and $\mathbf{C}(k)$, and the noises $\mathbf{w}(k)$ and $\mathbf{v}(k)$.

12.7.3 Sensor Fusion

Sensor fusion is the process of merging data from multiple sensors such that to reduce the amount of uncertainty that may be involved in a robot navigation motion or task performing. Sensor fusion helps in building a more accurate world model in order for the robot to navigate and behave more successfully. The three fundamental ways of combining sensor data are the following:

- *Redundant sensors*: All sensors give the same information for the world.
- *Complementary sensors*: The sensors provide independent (disjoint) types of information about the world.
- *Coordinated sensors*: The sensors collect information about the world sequentially.

The three basic sensor communication schemes are [14]:

- *Decentralized*: No communication exists between the sensor nodes.
- *Centralized*: All sensors provide measurements to a central node.
- *Distributed*: The nodes interchange information at a given communication rate (e.g., every five scans, i.e., one-fifth communication rate).

The centralized scheme can be regarded as a special case of the distributed scheme where the sensors communicate to each other every scan. A pictorial representation of the fusion process is given in Figure 12.8.

Consider, for simplicity, the case of two redundant local sensors ($N=2$). Each sensor processor i provides its own prior and updated estimates and covariances $\hat{x}^i(k+1|k)$, $\Sigma^i(k+1|k)$ and $\hat{x}^i(k+1|k+1)$, $\Sigma^i(k+1|k+1)$, $i=1,2$. Assume that the fusion processor has its own total prior estimate $\hat{x}(k+1|k)$ and $\Sigma(k+1|k)$. The fusion problem is to compute the total estimate $\hat{x}(k+1|k+1)$ and covariance matrix $\Sigma(k+1|k+1)$ using only those local estimates and the total prior estimate. The total updated estimate can be obtained using linear operations on the local estimates, as:

$$\begin{aligned}\hat{x}(k+1|k+1) = \Sigma(k+1|k+1)[&\Sigma^1(k+1|k+1)^{-1}\hat{x}^1(k+1|k+1)\\ &+\Sigma^2(k+1|k+1)^{-1}\hat{x}^2(k+1|k+1)\\ &-\Sigma^1(k+1|k)^{-1}\hat{x}^1(k+1|k)-\Sigma^2(k+1|k)^{-1}\hat{x}^2(k+1|k)\\ &+\Sigma(k+1|1)^{-1}\hat{x}(k+1|k)]\end{aligned}$$

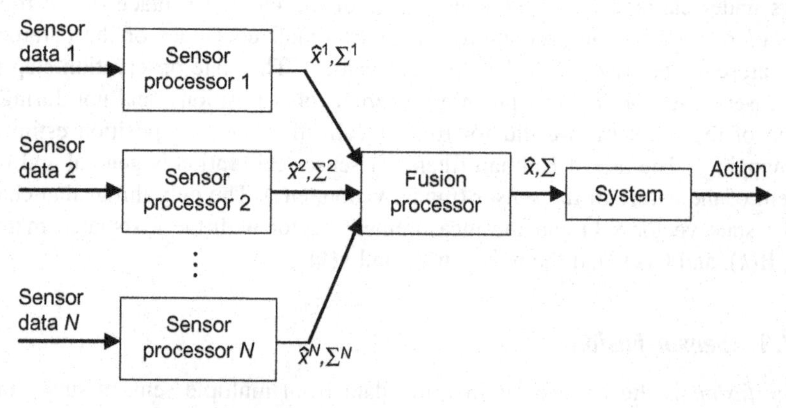

Figure 12.8 Illustration of the sensor fusion process.

where:

$$\Sigma(k+1|k+1) = [\Sigma^1(k+1|k+1)^{-1} + \Sigma^2(k+1|k+1)^{-1} - \Sigma^1(k+1|k)^{-1} \\ -\Sigma^2(k+1|k)^{-1} + \Sigma(k+1|k)^{-1}]^{-1}$$

When the local processors and the fusion processor have the same prior estimates, the above fusion equations can be simplified as:

$$\hat{x}(k+1|k+1) = \Sigma(k+1|k+1)[\Sigma^1(k+1|k+1)^{-1}\hat{x}^1(k+1|k+1) \\ +\Sigma^2(k+1|k+1)^{-1}\hat{x}^2(k+1|k+1) \\ -\Sigma(k+1|k)^{-1}\hat{x}(k+1|k)] \tag{12.31a}$$

$$\Sigma(k+1|k+1) = [\Sigma^1(k+1|k+1)^{-1} + \Sigma^2(k+1|k+1)^{-1} - \Sigma(k+1|k)^{-1}]^{-1} \tag{12.31b}$$

This means that the common prior estimates (i.e., the redundant information) are subtracted in the linear fusion operation. The above equations constitute the fusion processor shown in Figure 12.8. Typically, the sensor processors $1, 2, \ldots, N$ are Kalman filters.

Example 12.2

We will derive the formula for fusing the measurements x_k provided for a quantity x (e.g., the position of a WMR) by m independent sensors. It is assumed that the measurement x_k of the kth sensor is normally distributed (Gaussian) with variance σ_k^2.

We will apply the maximum likelihood estimation method in which the joint probability distribution $p(x_1, x_2, \ldots x_m | x, \sigma)$ of $x_1, x_2, \ldots, x_m$ is maximized with respect to the fused value x and fused variance σ^2. The joint Gaussian distribution $p(x_1, x_2, \ldots, x_m | x, \sigma)$ is:

$$p(x_1, x_2, \ldots, x_m | x, \sigma) = \prod_{k=1}^{m} \frac{1}{\sigma_k \sqrt{2\pi}} e^{-(x_k - x)^2 / 2\sigma_k^2}$$

The likelihood function L is defined as the logarithm of $p(x_1, x_2, \ldots, x_m | x, \sigma)$, that is:

$$L(x_1, x_2, \ldots, x_m | x, \sigma) = -\frac{1}{2} m \ln(2\pi) - m \sum_{k=1}^{m} \ln \sigma_k - \sum_{k=1}^{m} (x_k - x)^2 / (2\sigma_k^2)$$

To maximize L, we equate to zero the derivative of L with respect to x, that is:

$$\frac{\partial L}{\partial x} = \sum_{k=1}^{m} \frac{(x_k - x)}{\sigma_k^2} \\ = \sum_{k=1}^{m} \frac{x_k}{\sigma_k^2} - x\left(\sum_{k=1}^{m} \frac{1}{\sigma_k^2}\right) = 0$$

Thus, the fused estimate $\hat{x}$ of the m sensors is equal to:

$$\hat{x} = \left(\sum_{k=1}^{m} \frac{x_k}{\sigma_k^2}\right) \Big/ \left(\sum_{k=1}^{m} \frac{1}{\sigma_k^2}\right)$$

The variance $\sigma_{\hat{x}}$ of $\hat{x}$ is given by:

$$\sigma_{\hat{x}}^2 = \sum_{k=1}^{m} \sigma_k^2 \left(\frac{\partial \hat{x}}{\partial x_k}\right)^2$$

The partial derivative $\partial \hat{x}/\partial x_k$ is:

$$\frac{\partial \hat{x}}{\partial x_k} = \frac{\partial}{\partial x_k} \frac{\sum_{k=1}^{m}(x_k/\sigma_k^2)}{\sum_{k=1}^{m}(1/\sigma_k^2)} = \frac{1/\sigma_k^2}{\sum_{k=1}^{m}(1/\sigma_k^2)}$$

Therefore:

$$\sigma_{\hat{x}}^2 = \sum_{k=1}^{m} \sigma_k^2 \left[\frac{1/\sigma_k^2}{\sum_{k=1}^{m}(1/\sigma_k^2)}\right]^2 = \sum_{k=1}^{m} \frac{1/\sigma_k^2}{[\sum_{k=1}^{m}(1/\sigma_k^2)]^2} = \frac{1}{\sum_{k=1}^{m}(1/\sigma_k^2)}$$

or

$$\frac{1}{\sigma_{\hat{x}}^2} = \sum_{k=1}^{m} \frac{1}{\sigma_k^2}$$

To illustrate the meaning of the above formulas for $\hat{x}$ and $\sigma_{\hat{x}}^2$, we consider the case of two sensors ($m = 2$), for example, a laser range finder sensor and an ultrasonic range sensor, namely:

$$\hat{x} = \left(\frac{x_1}{\sigma_1^2} + \frac{x_2}{\sigma_2^2}\right) \Big/ \left(\frac{1}{\sigma_1^2} + \frac{1}{\sigma_2^2}\right) = \left(\frac{\sigma_2^2}{\sigma_1^2 + \sigma_2^2}\right) x_1 + \left(\frac{\sigma_1^2}{\sigma_1^2 + \sigma_2^2}\right) x_2$$

$$\frac{1}{\sigma_{\hat{x}}^2} = \frac{1}{\sigma_1^2} + \frac{1}{\sigma_2^2} = \frac{\sigma_1^2 + \sigma_2^2}{\sigma_1^2 \sigma_2^2}$$

or

$$\sigma_{\hat{x}}^2 = \sigma_1^2 \sigma_2^2 / (\sigma_1^2 + \sigma_2^2)$$

We see that the variance of the fused estimate is smaller than all the variances of the individual sensor measurements (similarly to the connection of pure resistances in parallel) (see Figure 12.9). The formula for $\hat{x}$ can be written as:

$$\hat{x} = x_1 + [\sigma_1^2/(\sigma_1^2 + \sigma_2^2)](x_2 - x_1)$$

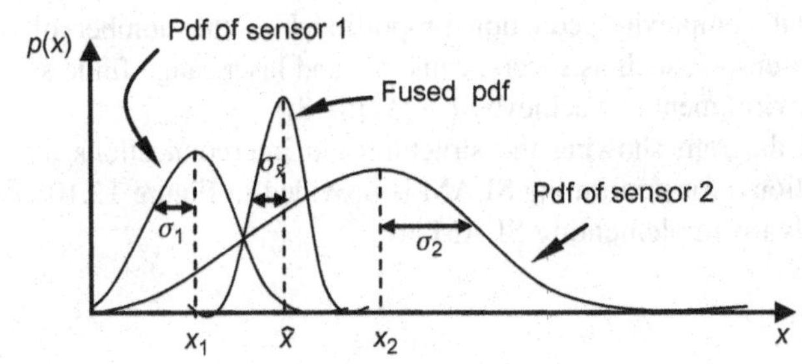

Figure 12.9 The variance $\sigma_{\hat{x}}$ of the fused estimate $\hat{x}$ is smaller than both σ_1 and σ_2.

Thus, we get the following sensor estimate updating formula:

$$\hat{x}_{k+1} = \hat{x}_k + \Sigma_{k+1}(y_{k+1} - \hat{x}_k), \quad y_{k+1} = x_2, \quad \hat{x}_k = x_1$$

$$\Sigma_{k+1} = \sigma_k^2/(\sigma_k^2 + \sigma_y^2)$$

with $\sigma_k^2 = \sigma_1^2$ and $\sigma_y^2 = \sigma_2^2$. The updated variance σ_{k+1} of $\hat{x}_{k+1}$ is found to be:

$$\sigma_{k+1}^2 = \sigma_k^2 - \Sigma_{k+1}\sigma_k^2$$

The above sequential equations for $\hat{x}_{k+1}$ and σ_{k+1}^2 represent a discrete Kalman filter (see Section 12.2.3 and Example 12.1) that can be used when the sensors collect and provide their data sequentially.

In the above analysis, the variable $\hat{x}$ was assumed to have a fixed value (e.g., when the WMR is not moving). If the variable x changes, and the change can be expressed by a dynamic stochastic system, then the dynamic Kalman filter should be used and σ_{k+1} is decreasing with time (see Table 12.1).

12.8 Simultaneous Localization and Mapping

12.8.1 General Issues

The SLAM problem deals with the question if it is possible for a WMR to be placed at an unknown location in an unknown environment, and for the robot to incrementally build a consistent map of this environment while simultaneously determining its location within this map.

The foundations for the study of SLAM were made by Durrant-Whyte [15] by establishing a statistical basis for describing relationships between landmarks and manipulating geometric uncertainty. The key element was the observation that there must be a high degree of correlation between estimates of the location of several landmarks in a map and that these correlations are strengthened with successive observations. A complete solution to the SLAM problem needs a joint state, involving the WMR's pose and every landmark position, to be updated following each landmark observation. Of course, in practice this would need the estimator to employ a very high dimensional state vector (of the order of the number of landmarks maintained in the map) with computational complexity reduction proportional to the number of landmarks. A wide range of sensors, such as sonars, cameras, and laser range finders, has been used to sense the environment and achieve SLAM ([5–8]).

A generic self-explained diagram showing the structure and interconnections of the various functions (operations) for performing SLAM is provided in Figure 12.10.

The three typical methods for implementing SLAM are:

- *EKF*
- *Bayesian estimator*
- *PF*

EKF is an extension of Kalman filter to cover nonlinear stochastic models, which allows to study observability, controllability, and stability of the filtered system. Bayesian estimators describe the WMR motion and feature observations directly using the underlying probability density functions and the Bayes theorem for probability updating. The Bayesian approach has shown a good success in many challenging environments. The PF (or as otherwise called, sequential *Monte Carlo method*) is based on simulation [15]. In the following sections, all these SLAM methods will be outlined.

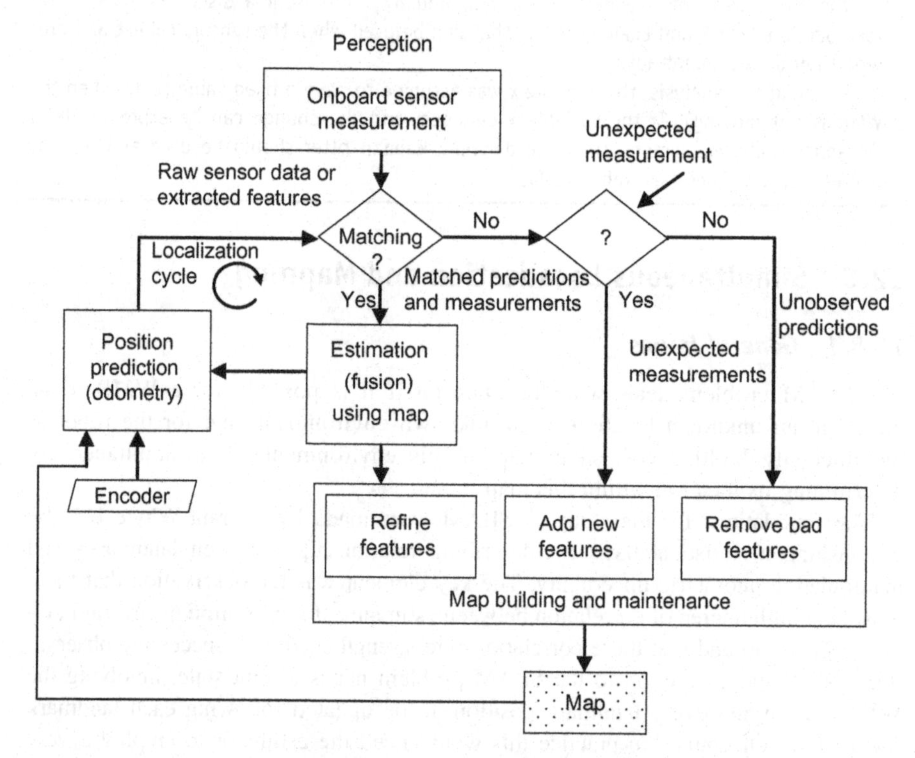

Figure 12.10 Interconnection diagram and flow chart of the SLAM functions.

12.8.2 EKF SLAM

The motion of the WMR and the measurement of the map features are described by the following stochastic nonlinear model [16]:

$$\mathbf{x}_{k+1} = \mathbf{f}(\mathbf{x}_k, \mathbf{u}_k, \mathbf{w}_k) \tag{12.32a}$$

$$\mathbf{z}_k = \mathbf{h}(\mathbf{x}_k) + \mathbf{v}_k \tag{12.32b}$$

where the state vector $\mathbf{x}_k$ contains the m-dimensional position $\mathbf{x}_{\mathrm{r},k}$ of the vehicle, and a vector $\mathbf{x}_\mathrm{f}$ of n stationary d-dimensional map features, that is, $\mathbf{x}_k$ has dimensionality $m + d \cdot n$:

$$\mathbf{x}_k = \begin{bmatrix} \mathbf{x}_{\mathrm{r},k} \\ \mathbf{x}_\mathrm{f} \end{bmatrix} \tag{12.33}$$

The l-dimensional input vector $\mathbf{u}_k$ is the robot's control command, and the l-dimensional process $\mathbf{w}_k$ is a Gaussian stochastic zero-mean process with constant covariance matrix $\mathbf{Q}$. The function $\mathbf{h}(\mathbf{x}_k)$ represents the sensor model, and $\mathbf{v}_k$ represents the inaccuracies and noise. It is also assumed that $\mathbf{v}_k$ is a zero-mean Gaussian process.

Given a set of measurements $\mathbf{Z}_k = \{\mathbf{z}_1, \mathbf{z}_2, \ldots, \mathbf{z}_k\}$ for the current map estimate $\mathbf{x}_{k|k}$ the expression:

$$\mathbf{x}_{k+1} = \mathbf{f}(\mathbf{x}_{k|k}, \mathbf{u}_k, \mathbf{0}) \tag{12.34a}$$

gives an *a priori* noise-free estimate of the new locations of the robot and map features after the application of the control input $\mathbf{u}_k$. In the same way:

$$\mathbf{z}_{k+1|k} = \mathbf{h}(\mathbf{x}_{k+1|k}) + \mathbf{0} \tag{12.34b}$$

provides a noise-free *a priori* estimate of the sensor measurements.

If $\mathbf{f}(\cdot,\cdot,\cdot)$ and $\mathbf{h}(\cdot)$ are linear, then the Kalman filter, Eqs. (12.5)–(12.10), can be directly applied. But, in general $\mathbf{f}(\cdot,\cdot,\cdot)$ and $\mathbf{h}(\cdot)$ are nonlinear in which case we use the following *linearizations* (first-order Taylor approximations):

$$\mathbf{x}_{k+1} \approx \mathbf{x}_{k+1|k} + \mathbf{F}(\mathbf{x}_k - \mathbf{x}_{k|k}) + \mathbf{G}\mathbf{w}_k \tag{12.35a}$$

$$\mathbf{z}_{k+1} \approx \mathbf{z}_{k+1|k} + \mathbf{H}(\mathbf{x}_{k+1} - \mathbf{x}_{k+1|k}) + \mathbf{v}_k \tag{12.35b}$$

where $\mathbf{F}$, $\mathbf{G}$, and $\mathbf{H}$ are the Jacobian matrices:

$$\mathbf{F} = \left.\frac{\partial \mathbf{f}}{\partial \mathbf{x}}\right]_{(\mathbf{x}_{k|k}, \mathbf{u}_k, \mathbf{0})}, \quad \mathbf{G} = \left.\frac{\partial \mathbf{f}}{\partial \mathbf{w}}\right]_{(\mathbf{x}_{k|k}, \mathbf{u}_k, \mathbf{0})}, \quad \mathbf{H} = \left.\frac{\partial \mathbf{h}}{\partial \mathbf{x}}\right]_{(\mathbf{x}_{k+1|k}, \mathbf{0})}$$

Given that the landmarks are assumed stationary, their *a priori* estimate is:

$$\mathbf{x}_{\mathrm{f},k+1|k} = \mathbf{x}_{\mathrm{f},k|k}$$

Thus, the overall state model of the vehicle and map dynamics is:

$$\mathbf{x}_{k+1} \approx \begin{bmatrix} \mathbf{x}_{\mathrm{r},k+1|k} \\ \mathbf{x}_{\mathrm{f},k|k} \end{bmatrix} + \begin{bmatrix} \mathbf{F}_{\mathrm{r}} & \mathbf{0} \\ \mathbf{0} & \mathbf{I} \end{bmatrix} \begin{bmatrix} \tilde{\mathbf{x}}_{\mathrm{r},k|k} \\ \tilde{\mathbf{x}}_{\mathrm{f},k|k} \end{bmatrix} + \begin{bmatrix} \mathbf{G}_{\mathrm{r}} \\ \mathbf{0} \end{bmatrix} \begin{bmatrix} \mathbf{w}_k \\ \mathbf{0} \end{bmatrix}$$

$$\mathbf{z}_{k+1} \approx \mathbf{z}_{k+1|k} + \begin{bmatrix} \mathbf{H}_{\mathrm{r}} & \mathbf{H}_{\mathrm{f}} \end{bmatrix} \begin{bmatrix} \tilde{\mathbf{x}}_{\mathrm{r},k+1|k} \\ \tilde{\mathbf{x}}_{\mathrm{f},k+1|k} \end{bmatrix} + \mathbf{v}_{k+1}$$

or

$$\mathbf{x}_{k+1} = \mathbf{x}_{k+1|k} + \mathbf{A}\tilde{\mathbf{x}}_{k|k} + \mathbf{B}\mathbf{w}_k^* \tag{12.36}$$

$$\mathbf{z}_{k+1} = \mathbf{z}_{k+1|k} + \mathbf{C}\tilde{\mathbf{x}}_{k|k} + \mathbf{v}_{k+1} \tag{12.37}$$

where:

$$\mathbf{A} = \begin{bmatrix} \mathbf{F}_{\mathrm{r}} & \mathbf{0} \\ \mathbf{0} & \mathbf{I} \end{bmatrix}, \quad \mathbf{B} = \begin{bmatrix} \mathbf{G}_{\mathrm{r}} \\ \mathbf{0} \end{bmatrix}, \quad \mathbf{C} = \begin{bmatrix} \mathbf{H}_{\mathrm{r}} & \mathbf{H}_{\mathrm{f}} \end{bmatrix}, \quad \mathbf{w}_k^* = \begin{bmatrix} \mathbf{w}_k \\ \mathbf{0} \end{bmatrix} \tag{12.38a}$$

$$\mathbf{x}_{k+1|k} = [\mathbf{x}_{\mathrm{r},k+1|k}^{\mathrm{T}} \quad \mathbf{x}_{\mathrm{f},k|k}^{\mathrm{T}}]^{\mathrm{T}}, \quad \tilde{\mathbf{x}}_{k|k} = [\tilde{\mathbf{x}}_{\mathrm{r},k|k}^{\mathrm{T}} \quad \tilde{\mathbf{x}}_{\mathrm{f},k|k}^{\mathrm{T}}]^{\mathrm{T}} \tag{12.38b}$$

Using the augmented linearized model for the dynamics of the position and environmental landmarks, we can apply directly the same steps as those given in Section 12.7 for the linear localization problem.

For convenience, we repeat here the sequence of the covariances:

$$\begin{aligned} \text{Prediction covariance: } \boldsymbol{\Sigma}_{k+1|k} &= \mathbf{A}\boldsymbol{\Sigma}_{k|k}\mathbf{A}^{\mathrm{T}} + \mathbf{B}\mathbf{Q}^*\mathbf{B}^{\mathrm{T}} \\ \text{Innovation covariance: } \mathbf{S}_{k+1|k} &= \mathbf{C}\boldsymbol{\Sigma}_{k+1|k}\mathbf{C}^{\mathrm{T}} + \mathbf{R} \\ \text{Filter gain: } \mathbf{K}_{k+1} &= \boldsymbol{\Sigma}_{k+1|k+1}\mathbf{C}^{\mathrm{T}}\mathbf{R}^{-1} \\ \text{Filter covariance: } \boldsymbol{\Sigma}_{k+1|k+1} &= \boldsymbol{\Sigma}_{k+1|k} - \mathbf{K}_{k+1}\mathbf{C}\boldsymbol{\Sigma}_{k+1|k} \end{aligned} \tag{12.39}$$

Under the condition that the pair $(\mathbf{A}, \mathbf{C})$ is completely observable, the filter covariance tends to a constant matrix $\boldsymbol{\Sigma}$ which is given by the following steady-state (algebraic) Riccati equation:

$$\boldsymbol{\Sigma} = \mathbf{A}[\boldsymbol{\Sigma} - \boldsymbol{\Sigma}\mathbf{C}^{\mathrm{T}}(\mathbf{C}\boldsymbol{\Sigma}\mathbf{C}^{\mathrm{T}} + \mathbf{R})^{-1}\mathbf{C}\boldsymbol{\Sigma}]\mathbf{A}^{\mathrm{T}} + \mathbf{Q} \tag{12.40}$$

It is noted that in SLAM the complete observability condition is not satisfied in all cases. Clearly, the steady-state covariance matrix $\boldsymbol{\Sigma}$ depends on the values of $\boldsymbol{\Sigma}_{r,0|0}$, $\mathbf{Q}$ and $\mathbf{R}$, as well as on the total number n of landmarks. The solution of (12.40) can be found by standard computer packages (e.g., Matlab).

The EKF-SLAM solution is well developed and possesses many of the benefits of the application of the EKF technique to navigation or tracking control problems. However, since EKF-SLAM uses linearized models of nonlinear dynamics and observation models, it leads sometimes to inevitable and critical inconsistencies. Convergence and consistence can only by assured in the linear case as illustrated in the following example.

Example 12.3

Consider a one-dimensional WMR (monorobot) with state $x_{r,k}$, and a one-dimensional landmark x_f as shown in Figure 12.11 [16].

The WMR's position error dynamics is:

$$\mathbf{x}_{r,k+1} = \mathbf{x}_{r,k} + \mathbf{u}_k + \mathbf{w}_k \tag{12.41}$$

and the landmarks dynamics is:

$$\mathbf{x}_{f,k+1} = \mathbf{x}_{f,k} \tag{12.42}$$

The map generated by this system is a single static landmark x_f. The observation (measurement) model for this landmark is:

$$\mathbf{z}_{k+1} = \mathbf{x}_{f,k+1} - \mathbf{x}_{r,k+1|k} + \mathbf{v}_k \tag{12.43}$$

where $\mathbf{v}_k$ is the landmark's measurement error. Equations (12.41)–(12.43) can be written in the standard form of Eqs. (12.2a) and (12.2b), that is:

$$\mathbf{x}_{k+1} = \mathbf{A}\mathbf{x}_k + \mathbf{B}'\mathbf{u}_k + \mathbf{B}\mathbf{w}_k \tag{12.44}$$

$$\mathbf{z}_{k+1} = \mathbf{C}\mathbf{x}_k + \mathbf{v}_k \tag{12.45}$$

where:

$$\mathbf{x}_k = \begin{bmatrix} x_{r,k} \\ x_{f,k} \end{bmatrix}, \quad \mathbf{A} = \begin{bmatrix} 1 & 0 \\ 0 & 1 \end{bmatrix}, \quad \mathbf{B}' = \begin{bmatrix} 1 \\ 0 \end{bmatrix}, \quad \mathbf{B} = \begin{bmatrix} 1 \\ 0 \end{bmatrix}, \quad \mathbf{C} = \begin{bmatrix} -1 & 1 \end{bmatrix}, \quad \mathbf{w}_k = \begin{bmatrix} w_k \\ 0 \end{bmatrix}$$

For the filtering problem, the term $\mathbf{B}'\mathbf{u}_k$ is just an additive term, and has no influence on the filtered estimate. The filtered estimate is given by:

$$\hat{\mathbf{x}}_{k+1|k+1} = \mathbf{A}\hat{\mathbf{x}}_{k|k} + \mathbf{B}'\mathbf{u}_k + \mathbf{K}(k+1)[\mathbf{z}_{k+1} - \mathbf{C}\mathbf{A}\hat{\mathbf{x}}_{k|k}] = (\mathbf{A} - \mathbf{KCA})\hat{\mathbf{x}}_{k|k} + \mathbf{B}'\mathbf{u}_k$$

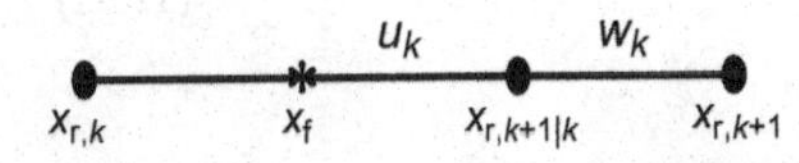

Figure 12.11 A single-dimensional robot with a one-dimensional landmark.

Here, the matrix $\mathbf{A}' = \mathbf{A} - \mathbf{KCA}$ is equal to:

$$\mathbf{A}' = \begin{bmatrix} k_1 + 1 & -k_1 \\ k_2 & -k_2 + 1 \end{bmatrix} \text{ where, } \mathbf{K} = \begin{bmatrix} k_1 \\ k_2 \end{bmatrix} \tag{12.46}$$

which has the eigenvalues $\lambda_1 = 1$ and $\lambda_2 = k_1 - k_2 + 1$. We see that λ_1 is always equal to 1 regardless of the filter gain matrix **K**. Therefore, the filter in this case is only *marginally stable* (i.e., it performs constant amplitude oscillations). This can easily be verified via simulation. To overcome this problem which is caused by a partially observable system, we can use several techniques to assure *complete observability* of the pair $(\mathbf{A}, \mathbf{C})$, that is, the matrix:

$$\mathbf{P} = \begin{bmatrix} \mathbf{C} \\ \mathbf{CA} \\ \vdots \\ \mathbf{CA}^{N-1} \end{bmatrix} \quad (N = \text{dimensionality of } \mathbf{A}) \tag{12.47}$$

is of full rank N, in which case it is invertible. These methods include [17]:

- use of anchors or markers,
- use of fixed global references,
- use of an external sensor,
- use of the relative position of the landmark with respect to the position of the robot instead of global positioning.

Therefore, let us assume that we add an anchor in the measurement process of Eq. (12.43) with state $z_k^{(0)}$ and noise $v_k^{(0)}$. Then, the measurement model of Eq. (12.45) has:

$$\mathbf{C} = \begin{bmatrix} -1 & 0 \\ -1 & 1 \end{bmatrix}, \quad \mathbf{z}_k = \begin{bmatrix} z_k^{(0)} \\ z_k \end{bmatrix}, \quad \mathbf{v}_k = \begin{bmatrix} v_k^{(0)} \\ v_k \end{bmatrix} \tag{12.48}$$

Now, the gain matrix **K** becomes:

$$\mathbf{K} = \begin{bmatrix} k_{11} & k_{12} \\ k_{21} & k_{22} \end{bmatrix} \tag{12.49}$$

and the matrix $\mathbf{A}' = \mathbf{A} - \mathbf{KCA}$ is:

$$\mathbf{A}' = \begin{bmatrix} 1 + k_{11} + k_{12} & -k_{12} \\ k_{21} + k_{22} & 1 - k_{22} \end{bmatrix} \tag{12.50}$$

The matrix gain **K** is given by Eqs. (12.10), (12.7), and (12.8). If σ_w and σ_v are the standard deviations of w and v, respectively, then using the present form of the matrices **A**,**B**, and **C** (see Eq. (12.48)), we can verify that:

$$\begin{aligned} k_{11} &= -(\sigma_v^4 + \sigma_w^2\sigma_v^4 + 4\sigma_w\sigma_v^2 + 2\sigma_v^2 + 3\sigma_w^2)/\mu \\ k_{12} &= -(\sigma_v^4 + \sigma_w^2 + 3\sigma_w^2\sigma_v^2 + \sigma_w^2\sigma_v^4)/\mu \\ k_{21} &= -(\sigma_v^2 + \sigma_w^2\sigma_v^2 + 2\sigma_w^2)/\mu \\ k_{22} &= (\sigma_v^4 + 2\sigma_v^2 + \sigma_w^2\sigma_v^2 + 2\sigma_w^2)/\mu \end{aligned} \tag{12.51}$$

where:

$$\mu = \sigma_v^6 + 2\sigma_w^2\sigma_v^4 + 6\sigma_v^4 + 7\sigma_w^2\sigma_v^2 + 4\sigma_v^2 + 4\sigma_w^2 \tag{12.52}$$

Now, we can see that the eigenvalues of $\mathbf{A}' = \mathbf{A} - \mathbf{KCA}$ given by Eq. (12.50) always lie inside the unit circle of the complex z plane, and so *adding the "anchor" the filter becomes stable.*

12.8.3 Bayesian Estimator SLAM

In probabilistic (Bayesian) form, the SLAM problem is to compute the conditional probability [18–21]:

$$P(\mathbf{x}_{r,k}, \mathbf{x}_f | \mathbf{Z}_{0,k}, \mathbf{U}_{0,k}, \mathbf{x}_{r,0}) \tag{12.53}$$

where:

$$\mathbf{x}_f = \{\mathbf{x}_{f1}, \mathbf{x}_{f2}, \ldots, \mathbf{x}_{fn}\}$$

$$\mathbf{Z}_{0,k} = \{\mathbf{z}_1, \mathbf{z}_2, \ldots, \mathbf{z}_k\} = \{\mathbf{Z}_{0,k-1}, \mathbf{z}_k\}$$

$$\mathbf{U}_{0,k} = \{\mathbf{u}_1, \mathbf{u}_2, \ldots, \mathbf{u}_k\} = \{\mathbf{U}_{0,k-1}, \mathbf{u}_k\}$$

represent the history of landmark observations $\mathbf{z}_k$ and the history of control inputs $\mathbf{u}_k$, respectively. This probability distribution represents the *joint posterior density* of the landmark locations and WMR state (at time k) given the recorded observations and control inputs up to and including the time k together with the initial state $\mathbf{x}_{r,0}$ of the robot. The Bayesian SLAM is based on Bayesian learning, discussed in Section 12.2.4. Starting with an initial estimate for the distribution:

$$P(\mathbf{x}_{r,k-1}, \mathbf{x}_f | \mathbf{Z}_{0,k-1}, \mathbf{U}_{0,k-1}, \mathbf{x}_{r,0}) \tag{12.54}$$

at time $k-1$ the joint posterior probability, following a control $\mathbf{u}_k$ and observation $\mathbf{z}_k$ is computed by Bayes learning (updating) formula (12.15b). The steps of the algorithm are as follows:

Step 1—Observation Model

Determine the observation model that describes the probability of making an observation $\mathbf{z}_k$ when the WMR location and landmark locations are known. In general, this model has the form:

$$P(\mathbf{z}_k | \mathbf{x}_{r,k}, \mathbf{x}_f) \tag{12.55}$$

Of course, we tacitly assume that once the robot's location and the map are defined, the observations are conditionally independent given the map and the current WMR state.

Step 2—WMR Motion Model

Determine the motion model of the vehicle which is described by the Markovian conditional probability:

$$P(\mathbf{x}_{r,k}|\mathbf{x}_{r,k-1},\mathbf{u}_k) \tag{12.56}$$

This probability indicates that the state $\mathbf{x}_{r,k}$ at time k depends only on the state $\mathbf{x}_{r,k-1}$ at time $k-1$ and the exerted control $\mathbf{u}_k$, and does not depend on the observations and the map.

Step 3—Time Update

We have:

$$\begin{aligned} &P(\mathbf{x}_{r,k},\mathbf{x}_f|\mathbf{Z}_{0,k-1},\mathbf{U}_{0,k},\mathbf{x}_{r,0}) \\ &= \int P(\mathbf{x}_{r,k}|\mathbf{x}_{r,k-1},\mathbf{u}_k)P(\mathbf{x}_{r,k-1},\mathbf{x}_f|\mathbf{Z}_{0,k-1},\mathbf{U}_{0,k-1},\mathbf{x}_{r,0})d\mathbf{x}_{r,k-1} \end{aligned} \tag{12.57}$$

Step 4—Measurement Update

According to Bayes update law:

$$\begin{aligned} &P(\mathbf{x}_{r,k},\mathbf{x}_f|\mathbf{Z}_{0,k},\mathbf{U}_{0,k},\mathbf{x}_{r,0}) \\ &= \frac{P(\mathbf{z}_k|\mathbf{x}_{r,k},\mathbf{x}_f)P(\mathbf{x}_{r,k},\mathbf{x}_f|\mathbf{Z}_{0,k-1},\mathbf{U}_{0,k},\mathbf{x}_{r,0})}{P(\mathbf{z}_k|\mathbf{Z}_{0,k-1},\mathbf{U}_{0,k})} \end{aligned} \tag{12.58}$$

Equations (12.57) and (12.58) provide a recursive algorithm for calculating the joint posterior probability $P(\mathbf{x}_{r,k},\mathbf{x}_f|\mathbf{Z}_{0,k},\mathbf{U}_{0,k},\mathbf{x}_{r,0})$ for the WMR state $\mathbf{x}_{r,k}$ and map $\mathbf{x}_f$ at time k using all observations $\mathbf{Z}_{0,k}$ and all control inputs $\mathbf{U}_{0,k}$ up to and including time k. This recursion uses the WMR model $\mathbf{P}(\mathbf{x}_{r,k}|\mathbf{x}_{r,k-1},\mathbf{u}_k)$ and the observation model $P(\mathbf{z}_k|\mathbf{x}_{r,k},\mathbf{x}_f)$. We remark that here the map building can be formulated as computing the conditional density $P(\mathbf{x}_f|\mathbf{X}_{0,k},\mathbf{Z}_{0,k},\mathbf{U}_{0,k})$. This needs the location $\mathbf{x}_{r,k}$ of the WMR to be known at all times, under the condition that the initial location is known. The map is then produced through the fusion of observations from different positions. On the other hand, the localization problem can be formulated as the problem of computing the probability distribution $P(\mathbf{x}_{r,k}|\mathbf{Z}_{0,k},\mathbf{U}_{0,k},\mathbf{x}_f)$. This requires the landmark locations to be known with certainty. The goal is to compute an estimate of the WMR location with respect to these landmarks.

The above formulation can be simplified by dropping the conditioning on historical variables in Eq. (12.54) and write the joint posterior probability as $P(\mathbf{x}_{r,k-1},\mathbf{x}_f|\mathbf{z}_k)$. Similarly, the observation model $P(\mathbf{z}_k|\mathbf{x}_{r,k},\mathbf{x}_f)$ makes explicit the dependence of observations on both the vehicle and landmark locations. But, here the joint posterior probability cannot be partitioned in the standard way, that is, we have:

$$P(\mathbf{x}_{r,k},\mathbf{x}_f|\mathbf{z}_k) \neq P(\mathbf{x}_{r,k}|\mathbf{z}_k)P(\mathbf{x}_f|\mathbf{z}_k) \tag{12.59}$$

Thus, care should be taken not to use the partition shown in Eq. (12.59) because this could lead to inconsistencies.

However, the SLAM problem has more intrinsic structure not visible from the above discussion. The most important issue is that the errors in landmark location estimates are highly correlated, for example, the joint probability density of a pair of landmarks, $P(\mathbf{x}_{fi},\mathbf{x}_{fj})$, is highly peaked even when the independent densities $P(\mathbf{x}_{fi})$ may be very dispersed. Practically, this means that the relative location $\mathbf{x}_{fi}-\mathbf{x}_{fj}$ of any two landmarks $\mathbf{x}_{fi},\mathbf{x}_{fj}$ may be estimated much more accurately than their individual positions where $\mathbf{x}_{fi}, \mathbf{x}_{fj}$ may be quite uncertain. In other words, the relative location of landmarks always improves and never diverges, regardless of the WMR's motion. Probabilistically, this implies that the joint probability density on all landmarks $P(\mathbf{x}_f)$ becomes monotonically more peaked as more observations are made.

Example 12.4

We consider a differential drive WMR mapping two-dimensional landmarks $\mathbf{x}_f = [x_f^i, y_f^i]^T$, $i = 1, 2, \ldots, m$ (Figure 12.12) [16]. The state space of the robot is three dimensional, where the state vector is:

$$\mathbf{x}_k = \begin{bmatrix} x_k & y_k & \phi_k \end{bmatrix}^T \tag{12.60}$$

The robot is controlled by a linear velocity v and an angular velocity ω.

Let l be the distance from the center of the wheel axis to the location of the center of projection for any given sensor, and Δt the discrete-time step. The dynamic model of the trajectory of the center of projection of the sensor, which includes the noises $w_{v,k}$ and $w_{\omega,k}$, is:

$$\mathbf{x}_{k+1} = \mathbf{f}_r(\mathbf{x}_k, \mathbf{u}_k, \mathbf{w}_k), \quad \mathbf{w}_k = [w_{v,k} w_{\omega,k}]^T \tag{12.61}$$

or in detailed form:

$$\begin{bmatrix} x_{k+1} \\ y_{k+1} \\ \phi_{k+1} \end{bmatrix} = \begin{bmatrix} x_k + [(v_k + w_{v,k})\cos\phi_k - l(\omega_k + w_{\omega,k})\sin\phi_k]\Delta t \\ y_k + [(v_k + w_{v,k})\sin\phi_k + l(\omega_k + w_{\omega,k})\cos\phi_k]\Delta t \\ \phi_k + (\omega_k + w_{\omega,k})\Delta t \end{bmatrix} \tag{12.62}$$

Differentiating Eq. (12.62) with respect to $\mathbf{x}_k$ and $\mathbf{w}_k = [w_{v,k}, w_{\omega,k}]^T$, we find the Jacobian matrices:

$$\mathbf{A}_r = \frac{\partial \mathbf{f}_r}{\partial \mathbf{x}_k} = \begin{bmatrix} 1 & 0 & -(v_k \sin\phi_k - l\omega_k \cos\phi_k)\Delta t \\ 0 & 1 & (v_k \cos\phi_k - l\omega_k \sin\phi_k)\Delta t \\ 0 & 0 & 1 \end{bmatrix}$$

$$\mathbf{G}_r = \frac{\partial \mathbf{f}_r}{\partial_w} = \begin{bmatrix} (\cos\phi_k)\Delta t & -(l\sin\phi_k)\Delta t \\ (\sin\phi_k)\Delta t & (l\cos\phi_k)\Delta t \\ 0 & \Delta t \end{bmatrix} \tag{12.63}$$

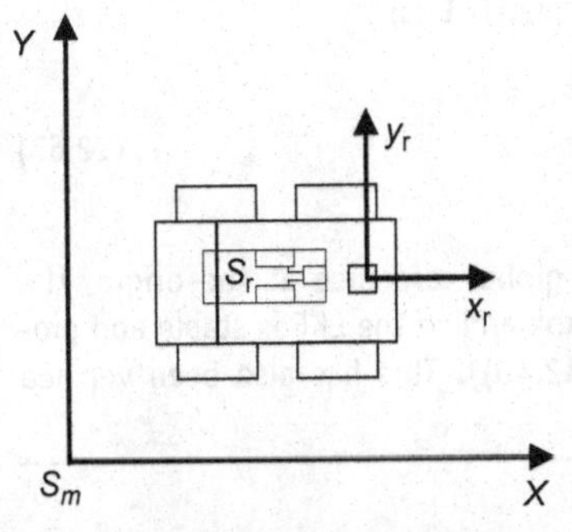

Figure 12.12 Differential drive WMR on the plane $X - Y$.

The measurement model of the sensor (here a laser range scanner) is:

$$\mathbf{z}_k = \begin{bmatrix} z_{r,k} \\ z_{\beta,k} \end{bmatrix} = \begin{bmatrix} \sqrt{(x_f^i - x_k)^2 + (y_f^i - y_k)^2} + v_{r,k} \\ tg^{-1}\left(\dfrac{y_f^i - y_k}{x_f^i - x_k}\right) - \phi_k + \dfrac{\pi}{2} + v_{\beta,k} \end{bmatrix} \quad (i = 1, 2, \ldots, m) \tag{12.64}$$

where $z_{r,k}$ and $z_{\beta,k}$ are the *range* and *bearing* of an observed point landmark with respect to the laser center of projection. The position of the ith landmark is (x_f^i, y_f^i) and the measurement noises are $v_{r,k}$ and $v_{\beta,k}$. The Jacobian matrix of this nonlinear model is:

$$\mathbf{H}_i = \begin{bmatrix} -\dfrac{x_f^1 - x_k}{d_1} & -\dfrac{y_f^1 - y_k}{d_1} & 0 & \cdots & \dfrac{x_f^m - x_k}{d_m} & \dfrac{y_f^m - y_k}{d_m} & 0 \\ \dfrac{y_f^1 - y_k}{d_1^2} & -\dfrac{x_f^1 - x_k}{d_1^2} & -1 & \cdots & -\dfrac{y_f^m - y_k}{d_m^2} & \dfrac{x_f^m - x_k}{d_m^2} & -1 \end{bmatrix} \tag{12.65a}$$

where:

$$d_i = \sqrt{(x_f^i - x_k)^2 + (y_f^i - y_k)^2}, \quad i = 1, 2, \ldots, m \tag{12.65b}$$

The measurement model of a global reference, fixed at the origin, for the nonlinear vehicle is:

$$\mathbf{h}^{(0)} = \begin{bmatrix} \sqrt{x_k^2 + y_k^2} + v_{r,k} \\ tg^{-1}(y_k/x_k) - \phi_k + \dfrac{\pi}{2} + v_{\beta,k} \end{bmatrix}$$

which has the Jacobian matrix:

$$\mathbf{H}_0 = \begin{bmatrix} x_k/q_k & y_k/q_k & 0 & 0 & \cdots \\ -y_k/q_k^2 & x_k/q_k^2 & -1 & 0 & \cdots \end{bmatrix} \tag{12.66}$$

where $q_k = (x_k^2 + y_k^2)^{1/2}$. Thus, the overall measurement matrix $\mathbf{C}$ is:

$$\mathbf{C} = \begin{bmatrix} \mathbf{H}_0 \\ \mathbf{H}_i \end{bmatrix} \tag{12.67}$$

It can be verified that with the addition of the global reference at the origin, the observability matrix $\mathbf{P}$ given by Eq. (12.47) is of full rank and so the EKF is stable and provides a steady-state covariance matrix $\mathbf{\Sigma}$ (see Eq. (12.40)). This has also been verified experimentally [16].

12.8.4 PF SLAM

The purpose of the PF is to estimate the robot's position and map parameters $\mathbf{x}_k$ (see Eq. (12.53)) for $k = 0, 1, 2, \ldots$ on the basis of the data $\mathbf{z}_k$, $k = 0, 1, 2, 3, \ldots$. The Bayes method computes the $\mathbf{x}_k$ using the posterior probability $p(\mathbf{x}_k|\mathbf{z}_0, \mathbf{z}_1, \ldots, \mathbf{z}_k; \mathbf{u}_0, \mathbf{u}_1, \ldots, \mathbf{u}_k)$. The *Markov sequential Monte Carlo* (MSMC) method (PF) is based on the total probability distribution $p(\mathbf{x}_0, \mathbf{x}_1, \ldots, \mathbf{x}_k|\mathbf{z}_0, \mathbf{z}_1, \ldots, \mathbf{z}_k; \mathbf{u}_0, \mathbf{u}_1, \ldots, \mathbf{u}_k)$ [18,22,23].

Here, the Markov stochastic model given by Eqs. (12.32a) and (12.32b) of the system is described in a probabilistic way as follows [18]:

1. $\mathbf{x}_0, \mathbf{x}_1, \ldots, \mathbf{x}_k$ is a first-order Markov process such that:
   $\mathbf{x}_k|\mathbf{x}_{k-1}$ corresponds to $P_{\mathbf{x}|\mathbf{x}_{k-1}}(\mathbf{x}|\mathbf{x}_{k-1})$ with initial distribution $P(\mathbf{x}_0)$.
2. Assuming that $\mathbf{x}_0, \mathbf{x}_1, \ldots, \mathbf{x}_k$ are known, the observations $\mathbf{z}_0, \mathbf{z}_1, \mathbf{z}_2, \ldots$ are conditionally independent, that is, $\mathbf{z}_k|\mathbf{x}_k$ is described by $P_{\mathbf{z}|\mathbf{x}}(\mathbf{z}|\mathbf{x}_k)$.

Particle methods belong to the *sampling statistical methods* which generate a set of samples that approximate the *filtering probability distribution* $p(\mathbf{x}_k|\mathbf{z}_0, \mathbf{z}_1, \ldots, \mathbf{z}_k)$.[2] Therefore, with M samples, the expectation with respect to the filtering distribution is approximately given by:

$$\int f(\mathbf{x}_k)p(\mathbf{x}_k|\mathbf{z}_0, \mathbf{z}_1, \ldots, \mathbf{z}_k)d\mathbf{x}_k \approx \frac{1}{M}\sum_{m=1}^{M} f(\mathbf{x}_k^m)$$

where $f(\mathbf{x}_k)$ can give, by the Monte Carlo technique, all moments of the distribution up to a desired degree. The particle method which is mostly used is the so-called *sequential importance resampling* (SIR) method proposed by Gordon and colleagues [18]. In this method, the filtering distribution $p(\mathbf{x}_k|\mathbf{z}_0, \mathbf{z}_1, \ldots, \mathbf{z}_k)$ is approximated by a set of M *particles (multiple copies)* of the variable of interest:

$$\{(\mathbf{x}_k^m, w_k^m), m = 1, 2, \ldots, M\}$$

where $w_k^m (m = 1, 2, \ldots, M; k = 0, 1, 2, \ldots)$ are weights that signify the relative quality of the particles, that is, they are approximations to the relative posterior probabilities (densities) of the particles such that:

$$w_k^1 + w_k^2 + \cdots + w_k^M = 1$$

SIR is a recursive (sequential, iterative) form of *importance sampling* where the expectation of a function $f(\cdot)$ is approximated by a weighted average:

$$\int f(\mathbf{x}_k)p(\mathbf{x}_k|\mathbf{z}_0, \ldots, \mathbf{z}_k)d\mathbf{x}_k \approx \sum_{m=1}^{M} w_k^m f(\mathbf{x}_k^m)$$

One of the problems encountered in PFs is the depletion of the particle population in some regions of space after some iterations. As most of the particles have

[2] For convenience, we drop the dependence on the inputs $\mathbf{u}_0, \mathbf{u}_1, \ldots, \mathbf{u}_k$.

drifted far enough, their weights become very small (close to zero) and they no longer contribute to estimates of $\mathbf{x}_k$ (i.e., they can be omitted).

The samples of $\mathbf{x}_k^m, m = 1, 2, \ldots, M$ are drawn using a *proposed probability distribution* $p_p(\mathbf{x}_k|\mathbf{x}_0, \mathbf{x}_1, \ldots, \mathbf{x}_{k-1}, \mathbf{z}_1, \ldots, \mathbf{z}_k)$. Usually, as proposed probability distribution we select the *transition prior distribution* $p(\mathbf{x}_k|\mathbf{x}_{k-1})$, which facilitates the computations.

The PF algorithm consists of several steps. At each step, for $m = 1, 2, \ldots, M$, we do the following:

1. Draw from the proposed distribution $p_p(\mathbf{x}_k|\cdots)$ samples $\mathbf{x}_k^m$:

$$\mathbf{x}_k^m \leftrightarrow p_p(\mathbf{x}_k|\mathbf{x}_0, \mathbf{x}_1, \ldots, \mathbf{x}_{k-1}; \mathbf{z}_0, \mathbf{z}_1, \ldots, \mathbf{z}_k)$$

2. Update the importance (or quality) weights up to a normalizing constant:

$$\hat{w}_k^m = w_{k-1}^m \frac{p(\mathbf{z}_k|\mathbf{x}_k^m)p(\mathbf{x}_k^m|\mathbf{x}_{k-1}^m)}{p_p(\mathbf{x}_k^m|\mathbf{x}_0^m, \mathbf{x}_1^m, \ldots, \mathbf{x}_{k-1}^m; \mathbf{z}_0, \mathbf{z}_1, \ldots, \mathbf{z}_k)} \tag{12.68a}$$

If as proposed distribution, the prior distribution $p(\mathbf{x}_k^m|\mathbf{x}_{k-1}^m)$ is used, that is, $p_p(\mathbf{x}_k^m|\mathbf{x}_0^m, \mathbf{x}_1^m, \ldots, \mathbf{x}_{k-1}^m; \mathbf{z}_0, \ldots, \mathbf{z}_k) = p(\mathbf{x}_k^m|\mathbf{x}_{k-1}^m)$, the above expression for $\hat{w}_k^m$ reduces to:

$$\hat{w}_k^m = w_{k-1}^m p(\mathbf{z}_k|\mathbf{x}_k^m) \tag{12.68b}$$

3. Compute the normalized weights as:

$$w_k^m = \hat{w}_k^m \Big/ \sum_{q=1}^{M} \hat{w}_k^q \tag{12.69}$$

4. Compute an estimate of the *effective sample size* (i.e., number of particles) ESS as:

$$\text{ESS} = 1 \Big/ \sum_{m=1}^{M} (w_k^m)^2 \tag{12.70}$$

5. If ESS $< N_{\max}$, where $N_{\max}$ is a maximum (threshold) number of particles, then perform population resampling as described below.

The PF method can be applied to the mobile robot localization problem. Actually, we have three phases:

1. *Prediction*—use a model for the simulation of the effect that a control action has on the set of particles with the noise added (see Eq. (12.32a)).
2. *Update*—use information obtained from sensors to update the weights such that to improve the probability distribution of the WMR's motion (see Eq. (12.32b)).
3. *Resampling*—draw M particles from the current set of particles with probabilities proportional to their weights and replace the current set of particles with this set selecting the weights as $w_k^m = 1/M (m = 1, 2, \ldots, M)$.

Note 1: Resampling is performed if ESS in Eq. (12.70) is ESS $< N_{\max}$ (see Section 13.13, Phase 3).

Note 2: A Matlab code for EKF- and PF-based SLAM can be found in: http://www.frc.ri.cmu.edu/projects/emergencyresponse/radioPos/index.html.

A schematic representation of the PF loop is as shown in Figure 12.13:

We recall that after a certain number of iterations, k, most weights become nearly zero and therefore the corresponding particles have negligible importance. This fact is indeed faced by the resampling process which replaces these low-importance particles with particles of higher importance.

12.8.5 Omnidirectional Vision-Based SLAM

The main issue in SLAM is that the robot can build a map of the environment and localize its position/posture in this map on the basis of noisy measurements of characteristic points (landmarks) from the images. The measurements of a catadioptric camera are particularly suited for use in conjunction with EKF in which every state variable or output variable is represented by its mean and covariance. The motion of the WMR and the measurements are described by Eqs. (12.32a) and (12.32b), where **f** and **h** are nonlinear functions of their arguments. The state vector $\mathbf{x}_k$ at time $t = kT$ $(k = 0, 1, 2, \ldots)$ contains the position vector $\mathbf{x}_{r,k}$ of the vehicle, and the vector $\mathbf{x}_f$ of the map features. The disturbances/noises $\mathbf{w}_k$ and $\mathbf{v}_k$ are assumed Gaussian with zero means and known covariances. The EKF equations involve the Jacobian matrices of **f** and **h**. For the catadioptric camera system, these

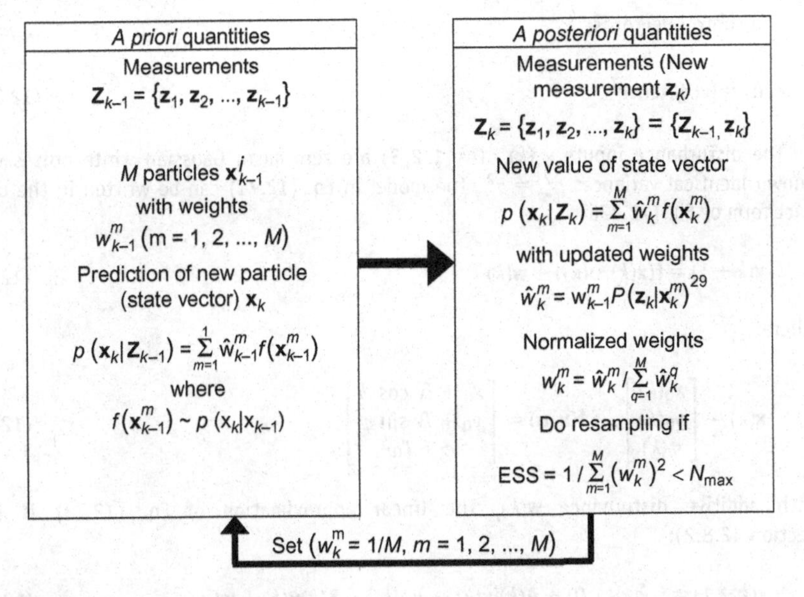

Figure 12.13 PF loop structure.

matrices have been derived in Example 9.4, and can be used to formulate the EKF equations in a straightforward way. Two examples of this application are provided in Refs. [24,25].

Example 12.5

In this example, we will develop the full EKF equations for the localization of a unicycle-type WMR using (fusing) two kinds of sensors [26]:

- Encoders on the powered wheels that provide a measure of the incremental rotation angles over a possibly varying sampling period Δt.
- A set of sonar sensors mounted on the platform of the WMR.

The WMR kinematic equations $\dot{x}_Q = v\cos\phi$, $\dot{y}_Q = v\sin\phi$, and $\dot{\phi} = \omega$ are discretized using the first-order approximation $\dot{x}(t) \simeq [x(k+1) - x(k)]/\Delta t$, giving the discrete-time model:

$$\begin{aligned} x_Q(k+1) &= x_Q(k) + Tv(k)\cos\phi(k) + w_1(k) \\ y_Q(k+1) &= y_Q(k) + Tv(k)\sin\phi(k) + w_2(k) \\ \phi(k+1) &= \phi(k) + T\omega(k) + w_3(k) \end{aligned} \tag{12.71}$$

where $t = kT$ (or $t = k$ for simplicity) and $T = \Delta t$ (constant). This model is nonlinear in the state vector:

$$\mathbf{x}(k) = [x_Q(k), y_Q(k), \phi(k)]^T \tag{12.72a}$$

The control vector is:

$$\mathbf{u}_k = [v(k), \omega(k)]^T \tag{12.72b}$$

The disturbance inputs $w_i(k)$ $(i = 1, 2, 3)$ are zero-mean Gaussian white noises with known identical variances $\sigma_{wi}^2 = \sigma_w^2$. The model in Eq. (12.71) can be written in the compact form of Eq. (12.32a):

$$\mathbf{x}(k+1) = \mathbf{f}(\mathbf{x}(k), \mathbf{u}(k)) + \mathbf{w}(k) \tag{12.73}$$

where:

$$\mathbf{x}(k) = \begin{bmatrix} x_Q(k) \\ y_Q(k) \\ \phi(k) \end{bmatrix}, \quad \mathbf{f}(\mathbf{x}, \mathbf{u}) = \begin{bmatrix} x_Q + Tv\cos\phi \\ y_Q + Tv\sin\phi \\ \phi + T\omega \end{bmatrix} \tag{12.74}$$

with additive disturbance $\mathbf{w}(k)$. The linear approximation of Eq. (12.73) is (see Section 12.8.2):

$$\mathbf{x}(k+1) = \mathbf{f}(\hat{\mathbf{x}}(k|k), \mathbf{0}) + \mathbf{A}(k)[\mathbf{x}(k) - \hat{\mathbf{x}}(k|k)] + \mathbf{B}(k)\mathbf{u}(k) + \mathbf{w}(k) \tag{12.75}$$

where $\hat{\mathbf{x}}(k|k)$ is the current estimate of $\mathbf{x}(k)$ (based on measurements up to time k), and:

$$\mathbf{f}(\hat{\mathbf{x}}(k|k),\mathbf{0})=\begin{bmatrix}\hat{x}_Q(k|k)\\ \hat{y}_Q(k|k)\\ \hat{\phi}(k|k)\end{bmatrix} \tag{12.76a}$$

$$\mathbf{A}(k)=\left[\frac{\partial\mathbf{f}}{\partial\mathbf{x}}\right]_{\hat{\mathbf{x}}(k|k),\mathbf{0}}=\begin{bmatrix}1&0&0\\0&1&0\\0&0&1\end{bmatrix} \tag{12.76b}$$

$$\mathbf{B}(k)=\left[\frac{\partial\mathbf{f}}{\partial\mathbf{u}}\right]_{\hat{\mathbf{x}}(k|k),\mathbf{0}}=\begin{bmatrix}T\cos\hat{\phi}(k|k)&0\\T\sin\hat{\phi}(k|k)&0\\0&T\end{bmatrix} \tag{12.76c}$$

Let $x_{r,i}, y_{r,i}$ be the coordinates of the ith sonar sensor in the vehicles coordinate frame Qx_ry_r, and $\phi_{r,i}$ the orientation angle of the ith sonar in Qx_ry_r as shown in Figure 12.14.

The discrete-time kinematic equations in Oxy of the ith sonar are:

$$\begin{aligned}x_i(k)&=x_Q(k)+x_{r,i}\sin\phi(k)+y_{r,i}\cos\phi(k)\\ y_i(k)&=y_Q(k)-x_{r,i}\cos\phi(k)+y_{r,i}\sin\phi(k)\\ \phi_i(k)&=\phi(k)+\phi_{r,i}(k)\end{aligned} \tag{12.77}$$

Now, consider a plane (surface) Π^j as shown in Figure 12.15, and a sonar i with *beam width* δ (all sonars are assumed to have the same beam width δ). Each plane Π^j can be represented in Oxy by p_n^j and θ_n^j, where:

- p_n^j is the (normal) distance of Π^j from the origin O of the world coordinate frame.
- θ_n^j is the angle between the normal line to the plane Π^j and the Ox direction.

The distance d_i^j of sonar i from the plane Π^j is (see Figure 12.15):

$$d_i^j=(p_n^j-x_i\cos\theta_n^j-y_i\sin\theta_n^j) \tag{12.78a}$$

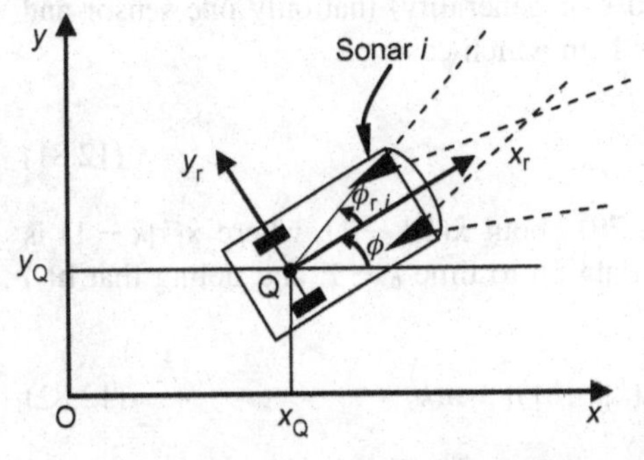

Figure 12.14 WMR with incremental encoders on the wheels and sonar sensors on its platform.

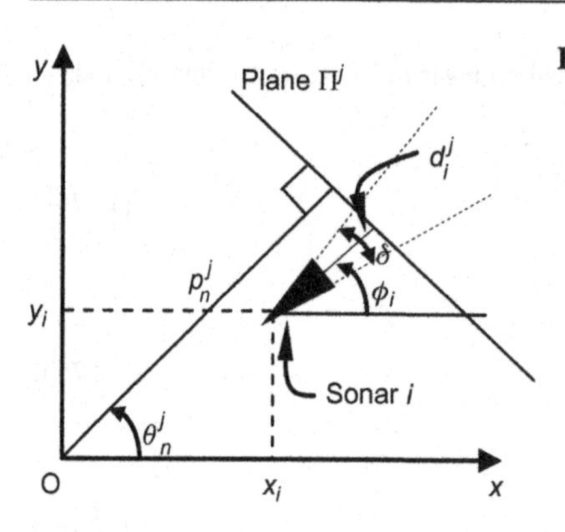

Figure 12.15 Geometry of sonar i.

for

$$\phi_i - \delta/2 \le \theta_n^j \le \phi_i + \delta/2 \tag{12.78b}$$

The measurement vector $\mathbf{z}(k)$ contains the encoder and sonar measurements and has the form:

$$\mathbf{z}(k) = \begin{bmatrix} \mathbf{z}_1(k) \\ \mathbf{z}_2(k) \end{bmatrix} = \mathbf{h}(\mathbf{x}(k)) + \mathbf{n}(k) \tag{12.79}$$

where $\mathbf{n}(k)$ is a Gaussian zero-mean white measurement noise with covariance matrix $\mathbf{R}(k) = \text{diag}[\sigma_n^2(k), \sigma_n^2(k), \ldots]$, and:

$$\mathbf{z}_1(k) = [x_Q(k) + n_1(k), y_Q(k) + n_2(k), \phi(k) + n_3(k)]^T \tag{12.80a}$$

$$\mathbf{z}_2(k) = [d_1^j(k) + n_4(k), d_2^j(k) + n_5(k), \ldots, d_{m_s}^j(k) + n_{3+m_s}(k)] \tag{12.80b}$$

with $i = 1, 2, \ldots, m_s$ (m_s is the number of sonars), $d_i^j(k)$ the distance measurement with respect to the plane $\mathbf{\Pi}^j$ provided by the ith sonar, $j = 1, 2, \ldots, m_p$ (m_p is the number of planes), and $\mathbf{h}(x(k)) = [x_Q(k), y_Q(k), \phi(k), d_1^1(k), d_2^1(k), \ldots, d_{m_s}^{m_p}]^T$

For simplicity, we assume (without loss of generality) that only one sensor and one plane are used (i.e., $m_s = 1$ and $m_p = 1$, in which case:

$$\mathbf{h}(\mathbf{x}(k)) = [x_Q(k), y_Q(k), \phi(k), d_1^1(k)]^T \tag{12.81}$$

Linearizing the measurement Eq. (12.79) about $\hat{\mathbf{x}}(k|k-1)$, where $\hat{\mathbf{x}}(k|k-1)$ is the estimate of $\mathbf{x}(k)$ using measurement data up to time $k-1$, and noting that $\mathbf{h}(\cdot)$ does not depend on $\mathbf{u}$, we get:

$$\mathbf{z}(k) = \mathbf{h}(\hat{\mathbf{x}}(k|k-1)) + \mathbf{C}(k)[\mathbf{x}(k) - \hat{\mathbf{x}}(k|k-1)] + \mathbf{n}(k) \tag{12.82}$$

where:

$$\mathbf{h}(\hat{\mathbf{x}}(k|k-1)) = \mathbf{z}(k|k-1) = [\hat{x}_Q(k|k-1), \hat{y}_Q(k|k-1), \hat{\phi}(k|k-1), d_1^1(k|k-1)]^T \tag{12.83a}$$

$$\mathbf{C}(k) = \left[\frac{\partial \mathbf{h}}{\partial \mathbf{x}}\right]_{\hat{x}(k|k-1)} = \begin{bmatrix} 1 & 0 & 0 \\ 0 & 1 & 0 \\ 0 & 0 & 1 \\ -\cos\theta_n^1 & -\sin\theta_n^1 & \begin{matrix} x_{r,1}\cos(\hat{\phi}(k|k-1)-\theta_n^1) \\ -y_{r,1}\sin(\hat{\phi}(k|k-1)-\theta_n^1) \end{matrix} \end{bmatrix} \tag{12.83b}$$

Now, having available the linearized model of Eqs. (12.75) and (12.82), we can use directly the linear Kalman filter Eqs. (12.5)–(12.12). For the robot localization we use the steps described in Section 12.7.1.

Therefore:

Step 1: One-Step Prediction

$$\begin{aligned} \hat{\mathbf{x}}(k+1|k) &= \mathbf{f}(\hat{\mathbf{x}}(k|k), \mathbf{0}) + \mathbf{B}(k)\mathbf{u}(k) \\ &= \hat{\mathbf{x}}(k|k) + \mathbf{B}(k)\mathbf{u}(k) \\ \boldsymbol{\Sigma}(k+1|k) &= \mathbf{A}(k)\boldsymbol{\Sigma}(k|k)\mathbf{A}^T(k) + \mathbf{Q}(k) \\ \mathbf{Q}(k) &= \text{diag}[\sigma_w^2, \sigma_w^2, \sigma_w^2] \end{aligned} \tag{12.84}$$

Step 2: Sensor Observation

A new set of sensors' measurements $\mathbf{z}(k+1)$ is taken at time $k+1$

Step 3: Matching

The measurement innovation process:

$$\begin{aligned} \tilde{\mathbf{z}}(k+1|k) &= \mathbf{z}(k+1) - \hat{\mathbf{z}}(k+1|k) \\ &= \mathbf{z}(k+1) - \mathbf{h}(\hat{\mathbf{x}}(k+1|k)) \end{aligned} \tag{12.85}$$

is constructed.

Step 4: Position Estimation

$$\mathbf{K}(k+1) = \boldsymbol{\Sigma}(k+1|k+1)\mathbf{C}^T(k+1)\mathbf{R}^{-1}(k+1) \tag{12.86a}$$

$$\hat{\mathbf{x}}(k+1|k+1) = \hat{\mathbf{x}}(k+1|k) + \mathbf{K}(k+1)\tilde{\mathbf{z}}(k+1|k), \mathbf{x}(0|0) = \mathbf{x}_0 \tag{12.86b}$$

$$\boldsymbol{\Sigma}(k+1|k+1) = \boldsymbol{\Sigma}(k+1|k) - \mathbf{K}(k+1)\mathbf{C}(k+1)\boldsymbol{\Sigma}(k+1|k), \quad \boldsymbol{\Sigma}(0|0) = \boldsymbol{\Sigma}_0 \tag{12.86c}$$

with $\boldsymbol{\Sigma}_0$ a given symmetric positive definite matrix.

Any valid controller that guarantees the stability of the closed-loop system can be used. Here, a dynamic state feedback linearizing and decoupling controller derived by the method of Example 6.7 will be used [26,27]:

We select the output vector:

$$\mathbf{y} = \begin{bmatrix} y_1 \\ y_2 \end{bmatrix} = \begin{bmatrix} x_Q \\ y_Q \end{bmatrix} \tag{12.87a}$$

and differentiate it to obtain:

$$\dot{\mathbf{y}} = \begin{bmatrix} \dot{y}_1 \\ \dot{y}_2 \end{bmatrix} = \mathbf{H}_1(\phi)\mathbf{u}, \quad \mathbf{H}_1(\phi) = \begin{bmatrix} \cos\phi & 0 \\ \sin\phi & 0 \end{bmatrix}, \quad \mathbf{u} = \begin{bmatrix} v \\ \omega \end{bmatrix} \tag{12.87b}$$

Clearly, ω does not influence $\dot{\mathbf{y}}$, and so we introduce a dynamic compensator

$$\dot{z} = \mu, \quad v = z \tag{12.88}$$

where μ is the linear WMR acceleration. Therefore, Eq. (12.87a) takes the form:

$$\dot{\mathbf{y}} = z\begin{bmatrix} \cos\phi \\ \sin\phi \end{bmatrix} \tag{12.89}$$

Differentiating Eq. (12.89) we get:

$$\ddot{\mathbf{y}} = \dot{z}\begin{bmatrix} \cos\phi \\ \sin\phi \end{bmatrix} + z\dot{\phi}\begin{bmatrix} -\sin\phi \\ \cos\phi \end{bmatrix} = \mathbf{H}_2(\phi)\begin{bmatrix} \mu \\ \omega \end{bmatrix} \tag{12.90}$$

where $\mathbf{H}_2(\phi)$ is the nonsingular decoupling matrix:

$$\mathbf{H}_2(\phi) = \begin{bmatrix} \cos\phi & -z\sin\phi \\ \sin\phi & z\cos\phi \end{bmatrix}, \quad \mathbf{H}_2^{-1}(\phi) = \begin{bmatrix} \cos\phi & \sin\phi \\ -(\sin\phi)/z & (\cos\phi)/z \end{bmatrix} \tag{12.91}$$

with $z = v \neq 0$.

Thus, defining new inputs w_1 and w_2 such that:

$$\ddot{y}_1 = w_1, \quad \ddot{y}_2 = w_2 \quad \text{(input} - \text{output decoupled system)} \tag{12.92}$$

and solving Eq. (12.90) for $[\mu, \omega]^{\mathrm{T}} = [\dot{z}, \omega]^{\mathrm{T}}$ we get:

$$\begin{bmatrix} \dot{z} \\ \omega \end{bmatrix} = \begin{bmatrix} \cos\phi & \sin\phi \\ -(\sin\phi)/z & (\cos\phi)/z \end{bmatrix}\begin{bmatrix} w_1 \\ w_2 \end{bmatrix}$$

which is the desired dynamic state feedback linearizing and decoupling controller:

$$\dot{z} = w_1\cos\phi + w_2\sin\phi \tag{12.93a}$$

$$v = z \tag{12.93b}$$

$$\omega = (w_2\cos\phi - w_1\sin\phi)/z \tag{12.93c}$$

with:

$$\ddot{y}_1 = w_1, \quad \ddot{y}_2 = w_2 \tag{12.93d}$$

Numerical Results—We consider the following values of the system parameters and initial conditions:

- *Initial WMR position:* $x_Q(0) = 1.4$ m, $y_Q(0) = 1.3$ m, $\phi(0) = 45°$
- *Sonar position in* $Q_{x_r y_r}$: $x_{r,1} = 0.5$ m, $y_{r,1} = 0.5$ m, $\phi_{r,1} = 0°$
- *Plane position:* $p_n^1 = 7.0$ m, $\theta_n^1 = 45°$
- *Disturbance/noise:*

$$\mathbf{Q} = \mathrm{diag}[0.1,\ 0.1,\ 0.1], \quad \mathbf{R} = \mathrm{diag}[10^{-3},\ 10^{-3},\ 10^{-3},\ 10^{-3}]$$

The desired trajectory starts at $y_{1,\mathrm{d}}(0) = x_{Q,\mathrm{d}}(0) = 1.5$ m, $y_{2,\mathrm{d}}(0) = x_{Q,\mathrm{d}}(0) = 1.5$ m and is a straight line that forms an angle of 45° with the world coordinate Ox axis as shown in Figure 12.16.

The new feedback controller $[w_1(t), w_2(t)]^{\mathrm{T}}$ is designed, as usual, using the linear PD algorithm:

$$w_1 = \ddot{y}_{1,\mathrm{d}} + K_{\mathrm{p}1}(y_{1,\mathrm{d}} - y_1) + K_{\mathrm{d}1}(\dot{y}_{1,\mathrm{d}} - \dot{y}_1)$$

$$w_2 = \ddot{y}_{2,\mathrm{d}} + K_{\mathrm{p}2}(y_{2,\mathrm{d}} - y_2) + K_{\mathrm{d}2}(\dot{y}_{2,\mathrm{d}} - \dot{y}_2)$$

Figure 12.17A shows the trajectory obtained using odometric and sonar measurements and the desired trajectory. Figure 12.17B shows the desired orientation ϕ_{d}, and the real orientation ϕ obtained using the EKF fusion. As we see, the performance of the EKF fusion is very satisfactory.

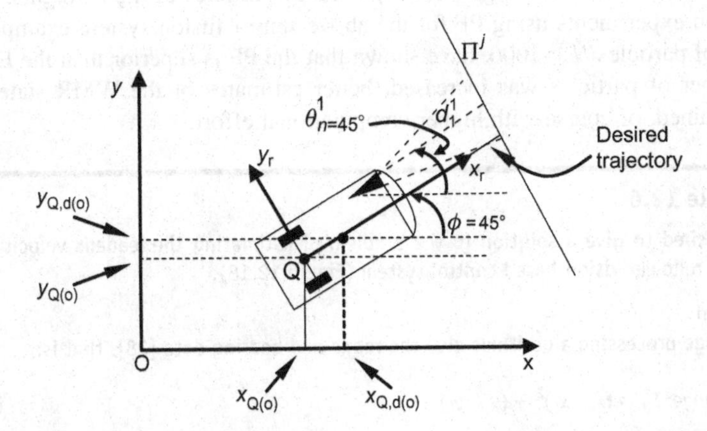

Figure 12.16 Desired trajectory of the WMR.

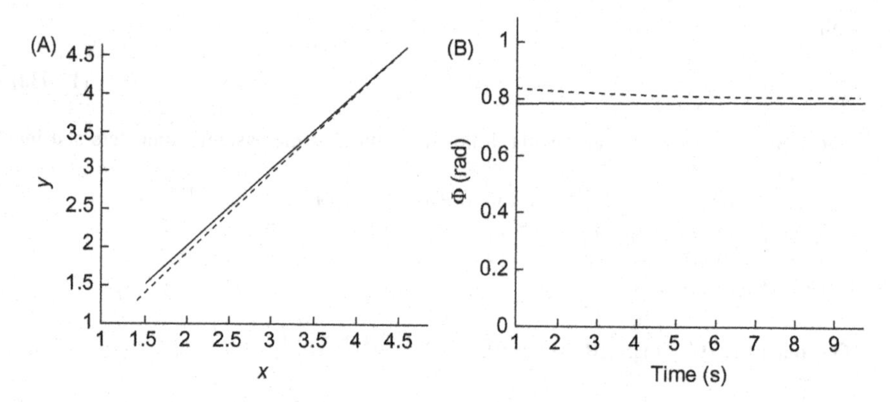

Figure 12.17 (A) Actual trajectory obtained using both kinds of sensors versus the desired trajectory (···) and (B) corresponding curves for the orientation ϕ_d (·····...) and ϕ (—).

However, the experiment showed that the trajectories obtained by pure odometric measurements, and by combined odometric and sonar data are similar. This is perhaps due to modeling reasons. Here, the case of using both sensors was treated jointly by a single EKF. A better picture of the improvement can be obtained by using the sensor fusion process of Figure 12.8 by first computing separately the (local) estimates provided by each individual kind of sensor, and then finding the combined estimate according to (12.31a) and (12.31b) which eliminates the effect of any redundant information.

Further improvement in the accuracy of the state vector estimation can be obtained using the PF approach. This is because the EKF assumes Gaussian disturbances and noise processes which in general is not so, whereas in the PF no assumptions are made about the probability distributions of the stochastic disturbances and noises. In PF, a set of weighted particles (state vector estimates) is employed, which evolve in parallel. Each iteration of the PF involves particle updating and weight updating, and convergence is assured through resampling by which particles with low weights are replaced by particles of high weights. Indeed, simulation experiments using PF for the above sensor fusion system example, with number of particles $N \geq 1000$, have shown that the PF is superior than the EKF. As the number of particles was increased, better estimates of the WMR state vector were obtained, of course with higher computational effort.

Example 12.6

It is desired to give a solution to the problem of estimating the leader's velocity v_l in a leader–follower vision-based control system (Figure 12.18).

Solution

The image processing algorithms give the range and bearing data [28], that is:

$$\textit{Range}: L_{lf}^2 = (x_l - x_f)^2 + (y_l - y_f)^2 \tag{12.94a}$$

$$\textit{Bearing}: \gamma_{lf} = \pi/2 - \phi_f + \theta_{lf}, \quad tg\theta_{lf} = (y_l - y_f)/(x_l - x_f) \tag{12.94b}$$

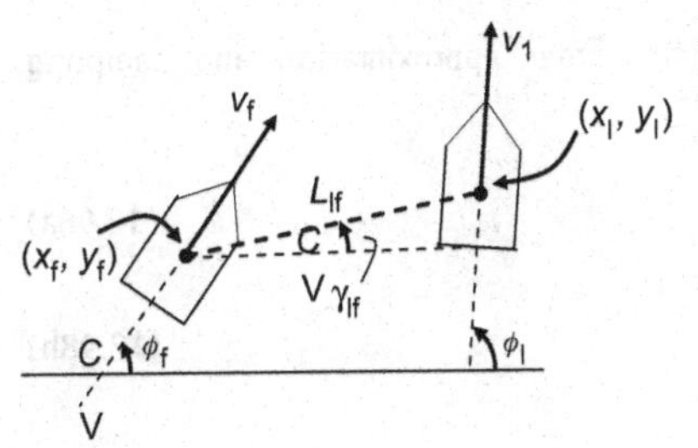

Figure 12.18 Leader–follower system with leader pose (x_l, y_l, ϕ_l) and follower pose (x_f, y_f, ϕ_f).

Differentiating L_{lf} and γ_{lf} we get:

$$\dot{L}_{lf} = (\tilde{x}_{lf}\dot{\tilde{x}}_{lf} + \tilde{y}_{lf}\dot{\tilde{y}}_{lf})/L_{lf} \tag{12.95a}$$

$$\dot{\gamma}_{lf} = (\tilde{x}_{lf}\dot{\tilde{y}}_{lf} - \tilde{y}_{lf}\dot{\tilde{x}}_{lf})/L_{lf}^2 - \dot{\phi}_f \tag{12.95b}$$

where:

$$\tilde{x}_{lf} = x_l - x_f, \quad \tilde{y}_{lf} = y_l - y_f \tag{12.95c}$$

Defining the angle ζ_{lf} as:

$$\zeta_{lf} = \gamma_{lf} + \phi_f - \phi_l \tag{12.96}$$

and assuming that $v_l = \text{const.}$ and $\omega_l = \text{const.}$, we find that the state vector under estimation is described by the nonlinear model:

$$\dot{\mathbf{X}}(t) = \mathbf{F}(\mathbf{X}, \mathbf{u}) + \mathbf{w}(t), \quad \mathbf{u} = [v_f, \omega_f]^T \tag{12.97a}$$

where $\mathbf{w}(t) \in R^6$ is a Gaussian stochastic disturbance input process with zero-mean and known covariance matrix and:

$$\mathbf{X} = \begin{bmatrix} \phi_l \\ v_l \\ \omega_l \\ L_{lf} \\ \gamma_{lf} \\ \phi_f \end{bmatrix}, \quad \mathbf{F}(\mathbf{X}, \mathbf{u}) = \begin{bmatrix} \omega_l \\ 0 \\ 0 \\ v_l \sin \zeta_{lf} - v_f \sin \gamma_{lf} \\ (v_l \cos \zeta_{lf} - v_f \cos \gamma_{lf})/L_{lf} - \omega_f \\ \omega_f \end{bmatrix} \tag{12.97b}$$

The measured output $\mathbf{z}(t)$ is:

$$\mathbf{z}(t) = \begin{bmatrix} L_{lf} \\ \gamma_{lf} \end{bmatrix} = \mathbf{H}(\mathbf{X}) + \mathbf{n}(t) \tag{12.97c}$$

where $\mathbf{n}(t)$ is the zero-mean sensor measurement Gaussian noise with known covariance matrix.

Discretizing Eqs. (12.97a)–(12.97c) with first-order approximation and sampling period T gives the nonlinear discrete-time model:

$$\mathbf{X}(k+1) = \mathbf{F}_d(\mathbf{X}(k), \mathbf{u}(k)) + \mathbf{w}(k) \tag{12.98a}$$

$$\mathbf{z}(k) = \mathbf{H}(\mathbf{X}(k)) + \mathbf{n}(k) \tag{12.98b}$$

where:

$$\mathbf{F}_d(\mathbf{X}(k), \mathbf{u}(k)) = \begin{bmatrix} \phi_l + T\omega_l \\ v_l \\ \omega_l \\ L_{lf} + T(v_l \sin \zeta_{lf} - v_f \sin \gamma_{lf}) \\ \gamma_{lf} + T\{(v_l \cos \zeta_{lf} - v_f \cos \gamma_{lf})/L_{lf}\} - \omega_f \\ \phi_f + T\omega_f \end{bmatrix}_{t=kT} \tag{12.98c}$$

We will apply the EKF technique to the nonlinear model given by Eqs. (12.98a)–(12.98c) which is linearized, as usual, about the current estimate $\hat{\mathbf{X}}(k|k)$ to give (see Eqs. (12.32a)–(12.35b)):

$$\mathbf{X}(k+1) = \mathbf{F}_d(\hat{\mathbf{X}}(k|k), \mathbf{0}) + \mathbf{A}(k)[\mathbf{X}(k) - \hat{\mathbf{X}}(k|k)] + \mathbf{B}(k)\mathbf{u}(k) + \mathbf{w}(k) \tag{12.99a}$$

where the estimate $\hat{\mathbf{X}}(k|k)$ is based on measurements $\mathbf{Z}(k) = \{\mathbf{z}(0), \mathbf{z}(1), \ldots, \mathbf{z}(k)\}$, and:

$$\mathbf{F}_d(\hat{\mathbf{X}}(k|k), \mathbf{0}) = \begin{bmatrix} \hat{\phi}_l(k|k) + T\hat{\omega}_l(k|k) \\ \hat{v}_l(k|k) \\ \hat{\omega}_l(k|k) \\ \hat{L}_{lf}(k|k) + T\hat{v}_l(k|k)\sin \hat{\zeta}_{lf}(k|k) \\ \hat{\gamma}_{lf}(k|k) + T[\hat{v}_l(k|k)\cos \hat{\zeta}_{lf}(k|k)]/L_{lf} \\ \hat{\phi}_f(k|k) \end{bmatrix} \tag{12.99b}$$

$$\mathbf{A}(k) = \left[\frac{\partial \mathbf{F}_d}{\partial \mathbf{X}}\right]_{\hat{\mathbf{X}}(k|k),0} = \left[\begin{array}{ccc|ccc} 1 & 0 & T & 0 & 0 & 0 \\ 0 & 1 & 0 & 0 & 0 & 0 \\ 0 & 0 & 1 & 0 & 0 & 0 \\ 0 & T \sin \hat{\zeta}_{lf}(k|k) & 0 & 1 & 0 & 0 \\ 0 & (T/\hat{L}_{lf}(k|k))\cos \hat{\zeta}_{lf}(k|k) & 0 & 0 & 1 & 0 \\ 0 & 0 & 0 & 0 & 0 & 1 \end{array}\right] \tag{12.99c}$$

$$\mathbf{B}(k)=\left[\frac{\partial \mathbf{F}_{\mathrm{d}}}{\partial \mathbf{u}}\right]_{\hat{\mathbf{x}}(k|k),\mathbf{0}}=\begin{bmatrix}0 & 0\\ 0 & 0\\ 0 & 0\\ -T\sin\hat{\gamma}_{\mathrm{lf}}(k|k) & 0\\ -(T/\hat{L}_{\mathrm{lf}}(k|k))\cos\hat{\gamma}_{\mathrm{lf}}(k|k) & -1\\ 0 & T\end{bmatrix} \tag{12.99d}$$

The measurement Eq. (12.97c) is linear by its own, and can be written as:

$$\mathbf{z}(k)=\begin{bmatrix}L_{\mathrm{lf}}\\ \gamma_{\mathrm{lf}}\end{bmatrix}=\mathbf{H}(\mathbf{X})+\mathbf{n}(t)=\mathbf{C}\mathbf{X}(k)+\mathbf{n}(t) \tag{12.100a}$$

where:

$$\mathbf{C}=\begin{bmatrix}0 & 0 & 0 & 1 & 0 & 0\\ 0 & 0 & 0 & 0 & 1 & 0\end{bmatrix} \tag{12.100b}$$

The linear state-space and measurement model given by Eqs. (12.99a)–(12.99d), (12.100a), and (12.100b) has the standard form of Eqs. (12.2a) and (12.2b) or Eqs. (12.75) and (12.82), and so the formulation of the EKF equations is straightforward. As an exercise the reader can write down these equations and write a computer program with proper values of the parameters.

Example 12.7

Given that a sensor is available to measure the range and the bearing of observed landmarks, it is desired to outline an algorithm that updates the *pose* (position and orientation) of a car-like WMR.

Solution

The algorithm is similar to the velocity estimation algorithm of the previous example. Here, the kinematic model of the car-like WMR is:

$$\begin{aligned}\dot{x}&=v_1\cos\phi\\ \dot{y}&=v_1\sin\phi\\ \dot{\phi}&=(1/Dv_1 tg\psi)\\ \dot{\psi}&=-a\psi+bv_2,\quad a>0\end{aligned} \tag{12.101}$$

where "a" is the decay parameter of the steering angle ψ, which is assumed to be constrained as $|\psi|<\psi_{\max}<90°$, and b is an input gain.

The control vector is:

$$\mathbf{u}=[u_1,u_2]^{\mathrm{T}}=[v_1,v_2]^{\mathrm{T}}$$

To use the EKF for updating the position and orientation of the WMR, we first discretize (as usual) this model and get (see Eqs. (12.98a) and (12.98b)):

$$\mathbf{X}(k+1) = \mathbf{F}_d(\mathbf{X}(k), \mathbf{u}(k)) + \mathbf{w}(k) \tag{12.102a}$$

$$\mathbf{z}(k) = \begin{bmatrix} z_1(k) \\ z_2(k) \\ \vdots \\ z_m(k) \end{bmatrix} = \begin{bmatrix} h_1(p_1, \mathbf{X}(k)) \\ h_2(p_2, \mathbf{X}(k)) \\ \vdots \\ h_m(p_m, \mathbf{X}(k)) \end{bmatrix} + \mathbf{n}(k) \tag{12.102b}$$

where $\mathbf{w}(k)$ and $\mathbf{n}(k) = [n_1(k), n_2(k), \ldots, n_m(k)]^\mathrm{T}$ are zero-mean Gaussian white processes with known covariance matrices $\mathbf{Q}(k)$ and $\mathbf{R}(k)$, respectively, and $z_i(k) = h_i(p_i, \mathbf{X}(k))$ is the position of the ith landmark:

$$p_i(k) = [p_{ix}(k), p_{iy}(k)]^\mathrm{T} \quad (i = 1, 2, \ldots, m) \tag{12.102c}$$

The output function $h_i(p_i, \mathbf{X}(k))$ is:

$$h_i(p_i, \mathbf{X}(k)) = \begin{bmatrix} [(p_{ix} - x(k))^2 + (p_{iy} - y(k))^2]^{1/2} \\ tg^{-1}[(p_{iy} - y(k))/(p_{ix} - x(k))] - \phi(k) \end{bmatrix} \tag{12.102d}$$

The function $\mathbf{F}_d(\mathbf{X}(k), \mathbf{u}(k))$ is given by:

$$\mathbf{F}_d(\mathbf{X}(k), \mathbf{u}(k)) = \begin{bmatrix} x(k) + Tu_1 \cos\phi(k) \\ y(k) + Tu_1 \sin\phi(k) \\ \phi(k) + (T/D)u_1 tg\psi(k) \\ \psi(k) - Ta\psi(k) + Tbu_2(k) \end{bmatrix} \tag{12.103a}$$

where:

$$\mathbf{X}(k) = [x(k), y(k), \phi(k), \psi(k)]^\mathrm{T}, \quad \mathbf{u}(k) = [u_1(k), u_2(k)]^\mathrm{T} \tag{12.103b}$$

The linearized state and measurement model is found to be:

$$\mathbf{X}(k+1) = \mathbf{F}_d(\hat{\mathbf{X}}(k|k), \mathbf{0}) + \mathbf{A}(k)[\mathbf{X}(k) - \hat{\mathbf{X}}(k|k)] + \mathbf{B}(k)\mathbf{u}(k) + \mathbf{w}(k) \tag{12.104a}$$

$$\mathbf{z}(k) = \mathbf{H}_d(\hat{\mathbf{X}}(k|k), \mathbf{0}) + \mathbf{C}(k)[\mathbf{X}(k) - \hat{\mathbf{X}}(k|k)] + \mathbf{v}(k) \tag{12.104b}$$

where:

$$F_d(\hat{\mathbf{X}}(k|k), \mathbf{0}) = \begin{bmatrix} \hat{x}(k|k) \\ \hat{y}(k|k) \\ \hat{\phi}(k|k) \\ (1 - Ta)\hat{\psi}(k|k) \end{bmatrix}$$

$$\mathbf{A}(k) = \left[\frac{\partial \mathbf{F}_d}{\partial \mathbf{X}}\right]_{\hat{\mathbf{X}}(k|k),\mathbf{0}} = \begin{bmatrix} 1 & 0 & -Tu_1 \sin\hat{\phi}(k|k) & 0 \\ 0 & 1 & Tu_1 \cos\hat{\phi}(k|k) & 0 \\ 0 & 0 & 1 & (T/D)u_1 \sec^2\hat{\phi}(k|k) \\ 0 & 0 & 0 & 1-Ta \end{bmatrix}$$

$$\mathbf{B}(k) = \left[\frac{\partial \mathbf{F}_d}{\partial \mathbf{u}}\right]_{\hat{\mathbf{X}}(k|k),\mathbf{0}} = \begin{bmatrix} T\cos\hat{\phi}(k|k) & 0 \\ T\sin\hat{\phi}(k|k) & 0 \\ (T/D)tg\hat{\psi}(k|k) & 0 \\ 0 & Tb \end{bmatrix}$$

$$\mathbf{H}_d(\hat{\mathbf{X}}(k|k),\mathbf{0}) = [h_1^T(p_1,\hat{\mathbf{X}}(k|k)),\ldots,h_m^T(p_m,\hat{\mathbf{X}}(k|k))]^T$$

$$h_i(p_i,\hat{\mathbf{X}}(k|k)) = \begin{bmatrix} [(p_{ix}-\hat{x}(k|k))^2+(p_{iy}-\hat{y}(k|k))^2]^{1/2} \\ tg^{-1}[(p_{iy}-\hat{y}(k|k))/(p_{ix}-\hat{x}(k|k))]-\phi(k) \end{bmatrix}$$

$$\mathbf{C}(k) = \left[\frac{\partial \mathbf{H}_d}{\partial \mathbf{X}}\right]_{\hat{\mathbf{X}}(k|k),\mathbf{0}} = \left[\left(\frac{\partial h_1^T}{\partial \mathbf{X}}\right),\left(\frac{\partial h_2}{\partial \mathbf{X}}\right)^T,\ldots,\left(\frac{\partial h_m}{\partial \mathbf{X}}\right)^T\right]_{\hat{\mathbf{X}}(k|k),\mathbf{0}}$$

$$\left[\frac{\partial h_i}{\partial \mathbf{X}}\right]_{\hat{\mathbf{X}}(k|k)} = \begin{bmatrix} \Delta\hat{x}(k|k)/\hat{\lambda}_p & \Delta\hat{y}(k|k)/\hat{\lambda}_p & 0 & 0 \\ -\Delta\hat{y}(k|k)/\hat{\lambda}_p^2 & \Delta\hat{x}(k|k)/\hat{\lambda}_p^2 & -1 & 0 \end{bmatrix}$$

where $\Delta\hat{x}_k(k|k) = \hat{x}(k|k) - p_{ix}$, $\Delta\hat{y}_k(k|k) = \hat{y}(k|k) - p_{iy}$, and $\hat{\lambda}_p = [\Delta\hat{x}^2(k|k)+\Delta\hat{y}^2(k|k)]^{1/2}$.

Again, the model given by Eqs. (12.104a) and (12.104b) is a standard linear time-varying stochastic model to which the EKF is applied directly. The four steps:

- One-step prediction
- Sensor observation
- Matching
- Pose estimation

provide the solution (see Section 12.7.1). In the matching step, we can use the χ^2 criterion to validate the matching (independence) of each landmark (sensor) measurement. Measurements that do not satisfy this criterion are ignored. It is remarked that the control input $\mathbf{u}(k)$ in the term $\mathbf{B}(k)\mathbf{u}(k)$ of Eq. (12.104a) is assumed to be known (since our aim here is purely the estimation of $\mathbf{X}(k)$). This input can be selected by a proper method (Chapters 5–9) for achieving a desired control objective as it was done in Example 12.5.

References

[1] Papoulis A. Probability, random variables and stochastic processes. New York, NY: Mc Graw-Hill; 1965.

[2] Meditch JS. Stochastic optimal linear estimation and control. New York, NY: Mc Graw-Hill; 1969.

[3] Anderson BDO, Moore JB. Optimal filtering. Prentice Hall, NJ: Englewood Cliffs; 1979.

[4] Borenstein J, Everett HR, Feng L. Navigating mobile robots: sensors and techniques. Wellesley, MA: A.K. Peters Ltd; 1999.

[5] Adams MD. Sensor modeling design and data processing for automation navigation. Singapore: World Scientific; 1999.

[6] Davies ER. Machine vision: theory, algorithms, practicalities. San Francisco, CA: Morgan Kaufmann; 2005.

[7] Bishop RH. Mechatronic systems, sensors and actuators: fundamentals and modeling. Boca Raton, FL: CRC Press; 2007.

[8] Leonard JL. Directed sonar sensing for mobile robot navigation. Berlin: Springer; 1992.

[9] Kleeman, L. Advanced sonar and odometry error modeling for simultaneous localization and map building. In: Proceedings of the 2004 IEEE/RSJ international conference on intelligent robots and systems, Sendai, Japan, 2004, p. 1866−71.

[10] Betke M, Gurvis L. Mobile robot localization using landmarks. IEEE Trans Rob Autom 1997;13(2):251−63.

[11] Andersen CS, Concalves JGM. Determining the pose of a mobile robot using triangulation: a vision based approach. Technical Report No I. 195-159, European Union Joint Research Center, December 1995.

[12] Hu H, Gu D. Landmark-based navigation of industrial mobile robots. Int J Ind Rob 2000;27(6):458−67.

[13] Castellanos JA, Tardos JD. Mobile robot localization and map building: a multisensor fusion approach. Berlin: Springer; 1999.

[14] Chang KC, Chong CY, Bar-Shalom Y. Joint probabilistic data association in distributed sensor networks. IEEE Trans Autom Control 1986;31:889.

[15] Durrant-Whyte HF. Uncertainty geometry in robotics. IEEE Trans Rob Autom 1988;4 (1):23−31.

[16] Vidal Calleja TA. Visual navigation in unknown environments. Ph.D. Thesis, IRI, Univ. Polit. de Catalunya, Barcelona, 2007.

[17] Guivant JE, Nebot EM. Optimization of the simultaneous localization and map-building algorithm for real-time implementation. IEEE Trans Rob Autom 2001;17 (3):242−57.

[18] Gordon NJ, Salmond DJ, Smith AFM. Novel approach to nonlinear/nonGaussian Bayesian estimation. Proc IEE Radar Signal Process 1993;140(2):107−13.

[19] Rekleitis I, Dudek G, Milios E. Probabilistic cooperative localization and mapping in practice. Proc IEEE Rob Autom Conf 2003;2:1907−12.

[20] Rekleitis I, Dudek G, Milios E. Multirobot collaboration for robust exploration. Ann Math Artif Intell 2001;31(1−4):7−40.

[21] Bailey T, Durrant−Whyte H. Simultaneous localization and mapping (SLAM), Part I. IEEE Rob Autom Mag 2006;13(2):99−110 Part II, *ibid*, (3):108−17.

[22] Doucet A, De Freitas N, Gordon NJ. Sequential Monte Carlo methods in practice. Berlin: Springer; 2001.

[23] Crisan D, Doucet A. A survey of convergence results on particle filtering methods for practitioners. IEEE Trans Signal Process 2002;50(3):736–46.

[24] Rituerto A, Puig L, Guerrero JJ. Visual SLAM with an omnidirectional camera. In: Proceedings of twentieth international conference on pattern recognition (ICPR), Istanbul, Turkey, 23–26 August, 2010, p. 348–51.

[25] Kim JM, Chung MJ. SLAM with omnidirectional stereo vision sensor. In: Proceedings of 2003 IEEE/RSJ international conference on intelligent robots and systems, Las Vegas, NV, October, 2003, p. 442–47.

[26] Rigatos GG, Tzafestas SG. Extended Kalman filtering for fuzzy modeling and multi-sensor fusion. Math Comput Model Dyn Sys 2007;13(3):251–66.

[27] Oriolo G, DeLuca A, Venditteli M. WMR control via dynamic feedback linearization: design implementation and experimental validation. IEEE Trans Control Sys Technol 2002;10(6):835–52.

[28] Das AK, Fierro R, Kumar V, Southall B, Spletzer J, Taylor CJ. Real-time mobile robot. In: Proceedings of 2001 international conference on robotics and automation, Seoul, Korea, 2001, p. 1714–19.

13 Experimental Studies

13.1 Introduction

In this book, we have presented fundamental analytic methodologies for the derivation of wheeled mobile robot (WMR) kinematic and dynamic models, and the design of several controllers. Unavoidably, these methodologies represent a small subset of the variations and extensions available in the literature, but the material included in the book is over sufficient for its introductory purposes, taking into account the required limited size of the book. All methods reported in the open literature are supported by simulation experimental results, and in many cases, the methods were applied and tested in real research mobile robots and manipulators.

The objective of this chapter is to present a collection of experimental simulation and physical results drawn from the open literature for most of the methodologies considered in the book. Specifically, the book provides sample results obtained under various artificial and realistic conditions. In most cases, the desired paths to be followed and the trajectories to be tracked are straight lines, curved lines, or circles or appropriate combinations of them. The experiments presented in the chapter cover the following problems treated in the book:

- Lyapunov-based model-based adaptive and robust control
- Pose stabilization and parking control using polar, chained, and Brockett-type integrator models
- Deterministic and fuzzy sliding mode control
- Vision-based control of mobile robots and mobile manipulators
- Fuzzy path planning in unknown environments (local, global, and integrated global–local path planning)
- Fuzzy tracking control of differential drive WMR
- Neural network-based tracking control and obstacle avoiding navigation
- Simultaneous localization and mapping (SLAM) using extended Kalman filters (EKFs) and particle filters (PFs)

By necessity, hardware, software, or numerical details are not included in the chapter. But, most of the simulation results were obtained using Matlab/Simulink functions. The reader is advised to reproduce some of the results with personal simulation or physical experiments.

Introduction to Mobile Robot Control. DOI: http://dx.doi.org/10.1016/B978-0-12-417049-0.00013-4

13.2 Model Reference Adaptive Control

Model reference adaptive control was studied in Chapter 7. Two equivalent controllers were derived. The first of them is given by Eqs. (7.40a), (7.40b), and (7.41), and the second is given by Eq. (7.77). In both cases, the mass m and the moment of inertia I of the WMR were considered to be unknown constants or slowly varying parameters. Starting the operation of the controllers using available initial guessed values of these parameters, the adaptive controllers perform simultaneously two tasks, namely, updating (adapting) the parameter values and stabilizing the tracking of the desired trajectories. As the time passes, the parameters approach their true values, and the tracking performance is improved. This fact has been verified by simulation of the controllers [1,2]. Figure 13.1A–C shows the performance of the first controller achieved with gain values $K_x = K_\phi = K_y = 5$ and adaptation parameters $\gamma_1 = \gamma_2 = 10$. The true parameter values are $m = 1$ and $I = 0.5$. The mobile robot initial pose is $x(0) = 0$, $y(0) = 0$, and $\phi(0) = 0$. Figure 13.1A depicts the convergence of the errors $\tilde{x} = x_d - x$, $\tilde{y} = y_d - y$, and $\tilde{\phi}_d = \phi_d - y$ using the true

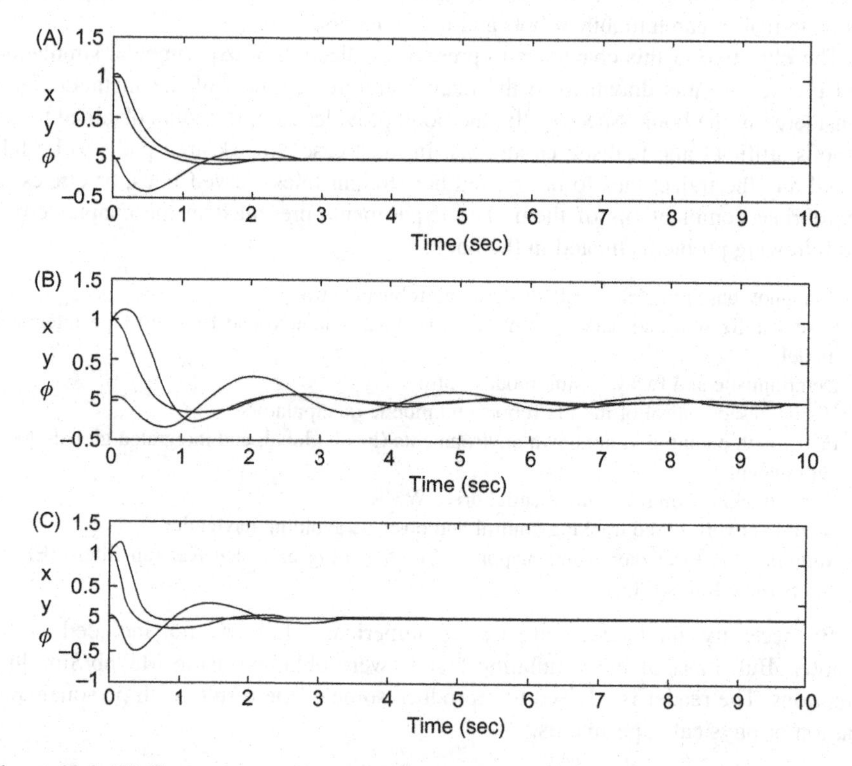

Figure 13.1 (A) Convergence with the true m and I values. (B) Convergence of the nonadaptive controller when the initial parameters are $m = 4$ and I = 2. (C) Convergence of the adaptive controller with the same initial parameter values.
Source: Reprinted from Ref. [1], with permission from European Union Control Association.

values. Figure 13.1B and C shows the error convergence performance when the initial parameter values of the nonadaptive and adaptive controller are $m=4$ and $I=2$, respectively.

The convergence when the true parameter values are used in the controller is achieved at 3 s (Figure 13.1A). The convergence of the nonadaptive controller when the parameter values are different from the true ones is achieved after 10 s (Figure 13.1B), and the convergence of the adaptive controller is achieved at about 4 s. The above results illustrate the robustness of the adaptive controller against parameter uncertainties. Analogous performance was obtained with the second adaptive controller [2]. The WMR parameters used in Ref. [2] are $\beta_1=\beta_2=0.5$ (true values of unknown dynamic parameters with known signs), $\gamma_1=\gamma_2=10$, and $K_4=K_5=100$. Two cases were studied, namely, (i) a 45° straight-line desired trajectory $x_d(t)=0.5t$, $y_d(t)=0.5t$, $\phi_d(t)=\pi/4$ rad and (ii) a unity radius circular trajectory, centered at the origin, produced by a moving point with constant linear velocity 0.5 m/s and initial pose $[x_d(0), y_d(0), \phi_d(0)]^T=[1,0,\pi/2]^T$. In the first case, the initial pose of the robot was $[x(0), y(0), \phi(0)]^T=[1,0,0]^T$. Clearly, $\phi(0)=0$ means that the robot was initially directed toward the positive x-axis. In the second case, the initial pose of the robot was $[x(0), y(0), \phi(0)]^T=[0,0,0]^T$. The results obtained in these two cases are shown in Figure 13.2 [2]. We see that in the first case, the robot is initially moving backward, and maneuvers to track the desired linear trajectory, whereas in the second case, the robot moves immediately toward the circular trajectory in order to track it. From Figure 13.2B and D, we see that the tracking (zero error reaching) times in the two cases are 2 and 1.5 s, respectively.

13.3 Lyapunov-Based Robust Control

Several simulation experiments were carried out for the differential drive WMR using the Lyapunov-based robust controller of Section 7.6, which involves the nonrobust linear controller part (7.109) and the robust controller part (7.112) [3]. In one of them, a circular reference trajectory was considered. The parameters of the nonrobust proportional feedback control part used are $K_{nrob}=\text{diag}[0.16, 0.16]$. The corresponding parameters for the robust controller are $K_{rob}=\text{diag}[0.96, 0.96]$. The simulation started with zero initial errors. The resulting (x, y) trajectories and the $\phi(t)$ trajectory are depicted in Figure 13.3. We see that the trajectories obtained with the nonrobust controller possess a deviation from the desired (reference) ones, which is eliminated by the robust controller. (Note that the dashed curves represent the desired trajectory and the solid lines the actual trajectory.)

A second simulation experiment was performed including an external disturbance pushing force ($F=-200$ N). The resulting (x, y) and $\phi(t)$ trajectories are shown in Figure 13.4, where again dotted lines indicate the reference trajectories and solid lines the actual ones.

One can see that the nonrobust controller cannot face the disturbance and leads to an unstable system. However, the robust controller leads to excellent tracking performance despite the existence of the large disturbancc.

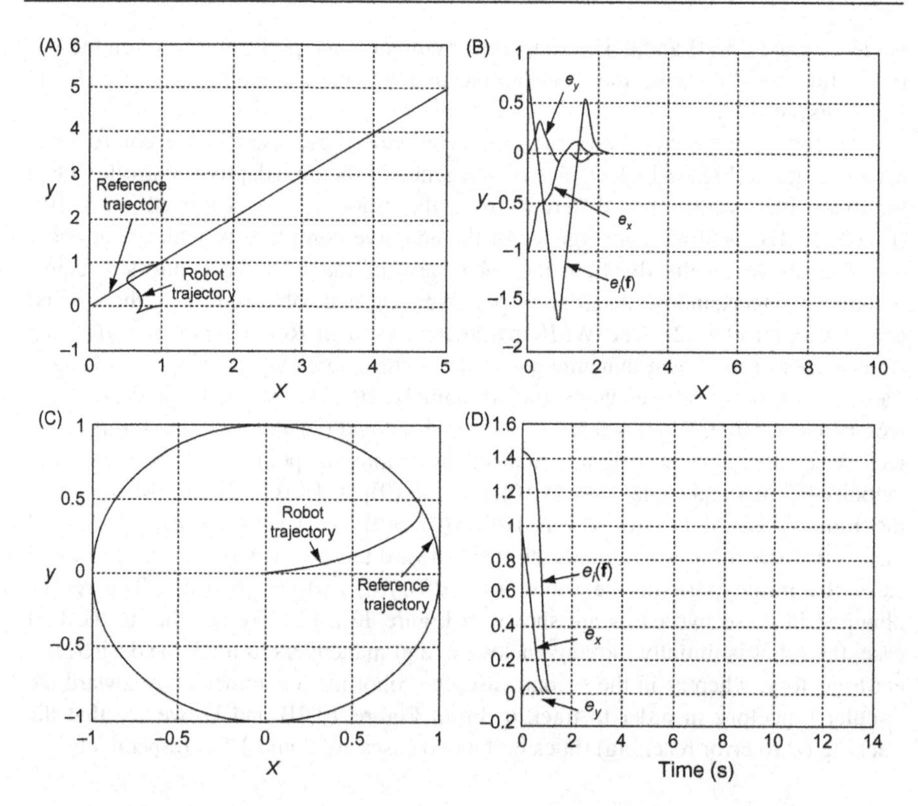

Figure 13.2 WMR trajectories and tracking errors. (A) Desired and actual robot trajectory for the straight-line case. (B) Corresponding trajectory tracking errors. (C) Desired and actual robot trajectory in the circular case. (D) Tracking errors corresponding to (C). *Source*: Reprinted from Ref. [2], with permission from Elsevier Science Ltd.

13.4 Pose Stabilizing/Parking Control by a Polar-Based Controller

The simulation was carried out using the WMR polar model (5.73a)–(5.73c) and the v, ω controllers (5.77) and (5.79):

$$v = lK_1(\cos\zeta)l, \quad K_1 > 0$$

$$\omega = K_2\zeta + K_1(\cos\zeta)(\sin\zeta)(\zeta + q_2\psi)/\zeta, \quad K_2 > 0$$

where l is the distance (position error) of the robot from the goal and ζ is the steering angle $\zeta = \psi - \phi$ (see Figure 5.11). Several cases were studied in Ref. [4]. Figure 13.5A shows the evolution of the robot's parking maneuver from a start pose until it goes to the desired goal pose.

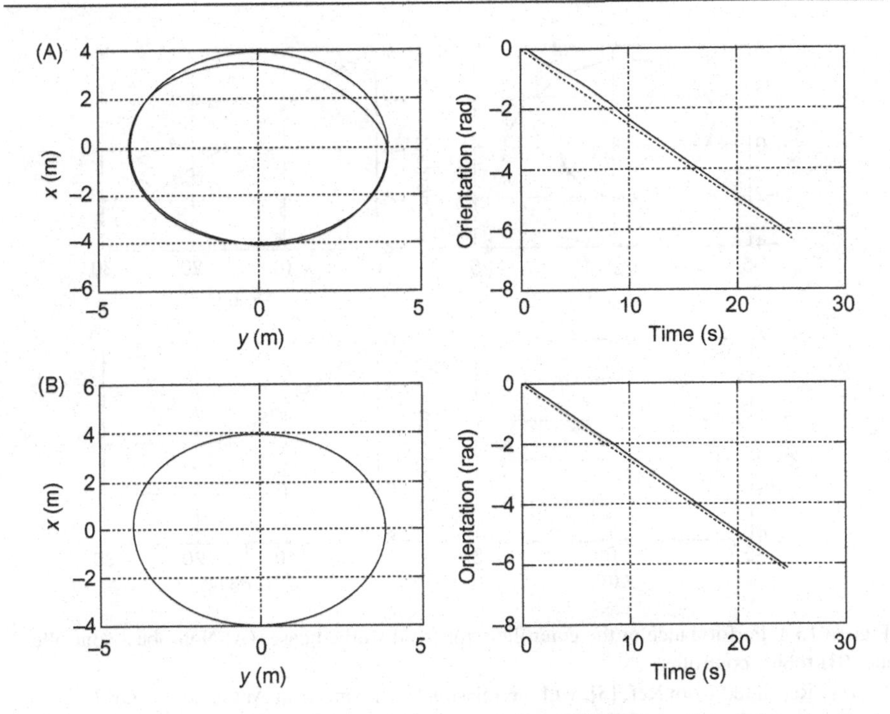

Figure 13.3 (A) Performance of the nonrobust controller. (B) Performance of the robust controller.
Source: Reprinted from Ref. [3], with permission from American Automatic Control Council.

We see that the starting pose is $(x, y, \phi) = (-1, 1, 3\pi/4)$ and the target pose is $(x, y, \phi) = (0, 0, 0)$. The maneuver trajectory shown in Figure 13.5A was obtained with gain values $K_1 = 3$, $K_2 = 6$, and $q_2 = 1$. The initial error in polar form corresponding to the initial pose is $(l, \zeta, \psi) = (\sqrt{2}, -\pi, -\pi/4)$ where $\zeta(0) = \psi(0) - \phi(0) = -\pi/4 - 3\pi/4 = -\pi$.

Figure 13.5B shows the resulting robot maneuvering for a different set of starting poses where the initial robot's orientation is always $\phi(0) = \pi/2$.

Note that the robot always approaches the parking pose with positive velocity, which is needed because $\zeta \rightarrow 0$ as the controller operates.

13.5 Stabilization Using Invariant Manifold-Based Controllers

The results to be presented here were obtained using the extended (double) Brockett integrator model (6.89) of the full differential drive WMR described by the kinematic and dynamic Eqs. (6.85a)–(6.85e). A typical "parallel parking" problem was considered with initial pose $(x, y, \phi)_0 = (0, 2, 0)$ and final

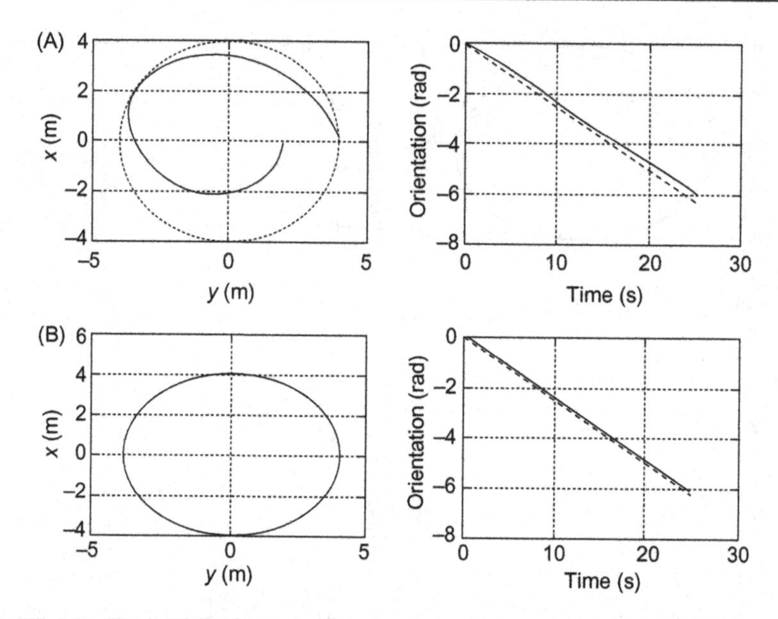

Figure 13.4 Performance of the controllers for the disturbed case. (A) Nonrobust controller and (B) robust controller.
Source: Reprinted from Ref. [3], with permission from American Automatic Control Council.

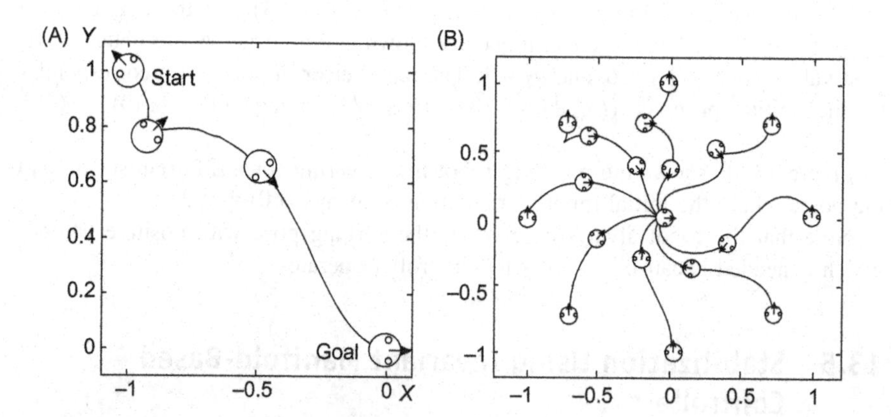

Figure 13.5 Parking movement using a polar coordinate stabilizing controller. (A) Robot parking maneuvering and (B) parking maneuver of the robot starting from a different initial pose ($\phi(0) = \pi/2$).
Source: Reprinted from Ref. [4], with permission from Institute of Electrical and Electronic Engineers.

pose $(x, y, \phi)_f = (0, 0, 0)$ [5]. The controller (6.92a) and (6.92b), derived using the invariant-attractive manifold method, has been applied, namely:

$$u_1 = -k_1 x_1 - k_2 \dot{x}_1 + k_3 x_3 x_2/(x_1^2 + x_2^2), \quad x_1^2 + x_2^2 \neq 0$$
$$u_2 = -k_1 x_2 - k_2 \dot{x}_1 - k_3 x_3 x_1/(x_1^2 + x_2^2), \quad x_1^2 + x_2^2 \neq 0$$

with gains $k_1 = 0.25$, $k_2 = 0.75$, and $k_3 = 0.25$. The parameters of the robot are $m = 10$ kg and $I = 15$ kg m^2. The control inputs are the pushing force F and the steering torque N. The resulting time evolution of these inputs is shown in Figure 13.6. When the initial conditions $x_1(0)$ and $x_2(0)$ do not violate the controller singularity condition, that is, when $x_1^2(0) + x_2^2(0) \neq 0$, the state converges to an invariant manifold, and once on the invariant manifold, no switching takes place. When $x_1^2(0) + x_2^2(0) = 0$, we use initially the controller (6.132):

$$u_2 = b\ \mathrm{sgn}(s), \quad u_1 = 0$$

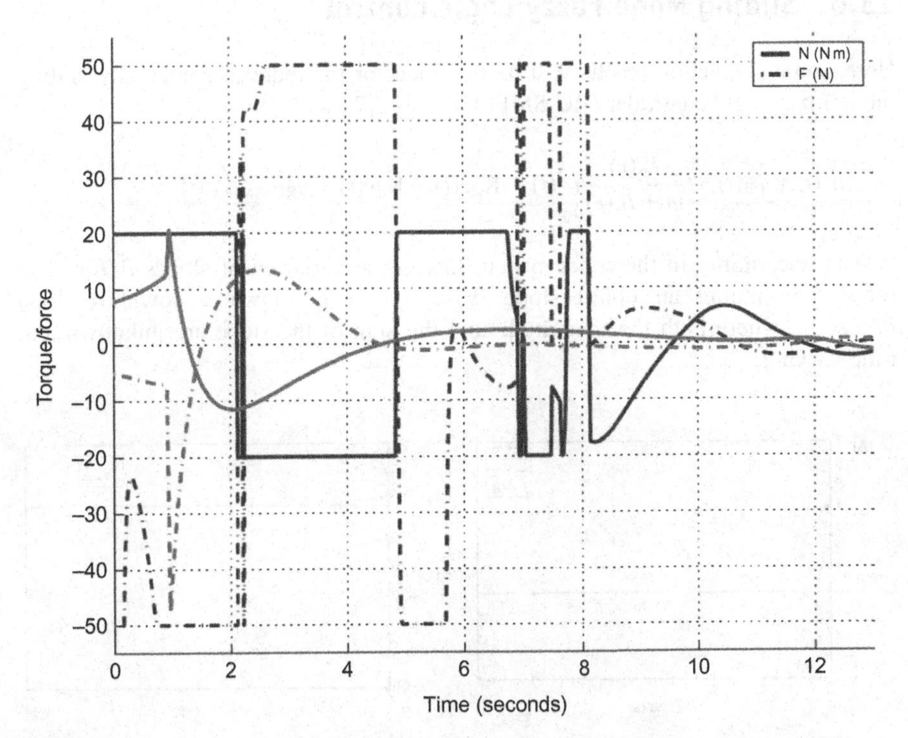

Figure 13.6 Control inputs (force and torque) for the dynamic and constrained kinematic stabilizing control.
Source: Reprinted from Ref. [5], with permission from Institute of Electrical and Electronic Engineers.

derived in Example 6.10, which drives the system out of the singularity region, and then the above controller which stabilizes to zero the system [5].

When the force and torque are constrained, the system cannot track the reference velocities provided by the kinematic controller.

The invariant manifold controllers in Eqs. (6.117a) and (6.117b) for the double integrator (kinematic) WMR model and Eq. (6.130b) for the extended Brockett integrator (full kinematic and dynamic) model were also simulated [6]. Figure 13.7A and B shows the time performance of the kinematic controller in Eqs. (6.117a) and (6.117b) with $x(0) = -1.5$ m, $y(0) = 4$ m, and $\phi(0) = -2.3$ rad, sampling period $\Delta t = 0.01$ s, and control gains $k_1 = 4$, $k_2 = 10$. The very quick convergence of $x(t), y(t)$, and $\phi(t)$ to zero is easily seen.

The performance of the full dynamic controller (6.130b) was studied for the case where $x(0) = -1.5$ m, $y(0) = 4$ m, $\phi(0) = -2.3$ rad, $v(0) = -1$ m/s, $\omega(0) = 1$ rad/s, $\Delta t = 0.01$ s, $k_1 = 1.5$, and $k_2 = 9$. The physical parameters of the WMR used are $m = 10$ kg, $I = 2$ kg m^2, $r = 0.03$ m, and $2a = 0.06$ m. A slightly better performance was observed.

13.6 Sliding Mode Fuzzy Logic Control

Here, some simulation results will be provided for the reduced complexity sliding mode fuzzy logic controller (RC-SMFLC) (8.39) [7,8]:

$$u^*(t) = \dot{v}_d(t) + \frac{I'_R(r)}{m + I_R(r)} v^2(t) + K_p^* e(t) + K_d \dot{e}(t) + u_{\text{RC-SMFLC}}(t)$$

The uncertainty in the robot system lies in the variation of slopes $\partial z/\partial r$. The robot may mount an uphill slope $\partial z/\partial r > 0$ or go down a downhill slope $\partial z/\partial r < 0$, where both the magnitude and the sign of the slope are unknown and time varying.

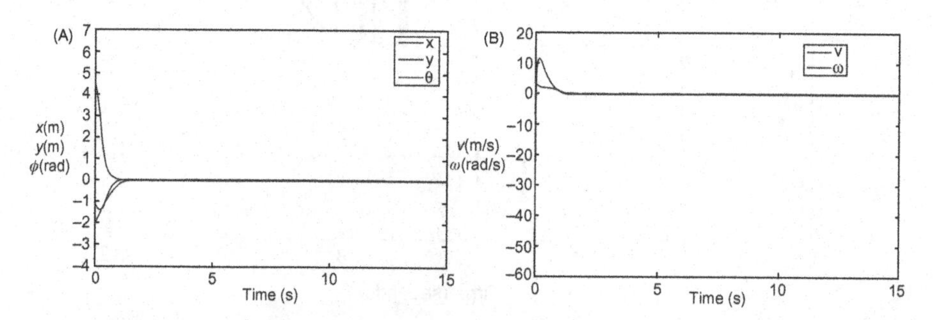

Figure 13.7 (A,B) Performance of the kinematic controller (6.117). (A) Trajectories of the states x, y, and ϕ. (B) Time evolution of linear velocity $v(t)$ and angular velocity $\omega(t)$. *Source*: Reprinted from Ref. [6], with permission from Institute of Electrical and Electronic Engineers.

The case of moving on an uphill was first considered. The fuzzy controller imitates the action of a human driver. If the slope increases, then the driver has to press more the acceleration pedal to maintain the velocity at the desired value, by compensating the increased effect of the gravitational term $[mg/(m + I_R(t))]z'(r)$. If the robot acceleration exceeds the desired value, the pressure on the acceleration must be reduced. These "increase–decrease" acceleration actions, with gradually reducing amplitude, continue until the desired action is reached. The operation of the RC-SMFLC controller is similar when the robot moves on a downhill slope. Figure 13.8 shows the velocity fluctuation and the corresponding robot trajectory for a robot moving on an uphill with slope varying between 5% and 10%. The desired velocity was 4.2 m/s. Figure 13.9 shows the results of the robot when the robot descends a downhill with slope in the interval [−10%, −5%].

The RC-SMFLC controller was also tested in the car-parking problem at a certain position of the uphill slope (here, at $x_d = 9$ m), which is in the middle of the uphill slope. Clearly, this is a simplified form of the backing up control of a truck on an uphill slope, which is a difficult task for all but the most skilled drivers. To achieve backing up the desired position, the driver has to attempt backing, move forward, back again, go forward again, and so on. Figure 13.10 shows the performance of the position controller when the robot mounts an uphill (Figure 13.10A and B) and when the robot goes down a downhill slope (Figure 13.10C and D).

13.7 Vision-Based Control

Here, a number of simulation studies carried out using vision-based controllers will be presented [9–11].

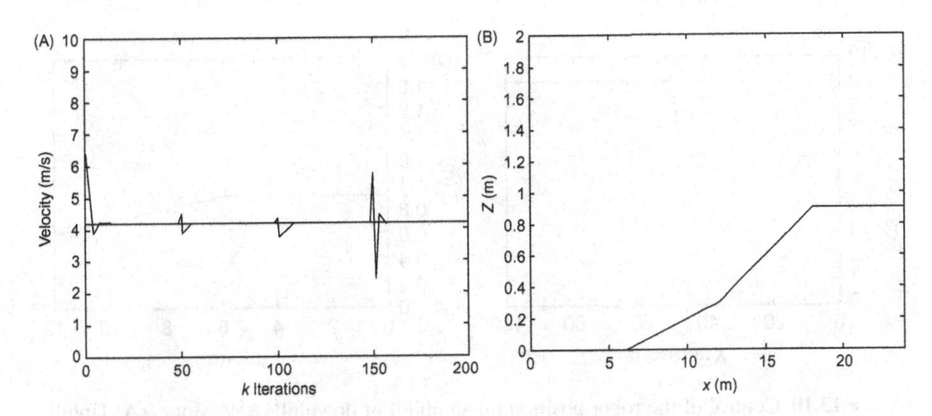

Figure 13.8 Robot performance ascending an uphill of uncertain slope. (A) Velocity fluctuation and (B) robot trajectory.
Source: Reprinted from Ref. [7], with permission from Elsevier Science Ltd.

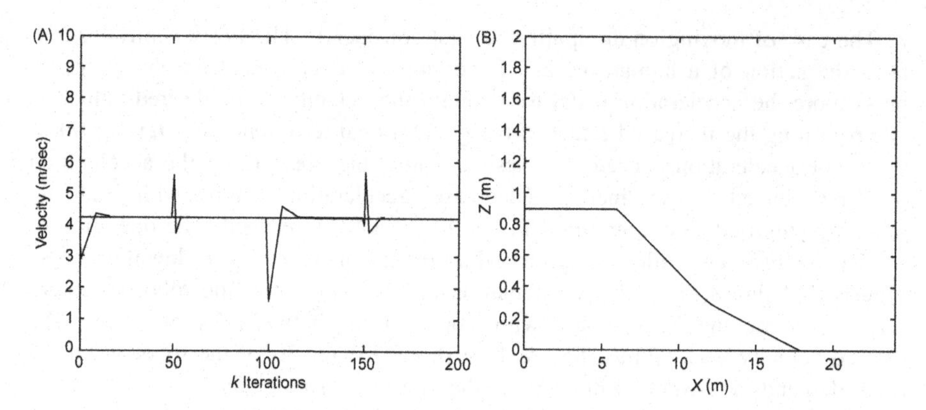

Figure 13.9 Robot performance descending a downhill of uncertain slope. (A) Velocity fluctuation and (B) robot trajectory.
Source: Reprinted from Ref. [7], with permission from Elsevier Science Ltd.

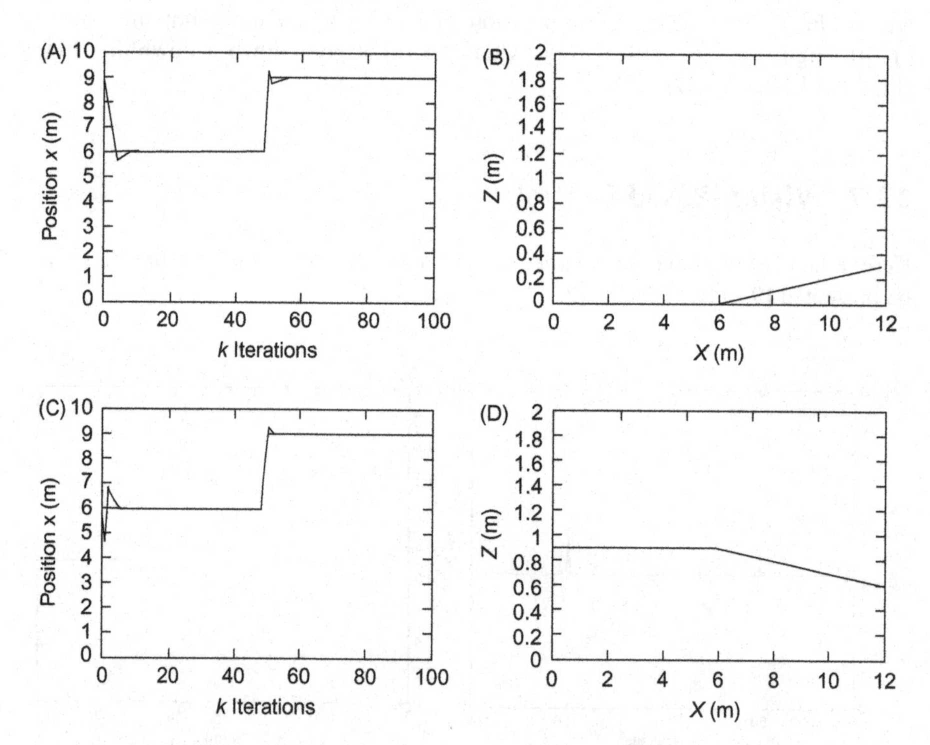

Figure 13.10 Control of the robot position on an uphill or downhill ±5% slope. (A) Uphill position evolution. (B) Robot trajectory corresponding to (A). (C) Downhill position evolution. (D) Robot trajectory corresponding to (C).
Source: Reprinted from Ref. [7], with permission from Elsevier Science Ltd.

13.7.1 Leader–Follower Control

First, some experimental results, obtained by the vision-based leader–follower tracking controller in Eqs. (9.75a) and (9.75b), with the linear velocity being estimated by Eq. (9.66), will be given [9]. These results were obtained for two Pioneer 2DX mobile robots, one following the other (Figure 13.11). The system hardware involved a frame grabber PXC200 in the vision part, which allowed image capturing by a SONY-EV-D30 camera mounted on the follower robot, and an image processor in charge of the calculation of the control actions (Pentium II-400 MHzPC).

Figure 13.12 shows the evolution of the distance to the leader. The task of the follower was to follow the leader robot maintaining a desired distance $l_d = 0.50$ m

Figure 13.11 Pioneer 2DX robots used in the experiment.
Source: Courtesy of R. Carelli.

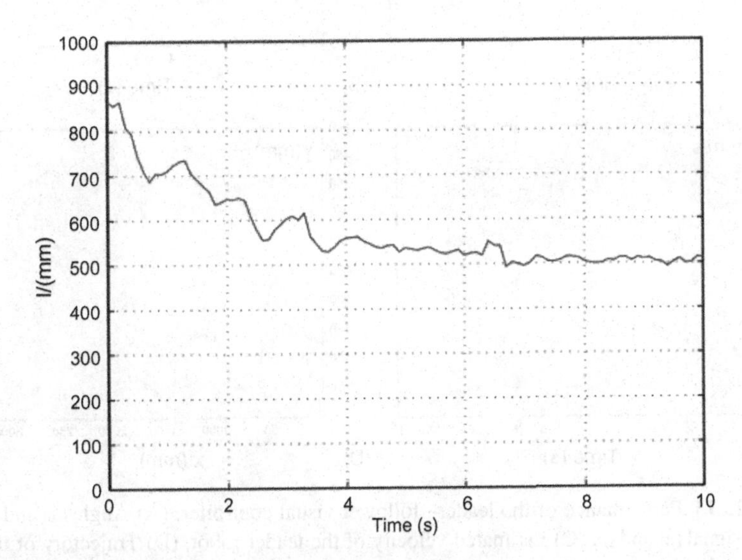

Figure 13.12 Time evolution of the leader–follower distance.
Source: Courtesy of R. Carelli.

and an angle $\varphi_d = 0°$. The desired control gains used are $K_l = 200$ and $K_\varphi = 10$, and the values of μ_l and μ_φ were $\mu_l = 0.005$ and $\mu_\varphi = 0.1$.

Figure 13.13 shows the time profiles of the angles φ and θ, the control signals v and ω of the follower robot, the estimated velocity of the leader, and the trajectory of the follower robot.

From Figures 13.12 and 13.13 we see that the visual control scheme with variable gains according to Eq. (9.75c) assures the achievement of the desired overall objectives, while avoiding possible control saturation.

13.7.2 Coordinated Open/Closed-Loop Control

Second, some results obtained using the coordinated hybrid open/closed-loop controller of Example 10.2 will be given for a mobile manipulator [10]. The MM consists of a differential drive mobile platform and a 3-DOF manipulator. The manipulator Jacobian is found analogously to the 2-DOF manipulator, by adding to Eq. (10.10) an extra term to each element of $\mathbf{J}$ corresponding to the third link (e.g., J_{11} is equal to $-l_1 \sin\theta_1 - l_2 \sin(\theta_1 + \theta_2) - l_3 \sin(\theta_1 + \theta_2 + \theta_3)$). The link lengths used in the

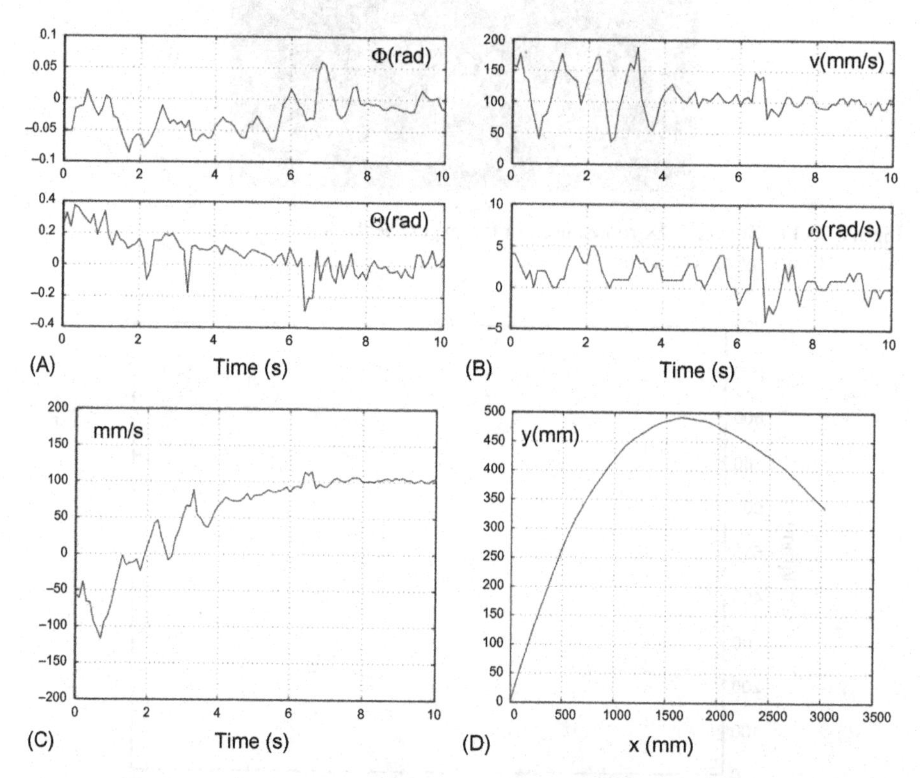

Figure 13.13 Performance of the leader–follower visual controller. (A) Angles ϕ and θ. (B) Control signals v and ω. (C) Estimated velocity of the leader robot. (D) Trajectory of the follower robot.
Source: Courtesy of R. Carelli.

simulation are $l_1 = 0.8$ m, $l_2 = 0.5$ m, and $l_3 = 0.3$ m. To maximize the manipulability index $w = |\det \mathbf{J}| = l|\sin \theta_2|$ of the manipulator, its desired configuration was selected as $\mathbf{q}_d = [0, \pi, 0]^T$. The kinematic model of the MM from the input velocities to the variation rates of the features is given in Eq. (10.62). The combination of controllers (10.63) and (10.64) for the feedback controller part was used with $k_f(\mathbf{q}_m)$, being given by Eq. (10.66). The hybrid (open-loop/closed-loop) controller scheme (10.68) with parameters $\sigma_0 = 0.15, \mu = 1$, and $K_p = 0.5$ was used (see Eqs. (10.65) and (10.67)). The times T_1 and T_2 for the open-loop controller (10.71) were selected as $T_1 = 5$ s and $T_2 = 10$ s. Three landmarks $[x, y]^T$, that is, $[1.7, 7]^T$, $[2.4, 9]^T$, and $[1, 7]^T$, were used. The initial pose of the mobile platform was at the origin, and the manipulator's initial configuration was selected close to the desired configuration. Figure 13.14 shows the features' camera measurement errors $\mathbf{e}_i(\mathbf{f})$, $i = 1, 2, 3$, the mobile base pose errors $e_x(t), e_y(t)$, the mobile platform (x, y) trajectory, and the platform's control inputs $v(t)$ and $\omega(t)$. The manipulability index value during the closed-loop control phase was $w = 0.4$, but it was sharply decreasing during the open-loop control phase. The camera

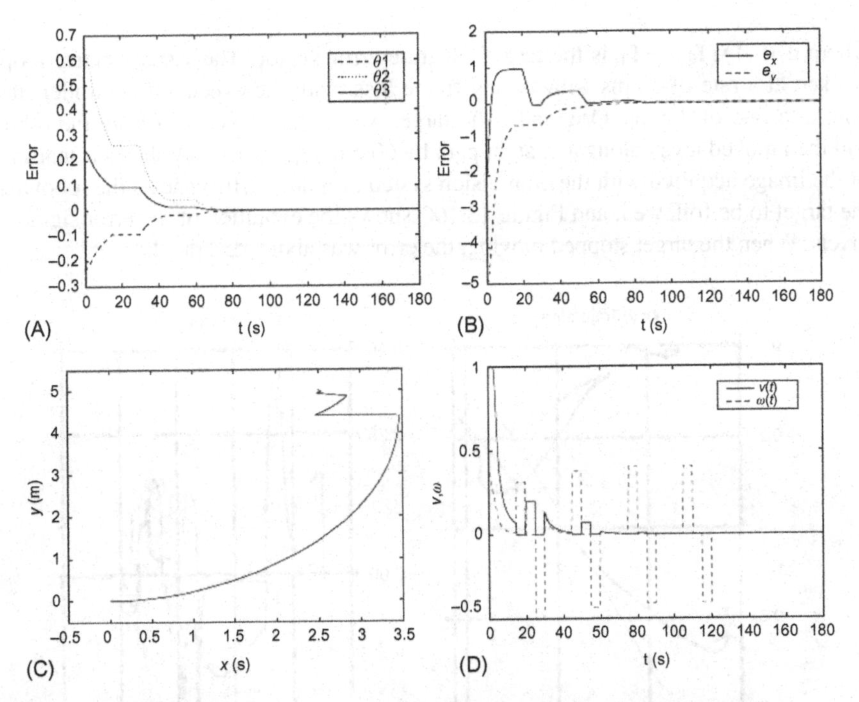

Figure 13.14 Performance of the coordinated hybrid visual controller of the MM. (A) Feature camera measurement errors. (B) Mobile platform pose errors. (C) Mobile Cartesian space trajectory. (D) Mobile platform's control inputs v and ω, which, in order to avoid the occurrence of limit cycles (due to the open-loop action), were applied only when the platform's pose errors were larger than a selected threshold.
Source: Reprinted from Ref. [10], with permission from Institute of Electrical and Electronic Engineers.

measurement errors $\mathbf{e}_i(\mathbf{f})$ were strictly decreasing during the closed-loop control phase but remained constant during the open-loop control phase.

Finally, some results obtained using a full-state image-based controller for the simultaneous mobile platform and end effector (camera) pose stabilization are depicted in Figure 13.15. This controller was designed via a merging of the results of Sections 9.2, 9.4, 9.5, and 10.3, as described in Section 10.5. The distance d of the target S along the x_w-axis of the world coordinate frame was selected as $d = 2.95$ m (see Figure 9.4), the camera focal length is $l_f = 1$ m, and the manipulator's link lengths are $l_1 = 0.51$ m and $l_2 = 0.11$ m [11].

13.7.3 Omnidirectional Vision-Based Control

Experimental results through the two-step visual tracking control procedure of Section 9.9.3 were derived in Ref. [12] using a synchro-drive WMR and a PD controller:

$$\dot{\mathbf{r}} = \mathbf{J}_{\text{im}}^{\dagger}[\mathbf{K}_p \mathbf{e}_T(\mathbf{f}_T) + \mathbf{K}_d \dot{\mathbf{e}}_T(\mathbf{f}_T)]$$

where $\mathbf{e}_T(\mathbf{f}_T) = \mathbf{f}_{T,d} - \mathbf{f}_T$ is the target's feature error vector. The visual control loops worked at a rate of 33 ms without any frame loss, with the execution controller running at a rate of 7.5 ms. Originally, the target was at a distance 1 m from the robot, and then moved away along a straight path by 60 cm. Figure 13.16A shows a snapshot of the image acquired with the omnivision system, Figure 13.16B shows the robot and the target to be followed, and Figure 13.16C shows the evolution of the error signal in pixels. When the target stopped moving, the error was about ± 2 pixels.

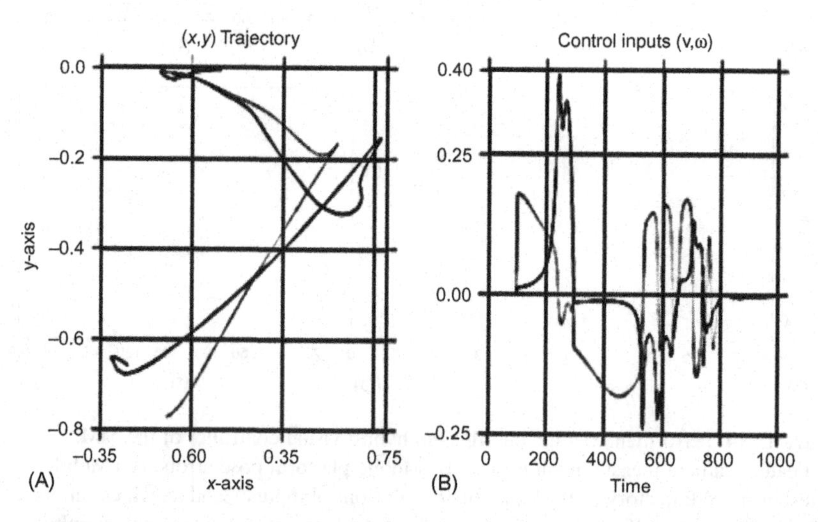

Figure 13.15 (A) Platform (x, y) trajectory from an initial to a final pose. (B) Control inputs (v, ω) profiles.
Source: Reprinted from Ref. [11], with permission from Springer Science+Business BV.

We will now present a method and some experimental results derived for a car-like WMR:

$$\dot{x} = v\cos\phi, \quad \dot{y} = v\sin\phi, \quad \dot{\phi} = (v/D)tg\psi$$

using a calibrated catadioptric vision system, modeled as shown in Figure 9.26, in which case, $\mathbf{A} = \mathbf{I}$, $\mathbf{G}_c = \mathbf{I}$, and:

$$\mathbf{x}_{im} = \mathbf{F}(\mathbf{x})$$

where $\mathbf{F}(\mathbf{x})$ is given in Eq. (9.121b). The coordinate frame F_r attached to the robot is assumed to coincide with the mirror frame F_m which implies that the camera frame and the WMR are subjected to identical kinematic constraints (Figure 13.17A).

The problem is to drive the x-axis of the WMR (control) frame parallel to a given 3D straight line L, while keeping a desired constant distance $y_{c,d}$ to the line (Figure 13.17B and C) [13]. The line L is specified by the position vector of a point Σ on the line and its cross-product with the direction vector $\mathbf{u}_L = [u_{Lx}, u_{Ly}, u_{Lz}]^T$

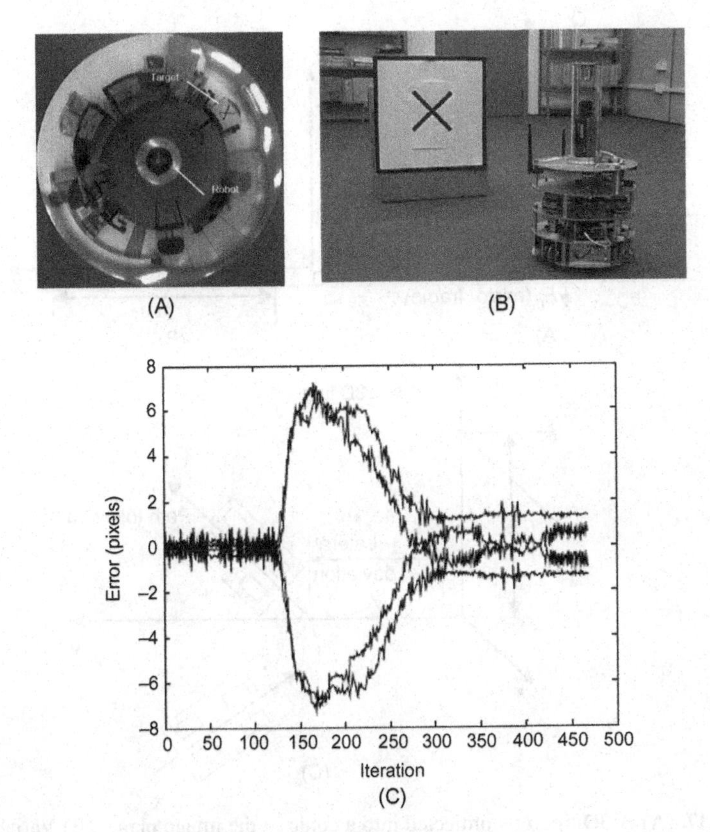

Figure 13.16 (A) Omnidirectional image showing the robot and target. (B) The robot and target. (C) Error signal. The robot started its movement at iteration 125.
Source: Courtesy of J. Okamoto Jr [12].

of the line, that is, $(\mathbf{u}_L; \overrightarrow{\mathrm{O_m}\Sigma} \times \mathbf{u}_L)$ in the mirror frame. Defining the vector $\mathbf{n} = \overrightarrow{\mathrm{O_m}\Sigma} \times \mathbf{u}_L / \| \overrightarrow{\mathrm{O}\Sigma} \times \mathbf{u}_L \| = [n_x, n_y, n_z]^{\mathrm{T}}$, the line L is expressed in the so-called *Plücker coordinates* $[\mathbf{u}_L^{\mathrm{T}}; \mathbf{n}^{\mathrm{T}}]^{\mathrm{T}}$ with $\mathbf{u}_L^{\mathrm{T}}\mathbf{n} = 0$. Let S be the intersection between the interpretation plane Π (defined by the line L and the mirror focal point $\mathrm{O_m}$) and the mirror surface. Clearly, S is the line's projection in the mirror surface. The projection S of the line L in the catadioptric image plane is obtained using the relation $\mathbf{x}_{\mathrm{im}} = \mathbf{f}(\mathbf{x})$, where $\mathbf{x}$ is an arbitrary point $\mathbf{x} = [x, y, z]^{\mathrm{T}}$ on L, and:

$$\mathbf{f}(\mathbf{x}) = \begin{bmatrix} x \\ y \\ z + \zeta\sqrt{x^2 + y^2 + z^2} \end{bmatrix} \qquad (\zeta = d')$$

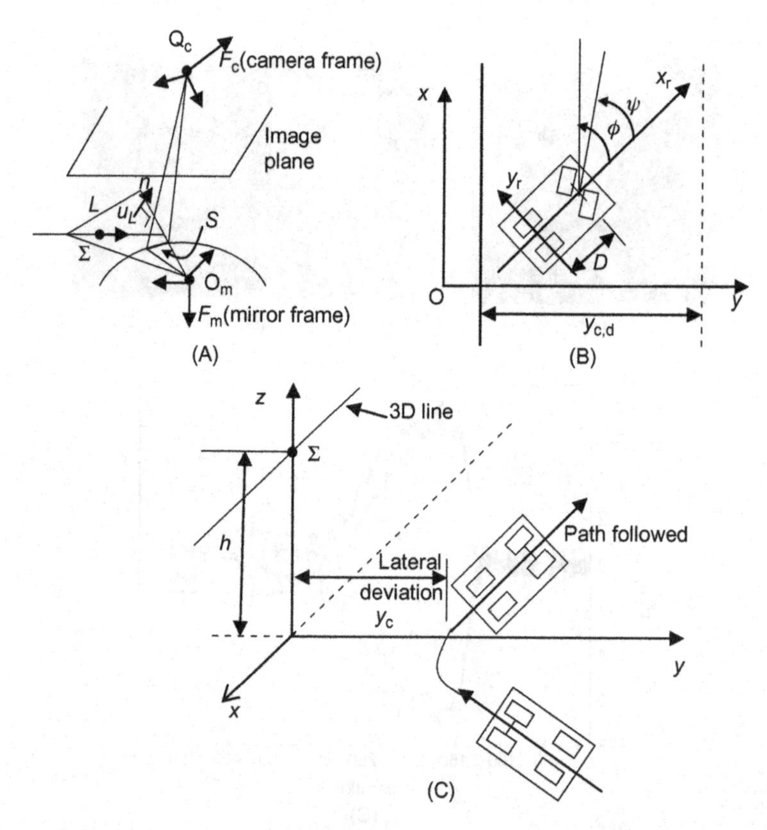

Figure 13.17 (A) A 3D line L is projected into a conic in the image plane. (B) Variables and parameters of the robot and the problem. (C) Pictorial representation of the task to be performed.
Source: Adapted from Refs. [13,14].

under the condition:

$$\mathbf{n}^{\mathrm{T}}\mathbf{x} = n_x x + n_y y + n_z z = 0$$

which expresses the fact that **n** is orthogonal to the interpretation plane. Inverting the relation $\mathbf{x}_{\mathrm{im}} = \mathbf{f}(\mathbf{x})$, we get:

$$\mathbf{x} = \mathbf{f}^{-1}(\mathbf{x}_{\mathrm{im}}), \quad \mathbf{x}_{\mathrm{im}} = [x_{\mathrm{im}}, y_{\mathrm{im}}, z_{\mathrm{im}}]^{\mathrm{T}}$$

where:

$$\mathbf{f}^{-1}(\mathbf{x}_{\mathrm{im}}) = \begin{bmatrix} m(\mathbf{x}_{\mathrm{im}})x_{\mathrm{im}} \\ m(\mathbf{x}_{\mathrm{im}})y_{\mathrm{im}} \\ m(\mathbf{x}_{\mathrm{im}})z_{\mathrm{im}} - \zeta \end{bmatrix}$$

with:

$$m(\mathbf{x}_{\mathrm{im}}) = \frac{z_{\mathrm{im}}\zeta + \sqrt{z_{\mathrm{im}}^2 + (1-\zeta^2)(x_{\mathrm{im}}^2 + y_{\mathrm{im}}^2)}}{x_{\mathrm{im}}^2 + y_{\mathrm{im}}^2 + z_{\mathrm{im}}^2}$$

Introducing the above expression for **x** into the condition $\mathbf{n}^{\mathrm{T}}\mathbf{x} = 0$, we get:

$$\mathbf{n}^{\mathrm{T}}\mathbf{f}^{-1}(\mathbf{x}_{\mathrm{im}}) = [n_x, n_y, n_z]\begin{bmatrix} m(\mathbf{x}_{\mathrm{im}})x_{\mathrm{im}} \\ m(\mathbf{x}_{\mathrm{im}})y_{\mathrm{im}} \\ m(\mathbf{x}_{\mathrm{im}})z_{\mathrm{im}} - \zeta \end{bmatrix} = 0$$

which after some algebraic manipulation can be written in the standard quadratic form [13,14]:

$$\mathbf{x}_{\mathrm{im}}^{\mathrm{T}}\mathbf{H}\mathbf{x}_{\mathrm{im}} = 0$$

$$\mathbf{H} = \begin{bmatrix} n_x^2(1-\zeta^2) - n_x^m\zeta^2 & n_x n_y(1-\zeta^2) & (2m-3)n_x n_z^{m-1} \\ n_x n_y(1-\zeta^2) & n_y^2(1-\zeta^2) - n_z^m\zeta^2 & (2m-3)n_y n_z^{m-1} \\ (2m-3)n_x n_z^{m-1} & (2m-3)n_y n_z^{m-1} & n_z^m \end{bmatrix}$$

where $m = 2$ in the general case, and $m = 1$ in the parabolic mirror–orthographic camera case. In expanded form, the above quadratic equation can be expressed in the following normalized form:

$$h_0 x_{\mathrm{im}}^2 + h_1 y_{\mathrm{im}}^2 + 2h_2 x_{\mathrm{im}} y_{\mathrm{im}} + 2h_3 x_{\mathrm{im}} + 2h_4 y_{\mathrm{im}} + 1 = 0$$

where $h_i = a_i/a_5 \quad (i = 1, 2, 3, 4, 5)$, and:

$$a_0 = \lambda[n_x^2(1-\zeta^2) - n_z^m \zeta^2], \quad a_1 = \lambda[n_y^2(1-\zeta^2) - n_z^m \zeta^2]$$
$$a_2 = \lambda n_x n_y (1-\zeta^2), \quad a_3 = \lambda(2m-3) n_x n_z^{m-1}$$
$$a_4 = \lambda(2m-3) n_y n_z^{m-1}, \quad a_5 = \lambda n_z$$

where λ is a scaling factor. The nondegenerate case $n_z \neq 0$ is considered. The state of the WMR (and camera) is:

$$\mathbf{x}_c = [x_c, y_c, \phi]^T$$

where x_c, y_c are the world coordinates of the camera and ϕ the angular deviation with respect to the straight line. The task is achieved when the lateral deviation y_c is equal to the desired deviation $y_{c,d}$, and the angular deviation is zero. These deviations are expressed in terms of image features as [14]:

$$y_c = h/\sqrt{h_3^2 + h_4^2}, \quad \phi = tg^{-1}\, h_3/h_4$$

assuming $h_3^2 + h_4^2 \neq 0$ and $h_3/h_4 \neq 0$ (i.e., that the line L does not lie on the xy plane of the camera frame F_c, and the x-axis of the mirror is not perpendicular to the line L). The parameter h is shown in Figure 13.17C. Choosing $z_1 = x_c$ in the WMR chained model we get:

$$u_1 = v \cos \phi$$

Further, choosing $z_2 = y_c$ we have $\dot{z}_2 = v \sin \phi$, and so $z_3 = tg\phi$ (for $\phi \neq \pi/2$). Now:

$$\begin{aligned} u_2 = \dot{z}_3 &= \dot{\phi}/\cos^2 \phi \\ &= (v/D) tg\psi / \cos^2 \phi \end{aligned}$$

Dividing the equations of the chained model by u_1 we get:

$$z_1' = 1, \quad z_2' = z_3, \quad z_3' = u_3$$

where $u_3 = u_2/u_1$. Selecting the state feedback controller u_3 for the above linear system as:

$$u_3 = -K_d z_3 - K_p z_2$$

we get the closed-loop error (deviation) equation:

$$\dot{z}_2' + K_d z_2' + K_p z_2 = 0$$

Therefore, selecting proper values for the gains $K_{\rm d}$ and $K_{\rm p}$ assures that $z_2 \to 0$ and $z_3 \to 0$, as $t \to \infty$ (with desired convergence rates) independently of the longitudinal velocity as long as $v \neq 0$. From the relations $z_2 = y_{\rm c}$ and $z_3 = tg\phi$, it follows that also $y_{\rm c} \to 0$ and $\phi \to 0$. The expression for the feedback control steering angle ψ is found from:

$$\begin{aligned} u_3 = u_2/u_1 &= \frac{(v/D)tg\psi/\cos^2\phi}{v\cos\phi} \\ &= (1/D)tg\psi/\cos^3\phi \end{aligned}$$

that is:

$$\begin{aligned} tg\psi &= D(\cos^3\phi)u_3 \\ &= -D(\cos^3\phi)[K_{\rm d}z_3 + K_{\rm p}z_2] \\ &= -D(\cos^3\phi)[K_{\rm d}tg\phi + K_{\rm p}y_{\rm c}] \end{aligned}$$

In terms of the measured features h_3 and h_4, this controller is expressed as:

$$tg\psi = -D\left[\cos^3\left(tg^{-1}\frac{h_3}{h_4}\right)\right]\left[K_{\rm d}\frac{h_3}{h_4} + K_{\rm p}\frac{h}{\sqrt{h_3^2+h_4^2}}\right]$$

The parameter h is obtained from proper measurement.

The above controller was tested by simulation for both a *paracatadioptric sensor* (parabolic mirror plus orthographic camera) and a *hypercatadioptric sensor* (hyperbolic mirror plus perspective camera) [13,14]. In the first case, the image of a line is a circle and in the second case is a part of a conic. Figure 13.18A and B shows the line configuration and actual robot trajectory in the world coordinate frame, and the profiles of the lateral and angular deviations provided by the simulation.

Similar results were reported for the case of hypercatadioptric system [13].

13.8 Sliding Mode Control of Omnidirectional Mobile Robot

Here, some simulation results for the omnidirectional mobile manipulator of Section 10.3.4, with three links ($n_{\rm m} = 3$) will be given. The sliding mode controller (10.53b) has been applied, replacing the switching term sgn(s) by the saturation function approximation sat(s_i/ϕ_i) as described in Section 7.2.2. The parameter values of the MM are given in Table 13.1. The masses and moments of inertia were assumed to contain uncertainties in intervals between $\pm 5\%$ and $\pm 30\%$ of

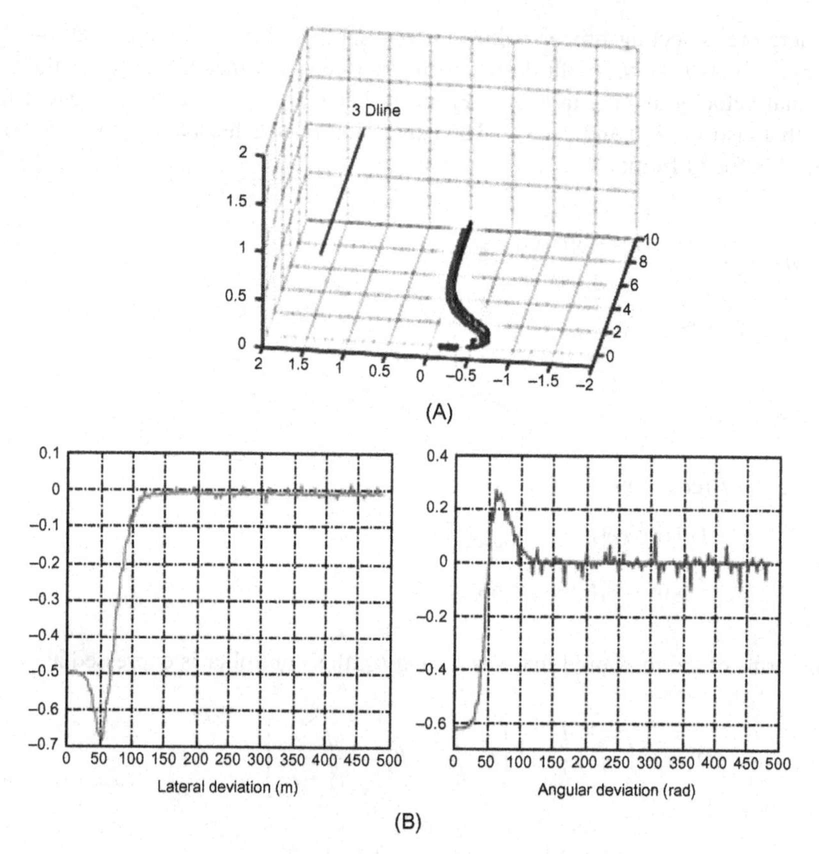

Figure 13.18 Omnidirectional visual servoing results with $(K_p, K_d) = (1, 2)$. (A) Desired line and actual robot trajectory, (B) Evolution of lateral and angular deviations.
Reprinted from [13], with permission from Institute of Electrical and Electronic Engineers.

their nominal values (Table 13.1). The simulation results were obtained using the extreme values as the real ones, and the mean values of the true and extreme values as the estimated values for the computation of the control signals.

Two cases were studied:

1. Linear desired trajectory of the platform's center of gravity
2. Circular desired trajectory of the platform's center of gravity

In both cases, an angular platform velocity of 1°/s with respect to the world coordinate frame was applied. For the joint variables, a ramp reference was used. The results obtained via simulation are shown in Figures 13.19 and 13.20 [15]. The results obtained using the exact values in the computed torque method showed that the sliding mode controller (with the "sat" function approximation) can indeed face large parametric uncertainties and follow the desired trajectories successfully.

Table 13.1 Parameter Values of the Robotic Manipulator

Physical Variable	True Value	Extreme Uncertain Values
Mass of each wheel	0.5 kg	0.525 kg (+5%)
Radius of each wheel	0.0245 m	0.0245 m
Mass of the platform	30 kg	33 kg (+10%)
Wheel distance from the platform COG	0.178 m	0.178 m
Platform moment of inertia with respect to axis z	0.93750 kg m^2	0.98435 kg m^2 (+5%)
Mass of link 1	1.25 kg	1.375 kg (+10%)
Length of link 1	0.11 m	0.11 m
Link 1 moment of inertia with respect to axis x	0.01004 kg m^2	0.010542 kg m^2 (+5%)
Mass of link 2	4.17 kg	5.421 kg (+30%)
Length of link 2	0.5 m	0.5 m
Position of link's COG along its axis	0.25 m	0.25 m
Link 2 moment of inertia with respect to axis x	0.34972 kg m^2	0.367206 kg m^2 (+5%)
Link-2 moment of inertia with respect to axis z	0.00445 kg m^2	0.0046725 kg m^2 (+5%)
Mass of link 3	0.83 kg	1.0790 kg (+30%)
Length of link 3	0.10 m	0.10 m
Position of link's 3 COG along its axis	0.05 m	0.05 m
Link 3 moment of inertia with respect to axis x	0.00321 kg m^2	0.0033705 kg m^2 (+5%)
Link 3 moment of inertia with respect to axis z	0.00089 kg m^2	0.0009345 kg m^2 (+5%)
Linear friction coefficient for all rotating joints	0.1 N m s	0.13 (+30%)

13.9 Control of Differential Drive Mobile Manipulator

13.9.1 Computed Torque Control

In the following discussion, some simulation results obtained by the application of the computed torque controller ((10.49a) and (10.49b)) to the 5-DOF MM of Figure 10.8 will be provided. The parameters of the MM are given in Table 13.2 [16].

The desired performance specifications for the error dynamics were selected to be $\zeta = 1$ and $\omega_n = 4$ (settling time $T_s = 1$ s). Then, the gain matrices $\mathbf{K}_p$ and $\mathbf{K}_\upsilon$ in Eq. (10.49b) are:

$$\mathbf{K}_p = \text{diag}[36, 36, 36, 36], \quad \mathbf{K}_\upsilon = \text{diag}[12, 12, 12, 12]$$

The control torques of the left and right wheel and the forearms and upper arms that result with the above gains are shown in Figure 13.21A and B. The desired trajectories of the manipulator front point O_b and the end effector tip E are shown in Figure 13.21C. The corresponding actual animated paths are shown in Figure 13.21D.

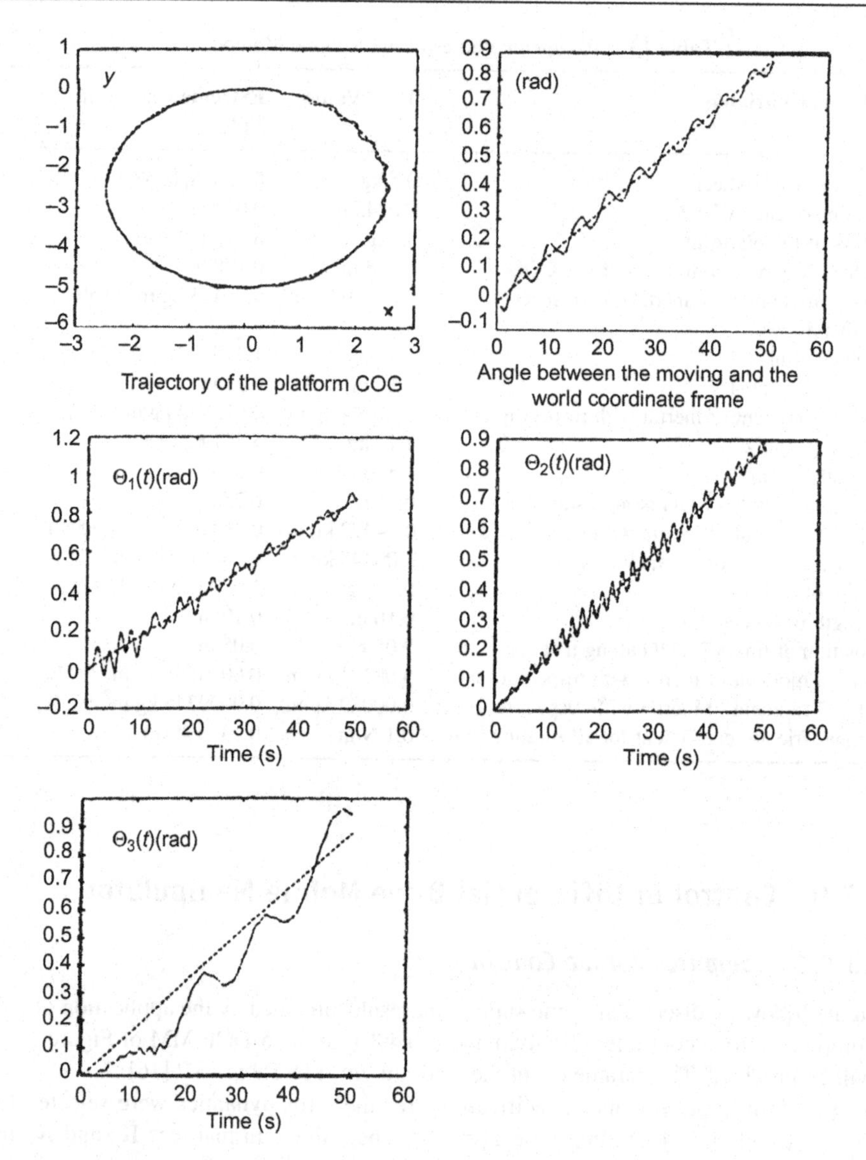

Figure 13.19 Control performance for the tracking of the circular trajectory of the platform's center of gravity using computed torque control.

13.9.2 Control with Maximum Manipulability

Here, we will present a few representative simulation results obtained in Refs. [17,18] for the MM of Figure 10.11. The platform parameters used are those of the LABMATE platform (Transition Research Corporation) and the manipulator parameters are $m_1 = m_2 = 4$ kg, $l_1 = l_2 = l = 0.4$ m, and $I_1 = I_2 = 0.0533$ kg m^2.

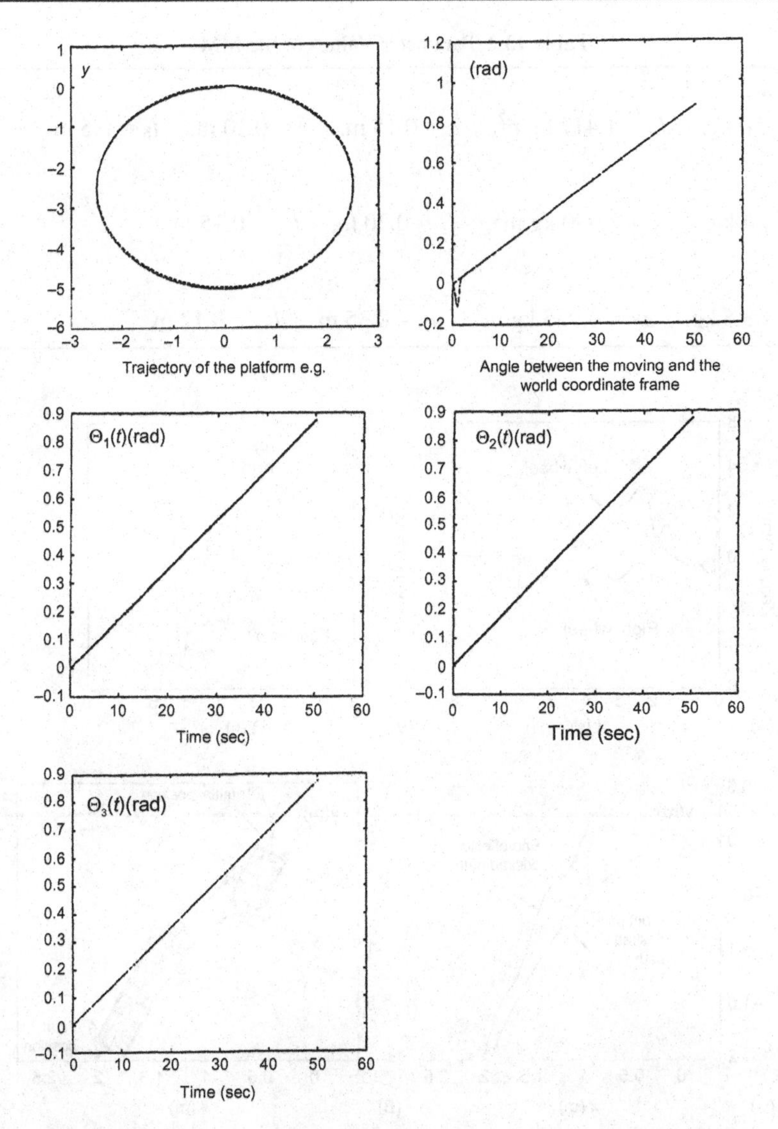

Figure 13.20 Control performance for the tracking of the circular trajectory of the platform's center of gravity using sliding mode control.

The center of gravity of each link was assumed to be at the midpoint of the link, and the sampling period was set to $T = 0.01$ s.

13.9.2.1 Problem (a)

The controller ((10.55a) and (10.55b)) was applied leading to the decoupled system (10.55c). This system was controlled by a diagonal (decoupled) PD tracking

Table 13.2 Parameter Values of the MM

Platform
$m = 50$ kg, $J_0 = 1.417$ kgm^2, $a = 0.15$ m, $r = 0.10$ m, $l_b = 0.5$ m
Link 1
$m_1 = 4$ kg, $J_1 = 0.030$ kg m^2, $l_1 = 0.30$ m, $l_{c1} = 0.15$ m
Link 2
$m_2 = 3.5$ kg, $J_2 = 0.035$ kg m^2, $l_2 = 0.35$ m, $l_{c2} = 0.12$ m

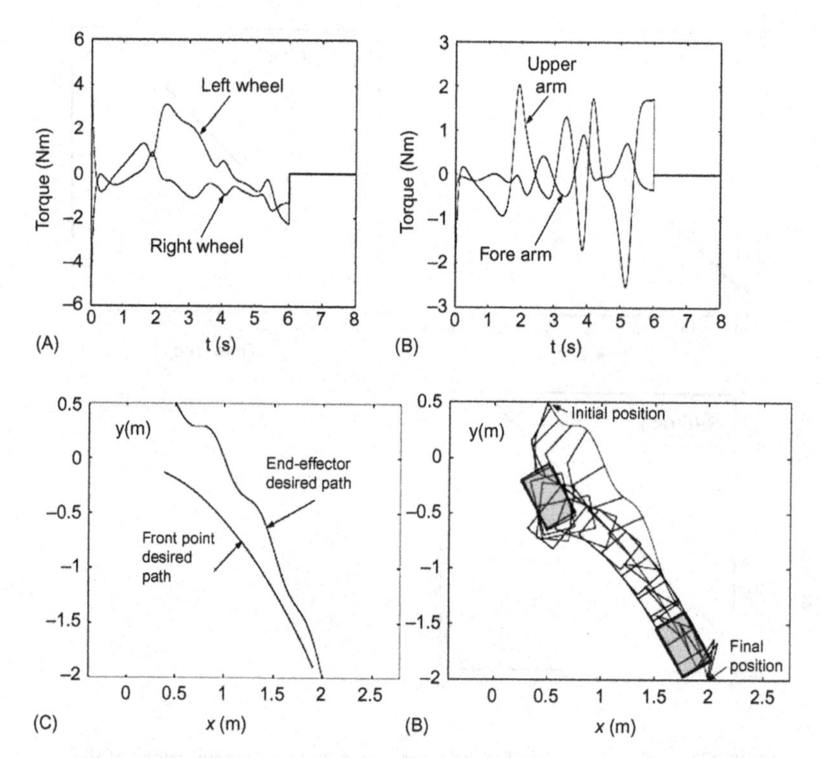

Figure 13.21 (A) Torques of driving wheels. (B) Forearm and upper arm torques. (C) Desired paths of the platform's front point O_b and end effector. (D) Actual animated paths of the platform and the end effector.
Source: Courtesy of E. Papadopoulos [16].

controller for the desired specifications $\omega_n = 2.0$ and $\zeta = 1.2$. The overdamping $\zeta = 1.2$ was selected to match the slow response of the platform. The velocity along the path was assumed constant.

Figure 13.22A and B shows the desired and actual trajectories of the wheel midpoint Q and the reference (end effector) point for the case where the desired path

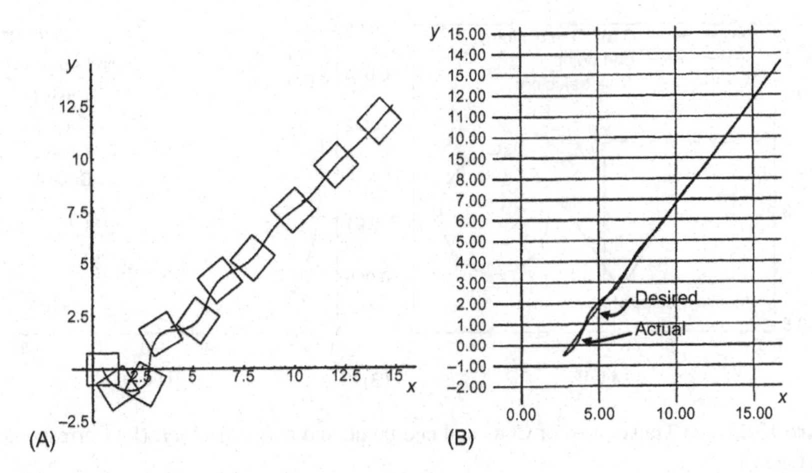

Figure 13.22 (A) Trajectory of the point Q for the 45° straight path. (B) Desired and actual trajectories of the end effector for the path (A).
Source: Reprinted from Ref. [17], with permission from Institute of Electrical and Electronic Engineers.

was a straight line 45° from the *x*-axis. The notch on one side of each square platform shows the forward direction of motion.

The manipulability was kept at its maximum except for the initial maneuvering period of the platform (about 5 s).

13.9.2.2 Problem (b)

Four cases were simulated with or without compensation as follows:

Case 1: Compensation of both the arm and the platform
Case 2: Compensation of the arm only
Case 3: Compensation of the platform only
Case 4: No compensation of the dynamic interaction

Figure 13.23A and B shows representative results for the case of a circular desired path.

13.10 Integrated Global and Local Fuzzy Logic-Based Path Planner

Here, simulation results of an integrated navigation (global planning and local/reactive planning) scheme for omnidirectional robots using fuzzy logic rules will be presented [19,20]. A description of this two-level model (Figure 13.24) will be first provided.

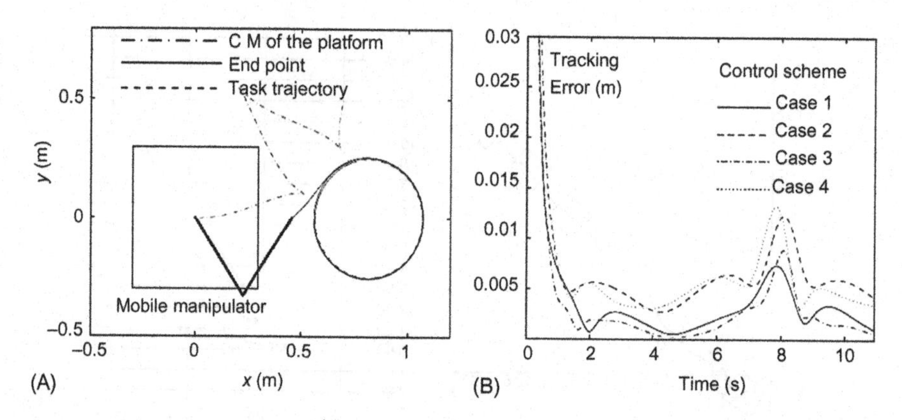

Figure 13.23 (A) Trajectories of COG and end point, and task trajectory. (B) Corresponding tracking errors.
Source: Reprinted from Ref. [18], with permission from Electrical and Electronic Engineers.

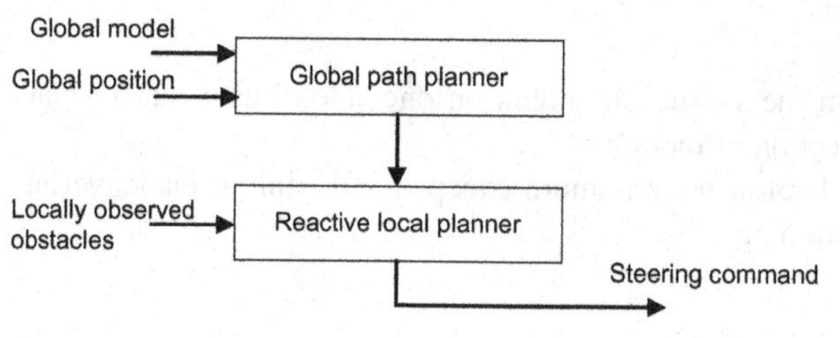

Figure 13.24 Two-level navigation control architecture.

The global path planner is based on the potential field method and provides a nominal path between the current configuration of the robot and its goal. The local reactive planner generates the appropriate commands for the robot actuators to follow the global path as close as possible, while reacting in real time to unexpected events by locally adapting the robot's movements, so as to avoid collision with unpredicted or moving obstacles. This local planner consists of two separate fuzzy controllers for path following and obstacle avoidance (Figure 13.25).

The inputs to the local planner are the path provided by the global path planner and the locally observed obstacles. The output is the steering command for the actuators of the robot. This steering command is a weighed sum of the outputs of the two fuzzy controllers as shown in Figure 13.26.

The purpose of the *fuzzy path following* module is to generate the appropriate commands for the actuators of the robot, so as to follow the global path as close as possible, while minimizing the error between the current heading of the robot and the desired one, obtained from the path calculated by the global planner (Figure 13.26). It has one input, the difference $\Delta\alpha$ between the desired and the actual heading of the robot, and one output, a steering command $\Delta\theta$.

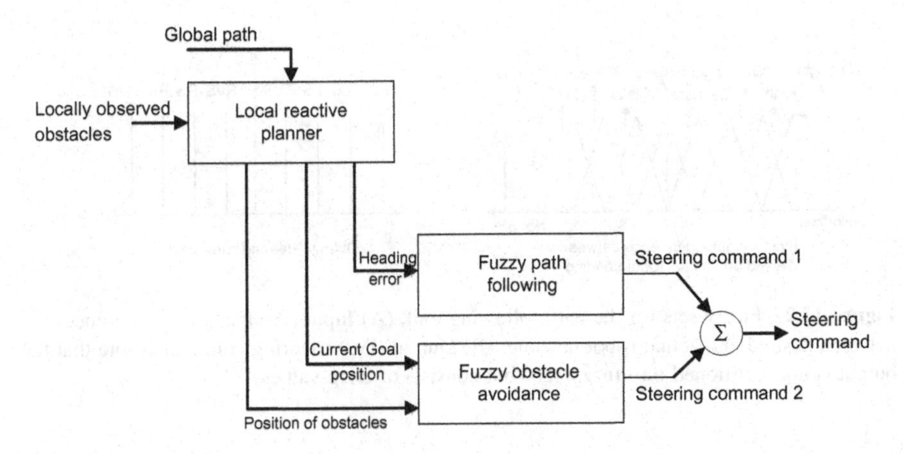

Figure 13.25 Structure of the fuzzy reactive local planner.

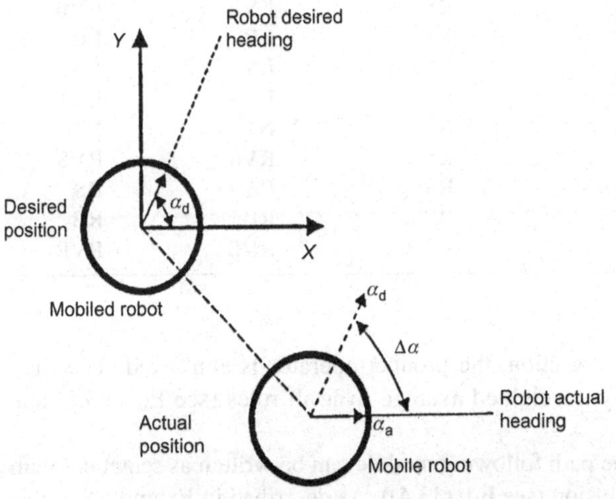

Figure 13.26 Actual and desired configurations of the mobile robot. α_d denotes the desired heading, α_a denotes the actual heading, and $\Delta\alpha$ denotes the difference between the actual and the desired heading of the mobile robot.

The fuzzy module/controller is based on the Takagi–Sugeno method.[1] The input space is partitioned by fuzzy sets as shown in Figure 13.27. Here, asymmetrical triangular and trapezoidal functions which allow a fast computation, essential under real-time conditions, are utilized to describe each fuzzy set (see Eqs. (13.1–13.3)) below.

[1] In the Takagi–Sugeno method, the rules have the form IF x_1 is A_1^i AND ... AND $x_n = A_n^i$, THEN $y^i = c_0^i + c_1^i x_1 + ... + c_n^i x_n$ where $c_0^i, c_1^i, ..., c_n^i$ are crisp-valued coefficients. The total conclusion of this fuzzy system is $y = \Sigma_{i=1}^m w^i y^i / \Sigma_{i=1}^m w^i$, where $w^i = \Pi_{k=1}^n \mu_{A_k^i}(x_k)$ are the weights of the averaging process.

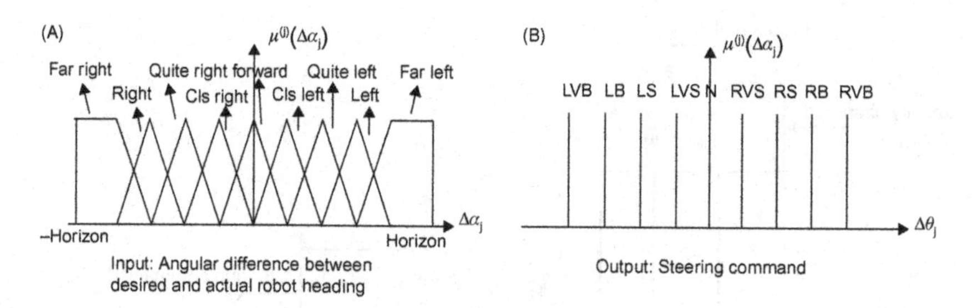

Figure 13.27 Fuzzy sets for the path following unit. (A) Input ($\Delta\alpha$): angular difference between desired and actual robot heading. (B) Output ($\Delta\theta$): steering command (note that the output is not partitioned into fuzzy sets, but consists of crisp values).

Table 13.3 A Fuzzy Rule Base for the Mobile Robot Navigation

v_i/μ_j	Very Close	Close	Far	Very Far
Far right	RB	RS	LVB	LVB
Right	RB	RS	LB	LB
Quite right	RB	RB	LS	LS
Close right	RB	RB	LVS	LVS
Forward	N	N	N	N
Close left	RVS	RVS	RVS	RVS
Quite left	RS	RS	RS	RS
Left	RB	RB	RB	RB
Far left	RVB	RVB	RVB	RVB

To calculate the fuzzy intersection, the product operator is employed. The final output of the unit is given by a weighted average over all rules (see Eq. (13.5) and Figure 13.27).

Intuitively, the rules for the path following module can be written as sentences with one antecedent and one conclusion (see Eq. (13.4)). As described in Example 8.3, this structure lends itself to a tabular representation. This representation is called *fuzzy associative matrix* (FAM) and represents the prior knowledge of the problem domain (Table 13.3).

The symbols in Table 13.3 have the following meaning: LVB, left very big; LB, left big; LS, left small; LVS, left very small; N, neutral (zero); RVB, right very big; RB, right big; RS, right small; RVS, right very small.

The rows represent the fuzzy (linguistic) values of the distance to an obstacle, the columns are the fuzzy values of angles to the goal, and the elements of the matrix are the torque commands of the motor. The tools of fuzzy logic allow us to translate this intuitive knowledge into a control system. The fuzzy set $\mu^{(j)}_{\tilde{p}_j}, \tilde{p}_j = 1, \ldots, p_j$ is described by asymmetrical triangular and trapezoidal functions.

Defining the parameters $ml_{\tilde{p}_j}^{(j)}$ and $mr_{\tilde{p}_j}^{(j)}$ as the x-coordinates of the left and right zero crossing, respectively, and $mcl_{\tilde{p}_j}^{(j)}$ and $mcr_{\tilde{p}_j}^{(j)}$ as the x-coordinates of the left and right side of the trapezoid's plateau, the trapezoidal functions can be written as:

$$\mu_{\tilde{p}_j}^{(j)}(\Delta\alpha_j) = \begin{cases} \max((\Delta\alpha_j - ml_{\tilde{p}_j}^{(j)})/(mcl_{\tilde{p}_j}^{(j)}) - ml_{\tilde{p}_j}^{(j)}, 0) & \text{if } \Delta\alpha_j < mcl_{\tilde{p}_j}^{(j)} \\ 1 & \text{if } mcl_{\tilde{p}_j}^{(j)} \le \Delta\alpha_j < mcr_{\tilde{p}_j}^{(j)} \\ \max((\Delta\alpha_j - mr_{\tilde{p}_j}^{(j)})/(mcr_{\tilde{p}_j}^{(j)} - mr_{\tilde{p}_j}^{(j)}), 0) & \text{if } \Delta\alpha_j > mcr_{\tilde{p}_j}^{(j)} \end{cases} \tag{13.1}$$

with $\tilde{p}_j = 1, 2, \ldots, p_j$. Triangular functions can be achieved by setting $mcl_{\tilde{p}_j}^{(j)} = mcr_{\tilde{p}_j}^{(j)}$. On the left and right side of the interval, the functions are continued as constant values of magnitude 1, that is:

$$\mu_1^{(j)}(\Delta\alpha_j) = \begin{cases} 1 & \text{if } \Delta\alpha_j \le mcr_1^{(j)} \\ \max((\Delta\alpha_j - mr_1^{(j)})/(mcr_1^{(j)} - mr_1^{(j)}), 0) & \text{if } \Delta\alpha_j > mcr_1^{(j)} \end{cases} \tag{13.2}$$

and

$$\mu_{p_j}^{(j)}(\Delta\alpha_j) = \begin{cases} \max((\Delta\alpha_j - ml_{p_j}^{(j)})/(mcl_{p_j}^{(j)} - ml_{p_j}^{(j)}), 0) & \text{if } \Delta\alpha_j \le mcl_{p_j}^{(j)} \\ 1 & \text{if } \Delta\alpha_j > mcl_{p_j}^{(j)} \end{cases} \tag{13.3}$$

The fuzzy set $\mu_{\tilde{q}_j}^{(j)}$ is associated with linguistic terms $A_{\tilde{q}_j}^{(j)}$. Thus, for the mobile robot, the linguistic control rules $R_1^{(j)}, \ldots, R_{r_j}^{(j)}$, which constitute the rule base, can be defined as:

$$R_{\tilde{r}_j}^{(j)}\text{:IF} \quad \Delta\alpha_j \quad \text{is} \quad A_{\tilde{p}_j}^{(j)}, \quad \text{THEN} \quad f(\Delta\theta_{\tilde{r}_j}) \quad (\tilde{r}_j = 1, 2, \ldots, r_j) \tag{13.4}$$

Finally, the output of the unit is given by the weighted average over all rules:

$$\Delta\theta_j = \sum_{\tilde{r}_j=1}^{r_j} \sigma_{\tilde{r}_j} \cdot \Delta\theta_{\tilde{r}_j} / \sum_{\tilde{r}_j=1}^{r_j} \sigma_{\tilde{r}_j} \tag{13.5}$$

Equation (13.4) together with Eq. (13.5) defines how to translate the intuitive knowledge reflected in the FAM into a fuzzy rule base. The details of this translation can be modified by changing the number of fuzzy sets, the shape of the sets (by choosing the parameters $ml_{\tilde{p}_j}^{(j)}$, $mr_{\tilde{p}_j}^{(j)}$, $mcl_{\tilde{p}_j}^{(j)}$, $mcr_{\tilde{p}_j}^{(j)}$) as well as the value $\Delta\theta_{\tilde{r}_j}$ of each of the rules in Eq. (13.5). As an example, in this application, the number of fuzzy sets that fuzzifies the angular difference in heading $\Delta\alpha_j$ is chosen to be nine. All the other parameters were refined by trial and error. The fuzzy-based obstacle avoidance unit, which controls the mobile robot, has three principal inputs:

1. The distance d_j between the robot and the nearest obstacle.
2. The angle γ_j between the robot and the nearest obstacle.
3. The angle $\theta_j = \alpha_j - \beta_j$ between the robot's direction and the straight line connecting the current position of the robot and the goal configuration, where β_j is the direction of the straight line connecting the robot's current position and the goal position, and α_j is the current direction of the robot (Figure 13.28).

The output variable of the unit is the motor torque command τ_j. All these variables can be positive or negative, that is, they do not only inform about the magnitude but also about the sign of displacement relative to the robot left or right. The motor command which can be interpreted as an actuation for the robot's direction motors is fed to the mobile platform at each iteration. It is assumed that the robot is moving with constant velocity, and no attempt is being made to control it.

For the calculation of the distance, the only obstacles considered are those which fall into a bounded area surrounding the robot and moving along with it. In this implementation, this area is chosen to be a cylindrical volume around the mobile platform and reaches up to a predefined horizon. This area can be seen as a simplified model for the space scanned by ranging sensors (e.g., ultrasonic sensors) attached to the sides of the robot. Besides an input from ultrasonic sensors, a camera can also be used to acquire the environment. Mobile robots are usually equipped with a pan/tilt platform where a camera is mounted. This camera can also be utilized. If no obstacle is detected inside the scan area, the fuzzy unit is informed of an obstacle in the far distance.

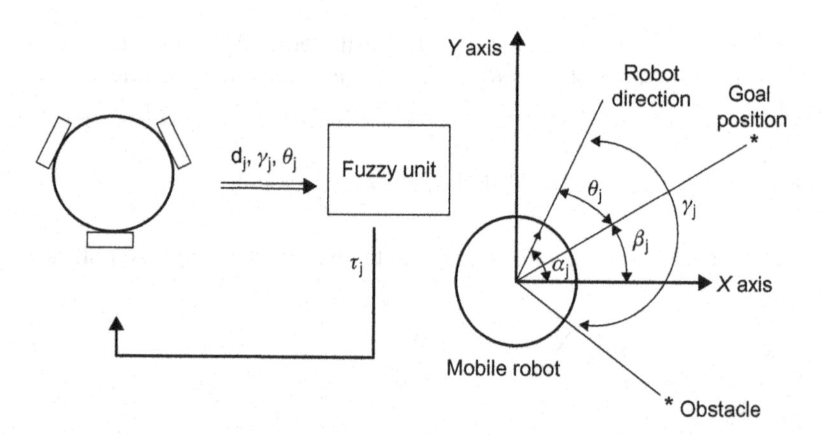

Figure 13.28 The omnidirectional mobile robot connected to the corresponding fuzzy-based obstacle avoidance unit. This unit receives via an input the angle $\theta_j = \alpha_j - \beta_j$ between the robot's direction and the straight line connecting the current position of the robot and the goal configuration, and the distance and the angle of the nearest obstacle (d_j, γ_j). The output variable of the unit is the motor command τ_j.

13.10.1 *Experimental Results*

The functioning of the proposed system, applied to an omnidirectional mobile robot, was evaluated through simulations using Matlab. In all cases, the proposed planner provided the mobile robot with a collision-free path to the goal position. Simulation results, obtained in three different working scenarios, are shown inFigure 13.29 [20].

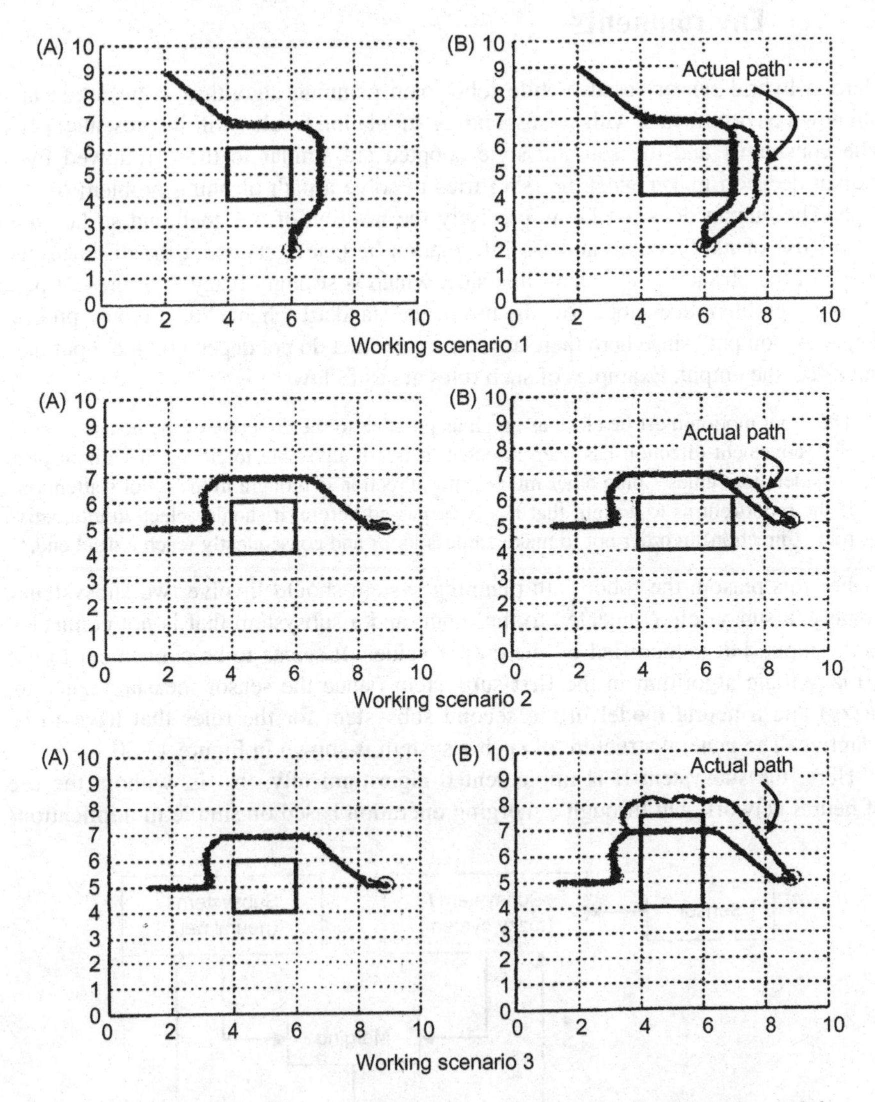

Figure 13.29 Three scenarios of simulation results. (A) Actual path of the robot and the global path coincide since the world model was accurate. (B) Actual path of the robot is different from the path provided by the global path planner since the position/size of the obstacle was not very accurate or the obstacle moved to a new position.

The present two-level fuzzy global and local path planning control method can also be directly applied to fixed industrial robots [19], as well as to mobile manipulators.

13.11 Hybrid Fuzzy Neural Path Planning in Uncertain Environments

Here, a hybrid sensor-based mobile robot path planning algorithm in uncertain and unknown environments, with corresponding simulation results, will be presented [21]. The constraints and the solution style adopted are similar to those followed by a human decision maker when he (she) tries to solve a path planning problem of this type. The human does not know precisely the position of the goal, and so he (she) makes use of its direction, and the information he (she) gets about the distance, the shape of the obstacles, at their visible side, which is strongly fuzzy. The present path planning problem does not allow the use of the standard scheme "data input–processing–data output" since here there are some rules that do not depend on the input data but affect the output. Examples of such rules are as follows:

- The robot motion must be close as much as possible to the direction of the goal.
- If a movement direction has been selected, it is not advisable to change it without prior consideration, unless some other much better direction is brought to the robot's attention.
- If the robot returns to a point that has been passed before, it should select an alternative route (direction) in order not to make endless loops and consequently reach a dead end.

For this reason, the robot path planning system should involve two subsystems, namely, a subsystem connected to the input, and a subsystem that is not connected to the input, but is modified by the output values. It seems to be convenient to use a fuzzy logic algorithm in the first subsystem (since the sensor measurements are fuzzy) and a neural model in the second subsystem for the rules that have to be adaptive. The general structure of such a system is shown in Figure 13.30.

Here, the subsystem II is implemented algorithmically, that is, without the use of neural network, but through a merging operation based on simple multiplication.

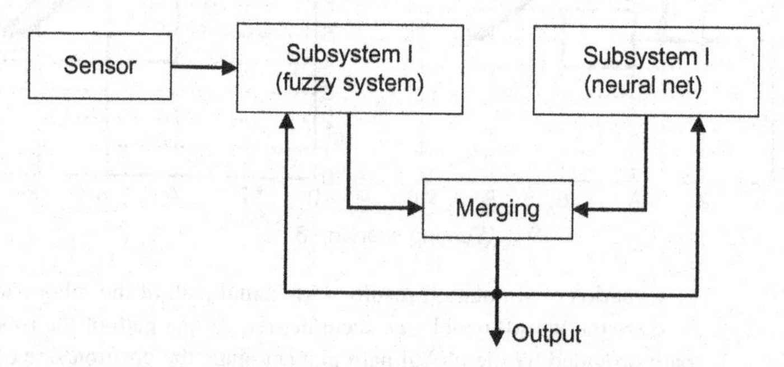

Figure 13.30 General structure of a WMR path planner in unknown terrain.

13.11.1 The Path Planning Algorithm

It will be assumed that the autonomous robot (represented by a point in the space) has to move from an initial position S_0 (start) to a final position G (goal). The available data are measurements from the sensors about all required quantities. These measurements are obtained from the current position S_i of the robot. The direction of the goal G, that is, the direction of the straight line S_iG is also assumed to be measured, but the length (distance) (S_iG) is not measured. Then, the algorithm works as follows: we first define a coordinate frame attached to the robot system with Ox-axis the straight line S_iG, and divide the space into K directions with reference the S_iG (or Ox) direction. Obviously, two consecutive directions have an angular distance of $360/K$ degrees. It is convenient to select the number K of directions such that the axes Ox, Ox', Oy, and Oy' are involved in them. Now, from the sensors' measurements and the fuzzy inference on the database, we determine and associate to each direction a priority. Then, we select the direction with the highest priority, and the new point S_{i+1} is placed in this direction with $(S_iS_{i+1}) = b$, where b is the step length that we select.

On the basis of the above, the path planning algorithm is implemented by the following steps:

Step 1: Obtain the measurements from the position S_i.
Step 2: Fuzzify the results of these measurements.
Step 3: Introduce the fuzzy results into the rules adopted.
Step 4: Determine a priority value for each direction.
Step 5: Multiply the above priority value by a given *a priori* weight for each direction.
Step 6: Determine the direction with the highest priority, move by a step length in this direction to find S_{i+1}.
Step 7: Repeat the algorithm for the position S_{i+1}.

The fuzzy database contains suitable rules, the inputs of which are obtained from the sensors. The *a priori* knowledge used for the robot motion is expressed in the form of a weight for each direction, and there is no need to be determined each time from the knowledge base. For example, the direction from the current position to the goal must have the highest priority. Similarly, the directions that approach the goal must have higher priorities than the directions that depart from the goal. The representation of this *a priori* knowledge in the form of weights suggests the use of a neural network which accepts as inputs the priorities of the K directions, and has as weights the weights of the *a priori* knowledge (Figure 13.30). These weights can be predetermined by trial and error work and can be updated periodically after a certain number of steps. In this way, we ensure the learning of the relationship between the *a priori* knowledge and the output of the fuzzy inference (not the learning of the *a priori* knowledge itself). An algorithm is also required to overcome the dead ends that may appear when a cycle occurs in which the weights of the *a priori* knowledge do not change. This algorithm varies from case to case. In the simulation example that follows, we used a fast algorithm for this purpose which runs over each cycle of the path planning algorithm.

The present algorithm has a very general structure and can be used to any path planning problem in an unknown environment where optimality is not required. The algorithm needs only to have available the rules of the fuzzy database, the length b of the motion step, and the number K of the directions that are examined.

13.11.2 Simulation Results

The simulation robot path planning example was designed to have the following particular features:

- The distance D from the point S_i to some obstacle point (e.g., an obstacle or the room wall) is assumed to be measured by a sensor.
- The measurement results are quantized in 11 intervals and fuzzified as given in Table 13.4

  The entries of this table are the values of membership functions $\mu_j(D)$ of the distance in the various ranges $j = 0, 1, \ldots, 10$ as distributed among the linguistic values L, VL, MORLL, NL, ME, MM, NM, HI, VH, MH, NH, and UN. The respective code of these values is described as follows: L, low; VL, very low; MORLL, more or less low; ME, medium; MM, more or less medium; NM, not medium; H, high; VH, very high; MH, more or less high; NH, not high; UN, unknown.
- The rules employed are the following:

  R1 = IF D_i is low, THEN P_i is high

  R2 = IF D_i is medium, THEN P_i is medium

  R3 = IF D_i is high, THEN P_i is low

  where P_i is the priority of the ith direction quantized in 11 intervals and fuzzified according to Table 13.4.
- The knowledge is represented by fuzzy matrices where Zadeh's max-min inference rule is applied.
- The fuzzification is performed using the singleton method, and the defuzzification using the center of gravity method.
- The number of directions is selected to be $K = 16$ and the step length to be $b = 5$ pixels.

Table 13.4 Distance Quantization

	Range	L	VL	MORLL	NL	ME	MM	NM	HI	VH	MH	NH
0	$D<15$	1.00	1.00	1.00	0.00	0.00	0.00	1.00	0.00	0.00	0.00	1.00
1	$15<D<60$	0.67	0.45	0.82	0.33	0.00	0.00	1.00	0.00	0.00	0.00	1.00
2	$60<D<105$	0.33	0.11	0.57	0.67	0.25	0.50	0.75	0.00	0.00	0.00	1.00
3	$105<D<150$	0.00	0.00	0.00	1.00	0.50	0.71	0.50	0.00	0.00	0.00	1.00
4	$150DD<195$	0.00	0.00	0.00	1.00	0.75	0.87	0.25	0.00	0.00	0.00	1.00
5	$195<D<240$	0.00	0.00	0.00	1.00	1.00	1.00	0.00	0.00	0.00	0.00	1.00
6	$240<D<285$	0.00	0.00	0.00	1.00	0.75	0.87	0.25	0.20	0.04	0.45	0.80
7	$285<D<330$	0.00	0.00	0.00	1.00	0.50	0.71	0.50	0.40	0.16	0.63	0.60
8	$330<D<375$	0.00	0.00	0.00	1.00	0.25	0.50	0.75	0.60	0.36	0.77	0.40
9	$375<D<420$	0.00	0.00	0.00	1.00	0.00	0.00	1.00	0.80	0.64	0.89	0.20
10	$D>420$	0.00	0.00	0.00	1.00	0.00	0.00	1.00	1.00	1.00	1.00	0.00

- The weights in the network selecting the maximum direction priority (i.e., the weights of the *a priori* knowledge) are predetermined as:

  $(10^{10}, 9000, 800, 70, 6, 5, 4, 3, 2, 3, 4, 5, 6, 70, 800, 9000)$

The robot environment used in the present example was obtained on a VGA640x480 screen. During the motion, the "mouse" was regarded as a moving obstacle. The results of four experiments with different starting and goal points are shown in Figure 13.31A–D [21].

13.12 Extended Kalman Filter-Based Mobile Robot SLAM

Here, some results obtained for the SLAM problem using EKFs will be presented (Section 12.8.2). The differential drive WMR SLAM problem of Example 12.3 (Figure 12.11) was solved using a global reference at the origin, which assures that the EKF is stable [22]. Actually, the robot and landmark covariance estimates were proved to be significantly reduced, and the actual robot path and landmark estimates showed a significant error reduction. This is shown in Figure 13.32 where the results are compared to the GPS measurements for two anchors located at (2.8953, −4.0353) and (9.9489, 6.9239).

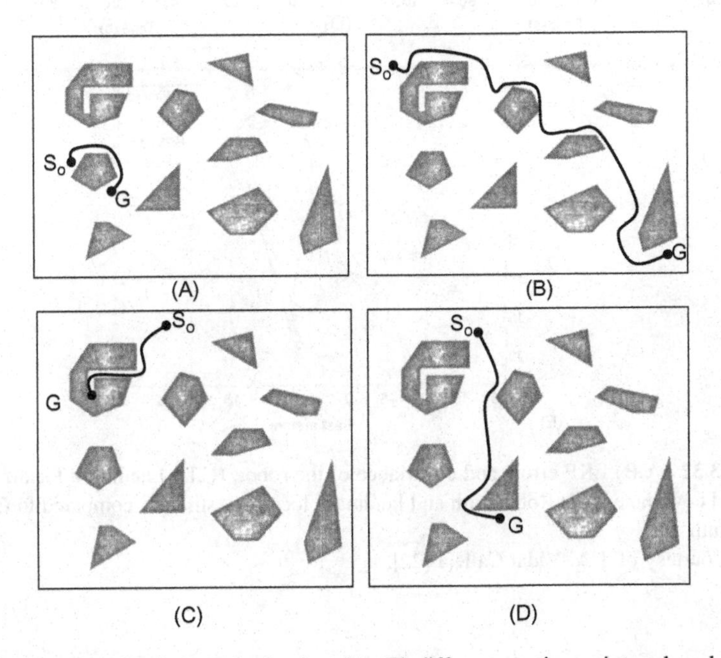

Figure 13.31 Path planning results for four (A–D) different starting point and goal configurations.

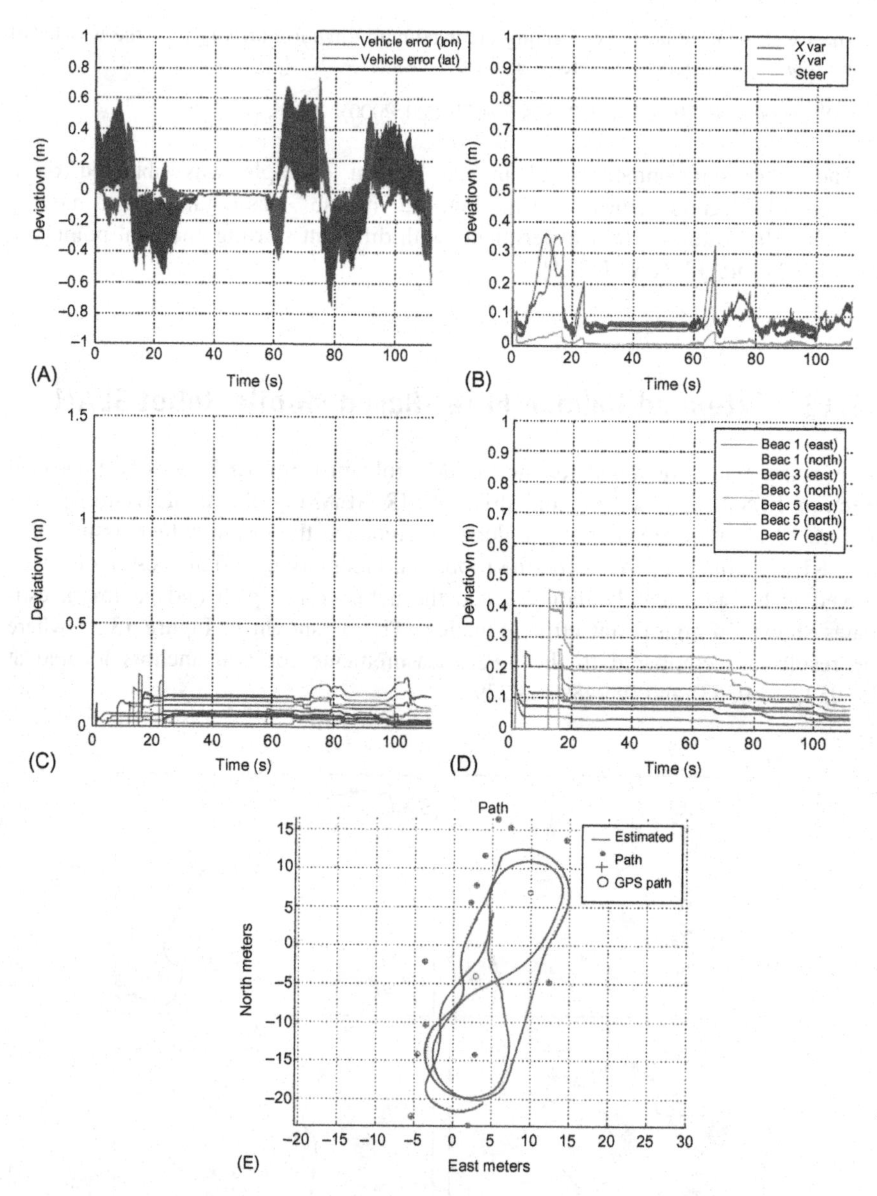

Figure 13.32 (A,B) EKF errors and covariance of the robot. (C,D) Landmark localization errors and covariance. (E) Robot path and landmark location estimates compared to GPS ground truth.
Source: Courtesy of T.A. Vidal Calleja [22].

13.13 Particle Filter-Based SLAM for the Cooperation of Two Robots

Here, some results obtained by applying the particle filter approach of Section 12.8.4 to WMR SLAM for a pair of robots performing collaborative exploration will be presented [23,24]. The parameters x_k used allow the modeling of the poses and uncertainties of the robots. The strategy for cooperative exploration of the two robots is to allow them to take turns moving such that at each time one robot is stationary and can be considered as a fixed reference point (Figure 13.33).

The positions of the two robots are estimated via the particle filter combining an open-loop estimate of odometry error with data from a range finder on one robot and a three-plane target mounted on top of the other robot.

13.13.1 Phase 1 Prediction

Here, the variable of interest is the pose $[x, y, \phi]^{\mathrm{T}}$ of the moving robot. The robot motion is performed as a rotation followed by a translation. The rotation angle is equal to $\Delta\phi = \phi_k - \phi$, where $\phi_k = \mathrm{arc}tg(\Delta y/\Delta x)$, and the forward translation distance is $l = \sqrt{(\Delta x^2) + (\Delta y^2)}$. If the starting pose is $[x_k, y_k, \phi_k]^{\mathrm{T}}$, then the resulting pose is $[x_{k+1}, y_{k+1}, \phi_{k+1}]^{\mathrm{T}}$ given by:

$$\mathbf{x}_{k+1} = \begin{bmatrix} x_{k+1} \\ y_{k+1} \\ \phi_{k+1} \end{bmatrix} = \begin{bmatrix} x_k + l\cos\phi_k \\ y_k + l\sin\phi_k \\ \phi_k \end{bmatrix}$$

The rotation error (noise) $\Delta\phi$ due to odometric error is assumed to be a Gaussian process with mean value μ_{rot} and standard deviation proportional to $\Delta\phi$, that is, $\sigma_{\mathrm{rot}} = \sigma\Delta\phi$. The translation noise has two components: the first is related to

Figure 13.33 The two cooperating WMRs exploring one side of a room.
Source: Reprinted from Ref. [24], with permission from Institute of Electrical and Electronic Engineers.

the actual distance travelled (*pure translation noise*), and the second is due to changes in orientation during the forward translation. This component is called *drift*. The mean values μ_{trans}, μ_{drift} and the standard deviations σ_{trans}, σ_{drift} of these two components can be determined experimentally by discretizing the simulated motion in L steps. The standard deviations per step are equal to:

$$\sigma_{\text{trs}} = \sigma_{\text{trans}}\sqrt{L} \quad \text{and} \quad \sigma_{\text{drft}} = \sigma_{\text{drift}}\sqrt{L/2}$$

Therefore, the WMR stochastic prediction model is:

$$\mathbf{x}_{k+1} = \begin{bmatrix} x_{k+1} \\ y_{k+1} \\ \phi_{k+1} \end{bmatrix} = \begin{bmatrix} x_k \\ y_k \\ \phi_k \end{bmatrix} + \begin{bmatrix} (\Delta l + \varepsilon_{\Delta l})\cos(\phi_k + \varepsilon_{\phi_1}) \\ (\Delta l + \varepsilon_{\Delta l})\sin(\phi_k + \varepsilon_{\phi_2}) \\ \varepsilon_{\phi_1} + \varepsilon_{\phi_2} \end{bmatrix}$$

where $\varepsilon_{\Delta l}, \varepsilon_{\phi_1}$, and ε_{ϕ_2} are Gaussian noises with mean values μ_{transl}, $\mu_{\phi_1} = \mu_{\phi_2} = \mu_{\text{drift}}/2$ and standard deviations $\sigma_{\Delta l} = \sigma_{\text{trans}}\sqrt{L\Delta l}$, $\sigma_{\phi_1} = \sigma_{\phi_2} = (\sqrt{\text{N}}/\sqrt{2})\sigma_{\text{drift}}\Delta l$.

An experiment where the robot moved forward three times (horizontally to the right), rotated by 90°, then moved forward three more times, after that rotated again by 90°, and translated forward five times showed that the uncertainty grows without bound.

13.13.2 Phase 2 Update

After each motion, the robot range finder (RF) sensor sends a measurement vector $\mathbf{z} = [l, \phi, \theta]^{\text{T}}$, where l, ϕ, and θ are as shown in Figure 13.34.

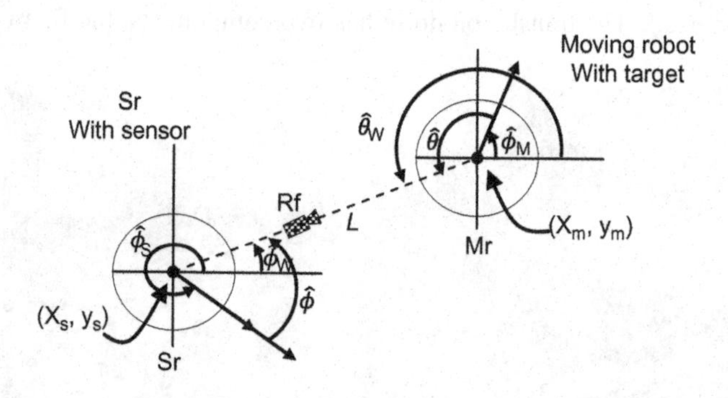

Figure 13.34 The range finder (RF) sensor mounted on the stationary robot (SR) observes the moving robot (MR) that carries the target. The tracker gives $l, \hat{\phi}$, and $\hat{\theta}$ ($\phi_{\text{w}} = \hat{\phi} + \hat{\phi}_{\text{s}}$, $\theta_{\text{w}} = \hat{\theta} + \hat{\phi}_{\text{m}}$).
Source: Adapted from Ref. [24].

The laser range finder can always detect at least two planes of the three-plane target from any position around the moving robot. The sensor output $[l, \phi, \theta]^{\mathrm{T}}$ is given by:

$$\mathbf{z} = \begin{bmatrix} l \\ \phi \\ \theta \end{bmatrix} = \begin{bmatrix} ((\mathrm{d}x)^2 + (\mathrm{d}y)^2)^{1/2} \\ \mathrm{atan}(dy, \mathrm{d}x) - \theta_{\mathrm{s}} \\ \mathrm{atan}(-\mathrm{d}y, \; -\mathrm{d}x) - \theta_{\mathrm{m}} \end{bmatrix}$$

where $\mathbf{x}_{\mathrm{m}} = [x_{\mathrm{m}}, y_{\mathrm{m}}, \phi_{\mathrm{m}}]^{\mathrm{T}}$ is the pose of the moving robot and $\mathbf{x}_{\mathrm{s}} = [x_{\mathrm{s}}, y_{\mathrm{s}}, \phi_{\mathrm{s}}]^{\mathrm{T}}$ is the pose of the stationary robot, $\mathrm{d}x = x_{\mathrm{m}} - x_{\mathrm{s}}$ and $\mathrm{d}y = y_{\mathrm{m}} - y_{\mathrm{s}}$.

Using the sensor measurement $\mathbf{z}$, the weights are updated as shown in Eqs. (12.68b) and (12.69):

$$w_{k+1}^{m} = \hat{w}_{k+1}^{m} / \left(\sum_{q=1}^{M} \hat{w}_{k+1}^{q} \right), \quad \hat{w}_{k+1}^{m} = w_{k}^{m} p(\mathbf{z}/\mathbf{x}_k)$$

where $p(\mathbf{z}/\mathbf{x}_k)$ is the following Gaussian distribution:

$$p(\mathbf{z}/\mathbf{x}_k) \equiv \frac{\mathrm{e}^{-\frac{(l-l_k)^2}{\sigma_l^2}}}{\sqrt{2\pi}\sigma_l} \cdot \frac{\mathrm{e}^{-\frac{(\phi-\phi_k)^2}{\sigma_\phi^2}}}{\sqrt{2\pi}\sigma_\phi} \cdot \frac{\mathrm{e}^{-\frac{(\theta-\theta_k)^2}{\sigma_\theta^2}}}{\sqrt{2\pi}\sigma_\theta}$$

assuming that the processes l, ϕ, and θ are statistically independent.

13.13.3 Phase 3 Resampling

When the effective sampling size ESS (see Eq. (12.70)) is less than the threshold $\mathrm{N}_{\mathrm{max}}$, the particle population is resampled to eliminate (probabilistically) the ones with small weights.

13.13.4 Experimental Work

The mapping algorithm of the room is based on triangulation of free space by the two robots (see Section 12.6.3). The "sweep" of the space is performed using the line of visual contact of the two robots, that is, if the two robots can see each other, there is no obstacle in the space between them. If a robot is located at a corner (stationary) and the other moves along a wall (always without losing visual contact), then a triangle of free space is mapped. The environment is completely mapped by the robots using an online triangulation of the free space [23,24]. Figure 13.35 shows experimental results with two robots exploring the convex area of the corridors of a building. Both subfigures show a spatial integration of the particles during the full trajectory.

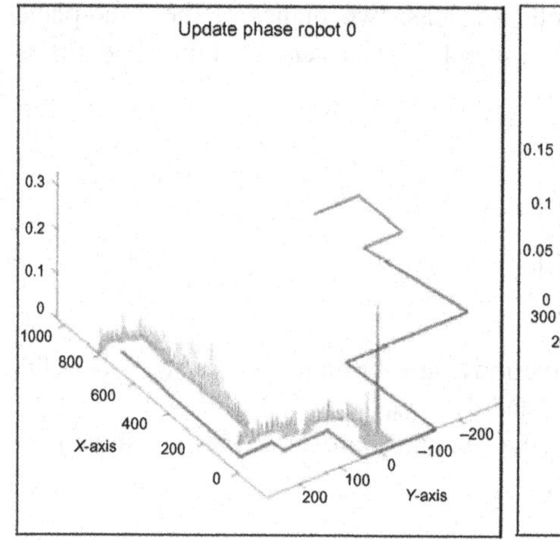

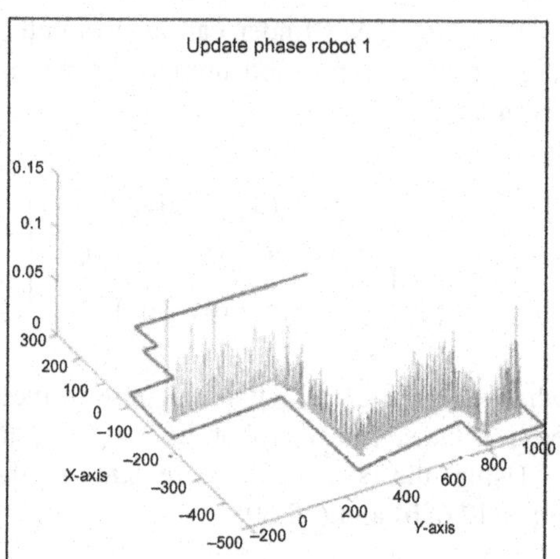

Figure 13.35 Convex area exploration by the two robots: (A) trajectory of robot 0; (B) trajectory of robot 1. The peak heights represent accuracy (the higher a peak the more accurate the estimation).
Source: Reprinted from Ref. [24], with permission from Institute of Electrical and Electronic Engineers.

13.14 Neural Network Mobile Robot Control and Navigation

Here, simulation results will be presented concerning two problems:

1. Neural network-based mobile robot trajectory tracking
2. Neural network-based mobile robot obstacle avoiding navigation.

13.14.1 Trajectory Tracking

The general method of Section 8.5 was applied to a differential drive WMR using a PD velocity controller to train a multilayer perceptron (MLP), as shown in Figure 8.11. The PD teacher controller has the form:

$$\mathbf{v} = \mathbf{K}(\mathbf{x}_d - \mathbf{x}) + \dot{\mathbf{x}}_d$$

where $\mathbf{x} = [x, y]^T$, and has been designed to guarantee exponential convergence of $\mathbf{x}(t)$ to the desired trajectory $\mathbf{x}_d(t)$. The desired trajectory was generated by a reference (virtual) WMR so as to assure that it is compatible with the kinematic nonholonomic constraint of the robot under control. The gain matrix $\mathbf{K}$ was chosen such that:

$$de/dt = -\mathbf{K}\mathbf{e}, \quad \mathbf{e} = \mathbf{x}_d - \mathbf{x}, \quad \mathbf{K} > 0$$

A specific value of $\mathbf{K}$ that assured good exponential trajectory convergence is $\mathbf{K} = \text{diag}[k_1, k_2] = [102.9822, 1.3536]$, and was used in the experiment. A two-layer

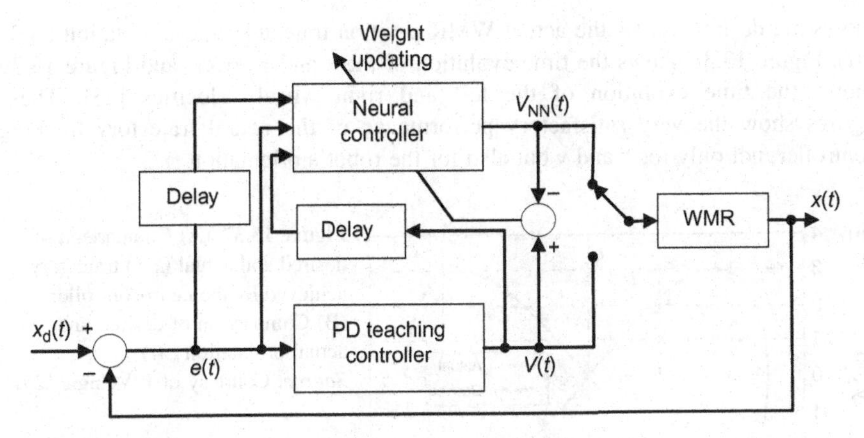

Figure 13.36 WMR supervised tracking neurocontroller.
Source: Adapted from Ref. [25].

MLP NN was used with 10 nodes in the hidden layer and 1 node in the output layer. This NN was trained by the BP algorithm with input–output data the velocity $\mathbf{v}(t)$ and the position $\mathbf{x}(t)$, respectively. The structure of the neurocontroller is shown in Figure 13.36 [25].

In this scheme the NN, instead of learning explicit trajectories, is trained to learn the relationship between the linear velocity $\mathbf{v}(t)$ and the position error $\mathbf{e}(t)$. The learning rate is adjusted by the rule:

$$\gamma(k+1) = \gamma(k)[1 - \mu e^{-\eta(k)}\mathrm{sgn}(\eta/k)], \quad \mu \in [0, 1]$$

where $\eta(t)$ is the normalized ratio:

$$\eta(k) = \frac{\Delta V(\mathbf{w}, k)}{V(\mathbf{w}, k)} = \frac{V(\mathbf{w}, k) - V(\mathbf{w}, k-1)}{V(\mathbf{w}, k)}$$

of the total squared error:

$$V(\mathbf{w}, t) = \sum_{k=0}^{N} ||\mathbf{v}(k) - \mathbf{v}_{\mathrm{NN}}(k)||^2$$

which was actually minimized by updating the NN weights $\mathbf{w} = [w_1, w_2, \ldots]^{\mathrm{T}}$. The experiment was performed for several values of $\mu \in [0, 1]$. The value of μ that leads to the best learning convergence was used in the NN training. In the experiments, the initial value $\gamma(0) = 0.02$ was used for both layers. The best results were obtained with $\mu = 0.71$.

After training, the neurocontroller was used to control the robot. The results obtained with the neurocontroller are shown in Figures 13.37–13.39. Figure 13.37

shows the desired versus the actual WMR position trajectory and orientation angle $\phi(t)$. Figure 13.38 shows the time evolution of the x and y errors, and Figure 13.39 shows the time evolution of the left and right wheel velocities [25]. These figures show the very satisfactory performance of the neural trajectory tracking controller, not only for x and y but also for the robot's orientation ϕ.

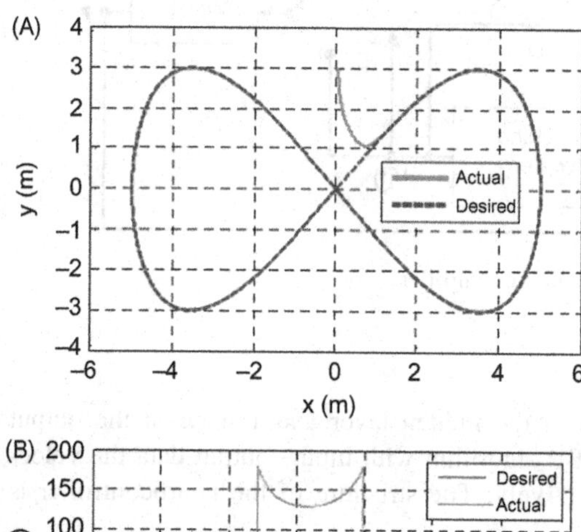

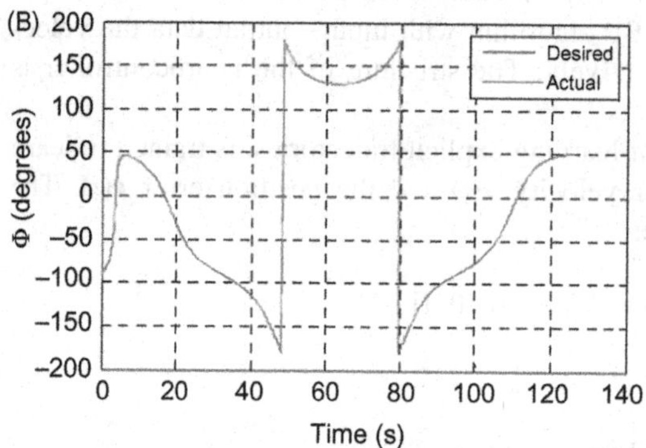

Figure 13.37 (A) Comparison of desired and actual (x, y) trajectory achieved by the neurocontroller. (B) Comparison of desired and actual orientation $\phi(t)$.
Source: Courtesy of J. Velagic [25].

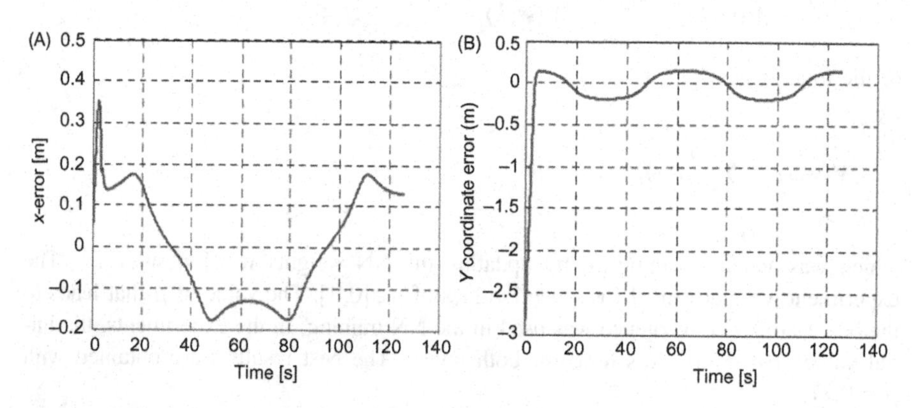

Figure 13.38 Coordinated errors: (A) x error; (B) y error.
Source: Courtesy of J. Velagic [25].

13.14.2 Navigation for Obstacle Avoidance

The neural nets can be used and have been used for WMR navigation purposes, besides the control purposes. The results to be provided here were obtained by combining MLPs for local modeling of the WMR, and RBFs for the activation of the local models. The division of the input space to subregions was performed using the fuzzy C-means method. For full comprehension of the present mobile robot navigation method, we first describe briefly the concepts of *local model NNs* (LMNs) and the *fuzzy C-means segmentation* (FCM) algorithm.

13.14.2.1 Local Model Networks

In this approach, the operation regime is divided in subareas, and each one of them is associated to a local neural model that approximates the system behavior within this subarea. The operating regime includes all the operating points where the system is capable of functioning. Usually, linear neural networks are used. To use the LMNs for system modeling, the parameters that must be included in the operating point (OP) must be determined. The structure of LMN modeling is shown in Figure 13.40 [26–28], where k is the serial number of the local model.

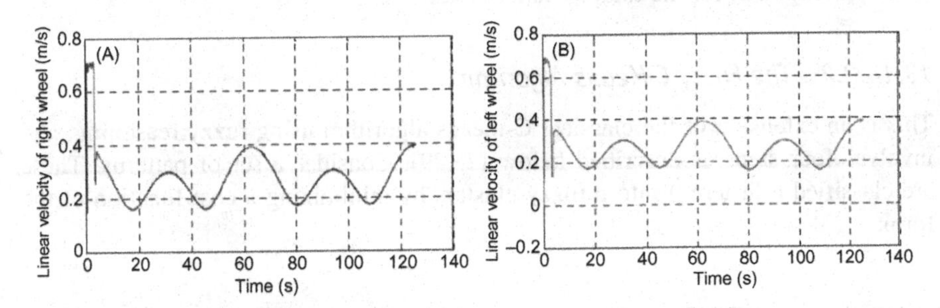

Figure 13.39 Linear velocities of the two wheels: (A) left wheel; (B) right wheel. *Source*: Courtesy of J. Velagic [25].

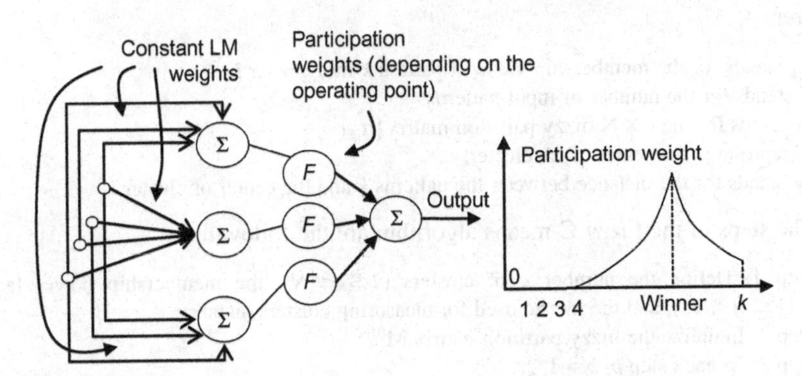

Figure 13.40 LMN structure (A), with participation weights (B).

Each linear model is actually a linear predictor, giving a linear estimate of the next value of the system output depending on the previous ones as well as on the past values, namely:

$$\hat{y}(k) = \alpha_1 y(k-1) + \cdots + \alpha_n y(k-n) + b_1 u(k-1) + \cdots + b_m u(k-m)$$

where n and m are the model's orders with respect to the output $\mathbf{y}(k) \in R^p$ and the input $\mathbf{u}(k) \in R^q$, respectively. In the special case where $m = n = 1$, the model reduces to:

$$\hat{\mathbf{y}}(k) = \boldsymbol{\alpha}\mathbf{y}(k-1) + \mathbf{bu}(k-1)$$

The estimator (predictor) $\hat{\mathbf{y}}$ is trained within each specific part of the operating regime, giving the corresponding coefficients $\boldsymbol{\alpha}_i (i = 1, 2, \ldots, n)$ and $\mathbf{b}_i (i = 1, 2, \ldots, m)$. Once there is an LMN trained for every part of the operating regime, the system modeling is complete. The values of its coefficients are given by the estimated values $\hat{\boldsymbol{\alpha}}_i (i = 1, 2, \ldots, n)$ and $\hat{\mathbf{b}}_i (i = 1, 2, \ldots, m)$. In Figure 13.40, the winner-take-all strategy is illustrated, but it is also possible to have a multi-LMN participation for the total system output.

13.14.2.2 The Fuzzy C-Means Algorithm

This is an extension of the classical C-means algorithm using fuzzy reasoning, and involves four steps as described below [26,29]. Consider a set of patterns. These are classified (clustered) into c fuzzy clusters by minimizing a cost function of the form:

$$J(\mathbf{M}, \upsilon) = \sum_{i=1}^{c} \sum_{k=1}^{N} (\mu_{ik})^m d_{ik}$$

where:

- μ_{ik} stands for the membership value of pattern k into cluster i;
- N stands for the number of input patterns;
- $\mathbf{M}$ stands for the $c \times N$ fuzzy partition matrix $[\mu_{ik}]$;
- υ_i represents the center of ith cluster;
- d_{ik} stands for the distance between the patterns k and the center of cluster i.

The steps of the fuzzy C-means algorithm are the following:

Step 1: Define the number c of clusters $(2 \leq c \leq N)$, the membership power factor m $(1 \leq m < \infty)$, and the metric used for measuring clusters distance.
Step 2: Initialize the fuzzy partition matrix $\mathbf{M}^{(0)}$
Step 3: At each step b, $b = 1, 2, \ldots$
(i) Calculate the vectors of the clusters' centers as $\upsilon_i = \sum_{k=1}^{N} (\mu_{ik})^m \mathbf{x}_k / \sum_{k=1}^{N} (\mu_{ik})^m$

(ii) Update the fuzzy partition matrix $\mathbf{M}^{(b)}$ and find the next one $\mathbf{M}^{(b+1)}$ as follows: Calculate for each pattern $\mathbf{x}_k$ the number of clusters that pattern is associated with: $I_k = \{i|1 \leq i \leq c,\ d_{ik} = ||\mathbf{x}_k - \upsilon_i|| = 0\}$, and the set of clusters $T_k, k = 1, 2, \ldots, c - I_k$ to which the pattern is not associated, with the rules:

IF $I_k \neq 0$, THEN $\mu_{ik} = 1/\sum_{j \in I_k} \left(\frac{d_{ik}}{d_{jk}}\right)^{2/(m-1)}$

IF $I_k = 0$, THEN $\mu_{ik} = 0$ for all $i \in T_k$ and $\sum_{j \in I_k} \mu_{jk} = 1$

Step 4: Compare $\mathbf{M}^{(b)}$ and $\mathbf{M}^{(b+1)}$ using the selected measure and if $||\mathbf{M}^{(b)} - \mathbf{M}^{(b+1)}|| \leq \varepsilon_L$ (convergence threshold), terminate the algorithm; otherwise return to step 3.

13.14.2.3 Experimental Results

The training process termination criterion of each LMN used is the root mean square (RMS):

$$y_{\mathrm{RMS}} = \sqrt{(1/N)\sum_{i=1}^{N} (y_{\mathrm{d}}(k) - y_{\mathrm{NN}}(k))^2}$$

where $y_{\mathrm{d}}(k)$ is the desired output and y_{NN} is the LMN output:

$$y_{\mathrm{NN}} = \sum_{i=1}^{M} \hat{y}_i \phi_i(||\mathbf{x} - \mathbf{c}_i||)$$

with $\hat{y}_i$ being the output estimate of the ith local submodel, and $\phi_i(||\cdot||)$ the ith RBF function defined by Eq. (8.62). The results were obtained using 500 learning datasets, each dataset involving five parameters:

1. LD, left distance
2. RD, right distance
3. FD, front distance
4. TB, target bearing
5. SA, change in steering angle

The experiment used one mobile robot, one target, and four obstacles (if more than one robot exists, each mobile robot is considered as an obstacle to the other robots). The MLP NNs used for local modeling involved three layers (input, hidden, output layer). The input layer has four nodes, that is, three nodes for the input values of the front, left and right distances from obstacles, and one node for the target bearing (if no target is detected, the input to the fourth node is zero). The output layer has a single node which computes the robot steering angle. Figure 13.41A–C shows the path of a single robot followed under the control of an MLP, an RBF, and an LMN (based on MLP and RBF), respectively. Details and statistics of the method are provided in Ref. [26].

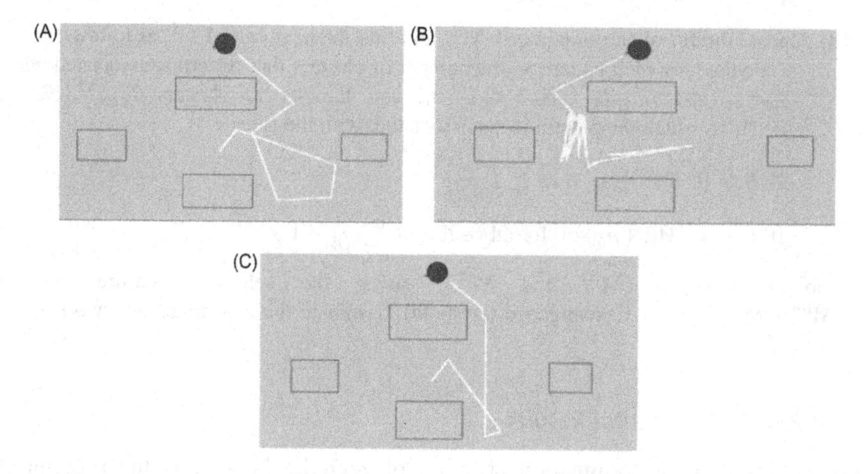

Figure 13.41 Performance of neural network-based navigation: (A) MLP, (B) RBF, and (C) full LMN navigator.
Source: Courtesy of H.A. Awad [26].

13.15 Fuzzy Tracking Control of Differential Drive Robot

Here, performance results of the fuzzy logic controller designed by the Mamdani model, as discussed in Section 8.4.1, will be presented. The kinematic model is:

$$\dot{x} = v\cos\phi, \quad \dot{y} = v\sin\phi, \quad \dot{\phi} = \omega, \quad \dot{x}\sin\phi = \dot{y}\cos\phi$$

and the dynamic model in the unconstrained form (3.19a):

$$\overline{\mathbf{D}}(\mathbf{q})\dot{\mathbf{v}} + \overline{\mathbf{C}}(\mathbf{q},\dot{\mathbf{q}})\mathbf{v} + \mathbf{F}\mathbf{v} = \boldsymbol{\tau} + \mathbf{d}(t)$$

where $\mathbf{v} = [v, \omega]^{\mathrm{T}}$, $\mathbf{q} = [x, y, \phi]^{\mathrm{T}} = \mathbf{x}$, $\mathbf{F}$ is the linear friction matrix, and $\mathbf{d}(t)$ is an external disturbance, has the parameters [30]:

$$\overline{\mathbf{D}}(\mathbf{q}) = \begin{bmatrix} 0.3749 & -0.0202 \\ -0.0202 & 0.3739 \end{bmatrix}, \quad \mathbf{F} = \begin{bmatrix} 10 & 0 \\ 0 & 10 \end{bmatrix}$$

$$\overline{\mathbf{C}}(\mathbf{q},\dot{\mathbf{q}}) = \begin{bmatrix} 0 & 0.1350\dot{\phi} \\ -0.1350\dot{\phi} & 0 \end{bmatrix}, \quad \mathbf{u} = \boldsymbol{\tau} = \begin{bmatrix} u_1 \\ u_2 \end{bmatrix}$$

The disturbance vector $\mathbf{d}(t)$ is:

$$\mathbf{d}(t) = \begin{bmatrix} \delta(t-t_k) \\ \delta(t-t_k) \end{bmatrix}, \quad t_k = 2, 4, \ldots$$

where the delta function $\delta(\cdot)$ is applied every 2 s. The desired velocity vector $\mathbf{v}_\mathrm{d} = [v_\mathrm{d}, \omega_\mathrm{d}]^\mathrm{T}$ is assumed to be:

$$\mathbf{v}_\mathrm{d} = \begin{bmatrix} v_\mathrm{d}(t) \\ \omega_\mathrm{d}(t) \end{bmatrix} = \begin{bmatrix} 0.25 - 0.25\cos(2\pi t/5) \\ 0 \end{bmatrix}$$

and the initial conditions are:

$$\mathbf{q}(0) = [0.1, 0.1, 0]^\mathrm{T}, \mathbf{v}(0) = [v(0), \omega(0)]^\mathrm{T} = [0, 0]^\mathrm{T}$$

The crisp kinematic controller used is (see Eq. (8.18)):

$$v_\mathrm{m} = v_\mathrm{d}\cos\varepsilon_3 + K_1\varepsilon_1, \quad \omega_\mathrm{m} = \omega_\mathrm{d} + K_2 v_\mathrm{d}\varepsilon_2 + K_3\sin\varepsilon_3$$

with gains $K_1 = K_2 = K_3 = 5$. The membership functions of the linguistic (fuzzy) variables $\tilde{\mathbf{v}} = [\tilde{v}, \tilde{\omega}]^\mathrm{T}, u_1, u_2$, where $\tilde{\mathbf{v}} = \mathbf{v} - \tilde{\mathbf{v}}_\mathrm{d}$, have the triangular and trapezoidal forms shown in Figure 8.15. The dynamic fuzzy controller has the rule base of Table 8.2 which represents $3 \times 3 = 9$ fuzzy rules as described in Section 8.4.1. The fuzzy values of u_1 and u_2 are converted to crisp values with the COG defuzzification method and are represented by the input–output surfaces shown in Figure 13.42.

The performance of the full controller was tested by simulation on the Matlab (Simulink) in Ref. [30]. It was verified that the error $\tilde{x}$ goes to zero at $t = 0.5$ s, $\tilde{y}$ goes to zero at $t = 1.0$ s, and $\tilde{\phi}$ goes to zero at $t = 1.2$ s. The velocity errors were brought to zero at $t = 0.25$ s. The switching behavior of the controller was very fast. The simulation plots of the unperturbed closed-loop control system are shown in Figure 13.43.

The simulation results for the perturbed system showed again a very good robustness of the controller. Both position and orientation errors converged quickly to zero despite the external disturbances.

13.16 Vision-Based Adaptive Robust Tracking Control of Differential Drive Robot

Here, some results obtained by simulation in Refs. [31,32], using the adaptive robust trajectory tracking controller (9.158a)–(9.158e), will be presented. The system and controller parameters used are given in Table 13.5.

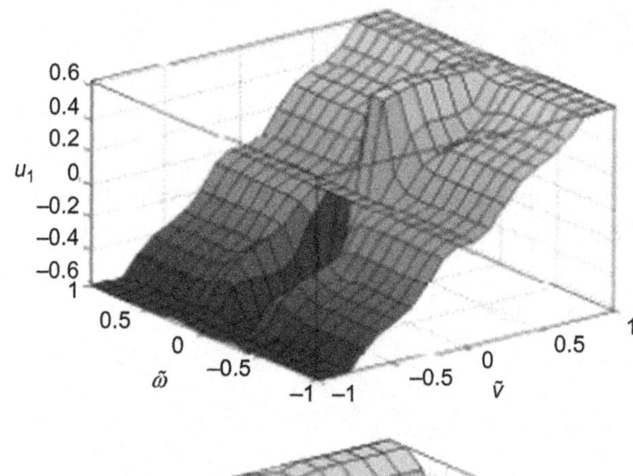

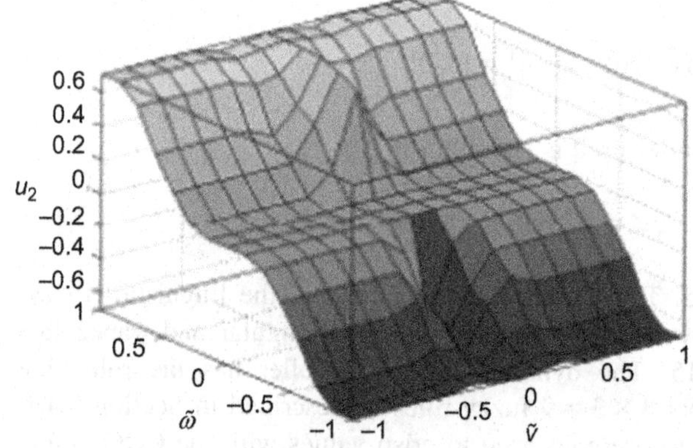

Figure 13.42 Overall input–output surfaces of the fuzzy controller $\{\tilde{v}, \tilde{\omega}) \rightarrow u_{1,\mathrm{COG}}, (\tilde{v}, \tilde{\omega}) \rightarrow u_{2,\mathrm{COG}}\}$. *Source*: Courtesy of O. Castillo [30].

The robot initial pose was $[0, 0, 0]$ and the desired trajectory started from $[3.9, 4.1, 0.2]$. The components d_1 and d_2 of $\mathbf{d}$ were random in the interval $[-1, 1]$ and $d_{\max} = 2$. The values of v_{d} and ω_{d} were selected as $v_{\mathrm{d}} = 1$ m/s and $\omega_{\mathrm{d}} = 0.5$ rad/s. Figure 13.44 shows the convergence of the errors $\varepsilon_1, \varepsilon_2, \varepsilon_3, \tilde{v}$, and $\tilde{\omega}$ to zero, despite the existence of the external disturbance.

Figure 13.45A shows the torques τ_1, τ_2 which, due to the chattering effect, converge to a small area around zero. Figure 13.45B shows the robot desired and actual trajectory in the world coordinate frame.

The experiments were repeated with the saturation-type controller using a boundary layer width $U(t) = 1/(1+t)^3$. Now, the convergence of the errors and the actual trajectory are about the same as in Figures 13.44A and B and 13.45B, but the chattering in the control signals τ_1 and τ_2 disappeared as shown in Figure 13.46.

The estimates of the parameters $\beta_1 = m, \beta_2 = I$, and λ converged to constant values $\hat{\beta}_1 \rightarrow 10^-, \beta_2 \rightarrow 6^+$, and $\lambda \rightarrow 0.97$. For comparison, the reader is advised to perform a simulation experiment using the alternative controller (9.160) with the same WMR vision system values, desired trajectory, and $v_{\mathrm{d}}, \omega_{\mathrm{d}}$ values.

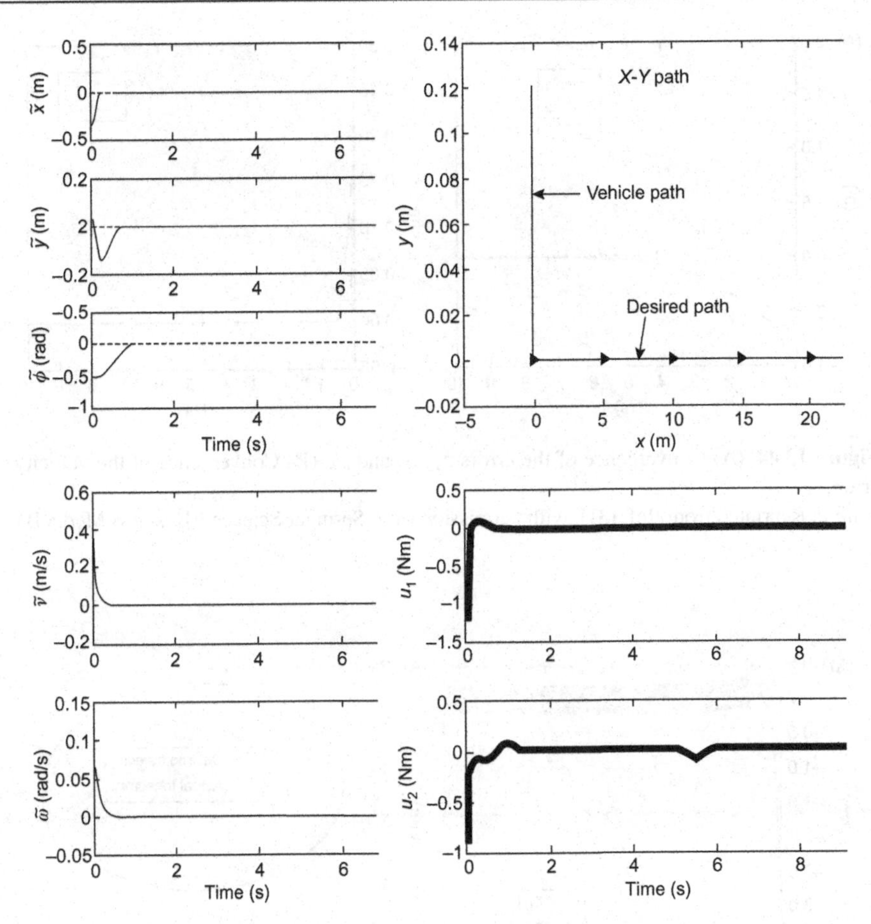

Figure 13.43 Closed-loop performance of the fuzzy controller when $\mathbf{d}(t) = \mathbf{0}$. (A) Errors $\tilde{x}, \tilde{y}, \tilde{\phi}, \tilde{v}$, and $\tilde{\omega}$. (B) xy path and control input (torques) u_1 and u_2.
Source: Courtesy of O. Castillo [30].

Table 13.5 System and Controller Parameters

WMR Vision System Parameters							
λ	$\lambda_{\max}$	$\lambda_{\min}$	r	a	m	I	ϕ_0
1	2	0.5	0.1 m	0.5 m	10 kg	5 kg m^2	$-\pi/2$ rad

Controller Parameters						
k_1	k_3	k_{a1}	k_{a2}	γ_1	γ_2	γ_3
2	2	20	20	1	1	5

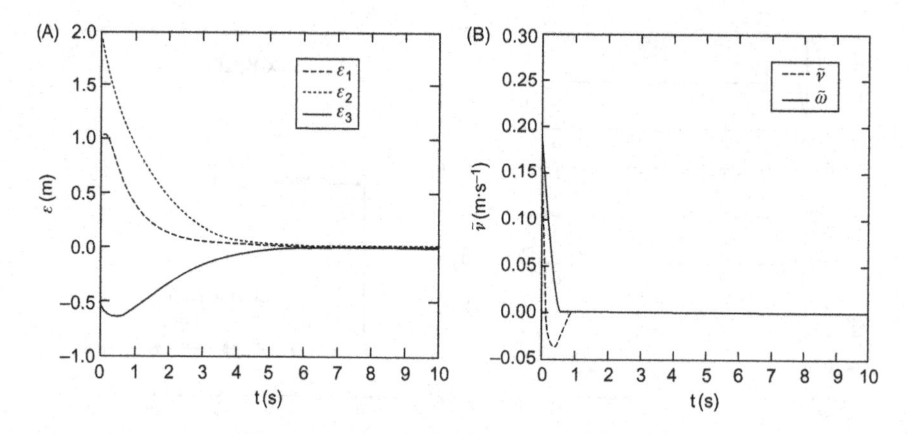

Figure 13.44 (A) Convergence of the errors $\varepsilon_1, \varepsilon_2$, and ε_3. (B) Convergence of the velocity errors.
Source: Reprinted from Ref. [31], with permission from Springer Science+Business Media BV.

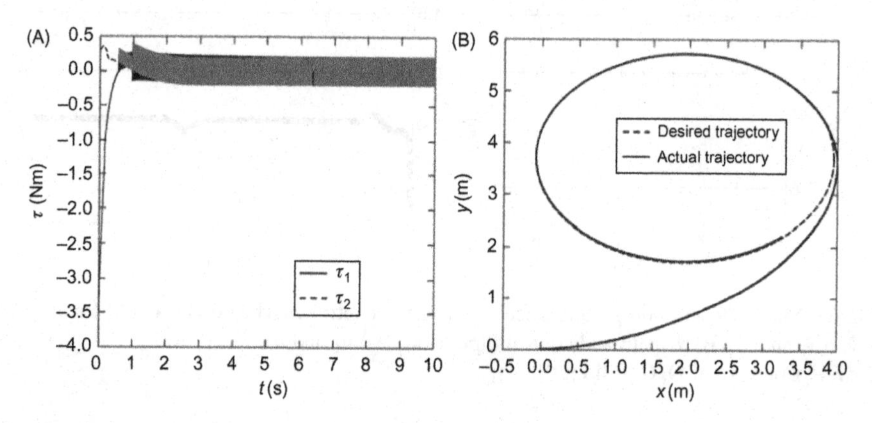

Figure 13.45 (A) The controls τ_1 and τ_2 suffer from chattering. (B) Actual and desired trajectories.
Source: Reprinted from Ref. [31], with permission from Springer Science+Business Media BV.

13.17 Mobile Manipulator Spherical Catadioptric Visual Control

The problem discussed in Example 10.4 was thoroughly studied in Ref. [33]. The resolved-rate visual servo controller (10.83) was applied using an estimate $\hat{\mathbf{J}}_{\mathrm{im}}$ of the image Jacobian computed using the EKF technique. Since the parameters

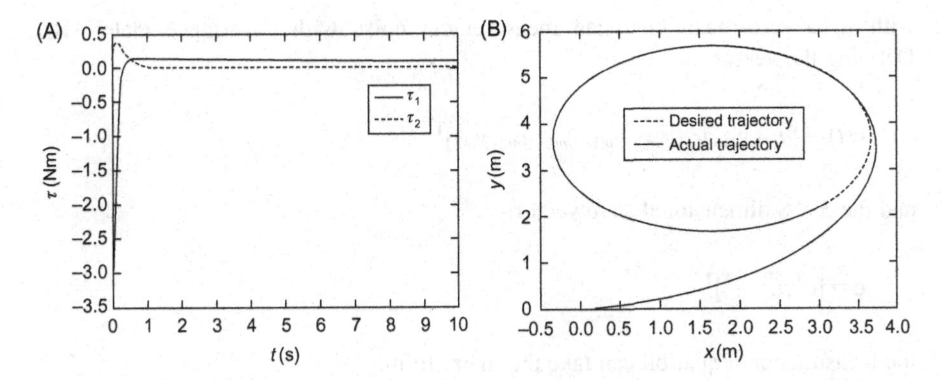

Figure 13.46 (A) Time evolution of τ_1 and τ_2 in the case of saturation-type robustifying control term. (B) Resulting actual trajectory versus desired trajectory.
Source: Reprinted from Ref. [31], with permission from Springer Science+Business Media BV.

involved in the vector $\boldsymbol{\xi}$ are constant, the state transition model has the linear form:

$$\boldsymbol{\xi}_{k+1} = \boldsymbol{\xi}_k + \mathbf{w}_k \tag{13.6}$$

where $\mathbf{w}_k$ is a Gaussian zero-mean disturbance process with covariance matrix $\mathbf{Q}_k$. The measurement vector is:

$$\mathbf{z}_k = [u_{m1}, v_{m1}, u_{r1}, v_{r1}; u_{m2}, v_{m2}, u_{r2}, v_{r2}]^{\mathrm{T}}$$

and the measurement equation is

$$\mathbf{z}_k = \mathbf{h}(\boldsymbol{\xi}_k, \mathbf{n}_k)$$

where:

$$\mathbf{h} = [\mathbf{h}_1 \vdots \mathbf{h}_2]^{\mathrm{T}}$$

$$\mathbf{h}_1 = \left[u_{\text{offset}} + \frac{x_{m1}^m \lambda_u}{z_{m1}^m - d}, v_{\text{offset}} + \frac{y_{m1}^m \lambda_v}{z_{m1}^m - d}, u_{\text{offset}} + \frac{x_{r1}^m}{z_{r1}^m}\lambda_u, v_{\text{offset}} + \frac{y_{r1}^m}{z_{r1}^m}\lambda_v \right]$$

$$\mathbf{h}_2 = \left[u_{\text{offset}} + \frac{x_{m2}^m}{z_{m2}^m - d}\lambda_u, v_{\text{offset}} + \frac{y_{m2}^m}{z_{m2}^m - d}\lambda_v, u_{\text{offset}} + \frac{x_{r2}^m}{z_{r2}^m - d}\lambda_u, v_{\text{offset}} + \frac{y_{r2}^m}{z_{r2}^m - d}\lambda_v \right]$$

with η_k a zero-mean Gaussian measurement noise with covariance matrix $\mathbf{R}_k$. Defining the vector:

$$\varepsilon(t) = [u_{r1}, v_{r1}, u_{r2}, v_{r2}, u_{m1}, v_{m1}, u_{m2}, v_{m2}]^{\mathrm{T}}$$

and the 3×8 dimensional state vector:

$$\mathbf{e} = [\varepsilon^{\mathrm{T}}, \dot{\varepsilon}^{\mathrm{T}}, \ddot{\varepsilon}^{\mathrm{T}}]^{\mathrm{T}}$$

the measurement equation can take the linear form:

$$\mathbf{z}_{k+1} = \mathbf{H}\mathbf{e}_k + \eta_k \tag{13.7}$$

with:

$$\mathbf{z}_k = [u_{r1}, v_{r1}, u_{r2}, v_{r2}, u_{m1}, v_{m1}, u_{m2}, v_{m2}]^{\mathrm{T}}$$

where, under the assumption that all the components of the state vector $\mathbf{e}$ are measurable:

$$\mathbf{H} = [\mathbf{I}_{8\times 8} \vdots \mathbf{O}_{8\times 8} \vdots \mathbf{O}_{8\times 8}]^{\mathrm{T}}$$

In the following, a small representative set of simulation results are provided with the following data [33,34]:

- Robot base location in the camera frame: $(400, 200, -1900)$ mm
- Mirror frame location (with origin at sphere's center) in the camera frame: $(0, 0, -4000)$ mm
- Spherical mirror radius: 500 mm
- Camera intrinsic parameters: $\lambda_u = -998.97$, $\lambda_v = -916.23$, $u_{\text{offset}} = 342.70$, $v_{\text{offset}} = 236.88$
- Landmarks' midpoint in robot frame: $(400, 500, 250)$ mm
- Landmarks' connecting line initial orientations: $\phi = \pi/6$ rad, $\theta = \pi/2$ rad
- Desired landmarks' midpoint in robot frame: $(200, -50, 200)$ mm
- Landmarks' connecting line desired orientations: $\phi = \pi/3$ rad, $\theta = \pi/6$ rad
- Deviation of camera intrinsic and extrinsic parameters from their real values: average +30%
- Variance of zero-mean Gaussian noise: 2 pixels

Figure 13.47A–D shows a set of results.

These results verify the decay over time of the features' errors, the accuracy of the EKF estimates of the camera parameters and image features, and show the actual trajectories of the landmarks in the 3D workspace.

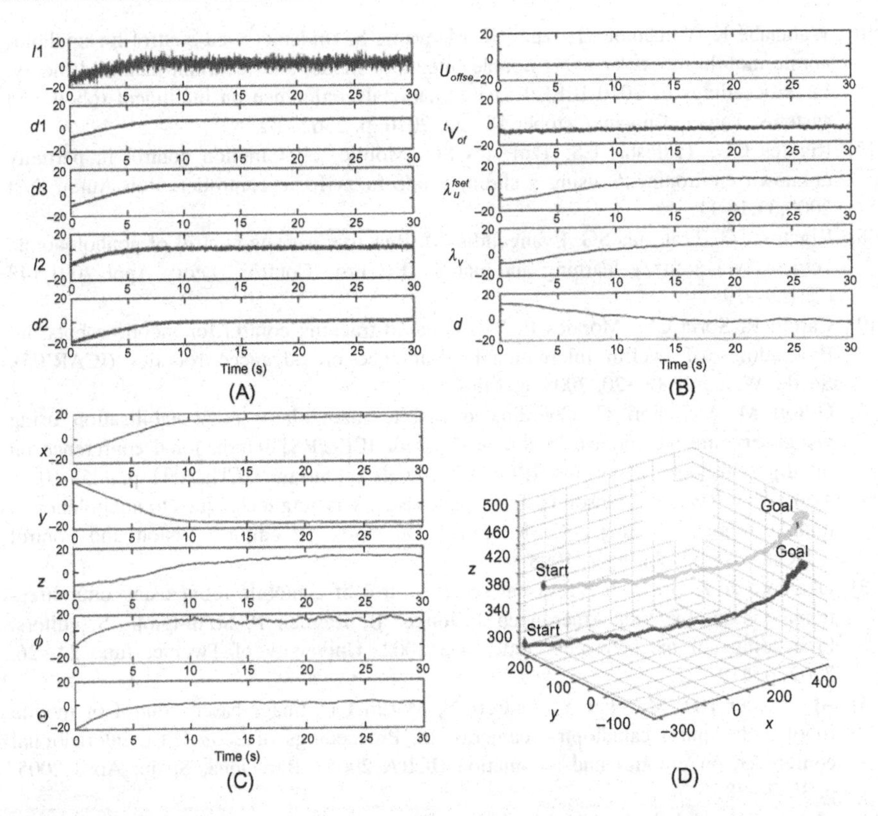

Figure 13.47 (A) Time evolution of the image features' errors. (B) Camera parameters' estimates. (C) Midpoint frame pose. (D) 3D trajectories of the landmarks.
Source: Reprinted from Ref. [34], with permission from International Society for Optical Engineering (SPIE).

References

[1] Gholipour A, Yazdanpanah MJ. Dynamic tracking control of nonholonomic mobile robot with model reference adaptation for uncertain parameters. In: Proceedings of 2003 European control conference (ECC'2003), Cambridge, UK; September 1–4, 2003.

[2] Pourboghrat F, Karlsson MP. Adaptive control of dynamic mobile robots with nonholonomic constraints. Comput Electr Eng 2002;28:241–53.

[3] Zhang Y, Hong D, Chung JA, Velinky SA. Dynamic model based robust tracking control of a differentially steered wheeled mobile robot. In: Proceedings of American control conference. Philadelphia, PA; June 1988. p. 850–55.

[4] Aicardi M, Casalino G, Bicchi A, Balestrino A. Closed-loop steering of unicycle vehicles via Lyapunov techniques. In: IEEE robotics and automation magazine. March 1995. p. 27–33.

[5] Devon D, Bretl T. Kinematic and dynamic control of a wheeled mobile robot. In: Proceedings of 2007 IEEE/RSJ international conference on intelligent robots and systems. San Diego, CA; October 29–November 2, 2007. p. 4065–70.

[6] Watanabe K, Yamamoto T, Izumi K, Maeyama S. Underactuated control for nonholonomic mobile robots by using double integrator model and invariant manifold theory. In: Proceedings of 2010 IEEE/RSJ international conference on intelligent robots and systems. Taipei, Taiwan; October 18–22, 2010. p. 2862–67.

[7] Rigatos GG, Tzafestas CS, Tzafestas SG. Mobile robot motion control in partially unknown environments using a sliding-mode fuzzy-logic controller. Rob Auton Syst 2000;33:1–11.

[8] Rigatos GG, Tzafestas SG, Evangelidis GJ. Reactive parking control of nonholonomic vehicles via a fuzzy learning automaton. IEE Proc Control Theory Appl 2001;148 (2):169–79.

[9] Carelli R, Soria CM, Morales B. Vision-based tracking control for mobile robots. In: Proceedings of twelfth international conference on advanced robotics (ICAR'05). Seatle, WA; July 18–20, 2005. p. 148–52.

[10] Gilioli M, Melchiori C. Coordinated mobile manipulator point-stabilization using visual-servoing techniques. In: Proceedings of IEEE/RSJ international conference on intelligent robots and systems (IROS'2002). vol. 1. Lausanne, CH; 2002. p. 305–10.

[11] Tsakiris D, Rives P, Samson C. Extending visual servoing techniques to nonholonomic mobile robots. In: Hager G, Kriegman D, Morse S, editors. Vision and control (LNCIS). Berlin: Springer; 1998.

[12] Okamoto Jr J, Grassi Jr V. Visual servo control of a mobile robot using omnidirectional vision. In: Van Amerongen J, Jonker B, Regtien P, Stramigiolis S, editors. Proceedings of mechatronics conference 2002. University of Twente; June 24–26, 2002. p. 413–22.

[13] Abdelkader HH, Mezouar Y, Andreff N, Martinet P. Image-based control of mobile robot with central catadioptric cameras. In: Proceedings of 2005 IEEE international conference on robotics and automation (ICRA 2005). Barcelona, Spain; April 2005. p. 3533–38.

[14] Mezouar Y, Abdelkader HH, Martinet P, Chaumette F. Central catadioptric visual servoing from 3D straight lines. In: Proceedings of IEEE/RS international conference on intelligent robots and systems (IROS'04). Sendai, Japan; 2004. p. 343–49.

[15] Tzafestas SG, Melfi A, Krikochoritis T. Kinematic/dynamic modeling and control of an omnidirectional mobile manipulator. In: Proceedings of fourth IEEE/IFIP international conference on information technology for balanced automation systems in production and transportation (BASYS2000). Berlin, Germany; 2000.

[16] Papadopoulos E, Poulakakis J. Trajectory planning and control for mobile manipulator systems. In: Proceedings of eighth IEEE Mediterranean conference on control and automation (MED'00). Patras, Greece; July 17–19, 2000.

[17] Yamamoto Y, Yun X. Coordinating locomotion and manipulation of a mobile manipulator. In: Proceedings of thirty-first IEEE conference on decision and control. Tucson, AZ; 1992. p. 2643–48.

[18] Yamamoto Y, Yun X. Modeling and compensation of the dynamic interaction of a mobile manipulator. In: Proceedings of IEEE conference on robotics and automation. San Diego, CA; 1994. p. 2187–92.

[19] Zavlangas PG, Tzafestas SG, Althoefer K. Navigation for robotic manipulators employing fuzzy logic. In: Proceedings of third world conference on integrated design and process technology (IDPT'98). vol. 6. Berlin, Germany; 1998. p. 278–83.

[20] Zavlagas PG, Tzafestas SG. Integrated fuzzy global path planning and obstacle avoidance for mobile robots. In: Proceedings of European workshop on service and humanoid robots (SERVICEROB'2001). Santorini, Greece; 2001.

[21] Tzafestas SG, Stamou G. A fuzzy path planning algorithm for autonomous robots moving in an unknown and uncertain environment. In: Proceedings of European robotics and intelligent systems conference (EURISCON'94). Malaga, Spain; August 1994. p. 140–49.
[22] Vidal Calleja TA. Visual navigation in unknown environments. PhD thesis. Barcelona: IRI, The Universitat Politècnica de Catalunya; 2007.
[23] Rekleitis I, Dudek G, Milios E. Multirobot collaboration for robust exploration. Ann Math Artif Intell 2001;31(1–4):7–40.
[24] Rekleitis I, Dudek G, Milios E. Probabilistic cooperative localization and mapping in practice. In: Proceedings of IEEE international conference on robotics and automation. vol. 2. Taipei, Taiwan; 2003. p. 1907–12.
[25] Velagic J, Osmic N, Lacevic B. Neural network controller for mobile robot motion control. World Acad Sci Eng 2008;23:193–8.
[26] Awad HA, Al-zorkany M. Mobile robot navigation using local model network. Trans Eng Comput Technol 2004;VI:326–31.
[27] Skoundrianos EN, Tzafestas SG. Fault diagnosis via local neural networks. Math Comput Simul 2002;60:169–80.
[28] Skoundrianos EN, Tzafestas SG. Finding fault: diagnosis on the wheels of a mobile robot using local model neural networks. IEEE Rob Autom Mag 2004;11(3):83–90.
[29] Tzafestas SG, Raptis S. Fuzzy image processing: a review and comparison of methods. Image Process Commun 1999;5(1):3–24.
[30] Castillo O, Aguilar LT, Cardenas S. Fuzzy logic tracking control for unicycle mobile robots. Eng Lett 2006;13(2):73–7 [EL.13-2-4].
[31] Yang F, Wang C. Adaptive tracking control for uncertain dynamic nonholonomic mobile robots based on visual servoing. J Control Theory Appl 2012;10(1):56–63.
[32] Wang C, Liang Z, Du J, Liang S. Robust stabilization of nonholonomic moving robots with uncalibrated visual parameters. In: Proceedings of 2009 American control conference. Hyatt Regency River Front, St. Louis, MO; 2009. p.1347–51.
[33] Zhang Y. Visual servoing of a 5-DOF mobile manipulator using panoramic vision system. M.A.Sc. thesis. Regina, Canada: Faculty of Engineering, University of Regina; 2007.
[34] Zhang Y, Mehrandezh M. Visual servoing of a 5-DOF mobile manipulator using a catadioptric vision system. Proc SPIE Conf Optomechatronic Syst Control III 2007; 6719:6–17.

14 Generic Systemic and Software Architectures for Mobile Robot Intelligent Control

14.1 Introduction

In this book we have studied several controllers of mobile robots, and provided a set of fundamental issues regarding the high-level functions of path/motion planning, task planning, localization, and mapping. In Chapter 13, we have presented a host of experimental results (most of which are simulation results) of the above methods and controllers. In this chapter, we will be concerned with systemic and software architectures which can be used for integrating controllers and high-level functional units to achieve overall intelligent performance of a mobile robot. Software architectures help to deal with the high degree of heterogeneity among the subsystems involved, to face strict operational requirements posed by real-time interactions with the robot's environment, and to treat the system's complexity that goes beyond the capabilities of a single designer. In general, robot software is the coded commands that tell the robot what functions to perform in order to carry out and control its actions. Actually, the development of robot software is a nontrivial task. Many software frameworks and systems have been developed to make robot programming easier, to achieve a proper run-time for fault tolerant execution, and to obtain systems that are portable to other robotic applications.

Unfortunately, there is no widely accepted software standard for developing mobile robot applications. Robot companies provide their own development frameworks such as ERSP from Evolution Robotics, Open-R from Sony, ARIA from ActivMedia, and Saphira from SRI International. On the other hand, university research groups have developed particular software platforms such as:

- Miro [1]
- CLARAty [2]
- Marie [3]
- Player/Stage [4]
- CARMEN [5]

The purpose of this chapter is to provide an overview of fundamental concepts and architectures that integrate low-, medium-, and high-level control and planning functions of mobile robotic systems. Specifically, the following issues are considered:

- Hierarchical, multiresolutional, reference model, and behavior-based intelligent control system architectures

Introduction to Mobile Robot Control. DOI: http://dx.doi.org/10.1016/B978-0-12-417049-0.00014-6

- Basic characteristics of mobile robot control software architectures
- Two examples of mobile robot control software architectures
- Comparative evaluation of two mobile robot control software architectures
- Intelligent human–robot interfaces
- Two integrated intelligent mobile robot research prototypes
- Design for heterogeneity and modularity

14.2 Generic Intelligent Control Architectures

14.2.1 General Issues

Intelligent control (IC) has an age of over 40 years and constitutes a generalization of traditional control of the 1940s and 1950s and modern control of the 1960s and 1970s to incorporate autonomous human-like interactive behavior of the controller with the environment. The term "intelligent control" was coined by Fu [6] to embrace the area beyond adaptive and learning control, and according to Saridis [7,8] represents the field which merges control, artificial intelligence (AI), and operational research (OR).

Intelligent control provides the means to achieve autonomous behavior such as planning at different levels of detail, imitation of human behavior, learning from past experience, integration and fusion of sensor information, identification of abrupt changes in system operation, and proper interaction with a changing environment [9].

The field of intelligent control started with the development of generic *IC architectures* (ICAs), which are mainly the following:

- Hierarchical ICA (Saridis)
- Multiresolutional/nested ICA (Meystel)
- Reference model ICA (Albus)
- Behavior-based ICAs, namely, subsumption ICA (Brooks) and motor schemas ICA (Arkin)

These architectures were expanded, enriched, or combined over the years in several ways [10,11]. In the following, we give a brief overview of these architectures in their original abstract form. Most of the software systems and integrated hardware–software systems developed for intelligent mobile robot control follow in one or the other way one of these generic architectures or suitable combinations of them. This will be clear from the discussions of Sections 14.4, 14.5, and 14.7.

14.2.2 Hierarchical Intelligent Control Architecture

This architecture has three main levels, namely (Figure 14.1):

1. Organization level
2. Coordination level
3. Execution level

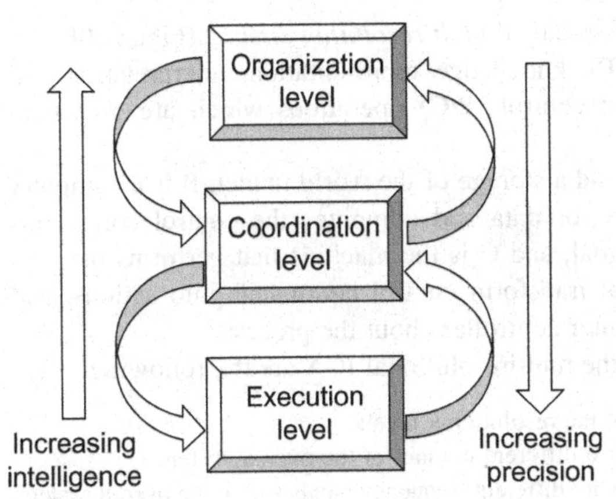

Figure 14.1 Hierarchical control architecture.

These levels may involve several layers within them that follow the human control mode of interaction between a director (supervisor) with his subordinate employees [7,8].

The *organization* level implements the higher level functions (e.g., learning, decision making) which imitate functions of human performance and can normally be represented and treaded by artificial intelligence techniques. This level receives and interprets feedback information from the lower levels, defines the planning/sequencing and decision making strategies to be executed in real time, and processes large amounts of knowledge/information with little or no precision. Here, long-term memory exchange is taking place.

The *coordination* level consists of several coordinators (each implemented by a piece of S/W or a dedicated microprocessor) which receive the task(s) from the organization level. All necessary details must be provided so that the chosen task's plan is successfully executed.

The *execution level* involves the actuators, the hardware controllers, and the sensing devices (visual, sonar, etc.) and executes the action programs issued by the coordination level.

Saridis has developed a complete analytic theory for this architecture, formulating and exploiting the *Principle of Increasing Precision with Decreasing Intelligence* using the information entropy concept. Neural networks, fuzzy systems, Petri nets, and optimal control have been used in these hierarchical levels [12].

14.2.3 Multiresolutional Intelligent Control Architecture

This architecture was developed by Meystel [13–15] and first applied to intelligent mobile robots. It follows the commonsense model *Planner-Navigator-Pilot-Execution Controller. The Planner* delivers a rough plan. The *Navigator* computes a more precise trajectory of the motion to be executed. The *Pilot* develops online tracking open-loop control. Finally, the *Execution Controller* executes plans and compensations computed by the planner, the navigator, and the pilot. This scheme is implemented in the form of the so-called *multiresolution six-box* (Figure 14.2A). Each level contains perception (P), knowledge representation, interpretation and processing (K), and planning and control (P/C) operations which are shown in more detail in Figure 14.2B.

In Figure 14.2B, A is a source and a storage of the world model, B is a computer controller which processes the sensor data and computes the control commands required to achieve the system's goal, and C is the machine that performs the process of interest with actuators that transform control commands into actions, and with sensors that inform the computer controller about the process.

The fundamental properties of the multiresolutional ICA are the following:

P1: Computational independence of the resolutional levels.
P2: Each resolution level represents a different domain of the overall system.
P3: Different resolution levels deal with different frequency bands within the overall system.
P4: Loops at different levels are six-box diagrams nested in each other.
P5: The upper and lower parts of the loop correspond to each other.

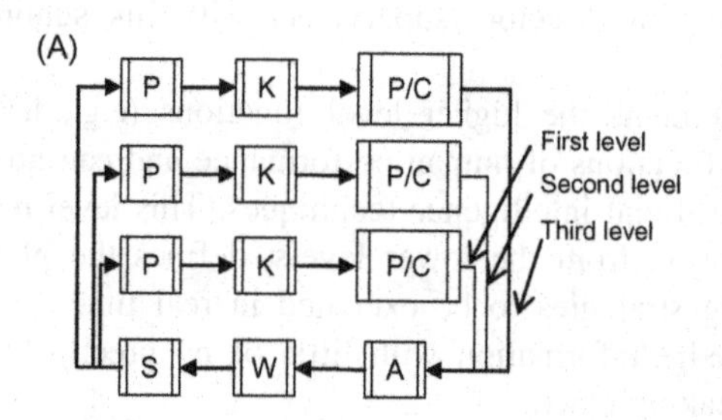

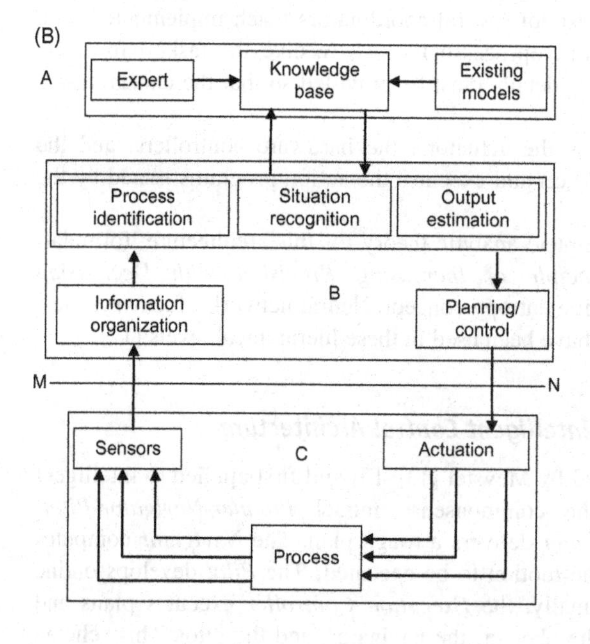

Figure 14.2 (A) Three-level multiresolution architecture (6-box-representation). (B) Each level has its own feedback loop.

P6: The system behavior is the result of superposition of the behaviors generated by the actions at each resolution level.
P7: The algorithms of behavior generation are similar at all levels.
P8: The hierarchy of representation evolves from linguistic at the top to analytical at the bottom.
P9: The subsystems of the representation are relatively independent.

14.2.4 Reference Model Intelligent Control Architecture

This architecture (RMA) was developed and expanded at the *National Institute of Standards* (NIST) by Albus and colleagues [16–18]. It is suitable for modular expansion (Figure 14.3).

The control problem in the reference model ICA is decomposed in the following subproblems:

- Task decomposition
- World modeling
- Sensory processing
- Value judgment

The various control elements are clustered into computational nodes arranged in hierarchical layers, each one of which has a particular function and a specific

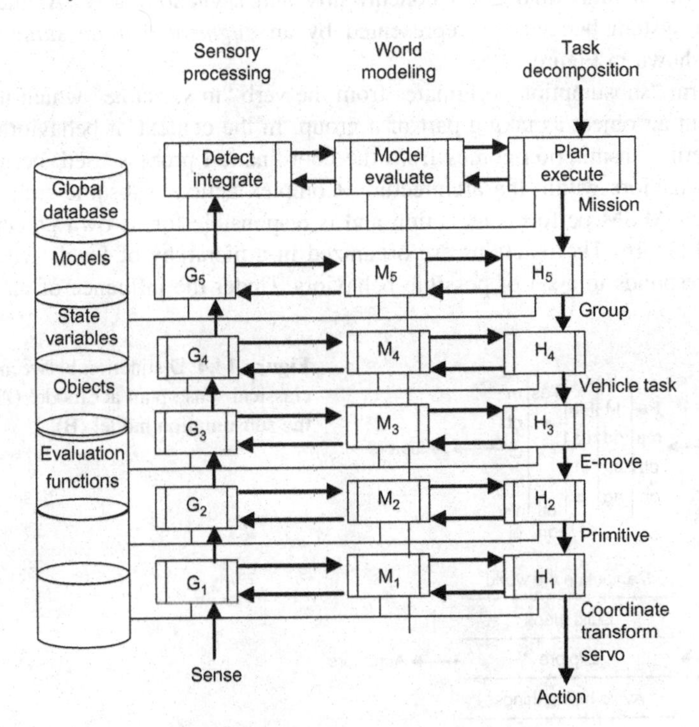

Figure 14.3 The NIST RMA hierarchical control architecture.

timing behavior. The NIST paradigm has been refined many times. Starting with the cerebellar model in the 1970s, it was evolved to the RCS-4 (Real-time Control System-4) in the 1990s and was applied to automated manufacturing systems, army field material handling systems, multiple underwater mobile robots, and telerobotic service systems. The main design issues addressed by RMA are the following:

- Real-time task and software execution
- Smart interface/communication methods
- Information/knowledge base management
- Optimized allocation of resources

14.2.5 Behavior-Based Intelligent Control Architectures

These architectures are based on the concept of *agent* and can be implemented using knowledge-based systems, neural, fuzzy or neurofuzzy structures [14–19]. The two most common behavior-based architectures are the *subsumption* architecture developed by Brooks [20,21] and the *motor schema* architecture developed by Arkin [22–24]. The subsumption architecture follows the decomposition of the behavior paradigm (Figure 14.4B) and was first employed in the autonomous mobile robot Shakey.

The tasks, achieving behavior, are represented as separate layers. Individual layers work on individual goals concurrently and asynchronously. At the lowest level, the system behavior is represented by an *augmented finite state machine* (AFSM) shown in Figure 14.5.

The term "subsumption" originates from the verb "to subsume" which means to think about an object as taking part of a group. In the context of behavioral robotics, the term subsumption comes from the coordination process used, between the layered behaviors within the architecture. Complex actions subsume simple behaviors. Each AFSM performs an action and is responsible for its own perception of the world [15,16].The reactions are organized in a hierarchy of levels where each level corresponds to a set of possible behaviors. Under the influence of an internal

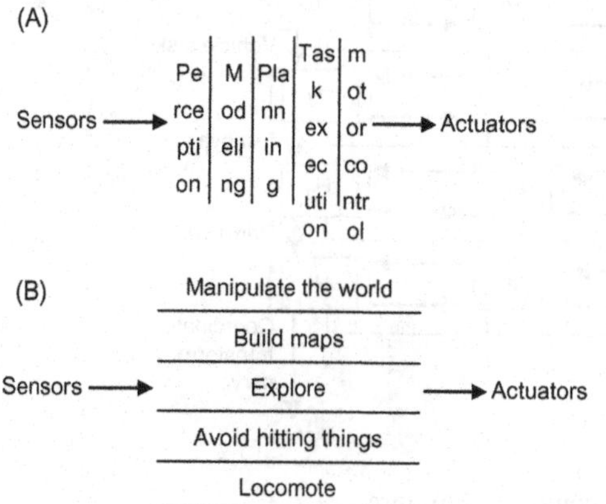

Figure 14.4 Distinction between the classical sense-plan-act model (A) and the subsumption model (B).

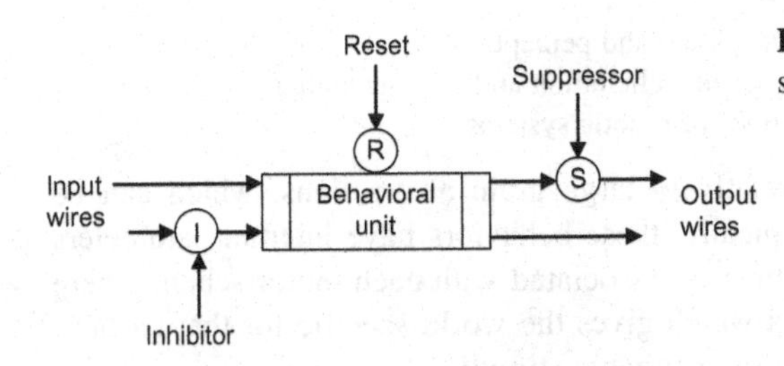

Figure 14.5 AFSM employed in the subsumption architecture.

or external stimulation, a particular behavior is required. Then, it emits an influx toward the inferior level. At this level, another behavior arises as a result of simultaneous action of the influx and other stimuli. The process continues until terminal behaviors are activated. A priority hierarchy fixes the topology. The lower levels in the architecture have no awareness of higher levels. This allows the use of incremental design. That is, higher level competencies are added on top of an already working control system without any modification of those lower levels.

The motor schemas architecture was more strongly motivated by biological sciences and uses the theory of schemas, the origin of which goes back to the eighteenth century (Immanuel Kant). Schemas represent a means by which understanding is able to categorize sensory perception in the process of realizing knowledge of experience. The first applications of schema theory include an effort to explain postural control mechanisms in humans, a mechanism for expressing models of memory and learning, a cognitive model of interaction between motor behaviors in the form of schemas interlocking with perception in the context of the perceptual cycle, and a means for cooperation and competition between behaviors.

From among the various definitions of the *schema concept* available in the literature, we give here the following representative ones [17,18]:

- A pattern of action or a pattern for action
- An adaptive controller which is based on an identification procedure for updating the representation of the object under control
- A perceptual entity corresponding to a mental entity
- A functional unit that receives special information, anticipates a possible perceptual content, and matches itself to the perceived information

A convenient working definition is the following [19]: "A schema is the fundamental entity of behavior from which complex actions can be constructed, and which consists of the knowledge how to act or perceive, as well as the computational process by which it is enacted."

Using schemas, robot behavior can be encoded at a coarser granularity than neural networks while maintaining the features of concurrent cooperative–competitive control involved in neuroscientific models. More specifically, schema theory-based analysis and design of behavior-based systems possesses the following capabilities:

- It can explain motor behavior in terms of the concurrent control of several different activities.
- It can store both how to react and how to realize this reaction.
- It can be used as a distributed model of computation.

- It provides a language for connecting action and perception.
- It provides a learning approach via schema elicitation and schema tuning.
- It can explain the intelligence functions of robotic systems.

Motor schema behaviors are relatively large grain abstractions, which can be used in a wide class of cases. Typically, these behaviors have internal parameters which offer extra flexibility in their use. Associated with each motor schema there is an embedded perceptual schema which gives the world specific for that particular behavior and is capable of providing suitable stimuli.

Three ways in which *planning* (*deliberative*) and *reactive* behavior can be merged are [24] as follows:

- Hierarchical integration of planning and reaction (Figure 14.6A)
- Planning to guide reaction, that is, permitting planning to select and set parameters for the reactive control (Figure 14.6B)
- Coupled planning–reacting, where these two concurrent activities, each guides the other (Figure 14.6C)

One of the first robotic control schemes that were designed using the hybrid deliberative (hierarchical) and reactive (schema-based) is the *autonomous robot architecture* (AuRA) [24]. AuRA incorporated a traditional planner that could reason over a modular and flexible behavior-based control system (Figure 14.7).

14.3 Design Characteristics of Mobile Robot Control Software Architectures

To design or evaluate a robot control software architecture, the following desirable key characteristics should be considered [25]:

- Robot hardware abstraction
- Extendibility–Scalability

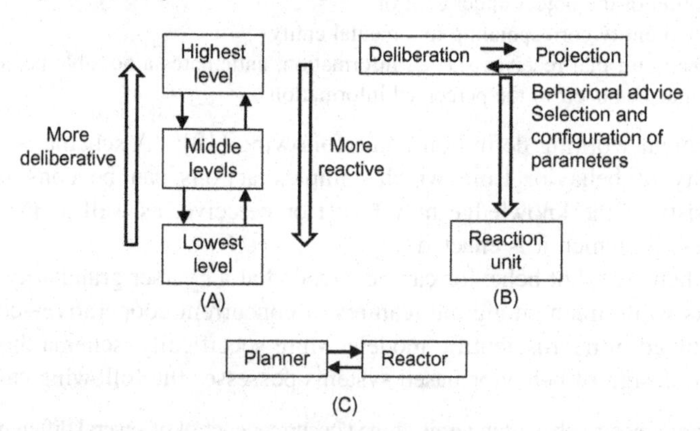

Figure 14.6 (A) Hierarchical hybrid deliberative–reactive structure. (B) Planning to guide reaction scheme. (C) Coupled planning and reacting scheme.

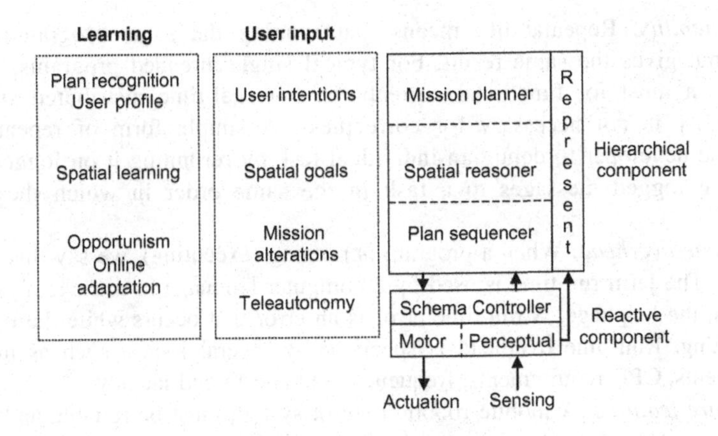

Figure 14.7 General AuRA structure.

- Reusability
- Repeatability
- Run-time overhead
- Software features
- Tools and techniques
- Documentation

A brief description of each of them is as follows.

Robot hardware abstraction: A primary design goal of hardware design is *portability* because robot hardware is generally changing. Abstraction of hardware such as actuators and sensors must be accommodated in a portable architecture. Typically, the hardware provided by manufacturers involve hardware-specific commands (such as to move in absolute or velocity mode) which are encapsulated into a generalized set of commands. It is highly desirable to keep the hardware characteristics in a single file of the software source. This file must be the only place where changes have to be performed when moving the system to a new hardware.

Extendibility–scalability: Extendibility is the capability to add new software components and new hardware modules to the system. This is a very important characteristic since robotic systems in research and development environments evolved in terms of both hardware and software. For example, the addition of new sensors is a typical process in these environments. Scalability can be achieved using dynamic objects and process invocation on a live-when-needed basis. Modern software tools have factory patterns that help in this.

Reusability: Reusing existing knowledge from previous designs can speed up the mobile robot software development. A popular approach to this is through software reuse of components, structure, framework, and software patterns [26,27]. Software patterns are classified according to three software development levels, namely, analysis or conceptual patterns for the analysis level, design patterns for the design level, and programming patterns for the implementation level.

Repeatability: Repeatability means that running the same program on the same input gives the same result. For typical single-threaded programs, repeatability is a must for functional correctness. For real-time distributed systems, repeatability is not necessary for correctness. A simple form of repeatability allows the developer to debug an individual task by rerunning it on logged data, presenting logged messages to a task in the same order in which they were received.

Run-time overhead: When a program is running (executing), we say that it is in run-time. The term run-time is used by a computer language to manage a program written in the language. A run-time error is an error that occurs while the program is executing. Run-time overhead is specified by several issues, such as memory requirements, CPU requirements, frequency, and end-to-end latency.

Software features: A mobile robotic control system must be reliable and robust to unexpected events. The framework for robust integration should integrate all skills. For research and development purposes, software architecture, besides reusability and repeatability, should provide the means for the following:

- Simple integration of new devices and units
- Clear distinction between levels of competence
- Prototyping
- Simple debugging

For a general software system, the following features are important:

- Design simplicity (in both the implementation and interface)
- Design correctness (in all aspects)
- Design consistency (i.e., absence of inconsistencies)
- Design completeness (i.e., coverage of as many important aspects as it is practical)

A good architecture must be based on a formal theory which is adhered by the software developers.

Tools and methods: Today, several tools for constructing software architectures, standardized by international bodies (ISO, ANSI, OMG, etc.), are available. The hardware providers offer the basic interface for evaluating the hardware. This may be a C language API. Early software systems for mobile robots were almost exclusively programmed in C. Artificial intelligence workers were using LISP. Now, we use popular OO-based languages such as C+ + and Java or the component technology CORBA (*common object request broker architecture*). Very popular are also the LabVIEW (*laboratory virtual instrumentation engineering workbench*) of national instruments (NI), and the UML (*unified modeling language*) which is supported by tools that allow the automatic synthesis, analysis, and code generation.

A key component in a robotic system is a reliable and efficient communication mechanism for exchanging and transmitting data and events. Data transfer can be initiated in either a *pull* or a *push* way. The interaction of tools such as CORBA, Microsoft Active X, and enterprise Java beans (EJB) can be achieved through interface description language (IDL).

Documentation: Software architecture should be accompanied with proper and rigorous documentation, which can be used nonlocally, and includes the following:

- The architecture's philosophy
- A programmer's guide
- A user's guide
- A reference manual
- Code documentation

Obviously, the philosophy of an architecture remains the same throughout, but the other elements of documentation should be refreshed to reflect any changes or improvements performed over the time. Documentation can be made in the following ways:

- Printed manuals
- Web-based documentation
- UML/class diagrams
- Comments in the source

A combination of them is very useful and desirable. Today, there are available *JavaDoc/Doxygen* utilities that are embedded in the code.

14.4 Brief Description of Two Mobile Robot Control Software Architectures

14.4.1 The Jde Component-Oriented Architecture

The *Jde* architecture uses *schemas* which are combined in dynamic hierarchies to unfold the global behavior [28,29]. Each schema is built separately into a plug-in and linked to the framework dynamically when required. The *Jde* software architecture follows the hierarchical scheme shown in Figure 14.6A, that is, it combines deliberation and reactiveness in a proper and successful way. Each schema is a task-oriented piece of software which is executed independently. At any time, there may be in execution several schemas, each one designed to achieve a goal or complete a particular task. In *Jde,* a schema:

- is *tunable* (i.e., it can modulate its behavior accepting continuously some parameters);
- is an *iterative process* (i.e., it performs its work via periodical iterations giving an output when each iteration is finished);
- can be stopped or resumed at the end of any iteration.

In Jde, hierarchy is considered as a co-activation that only means predisposition. A parent can coactivate several children at the same time, but this does not mean that all children gain control of the robot. The children's real activation is determined by an *action–selection* mechanism that continuously selects which one gets the control at each iteration, given the current goal and environment condition. Three advantages of this hierarchical scheme are reduced complexity for action–selection, action–perception coupling, and distributed monitoring. A motor

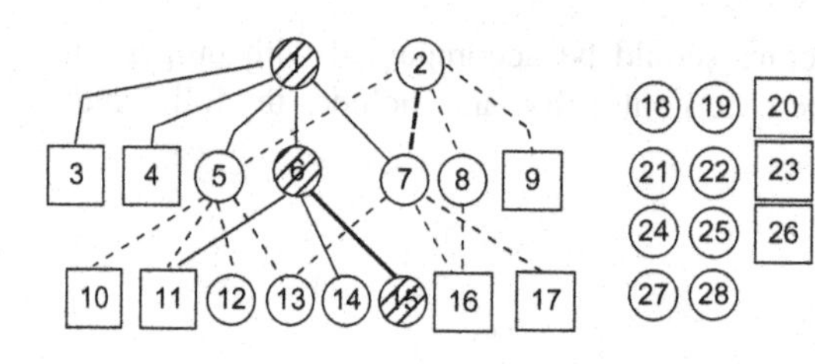

Figure 14.8 Jde hierarchical architecture where there is one winner (denoted in shaded) at each level.

schema may control the actuators directly or may awake a set of new child schemas. The sequence of activations creates a specific hierarchy of schemas for producing a particular global behavior (Figure 14.8) [28]. All awake schemas (checking, ready, winner) run concurrently. Hierarchies are specific to each global behavior. Once the parent has awaked its children schemas, it continues executing and checking its own preconditions, monitoring the actions of its children, and modulating them appropriately. All the active schemas, except the winner one, are deactivated, and a new tree is generated under the new winner.

In Figure 14.8, circles represent *motor schemas*, and squares represent *perceptual schemas*. The *Jde* architecture was implemented in the *Jdec* software platform in the language C. The Jdec platform supports the hardware of Pioneer robot equipped with additional vision sensors. Several schema-based behaviors were constructed, namely, person following, laser-based and vision-based localization, virtual force field-based local navigation, and gradient-based deliberative global navigation. In the hierarchy, each schema provides a set of shared variables for communication with other schemas which is performed by shared memory. When in winner state, a schema specifies and updates continuously its output. Perceptive schemas do not take part in the action–selection process and always gain easily the winner state.

The pseudocode of a Jdec schema is [29]:

Initialization code
Loop
If (slept) stop-the-schema
Action_selection
    Check preconditions
    Check brother's state
  If (collision OR absence)
      father_arbitrates
If (winner) then schema_iteration
msleep
End loop

Due to the iterative execution style, CPU consumption is moderate and facilitates the design of an application in a reactive fashion. Each schema is written in two separate C files, namely:

1. *myschema.h* (with the declaration of shared variables)
2. *myschema.c*

They are both compiled jointly in a single C module. All schemas of an application are statically linked together in the executable.

In the *enhanced Jde* architecture (called *Jde-neoc*), several new tools were developed and added. These are as follows:

- A *visualization tool* for the visualization of sensors, actuators, and other elements
- A *management tool* which allows the manual activation and deactivation of schemas, as well as their graphical user interface (GUI). This helps very much the debugging of any set of schemas.

In *Jde-neoc*, the perception and control are distributed among a set of schemas which are software elements, with a clear API each built as a plug-in on a separate file. Details on the design and implementation of Jde and Jde-neoc can be found in Refs. [28,29].

14.4.2 Layered Mobile Robot Control Software Architecture

This is simple software architecture of the type shown in Figure 14.1, which involves three or four hierarchical layers, each of which depends only on the specific hardware platform used, and is not informed on the contents of layer above or below it [30]. This layered architecture is depicted in Figure 14.9 for a mobile robot or manipulator, which includes all necessary high-level and low-level functions from task definition, path planning, and sensor fusion to sensor/actuator interfacing and motor control.

For fully autonomous operation, the layers 2 through 4 are needed; the layer 1 is not always required.

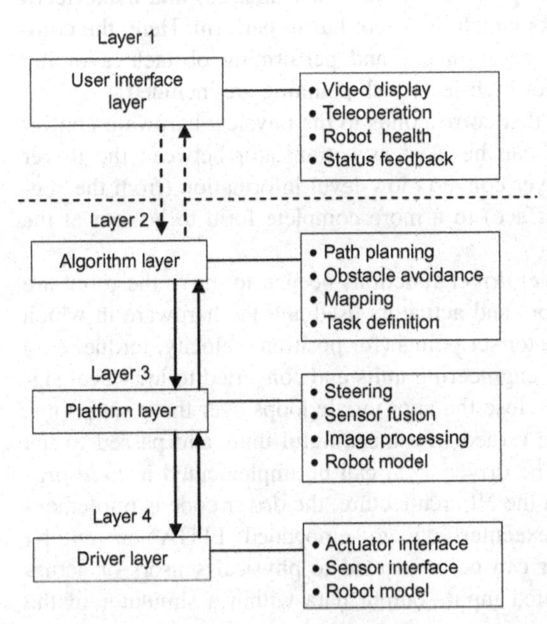

Figure 14.9 Layered architecture of a mobile robot/manipulator.

This software architecture was designed and developed for implementation on the NI *CompactRIO* platform for robotics combined with the *LabVIEW* graphical development environment.

LabVIEW is typically used for data acquisition, instrument control, and industrial automation on a variety of platforms including Microsoft Windows, and several versions of UNIX, Linux, and MacOS [31]. The programming language used in LabVIEW is called G, and is a data flow programming language. The execution sequence of the LabVIEW graphical syntax is as well defined as with any textually coded language (C, Visual BASIC, etc.). LabVIEW embeds into the development cycle, the construction of user interfaces (called "front panels"). LabVIEW programs and subroutines are called *virtual instruments* (VIs), each one of which has three elements (*block diagram, front panel, connector panel*). The front panel exhibits controls and indicators that allow a user to input data into or extract data from a running VI. In addition, the front panel can also act as a programming interface. A benefit of LabVIEW over other development environments is the extensive support for accessing instrumentation hardware (http://ni.com//labview).

A short description of the functions performed in each layer of the architecture is as follows.

User interface layer: The *user interface* (UI) allows a human operator to interact physically with the robot via relevant information provided on the host PC. To display live data from an on-board or fixed camera or the *xy* coordinates of nearby obstacles on a map, a GUI should be employed. This layer can also be used for reading input data from a mouse or joystick or to drive a simple display. An emergency (high priority) stop must also be included in this layer.

Algorithm layer: This layer involves the high-level control algorithms of the robot. Its units obtain information (position, speed, video images) and make feedback control decisions for the tasks which the robot has to perform. Here, the components for mapping the robot's environment and performing obstacle avoiding path/motion planning, as well as for high-level task planning are included.

Platform layer: Here, the code that corresponds to the physical hardware configuration is contained. Actually, it can be used as a translator between the driver layer and algorithm layer. This layer converts low-level information (from the sensors' interface and actuators' interface) to a more complete form to be sent at the algorithm layer, and vice versa.

Driver layer: Here, the low-level driver functions needed to move the robot are generated, depending on the sensors and actuators used and the hardware in which the driver software runs. The actuator set points (for position, velocity, torque, etc.) are received by the driver layer in engineering units and converted to low-level signals, potentially including code to close the appropriate loops over those set points. Similarly, the raw sensor data are turned into meaningful units and passed to the other layers of the architecture. The driver level can be implemented in *field programmable gate array* (FPGA). In the NI architecture, the driver code is implemented in *LabVIEW FPGA* and executes on an embedded FPGA on an NI *CompactRIO* platform. The driver can be connected to physical sensors or actuators, or it can interface to simulated input–output data within a simulator of the environment. For research and development purposes, a switch between simulation and actual hardware must be provided, which operates without affecting the other layers. An overall pictorial representation of the above mobile robot reference control software architecture overlayed on an NI CompactRIO or NI Single-Board RIO embedded system is shown in Figure 14.10 [30].

The above architecture is similar to that used in the NASA mobile manipulators designed by "Superdroid Robots" (see Figure 1.30) (http://superdroid.com/#customized-robots-and-robot-parts).

14.5 Comparative Evaluation of Two Mobile Robot Control Software Architectures

14.5.1 Preliminary Issues

Here, a comparative evaluation of two mobile robot control software systems drawn from Ref. [25] will be summarized. The two systems evaluated are the *Saphira* architecture, which was developed at SRI International Artificial Center [32], and *behavior-based robot research architecture* (BERRA) [33].

Saphira architecture was developed to exert intelligent control to the *Flakey* mobile robot [34] following the perception–action cycle scheme. The software runs a fuzzy logic-based reactive planning and a behavior sequencer. The system includes integrated modules for sonar sensor interpretation, mapping, and navigation. The core of the system consists of a server that manages the hardware, and Saphira software as a client to this server. The Saphira architecture is depicted in Figure 14.11A and B [32].

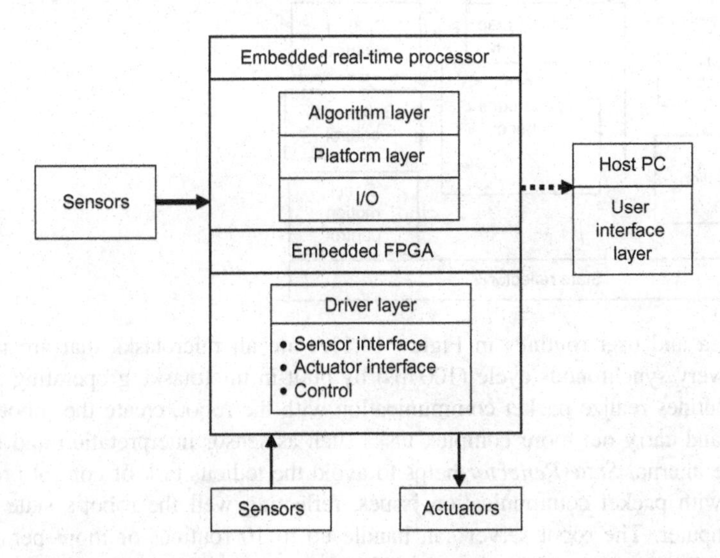

Figure 14.10 Mobile robot reference control software architecture overlayed onto an embedded real-time processor and FPGA.

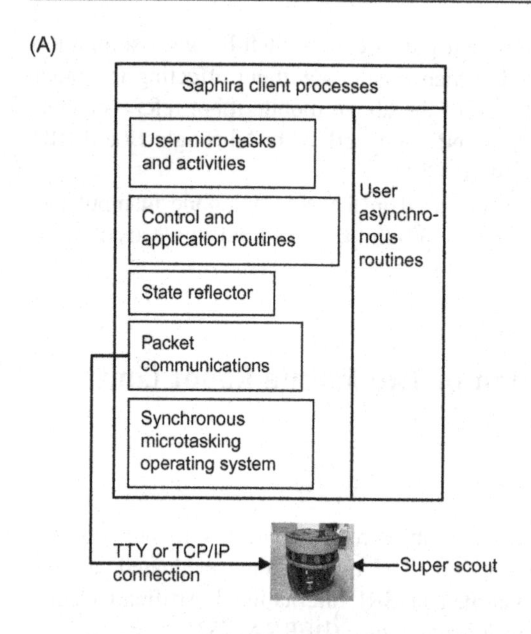

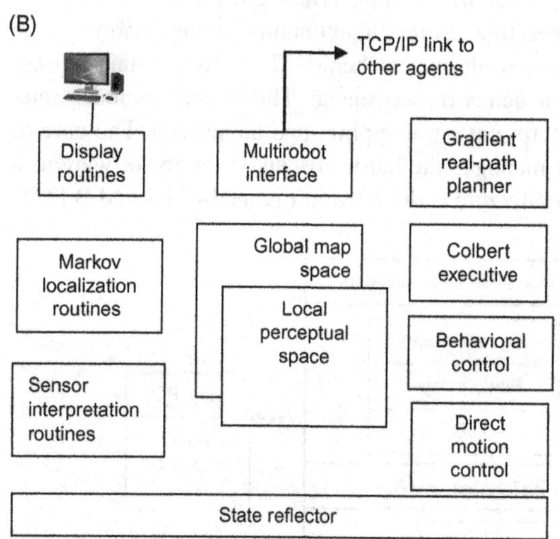

Figure 14.11 (A) Saphira system architecture. (B) Saphira control architecture.

Saphira and user routines in Figure 14.11A are all microtasks that are invoked during every synchronous cycle (100 ms) by built-in microtasking operating system. These routines realize packet communication with the robot, create the robot's state picture, and carry out more complex tasks such as sensor interpretation and navigation. The internal *State Reflector* helps to avoid the tedious task of control programs to deal with packet communication issues, reflecting well the robot's state on the host computer. The robot server can handle up to 10 routines or more per 100 ms cycle. Additional user routines on the left subsystem of Figure 14.11A can be executed asynchronously as separate threads sharing the same address space.

As shown in Figure 14.11B, the *control architecture* is built on the top of the state reflector, and consists of a set of microtasks/asynchronous tasks that implement all navigation functions, interpreting the sensor readings about a geometric world model, and mapping the robot states to control functions.

Localization is performed using Markov-type routines that connect the robot's local sensor readings to its map of the world. The *multirobot interface* links the robot to other robots via TCP/IP connections [35]. The path planning is performed combining the two available geometric representations for local path planning (*local perceptual space*, LPS) and global path planning (*global map space*, GMS).

BERRA was developed with primary design goals the flexibility and scalability. It was implemented using the *adaptive communication environment* (ACE) package [36]. The use of ACE allows the portability of the system across a large class of operating systems and provides powerful means for server/client interaction and service functioning.

The evaluation presented in Ref. [25] was carried out using a test-case service agent, where the operator could command the robot to navigate in an office environment for which a map was known *a priori*. The two architectures were ported to the Nomadic *Super Scout* robot shown in Figure 14.12.

This is a small robot having 16 ultrasonic sensors (Polaroid) and 6 tactile sensors. A serial port is used for the communication of the motherboard and the controller board. The Scout robot is equipped with the *Red Hat Linux* operating system and a C language API.

14.5.2 The Comparative Evaluation

14.5.2.1 Operating System and Language Support

Saphira: Most operating systems are supported (UNIX, MS Windows, etc.). The GUI is based on Motif. The core of the system is programmed in C. It has a C-like syntax with finite state machine-based semantics, and part of Saphira is written in LISP.

Figure 14.12 The Nomadics super scout mobile robot.
Source: http://ubirobot.ucd.ie/content/nomad-scout-2-and-nomad-super-scout.

BERRA: Linux and Solaris Operating Systems and all ACE platforms are supported. The *Esmeralda* speech recognition system is used [37], and vision functions employ Blitz++ [38].

14.5.2.2 Communication Facilities

Only BERRA is of the multiprocess communication. BERRA uses sockets based on ACE, and support UNIX and INET socket protocols.

14.5.2.3 Hardware Abstraction

In *Saphira*, hardware abstraction is performed in the robot server (i.e., there is only one abstraction level). Client processes cannot address the lower level hardware. *BERRA* also has one high-level abstraction, but control of lower level hardware can be achieved by parameterizing the high-level commands using a difficult syntax.

14.5.2.4 Porting and Application Building

Saphira porting of the hardware level code needs major effort. In the study described in Ref. [20], only the source code of the Pioneer platform server was available (actually a number of C files with no clear interdependencies). All relevant behaviors were provided with the system, and the construction of the map and its incorporation into the LPS was easy. So, localization performed well out of the box. Originally, BERRA ran on Nomad 2000 and Nomad 4000. Although the hardware of them can accept calls from multiple clients, the Scout can only be accessed by one client and created a problem for BERRA. A newer version of BERRA allowed only one of its processes to access and control the robot hardware.

14.5.2.5 Run-Time Consideration

Saphira can be easily started by first starting the robot server and then Saphira which then is connected to the server. Then, behaviors and tasks can be directly started and stopped by the operator in the GUI with the Colbert interpreter. Libraries can be dynamically loaded in run-time. Using GUI the robot's position in the map of the environment can be updated. In Saphira, 10 MB of memory is sufficient, but the response time ($\sim$0.6 s) is high. The major drawback of Saphira is the inaccuracy of the localization system (position track is lost after approximately 10 m).

BERRA can be started by a shell-script as long as everything goes as planned. If for any reason a process goes wrong, the system needs to be totally restarted. The time needed from a sensor reading to a corresponding actuator control signal is very low (about 0.17 s). BERRA needs 36 MB of run-time memory, but does not have a GUI. In the tests, BERRA performed very well (the scout could traverse a department for hours according to navigation requests).

14.5.2.6 Documentation

Saphira is very well documented and supported by many publications and a complete manual that includes a user's guide. But the code is not so well documented. *BERRA* is supported by many publications and a web-based documentation. Users and programmers have at their disposal short guides.

Details of the evaluation together with problems that must be avoided, and guidelines on how to choose among available commercial or research mobile robot platforms in order to meet specific goals and requirements can be found in Ref. [25]. A survey of nine open sources, freely available, *robotic development environments* (RDEs) is provided in Ref. [39]. This survey compares and evaluates these RDEs by establishing and using a comprehensive list of evaluation criteria, which includes the criteria presented in Section 14.3. First, a conceptual framework of four broad categories is presented based on the characteristics and capabilities of RDEs. The evaluation and comparison of these nine RDEs conclude with guidelines on how to use profitably its results. A comprehensive book dedicated to software architecture with deep and illuminated design issues is Ref. [40].

14.6 Intelligent Human–Robot Interfaces

14.6.1 Structure of an Intelligent Human–Robot Interface

Interfaces play a key role for the successful and efficient operation of an intelligent robot such that to fulfill its goals with the aid of multisensors and shared autonomy [41]. Here, the basic design principles of intelligent *human–robot interfaces* (HRIs) will be outlined. An intelligent HRI has the general self-explained structure shown in Figure 14.13.

The robotic system involves a supervisor, a planner, and a controller, and sometimes, if required, a decision support component which contributes to the realization of cooperative human–robot decision making and control.

The three main types of users are operators, engineers, and maintenance specialists. These users interact with the robotic system via the HRI. Users have in general different but overlapping needs with respect to depth and quantity.

14.6.2 Principal Functions of Robotic HRIs

The principal functions of HRIs are the following [42]:

- Input handling
- Perception and action
- Dialogue handling
- Tracking interaction
- Explanation
- Output generation

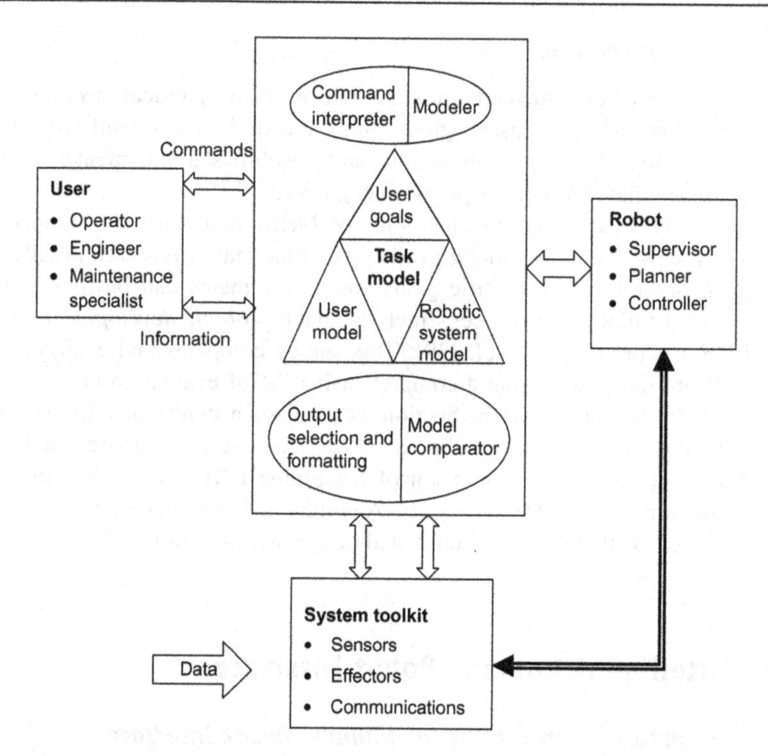

Figure 14.13 General structure of an intelligent HRI.

The *input handling* function provides the means to utilize the type of inputs received by the system which may be analog, digital, probabilistic, linguistic, and fuzzy.

The *perception and action* function is fundamental for the overall HRI performance and is supported by the presentation level of the HRI which determines how to present the information to the user and how to transform his/her control inputs.

Dialogue handling (control) undertakes the task of determining what information to treat and when. A dialogue is any logically coherent sequence of actions and reactions exchanged between the user and the HRI. Human–machine dialogues are necessary for many robotic operations (e.g., scheduling, supervision, planning, control).

Tracking interaction is concerned with tracking the entire interaction between the HRI and the human user, as well as between the HRI and the robotic system at hand.

The *explanation* function needs a model of the robotic system to be available. Its role is to explain to the user the meaning of the various aspects and components of the robotic system, and sometimes of the HRI itself. It should be also capable of explaining how the various parts of the system operate.

Output generation is realized using graphical editors and typically offers appropriate graphical and textual pictures which are dynamically changing. In more recent applications, multimedia presentations are also provided. If the HRI is required to be able to adapt to different users or user classes, a *user model* is also needed. To design a user model, it is necessary to use our knowledge on human processing behavior and represent the cognitive strategies, via rules, algorithms, and reasoning mechanisms. A more complete user model must also include a model of the robotic system in order to incorporate the user's view with respect to the robotic system.

14.6.3 Natural Language Human–Robot Interfaces

A special very popular class of HRIs is the class of *natural language interfaces* (NLIs). NLIs possess humanized properties since the user can communicate with the robot through a kind of verbal language (e.g., a small subset of English). Actually, NLIs are not the best interfaces in all cases. Thus, to decide whether to use a NLI or not, one has to consider several factors, of which some examples are the following:

- *Ease of learning*: If a full natural language (NL) is used, no human effort is necessary to learn it. This is not so if a restricted language with legal statements is used.
- *Conciseness*: The desire for conciseness is usually in conflict with the user friendliness.
- *Precision*: Many English sentences are ambiguous. It is so natural that English does not use parentheses as artificial logical languages do.
- *Need for pictures*: Words are not the best way to describe shapes, positions, and curves. A picture is worth many words. However, programs that handle graphical objects (e.g., CAD systems) are still good candidates for NLIs and other linguistic interfaces.
- *Semantic complexity*: NLs are concise and efficient when the universe of possible messages is large. Actually, no trivial language can perform the interfacing job, since the number of different messages that have to be handled is extremely large.
- *Cost*: The cost of NLIs used is higher than that of standard HRIs.

The components of an NL understanding system, that is, a system that transforms statements from the language in which they were made in a program-specific form that initiates appropriate actions, are as follows:

- Words and lexicons
- Grammar and sentence structure
- Semantics and sentence interpretation

The above primary components can be merged into an integrated understanding system in the following three ways:

1. *Interactive selection*: The system displays the options to the user who chooses among them to gradually construct a complete statement, which corresponds to actions that the target program can perform.
2. *Semantic grammars*: The window-based approach does not allow the user to control interactions or compose free-form statements that the system has to understand. The alternative is for the user to compose entire statements. A semantic grammar provides one implementation of this alternative approach but is appropriate when a relatively small subset of an NL has to be recognized.

3. *Syntactic grammars*: If a large part of NL is used as HRI, the capture of as much of the language regularity as possible is required. To this end, it is necessary to capture the syntactic regularity of the NL at hand. Thus, one needs to use a syntactically motivated grammar.

In the literature, several tools are described that assist in the building process of the lexicon, the grammar, the semantic rules, and the code that uses all of them. Also, some programs exist that do most of the understanding in all three approaches discussed above.

NLIs in robotics have been considered and used by many researchers, for example:

- *Nilsson* [43], where the mobile robot Shakey, capable of understanding simple NL commands, is presented.
- *Sato and Hirai* [44], where NL instructions are employed for teleoperation control.
- *Torrance* [45], where an NL interface is used to navigate an indoor mobile robot.

14.6.4 Graphical Human–Robot Interfaces

Graphical HRIs (GHRIs) represent a very large area of information technology. GHRIs are used for task analysis, online monitoring, and direct control. For example, to teleoperate a mobile robot in a critical workspace, a considerable effort must be devoted to preparing the task, training the operator, and finding the optimal cooperation modes in various situations. Before actually executing a task, a GHRI can help the user to specify his intention, display the commands, and the expected consequences on the monitor. In this way, the user can interactively generate and modify a plan.

On a GHRI, an operator can define a series of movements and actions by clicking or dragging a mouse on the screen. The available task and geometric planners can then find a sequence of motions and actions that implement the task. A simulation system is usually designed and used to animate the robot's motion on a 2D or 3D workspace, where several viewpoints can be set to monitor and observe the robot's behavior and its relation to the world. Possible collisions with obstacles, robots, and other objects are avoided. Here, the optimal utilization of various sensors is a fundamental prerequisite. As an additional aid, a *task editor* is used to support the task specification by interactively modifying a plan. It is useful if with this task editor, the operator can also define a sequence of actions as a macro. The macros can be retrieved and used to represent and implement an entire task plan. A useful concept that can be used in task analysis is the concept of *telesensor programming* (Hirzinger, [10]). Due to the unavoidable errors in the dead-reckoning and world models, the sensor patterns have to be employed by the robot to ensure an accurate relation with the world.

Graphical interfaces are very often combined with animation and virtual reality (VR) tools. Examples of this type are the works of Heinzmann [46], Rossmann [47], and Wang et.al [48].

In Ref. [46], the HRI of the robot consists of a visual face tracking system. The system employs a monocular camera and a hardware vision system to track several facial features (eyes, eye brows, ears, mouth, etc.). The 3D pose and orientation of the head are computed using this information. The solution to the design of human-friendly robots, provided, satisfies two safety goals.

Safety goal 1: A human-friendly robot should be able to operate without posing a threat when humans are inside the robot's workspace.

Safety goal 2: In an unstructured environment which may involve humans, any action autonomously taken by the robot must be safe even when the robot's sensor information about the environment is uncertain or false.

In Ref. [48] the human–machine system includes a virtual tools system, an automatic path planner, and a collision detection simulator. Tests on the performance of the path planner are also discussed. A virtual tools HRI for point specifications of tasks, which interweaves virtual robot end/effector representations with physical reality to immerse the human in the scene using simple hand gestures, needs to be developed for flexibly designating where the robot should grasp as an incoming part. The virtual tools system is displayed in four quadrants on a Silicon Graphics workstation with Galileo video. The virtual gripper is displayed on the two left quadrants display, superimposed on two camera views and blended with live video, to create the illusion of a real gripper in two views in the physical scene. The top right quadrant is occupied by the toolbox of graphic icons representing various tools available for use by the robot. The bottom right quadrant displays homogeneous transformation matrix information such as graphic object models and views from the robot camera. This Pennsylvania University system which is based on the virtual tool concept allows the operator to direct robot tasks in a natural way in almost real-time.

In Ref. [47] a multirobot system (called CIROS) is designed that implements the capability to derive robot operations from tasks performed in the virtual reality environment. To this end, two appropriate components are used: the change-detection component and the change-interpretation component. The VR system employed is based on a special simulation system (COSIMIR, cell-oriented simulation of industrial robots) developed at the Institute of Robotics Research in Dortmund, Germany. In CIROS, a new VR concept is used. This is called *projective virtual reality* (PVR), because the actions carried out by humans in the VR are projected on to robots to carry out the task in the physical environment. The intelligent controller implements PVR-based control by adding the levels for *online collision avoidance*, multirobot coordination, and automatic action planning.

As an example, we describe briefly the NL HRI of the KAMRO intelligent mobile robot, which was designed and built at the University of Karlsrue [49,50], and uses a multiagent architecture.

A fundamental problem in such *multiagent system* (MAS) is the negotiation among the agents that compete for a given task. This negotiation process can be performed by a *centralized mediator* or a *selected candidate* or by *many* (or *all*) *candidates*. All agents should be able to negotiate with the competing agents. One way to manage the communication among agents is via a *blackboard system*. Possible in a MAS are the following:

- Deadlocks caused by agent bodies/external resources
- Deadlocks caused by special agents
- Deadlocks caused by agent teams

These deadlock situations are successfully translated into corresponding mechanisms in the KAMRO robot.

The NL HRI of KAMRO performs the following functions:

- Task specification/representation, that is, analysis of instructions related to the implicit robot operations (e.g., "pick-and-place," bin picking)
- Execution representation
- Explanation of error recovery
- Updating and describing the environment representation

The KAMRO NL HRI architecture is as shown in Figure 14.14.

The robot and the NL HRI have permanent access to the correct environment representation via an overhead camera. This information is stored in a common database. Since the world representation changes over time, a timestamp of the snapshot is used which allows merging older and newer knowledge about the environment. The processing of NL instructions is illustrated in Figure 14.15.

An edited book with outstanding contributions covering a wide repertory of concepts, techniques, and applications of HRIs is Ref. [51].

14.7 Two Intelligent Mobile Robot Research Prototypes

Here, two working research prototypes of integrated mobile robots will be briefly presented. These are the robotic wheelchair SENARIO, developed within the frame

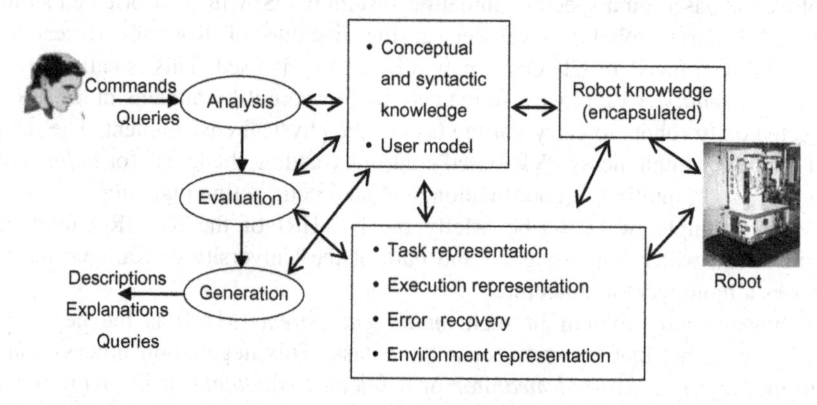

Figure 14.14 NL HRI architecture of the KAMRO (Karlsruhe autonomous intelligent mobile robot).

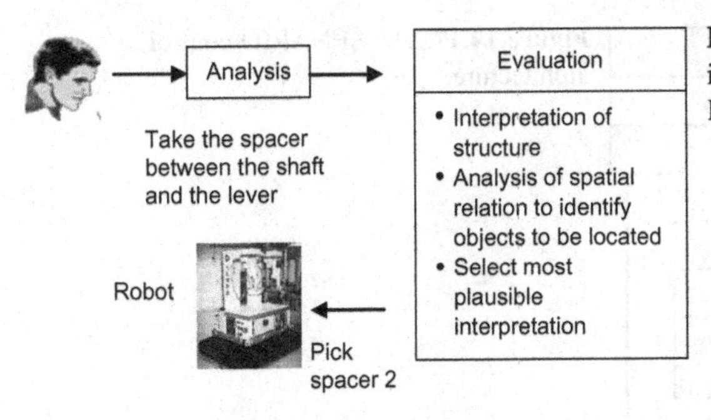

Figure 14.15 Structure of NL instruction processing in the KAMRO.

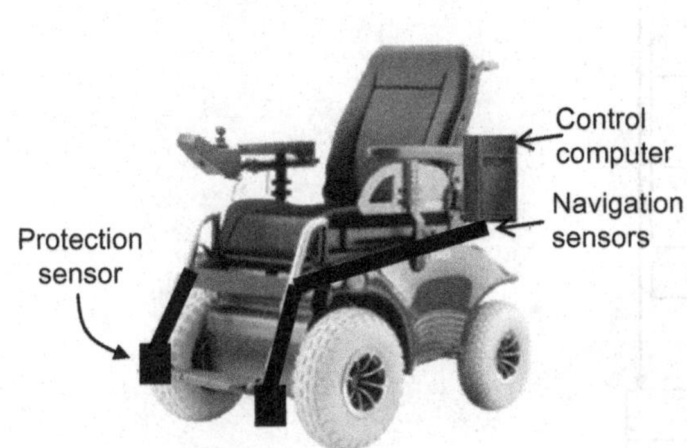

Figure 14.16 The SENARIO wheelchair intelligent mobile robot system was integrated on a commercial MEYRA platform. The components of the system include a computer, an orientation (encoder) sensor, and an ultrasonic sensor array (eight sensors for navigation and three sensors for protection; two in the front and one in the back).

of the European Union TIDE project SENARIO: *sensor-aided intelligent navigation system for powered wheelchairs*[1] [52,53], and the ROMAN service MM, designed and built at the Technical University of Munich [54–56].

14.7.1 The SENARIO Intelligent Wheelchair

This intelligent mobile robot system was implemented and tested on a Meyra powered wheelchair as shown in Figure 14.16.

The architecture of the SENARIO mobile robot, which is a *virtually centralized hierarchical control architecture*, is shown in Figure 14.17. Two functional alternatives are possible in SENARIO:

1. *Semi-autonomous*
2. *Fully autonomous*

The following four processes determine the autonomous behavior actions of SENARIO (Figure 14.17):

1. *Task planning*, which schedules the execution of all the other processes and is responsible for overall system control. The planning, path following, and goal monitoring procedures

[1] SENARIO Consortium: Zenon S.A. (GR), National Technical University of Athens (GR), Microsonic GmbH (DE), Reading University (UK), and Montpellier University (FR).

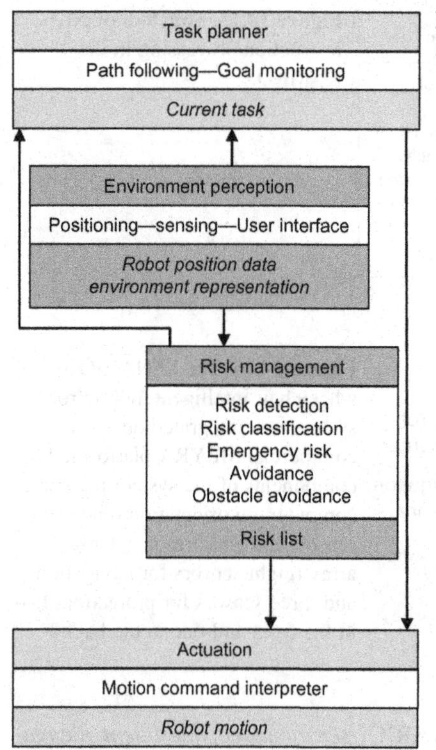

Figure 14.17 The SENARIO control architecture.

reside in the task planner. Task planning is on the top of the control hierarchy of the robot.

2. *Environment perception*, where a set of sensors is used to truck the interactions of the robot with its environment. Sensor data processing, environment features extraction, and positioning, as well as user input, are grouped together as environment perception processes.
3. *Risk management*, which is responsible for both risk detection and avoidance. This process uses the output of environment perception to detect potential risks and determine their importance. Then, it computes the necessary sequence of actions for risk avoidance.
4. *Actuation*, which handles the interface with the robot actuators.

This organization is an instance of a *centralized intelligent control scheme* (Figure 14.1) because task planning assumes the responsibility for the coordination of all the other processes in the system (*organization level*). However, some of the processes can override task planning and can communicate with each other directly, in cases of emergency. This option offers a distributed alternative to the centralization organization. This hybrid solution is called *virtually centralized control* [24]. Virtually centralized control combines the reactivity of the distributed approach, along with the high-level control features of the centralized one.

Semi-autonomous mode: The system receives commands to move on a direction, or to take an action (e.g., go ahead, turn left, stop). The system realizes the instructed action, while preventing risk conditions and avoiding obstacles during execution. Each time a risk is detected, SENARIO informs the user and takes appropriate corrective measures to preserve safety and continues with the execution of the instruction. In semi-autonomous mode, the user can override system actions, for example, approach closer to a wall than the system's alarm distance. In these cases, the system applies minimum speed limit in all instructed commands. In any case, if SENARIO detects an emergency situation, it stops moving and asks for user instruction to recover. So, the responsibility of actions is shared between the system and the user. This mode requires *risk detection* and *risk avoidance* functionality.

Fully autonomous mode: This is a superset of semi-autonomous mode. The system receives all the commands of semi-autonomous mode, along with "go to goal" commands. For example, the user can issue commands such as "go to living room." In this case, the system locates itself and then the target position in the environment map. It then plans and executes a path to the specified destination, avoiding all obstacles and risks on the way. During goal execution, the user can interfere with the system, as in semi-autonomous mode, or he/she can specify a new destination at any point of the goals set. In this mode, the system takes full responsibility for execution. Full autonomy requires *path planning*, along with *risk detection* and *risk avoidance* functionality.

Each process of the system consists of a series of executive tasks. These are specialized procedures computing the parameters that characterize each process. In particular, the task planner monitors the overall system functionality through the *path following* and *goal monitoring* tasks. The task planner computes the *current task* of the robot based on the risk management and environment perception outputs. Environment perception consists of the *positioning*, *sensing*, and *user interfacing* executive tasks. Similarly, risk management is split into *risk detection*, *risk classification*, *emergency risk avoidance*, and *obstacle avoidance tasks*.

The positioning task is responsible for reporting the robot's position, whenever it is asked to report, while running individually in its loop. As a task belonging to environment perception, it employs sensors and processes their information. The output of this task is referred to as the *position estimation* of the robot or, equivalently, as the *robot position data*.

Supplementary to the action of positioning is the sensing task. Both tasks employ sensors, and occasionally, share the same environmental information. We refer to the output of the sensing task as the *environment representation*. SENARIO supports multiple environment representations ranging from simple combinations of sensor data to occupancy grid representations for the environment.

Risk detection is responsible for the detection of both *external risks* threatening the robot and originating from the environment and *risks internal* to the control system, such as malfunctions. The detection and reaction methods to these risks are different. The former, allows the subsequent use of risk classification, while the latter is implemented by low-level, robust reliable, and fast components, that do not require additional processing and that directly react on the actuators through emergency lines.

Risk classification employs a default set of criteria for classifying risks according to their emergency. All the risks identified during the detection process are classified according to the criteria used in the risk classification task. The outcome of these tasks is a *risk list* sorted in decreasing order of emergency. This risk list is further processed, either by task planning, or by risk avoidance (emergency risk avoidance task), or directly by the actuation processes, thus supporting the virtually centralized control scheme specified above.

Obstacle avoidance receives input from three processes: environment perception (positioning–position estimation, and sensing–environment representation), risk detection (risk classification–risk list) and task planning (current target position or direction).

The task needs only one of the three input sources in order to maintain reliable operation, namely, the environment representation. Any additional information is affecting the robot route according to the needs of either task planning or risk detection with varying priority. This scheme is another instance of virtually centralized control, due to the fact that either risk avoidance or task planning has absolute control of the system, but there is a *supervised distribution of control* [24] among the subsystems based on dynamic priority ordering.

Clearly, the interactions between risk detection and risk avoidance, and the difference between emergency risk avoidance and obstacle avoidance tasks should be discriminated. Emergency risk avoidance is triggered by an emergency risk situation (i.e., a risk with a higher emergency classification in the risk list), while obstacle avoidance covers the rest of the cases. Obstacle avoidance uses a local path planning module based on the *vector field histogram* (VHF) method extended such as to hold for nonpoint WMRs. This extension is known as *active kinematic histogram* (AKH) [57].

Actuation realizes the commands of the rest of the supervised distributed control scheme described above. It consists of a motion command interpreter task, which receives commands by the risk management and task planner tasks in a common format, and translates these instructions into motion commands for the actuators. The output of the actuation task is identical to the output of the overall system and is referred to as the *robot motion*.

Implementation: The configuration used and some of the results obtained are as follows.

Sensing task: The sensing task is multimodal, that is, it uses a combination of sensing principles (ultrasound, infrared light, etc.). The wheelchair is equipped with proximity (ultrasonic) and positioning (encoders-infrared scanners) sensors. A minimum number of sensors were used to achieve the required functionality in order to keep the trade-off between functionality and cost in balance. Specifically, there are 11 ultrasonic sensors in total, supplied by Microsonic GmbH-Dortmund. The ultrasonic sensors are divided into two clusters: *navigation* and *protection*.

The difference in the two sensor's clusters is that protection sensors supply human protection functionality, working in fail-safe mode. Both protection and navigation sensors cover a range of 250 cm, while the robot dimensions are 132 cm in length and 82 cm in width. The ultrasonics sensors are mounted on the robot as

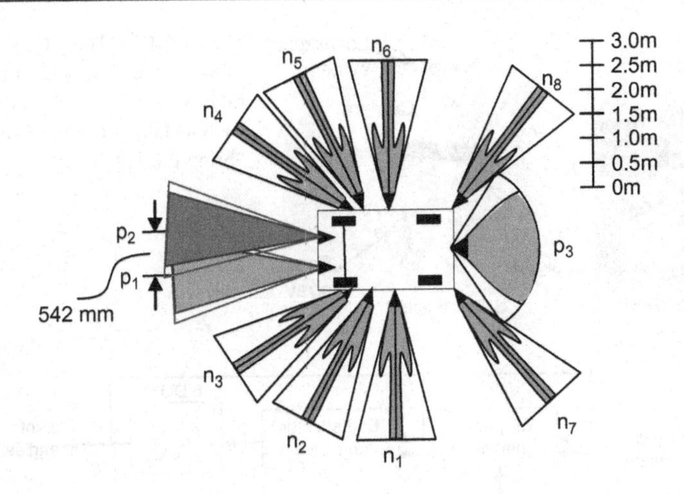

Figure 14.18 Field of view of navigation and protection ultrasonic sensors (p_1, p_2 are the front protection sensors).

shown in Figure 14.18. The letter "n" denotes the navigation sensors, while "p" denotes the protection ones.

Man–machine interface: The user interface supports speech recognition using an automatic speech recognition module. The module can record and interpret user commands providing the appropriate control signals to the task planner. The emphasis is on user-dependent speech recognition, to avoid accidental command initiation due to other people talking nearby. During this phase, the speech recognition unit maps the acoustic signal (voiceprints) in its input to a set of commands for the actuation into the range position in less than 1 min.

The system was tested in several environments with complexity below and above average, involving points in adjacent rooms acting in real time, and showed a success performance. All tests were performed with an average speed of 0.2 m/s.

Typical tasks performed successfully by the system are avoidance of furniture, avoidance of persons seating or standing on its way, location of a door besides a glass wall, passing through the door, and arriving at the desired goal destination.

14.7.2 The ROMAN Intelligent Service Mobile Manipulator

ROMAN involves the following subsystems (Figure 14.19):

- *Mobile platform*: A three-wheel (diameter 0.2 m) omnidirectional platform (0.63 m width × 0.64 m depth × 1.85 m height) equipped with a multisensoric system (a laser-based angle measurement system and an eye-safe laser beam for scanning the workspace in a horizontal plane and measuring the azimuth angle).
- *Robotic arm*: An anthropomorphic manipulator (maximum range 0.8 m, maximum payload 1.5 kg) for carrying out service tasks.

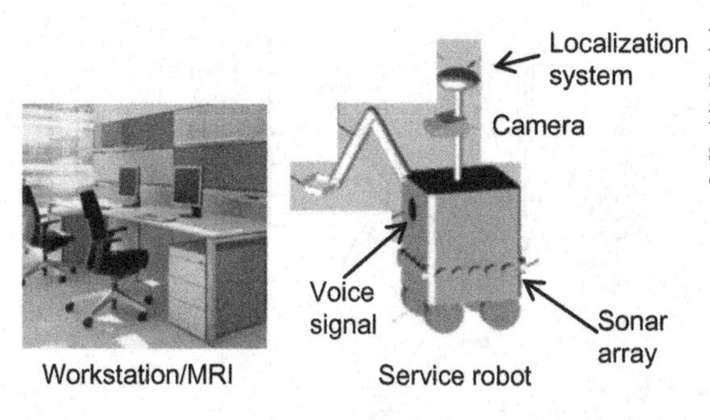

Figure 14.19 The ROMAN service mobile service robot (MRI includes an NLI with commands such as take the "cup away," and "bring the box").

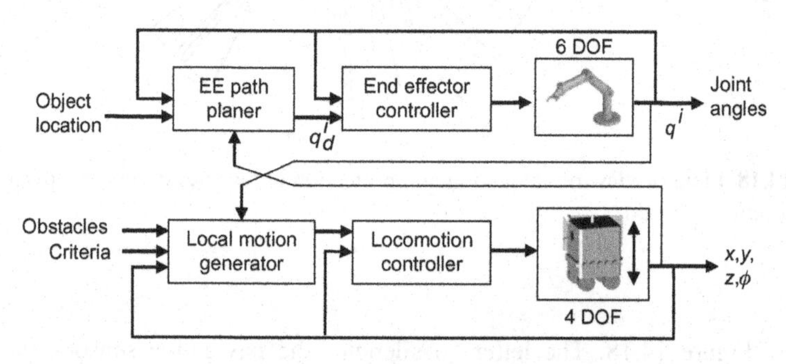

Figure 14.20 Coordinated control structure of ROMAN's platform and arm.

- *Task planner and coordinator*: The planner enables ROMAN to autonomously perform typical tasks such as finding its way to the goal position, opening the doors, or handling the desired objects.
- *Vision system*: The camera is mounted on a tilt-unit to allow object recognition over the entire workspace. Two object recognition techniques are used to handle a variety namely: a feature-based method for large objects, and an appearance-based method for small objects.
- *Multimodal human–robot interface* (MRI): This is used for natural voice/speech based dialogue between the human user and the robot.

The navigation of the platform is helped via a localization system that provides platform position and orientation data in real time. An ultrasonic sensor array helps to detect any obstacle(s) and works with the motion platform controller to avoid collisions with the obstacles.

The 6-DOF robotic arm of ROMAN is suitable for manipulating/handling light-weight geometrically simple objects (e.g., dishes, glasses, bottles, journals). The control strategy for motion coordination of the robotic arm and the mobile platform of ROMAN is shown in Figure 14.20.

The integrated processing and control architecture of ROMAN is depicted in Figure 14.21. It is implemented on a VME bus-based multiprocessor system, communicating with the environment via an Ethernet link (10 Mbit/s).

In ROMAN, the information exchange between the operator and the robot is performed in two stages: *task specification* and *task execution* (semi- or fully

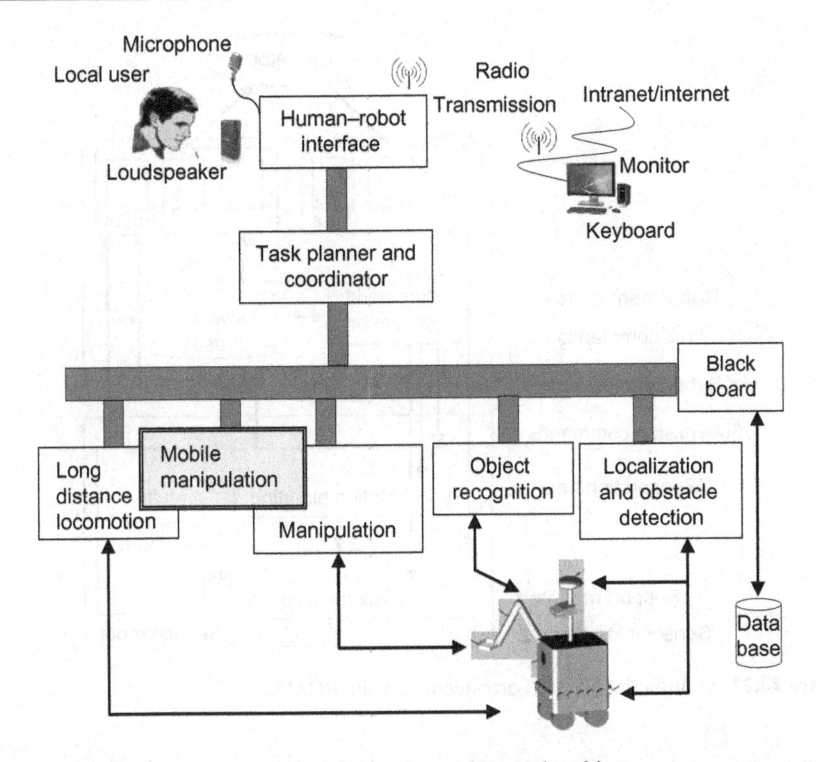

Figure 14.21 ROMAN's hierarchical processing and control architecture.

autonomous). The task specification requirements include task description. The task execution requirements involve approaching the goal area, object specification, object handling, and object handover. In addition, there are the monitoring and sensor support requirements. The human–robot dialogue involves the following:

- Dialogue-oriented natural speech (voice) command input
- Visual screen-based monitoring
- Tactile supervisory control during mobile handling
- Voice output during task operation

The NL HRI architecture of ROMAN is as shown in Figure 14.22.

The *task commands* consist of service task-specific actions, the *support commands* are the operators' responses to requests received from the motion planning level during task execution, and the *supervision commands* are initiated by the operator during task execution and immediately interrupt the current operation. The command language is able to represent both the user-defined service tasks and service robot-specific commands. The sensor information passed to the NL HRI from the planning level involves off-line environmental data, continuous sensor data, and abstract sensor data. Any problem arising during task execution initiates a request for support, which needs to be interpreted by the human operator.

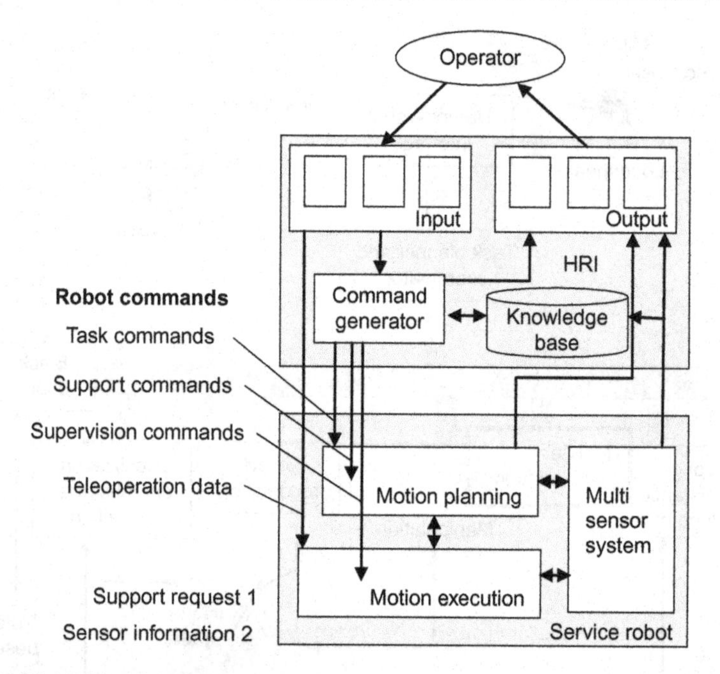

Figure 14.22 Multimodal NL HRI architecture of the ROMAN.

The command generator translates semantic structures into robot commands. The command generator of ROMAN receives the operator's instructions and performs the following functions:

- Translation
- Consistency check
- Completeness check
- Data expansion
- Macro separation

Its output is the corresponding robot command. When receiving an NL command from the user, ROMAN's interface converts the voice signal into an executable robot command, and splits it up via the task planner into a sequence of typical subtasks such as *open door*, *pass door*, or *travel along corridor*. These typical tasks are then executed by either the expert for long distance (e.g., room to room) or by the mobile manipulation expert. The task planner also coordinates the experts during task execution (Figure 14.21). For example, during the "open door" task execution, the object recognition expert is connected with the expert for mobile manipulation to locate the door handle.

Typical ROMAN's tasks are the following:

- Cleaning a table among several tables, by receiving the desired table specification by the user (via NL or mouse clicking), decomposing the command into a sequence of subtasks, planning a suitable path to the table, and starting its motion.

- Door opening, via a standard opening maneuver of its manipulator, and then passing it. If an obstacle is encountered on its way, ROMAN slows down and performs an obstacle avoidance maneuver, if possible. Upon arriving at the desired table, ROMAN looks for specified objects (e.g., a cup, a bottle), grasp it, and puts it (e.g., on the refrigerator).
- Drawer opening by determining the drawer's position (via the object recognition unit), taking out a box, and carrying it at the hands of the user.

14.8 Discussion of Some Further Issues

Two further important issues for the development and application of software architectures for intelligent mobile robot control (and other control systems) are the following:

1. Design for heterogeneity
2. Modular design

14.8.1 Design for Heterogeneity

The hardware heterogeneity that exists between the various components in commercial and research mobile robots is one of the most difficult problems to face. The heterogeneity is due to the variety of multiple processing units, communication components, central control computers and workstations, sensor and actuator hardware, and so on. In addition, on every processing node, *off-line* and *online* (real-time) operations coexist. To face this extended heterogeneity, the concept of *middleware* was developed [1]. Very broadly, middleware is a software layer that defines unified (standardized) interfaces and communication services according to the capabilities of each individual robot. In the mobile robotics field, the middleware must provide interfaces to the various types of actuators and sensors and encapsulate them such that advanced software can be easily ported from one robot (hardware) to another.

Commercially, several middleware platforms are available such as Miro [1], Marie [3], Player [4], ORCA 1/2 [58], and MCA [59].

A general approach for the development of middleware that faces high-degree heterogeneity involves the following three architectural abstractions [60–62]:

1. *Architecture design abstractions*, which enable the development of adaptive, reusable, and hierarchical subsystems and components
2. *Architecture modeling and analysis*, which permits early, integrated, and continuous evaluation of system behaviors
3. *Middleware architecture*, which permits self-adaptation in highly dynamic, changing, and heterogeneous environments

A general diagram that illustrates the middlewear layers between hardware and user is shown in Figure 14.23.

Design abstractions concern the representation and reasoning about complex systems at a high level. To this end, several canonical architectural constructs were

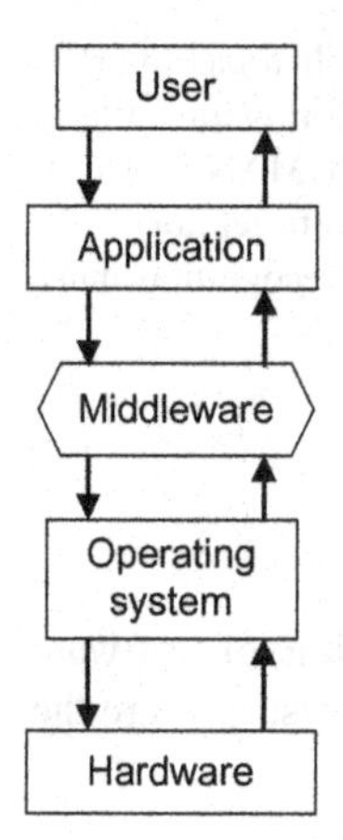

Figure 14.23 Structure of middlewear layers.

developed, namely, *components*, *connectors*, *interfaces*, *services*, *communication ports*, and *configuration*. The use of these constructs which are described by heuristics or constraints can be made in two styles: *client-to-server* and *peer-to-peer*. Traditional software architectures are layered, where components at a given layer need the services of components at the layer below. In an *adaptive-layered style*, components at a given layer monitor, manage, and adapt components of the lower layer. For a discussion of an adaptive-layered architecture, called PLASMA (plan-based layered architecture for software systems) the reader is referred to Ref. [60]. PLASMA possesses three bottom-up adaptive layers:

1. *Application layer* (components at this layer reside at the bottom layer)
2. *Middle layer* (adaptation layer), which monitors, manages, and adapts components of the bottom layer
3. *Top layer* (planning layer), which manages the adaptation layer and the generation of plans or user-supplied goals and component specifications

This architecture is an instance of the generic hierarchical architecture shown in Figure 14.1. Clearly, if a nonadaptive system is designed, only the application layer is needed.

Modeling and analysis of software for robotics is concerned with architectural modes and analyses to guide and direct design decisions about dynamic planning and adaptation. A software language that can be used for modeling and analysis is SADEL (Software Architecture Description and Evaluation Language) [60,63]. In SADEL, a model specifies the functional interfaces and application components, and another model deals with the management interfaces of components (deploy, connect, suspend, etc.). With SADEL, the implementation of tools that allow one to carry experiments with various system design decisions regarding nonfunctional features, policies for initiating replanning, and alternatives for reusing software components is feasible.

Middleware architectures available for robotics are not always effective, especially in cases where the systems are distributed across multiple, heterogeneous platforms.

A modified middleware solution that alleviates these shortcomings and can be used effectively in many mobile robotic platforms is *RoboPrism* [64]. This is achieved by providing the required low-level abstractions for interfacing with the operating system at hand, implementing software systems through the use of constructs (component, connector, etc.), offering a wide collection of metalevel services, and enabling the management and adaptation of the metalevel services to obtain in overall an adaptive-layered system. All the above can be achieved with low total cost (memory, CPU, network).

Another middleware solution with many important features is *Miro Architecture*[1]. *Miro* is a CORBA-based framework for programming robots, developed at the University of Ulm (Germany). Common object request broker architecture (CORBA) is an open vendor-independent architecture and infrastructure produced and offered by OMG [65]. Through the standard HOP protocol, a CORBA-based program from any vendor (on almost any computer, operating system, programming language, and network) can interoperate with any other CORBA-based program. CORBA is a *middleware* suitable for servers that have to handle reliably large number of users at high bit rates. CORBA takes successful care of issues such as load balancing, resource control, and fault tolerance on the server side. CORBA 2 and CORBA 3 represent complete releases of the entire CORBA specification [66].

The Miro Architecture (two representations) is shown in Figure 14.24. Miro manual is available in Ref. [68].

Miro provides an object-oriented middleware for robots which besides CORBA employs standard and widely used packages such as ACE, TAO, and Qt. It facilitates and improves the development process and the integration of system information processing frameworks.

Miro satisfies the following goals of robotics middleware:

- Object-oriented design
- Open architecture style
- Hardware and operating system abstraction
- Proper communication support and interoperability
- Client-server style
- Software design patterns that offer a high-quality framework for common well-understood functionalities

In general, the integration of new hardware devices on a given middleware falls in the following categories:

1. The hardware is already fully supported by the middleware.
2. The middleware already provides support for hardware services with equal functionality.
3. The middleware supports similar devices.
4. The middleware does not support the hardware at hand.

In case 1, the service offered can be used with no or moderate additional effort. In case 2, one may reuse the existing interfaces and so he/she has to implement only the hardware-specific parts. In case 3, the system designer must derive his/her

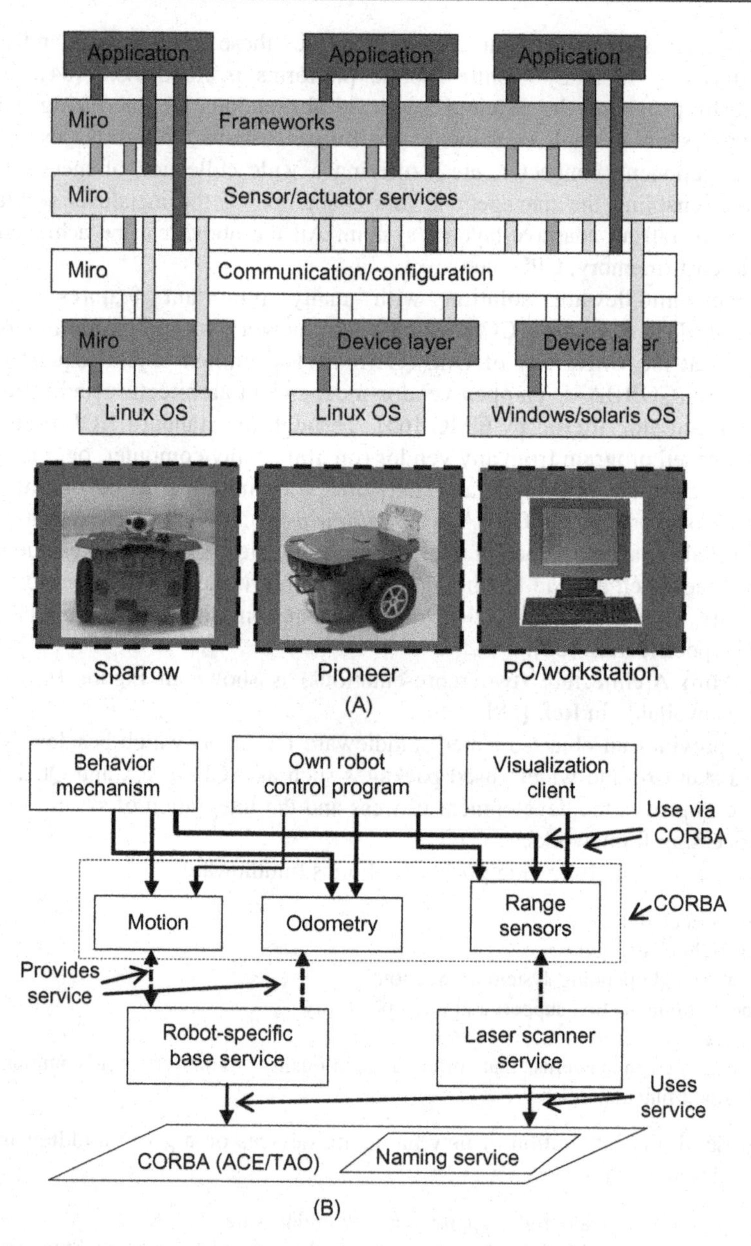

Figure 14.24 Two representations of the Miro architecture: (A) general abstraction layers and (B) the role of CORBA in the architecture.
Source: Adapted from Refs. [1,67].

own interfaces and add the missing functionality. Finally, in case 4, the implementation of new services has to be made almost from the beginning. A successful example of porting new robots to the Miro middleware is reported in Ref. [69]. A discussion on reusing software for robotic applications employing analysis patterns is provided in Ref. [27].

14.8.2 Modular Design

The modular design of software can be based on the following:

- Software design
- Software architecture

The *software design* is concerned with the decomposition of functionality into layers with increasing degree of abstraction. For complex environments pure reactive control is likely not to be successful. Therefore, high-level AI functions and reactive behavior must be suitably merged.

The *software architecture* deals with the implementation details. It is based on available middleware such as the *middleware for cooperative robotics* (MIRC) [1].

The hardware may be split into the following modules (layers) [70]:

- *Driving module*, which is responsible for handling the motion of the robot
- *Actuator module*, which performs all active interactions with the environment
- *Sensor module*, which is responsible for the entire sensing of the environment
- *Control module*, which performs the more complex information processing for the robot control

A modular software architecture that achieves the goals of:

- Flexibility
- Maintainability
- Testability
- Modifiability

is reported in Ref. [71]. This architecture is based on an asynchronous *publish–subscribe* mechanism and a *blackboard* object that handles synchronized access to shared data. The publish–subscribe mechanism with a blackboard decouples the sender and receiver and reduces modules' dependencies to a very large extent.

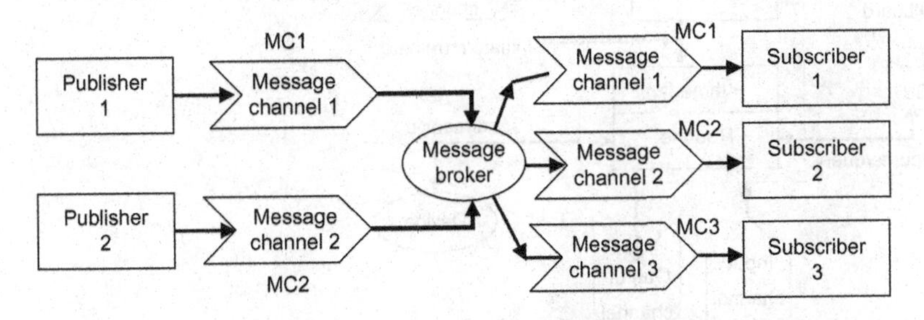

Figure 14.25 Structure of the publish–subscribe messaging pattern.

The publish–subscribe messaging pattern has the structure shown in Figure 14.25, and involves three main components: the *publisher*, the *broker*, and the *subscriber* [71].

The publisher generates and publishes signals (messages) which are used by the subscriber. The broker is a signal router that monitors every module's output channel signal and according to the signal type passes it to the input channel of each subscriber. Publishers and subscribers are actually decoupled via the message channels that are configured during the system's initialization.

To achieve the desired (high) flexibility, implementability, and testability, the perception, planning, localization, and control tasks are decomposed into a set of simple modules. For example, localization using a GPS sensor can be split into the following two modules:

1. *GPS reader* module, which receives and processes messages from the GPS sensor
2. *Localization* module, which filters raw sensor data and updates the robot's state

In this way, there is no need for the localization module to know how to connect and get data from the GPS sensor (i.e., changing the sensor or communication protocol does not influence localization). A typical module has the structure shown in Figure 14.26 [71].

An overall generic software architecture that uses the above concepts is shown in Figure 14.27 [71].

The high-level modules A, B, C perform task-specific perception, planning, and control. The low-level modules perform the execution of the commands issued by the high-level modules, accept and process sensor data sending the processed data to high-level modules, and send the proper commands to the robot's actuators. The data synchronization with each other is achieved through the publish–subscribe pattern and the shared blackboard.

In Ref. [71], the flexibility, extensibility, and testability of the above architecture was tested by constructing and applying a controller for a real automobile followed

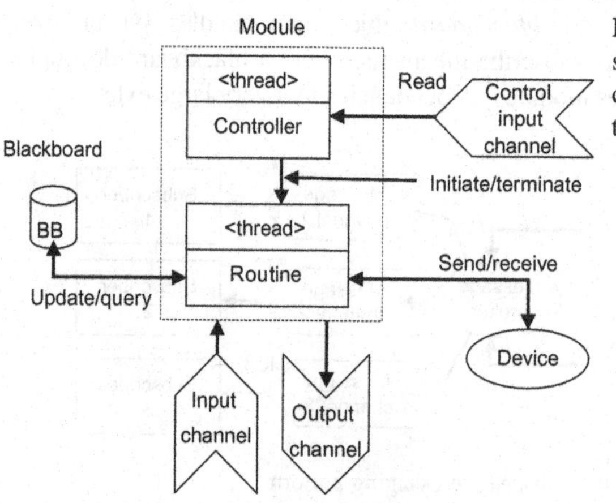

Figure 14.26 The standard module structure (which involves a controller thread and a routine thread).

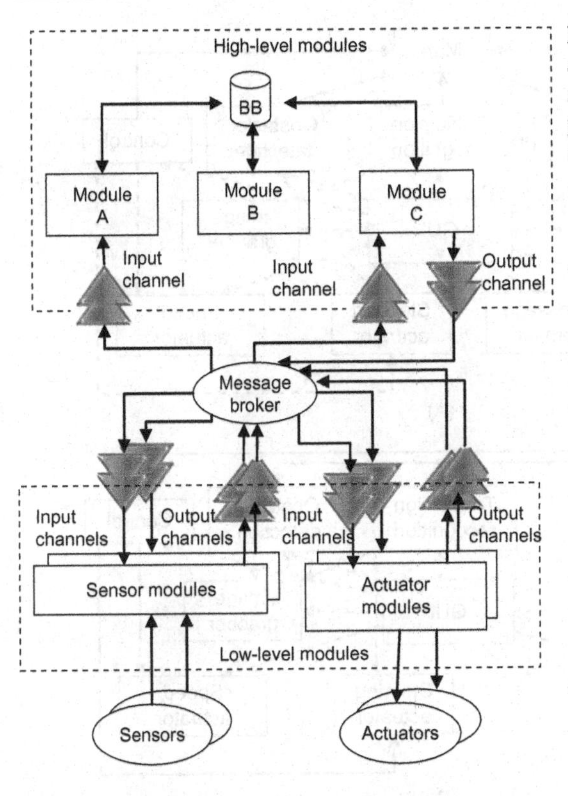

Figure 14.27 A generic architecture for a mobile robotic system based on publish–subscribe paradigm with a message broker.

by a virtual reality model of the car. To this end, the design was based on five principal high-level modules, namely:

1. *Localization module* (based on a GPS sensor and an electronic compass). An extended Kalman filter was used for sensor fusion and state estimation.
2. *Obstacle detection module* (using a front-facing monocular camera).
3. *Traffic recognition module* (using a second monocular camera for the detection of traffic signs via color and shape information).
4. *Planning module* (which updates the status of wave-points that have been reached and decides what movement to do next).
5. *Control module* (with steering input set-point found using a fuzzy logic control algorithm).

The simulation experiments were based on the bicycle model of the car and a camera simulator based on the Open GL library for rendering 3D scenes. The car module is an active blackboard storing the current position, orientation, and speed of the car-like robot, while the simulator is running with 1 ms updating period. The direct communication and the publish–subscribe schemes were applied and compared. The dependency diagrams of the above two schemes are shown in Figure 14.28 [71].

The overhead of the publish–subscribe scheme for message payload from 50 up to 1000 bytes was found to be 100 μs. Although it is much higher than the 10-μs overhead of the direct communication scheme, it was proved to be quite acceptable. This is so because the sensor, actuator, and control loop period is much slower.

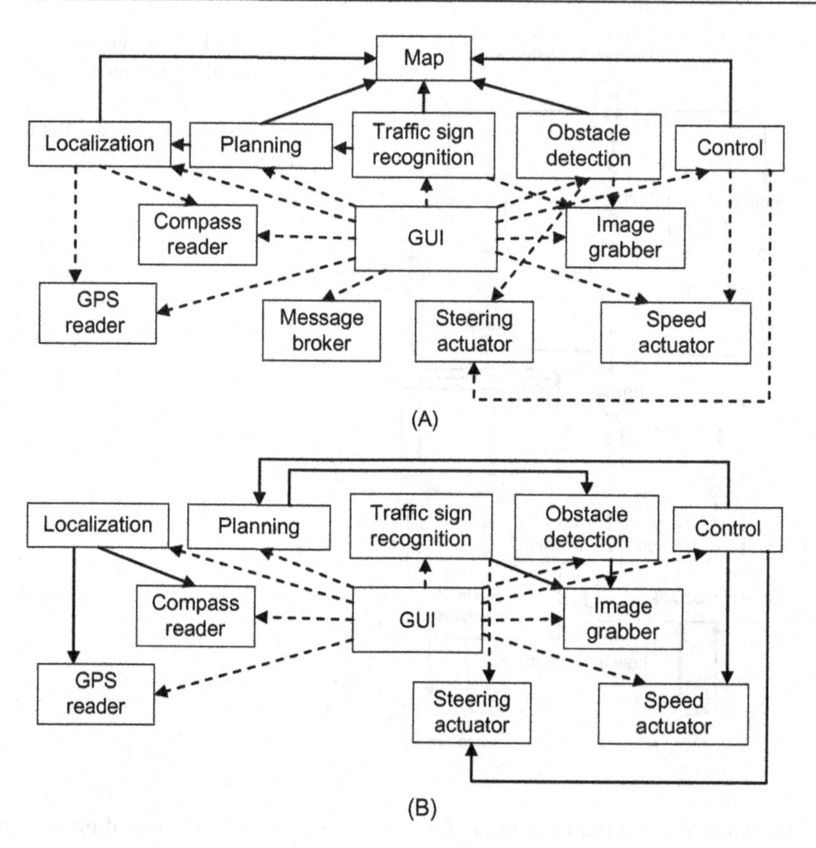

Figure 14.28 (A) Dependency diagram of the publish–subscribe scheme. (B) Dependency diagram in case of direct communication.

Two other works devoted to software/middleware architectures using the blackboard concept are described in Refs. [72,73]. A comprehensive review of the middleware work on networked robots, up to 2009, is provided in Ref. [74], and an annotated literature survey with rich comments up to 2012 is given in Ref. [67].

A global picture of the features of 15 robotics middleware frameworks, as summarized in Ref. [67], can be obtained by the following list:

- *OpenRTMaist*, a modular software structure platform that simplifies the process of building robots by simply combining selected modules
- *ASEBA*, which allows distributed control and efficient resources utilization of robots with multiprocessors
- *MARIE*, which creates flexible distributed components that allow sharing, reusing, and integrating new or existing software programs for rapid robotic application development
- *RSCA*, a real-time support for robotic applications with abstractions that make them both portable and reusable on different hardware platforms
- *MRDS*, a robotic software platform that supports a wide variety of hardware devices and a set of useful tools facilitating programming and debugging

- *OPROS*, a component-based generic software platform that enables complicated functions to be developed easily by using the standardized components in the heterogeneous communication network
- *CLARAty*, a reusable robotic framework that enables integration, maturation, and demonstration of advanced robotic technologies, from multiple institutions on NASA's rover platforms
- *ROS*, which provides OS services such as hardware abstraction, low-level device control, message-passing between processes, and package management
- *OROCOS*, a general-purpose modular framework for robot and machine control
- *PYRO*, a programming environment for easily exploring advanced topics in artificial intelligence and robotics without having to worry about the low-level details of the underlying hardware
- *PLAYER*, a development framework that supports different hardware devices and common services needed by different robotic applications
- *ORCA*, a framework that enables software reuse in robotics using component-based development
- *ERSP*, which provides cutting edge technologies for vision, navigation, and system development
- *WEBOTS*, a rapid prototyping environment for modeling, programming, and simulating mobile robots
- *ROBOFRAME*, a system framework that covers the special needs of autonomous lightweight robots such as dynamical locomotion and stability

Obviously, the selection of a commercial software/middleware platform for a given robotic system or application is a complex problem and needs deep and careful considerations.

References

[1] Utz H, Sablantnög S, Euderde E, Kraetzschmar G. Miro-middleware for mobile robot applications. IEEE Trans Robot Autom 2002;18(4):493−7.

[2] Nesnas A, Wright A, Bajiracharya M, Simmons R, Estlin T. CLARAty and challenges of developing interoperable robotic software. Proceedings IEEE/RSJ international conference on intelligent robots and systems (IROS'2003), vol. 3. Las Vegas, NV; 2003. p. 2428−35.

[3] Cote C, Brosseau Y, Letourneau D, Raivesty C, Michand F. Robotic software integration using MARIE. Int J Adv Robot Syst 2006;3(1):55−60.

[4] Gerkey B, Vaugham R, Howard A. The player/stage project: tools for multi-robot and distributed sensor systems. Proceeding of eleventh international conference on advanced robotics (ICAR'2003). Coimbra, Portugal; 2003. p. 317−323.

[5] Montemerlo M, Roy N, Thrun S. Perspectives on standardization in mobile robot programming: The Carnegie Mellon Navigation (CARMEN) toolkit. Proceedings of IEEE/RSJ international conference on intelligent robotics and systems. Las Vegas, NV; 2003. p. 2436−2441.

[6] Fu K-S. Learning control systems and intelligent control systems: an intersection of artificial intelligence and automatic control. IEEE Trans Autom Control 1971;AC-16(1):70−2.

[7] Saridis GN. Toward the realization of intelligent controls. Proc IEEE 1979;67 (8):1115−33.

[8] Saridis GN. Foundations of intelligent controls. Proceedings of IEEE workshop on intelligent control. Troy, NY; 1985. p. 23–8.
[9] Antsaklis P, Passino KM, editors. An introduction to intelligent and autonomous systems. Berlin: Kluwer/Springer; 1993.
[10] Coste-Maniére E, Simmons R. Architecture, the backbone of robotic systems. Proceedings of IEEE international conference on robotics and automation (ICRA'2000). San Francisco, CA; 2000. p. 67–72.
[11] Xie W, Ma J, Yang M, Zhang Q. Research on classification of intelligent robotic architecture. J Comput 2012;7(2):450–7.
[12] Saridis GN. Analytical formulation of the principle of increasing precision and decreasing intelligence for intelligent machines. Automatica 1989;25(3):461–7.
[13] Meystel AM. Architectures of intelligent control. The science of autonomous intelligence. Proceedings of IEEE international symposium on intelligent control. Chicago, IL; 1993. p. 42–8.
[14] Meystel AM. Autonomous mobile robots: vehicles with cognitive control. Singapore: World Scientific Singapore; 1991.
[15] Meystel AM. Multiresolutional hierarchical decision support systems. IEEE Trans Syst Man Cybern-Part C: Appl Rev 2003;SMR-AR 33:86–101.
[16] Albus JS. System description and design architecture for multiple autonomous undersea vehicles. NIST Tech. Note 1251, Washington, D.C.; September 1988.
[17] Albus JS, Quintero R. Towards a reference model architecture for real-time intelligent control systems. Robotics and manufacturing, vol. 3. New York, NY: ASME; 1990.
[18] Albus JM. Outline for a theory of intelligence. IEEE Trans Syst Man Cybern 1991; SMC-21(3):473–509.
[19] Ayari I, Chatti A. Reactive control using behavior modeling of a mobile robot. Int J Comput, Commun Control 2007;2(3):217–28.
[20] Brooks RA. A robust layered control system for a mobile robot. IEEE J Robot Automn 1986;RA-2:14–23.
[21] Brooks RA. Intelligence without reason. AI Memo. No. 1293, AI Laboratory, MIT; 1991.
[22] Arkin RC. Motor schema-based mobile robot navigation. Int J Robot Res 1989;8 (4):92–112.
[23] Arkin RC. Cooperation without communication: multi-agent schema based robot navigation. J Robot Syst 1992;9(2):351–64.
[24] Arkin RC. Behavior-based robotics. Cambridge, MA: The MIT Press; 1998.
[25] Oreback A, Christensen HL. Evaluation of architectures for mobile robots. Auton Robots 2003;14:33–49.
[26] Riehle D, Zullighoven H. Understanding and using patterns in software development. Theory Pract Object Syst 1996;2(1):3–13.
[27] Jawawi D, Deris S, Mamat R. Software reuse for mobile robot applications through analysis patterns. Int Arab J Inf Technol 2007;4(3):220–8.
[28] Canas JM, Matellan V. Integrating behaviors for mobile robots: an ethological approach. Cutting edge robotics. Pro Literature Verlag/ARS; 2005. p. 311–50.
[29] Canas JM, Ruiz-Ayucar J, Aguero C, Martin F. Jde-neoc: component oriented software architecture for robotics. J Phys Agents 2007;1(1):1–6.
[30] Kerry M. Simplifying robot software design layer by layer. National Instruments RTC Magazine. <http://rtcmagazine.com/articles/view/102283>; 2013 [20 AUGUST].
[31] Travis J, Kring J. LabVIEW for everyone: graphical programming made easy and fun. Upper Saddle River; NJ: Prentice-Hall; 2006.

[32] Konolige K., Myers K. The Saphira architecture for autonomous mobile robots. SRI International. <http://www.wv.inf.tu-dresden.de/Teaching/MobileRoboticsLab/Download/Saphira-5.3-Manual.pdf, http://www.cs.jhu.edu/~hager/Public/ICRAtutorial/Konolige-Saphira/saphira.pdf>; 2013 [20 AUGUST].

[33] Lindstrom M, Oreback A, Christensen H. BERRA: a research architecture for service robots. Proceedings of IEEE international conference robotics and automation (ICRA'2000). San Francisco, CA; April 24–28, 2000. p. 3278–83.

[34] Saffioti A, Ruspini E, Konolige K. Blending reactivity and goal-directness in a fuzzy controller. Proceedings of 2nd IEEE international conference on fuzzy systems. San Francisco, CA; 1993. p. 134–9.

[35] Guzzoni D, Cheyer A, Julia A, Konolige K. Many robots make short work. AI Mag 1997;18(1):55–64.

[36] Schmidt DC. The ADAPTIVE communication environment: object-oriented network programming components for developing client/server applications. Proceedings of eleventh and twelveth Sun Users Group conference. San Jose, CA; June 14–17, December 7–9, 1993.

[37] Fink GA. Developing HMM-based recognizers with ESMERALDA. Lecture notes in artificial intelligence, vol. 1692. Berlin: Springer; 1999. p. 229–34

[38] Veldhuizen TL. Arrays in Blitz+ +. Proceedings of second international scientific computing in object-oriented parallel environments: ISCOPE' 98. Santa Fe. NM, Berlin: Springer; 1998.

[39] Kramer J, Scheutz M. Development environments for autonomous mobile robots: a survey. Auton Robots 2007;22(2):101–32.

[40] Qian K, Fu X, Tao L, Xu C-W. Software architecture and design illuminated. Burlington, MA: Jones and Bartlett Publishers; 2009.

[41] Hirzinger G. Multisensory shared autonomy and telesensor programming: key issues in space robotics. Robot Auton Syst 1993;11:141–62.

[42] Tzafestas SG, Tzafestas ES. Human–machine interaction in intelligent robotic systems: a unifying consideration with implementation examples. J Intell Robot Syst 2001;32(2):119–41.

[43] Nilsson NJ. Shakey the robot. Technical Note No. 323, Al Center. Menlo Park, CA: SRI International; 1984.

[44] Sato T, Hirai S. Language-aided robotic teleoperation system (LARTS) for advanced teleoperation. IEEE J Robot Autom 1987;3(5):476–80.

[45] Torrance MC. Natural communication with robots. MScthesis, DEEC. MA: MIT Press; 1994.

[46] Heinzmann J. A safe control paradigm for human–robot interaction. J Intell Robot Syst 1999;25:295–310.

[47] Rossmann J. Virtual reality as a control and supervision tool for autonomous systems. In: Remboldt U, editor. Intell Auton Syst. Amsterdam: IOS Press; 1995. p. 344–51.

[48] Wang C, Ma H, Cannon DJ. Human–machine collaboration in robotics: integrating virtual tools with a collision avoidance concept using conglomerates of spheres. J Intell Robot Syst 1997;18:367–97.

[49] Laengle T, Remboldt U. Distributed control architecture for intelligent systems. Proceedings of international symposium on intelligent systems and advanced manufacturing. Boston, MA; November 18–22, 1996. p. 52–61.

[50] Laengle T, Lueth TC, Remboldt U, Woern H. A distributed control architecture for autonomous mobile robots—implementation of the Karlsruhe multi-agent robot architecture. Adv Robot 1998;12(4):411–31.

[51] Sarkar M. Human robot interaction. In-Tech, e-Books; 2008.

[52] Katevas NI, Sgouros NM, Tzafestas SG, Papakonstantinou G, Beatie G, Bishop G, et al. The autonomous mobile robot SENARIO: A sensor-aided intelligent navigation system for powered wheelchairs. IEEE Robot Autom Mag 1997;4(4):60–70.

[53] Katevas NI, Tzafestas SG, Koutsouris DG, Pnevmatikatos CG. The SENARIO autonomous navigation system. In: Tzafestas SG, editors. Mobile robotics technology for health care services. Proceedings of first MobiNet symposium. Athens; 1997. p. 87–99.

[54] Ettelt E, Furtwangler R, Hanbeck UD, Schmidt G. Design issues of a semi-autonomous robotic assistant for the health care environment. J Intell Robot Syst 1998;22 (3–4):191–209.

[55] Fisher C, Buss M, Schmidt G. Hierarchical supervisory control of service robot using human-robot interface. Proceedings of international conference on intelligent robots and systems (IROS'96). Osaka, Japan; 1996. p. 1408–16.

[56] Fischer C, Schmidt G. Multi-modal human-robot interface for interaction with a remotely operating mobile service robot. Adv Robot 1998;12(4):397–409.

[57] Katevas NI, Tzafestas SG. The active kinematic histogram method for path planning of non-point non-holonomically constrained robots. Adv Robot 1998;12(4):375–95.

[58] Brooks A. Toward component-based robotics. Proceedings of IEEE/RSJ international conference on intelligent robots and systems (IROS 2005). Edmonton, Alberta, Canada; August 2–6, 2005. p. 163–8. <http://orca-robotics.sourceforge.net>; 2013 [20 AUGUST].

[59] Scholl KU. MCA2-modular controller architecture. <http://mac2.sourceforge.net>; 2013 [20 AUGUST].

[60] Brun Y, Edwards G. Engineering heterogeneous robotic systems. Computer 2011; May:61–70.

[61] Taylor RN, Medvidovic N, Dashofy EM. Software architecture: foundations, theory and practice. New York, NY: John Wiley & Sons; 2009.

[62] Edwards G, Garcia J, Tajalli H, Popescu D, Medvidovic N, Sukhatme G, et al. Architecture-driven self-adaption and self-management in robotics systems. Proceedings of international workshop on software engineering for adaptive and self-managing systems (SEAMS'09). Los Angeles, CA: IEEE Computer Society Press; March 2009. p. 142–51.

[63] Medvidovic N. A language and environment for architecture based software development and evolution. Proceedings of twentyfirst international conference on software engineering (ICSE'99). IEEE Computer Science Press; 1999. p. 44–53.

[64] Available from: http://sunset.usc.edu/~softarch/Prism; 2013 [20 AUGUST].

[65] Available from: http://omg.org; 2013 [20 AUGUST].

[66] Available from: http://omg.org/getingstarted/corba.faq.htm.

[67] Elkady A, Sobh T. Robotics middleware: a comprehensive literature survey and attributed-based bibliography. J Robot 2012 [Open Access].

[68] Miro-middleware for robots. <http://orcarobotics.sourceforge.net>; 2013 [20 AUGUST].

[69] Kruger D, Van Lil I, Sunderhauf N, Baumgartl, Protzel P. Using and extending the Miro middleware for autonomous mobile robots. Proceedings of international conference on towards autonomous robotic Systems (TAROS 06). Survey, UK; September 4–6, 2006. p. 90–5.

[70] Steinbauer G, Fraser G, Muhlenfeld A, Wotawa A. A modular architecture for a multi-purpose mobile robot. Proceedings of seventeenth conference on industrial and

engineering applications of AI and ES (IEA/AIE): innovations of artificial intelligence. Ottawa, Canada; 2004. p. 1007–15.

[71] Limsoonthrakul S, Dailey ML, Sirsupundit M. A modular system architecture for autonomous robots based on blackboard and publish-subscribe mechanisms. Proceedings of IEEE international conference on robotics and biomimetics (ROBIO 2009). Bangkok, February, 22–25, 2009. p. 633–38.

[72] Schneider S, Ullman M, Chen V. Controlshell: a real-time software framework. Proceedings of IEEE international conference on systems engineering. Fairborn, OH; 1991. p. 129–34.

[73] Shafer S, Stentz A, Thorpe C. An architecture for sensor fusion in a mobile robot. Proceedings of IEEE international conference on robotics and automation. San Francisco, CA; April 1986. p. 2002–11.

[74] Mohamed N, Al-Jaroodi J, Jawhar I. A review of middleware for networked robots. Int J Comput Sci Netw Secur 2009;9(5):139–43.

15 Mobile Robots at Work

15.1 Introduction

Robots of all types constitute one of the keys to human social and economic development. The first robots were devices with one or more arms making human-like motions. In our days, the shape of robots covers a very large repertory of forms including wheeled platforms, mobile manipulators, legged bodies, animal-like robots, and so on. All these robots contribute in one or the other way to human, industrial, agricultural, technical, and social life improvements. The benefits of using autonomous and intelligent mobile robots (wheeled or legged) in medical, assistive, and service applications are numerous, and their positive impact in modern society is continuously increasing.

The purpose of the present chapter is purely encyclopedic and aims to provide the reader a small set of mobile robots of various types that are actually used in modern industry and society [1–25]. Specifically, the following robot categories are discussed with relevant photos included:

- mobile robots and manipulators in the factory and industry,
- mobile robots in the society (rescue, guidance, hospital),
- mobile robots for home services (cleaning, other services),
- assistive mobile robots (autonomous wheelchairs, service mobile manipulators for the impaired),
- mobile telerobots and web robots,
- other mobile robot applications.

The above applications illustrate the importance and value of mobile robots for achieving a better quality of human life in a wide variety of directions.

15.2 Mobile Robots in the Factory and Industry

Mobile robots used on the factory shop floor are typically called *automated* or *autonomous guided vehicles* (AGVs) and move following markers or wires on the floor or employ sensory systems (lasers, vision). AGVs are used in factory for material handling, product transfer from one place to another, for inspection and quality control, etc. They work around the clock contributing to better continuous

Introduction to Mobile Robot Control. DOI: http://dx.doi.org/10.1016/B978-0-12-417049-0.00015-8

floor and just-in-time delivery. An AGV can tow objects behind them in trailers to which they can autonomously attach. The first AGV was commercially available in the 1950s by Barrett Electronics (Northbrook, IL), actually being a tow truck following a wire in the floor instead of a rail. Today, AGVs are primarily *laser navigated*, and can be programmed to communicate with other robots to assure smooth product movement and storage in the industrial premises. All modern flexible manufacturing systems use AGVs for achieving their goals and assure fast and high-quality production 24 h a day. The navigation of AGVs can be performed in the following ways (Wikipedia article: Automated-Guided -Vehicles#Wired).

- *Wired navigation*—A wired sensor placed on the bottom of the platform (facing the ground) detects the radio frequency which is transmitted from a wire placed in a slot about 1 in. below the ground. The AGV is then moved following the wire.
- *Guide tape-based navigation*—This navigation type is suitable for automated guided carts (i.e., light-duty AGVs). A tape is used to guide the vehicle, which is equipped with a suitable guide sensor enabling it to follow the path of the tape.
- *Laser-based navigation*—This is a wireless navigation process which is performed by mounting retroreflective tape on walls, poles, and machines. The AGV has a rotating turret carrying a laser transmitter and receiver. The laser is sent off and then received again allowing the computation of the vehicle's orientation (and in some cases the distance) automatically. The AGV is equipped with a reflector map stored in memory, and can correct its position by a feedback control technique (using the error between the desired and received measurement).
- *Gyroscope-based navigation*—This is an inertial guidance system. Transporters are embedded in the floor of the work environment. The AGV employs these transporters to detect the correctness of its route, and a gyroscope measures the change in the direction of the vehicle. This change (error) is used to correct the vehicle's motion and return it on the path.
- *Natural features-based navigation*—This allows the AGV navigation with no workspace retrofitting. Typically, one or more range finder sensors (laser, gyroscope, etc) are used, and the extended Kalman filter or particle/Monte Carlo filters are used for localization and mapping. The most common steering type in factory AGVs is the differential drive. When there are several AGVs working on the shop floor, some kind of traffic control is required (forward sensing control, zone control, etc.).

The applications of factory and industrial AGVs include the following:

- *Trailer loading*—Pick up of pallets from conveyors, or staging lanes and deliver into a trailer according to a loading pattern.
- *Raw material handling*—Transportation of raw materials (metal, plastic, rubber, paper, etc.) received in the warehouse, and delivery to the proper production line.
- *Finished product handling*—Transportation of finished goods from the production line to storage or shipping to customers.

The above applications of AGVs are needed in most modern industries—automotive industry, food and beverage industry, paper-and-print industry, chemical/pharmaceutical industry, manufacturing industry, etc.

Figures 15.1–15.5 show some AGVs used in factories and industries, including autonomous mobile manipulators.

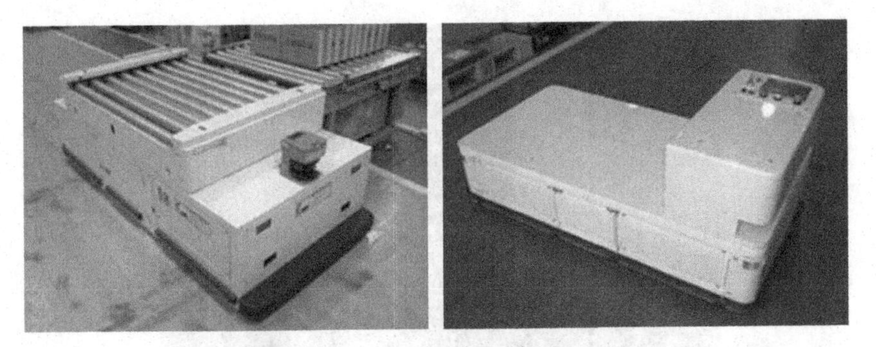

Figure 15.1 Two Hitachi autonomously guided vehicles for the factory.
Source: http://www.hitachi-pt.com/agv/intelligentcarry/index.html.

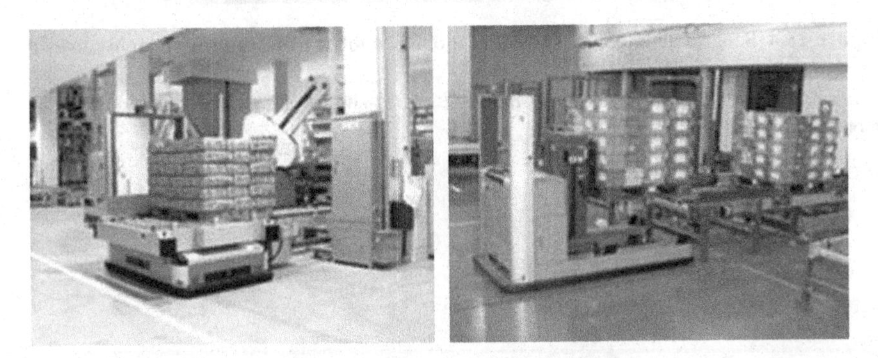

Figure 15.2: Two more industrial AGVs. (A) Adept handling AGV, (B) DTA Vehicles AGV pallet truck.
Source: http://www.directindustry.com/industrial-manufacturer/agv-80196.html/.

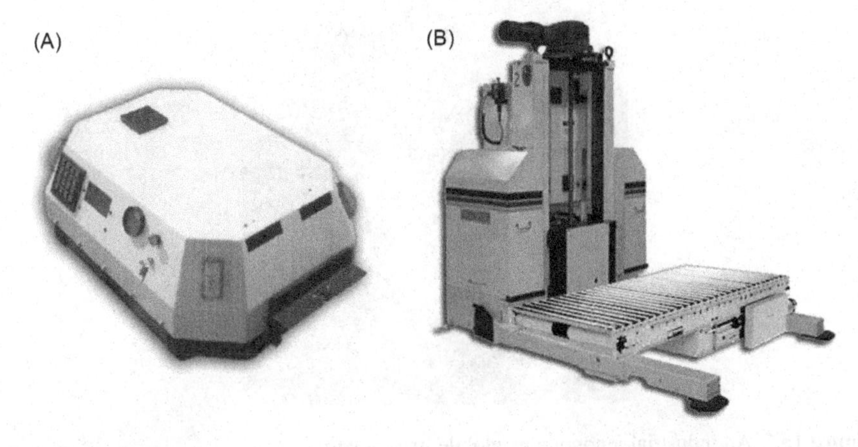

Figure 15.3 (A) Eagle series E200 cart transporter and (B) Falcon series F150 heavy-duty conveyor.
Source: http://www.coreconagvs.com/products.

Figure 15.4 Seegrid autonomous industrial mobile robot at work (moving a highly loaded passive vehicle).
Source: http://www.engadget.com/2008/06/04/seegrid-shows-off-autonomous-industrial-mobile-robot-system/.

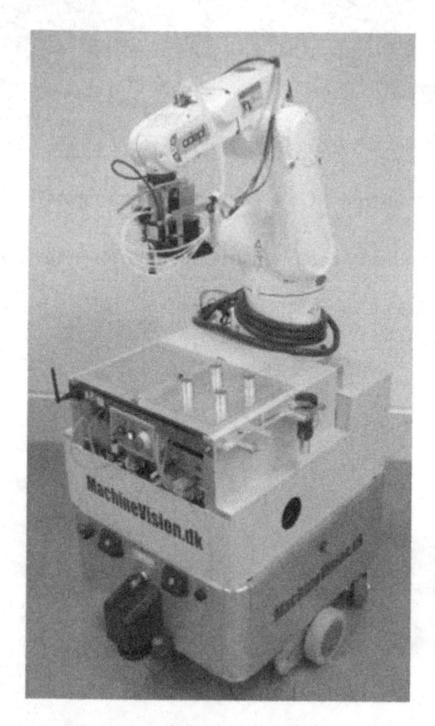

Figure 15.5 An industrial autonomous mobile manipulator.
Source: http://blog.robotiq.com/bid/32556/Hybrid-Robots-Autonomous-Industrial-Mobile-Manipulators; http://www.machinevision.dk/joomla/index.php?lang=en.

15.3 Mobile Robots in the Society

The applications of mobile robots in the society are numerous with new ones added continuously. Some of these applications are as follows:

- mobile manipulators for rescue,
- robotic canes and guiding assistants,
- mobile robots for home services,
- assistive mobile robots for the elderly and "*persons with special needs*" (PwSN),
- mobile telerobots and web robots.

15.3.1 Mobile Manipulators for Rescue

Natural and manmade disasters offer unique challenges for effective cooperation of robots and humans. The location of disasters are usually too dangerous for human intervention or cannot be reached. In many cases, there are additional difficulties such as extreme temperatures, radioactive levels, strong wind forces that do not allow a fast action of human rescuers. Lessons learned from past disaster experience have motivated extended research and development in many countries for the construction of suitable robotic rescuers. Due to the strong earthquake activity, Japan is one of the countries where strong and effective autonomous or semiautonomous robotic systems for rescue were developed. Modern robot rescuers are light, flexible, and durable. Many of them have cameras with 360° rotation that provide high resolution images, and other sensors that can detect body temperature and colored clothing. Figure 15.6 shows the search and rescue robot "Telemax" promoted at RoboCup2009, and Figure 15.7 shows the urban search and rescue robot (developed by NIST/DHS) at an actual exercise.

Figure 15.8 shows a rescue robot operated when needed by the Tokyo Fire Department.

Rescue robots are also of the *snake*-type as shown in Figure 15.9A and B.

Figure 15.6 The Telemax robot for rescue.
Source: http://www.gizmag.com/search-and-rescue-robots-at-robocup-2009-12144-12144/.

Figure 15.7 The search and rescue robot moving across a rubble pile in a NIST/DHS exercise. *Source*: http://www.science20.com/news/rescue_robots_are_on_the_way.

Figure 15.8 Rescue robot used by Tokyo Fire Department.
Source: http://web-japan.org/trends/09_sci-tech/sci100909.html.

15.3.2 Robotic Canes, Guiding Assistants, and Hospital Mobile Robots

Mobile robots and humanoid robots have been exploited for developing and constructing systems that help blind people to find their way around large buildings (such as supermarkets, museums, hospitals, airports). Two of them are:

- the eye-Robot,
- the robotic shopping assistant.

The *eye-Robot* (Figure 15.10) is a system that can be regarded as being between a white cane and a seeing eye dog.

Figure 15.9 Two examples of snake-type rescue robots.
Source: (A) http://dart2.arc.nasa.gov/Exercises/TMR2004/TMR-d2/images/23DSC00207.jpg and (B) http://www.elistmania.com/images/articles/21/Thumbnail/Snake_Robots.jpg.

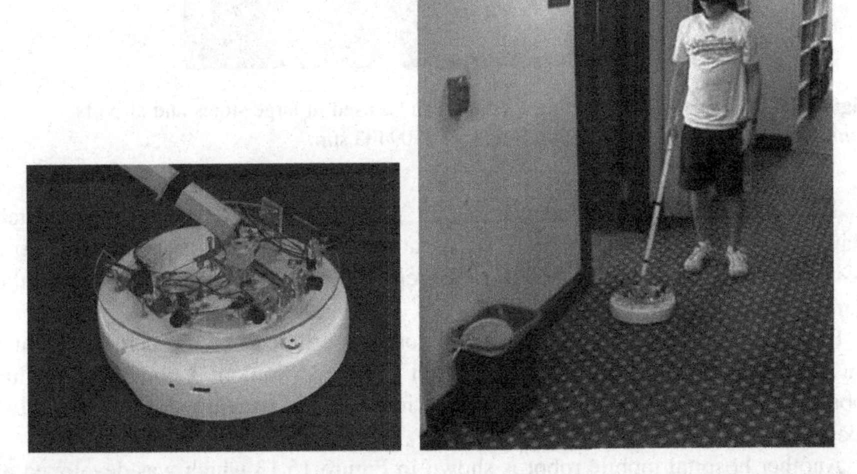

Figure 15.10 The eye-Robot cane for the blind.
Source: http://forums.trossenrobotics.com/showthread.php?1409-The-eyeRobot-Robot-Blind-Aid; http://www.instructables.com/id/eyeRobot---The-Robotic-White-Cane/.

The eye-Robot has been designed by using the Roomba robot as a base and can guide a blind or visually impaired user through cluttered and populated environments. The user indicates his/her desired motion by intuitively pushing on and twisting the handle. The robot uses this information and finds an obstacle-free route down a hallway or across a room, using sonar to steer the user in an appropriate direction. In actuality, the user can naturally follow behind the robot with little conscious thought. The eye-Robot has four ultrasonic range finders and two IR sensors facing out 90° to the right and left to enable the robot in wall following.

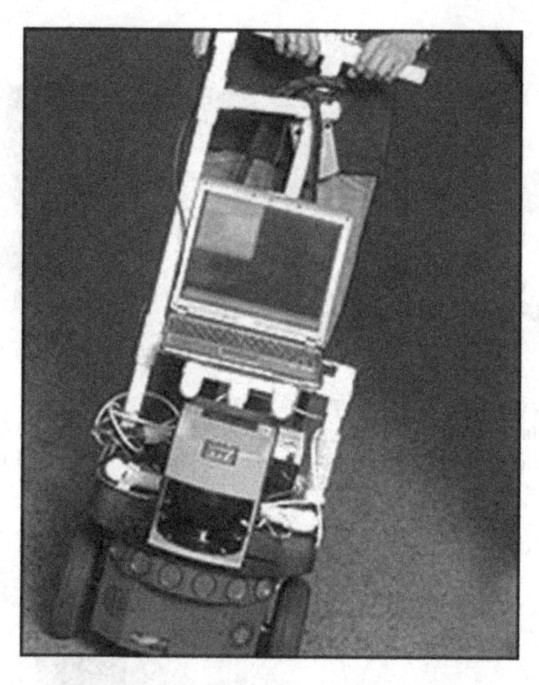

Figure 15.11 The robotic shopping assistant can be used in large stores and airports.
Source: http://news.bbc.co.uk/2/hi/technology/4509403.stm.

The robotic shopping assistant is shown in Figure 15.11. It was created at Utah State University. The user, when arriving at the grocery store, grabs the shopping assistant which leads him/her to the different products. When the user leaves the store he/she leaves the robot behind.

Figure 15.12 shows a mobile hospital robot developed by InTouch Health and put in operation at the Healdsburg District Hospital (Somona County, CA). This robot can implement online remote interviews of the patient with the specialist doctor.

Another hospital mobile robot is shown in Figure 15.13 which was developed at Keio University (Japan). This robot can automatically generate a map of its environment and carry medical supply in hospitals.

15.3.3 Mobile Robots for Home Services

Home or domestic robots are mobile robots and mobile manipulators designed for household tasks such as floor cleaning, pool cleaning, coffee making, serving. Also, robots suitable for helping elderly people and PwSN may be included in the class of home robots, although they can be regarded to belong to the more general class of *assistive robots* which will be discussed separately. Today, robots include also *humanoid robots* suitably created for helping in the house.

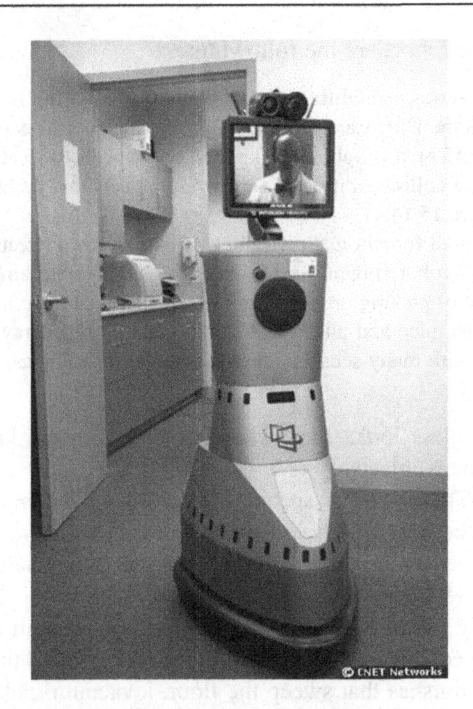

Figure 15.12 The InTouch Health RP-7 hospital robot.
Source: http://www.cnet.com/2300-11394_3-6184443-2.html.

Figure 15.13 The Keio University MKP003 hospital mobile robot.
Source: http://www.robotliving.com/robot-news/hospital-robot.

Examples of home robots are the following:

Dustbot—This is a series of multifunctional robots that can keep homes and cities clean. This series includes the Dust-Cart, a humanoid robot for door-to-door garbage collection. It has a height of 1.45 m, a weight of 70 kg, and two wheels for feet. It is appropriate for door-to-door garbage collection in very narrow city streets that traditional garbage trucks cannot access (Figure 15.14A).

Dust Clean can be used for automatic cleaning of narrow town streets (Figure 15.14).

Care-O-bot 3—This robot (Figure 15.15) has a highly flexible arm with a three-finger hand that is capable of picking up home items (a bottle, a cup, etc.). It can carefully pick up a bottle of orange juice and put it next to the glasses on the tray in front of it. To do this, it is equipped with many sensors (stereo vision, color cameras, laser scanners, and a 3D range camera).

It knows what a glass looks like, where to find it in the kitchen, and can be taught to recognize new objects.

ARI-100 robot—This robot is specialized to duct cleaning and inspection. It involves an articulated arm which can move in all directions allowing it to clean efficiently every corner of the duct, without the need to manually adjust the length of the brush (Figure 15.16).

Roomba discovery vacuum—This is a robotic floor vacuum capable of moving around home and sweeping up dirt while moving. It performs three types of cleaning via two rotating brushes that sweep the floor, a vacuum sucking dust and particles off the floor, and side sweeping brushes to clean baseboards and walls (Figure 15.17).

Roomba recognizes its position in the room and avoids possible risks and stairs. It returns for a self-charging home after the floor is clean or when it needs to recharge.

Figure 15.14 The humanoid garbage collector Dust-Cart (left) and the mobile robot automatic street-cleaner (right).
Source: http://www.gizmag.com/dustbot-multifunctional-robots-keep-town-tidy/12923/.

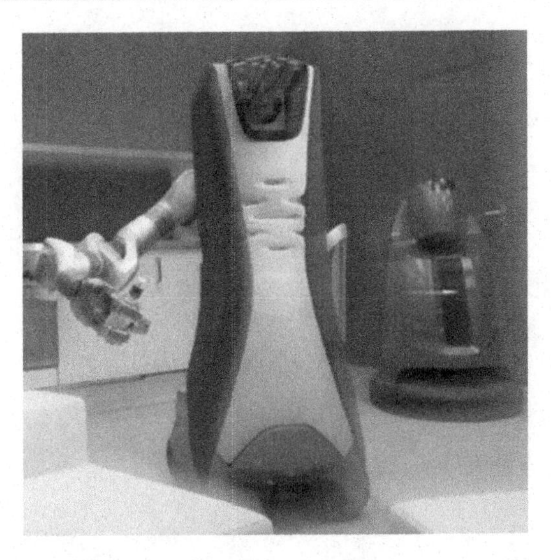

Figure 15.15 The Care-O-bot 3 home service robot (height 1.45 m).
Source: http://phys.org/news134145359.html.

Figure 15.16 The duct cleaning and inspection robot of Robotics Design Inc (Anatroller).
Source: http://roboticsdesign.en.ec21.com/Duct_Cleaning_and_Inspection_Robot−3113255_3113256.html.

15.4 Assistive Mobile Robots

Assistive mobile (and fixed) robots belong to *assistive technology* which in our times is a major field of research, given the aging of population and diminishing number of available caregivers. *Assistive robotics* (AR) includes all robotic systems that are developed for PwSN and attempt to enable disabled people to reach and

Figure 15.17 The Roomba Discovery vacuum.
Source: www.robotshop.com/robotics-floor-cleaners.html.

maintain their best physical and/or social functional level, improving their quality of life and work productivity [6,7].

The main categories of PwSN are as follows:

- PwSN with loss of lower limb control (paraplegic patients, spinal cord injury, tumor, degenerative disease)
- PwSN with loss of upper limb control (and associated locomotor disorders)
- PwSN with loss of spatiotemporal orientation (mental, neuropsychological impairments, brain injuries, stroke, ageing, etc.).

The field of AR was initiated in North America and Europe in the 1960s. A landmark assistive robot is the so-called *Golden Armo* developed in 1969, a seven degrees-of-freedom orthosis moving the arm in space (Rancho Los Amigos Hospital, CA). In 1970, the first robotic arm mounted on a wheelchair was designed. Today many smart AR systems are available, including the following:

i. *Smart-intelligent wheelchairs* that can eliminate the user's task to drive the wheelchair and can detect and avoid obstacles and other risks.
ii. *Wheelchair-mounted robots* (WMRs) which offer the best solution for people with motor disabilities increasing the user's mobility and the ability to handle objects. Today, WMRs can be operated in all alternative ways (manual, semiautomatic, automatic) through the use of proper interfaces.
iii. *Mobile autonomous manipulators*, that is, robotic arms mounted on mobile platforms, that can follow the user's (PwSNs) wheelchair in the environment, can perform tasks in open environments, and can be shared between several users.

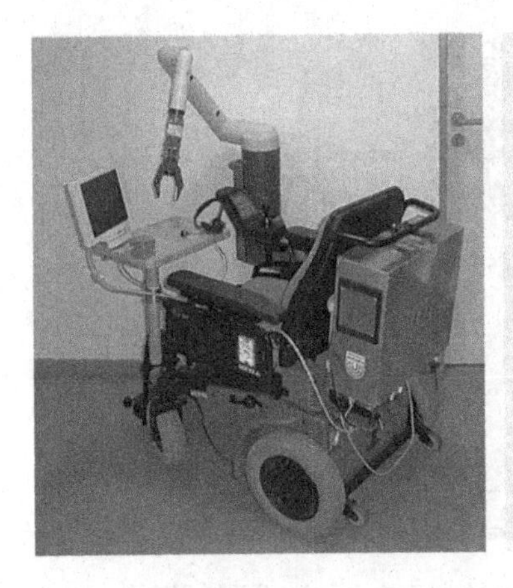
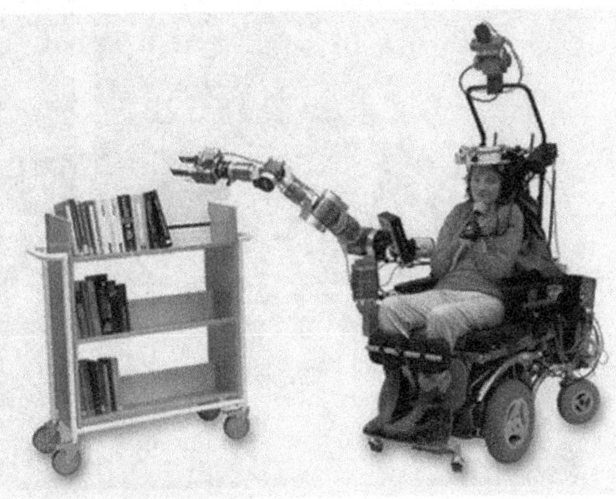

Figure 15.18 The wheelchair FRIEND armed with the MANUS robot arm.
Source: www.AMaRob.de.

Three well-known European assistive robots are the French *MASTER robot*, the Dutch *MANUS robot*, and the UK *RTX robot*. The European Union has launched in 1991 the "*Technology Initiative for Disabled and Elderly People*" (TIDE). During the pilot phase of TIDE the following robotic systems were developed: *MARCUS, M3S, RAID,* and *MECCS*. During the next phase (Bridge Phase) the following systems were created: *SENARIO, FOCUS, EPI-RAID, OMNI,* and *MOVAID* in the framework of respective R&D projects [8–15]. Figures 15.18 and 15.19 show the Bremen (IAT) wheelchair with the "functional robot arm with user friendly interface for disabled people (FRIEND)" and the service mobile manipulator MOVAID [11].

Typical tasks of MOVAID are to warm up some food in a microwave oven and serve it to the user's bed, clean the kitchen surface, remove dirty sheets from a bed, etc. [11].

15.5 Mobile Telerobots and Web Robots

Telerobots combine the capabilities of standard robots (rigid or mobile) and teleoperators. *Teleoperators* are operated by direct manual control and need an operator to work in real time for hours. Of course, due to the human supervision they can perform nonrepetitive tasks (as, e.g. it is required in nuclear environments) [4]. *Telerobots* have more capabilities than either a standard robot or a teleoperator, because they can carry out many more tasks that can be accomplished by each one of them alone. Therefore, the advantages of both are fruitfully exploited, and their limitations minimized. Telerobots can work with incomplete knowledge and models of the task space, being able to perform nonrepetitive tasks. The basic drawback

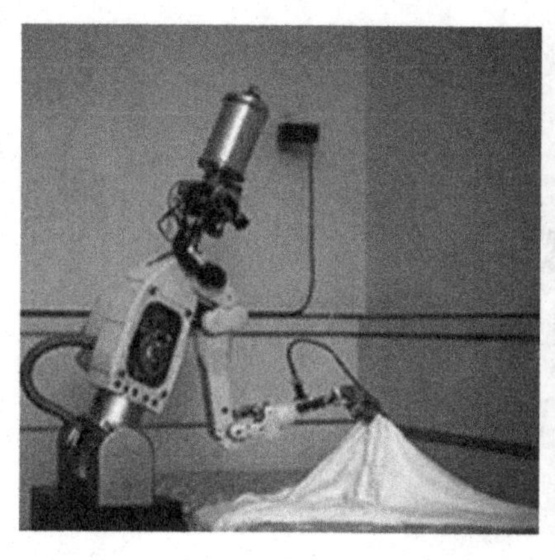
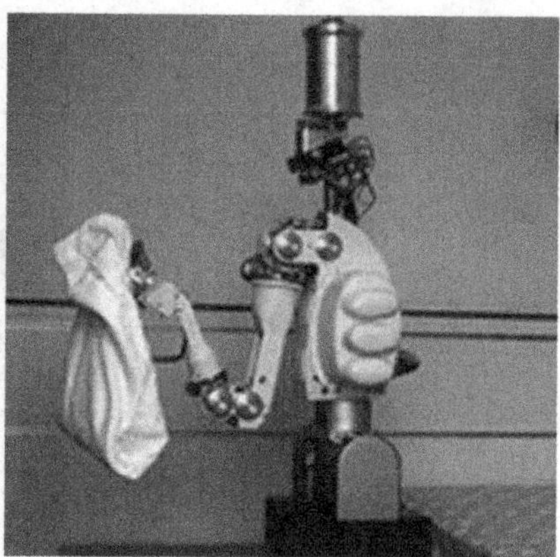

Figure 15.19 MOVAID—Mobility and activity assistance system for the disabled.
Source: http://www.robocasa.net/workshop/2007/pdf.laschi.pdf.

of telerobots is the occurrence of variable delays between the operator and the manipulator, especially in space applications.

Web robots, that is, robots taking their inputs and sending their outputs using as infrastructure the World Wide Web, are used for teleoperation, education, and entertainment purposes [26,27]. These systems can be remotely controlled via the Internet from any site having a typical web browser that incorporates the human operator control interface [25]. Telerobots find applications in space, terrestrial and deep sea exploration, as well as in remote intervention/surgery [4].

One of the major problems of telerobotic systems is that if there exist significant delays in the communication links, the system may exhibit instability. One way to face this problem is to use a shared supervisory scheme, where the control of the robots and other devices can be shared between a local control system and the human operator [4]. Two other techniques for solving this problem are the following:

- the scattering/wave variable teleoperation technique [28],
- the autoregressive integrated moving average (ARIMA) delay modeling/identification technique [29].

Human factors do not, in general, play a major role in robotic applications, but when a supervisory type of control is used with a human being at the center of the control loop they should be necessarily taken into account. That is, important information on how to improve the system can be acquired from the analysis of the human operator.

Figure 15.20 shows a pictorial representation of the evolution of teleoperation towards intelligent telerobotics [30].

Figure 15.21 shows a prototype mobile telerobot built at Sethu Institute of Technology which is capable of picking up objects with maximum degree of freedom with distant commands.

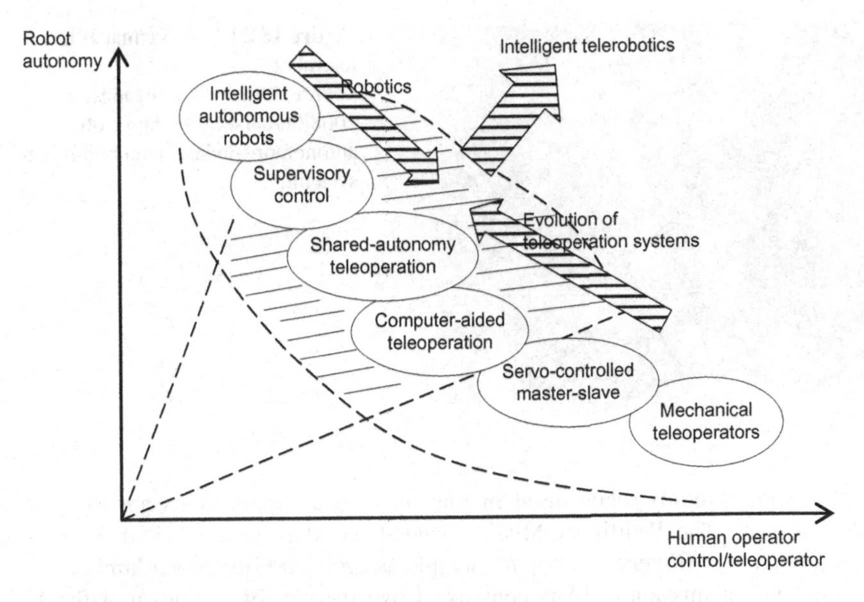

Figure 15.20 Evolution of teleoperation systems.

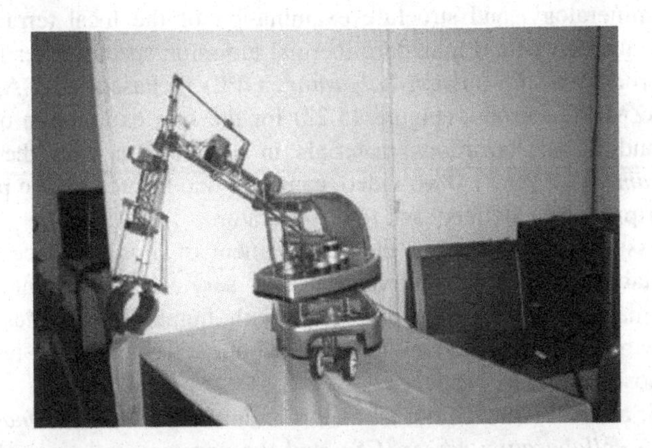

Figure 15.21: The Sethu mobile telerobotic manipulator.
Source: http://robots.net/robomenu/1195489702.html.

NASA has put a considerable research effort and investment in three fundamental areas [5]:

- Remote operations on planetary and lunar surfaces
- Satellite and space system servicing
- Robotics tending of scientific payloads

These areas required advance automation technology (to reduce crew interaction), hazardous material handling, robotic vision systems, collision avoidance algorithms, etc.

Figure 15.22 The Hazmat JPL telerobot.
Source: http://www.engadget.com/2008/06/04/seegrid-shows-off-autonomous-industrialmobile-robot-system/.

The most famous robots used in the outer space applications are the NASA *Mars Rovers*. The Pathfinder Mission landed on Mars in 1997, with its robotic rover, *Sojourner* (Figure 1.2, http://mars.jpl.nasa.gov/MPF/mpf/rover.html).

The current mission to Mars consists of two robotic rovers known as the *Mars Exploration Rovers*. These twin rovers have a panoramic camera suitable for texture, color, mineralogy, and structure examination of the local terrain. Also the rovers were equipped with a miniature thermal emission spectrometer for the identification of rocks. *Jet Propulsion Laboratory* (JPL) in Pasadena (CA) has developed the HAZMAT telerobot (Figure 15.22) for the safe exploration of dangerous sites and handling of hazardous materials in conjunction with the *Hazardous Materials Team* (HAZMAT). Two video, cameras, one located on the platform and one on the gripper, provide feedback to the operator.

The key issue that will lead to the development of more advanced telerobotic systems, suitable for precise intervention and services, is the interaction and merging of machine intelligence properties with human capabilities and skills. Web-based remote laboratories can be used for both actual operation tests and educational purposes.

In general, a web-based remote laboratory contains an *access management system* (AMS), a *collaboration server* (CS), and the *experimental server* (ES). These components can be implemented using any proper combination of available technological tools (Matlab, LabView, VRML, Java, etc). This architecture (Figure 15.23) uses the client/server scheme implemented in several ways. The AMS coordinates the accessibility of users (operators, students, etc) to the experiment, who can use any station equipped with a web browser and Java environment if the heterogeneity problem must be eliminated.

A few representative examples of the development and use of web robots, including their educational applications, are the following:

- the Mercury web robot [16],
- the Telegarden web robot [19],
- the Australian telerobot [17],

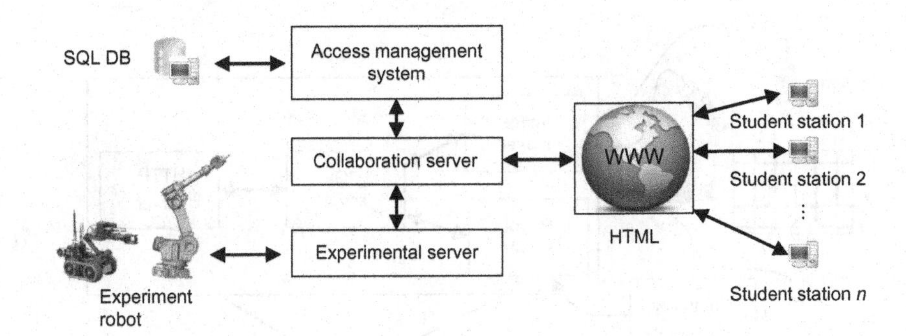

Figure 15.23 General architecture of a web-based telelab (experiment).

- the teleoperated multirobot platform (Teleworkbench) [18].
- the Swish open access web robot (Khep on the Web) [20].

The *Mercury web robot* (one of the early web-based robots, built in 1994) consists of an industrial robot arm fitted with a CCD camera and a pneumatic system (http://usc.edu/dept/raiders). All robots are accessible via standard point-and-click mouse commands, using a 2D workspace.

The *Telegarden web robot* was developed as a continuation of the Mercury robot at the University of Southern California, launched in June 1995 (http://www.usc.edu/dept/garden).

The *Australian telerobot* was invented by Ken Taylor and developed at the University of Western Australia in 1994. A 6-axis ABB robot is teleoperated over the web. Java and Java script is used which enables standard common gateway interface (CGI) communication with the web server while enabling moves to be programmed using a Java wireframe of the robot. The overall structure of the system is shown in Figure 15.24.

The *teleworkbench* was developed at the University of Paderborn (Germany) to ease the tasks of carrying out experiments with single or multiple mini (mobile) robots. The general architecture of Teleworkbench is shown in Figure 14.25A with many communicating minirobots, and the teleworkbench server structure is shown in Figure 15.25B.

Teleworkbench provides a standard environment in which algorithms and programs can be tested and validated on real robots. Two of the robots used in Teleworkbench are Khepera II and Bebot minirobot (Heinz Nixdorf Institute).

Khep on the Web—This system was built to remotely control Khepera mobile robots at the *LAMI* (Microprocessor and Interface Lab) of the Swiss Federal Institute of Technology of Lausanne (EPFL). The objective of *Khep on the Web* was to provide an open access web robotic platform for scientific research in mobile robotics. Khepera is a small cylindrical robot (diameter 55 mm, variable height) with a suspended cable for power and other signals supply without any disturbance on its motion. The camera used may not have wide-angle lenses. The system consists of a sensory-motor board (with eight infrared proximity sensors), a

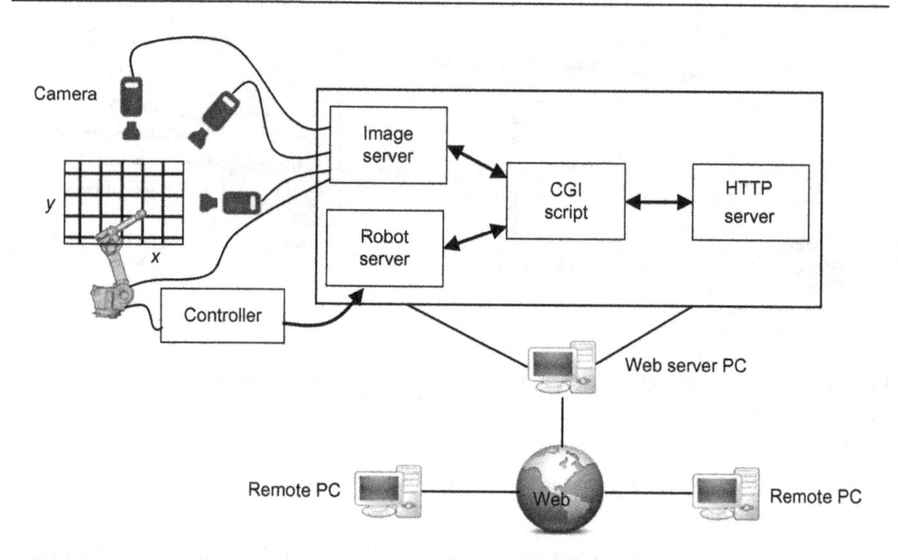

Figure 15.24 Architecture of the Australian UWA telerobot (http://telerobot.mech.uwa.edu.au).

CPU board (Motorola 68331 μC), and a video board having a color CCD camera (500 × 582 pixels). The web server launches a number of CGI scripts to perform the tasks that communicate via shared memory. A Java applet running on the client side sends several information requests, and a CGI script on the server replies to these requests by obtaining the information from the robot and the shared memory. The structure of the *Khep on the Web* system is shown in Figure 15.26.

Figure 15.27 shows the Khepera robot equipped with a video camera and looking to a mirror. The robot worked in a maze environment with walls higher than the robot such that the visitor has to move around to visit the maze (Figure 15.28). One of the walls is a mirror [20].

15.6 Other Mobile Robot Applications

Other applications of mobile robots include war robots and entertainment robots.

15.6.1 War Robots

The design, development, and construction of robots for the war has raised strong ethical questions. In general, the weaponized of robots (missile, unmanned combat air vehicles, unmanned terrestrial vehicles, etc.) has got a substantial portion of the investment in robotics research and development [24]. In general, military robots operate in geopolitically sensitive environments, and so a greater caution is certainly needed. For example, what happens if a unmanned aerial vehicle (UAV) mistakenly decides that a "friendly" is a target and then fires it? A small subset of

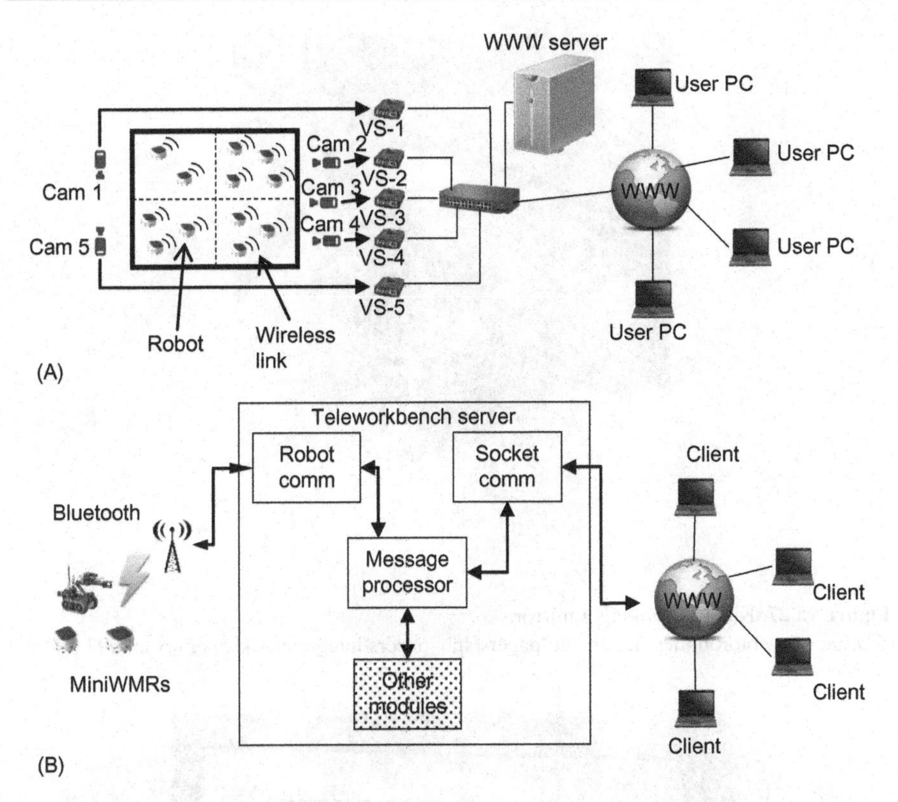

Figure 15.25 Teleworkbench system: (A) General architecture and (B) Teleworkbench server.

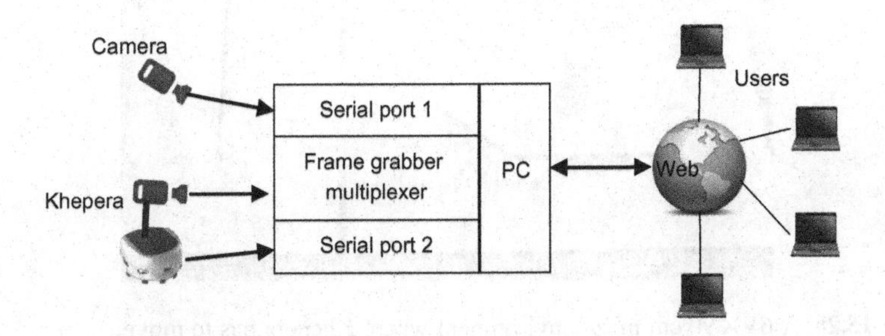

Figure 15.26 The architecture of *Khep on the Web* system.

land, aerial, and underwater robotized autonomous combat or exploration vehicles is the following:

Squat four-wheeled robot—This is driving itself through densely wooded terrain and it is too small to be threatening (Figure 15.29). It is called XU12 (for "experimental unmanned vehicle") and can navigate autonomously from point A to point B without

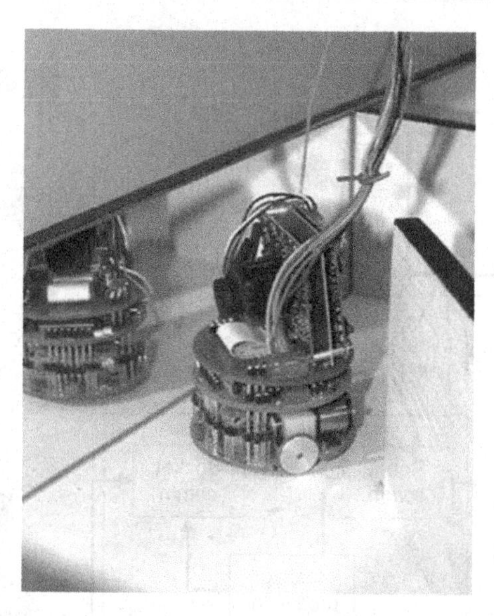

Figure 15.27 Khepera looking a mirror.
Source: www.biorobotics.ri.cmu.edu/papers/spb_papers/integrated1/khepera_vsmm97.pdf.

Figure 15.28 A 65 × 90 cm maze environment where Khepera has to move.
Source: www.biorobotics.ri.cmu.edu/papers/spb_papers/integrated1/khepera_vsmm97.pdf.

clobbering a boulder, hammering a tree, and so on. The robot is useful for inspection and surveillance purposes.

Mapping swarmbots—These have been developed jointly by Georgia Tech. University and JPL. They are collaborative mobile robots capable of autonomously mapping whole building for first responder and military purposes. They are equipped with a video camera and a laser scanner each (Figure 15.30).

Figure 15.29 The XU12 vehicle.
Source: http://www.popsci.com/scitech/article/2005-12/robots-go-war.

Figure 15.30 GaTech/JPL mapping swarmbots.

Voyeur-autonomous UAV—This is an autonomous UAV that can be used for military applications such as surveillance and target acquisition, as well as locating and detonating improvised explosive devices (IEDs) (Figure 15.31). It can be launched from an airplane or carried in a backpack and launched manually.

HAUV-N—This is an IED detection and neutralization version of the autonomous underwater vehicle (AUV) able to identify and destroy ship hull mines and underwater IEDs sparing human divers from this dangerous task (Figure 15.32).

X-47B—This is a stealth UAV that closely resembles a strike fighter (Figure 15.33). It can take off from and land on an aircraft carrier and support midair refueling. X-47B has a range of 3380 km, it can fly up to 40,000 ft at high subsonic speeds, and can carry up to 2000 kg of ordnance in two weapons bays.

Figure 15.31 The Voyeur-Autonomous UAV (Northrop Grumman Corp.).
Source: http://thefutureofthings.com/pod/1261/voyeur-autonomous-uav.html.

Figure 15.32 HAUV-N underwater IED detector and neutralizer (Bluefin Robotics Corp).
Source: www.militaryaerospace.com/index/display/mae-defense-executive-article-display/3856814793/articles/military-aerospace-electronics/executive-watch-2/2011/.

15.6.2 Entertainment Robots

These robots belong to the more general class of *social robots*. A social robot is a high-level intelligent autonomous robot that knows how to interact in a humanistic way with humans. Social robots are equipped with sensory, learning, social rules, and performance, etc. A basic requirement for a robot to be socialized is to be completely autonomous, capable of communication and cooperative interaction with everyday persons (nonprofessionals). Most of the available socialized and entertainment robots are mobile robots that have a legged-humanoid or wheeled-

Figure 15.33 The X-47B US Navy Stealth UAV (Northrop Grumman Corp.).
Source: http://thefutureofthings.com/pod/6239/x-47b-first-navy-stealth-uav-ready.html.

semihumanoid (the upper body part) appearance, that have the ability to write, play musical instruments, interact emotionally, dance, soccer, etc.

According to Dautenhahn [22], a list of basic social skills that an entertainment/ socialized robot must have is the following:

- Ability to contact with humans in a repeated and long-life setting. The robot should have the ability to become personalized, recognizing and adapting to its owner's preferences.
- Ability to negotiate tasks and preferences and provide "companionship."
- Ability to adapt, to learn, and to expand its skills, e.g., by being taught new performances by its owner.
- Ability to play a role of companion in a more human-like way (probably similarly to pets).
- *Social skills*—These are essential for a robot to be acceptable as a companion. For example, is good to have a robot that says "would you like me to bring a cup of coffee?". But, it may not be desirable to ask this question while we are watching our favorite TV.
- When moving in the same area as a human, the robot always changes its route to avoid getting too close to the human, especially if the human's back is turned.
- The robot turns its camera properly to indicate by its gaze that it was looking in order to participate or anticipate what is going on in the surrounding area.

Some functional entertainment and socialized robots are shown in Figures 15.34–15.37.

- *Kaspar*—This is a child-sized humanoid robot able to interact with children with autism, using gestures, expressions, synchronization, and imitation [23]. It has eight degrees of freedom in the head and neck and six degrees of freedom in the arms and hands. The face is a silicon-rubber mask, which is supported by an aluminum frame (Figure 15.34).
- *Wow Wee Roboscooper*—This entertainment robot has both functional capabilities and funny and cute looking (Figure 15.35).

Figure 15.34 The socialized robot Kaspar with open arms entertains a girl.
Source: http://kaspar.herts.ac.uk.

Figure 15.35 The entertainment robot *Wow Wee Roboscooper*.
Source: http://www.learnaboutrobots.com/entertainment.htm.

Figure 15.36 The bar service entertainment robot.

Figure 15.37 Bots testing their soccer skills in RoboCup 2010.
Source: www.youtube.com/watch?v=4wMSiKHPKX4.

- *Robot-Barman*—This robot is capable of opening beer bottles and other drink bottles and serve the bar's customers (Figure 15.36).
- *Humanoid robot soccer*—This is an adult-sized humanoid robot, fully autonomous, that can participate efficiently in a robot football championship (Figure 15.37). The ultimate goal of the designers is to develop by 2050 robot football player that can win against the human world champion team in soccer. Every robot on the field can handle different situations in the field (e.g., finding the ball, moving towards it, and carefully manipulating it towards the goal).

Eventually the skills developed for a soccer team will be transferred to other areas like domestic robotics, education, rescue, etc.

15.6.3 Research Robots

All the robots considered in the present chapter have passed through several stages of research and development, and most of them are still used in universities and institutions for further research. Just as a final example, Figure 15.38 shows two fully equipped P3-DX robots that can be used for research and actual cooperative operations in large areas such as workshop and office floors, logistic areas, hospitals. Each robot is equipped with an ultrasonic array, a SICK laser range finder, a pan-tilt camera, an onboard netbook, and a different bar code landmark for robot identification and distinction. Tasks that can be performed by groups of such robots include a large repertory of exploration and surveillance tasks, such as building and updating maps, locating particular objects on the map, tracing the product flow in a logistic space, etc. Actually, Section 13.13 discusses a case of cooperative robotic exploration of this kind.

Figure 15.38 Two P3-DX robots that can be used in cooperative exploration and surveillance tasks.
Source: http://areeweb.polito.it/ricerca/MacP4Log/index.php?option=com_content&view=article&id=28:cooproboteamsupervandmgmtlosspaces&catid=9.

15.7 Mobile Robot Safety

The application of mobile robots in industry and society removes the requirement for humans to perform several useful operations that may be difficult or dangerous for the human operator. Although fixed robot safety has long been a central issue of consideration, actually there has been less effort for the investigation and assurance of autonomous robots' safety. Safety is especially critical for mobile robots in home and health services where the humans (typically not accustomed to robots) might get in contact with them. Another point that creates the demand for safety, when using mobile robots, is the fact that these robots work autonomously with nonrepetitive or nonpredictable motions. These motions are due to that the robots operate in uncertain and unstructured environments, which need their ability to autonomously (and intelligently) change tasks, operating online on the basis of sensory information that may be not accurate enough, etc. Therefore, to secure safety for the people and the environment, the following safety measures are needed:

- Accurate, reliable, and redundant sensory systems
- Reliable software
- Low speed of operation

The maximum speed of operation that can be achieved depends on the following:

- Mechanical and dynamic limitations
- Computational and control limitations
- Unexpected dynamic changes of the environment

The first issue concerns the case where there is the possibility of longitudinal or lateral slippage when a certain velocity is exceeded. In practice, however this is not a major problem since maximum speed of motion is subject to other more strict limitations. The second issue has to do mainly with how quickly the robot can avoid obstacles, which depends on the computational speed, the control algorithm, and the sensing rate and accuracy. The third issue refers to unexpected collision and visibility problems. All these issues were studied following different approaches under different conditions and assumptions (see, e.g., Refs. [31–33]).

The implementation of safety measures, to avoid any risk both for the user and persons in the environment of the mobile robot, can be made in the following three levels:

- *Passive safety*—This depends on the electromechanical design (low inertia of moving parts, restricted motor speeds at maximum voltage, limited static motor torques at stall, possibility of passive elements, etc.).
- *Supervised safety*—This includes watch dog, safety monitor, velocity and force/torque limits, forbidden zones, etc.
- *Interactive safety*—This implies that the user is always able to switch off the system, that there exist proper (well distinguished) alarm signals, and that "control-blocking" conditions (i.e., conditions under which the user looses access to the control transducers where the transducers get out of reach) are not possible to occur.

The dominant means for mobile robots safety systems is *braking*. There must always be possible to stop the robot's motion in one of the following ways:

- *Emergency stop*—user initiated or automatic shutdown due to an unrecoverable hazard condition.
- *Intrusion pause*—initiated by a sensor when an external hazard occurs, in which case the user must be able to restart the system when the hazard is removed.
- *User pause*—initiated by the user whenever he/she detects a potential hazard or problem, in which case the user can restart or cancel the operation after the removal of the problem.

All these aspects are under extensive study by research institutes and mobile robot manufacturers, and today many alternative safety systems are available internationally. These systems are compatible with existing safety standards, such as the *US ASME B56.5-2004 Safety Standard* for driverless industrial vehicles [34], and the *British EN 1525-1998 Safety Standard* for driverless industrial vehicles allowing the use of noncontact safety sensors [35]. The *National Institute Standards and Technology* (NIST) is working continuously to enhance the safety standards for AGVs with the use of noncontact sensors, but also for providing evaluation standards of novel 3D real-time range sensor technology. A study where the above NIST enhancements are considered is provided in Ref. [36]. A discussion of the hazards associated with the use of industrial robots and the principles of guarding to assure human safety is provided in Ref. [37]. In this work, the hazards associated to robots are distinguished as follows:

- Control errors
- Mechanical hazards
- Environmental hazards
- Human errors
- Ancillary equipment

The guarding measures, while the robots are functioning, are classified as:

- Mechanical guards with safety interlock
- Presence sensing systems
- Trip devices
- Emergency stop equipment
- Screens between station
- Work envelope limit stop

References

[1] Nof S, editor. Handbook of industrial robotics. New York, NY: John Wiley & Sons; 1999.

[2] Schraft RD, Schmierer G. Service robots. London: Peter AK/CRC Press; 2000.

[3] Takahashi Y, editor. Service robot applications. In Tech/Read Online; 2008.

[4] Sheridan TB. Telerobotics, automation and human supervisory control. Cambridge, MA: MIT Press; 1992.

[5] Votaw B. Telerobotic applications. <http://www1.pacific.edu/eng/research/cvrg/members/bvotaw>; 2013 [accessed 20 August].

[6] Cook AM, Hussey SM. Assistive technologies: principles and practice. St. Louis, MO: Mosby; 2002.

[7] Reddy R. Robotics and intelligent systems in support of society. IEEE Intell Syst 2006;May–June:24–31.

[8] Dallaway JL, Jackson RD, Timmers PHA. Rehabilitation robotics in Europe. IEEE Trans Rehabil Eng 1995;23:35–45.

[9] Tzafestas SG, editor. Autonomous robotic wheelchair projects in Europe improve mobility and safety (Special Issue). IEEE Robot Autom Mag 2001;17(1):1–73.

[10] Tzafestas SG, editor. Autonomous mobile robots in health care services (Special Issue). J Intell Robot Syst 1998;22(3–4):177–350.

[11] Dario P, Guglielmelli E, Laschi C, Teti G. MOVAID: a personal robot in everyday life of disabled and elderly people. J Technol Disabl 1999;10:77–93.

[12] Pires G, Honorio N, Lopes C, Nunes U, Almeida AT. Autonomous wheelchair for disabled people. In: Proceedings of IEEE international symposium on industrial electronics (ISIE '97). Guimaraes; 1997. p. 797–801.

[13] Duffy BR. Social embodiment in autonomous mobile robotics. Int J Adv Robot Syst 2004;1(3):155–70.

[14] Tiwari P, Warren J, Day KJ, McDonald B. Some non-technology implications for wider application of robots assisting older people. In: Proceedings of HIMMS Asia Pacific 11. Melbourne, Australia; 20–23 September 2011. p. 1–15.

[15] Martens C, Prenzel O, Gräser A. The rehabilitation robots FRIEND I&II: daily life independence through semi-autonomous task execution. Vienna: I-Tech Education and Publishing; 2007.

[16] Goldberg K, Gentler S, Sutter C, Wiegley J. The Mercury project: a feasibility study for internet robots. IEEE Robot Autom Mag 2000;7(1):35–40.

[17] Trevelyan J. Lessons learned from 10 years experience with remote laboratories. in: Proceedings of international conference on engineering education and research. VSB-TUO, Ostrava; 2004. p. 1–10.

[18] Tanoto A, Rückert U, Witkowski U. Teleworkbench: a teleoperated platform for experiments in multirobotics. In: Tzafestas SG, editor. Web-based control and robotics education. Berlin/Dordrecht: Springer; 2009. p. 267–96.

[19] Goldberg K, editor. The robot in the garden: telerobotics and telepistemology in the age of internet. Cambridge, MA: MIT Press; 2000.

[20] Saucy P, Mondada F. Khep on the Web: open access to a mobile robot on the internet. IEEE Robot Autom Mag 2000;7(1):41–7.

[21] Prassler E, Scholz J, Fiorini P. Navigating a robotic wheelchair in railway station during rush hour. Int J Robot Res 1999;18:760–72.

[22] Dautenhahn K. Socially intelligent robots: dimensions of human–robot interaction. Philos Trans R Soc Lond B Biol Sci 2007;362:679–704.

[23] Dautenhahn K. Kaspar: kinesis and synchronization in personal assistant robotics. University of Hertfordshire, UK: Adaptive Research Group; <http://kaspar.feis.herts.ac.uk>.

[24] Zaloga S. Unmanned aerial vehicles: robotic air warfare 1917–2007. Oxford: Osprey Publishing; 2008.

[25] Taylor K, Dalton B. Internet robots: a robotics niche. IEEE Robot Autom Mag 2000;7(1):27–34.

[26] Tzafestas SG, editor. Web-based control and robotics education. Berlin: Springer; 2009.

[27] Tzafestas SG, Mantelos A-I. Time delay and uncertainty compensation in bilateral telerobotic systems: state-of-art with case studies. In: Habib M, Davim P, editors. Engineering creative design in robotics and mechatronics. Mershey, PA: IG Global; 2013. p. 208–38.

[28] Munir S, Book WJ. Internet-based teleoperation using wave variable with prediction. Proc IEEE/ASME Trans Mechatron 2002;7:124–33.

[29] Yang M, Li XR. Predicting end-to-end delay of the internet using time series analysis. Technical Report, University of New Orleans, Lake front; November 2003.

[30] Tzafestas CS. Web-based laboratory on robotics: remote vs virtual training in programming manipulators. In: Tzafestas SG, editor. Web-based control and robotics education. Berlin: Springer; 2009. p. 195–225.

[31] Hong T, Bostelman R, Madhavan R. Obstacle detection using a TOF range camera for indoor AGV navigation. Gaithersburg, MD: PerMIS; 2004.

[32] Pare C, Seward DW. A model for autonomous safety management in a mobile robot. Proceedings of CICMCA'05 international conference on computational intelligence for modeling control and automation. Washington, DC: IEEE Computer Society; 2005.

[33] Chung W, Kim S, Choi M, Choi J, Kim H, Moon CB, et al. Safe navigation of a mobile robot considering visibility of environment. IEEE Trans Ind Electron 2009;56(10):3941–9.

[34] American Society of Mechanical Engineers. Safety standard for guided industrial vehicle and automated functions manned industrial vehicle. Technical Report ASME B56.5, 1993.

[35] British standard safety of industrial trucks—driverless trucks and their systems. Technical Report BSEN-1525, 1998.

[36] Bostelman RV, Hong TH, Madhavan R, Chang TY. Safety standard advancement toward a mobile robot use near humans. Proceedings of RIA -SIAS'05 Conference, Chicago, IL, U.S.A.; 2005. www.et.byu.edu/~ered/ME486/Professional_Journal.pdf.

[37] Department of Labor, Robot safety. <http://www.osh.dol.govt.nz/order/catalogue/robotsafety.shtml>; 2013 [accessed 20 August].

Problems

A. Kinematics

1. Derive the formulas for $\sin(\theta + \phi)$ and $\cos(\theta + \phi)$ expanding symbolically two rotations of θ and ϕ via the concept of the rotation matrix $\mathbf{R}$.
2. Extending the kinematic analysis of the three-omni-wheel system of Section 2.4, derive the kinematic equations of a four-wheel and five-wheel omnidirectional WMR shown in Figure P.1, and study the velocity augmentation factor in each one of them. To this end, write a Matlab program for angular displacements $\phi \in [0, 180°]$.
3. Derive the kinematic model of a bicycle.
4. Derive the Jacobian matrix of a WMR with longitudinal and lateral wheel slip: (a) for differential drive, (b) for car-like, (c) for three- and four-wheel omnidirectional robots.
5. Study analytically the inaccessible area for a differential drive, a bicycle, and a car-like WMR.

B. Dynamics

6. Derive the Lagrange dynamic model of the nonholonomic two-wheel inverted pendulum robot shown in Figure P.2.

   The body of the pendulum is assumed a particle with its mass concentrated at its end. The wheels are conventional wheels.
7. Derive the Lagrange dynamic model of a conventional wheel with slip. Use this result to find the corresponding dynamic model for differential-drive WMR with slip.
8. Derive the kinematic and Lagrange dynamic model of the WMR shown in Figure P.3 with pure rolling nonslipping wheels. The two wheels are placed in the front of the robot and the third wheel is connected with a fixed structure and can rotate freely around its vertical axis. The generalized coordinates vector is $\mathbf{q}(t) = (x, y, \theta, \psi, \phi_1, \phi_2, \phi_3)$ as shown in Figure P.3. Formulate a least-squares parameter procedure in the following two cases:
   (a) Without the wheel motor's dynamics
   (b) Including the wheel motor's dynamics
9. **(a)** Derive the dynamic equation of the double pendulum shown in Figure P.4 where coordinates, parameters, and forces/torques are shown. Formulate the dynamic model using the Newton and the Lagrange method. Describe how to identify its parameters converting the dynamic model to a regressor form.
   (b) Derive the Lagrange dynamic model of a SCARA (selective compliance assembly robot arm) (see Figure 10.4E).
10. Propose a suitable optimization method for finding optimal persistently exciting trajectories by using either a Fourier series or a polynomial representation of the trajectory. Explain why these are good approximate persistently exciting trajectories.

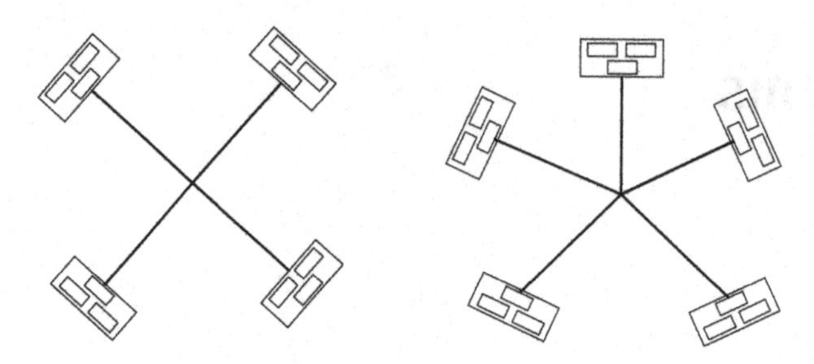

Figure P.1 Four- and five-wheel omnidirectional robots.

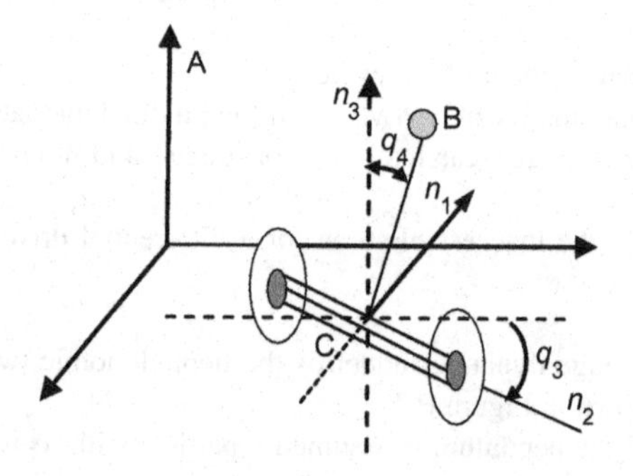

Figure P.2 Schematic of two-wheel inverted pendulum WMR.

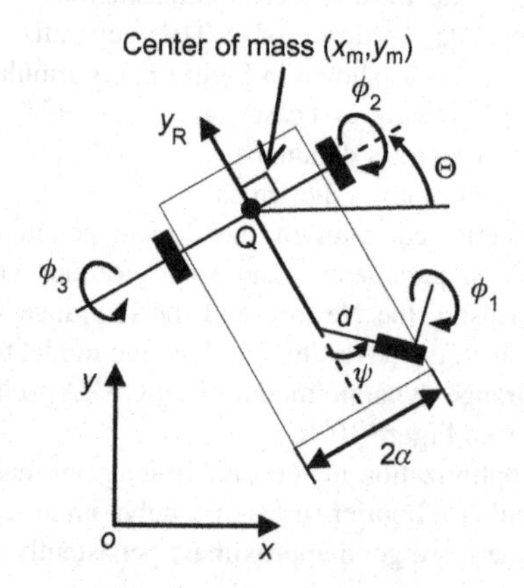

Figure P.3 Mobile robot for identification.

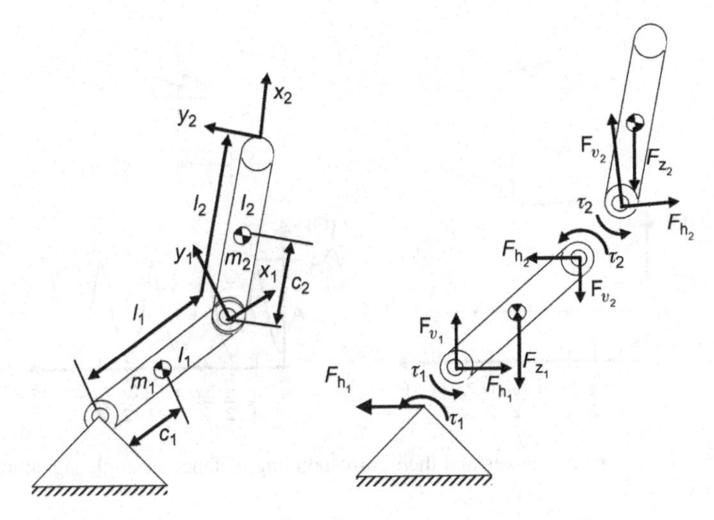

Figure P.4 The double pendulum model ($F_h = F_{horizontal}, F_v = F_{vertical}$).

C. Sensors

11. We are given an ultrasonic proximity sensor for detecting the presence of objects within 0.5 m of the device. At time $t = 0$, the sensor is pulsed for 0.1 ms. Suppose it takes 0.4 ms for resonances to decay out within the transducer, and 20 ms for echoes in the environment to die out.
  (a) Find the range of time needed as a window.
  (b) What is the minimum detectable distance?
  (The speed of sound is 344 m/s.)

12. The output of a laser finder is corrupted by a Gaussian zero-mean noise having a standard deviation $\sigma = 100$ cm.
  (a) How many measurements have to be averaged to obtain an accuracy of ± 0.5 cm with probability $p = 0.95$?
  (b) Show how to compensate the range measurements in case the noise has mean value 5 cm.

13. What is the upper limit on the frequency of a modulating sine wave to obtain a working distance of up to (but not including) 5 m using a continuous-beam laser range finder?

14. **(a)** How many bits are needed to store a 512×512 image in which each pixel can have 256 possible intensity values?
  (b) Propose a procedure for computing the median in an $n \times n$ neighborhood.
  (c) Propose a method which uses a single light sheet for determining the diameter of cylindrical objects (assuming that the distance between the camera and the centers of the cylinders is fixed, and that the array camera has a resolution of N pixels).

15. What is the minimum number of moment descriptors required to differentiate between the boundary shapes of Figure P.5?

16. What is the number of different shapes (called shape number) and the order of the shape number in each of the following 2D objects? Consider the clockwise direction (Figure P.6).

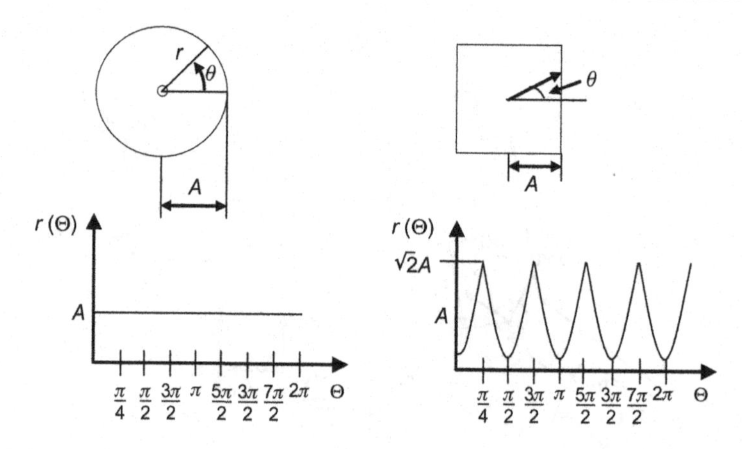

Figure P.5 Two boundary shapes and their corresponding distance vs angle signatures (a) $r(\theta)$ = const. (b) $r(\theta)$ = A sec θ.

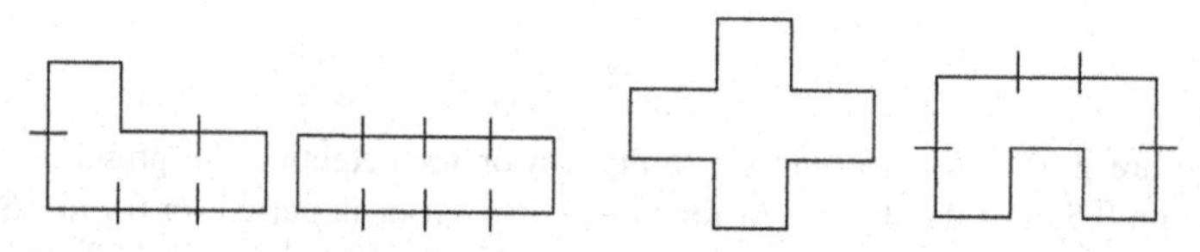

Figure P.6 Four planar objects specified by boundary primitives ($\rightarrow o_{\rightarrow}, \uparrow 1, \leftarrow 2_{\leftarrow}, \downarrow 3$).

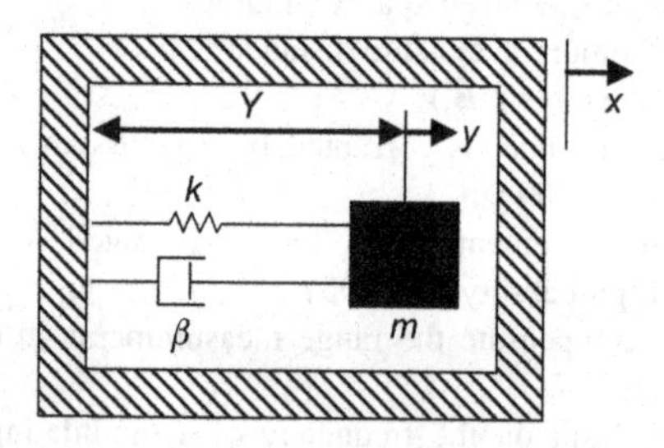

Figure P.7 Accelerometer.

17. Figure P.7 shows a linear accelerometer in which y is the displacement of the mass m with respect to the frame, and x is the displacement of the frame.

(a) Write down the differential equation of the accelerometer and find the transfer function $\overline{y}(s)/\overline{a}(s)$, where $a(t) = d^2x/dt^2$ is the frame acceleration. Explain why the output of the accelerometer is proportional to the acceleration and the condition under which this is true.

(b) Repeat the same for a rotational accelerometer.

18. We are given the 1-DOF gyroscope of Figure P.8.

A rotation of the frame about z generates a rotation of the gyroscope's disc about the axis y. This motion is affected by the spring parameter k and the friction β according to the relation $T_y = -\left(k\theta_y + \beta\frac{d\theta_y}{dt}\right)$

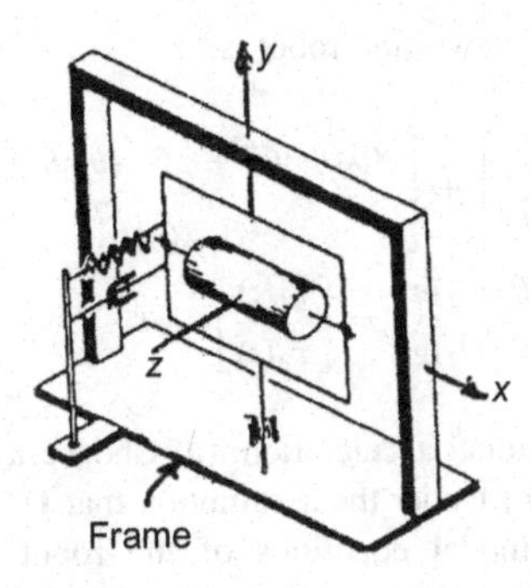

Figure P.8 One degree of freedom gyroscope.

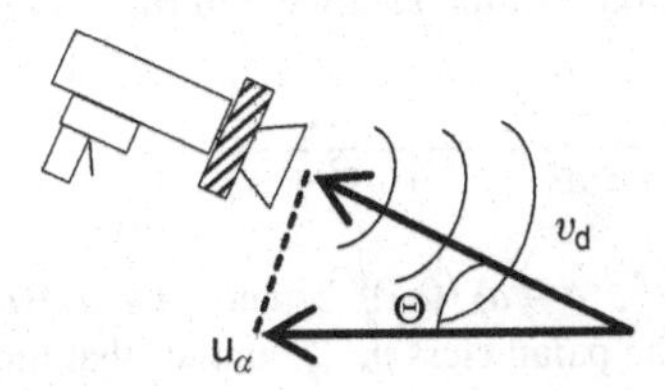

Figure P.9 A Doppler ground-speed sensor inclined at an angle θ.

(a) Using the differential equation:

$$T_y = J_y \frac{d^2\theta_y}{dt^2} + \beta_y \frac{d\theta_y}{dt} + J_0\omega_0 \frac{d\theta_z}{dt}$$

determine the equation that relates θ_y and θ_z (omit J_y and β_y).

(b) When only the spring exists, find the transfer function of the resulting rate gyroscope.

(c) When only the friction exists, find the transfer function of the resulting integrated rate gyroscope.

19. Doppler sensors are based on the Doppler shift in frequency for measuring velocity in maritime and aeronautical applications. Maritime systems use acoustical energy, reflected from the ocean floor, and airborne systems sense microwave radio frequency energy bounced off the surface of the earth. The microwave radar sensor is aimed downward at an angle θ (usually 45°) to sense ground velocity (Figure P.9).

Derive the formula that gives the true (actual) ground velocity v_a in terms of the measured Doppler velocity component v_d, the inclination angle θ, the speed of light c, the transmitted frequency f_0, and the observed Doppler shift frequency.

D. Control

20. Consider the control loop of a single robot axis. (a) Derive the transfer functions of the open-loop and the closed-loop system. (b) Study the position and velocity steady-state errors.

21. The dynamic model of a two-link robot is:

$$\begin{bmatrix} d_{11}(\theta_2) & d_{12}(\theta_2) \\ d_{12}(\theta_2) & d_{22} \end{bmatrix} \begin{bmatrix} \ddot{\theta}_1 \\ \ddot{\theta}_2 \end{bmatrix} + \begin{bmatrix} \beta_{12}(\theta_2)\dot{\theta}_2^2 + 2\beta_{12}(\theta_2)\dot{\theta}_1\dot{\theta}_2 \\ -\beta_{12}(\theta_2)\dot{\theta}_1^2 \end{bmatrix} + \begin{bmatrix} c_1(\theta_1, \theta_2)g \\ c_2(\theta_1, \theta_2)g \end{bmatrix} = \begin{bmatrix} \tau_1(t) \\ \tau_2(t) \end{bmatrix}$$

where g is the gravitational acceleration. (a) Choose a suitable state vector $\mathbf{x}(t)$ and a control vector $\mathbf{u}(t)$. (b) Under the assumption that $\mathbf{D}^{-1}(\boldsymbol{\theta})$ exists, express the corresponding state-space model equations of the robot in terms of d_{ij}, β_{ij}, and c_i. (c) Find a nonlinear input–output decoupling law when $\mathbf{y}(t) = \mathbf{x}(t)$. Determine the Jacobian matrix of the robot with respect to the base coordinate frame.

22. Describe two drawbacks for each of the following two robot control methods: (a) resolved rate control and (b) resolved acceleration control.

23. Find the unit step response of the following third-order system:

$$G(s) = \frac{a_0}{s^3 + a_2 s^2 + a_1 s + a_0} = \frac{1}{\tilde{s}^3 + a\tilde{s}^2 + \beta\tilde{s} + 1}$$

where $\tilde{s} = s/(a_0)^{1/3}$, $a = a_2/(a_0)^{1/3}$, and $\beta = a_1/(a_0)^{2/3}$. Express the transfer function in terms of the parameters p, ζ, and ω_c that are defined by:

$\tilde{s}^3 + a\tilde{s}^2 + \beta\tilde{s} + 1 = (\tilde{s} + p)(\tilde{s}^2 + 2\zeta\omega_c\tilde{s} + \omega_c^2)$. Explain why for $a = 1.3$ and $\beta = 2.0$ the step response involves a reverse motion before it reaches the steady-state value 1. Does this phenomenon occur in the second-order system? Where can this be used in the third-order systems?

24. Prove that the control law $\tau_j = -k_{jp}\tilde{q}_j - k_{jD}\dot{\tilde{q}}_j + k_{jI}\int_0^t \tilde{q}_j \, d\tau, \; j = 1, 2, \ldots, n$ assures asymptotic stability despite the presence of friction and gravity terms in the robot equation (3.11a).

25. Consider the two-link robot of Figure 10.6 and assume that there is an uncertainty in the load placed at its end point (tip). The load mass is less than 1. The smallest structural resonance frequency is 10 Hz. Compare via simulation (using Matlab) the performance of the controllers:

- Local PID controller
- Computed torque controller
- Sliding mode controller

in the following two cases:

(a)

$$\theta_{1d} = -\pi/3 + (\pi/3)[1 - \cos(\pi t/T)]$$
$$\theta_{2d} = 2\pi/3 - (\pi/3)[1 - \cos(\pi t/T)]$$

using $T = 1$, $T = 0.5$, and $T = 2$. What is the minimum sampling period T required for the implementation of the controller?

(b) The desired trajectory is a straight line from the initial position $(x, y) = (1, 0)$ to the final position $(x, y) = (0, 1.5)$ which must be traversed with constant velocity in 2 s time. Initially, the robot is stationary at the position $(x, y) = (1, 0)$ and the "elbow-down" configuration.

Hint: When the derivatives of the desired trajectories are not directly available, we generate q_d by smoothing the reference trajectory using a low-pass second-order filter with bandwidth Ω. This filter gives directly $\dot{\mathbf{q}}_d$ and $\ddot{\mathbf{q}}_d$.

26. We are given a two-link robotic arm of Figure P.10 which is attached to the ceiling and is under the influence of gravity. The reference coordinate frame is (x_0, y_0, z_0). The link masses are assumed to be lumped at their ends.

(a) Find the transformation matrices $\mathbf{A}_i^{i-1}$, $i = 1, 2$.

(b) Find the pseudo inverse matrix $\mathbf{J}_i$ of each link.

(c) Find the elements of the $\mathbf{D}(\boldsymbol{\theta})$, $\mathbf{h}(\boldsymbol{\theta}, \dot{\boldsymbol{\theta}})$, and $\mathbf{c}(\boldsymbol{\theta})$ matrices and write down the Lagrange dynamic model of the robot.

(d) Formulate the computed torque control problem.

(e) Choose the values for the robot and the controller parameters and simulate the system.

27. Consider a differential-drive WMR in the inertial coordinate frame (Figure P.11):

(a) Write the WMR's dynamic model in polar coordinates (ρ, α, β), where the error is given by the relations $\rho = \sqrt{(\Delta x)^2 + (\Delta y)^2}$, $\alpha = -\phi + \text{atan2}(\Delta y, \Delta x)$, and $\beta = -\phi - \alpha$.

(b) Show that the control law $v = K_\rho \rho$, $\omega = K_\alpha \alpha + K_\beta \beta$ gives an exponentially stable system when $K_\rho > 0$, $K_\beta < 0$, $K_\alpha - K_\rho > 0$.

Hint: For small $x = \alpha, \beta$ use the approximation $\cos x = 1$ and $\sin x = x$. Determine the values of K_ρ, K_α, and K_β which assure that the poles of the closed-loop characteristic polynomial have negative real parts.

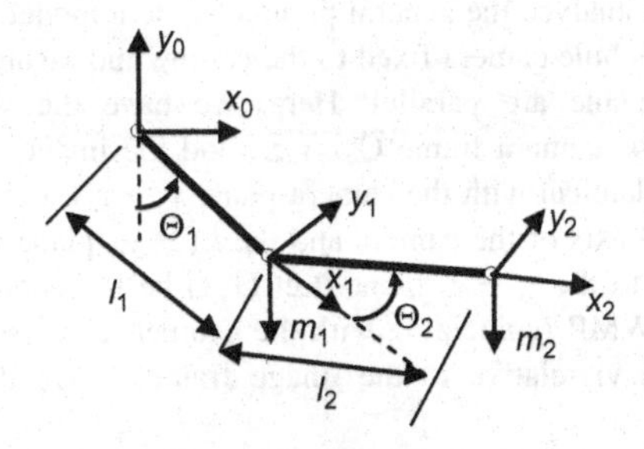

Figure P.10 A 2-linkrobot attached to the ceiling.

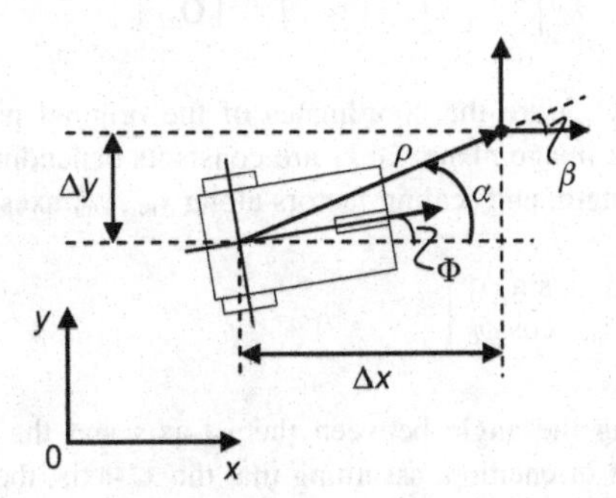

Figure P.11 Differential-drive robot in the inertial coordinate frame.

28. Formulate the control problem of a mobile manipulator, where the platform is an Ackerman's vehicle and the manipulator has two links. Write down the corresponding equations, and simulate them with parameter values of your choice.

29. **(a)** Apply the sliding mode control to a differential-drive WMR with uncertainty in the platform's mass.

    (b) Derive a discontinuous feedback control law with exponential convergence for the WMR with slip studied in Section 3.3 using the dynamic model ((3.47a)–(3.47c)). Choose an appropriate wheel surface traction model (see Ref. [5] of Chapter 3; Ward C, Iagnemma K. ICRA2007, p. 2724–9; Pota H, Katupitiya J, Eaton R. CDC2007, p. 596–601; Albagul A, Martono W, Muhida R. Cutting edge robotics. In: Kordic V et al., editors. ARS/pIV, Germany, www.intercho-pen.com/download/pdf/1. p. 784.

30. Formulate and solve the control problem of a four-wheel Mecanum vehicle in the following cases:

    (a) Lyapunov-based control

    (b) Adaptive control

    (c) Sliding mode control

E. Visual Servoing

31. **(a)** Describe the visual servo control following the typical classification of the problem.

    (b) Describe the camera calibration and the hand-eye (on-board) calibration problems.

32. Describe the general visual servoing problem and a solution when the vision system used is a central catadioptric mirror camera system.

33. **(a)** Describe and analyze the general pinhole camera model.

    (b) Consider a pinhole camera fixed to the ceiling and assume that the camera plane and WMR plane are parallel. Here, we have the world coordinate frame $O_w x_w y_w z_w$, the camera frame $O_c x_c y_c z_c$, and the image frame $O_{im} x_{im} y_{im}$ (which is assumed identical with the camera plane $x_c - y_c$). Let C be the crossing point of the optical axis of the camera and the $x_w - y_w$ plane with coordinates (x_p, y_p) with respect to the $x_c - y_c$ plane. Let (x, y) be the coordinates of the center of mass of the WMR (coinciding with the geometric center) and (x_m, y_m) the coordinates of (x, y) relative to the image frame. Prove that the pinhole camera model is:

$$\begin{bmatrix} x_m \\ y_m \end{bmatrix} = \begin{bmatrix} k_1 & 0 \\ 0 & k_2 \end{bmatrix} \mathbf{R} \left\{ \begin{bmatrix} x \\ y \end{bmatrix} - \begin{bmatrix} x_p \\ y_p \end{bmatrix} \right\} + \begin{bmatrix} O_{c1} \\ O_{c2} \end{bmatrix}$$

where (O_{c1}, O_{c2}) are the coordinates of the original point of the camera with respect to the image plane, k_1, k_2 are constants depending on the depth information, focal length, and scaling factors along x_{im}, y_{im} axes, and

$$\mathbf{R} = \begin{bmatrix} \cos\phi_0 & \sin\phi_0 \\ -\sin\phi_0 & \cos\phi_0 \end{bmatrix}$$

with ϕ_0 being the angle between the x_m-axis and the x_w-axis with a positive anticlockwise orientation, assuming that the x_w-axis, the x-axis, and the x_m-axis have the same orientation.

(c) Develop fully the general catadioptric camera model and express it in the form $\mathbf{x}_{\text{im}} = \mathbf{Kf}(\mathbf{x})$

where $\mathbf{x}_{\text{im}} = [x_{\text{im}}, y_{\text{im}}, 1]^{\text{T}}$ is the projection of the 3D world space point $\mathbf{x} = [x, y, z]^{\text{T}}$, $\mathbf{f}(\mathbf{x}) = (1/w)[x, y, 1]^{\text{T}}$, $w = z + \zeta\sqrt{x^2 + y^2 + z^2}$, $\mathbf{K}$ is the triangular calibration matrix of the catadioptric camera (containing the mirror-lens intrinsic parameters) and ζ is the mirror intrinsic parameter.

34. Define the camera robot system as a dynamic system:

$$\mathbf{x}_{k+1} = \mathbf{F}(\mathbf{x}_k, \mathbf{u}_k), \quad \mathbf{y}_k = \mathbf{H}(\mathbf{x}_k)$$

where $\mathbf{y}_k \in Y \subset R^m$ is the set of possible output values, $\mathbf{F}(\cdot)$ describes the dynamics, $\mathbf{H}(\cdot)$ is the output mapping, and $\mathbf{u}_k \in U \subset R^6$ contains the desired pose change of the camera coordinate system.

35. In the above model, define the mapping of robot movements to image changes (forward visual model) $\varphi_k(\mathbf{u}) = \varphi(\mathbf{x}_k, \mathbf{u}) = \mathbf{H}(\mathbf{F}(\mathbf{x}_k, \mathbf{u}))$, $\varphi: X \times U \to Y$ with X being system state space.

36. Derive the inverse model of the above forward model. Since the mapping φ is too complicated, describe the most popular method of linear approximation of $\mathbf{y}_{k+1} = \mathbf{H}(\mathbf{F}(\mathbf{x}_k, \mathbf{u}_k))$.

37. A more accurate approximation of the camera robot system model is to use a quadratic approximation. Describe and develop this "quadratic model."

38. Figure P.12 shows the structure of a point-to-point positioning system using a binocular camera.

Specifically, this positioning task may be one of the following: (i) to grasp an object, (ii) to execute an insertion task, (iii) to place an object held by the gripper to a pose defined in the image. In Figure P.12 the error function is defined as $\mathbf{e}_{\text{l}} = \mathbf{f}_l^c - \mathbf{f}_l^d$ for the left image, and $\mathbf{e}_{\text{r}} = \mathbf{f}_r^c - \mathbf{f}_r^d$ for the right image. This error can be driven to zero by estimating the image Jacobian via stacking the two monocular image Jacobians defined for each of the two cameras. To control three translational degrees of freedom of the robotic manipulator, it is sufficient to estimate the distance between two points in the image, that is, it is not needed an accurate estimation of the transformation between the robot and the camera coordinate systems (X_R^C must only roughly known).Develop an image-based visual servoing control algorithm for this task, assuming that an image-based tracking algorithm is available (that estimates 2D image positions of feature points), and the desired positions $\mathbf{f}_l^d$ and $\mathbf{f}_r^d$ are selected manually at the beginning of the servoing procedure.

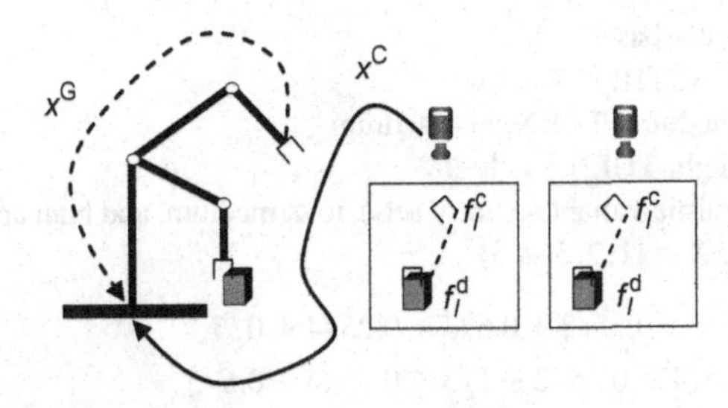

Figure P.12 Robotic positioning via a binocular camera.

F. Fuzzy and Neural Methods

39. Explain in your own words what we mean by the term intelligent control, describing its ingredients and primary architectures.

40. Describe briefly the three main components of computational intelligence.

41. Investigate using Matlab the membership function: $\mu_{big}(x) = 1/[1 + (x/F_2)^{-F_1}], x \in R$
The parameters F_1 and F_2 are known as "exponential fuzzifier" and "denominator fuzzifier" respectively.
Hint: Plot $\mu_{big}(x)$ in two cases:
(a) Constant F_2 and varying F_1 (e.g., $F_2 = 50; F_1 = 1, 2, 4, 100, 100$)
(b) Constant F_1 and varying F_2 (e.g., $F_1 = 4; F_2 = 30, 40, 50, 60, 70$)

42. We are given the fuzzy sets A and B, where A = {x greater than 15} and B = {x nearly 17} with membership functions:

$$\mu_A(x) = \begin{cases} \dfrac{1}{1+(x-15)^{-2}}, x > 15 \\ 0, x \leq 15 \end{cases} \qquad \mu_B(x) = \frac{1}{1+(x-17)^4}$$

(a) Find and draw the membership function of the fuzzy set $C = (x$ greater than 15) AND (x nearly 17)
(b) Similarly find and draw the membership function of $D = (x$ greater than 15) OR (x nearly 17)
(c) Finally draw the membership function of E = (x not greater than 15) AND (x nearly 17)

43. We are given the fuzzy rule: "IF x is A THEN y is B" where:

$$A = 0.33/6 + 0.67/7 + 1.00/8 + 0.67/9 + 0.33/10$$
$$B = 0.33/1 + 0.67/2 + 1.00/3 + 0.67/4 + 0.33/5$$

If we know that "x is A′" with:
$A' = 0.5/5 + 1.00/6 + 0.5/7$
Find the result B′ with the Mamdani rule.

44. Consider the bell-type membership function $\mu_A(x) = \text{bell}(x; 1.5, 2, 0.5)$ and the function:

$$f(x) = \begin{cases} (x-1)^2 - 1, & x \geq 0 \\ x, & x \leq 0 \end{cases}$$

Find the membership function of the fuzzy set B $= f(A)$.
Hint: Use the relation $f(A) = \Sigma \mu_A(x)/f(x)$, where Y is the superset Y of the mapping $y = f(x), y \in Y$.

45. Consider the rule base:
R_1: IF x is low, THEN y is low
R_2: IF x is medium, THEN y is medium
R_3: IF x is high, THEN y is high
where the linguistic variables (fuzzy sets), low, medium, and high are defined as follows:
Variablex: $X = \{1, 2, 3, 4, 5\}$

$$\text{Low} = 1/1 + 0.75/2 + 0.5/3 + 0.25/4 + 0/5$$
$$\text{Medium} = 0.5/1 + 0.75/2 + 1/3 + 0.75/4 + 0.5/5$$
$$\text{High} = 0/1 + 0.25/2 + 0.5/3 + 0.75/4 + 1/5$$

Variabley: $Y = \{6, 7, 8\}$

$$\text{Low} = 1/6 + 0.6/7 + 0.3/8$$
$$\text{Medium} = 0.6/6 + 1/7 + 0.6/8$$
$$\text{High} = 0.3/6 + 0.6/7 + 1/8$$

Find the equivalent relational matrix R of this rule database using (i) Mamdani's rule and (ii) Zadeh's rule.

46. In the above case, find the output:

$$B' = A' \circ R$$

using the max–min composition rule, where:
A′ = "nearly high"
$= 0.25/1 + 0.50/2 + 0.75/3 + 1/4 + 0.75/5$

and express the result linguistically.

47. Modify the BP propagation of Section 14.4.3.3 such that to hold for the hyperbolic tangent sigmoid function.

48. Prove the convergence of the BP algorithm.

49. Write the equations that describe the MLP shown in Figure P.13 in the matrix form (14.9). Assume that all neurons (except the neurons of the first-input layer) have the same activation function $\sigma(\cdot)$ and zero threshold value.

50. Describe the principal optimization algorithms that are based on the steepest descent learning method. How can we improve their rate of convergence?

51. Apply the BP training algorithm to the NN shown in Figure P.14. Give in analytic form the weight updating relations. Use the activation function:

$$\sigma(z) = [1 - \exp(z)]/[1 + \exp(z)]$$

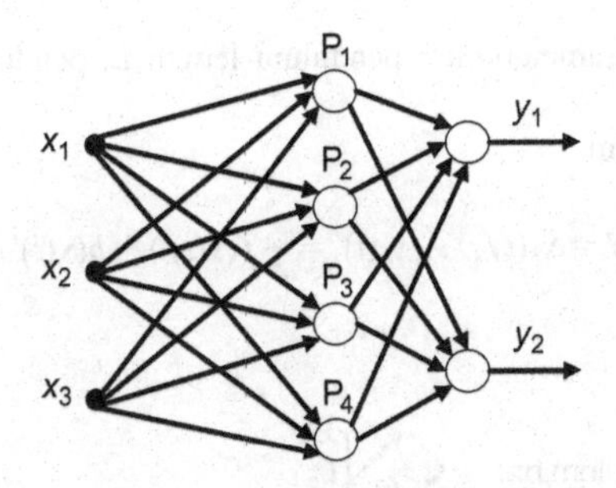

Figure P.13 An MLP with the same activation functions in the hidden and output layers.

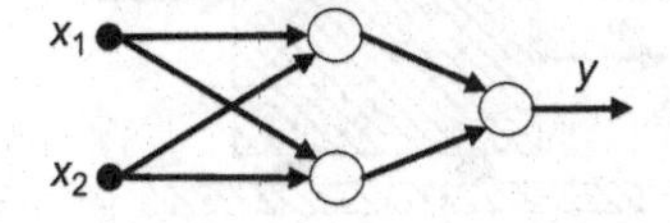

Figure P.14 A simple NN.

52. Describe the basic algorithms for selecting the centers $\mathbf{c}_i$ of the RBF basis functions. When the centers $\mathbf{c}_i$ have been selected, how can we select the parameters σ_i of the RBFs?

53. Consider an RBF NN with p inputs, one output, and m RBF basis functions. The basic functions are:

$$\phi(||\mathbf{x}-\mathbf{c}_i||) = 1/[1+\sigma_i(\mathbf{x}-\mathbf{c}_i)^2]$$

Give in detailed form the training equations for the parameters:

- The weights w_i of the output layer
- The widths σ_i of the RBFs
- The centers $\mathbf{c}_i$ of the RBFs

using the steepest algorithm.

54. **(a)** Explain the universal approximation property of an NN with three layers.

(b) Examine the use of BP learning method, with a sigmoid function, of the following forms:

(i) $\sigma(x) = 1/x, \quad 1 \leq x \leq 100$

(ii) $\sigma(x) = \log_{10}x, \quad 1 \leq x \leq 10$

(iii) $\sigma(x) = \exp(-x), \quad 1 \leq x \leq 10$

(iv) $\sigma(x) = \sin x, \quad 0 \leq x \leq \pi/2$

In each case use two data sets: training data and verification data. Using the training data and one hidden layer compute the synaptic weights. Evaluate the accuracy using the verification data.

55. **(a)** Construct an ANFIS two-rule two-input system Mamdani fuzzy system using the max–min composition and the COG deffuzification method. Describe fully the operation of the layers of this neuro-fuzzy system.

(b) Derive the data using the Matlab file "*tm-2in.m*" and the Matlab program "*tan-mip.m*" to learn this mapping. The NN must have about the same number of parameters as in (a). Plot the results.

56. Design a fuzzy controller for the stabilization of the inverted pendulum of Figure P.15.

The system parameters are: pendulum length L, pendulum bar mass m, vehicle mass M.

57. Consider the system

$$\dot{x}_1(t)^j = x_2(t)^j, \dot{x}_2(t)^j = x_3(t)^j, \ldots, \dot{x}_n(t)^j = -f(\mathbf{x}(t)^j) + b(\mathbf{x}(t)^j)u(t)^j$$

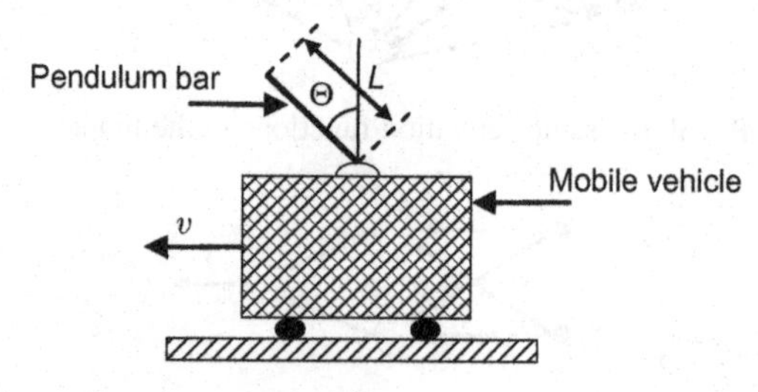

Figure P.15 Inverted pendulum-cart system.

where $\mathbf{x}(t)^j = [x_1(t)^j, x_2(t)^j, \ldots, x_n(t)^j]^T$ is the state vector, $u(t)^j$ is the control input, and $f(\mathbf{x}^j), b(\mathbf{x}^j)$ are unknown functions. The index j denotes the cycle (repetition) number and $t \in [0, T]$. The problem is to control $\mathbf{x}^j(t)$ such that to follow a desired state $\mathbf{x}_d(t) = [x_d, \dot{x}_d, \ldots, x_d^{(n-1)}]^T$ in $t \in [0, T]$. Assume that $f(\mathbf{x})$ and $b(\mathbf{x})$ are bounded, $\mathbf{x}_d(t)$ is measurable and bounded, $[\mathbf{e}^j(t)] = [\mathbf{x}^j(t) - \mathbf{x}_d^j(t)]_{t=0} = \mathbf{0}$ for each cycle for all $\mathbf{x}^j \in R^n$, and that the bound b_L: $0 < b_L < b(\mathbf{x}^j)$ is a positive constant.
Hint: Show that the controller u^j has the form:

$$u^j = U_m^j - \text{sgn}(s^j)[1 + 1/b_L]|U_m^j|$$

$$\text{sign}(s^j) = \begin{cases} 1, & s^j > 0 \\ 0, & s^j = 0 \\ -1, & s^j < 0 \end{cases}$$

58. Apply the above controller through simulation to: (a) a two-link robot (with $l_1 = l_2 = 1$ m, $m_1 = m_2 = 1$ kg), (b) a car-like WMR, and (c) to the MM consisting of the above.

59. Derive an adaptive neuro-fuzzy controller with Gaussian functions for controlling the differential-drive WMR shown in Figure P.16.
Hint: The world coordinate frame is *Oxy*. Using the symbols of Figure P.16, the WMRs dynamic equations are:

$$I_v \ddot{\phi} = D_r^l - D_l^l$$
$$M\dot{v} = D_r + D_l$$
$$I_\omega \ddot{\theta}_i + c\dot{\theta}_i = ku_i - rD_i \quad (i = l, r)$$
$$r\dot{\theta}_r = v + l\dot{\phi}$$
$$r\dot{\theta}_l = v - l\dot{\phi}$$

which, defining the state vector $\mathbf{x} = [v, \phi, \dot{\phi}]^T$, the control vector $\mathbf{u} = [u_r, u_l]^T$, and the output vector $\mathbf{y} = [v, \phi]^T$, can be written in the standard state-space form: $\dot{\mathbf{x}} = \mathbf{A}\mathbf{x} + \mathbf{B}\mathbf{u}$, $\mathbf{y} = \mathbf{C}\mathbf{x}$.

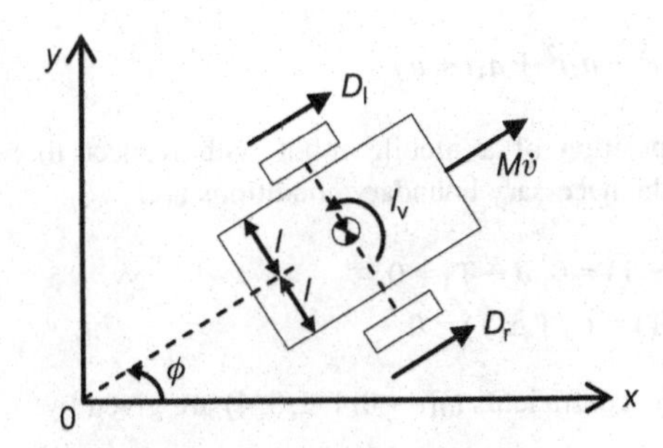

Figure P.16 Differential drive WMR for neuro-fuzzy control.

60. Derive a stable 2D visual servoing controller for MMs with planar robotic manipulators using RBF NNs under the following conditions: (i) the gravity and friction terms are unknown and (ii) there are modeling errors in the vision system.
Hint: Use RBFs with Gaussian base functions $\phi(x) = \mathrm{e}^{-(x-c)^2/b}(b>0)$, and the standard PD control law:

$$\mathbf{u} = -\mathbf{K}_\mathrm{p}(\mathbf{q} - \mathbf{q}_\mathrm{d}) - \mathbf{K}_\mathrm{d}\dot{\mathbf{q}}$$

assuming that the joint velocities $\dot{\mathbf{q}}$ are measurable. Express the NN in standard way: $\mathbf{y} = \mathbf{W}^\mathrm{T}\boldsymbol{\Phi}(\mathbf{x})$. Add the neural term in the above control law which is not accurate due to that gravity and friction are unknown, that is:

$$\mathbf{u} = \mathbf{K}_\mathrm{p}(\mathbf{q} - \mathbf{q}_\mathrm{d}) - \mathbf{K}_\mathrm{d}(\dot{\mathbf{q}} - \dot{\mathbf{q}}_\mathrm{d}) + \hat{\mathbf{W}}\boldsymbol{\Phi}(\hat{\mathbf{V}}\mathbf{s})$$

where $\hat{\mathbf{W}}$ is the weight matrix produced after the training of the NN. Since the robot is to be controlled by visual servoing the above PD control must be replaced by:

$$\boldsymbol{\tau} = \mathbf{J}^\mathrm{T}\mathbf{K}_\mathrm{p}\mathbf{R}^\mathrm{T}\tilde{\mathbf{x}}_\mathrm{s} - \mathbf{K}_\mathrm{d}\dot{\mathbf{q}} + \hat{\mathbf{W}}^\mathrm{T}\boldsymbol{\Phi}(\mathbf{s})$$

where $\mathbf{s} = [\mathbf{s}_0, \mathbf{q}^\mathrm{T}, \dot{\mathbf{q}}^\mathrm{T}_{s0}]$ is the input vector to the NN, $\mathbf{s}_0$ is the NN threshold, $\mathbf{W}$ is the weight matrix with N hidden-layer neurons, and $\tilde{\mathbf{x}}_\mathrm{s} \in R^n$ is the position error in the camera:

$$\tilde{\mathbf{x}}_\mathrm{s} = ah\mathbf{R}(\boldsymbol{\theta})[\mathbf{f}_\mathrm{d} - \mathbf{f}], h = l_\mathrm{f}/(l_\mathrm{f} - z)$$

Here, $\mathbf{R}(\boldsymbol{\theta})$ is the standard 2×2 rotation matrix, l_f the focal length, 'a' the scale factor, and $\mathbf{f}(\mathbf{q}), \mathbf{f}(\mathbf{q}_\mathrm{d})$ the image features (actual and desired). The term $\hat{\mathbf{W}}^\mathrm{T}\boldsymbol{\Phi}(\mathbf{s})$ approximates the gravity and the friction term $\mathrm{B}(\mathbf{s}) = \mathbf{g}(\mathbf{q}) + \mathbf{F}_\upsilon(\dot{\mathbf{q}})$. Treat first the case where the visual space can match the world frame exactly, and then the case where the vision system cannot provide exactly the world frame. Prove that the controller is asymptotically stable, and simulate the resulting close-loop system in the Matlab using parameter values of your choice.

G. Planning

61. How many 8-bit bytes are necessary for storing a continuous path of 1 min duration for a six degrees of freedom robot. Assume that for the storage of the position of each joint we need 16-bit words and that the sampling period is 16 ms.

62. Assume that the polynomial

$$x(t) = a_4t^4 + a_3t^3 + a_2t^2 + a_1t + a_0$$

describes the position of a mobile robot with respect to t in the time interval $[-\mathrm{T}, +\mathrm{T}]$. If the necessary boundary conditions are:

$$x(-\mathrm{T}) = 0, \dot{x}(-\mathrm{T}) = 0, \ddot{x}(-\mathrm{T}) = 0$$
$$x(\mathrm{T}) = C, \dot{x}(\mathrm{T}) = C/\mathrm{T}, \ddot{x}(\mathrm{T}) = 0$$

Show that the coefficients $a_i(i = 0, 1, 2, 3, 4)$ are given by:

$$a_0 = 3C/16, a_1 = C/2\mathrm{T}, a_2 = 3C/8\mathrm{T}^2, a_3 = 0, a_4 = -C/16\mathrm{T}^4$$

63. Write a computer program for implementing the paths:

$$x(t) = a_4t^4 + a_3t^3 + a_2t^2 + a_1t + a_0$$
$$y(t) = b_4t^4 + b_3t^3 + b_2t^2 + b_1t + b_0$$

with the following boundary conditions:
For $t = -3$: $x = 0, \dot{x} = 0, \ddot{x} = 0, y = 0, \dot{y} = 0, \ddot{y} = 0$
For $t = +3$: $x = 10, \dot{x} = 5, \ddot{x} = 0, y = 16, \dot{y} = 8, \ddot{y} = 0$
Plot the results with point spacing $\Delta t = 0.2$ s.

64. The motion planning technique described in Example 11.2 is called 2-1-2 because the path is described by a second-order polynomial of time (constant acceleration), followed by a first-order (linear) polynomial (constant velocity), and then again by a second-order polynomial (constant deceleration). Develop the analogous methods:
(a) 4-1-4 where the first and last path sections are described by fourth-order polynomials
(b) 4-3-4 where the second path section is described by a third-order polynomial.

65. A single link robot is to move from an initial angle $\theta(0) = 30°$ to a final angle $\theta(2) = 100°$ in 2 s. The velocity and acceleration of the joint at the initial and final positions are zero. Find the coefficients of a third-, fourth-, and fifth-order polynomial that realize this motion.

66. **(a)** Define and illustrate graphically the concept of shortest-path roadmaps.
(b) Define and illustrate graphically the concept of maximum clearance roadmaps.

67. The exact cell decomposition illustrated in Figure 11.7 is known as *vertical cell decomposition*, where the free configuration space is decomposed into a finite collection of 2-cells (trapezoids that have vertical sides or triangles which are degenerate trapezoids) and 1-cells (vertical segments that are the borders between two 2-cells).
(a) Extend the above basic vertical cell decomposition to treat the case where CS_{free} has two or more points that lie on the same vertical segments, without using random perturbations.
(b) Apply the vertical decomposition method for obstacle boundaries which are described as chains of circular arcs and line segments.

68. **(a)** For the obstacle environment and the initial-goal states shown in Figure P.17A draw the vertical decomposition.
(b) For the obstacle-start-goal state situation of Figure P.17B illustrate the visibility graph.
(c) Draw the shortest path for the cases shown in Figure P.18A and B. Explain your answer.

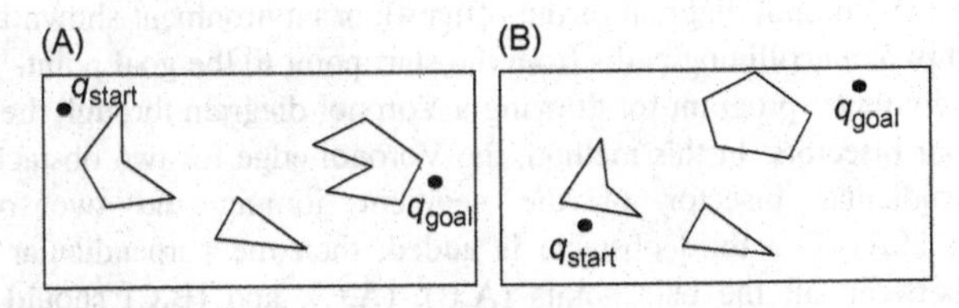

Figure P.17 Three-obstacle environments with different start-goal positions.

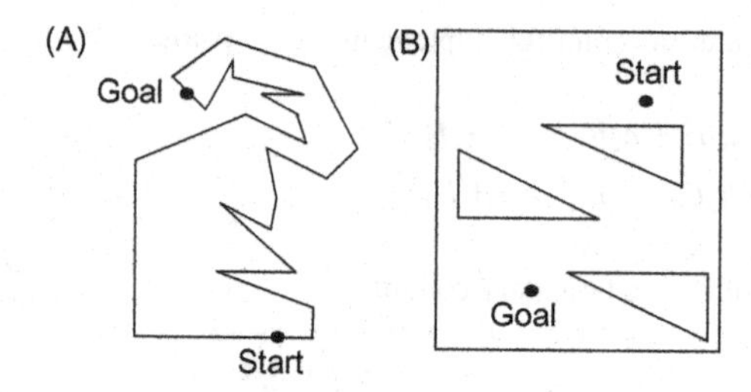

Figure P.18 Two environments for drawing the shortest start-goal paths.

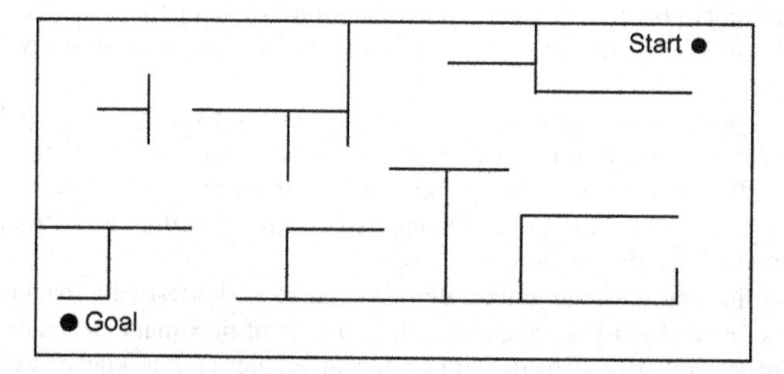

Figure P.19 The map of an office floor.

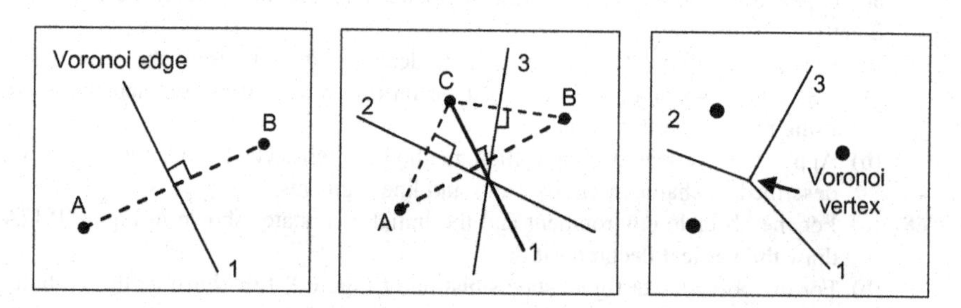

Figure P.20 (A) Bisection Voronoi edge (1) for two objects A and B, (B) bisections 1, 2, and 3 of the objects A, B, and C, and (C) bisection-based Voronoi vertex (path) for three objects.

69. Produce the Voronoi diagram of the office floor environment shown in Figure P.19, and find two noncolliding paths from the start point to the goal point.
70. Write a computer program for drawing a Voronoi diagram through the so-called perpendicular bisectors. In this method, the Voronoi edge for two obstacles A and B is a perpendicular bisector of the segment joining the two obstacles (see Figure P.20A). If a third obstacle is added, then the perpendicular bisectors 1,2, and 3 between all the pair points (A,B), (A,C), and (B,C) should be computed (Figure P.20B).

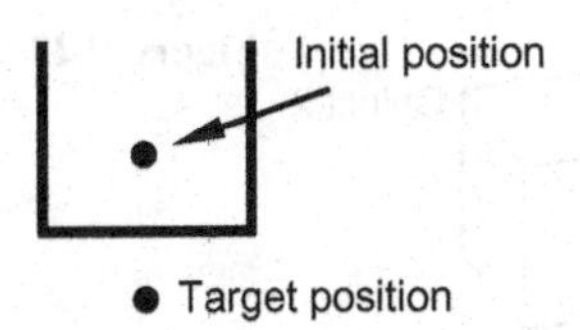

Figure P.21 A U-shaped obstacle.

71. We are given a U-shaped environment of the form shown in Figure P.21.
Derive the potential field equations and write a program that finds the respective path for going from the initial point to the target point showing by arrows the attractive repulsive field around the U-shaped wall.

72. Four significant drawbacks that are inherent to the potential field method (independently of the particular implementation) are the following:
(a) Trap situations due to local minima
(b) No passage between closely spaced obstacles
(c) Oscillations in the presence of obstacles
(d) Oscillations in narrow passages
Describe in your own way each of the above situations and justify their possibility.

73. Define the concepts of path planning, motion planning, and task planning and describe their similarities and differences.

74. Give a short description of the robot task planning in terms of the AI problem-solving methodology, that is, problem representation, problem reduction, and problem solution (solution space search methods).

75. Solve the well-known "cannibals" problem which is the following: Three missionaries and three cannibals want to cross a river from the right bank to the left bank by boat. The maximum capacity of the boat is two persons. If the missionaries are outnumbered at any time by the cannibals, the cannibals will eat the missionaries. Find a solution for the safe crossing of all six persons.
Hint: The state of the system is (N_{miss}, N_{cannib}) where N_{miss} and N_{cannib} is the number of missionaries and cannibals to the left bank respectively. The possible intermediate states are (0,1), (0,2), (0,3), (1,1), (2,2), (3,0), (3,1), and (3,2).

76. Describe the three steps of robot task planning, namely: world modeling, task specification, and program synthesis. The three principal types of 3D object representation schemes are: (i) boundary representation, (ii) sweep representation, and (iii) volumetric representation. Provide a short description of the volumetric representation, including the constructive solid geometry (using operations on primitive shapes and blocks).

77. We are given the objects as shown in Figure P.22, which is described as:
Put object face 1 (S_1 against S_3) and (S_2 against S_4). It is desired to find a set of relations that constrain the configuration of object 1 in relation to the known configuration of object 2.

78. Find a task plan for placing four protectors at the four nonadjacent edges of an object (product) which is to be packed into the box. The protectors are required for keeping the object (e.g., a fragile object) into the box. The initial state of the configuration is an empty box (Figure P.23A), and the goal state is shown in Figure P.23C with intermediate state as shown in Figure P.23B.

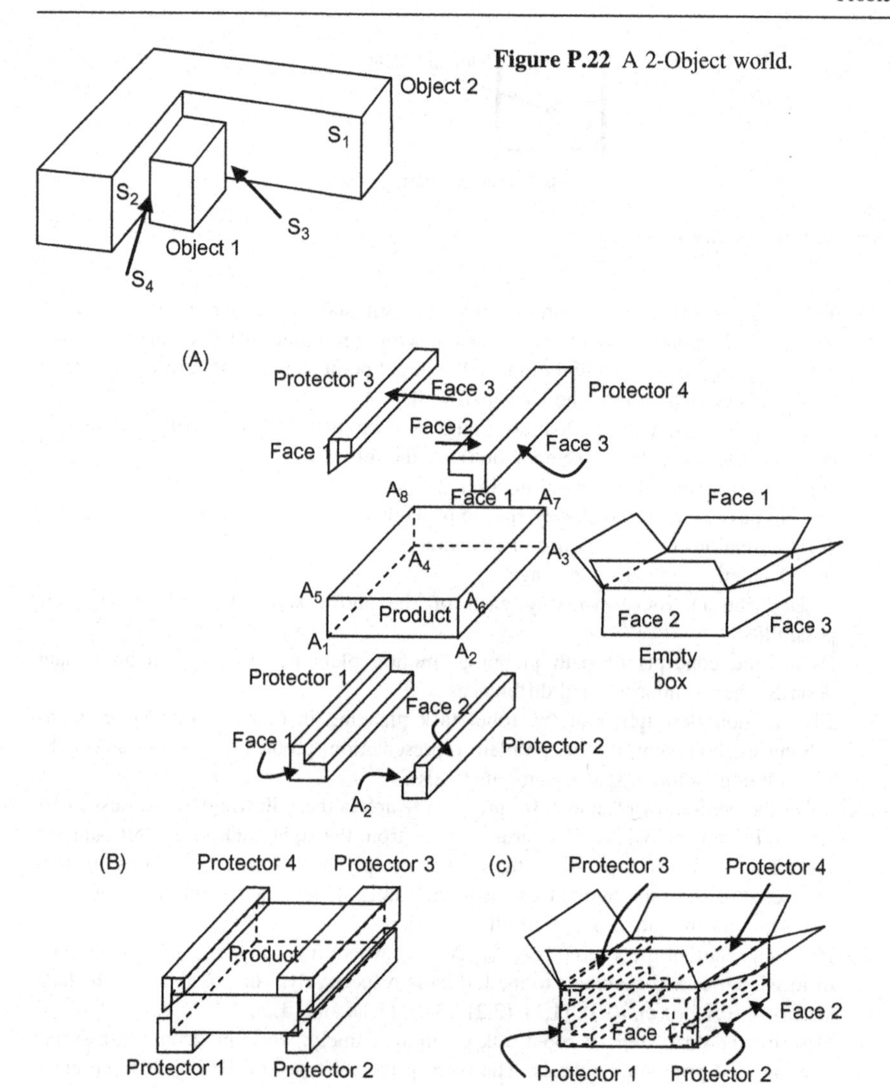

Figure P.22 A 2-Object world.

Figure P.23 (A) Initial configuration (empty box), (B) protectors placed on the product, and (C) protectors placed in the box.

79. Apply the method of triangle table to generate a task plan for transferring a box B_1 using a WMR into a nearby room as shown in Figure P.24 nearby (the robot is initially in room R_1, the box is in room R_2, and the box should be transferred in room R_3). The robot has available two motion operators:

GOTHRU—Movement from the room R_1 to the room R_2 through the respective door, etc.

PUSHTHRU—The robot pushes the box B_1 via the door D_2 from room R_2 to room R_3

The initial database is:

INROOM (ROBOT, R_1)

CONNECTS (D_1, R_1, R_2)

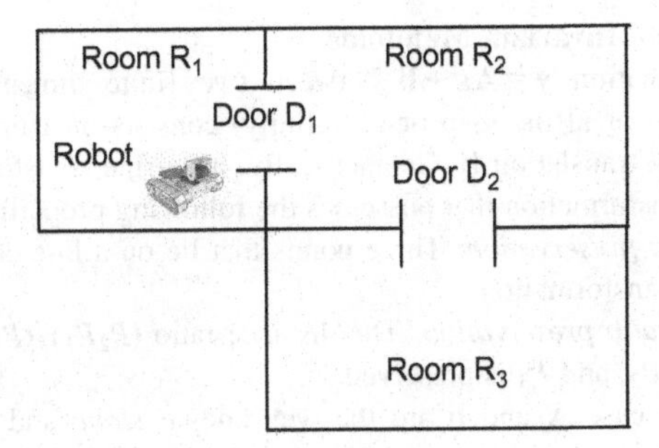

Figure P.24 A 3-room environment with two doors.

CONNECTS (D_2, R_2, R_3)
BOX (B_1)
INROOM (B_1, R_2)
The rule is:

$$(\forall x, \forall y, \forall z)[CONNECTS(X)(Y)(Z) \rightarrow CONNECTS(X)(Y)(Z)]$$

The goal state G_0 is:

$$(\exists X)[BOX(X) \wedge INROOM(X, R_2)]$$

80. Solve problem 79 using AND/OR graphs.

H. Localization and Mapping

81. Describe the following localization methods: (i) odometry, (ii) inertial navigation, (iii) active beacons, (iv) landmark recognition (artificial, natural), and (v) model matching.

82. Describe some ways of measuring and reducing nonsystematic odometry errors.

83. Describe how the global positioning system operates and give an example.

84. Describe a method for probabilistic localization.

85. Describe in your own words the boundary-following algorithm.

86. Describe a method for sensor array calibration via tracking with the EKF.

87. Formulate two ways of SLAM based on probabilities.

88. Formulate a method for localizing a WMR in an environment with landmarks.

89. Formulate the particle filter (sequential Monte Carlo) method for SLAM.

90. When the environment of a WMR involves moving objects we need to solve, in addition to the SLAM problem, the detection and tracking problem (DTM) of these dynamic objects. Derive the Bayesian formula for the combined SLAM−DTM problem. Propose a practical algorithm for performing DTM from a moving platform equipped with range sensors.

91. Show how omnidirectional images, provided by a single catadioptric camera, can be used to generate the representations required for topological navigation and visual path following. It is recalled that topological navigation is based on the robot's global position, estimated by a set of omnidirectional images obtained during the training stage.

92. Develop a method for integrating a catadioptric camera model with: (a) the extended Kalman filter SLAM and (b) the particle filter SLAM.

I. Affine Systems and Invariant Manifolds

93. The transformation $\mathbf{y} = \mathbf{A}\mathbf{x} + \mathbf{b}$ between two finite dimensional spaces (affine transformation or affine map or an affinity) consists of a linear transformation $\mathbf{A}$ followed by a translation $\mathbf{b}$. Geometrically, an affine transformation in Euclidean space is a transformation that possesses the following properties:

- *Colinearity preservation:* Three points that lie on a line continue to be colinear after the transformation.
- *Distance ratio preservation:* The distance ratio $(P_2P_1)/(P_3P_2)$ of three colinear points P_1, P_2, and P_3 is preserved.

In the 1D case $\mathbf{A}$ and $\mathbf{b}$ are the well-known *slope* and *intercept* of the plot $\mathbf{y} = \mathbf{A}\mathbf{x} + \mathbf{b}$. The matrix $\mathbf{A}$ represents a rotation (or shear) and $\mathbf{b}$ a "shift" (translation). So, homogeneous transformations are affine transformations:

$$\begin{bmatrix} \mathbf{y} \\ \cdots \\ 1 \end{bmatrix} = \left[\begin{array}{c|c} \mathbf{A} & \mathbf{b} \\ \hline \mathbf{0} & 1 \end{array}\right] \begin{bmatrix} \mathbf{x} \\ \cdots \\ 1 \end{bmatrix}, \quad \mathbf{T} = \left[\begin{array}{c|c} \mathbf{A} & \mathbf{b} \\ \hline \mathbf{0} & 1 \end{array}\right]$$

(a) Prove that the following propositions are equivalent:
  (i) $\mathbf{A} - \mathbf{I}$ is invertible.
  (ii) $\mathbf{A}$ does not have the number 1 as one of its eigenvalues.
  (iii) For all $\mathbf{b}$ the transformation has exactly one fixed point.
  (iv) There exists a $\mathbf{b}$ for which the transformation has exactly one fixed point.
  (v) Affine transformations with matrix $\mathbf{A}$ can be written as a linear transformation with some point as origin.

(b) Show that affine transformations in 2D, where $\mathbf{A}$ has an eigenvalue 1 (i.e., a 2D affine transformation without fixed point) is a pure translation. Determine when an affine transformation is invertible.

94. For the following nonlinear control system:

$$\dot{x}_1 = 3x_1 + x_2^2 - 2x_2 + u, \quad \dot{x}_2 = 3 \sin x_1 - x_2 - u$$

do the following: (a) For $u = 0$ plot the state-space trajectories in the neighborhood of the origin which is defined by $|x_1| \leq 1$ and $|x_2| \leq 2$. (b) Linearize the system around the point $x_1 = 0, x_2 = 0$, and $u = 0$, and show that the linearized system is unstable. (c) Design a linear state feedback controller for this system such that the closed-loop system has the specifications $\zeta = 0.5$ and $\omega_n = 3$. (d) Apply this controller to the original system and draw the trajectories of the resulting closed-loop system in state space.

95. A controller that leads the system $\dot{x}_1 = x_1 + u_1$, $\dot{x}_2 = x_2 + u_2$ to the origin $(x_1, x_2) = (0, 0)$ is $u_i = \begin{cases} -y_i & \text{for} \quad |y_i| \leq 5 \\ -5\text{sgn}(y_i) & \text{otherwise} \end{cases}$

with $y_i = 5x_i$.

(a) Determine analytically the region of attraction of the system by studying the derivative $\dot{V}$ of the Lyapunov function: $V = (1/2)(x_1^2 + x_2^2)$

(b) Determine the real region of attraction.

96. For the system: $\dot{x}_1 = \sin x_2, \dot{x}_2 = x_1^4 \cos x_2 + u$: (a) Design a controller for which the system can track the arbitrary trajectory $x_{d_1}(t)$ under the assumption that the state $[x_1(t), x_2(t)]^{\mathrm{T}}$ is measured exactly, the signals $x_{d_1}(t), \dot{x}_{d_1}(t)$ and $\ddot{x}_{d_1}(t)$ are all known, and the system does not involve any uncertainty, and (b) Verify using Matlab the ability of the closed-loop system to track the desired trajectory.

97. Find a controller $\upsilon(\mathbf{x})$ that stabilizes robustly the system: $\dot{x}_1 = x_2 + d_1(x,t), \dot{x}_2 = z + d_2(x,t), \dot{z} = \upsilon + d_3(x,t)$ when the disturbances are bounded: $|d_1| \le \rho_1(x_1), |d_2| \le \rho_2(x_1,x_2), |d_3| \le \rho_3(x_1,x_2,x_3)$. Use the Lyapunov-based robust control method.

98. (a) Determine whether the system

$$\dot{\mathbf{x}} = \mathbf{f}(\mathbf{x}) + \mathbf{b}u, \quad \mathbf{x} = [x_1, x_2, x_3]^{\mathrm{T}}$$
$$\mathbf{b} = [1,0,1]^{\mathrm{T}}, \quad \mathbf{f}(\mathbf{x}) = [x_2 + x_2^2 + x_3^2, x_3 + \sin(x_1 - x_3), x_3^2]^{\mathrm{T}}$$

is input-state linearizable, and (b) can the variables

$$z_1 = x_1 - x_3, z_2 = x_2 + x_2^2$$
$$z_3 = x_3 + \sin(x_1 - x_3) + 2x_2[x_3 + \sin(x_1 - x_3)]$$

be used as linearization state variables?

99. Examine the controllability of the differential drive, tricycle, car-like, and three- or four-wheel omnidirectional mobile robots using their affine models.

100. (a) Develop a steering method for a car-like WMR that provides smooth paths under curvature constraints and integrate it with a global motion planning technique for obstacle avoidance. (b) Develop a robust controller for a car-like robot using the "virtual" vehicle approach. Consider the case of existing errors and disturbances.

101. Develop a tracking controller with collision avoidance for a group of unicycle-type WMRs. Use a supervisory system that assigns to each robot its reference path, together with the desired velocity profiles of a function of the position along the path.

102. Develop an asymptotically stable control scheme for simultaneous position and torque tracking using the backstepping technique including the actuator dynamics.

103. Consider the m-input affine system $\dot{\mathbf{x}} = \mathbf{f}(\mathbf{x}) + \Sigma_{i=1}^{m} \mathbf{g}_i(\mathbf{x})u_i$ where $\mathbf{x} \in R^n$, $\mathbf{u} = [u_1, u_2, \ldots, u_m]^{\mathrm{T}} \in R^m$, and the nonlinear static state feedback control law:

$$\mathbf{u} = a(\mathbf{x}) + \beta(\mathbf{x})\upsilon(t), \quad \upsilon \in R^{m_0}, \quad m_0 < mu$$

We know that the above system is nonregular (static state) feedback linearizable if there exists a discontinuous state transformation $\mathbf{z} = \Phi(\mathbf{x}), \mathbf{z} \in R^n$ and a state feedback of the above form, such that the transformed system with state $\mathbf{z}$ and input υ is a controllable system. Show via the Frobenius theorem that the two-input driftless system: $\dot{\mathbf{x}} = \mathbf{g}_1(\mathbf{x})u_1 + \mathbf{g}_2(\mathbf{x})u_2$ is nonregular feedback linearizable if the nested distributions defined by:

$$\boldsymbol{\Delta}_i = \mathrm{span}\{\mathbf{g}_2\}, \quad \boldsymbol{\Delta}_i = \boldsymbol{\Delta}_{i-1} + \mathrm{ad}_{\mathbf{g}_1}\boldsymbol{\Delta}_{i-1}, \quad i = 1,2,\ldots,n-2$$

and $\boldsymbol{\Delta}_{n-1} = \boldsymbol{\Delta}_{n-2} + span\{\mathbf{g}_1\}$ have the following properties:

- $\boldsymbol{\Delta}_i$ is involutive and has constant rank for $i = 0,1,\ldots,n-1$
- rank $\boldsymbol{\Delta}_{n-1} = n$

Apply this result to the nonholonomic chained system:

$$\dot{x}_1 = u_1, \dot{x}_2 = u_2, \dot{x}_3 = x_2u_1, \ldots, \dot{x}_n = x_{n-1}u_1$$

104. Consider the chained WMR kinematic model:

$$\dot{x}_1 = u_1, \dot{x}_2 = u_2, \dot{x}_3 = x_2 u_1$$

with output $z = h(\mathbf{x}) = x_1$, and apply the nonsmooth transformation $\phi(\mathbf{x}) = x_3^{1/3}$. Use the result of problem 103 to derive the discontinuous state and input transformations that converts the system into the canonical form starting with the transformation $u_1 = x_3^{1/3}$. Prove that the resulting linear single-input canonical system is:

$$\dot{z}_1 = z_2, \dot{z}_2 = z_3, \dot{z}_3 = v$$

and find the feedback control law $u_1 = u_1(x), u_2 = u_2(x)$ which leads to a closed-loop system with eigenvalues $-\lambda_1, -\lambda_2, -\lambda_3 \quad (0 < \lambda_1 < \lambda_2 < \lambda_3)$, also giving the stability conditions.

105. Consider the chained WMR model:

$$\dot{x}_1 = u_1, \dot{x}_2 = u_2, \dot{x}_3 = x_1 u_2, v = u_1 + x_3 u_2, \omega = u_2$$

Determine the conditions, via the invariant and attractive manifold technique, under which the control law:

$$u_1 = -k_1 x_1 + k_2 \frac{s(x)}{x_1^2 + x_2^2} x_2, u_2 = -k_1 x_2 - k_2 \frac{s(x)}{x_1^2 + x_2^2}$$

with $s(\mathbf{x}) = x_3 - x_1 x_2 / 2$, stabilizes the system to the origin for any initial conditions $x_1 \neq 0, x_2 \neq 0$ if $k_1 > 0, k_2 > 0$. Furthermore, show that the control inputs u_1 and u_2 are bounded along the trajectories of the closed-loop system under the condition: $k_2 > 2k_1 > 0$.

106. Consider the $(n, 2)$ chained system: $\dot{x}_1 = u_1, \dot{x}_2 = u_2, \dot{x}_3 = x_2 u_1, \ldots, \dot{x}_n = x_{n-1} u_1$ which is completely controllable but not asymptotically stabilizable by smooth (or even continuous) static or dynamic feedback controls (Brockett's theorem). Find under what conditions the control law:

$$u_1 = -x_1$$

$$u_2 = k_2 x_2 + k_3 \left(\frac{x_3}{x_1} \right) + k_4 \left(\frac{x_4}{x_1^2} \right) + \cdots + k_n \left(\frac{x_n}{x_1^{n-2}} \right)$$

with $[x_i / x_1^{i-2}]_{(0,0)} = 0$, leads to a closed-loop system that possesses a unique forward solution for any initial condition $\mathbf{x}(0)$ such that $x_1(0) \neq 0$.

Hint: Use matrix A where:

$$\mathrm{A} = \begin{bmatrix} k_2 & k_3 & k_4 & k_5 & \ldots & k_{n-1} & k_n \\ -1 & 1 & 0 & 0 & \ldots & 0 & 0 \\ 0 & -1 & 2 & 0 & \ldots & 0 & 0 \\ 0 & 0 & -1 & 3 & \ldots & 0 & 0 \\ 0 & 0 & 0 & 0 & \ldots & -1 & n-2 \end{bmatrix}$$

107. **(a)** For the above $(n,2)$-chained system find the conditions under which we have $\dot{s} = -bs, b>0$, where $x_1(0) \neq 0$, and $s = x_2 + a_1\frac{x_3}{x_1} + a_2\frac{x_4}{x_1^2} + \ldots + a_{n-2}\frac{x_n}{x_1^{n-2}}$. This means that the manifold $s=0$ is invariant and attractive (exponentially).

(b) For the same system find the conditions under which the feedback controller:

$$u_1 = \begin{cases} u_{1A}(x) = -x_1, & |s| \leq \mu \\ u_{1B}(x) = \text{sgn}(x_1), & |s| > \mu \end{cases}$$

$$u_2 = \begin{cases} u_{2A}(x) = k_2x_2 + k_3\dfrac{x_3}{x_1} + \cdots + k_n\dfrac{x_n}{x_1^{n-2}}, & |s| \leq \mu \\ u_{2B}(x) = -\lambda x_2, & |s| > \mu \end{cases}$$

with $\lambda > 0, \mu > 0$, and $\text{sgn}(x_1) = \begin{cases} 1, x_1 \geq 0 \\ -1, x_1 < 0 \end{cases}$

assures a unique forward solution, for any initial condition $x(0)$, which converges exponentially to zero.

Note: A solution approach to problems 106 and 107 is provided by Astolfi A. and Valtolina. Discontinuous control of nonholonomic system. Systems Control Lett 1996;27:37–45; Local robust regulation of chained systems. Systems Control Lett 2003;49:231–8; and Global regulation and local robust stabilization of chained systems. In: Proceedings of the IEEE CDC, Sydney, Australia; December 12–15, 2000. p. 1637–42.

Robotics Web Sites*

1. General Robotics

1. www.robocommunity.com/article/10050/List-of-Other-Robot-Websites
2. www.robotstxt.org
3. www.dprg.org/robolinks.html
4. www.zerorobotics.org/web/zero-robotics/home-public
5. www.nasa.gov/audience/foreducators/robotics/home/index.html
6. www.msdn.microsoft.com/en-us/robotics/aa731517
7. http://researchguides.library.tufts.edu/content.php?pid=127295&sid=1118912
8. www.jafsoft.com/searchengines/webbots.html
9. http://www.seattlerobotics.org/
10. www.topsite.com/best/robotics
11. www.roboteers.com
12. www.cimwareukandusa.com/aRobotAdam.html
13. www.ryerson.ca/aferwon/courses/CPS607/CLASSES/CPS607CL.HTML
14. http://www.eventscope.org/es/index.shtml
15. www.docstoc.com/docs/35855596/Agrobots—Robots-in-Agriculture
16. www.universal-robots.com
17. www.robotics.org
18. www.rethinkrobotics.com
19. www.densorobotics.com/world
20. www.therobotreport.com/index.php/industrial_robots

2. Mobile Robotics

1. www.mrpt.org
2. www.mobosoft.com
3. www.ccsrobotics.com
4. http://mobots.epfl.ch/self-assembling-robots.html
5. www.davidbuckley.net/DB/HistoryMakers.htm
6. www.surveyor.com/SRV_info.html
7. www.wn.com/khepera_mobile_robot
8. www.k-team.com
9. www.arrickrobotics.com/arobot
10. www.hobbyengineering.com/H1937.html
11. https://researchspace.auckland.ac.nz/handle/2292/2725
12. www.robots.net/rcfaq.html

* All web sites provided in the book were valid on 20 AUGUST 2013. Perhaps, some of them may be changed or removed by their creators at a later time.

13. www.automation.com/content/rmt-robotics-announces-audio-feature-for-its-adam-mobile-robot
14. www.mobilerobot.ru
15. www.automation.hut.fi
16. www.roboticsclub.org/links.html
17. www.roboticsbusinessreview.com/rbr50/category
18. www.cresis.ku.edu/sites/default/files/TechRpt101.pdf
19. www.computerworld.com/s/article/9027523/Mobile_robots_aren_t_science_fiction_anymore
20. www.automation.com/product-showcase/rmt-robotics-makes-adam-mobile-robots-vocal
21. www.ri.cmu.edu/pub_files/pub1/simmons_reid_1999_1/simmons_reid_1999_1.pdf
22. www.dtic.mil/cgi-bin/GetTRDoc?AD=ADA433772
23. www.faculty.cooper.edu/mar/mar.htm
24. www.robots.net/rcfaq.html
25. www.arrickrobotics.com/arobot

3. Thirty Five Mobile Robot Companies

1. www.autopenhosting.org/robots/companies.html
2. http://stason.org/TULARC/science-engineering/robotics/35-Mobile-Robot-Companies.html#.Ufn2KNLTw2c
3. www.hotstockchat.com/mobile-robot-companies-geckosystems-us-and-zmp-japan-sign-mou/
4. www.k-team.com/kteam/index.php?rub=3&site=1&version=EN&page=3
5. www.mesa-robotics.com
6. http://robotics.sandia.gov/Roboticvehicles.html
7. http://www.barrett.com/robot/products/hand/handfrom.htm
8. www.aai.ca/robots
9. www.wifibot.com
10. www.esit.com/mobile-robots/
11. www.themachinelab.com/
12. www.ise.bc.ca/robotics.html
13. www.alibaba.com
14. www.motoman.com
15. www.seegrid.com
16. www.robotics.nasa.gov/links/industry.html
17. www.bastiansolutions.com
18. www.directindustry.com
19. www.wanyrobotics.com
20. www.botsinc.com/list-of-robot-companies
21. www.cyperbotics.com
22. www.personalrobots.com
23. www.robots.com
24. www.reisrobotics.com
25. www.rotundus.se
26. www.robotshop.com
27. www.pedsco.com
28. www.floorbotics.com
29. www.recce-robotics.com

30. www.robosoft.com
31. http://gizmodo.com/5966895/mitsubishis-remote-control-tankbot-is-yet-another-member-of-the-robot-clean+up-crew-army
32. www.irobot.com
33. www.destaco.com
34. www.mobilerobots.com
35. www.geckosystems.com

A Sample of Commercial Robots

No.	ROBOT NAME	COMPANY	DIMENSIONS	WEB SITE (See Mobile Robot Companies)
1.	Matilda III	MESA ROBOTICS...	30 in × 21 in × 40 in	[5]
2.	Ratler Marvin	SANDIA		[6]
3.	M2GAIA	Applied AI Systems	14.2 cm × 73 cm × 48 cm	[8]
4.	MR-2,MR-7	Engineering Services		[10]
5.	Rotundus	ROTUNDUS	60 cm × 90 cm × 80 cm	[25]
6.	Easy-Roller	ROBOTSHOP		[26]
7.	RMI-9XD	PEDSCO		[27]
8.	Micro VGTV	RECCE ROBOTICS	22 cm × 43 cm × 10 cm	[29]
9.	Robu-CAR-TT	ROBOSOFT		[30]
10.	TankBot	MITSUBISHI		[31]
11.	ATRVJr	i-RobotCorporation	75.5 cm × 49 cm × 55 cm	[32]
12.	PackBot	DESTACO		[33]
13.	SeekurJr	Adept Mobile Robots	119.8 cm × 83 cm × 50 cm	[34]
14.	PioneerLX	Adept Mobile Robotics	69.7 cm × 43.7 cm × 44.8 cm	[34]
15.	CareBot	Gecko Systems		[35]
16.	AmigoBot	Adept Mobile Robotics	33 cm × 29 cm × 15 cm	[34]
17.	Pioneer P3-DX	Adept Mobile Robotics	45.5 cm × 38.1 cm × 23.7 cm	[34]
18.	GP8 Pallet Track	SEEGRID	91in × 36in × 90.75in	[15]
19.	PeopleBot	Adept Mobile Robotics	41 cm × 49 cm × 112 cm	[34]
20.	AQUA2	Adept Mobile Robotics	63.8 cm × 46.7 cm × 12.7 cm	[34]

Printed in the United States
By Bookmasters